# INDUSTRIAL ORGANIC CHEMICALS

## Starting Materials and Intermediates

## VOLUME 2

Weinheim · New York · Chichester · Brisbane · Singapore · Toronto

# INDUSTRIAL ORGANIC CHEMICALS

### VOLUME 1
**Acetaldehyde** to **Aniline**

### VOLUME 2
**Anthracene** to **Cellulose Ethers**

### VOLUME 3
**Chlorinated Hydrocarbons** to **Dicarboxylic Acids, Aliphatic**

### VOLUME 4
**Dimethyl Ether** to **Fatty Acids**

### VOLUME 5
**Fatty Alcohols** to **Melamine and Guanamines**

### VOLUME 6
**Mercaptoacetic Acid and Derivatives** to **Phosphorus Compounds, Organic**

### VOLUME 7
**Phthalic Acid and Derivatives** to **Sulfones and Sulfoxides**

### VOLUME 8
**Sulfonic Acids, Aliphatic** to **Xylidines**

**Index**

# INDUSTRIAL ORGANIC CHEMICALS

## Starting Materials and Intermediates

### VOLUME 2
**Anthracene**
to **Cellulose Ethers**

Weinheim · New York · Chichester · Brisbane · Singapore · Toronto

> This book was carefully produced. Nevertheless, authors and publisher do not warrant the information contained therein to be free of errors. Readers are advised to keep in mind that statements, data, illustrations, procedural details or other items may inadvertently be inaccurate.

Library of Congress Card No.: Applied for.
British Library Cataloguing-in-Publication Data: A catalogue record for this book is available from the British Library.

Die Deutsche Bibliothek – CIP-Einheitsaufnahme
**Industrial organic chemicals** : starting materials and intermediates ;
an Ullmann's encyclopedia. – Weinheim ; New York ;
Chichester ; Brisbane ; Singapore ; Toronto : Wiley-VCH
   ISBN 3-527-29645-X
Vol. 2. Anthracene to Cellulose Ethers. – 1. Aufl. – 1999.

© WILEY-VCH Verlag GmbH, D-69469 Weinheim (Federal Republic of Germany), 1999
Printed on acid-free and chlorine-free paper.
All rights reserved (including those of translation in other languages). No part of this book may be reproduced in any form – by photoprinting, microfilm, or any other means – nor transmitted or translated into machine language without written permission from the publishers. Registered names, trademarks, etc. used in this book, even when not specifically marked as such, are not to be considered unprotected by law.

Composition and Printing: Rombach GmbH, Druck- und Verlagshaus, D-79115 Freiburg
Bookbinding: Wilhelm Osswald & Co., D-67433 Neustadt (Weinstraße)
Cover design: mmad, Michel Meyer, D-69469 Weinheim
Printed in the Federal Republic of Germany

# Contents

## 1 Anthracene

1. Physical Properties . . . . . . . . . . . . . 643
2. Chemical Properties . . . . . . . . . . . 643
3. Production . . . . . . . . . . . . . . . . . 644
4. Analysis . . . . . . . . . . . . . . . . . . . 644
5. Uses . . . . . . . . . . . . . . . . . . . . . 645
6. Toxicology . . . . . . . . . . . . . . . . . 645
7. References . . . . . . . . . . . . . . . . . 646

## 2 Anthraquinone

1. Introduction . . . . . . . . . . . . . . . . 649
2. Properties . . . . . . . . . . . . . . . . . 650
3. Production . . . . . . . . . . . . . . . . . 652
4. Purification, Quality Requirements, and Analysis . . . . . . . . . . . . . . . 659
5. Uses . . . . . . . . . . . . . . . . . . . . . 659
6. Economic Aspects . . . . . . . . . . . . 660
7. Toxicology . . . . . . . . . . . . . . . . . 661
8. References . . . . . . . . . . . . . . . . . 661

## 3 Aziridines

1. Introduction . . . . . . . . . . . . . . . . 663
2. Physical Properties . . . . . . . . . . . . 664
3. Chemical Properties . . . . . . . . . . . 664
4. Production . . . . . . . . . . . . . . . . . 666
5. Environmental Protection . . . . . . . 668
6. Uses . . . . . . . . . . . . . . . . . . . . . 668
7. Economic Aspects . . . . . . . . . . . . 669
8. Toxicology and Occupational Health 669
9. References . . . . . . . . . . . . . . . . . 671

## 4 Benzaldehyde

1. Introduction . . . . . . . . . . . . . . . . 673
2. Physical Properties . . . . . . . . . . . . 674
3. Chemical Properties . . . . . . . . . . . 674
4. Production . . . . . . . . . . . . . . . . . 677
5. Quality Specifications and Test Methods . . . . . . . . . . . . . . . . . . . . . 682
6. Storage, Transportation, and Safety Regulations . . . . . . . . . . . . . . . . 682
7. Uses . . . . . . . . . . . . . . . . . . . . . 683
8. Derivatives . . . . . . . . . . . . . . . . . 684
9. Economic Aspects . . . . . . . . . . . . 689
10. Toxicology . . . . . . . . . . . . . . . . . 689
11. References . . . . . . . . . . . . . . . . . 689

## 5 Benzene

1. Introduction . . . . . . . . . . . . . . . . 693
2. Physical Properties . . . . . . . . . . . . 694
3. Chemical Properties . . . . . . . . . . . 694
4. Raw Materials . . . . . . . . . . . . . . . 698
5. Production . . . . . . . . . . . . . . . . . 700
6. Separation and Purification . . . . . . 714
7. Quality Specifications . . . . . . . . . . 726
8. Storage and Transportation . . . . . . 728
9. Economic Aspects . . . . . . . . . . . . 728
10. Uses . . . . . . . . . . . . . . . . . . . . . 732
11. Toxicology and Occupational Health 733
12. References . . . . . . . . . . . . . . . . . 737

V

# 6 Benzenesulfonic Acids and Their Derivatives

1. General Aspects . . . . . . . . . . . . . . . . 742
2. Individual Benzenesulfonic Acids and Derivatives . . . . . . . . . . . . . . . . . . 755
3. References . . . . . . . . . . . . . . . . . . . 786

# 7 Benzidine and Benzidine Derivatives

1. Introduction . . . . . . . . . . . . . . . . . 793
2. Production . . . . . . . . . . . . . . . . . . 794
3. Storage and Transportation . . . . . . 799
4. Occupational Health . . . . . . . . . . . 799
5. Environmental Protection . . . . . . . 802
6. Characteristics of the Diphenyl Bases 803
7. Economic Aspects . . . . . . . . . . . . . 812
8. Toxicology . . . . . . . . . . . . . . . . . . 812
9. References . . . . . . . . . . . . . . . . . . 815

# 8 Benzoic Acid and Derivatives

1. Introduction . . . . . . . . . . . . . . . . . 819
2. Physical Properties . . . . . . . . . . . . 820
3. Chemical Properties . . . . . . . . . . . 821
4. Production . . . . . . . . . . . . . . . . . . 821
5. Quality Specifications . . . . . . . . . . 823
6. Storage, Transportation and Legal Aspects . . . . . . . . . . . . . . . . . . . . 823
7. Uses and Economic Aspects . . . . . . 825
8. Derivatives of Benzoic Acid . . . . . . 825
9. Toxicology . . . . . . . . . . . . . . . . . . 836
10. References . . . . . . . . . . . . . . . . . 838

# 9 Benzoquinone

1. Introduction . . . . . . . . . . . . . . . . . 841
2. Physical Properties . . . . . . . . . . . . 842
3. Chemical Properties . . . . . . . . . . . 842
4. Production of 1,4-Benzoquinone . . . 845
5. Uses of the Quinones . . . . . . . . . . 845
6. Toxicology of Benzoquinones . . . . . 846
7. References . . . . . . . . . . . . . . . . . . 846

# 10 Benzyl Alcohol

1. Introduction . . . . . . . . . . . . . . . . . 849
2. Physical Properties . . . . . . . . . . . . 850
3. Chemical Properties . . . . . . . . . . . 851
4. Production . . . . . . . . . . . . . . . . . . 852
5. Quality Standards and Test Methods 856
6. Storage, Transportation, and Safety Regulations . . . . . . . . . . . . . . . . . 857
7. Uses . . . . . . . . . . . . . . . . . . . . . . 857
8. Derivatives . . . . . . . . . . . . . . . . . 857
9. Economic Aspects . . . . . . . . . . . . . 860
10. Toxicology . . . . . . . . . . . . . . . . . 860
11. References . . . . . . . . . . . . . . . . . 861

## 11 Benzylamine

1. Physical Properties . . . . . . . . . . . . 863
2. Chemical Properties . . . . . . . . . . 863
3. Production . . . . . . . . . . . . . . . . . 864
4. Quality Specifications and Analysis . 864
5. Uses . . . . . . . . . . . . . . . . . . . . . . 865
6. Storage and Transportation . . . . . . 865
7. Ecology . . . . . . . . . . . . . . . . . . . 865
8. Toxicology . . . . . . . . . . . . . . . . . 866
9. References . . . . . . . . . . . . . . . . . 866

## 12 Bromine Compounds, Organic

1. Introduction . . . . . . . . . . . . . . . . 867
1.1. Physical Properties . . . . . . . . . . . . 868
1.2. Chemical Properties . . . . . . . . . . 868
1.3. Production . . . . . . . . . . . . . . . . . 871
1.4. Commercial Products . . . . . . . . . . 873
1.5. Manufacturers . . . . . . . . . . . . . . . 887
1.6. Toxicology and Occuptional Health . 887
2. References . . . . . . . . . . . . . . . . . 892

## 13 Butadiene

1. Introduction . . . . . . . . . . . . . . . . 899
2. Physical Properties . . . . . . . . . . . 900
3. Chemical Properties . . . . . . . . . . 900
4. Production . . . . . . . . . . . . . . . . . 906
5. Specifications of Butadiene . . . . . . . 914
6. Stabilization, Storage, and Transportation . . . . . . . . . . . . . . . . . . . . 914
7. Uses and Economic Importance . . . 916
8. Toxicology . . . . . . . . . . . . . . . . . 917
9. References . . . . . . . . . . . . . . . . . 921

## 14 Butanals

1. Introduction . . . . . . . . . . . . . . . . 925
2. Physical Properties . . . . . . . . . . . 925
3. Production . . . . . . . . . . . . . . . . . 926
4. Quality Specifications and Testing . 929
5. Handling, Storage, and Shipment . . 929
6. Chemical Reactions and Applications 930
7. Toxicology . . . . . . . . . . . . . . . . . 934
8. References . . . . . . . . . . . . . . . . . 934

## 15 Butanediols, Butenediol, and Butynediol

1. 1,4-Diols . . . . . . . . . . . . . . . . . . . 937
2. Other Butanediols . . . . . . . . . . . . 945
3. Toxicology . . . . . . . . . . . . . . . . . 947
4. References . . . . . . . . . . . . . . . . . 948

## 16 Butanols

1. Introduction . . . . . . . . . . . . . . . . 951
2. Physical Properties . . . . . . . . . . . 952
3. Chemical Properties . . . . . . . . . . 954
4. Production . . . . . . . . . . . . . . . . . 956
5. Quality Requirements and Control . 960
6. Storage and Transportation . . . . . . 961
7. Uses . . . . . . . . . . . . . . . . . . . . . . 961
8. Economic Aspects . . . . . . . . . . . . 963
9. Toxicology and Occupational Health 964
10. References . . . . . . . . . . . . . . . . . 967

## 17  2-Butanone

1. Introduction ................ 971
2. Physical Properties ........... 972
3. Chemical Properties .......... 973
4. Production ................ 975
5. Quality, Storage, Transportation ... 978
6. Uses ..................... 979
7. Economic Aspects ............ 980
8. Toxicology ................. 980
9. References ................. 981

## 18  Butenes

1. Introduction ................ 985
2. Physical Properties ........... 986
3. Chemical Properties .......... 986
4. Resources and Raw Materials .... 990
5. Upgrading of Butenes ......... 991
6. Quality Specifications ......... 995
7. Analysis ................... 996
8. Storage and Transportation ..... 996
9. Uses and Economic Data ....... 997
10. Toxicology ................. 1001
11. References ................. 1001

## 19  Butyrolactone

1. Introduction ................ 1005
2. Physical Properties ........... 1005
3. Chemical Properties .......... 1006
4. Production ................ 1007
5. Quality Specifications and Analysis. 1008
6. Storage and Transportation ..... 1008
7. Uses ..................... 1009
8. Toxicology and Occupational Health  1009
9. References ................. 1010

## 20  Caprolactam

1. History ................... 1013
2. Physical Properties ........... 1014
3. Chemical Properties .......... 1015
4. Production ................ 1015
5. Quality Specifications ......... 1037
6. Analysis ................... 1037
7. Storage and Transportation ..... 1039
8. Economic Aspects ............ 1039
9. Toxicology ................. 1040
10. References ................. 1041

## 21  Carbamates and Carbamoyl Chlorides

1. Introduction ................ 1045
2. Salts of Carbamic Acid ........ 1045
3. Esters of Carbamic Acids ....... 1046
4. Carbamoyl Chlorides .......... 1048
5. Toxicology and Occupational Health  1053
6. References ................. 1055

## 22 Carbazole

1. Physical Properties . . . . . . . . . . . . 1057
2. Chemical Properties . . . . . . . . . . 1057
3. Production . . . . . . . . . . . . . . . . . 1058
4. Analysis . . . . . . . . . . . . . . . . . . . 1058
5. Uses and Economic Aspects . . . . . . 1058
6. Derivatives . . . . . . . . . . . . . . . . . 1059
7. Toxicology. . . . . . . . . . . . . . . . . . 1059
8. References. . . . . . . . . . . . . . . . . . 1060

## 23 Carbohydrates

1. Introduction . . . . . . . . . . . . . . . 1061
2. Monosaccharides . . . . . . . . . . . . 1063
3. Oligosaccharides. . . . . . . . . . . . . 1066
4. Polysaccharides . . . . . . . . . . . . . 1068
5. Nomenclature . . . . . . . . . . . . . . 1069
6. Reactions of Carbohydrates . . . . . . 1070
7. References. . . . . . . . . . . . . . . . . . 1082

## 24 Carbonic Acid Esters

1. Introduction . . . . . . . . . . . . . . . 1085
2. Physical Properties . . . . . . . . . . . . 1085
3. Chemical Properties . . . . . . . . . . 1086
4. Production . . . . . . . . . . . . . . . . . 1087
5. Environmental Protection and Toxicology . . . . . . . . . . . . . . . . . . . . 1088
6. Quality Specifications . . . . . . . . . . 1089
7. Analysis . . . . . . . . . . . . . . . . . . . 1089
8. Storage and Transportation . . . . . . 1089
9. Uses. . . . . . . . . . . . . . . . . . . . . . 1089
10. Economic Aspects. . . . . . . . . . . . 1091
11. References. . . . . . . . . . . . . . . . . . 1091

## 25 Carboxylic Acids, Aliphatic

1. Introduction . . . . . . . . . . . . . . . 1096
2. Physical Properties . . . . . . . . . . . . 1096
3. Chemical Properties . . . . . . . . . . 1099
4. Natural Sources . . . . . . . . . . . . . . 1100
5. Production . . . . . . . . . . . . . . . . . 1100
6. Environmental Protection . . . . . . . 1104
7. Quality Specifications and Analysis . 1105
8. Storage and Transportation . . . . . . 1105
9. Uses. . . . . . . . . . . . . . . . . . . . . . 1105
10. Specific Aliphatic Carboxylic Acids . 1107
11. Trade Names, Economic Aspects. . . 1112
12. Toxicology and Occupational Health 1112
13. Derivatives . . . . . . . . . . . . . . . . . 1114
14. References. . . . . . . . . . . . . . . . . . 1117

## 26 Carboxylic Acids, Aromatic

1. Introduction . . . . . . . . . . . . . . . 1120
2. Physical and Chemical Properties . . 1120
3. Production . . . . . . . . . . . . . . . . . 1125
4. Polycarboxylic Acids and Anhydrides 1129
5. Economic Aspects. . . . . . . . . . . . 1134
6. Toxicology. . . . . . . . . . . . . . . . . . 1134
7. References. . . . . . . . . . . . . . . . . . 1134

## 27  Cellulose Esters

1. Inorganic Cellulose Esters . . . . . . . 1138
2. Organic Esters . . . . . . . . . . . . . . . 1167
3. References . . . . . . . . . . . . . . . . . . 1198

## 28  Cellulose Ethers

1. Introduction . . . . . . . . . . . . . . . . 1204
2. Production of Cellulose Ethers . . . . 1211
3. Methyl Cellulose (MC) and Mixed Methyl Cellulose Ethers . . . . . . . . . 1214
4. Ethyl Cellulose Ethers . . . . . . . . . . 1222
5. Hydroxyalkyl Cellulose Ethers . . . . . 1224
6. Carboxymethyl Cellulose (CMC) . . . 1230
7. Other Cellulose Ethers . . . . . . . . . . 1233
8. Analysis . . . . . . . . . . . . . . . . . . . 1236
9. Uses . . . . . . . . . . . . . . . . . . . . . . 1238
10. Economic Facts . . . . . . . . . . . . . . 1243
11. References . . . . . . . . . . . . . . . . . . 1245

# Anthracene

GERD COLLIN, DECHEMA e.V., Frankfurt/Main, Federal Republic of Germany
HARTMUT HÖKE, Weinheim, Federal Republic of Germany

1. Physical Properties ........ 643
2. Chemical Properties ....... 643
3. Production .............. 644
4. Analysis ................ 644
5. Uses ................... 645
6. Toxicology .............. 645
7. References .............. 646

Anthracene [120-12-7], was discovered in coal tar by J. DUMAS and H. A. LAURENT in 1832.

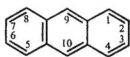

## 1. Physical Properties

Anthracene, $C_{14}H_{10}$, $M_r$ 178.24, mp 218 °C, bp 340 °C (at 101.3 kPa), $d_4^{25}$ 1.252, colorless plates with blue-violet fluorescence, sublimes readily and forms a continuous series of mixed crystals with phenanthrene and carbazole in binary and ternary systems. Anthracene is slightly soluble in benzene, chloroform, and carbon disulfide; less soluble in ether and alcohol; and insoluble in water.

| | |
|---|---|
| Specific heat capacity | 1164 J/kg (at 24 °C) |
| Heat of fusion | 162 kJ/kg |
| Heat of vaporization | 294 kJ/kg |
| Heat of combustion | 40110 kJ/kg (at 25 °C) |

## 2. Chemical Properties

Anthracene is converted to dianthracene by irradiation with UV light. Pyrocondensation gives 2,9-bianthryl and 9,10-dihydroanthracene [1]. Anthracene is readily hydrogenated to 9,10-dihydroanthracene. With homogeneous ruthenium catalysts, anthracene is hydrogenated to 1,2,3,4-tetrahydroanthracene [2]; further hydrogenation yields 1,2,3,4,5,6,7,8-octahydroanthracene. Oxidation of anthracene in the liquid or gas phase gives anthraquinone. Anthracene is primarily halogenated and nitrated in the 9 and 10 positions, and sulfonated in the 1 or 2 position, depending on the reaction conditions. Anthracene undergoes Diels – Alder addition of maleic anhydride and other dienophiles in the 9 and 10 positions.

# 3. Production

High-temperature coal tar contains, on average, 1.5% of anthracene. Continuous tar distillation concentrates the anthracene in the fraction boiling between 300 and 360 °C (anthracene oil I). This oil amounts to ca. 20% of the raw tar and contains about 7% anthracene; it is cooled to 20–30 °C and then centrifuged. About 10–15% of the crude anthracene oil I is obtained as a yellow-green crystalline material with an anthracene content of 20–35% (anthracene residues). These residues, which also contain 40–50% phenanthrene and 12–20% carbazole, are crystallized or distilled to yield a crude product containing 45–55% anthracene (40s anthracene). The crude anthracene then is purified by recrystallization from selective solvents and by distillation. Repeated recrystallization from a three- to fourfold excess of pyridine produces anthracene (95% purity) from the 40s anthracene in 80% yield relative to the anthracene content. Other selective solvents suitable for recrystallization are acetone [3]–[7], acetophenone [8], benzene–methanol [9], dialkyl sulfoxides [10], dialkylformamides [11], N-methylpyrrolidone [12], dimethylacetamide [13], and propylene carbonate [14]. Anthracene can be separated from the higher-boiling carbazole by distillation with lower-boiling hydrocarbon fractions as reflux medium [15], [16] or by azeotropic distillation with ethylene glycol [17], [18]. Azeotropic distillation also is used to separate the anthracene-accompanying tetracene and to obtain very pure anthracene for scintillation counting [19].

Carbazole-free crude anthracene can be prepared by distillation and crystallization of residues obtained from the pyrolysis of hydrocarbon fractions to olefins; however, the anthracene content of these pyrolysis oils generally is less than 1% [20]. The anthracene yield can be increased by selective hydrodealkylation of the aromatic fraction containing methylphenanthrenes and methylanthracenes [21].

Anthracene can be synthesized by hydrogenation of phenanthrene, isomerization of the resulting *sym*-octahydrophenanthrene to *sym*- octahydroanthracene, and subsequent dehydrogenation [22]–[25]. A mixture of *sym*-octahydroanthracene and *sym*-octahydrophenanthrene is obtained by catalytic disproportionation of tetralin [26]–[28]. In addition, anthracene is formed from diphenylmethane in the presence of $HF/BF_3$ at 80 °C [29], [30] and by thermal reaction of *o*-methyldiphenylmethane at ca. 600 °C [31]. These syntheses have no commercial importance to date because sufficient anthracene is obtained from coal tar.

# 4. Analysis

Anthracene is determined quantitatively in the presence of other coal tar constituents by oxidation with chromic acid in glacial acetic acid to produce anthraquinone; by Diels–Alder reaction with maleic anhydride [32], [33]; by gas chromatography [34]; by UV spectroscopy [35], [36]; or by luminescence spectroscopy [37].

# 5. Uses

On a commercial scale, about 30 000 t/a of anthracene are used, almost exclusively for the production of anthraquinone. The use of anthracene as a plasticizer for thermosetting resins and as a light stabilizer for polymers has been proposed. Anthracene is a crystalline organic photoconductor that can be used in electrophotography. In its purest form, anthracene is frequently employed as a scintillant to detect high-energy radiation, e.g., in nuclear physics.

# 6. Toxicology

As a polycyclic aromatic hydrocarbon, anthracene is suspected to be carcinogenic [38]–[45]. This earlier experience involving workers is based on crude anthracene that was contaminated with various other polycyclic aromatic hydrocarbons. In some reference books [40]–[42]—but not in others, e.g., [50]— and official statements [48]anthracene is therefore classified as a carcinogen. Pure anthracene, however, has no appreciable carcinogenic effect. Anthracene is classified as "not classifiable as to its carcinogenicity to humans" by IARC (Category 3) and by EPA (Group D) [51]. As early as 1924, epicutaneous tests on mice showed that the pure substance did not produce any skin tumors [46]. In a study lasting 33 months [43], rats were fed 5 to 15 mg of anthracene daily for 550 d, up to a total dose of 4.5 g per animal. Again, no tumorigenic reactions were observed. Only subcutaneous injections of an oily solution containing 20 mg of anthracene, given 33 times at the rate of one per week, resulted in local development of fibroma, to some extent with sarcoma-like excrescences [43], [44]. An epicutaneous tumor initiation test conducted over 35 weeks on mice with pure anthracene and phorbol ester as promoter resulted in papilloma in a few cases (4 out of 28 animals) [47].

Peroral application of 1.7 g/kg of pure anthracene has no lethal effect on mice [49]. Anthracene is absorbed percutaneously: after topical application of a $^{14}$C-labelled solution in hexane or acetone (ca. 9 µg/cm$^3$) to rat skin, some 50 % were absorbed in 6 d (ca. 29 % were recovered from the urine, ca. 22 % from the feces, and ca. 1 % from tissue, mainly the liver and kidneys); after 1 d 20 % of the dose was already present in the urine (ca. 17 %) and feces (ca. 3 %) [52]. Anthracene can sensitize the skin locally to light. GERARDE proposes a threshold limit value (TLV) of 0.1 mg/m$^3$ [45].

# 7. References

General References

*Beilstein,* **5**, 657; **5 (1)**, 321; **5 (2)**, 569; **5 (3)**, 2123.

H. J. V. Winkler: *Der Steinkohlenteer und seine Aufarbeitung,* Verlag Glückauf, Essen 1951, p. 174–181.

H.-G. Franck, G. Collin: *Steinkohlenteer,* Springer Verlag, Berlin – Heidelberg – New York 1968, p. 56–58, 136–137.

Rütgerswerke, *Erzeugnisse aus Steinkohlenteer,* Frankfurt / Main 1958.

Specific References

[1] S. E. Stein, L. L. Griffith, R. Billmers, R. H. Chen, *J. Org. Chem.* **52** (1987) 1582–1592.
[2] Rütgerswerke, DE 3303742, 1983 (B. Fell, G. Maletz).
[3] Ukrainian Scientific-Research Institute of Coal Chemistry, SU 386889, 1971 (V. E. Privalov, K. A. Belov, E. I. Vail, Y. T. Rezunenko).
[4] B. Karabon, A. Zin, K. Mnich, *Koks Smola Gaz* **19** (1974) 33–37.
[5] M. S. Litvinenko, L. D. Gluzman, A. A. Rok, S. N. Kipat, V. P. Bogunets, N. I. Zhuravskaya, R. M. Zil'berman, L. I. Didenko, *Koks Khim.* 1976, no. 3, 33–36; *Coke Chem.* 1976, no. 3, 38–42.
[6] L. S. Kuznetsova, V. I. Borodin, A. A. Stepanenko, A. A. Tolochko, V. N. Nazarov, *Koks Khim.* 1978, no. 1, 38–40, *Coke Chem.* 1978, no. 1, 50–52.
[7] B. Karabon, A. Zin, *Koks Smola Gaz* **26** (1980) 95–98.
[8] Rütgerswerke, DE 2020973, 1970 (H. Buffleb, J. Falenski, H.-G. Franck, J. Turowski, F. Melichar, G. Collin, M. Zander).
[9] I. Jurkiewicz, K. Wiszniowski, *Koks Smola Gaz* **5** (1960) 117–121.
[10] B. Marciniak, *Mol. Cryst. Liq. Cryst.* **162 B** (1988) 301–313.
[11] Union Rhein. Braunkohlen Kraftstoff, DE 1046002, 1957 (F. Hübenett, G. Altena).
[12] J. Polaczek, Z. Lisicki, T. Tecza, *Przem. Chem.* **60** (1981) 169–171.
[13] Charkovskij politechnitsheskij institut imeni V. T. Lenina, Ukrainskij nautshnoissledovatelskij Uglechimit sheskij institut, DE-OS 2626361, 1976 (V. J. Privalov, J. I. Vail, L. S. Kusnezova, K. A. Belov, I. M. Nosalevich, I. V. Romanov).
[14] G. Kesicka, S. Trybuta, *Przem. Chem.* **65** (1986) 145–148.
[15] A. Zboril, J. Ruzek, CS 150737, 1971.
[16] J. Komurka, I. Koropecky, V. Vlcek, J. Jelinek, J. Mixa, J. Ruzek, CS 164058, 1973.
[17] Soc. Chim. Gerland, FR 976 773, 1948 (A. J. M. Saunier).
[18] Chem. Systems, DE-OS 2123342, 1971 (G. M. Sugarman).
[19] R. Sizmann, *Angew. Chem.* **71** (1959) 243–245.
[20] Rütgerswerke, DE 2000351, 1970 (H. Buffleb, H.-G. Franck, J. Turowski, G. Collin, M. Zander).
[21] Rütgerswerke, DE 1906807, 1969 (H. Buffleb, H.-G. Franck, R. Oberkobusch, J. Turowski, G. Collin, M. Zander).
[22] Koppers Co., US 3389188, 1966 (W. A. Michalowicz).
[23] G. Kölling, *Erdöl Kohle Erdgas Petrochem.* **20** (1967) 726–729.
[24] K. Handrick, W. Hodek, F. Mensch, *DGMK-Compendium 78/79,* **2** (1978) 1089–1106.
[25] Bergwerksverband, DE 2952062, 1979 (K. Handrick, G. Kölling, F. Mensch).
[26] Tetralin, DE 333158, 1919.

[27] Sun Oil, US 3320332, 1964 (A. Schneider).
[28] Sun Oil, DE-OS 1493030, 1965 (R. D. Bushick, G. Mills).
[29] F. Meyer, D. Hausigk, G. Kölling, *Liebigs Ann. Chem.* **736** (1970) 140–141.
[30] Bergwerksverband, DE 1951127, 1969 (F. Meyer, G. Kölling, D. Hausigk).
[31] Bayer, DE-OS 2218004, 1972 (H. Wolz, R. Wenzel, M. Martin, G. Scharfe).
[32] H. P. Kaufmann, J. Baltes, *Fette Seifen* **43** (1936) 93–95.
[33] E. Funakubo, Y. Matsumoto, M. Fujiura, I. Kawanshi, S. Hiroike, *Brennst. Chem.* **40** (1959) 377–383.
[34] M. Zander, H. D. Sauerland, *Erdöl Kohle Erdgas Petrochem.* **25** (1972) 526–530.
[35] H. Schmidt, *Erdöl Kohle Erdgas Petrochem.* **19** (1966) 275–278.
[36] F. P. Hazlett, R. B. Hannan, J. H. Wells, *Anal. Chem.* **22** (1950) 1132–1136.
[37] M. Zander, *Angew. Chem. Int. Ed. Engl.* **4** (1965) 930–938.
[38] K. B. Lehmann, F. Flury: *Toxicology and Hygiene of Industrial Solvents,* Williams and Wilkins Co., Baltimore 1943.
[39] W. C. Hueper: *Occupational Tumors and Allied Diseases,* C. C. Thomas, Springfield 1942.
[40] National Institut for Occupational Safety and Health (NIOSH): *Registry of Toxic Effects of Chemical Substances,* US Government Printing Office, Washington, D.C., July 1982, p. 1277.
[41] N. I. Sax: *Cancer-Causing Chemicals,* Van Nostrand Reinhold Comp., New York 1981, p. 288.
[42] N. I. Sax: *Dangerous Properties of Industrial Materials.* 5th ed., Van Nostrand Reinhold Comp., New York 1981, p. 288.
[43] D. Schmähl, *Z. Krebsforsch.* **60** (1955) 697.
[44] H. Druckrey, D. Schmähl, *Naturwissenschaften* **42** (1955) 159.
[45] H. W. Gerarde: *Toxicology and Biochemistry of Aromatic Hydrocarbons.* Elsevier, Amsterdam 1960, p. 240–248.
[46] E. L. Kennaway, *J. Ind. Hyg.* **5** (1924) 462.
[47] J. D. Scribner, *J. Natl. Canc. Inst.* **50** (1973) 1717.
[48] Commission of the European Communities: "Memorandum on Occupational Diseases of 23 July 1962", *Off. J. Eur. Comm.* **1962,** no. 80, 2188–2193.
[49] P. A. Nagornyi, *Gig. Tr. Prof. Zabol.* **13** (1969) 59.
[50] M. O. Amdur et al. (Eds.): *Casarett and Doull's Toxicology,* 4th ed., Pergamon Press, 1991, p. 173
[51] M. Hassauer, F. Kalberlah, J. Oltmanns, K. Schneider : *Basisdaten Toxikologie,* UBA Report No. 102 03 443, E. Schmidt, Berlin 1993.
[52] J. J. Yang, T. A. Roy, C. R. Mackerer, *Toxicol. Ind. Health* **2** (1986) 79–84.

# Anthraquinone

AXEL VOGEL, Bayer AG, Leverkusen, Federal Republic of Germany

| | | | | | |
|---|---|---|---|---|---|
| 1. | Introduction | 649 | 3.5. | Styrene Process | 658 |
| 2. | Properties | 650 | 3.6. | Environmental Considerations | 658 |
| 3. | Production | 652 | 4. | Purification, Quality Requirements, and Analysis | 659 |
| 3.1. | Oxidation of Anthracene with Chromic Acid | 653 | 5. | Uses | 659 |
| 3.2. | Vapor-phase oxidation of Anthracene with air | 654 | 6. | Economic Aspects | 660 |
| 3.3. | Naphthalene Process | 655 | 7. | Toxicology | 661 |
| 3.4. | Synthesis from Phthalic Anhydride and Benzene | 656 | 8. | References | 661 |

# 1. Introduction

Anthraquinone [84-65-1] is the parent compound for a large palette of anthraquinone dyes and so is the most important starting material in their production. Furthermore, anthraquinone is gaining importance as a catalyst in the pulping of wood.

**1**

In 1835, anthraquinone was prepared for the first time by LAURENT, via oxidation of anthracene, which at first went largely unnoticed. Its special importance for the dye industry was recognized only in 1868 when GRAEBE and LIEBERMANN prepared anthracene from alizarin (1,2-dihydroxyanthraquinone) and, in turn, alizarin via anthraquinone. This laid the groundwork for the synthesis route of this – at the time – important dye. Since then, science and technology have extended the chemistry of anthraquinone swiftly, first by the discovery of new mordant dyes and acid dyes, later by the introduction of vat dyes and disperse dyes, and, since 1950, by the anthraquinone reactive dyes. At the same time, anthraquinone chemistry has acted as a stimulus for the rapid extension of the production processes of chemicals, such as oleum and chlorine.

Since 1970, the industrial chemistry of anthraquinone has been undergoing a revolution worldwide. Initiated by a growing shortage of anthracene accompanied by an increasing demand, new procedures of synthesis using naphthalene or styrene as starting materials were developed. Moreover, further processing of anthraquinone is changing; α-substituted anthraquinones should no longer be prepared by mercury-catalyzed sulfonation but by new nitration procedures.

A completely new use for anthraquinone was established by HOLTON in 1977, when he discovered the catalytic effect of anthraquinone on the alkaline pulping of wood. This branch of anthraquinone chemistry is experiencing a rapid development and has good chances of outdistancing the needs of the dye industry. The demand for anthraquinone has risen continuously since 1870. At present the annual demand is ≈ 30 000 t.

## 2. Properties

**Physical Properties.** Anthraquinone [*84-65-1*], $C_{14}H_8O_2$, $M_r$ 208.20, mp 287 °C, bp 377 °C, crystallizes as thin, light yellow to pale yellow needles; it is almost colorless if finely divided, as when precipitated from sulfuric acid. The faint colors of the crystals are effected strongly by the granular structure and in commercial products often are covered up by trace impurities.

Crystallographic class: monoclinic, space group P $2_1$/a. Dimensions of the unit cell: $a$ = 1.583 nm, $b$ = 0.397 nm, $c$ = 0.789 nm, $\beta$ = 102.5
Density of the crystals at 20 °C: 1.438 g/cm$^3$, of the melt (at 293 °C): 1.067 g/cm$^3$
Vapor pressure: 10.0 kPa at 286 °C, 20.0 kPa at 300 °C, 48.0 kPa at 340 °C
Molar heat capacity at constant pressure $c_p$ (solid): 265 J mol$^{-1}$ K$^{-1}$
Enthalpy of formation: −179 kJ/mol
Enthalpy of combustion at 25 °C: 6449 kJ/mol
Melting enthalpy at 287 °C: 32.57 kJ/mol
Flash point: 185 °C

The UV spectrum exhibits four bands – 251 nm ($\epsilon_{max}$ 54000), 279 nm ($\epsilon_{max}$ 17600), 321 nm ($\epsilon_{max}$ 4800), and 377 nm ($\epsilon_{max}$ 110) – each with pronounced vibrational structure. Unlike solutions of anthracene, most solutions of anthraquinone do not show fluorescence. The IR spectrum has very few bands (characteristic bands at $v_{C \to O}$ = 1680 cm$^{-1}$) because of the high symmetry of the compound.

Anthraquinone is sublimed easily without decomposition, a property exploited industrially for purification.

Anthraquinone is practically insoluble in water (0.006 g/L at 50 °C) and nearly insoluble to sparingly soluble in organic solvents at room temperature. However, solubility in organic solvents increases with temperature (in ethanol 0.05 g per 100 g ethanol dissolves at 18 °C, 2.25 g per 100 g at 78 °C; in toluene 0.19 g per 100 g toluene at 15 °C, 2.56 g at 100 °C). Therefore anthraquinone generally can be recrystallized from high-boiling solvents, particularly from the polar solvents pyridine, aniline, nitrobenzene, formamide, or dimethylformamide. Anthraquinone dissolves in 90% sulfuric acid giving a yellow to orange solution, and in oleum (20% $SO_3$), giving a red solution. It can be precipitated from these solutions in a finely divided form by dilution with water. The solubility increases with the concentration of the sulfuric acid. Anthraquinone behaves similarly in hydrofluoric acid; however, the solubility in almost 100% phosphoric acid is relatively small.

Anthraquinone can form explosive dust mixtures with air. Friction causes anthraquinone to become electrostatically charged.

**Chemical Properties.** Anthraquinone exhibits an extraordinary thermal stability, even in the presence of oxidizing agents. Therefore it can be prepared by oxidizing anthracene in the vapor phase. A yield of 90% is obtained if conversion is complete.

Anthraquinone is attacked by oxidizing agents only under vigorous conditions. For example, it can be converted to alizarin by fusion with sodium hydroxide and sodium chlorate. Anthraquinone can be converted to the polyhydroxyanthraquinones alizarin, quinizarin, anthrarufin, and purpurine by oxidation with concentrated sulfuric acid in the presence of boric acid, persulfates, or metaarsenic acid at temperatures generally above 200 °C. The oxidation can be carried out with oleum at a lower temperature that depends on the $SO_3$ content. Under the usual conditions, concentrated nitric acid does not oxidize anthraquinone, nitration taking place instead. Several mixtures of anthraquinone with 94 – 100% nitric acid with specific anthraquinone concentrations are explosivy.

On the other hand, anthraquinone can be reduced easily by ordinary reducing agents or by hydrogen in the presence of a catalyst. The standard potential (in aqueous ethanolic hydrochloric acid) for the anthraquinone–anthrahydroquinone equilibrium is only 0.155 V. Especially the quinone nucleus is reduced. Sodium hyposulfite in alkaline solution reduces anthraquinone to a blood-red sodium salt of anthrahydroquinone, which rearranges to its tautomer oxanthrone (9-keto-10-hydroxy-9,10-dihydroanthracene) in acidic solution. More vigorous reduction (tin – hydrochloric acid or catalytic hydrogenation at high pressure or high temperature) leads to anthrone (**2**) [*90-44-8*], which is in equilibrium with its tautomer anthrol (**3**) (9-hydroxyanthracene) [*529-86-2*], and then to anthracene and hydroanthracene.

If zinc or sodium hydroxide is used, dianthranol is obtained via dimerization.

The reduction can be restricted to the two outer nuclei by choice of reaction conditions. 1,2,3,4-Tetrahydroanthraquinone [*28758-94-3*] or octahydroanthraquinone

[23585-26-4] can be obtained. Reactions of anthraquinone with the usual ketone reagents take place with great difficulty or not at all.

If anthraquinone is heated with glycerol, iron powder, and sulfuric acid, condensation takes place at positions 1 and 9, providing the commercially important benzanthrone [82-05-3].

Chlorine halogenates anthraquinone in oleum in steps, depending on the proportion used, mainly at the α position. Heating anthraquinone with antimony pentachloride or a similar chlorine-releasing reagent produces chlorinated products, from monochloro- up to perchloroanthraquinone.

Sulfuric acid or oleum can sulfonate and/or hydroxylate anthraquinone. Concentrated sulfuric acid sulfonates only above 200 °C; usually, anthraquinone is sulfonated with 20–45% oleum at 140–160 °C. The major reaction products are β-sulfonic acids, i.e., anthraquinone-2-sulfonic acid [84-48-0], as well as anthraquinone-2,6-disulfonic acid [14486-58-9], and anthraquinone-2,7-disulfonic acid [84-49-1]. In the presence of mercury, however, the sulfonation takes place almost exclusively at the α position, providing anthraquinone-1-sulfonic acid [82-49-5], anthraquinone-1,5-disulfonic acid [117-14-6], and anthraquinone-1,8-disulfonic acid [82-48-4].

The nitration of anthraquinone requires vigorous conditions, and provides mainly α-nitroanthraquinones along with 10–20% β-nitro- and α,β-dinitroanthraquinones. Both mono- and dinitration take place, so that impure nitration products are obtained. These can be separated into 1-nitroanthraquinone, 1,5-dinitroanthraquinone, and 1,8-dinitroanthraquinone only by extensive processing. Trinitration of anthraquinone does not take place.

1-Nitroanthraquinone [82-34-8] can be prepared by nitrating anthraquinone in mixed acid ($H_2SO_4$-$HNO_3$) with excess nitric acid, in nearly 30 mol 94–98% nitric acid, or with an almost stoichiometric amount of the mixed acid in the presence of chlorinated aliphatic hydrocarbons. The crude 1-nitroanthraquinone is purified by recrystallization and, if necessary, by distillation; likewise it is processed directly to crude 1-aminoanthraquinone [82-45-1], which is then purified by distillation.

1,5- and 1,8-Dinitroanthraquinone are prepared by exhaustive nitration of anthraquinone in excess, nearly anhydrous mixed acid. Pure 1,5-dinitroanthraquinone [82-35-9] and 1,8dinitroanthraquinone [129-39-5] are obtained by recrystallizing the crude dinitroanthraquinone mixtures from nitrobenzene.

# 3. Production

Four processes are used in the industrial production of anthraquinone today:

1) Oxidation of anthracene with chromic acid
2) Vapor-phase oxidation of anthracene with air
3) Naphthalene process
4) Synthesis from phthalic anhydride and benzene

Anthraquinone is produced from anthracene, where this is available from coal tar, either by oxidation with chromic acid in 48% sulfuric acid or by oxidation with air in the vapor phase. The oxidation with chromic acid is competitive, provided that the chromium(III) sulfate lye formed can be processed to tanning agents. Anthracene with purity of $\geq 94\%$ is required for both oxidation processes; crude anthracene from coal tar must be purified by recrystallization.

About 85% of world production today is based on the oxidation of anthracene. Since the mid-1970s, anthracene production has fallen continuously, creating a supply shortage. Therefore, the complex naphthalene process is gaining in importance. There is an adequate supply of naphthalene in coal tar. If necessary, additional naphthalene can be isolated from the residual oils of gasoline reforming, a process common in the United States.

In the new naphthalene processes developed in Japan by Kawasaki and in Europe by Bayer anthraquinone is synthesized in three steps: Naphthoquinone is prepared by vapor-phase oxidation with air. Butadiene is added to this naphthoquinone in a Diels-Alder reaction. The tetrahydroanthraquinone formed is oxydehydrogenated. In this naphthalene process a significant amount of phthalic anhydride byproduct is produced.

In the synthesis of anthraquinone from phthalic anhydride and benzene approximately 1.4 t of aluminum chloride and 4–6 t of sulfuric acid per ton of anthraquinone are used. This procedure is used in areas where anthracene is not available in sufficient amounts from coal tar. It may become important again, as a result of the shortage of anthracene. Phthalic anhydride has become available at low prices and in sufficient quantities following the introduction of the economical oxidation of naphthalene or *o*-xylene with air.

In the early 1970s another interesting anthraquinone process was developed by BASF. Styrene is first dimerized to 1-methyl-3-phenylindane in the presence of an acid catalyst, which is then converted to anthraquinone in the vapor phase by oxidation with air.

## 3.1. Oxidation of Anthracene with Chromic Acid

Today the classic process of oxidizing anthracene with chromic acid is still carried out industrially on a large scale. In combination with the manufacture of tanning agents this process involves virtually no expense for the oxidizing agent. The batch process is still in use.

$$\text{anthracene} + Na_2Cr_2O_7 + 4\,H_2SO_4 \xrightarrow{105\,°C}$$

$$\text{anthraquinone} + Na_2SO_4 + Cr_2(SO_4)_3 + 5\,H_2O$$

Pure, finely pulverized 94–95% anthracene (2600 kg) is pasted with water (9000 kg) in a 30000-L mixing vessel. A wetting agent is added. For oxidation, portions of 48% sulfuric acid and 20% sodium dichromate solution are added alternately; the reaction begins at 50–60 °C. The temperature increases to 100–105 °C but should not be allowed to increase any further. The temperature is controlled by regulating the amount of reagents added. Foaming is severe. A total of 10 200 kg of 48% sulfuric acid and 23 500 kg of 20% sodium dichromate solution are used. At the end of the reaction there should be a small excess of oxidizing agent, which is reduced with hydrogen sulfite. The oxidation takes 30–36 h [1]. The reaction mixture is filtered through a suction filter while being stirred, washed, suspended with water, and filtered into a chambered rotary filter. A yield of ≈ 3000 kg of 95% crude anthraquinone is obtained. The product is dried, or it is recrystallized from nitrobenzene, in which case the retained water is evaporated first. A purity of 99% is achieved.

## 3.2. Vapor-phase oxidation of Anthracene with air

The vapor-phase oxidation of anthracene with air was described first by A. WOLF in 1916. The catalysts are similar to those for the oxidation of naphthalene or o-xylene to phthalic anhydride: vanadium compounds, mainly iron vanadate [2] or vanadic acid doped with substoichiometric amounts of alkali metal or alkaline-earth metal ions [3].

$$\text{anthracene} + 3/2\,O_2 \longrightarrow \text{anthraquinone} + H_2O$$

The conversion of the anthracene is nearly quantitative. The major byproduct is a small quantity of phthalic anhydride, which is separated easily. The advantage of this procedure is that it produces anthraquinone of 99% purity. It can be converted further without purification.

Details on a plant constructed by Aziendi Colori Nazionali Affini, Milan, were published [4]:

Anthracene is evaporated with a preheated air–water vapor mixture, and the gas current is mixed carefully with more air (danger of explosion!) and led into a catalytic furnace from the bottom at 325 °C. The catalytic furnace consists of 1400 L of catalyst piled in layers; temperature is controlled by tubes in which pressurized water circulates through these layers. The lower part of the furnace, where the reaction takes place, is cooled to 390 °C; the upper part is heated to 339 °C. The catalyst consists of iron vanadate prepared from ammonium vanadate and iron(III) chloride. The air flow rate is 2150 m³/h, each cubic meter containing 20 g of 94 % anthracene. The gases leaving the catalytic furnace first go through heat exchangers and then through cooling towers, cooling chambers, and dust filters. The anthraquinone produced has an average purity of 99.6 % The iron vanadate catalyst is prepared as follows [2]: 2.4 kg of red iron(III) oxide is dissolved by heating it in 9.1 L of 30 % hydrochloric acid, 9.6 L of water, and 120 g of potassium sulfate. The resulting solution is mixed with a solution of 30.1 kg of ammonium vanadate in 600 L of water at 60 °C. The solution is made alkaline with ammonia and the water evaporated until precipitation begins. The pregnant iron vanadate solution is evaporated onto pumice at 110–130 °C in a heatable ball mill. The catalyst is calcined at 330–350 °C. The catalyst should be yellow brown but not dark brown.

After a period of time the activity of the catalyst falls off, resulting in a decline in the yield of anthraquinone and an increase in phthalic anhydride. To keep the activity of the catalyst constant or to reactivate it, a small amount of ammonia is added to the oxidation air or the exhausted catalyst is washed with a dilute alkali solution [5].

## 3.3. Naphthalene Process

The naphthalene process developed by Kawasaki Kasei Chemicals [6] consists of three steps. In the first step naphthalene is oxidized in the gas phase with air on a vanadium pentoxide catalyst to produce the relatively unstable naphthoquinone (**4**). Large amounts of phthalic anhydride form at the same time. Because most of the naphthalene reacts, recycling of unreacted naphthaline is not necessary [7]–[10].

The hot reaction mixture is precooled in a gas cooler and then quenched and washed with water in a tower. The naphthoquinone is obtained mixed with phthalic acid anhydride (or phthalic acid) as a suspension in water. The naphthoquinone is extracted with an aromatic solvent, for example xylene or tetrahydroanthraquinone, to separate it from the phthalic acid. The small residue of acid is carefully removed by washing with dilute alkali [11]–[13].

In the second step the naphthoquinone is reacted with butadiene in a Diels-Alder reaction to give 1,4,4 a,10 a-tetrahydroanthraquinone (**5**). The Diels-Alder product is separated from unreacted naphthoquinone by extraction with aqueous alkali, which converts the quinone into a water-soluble alkali-metal salt. The organic phase, which contains naphthoquinone, is recycled to the first step [7], [8], [14].

In the third step the aqueous tetrahydroanthraquinone solution is reacted with air. The anthraquinone that forms is insoluble in aqueous alkali and can be isolated by filtration. The aqueous alkali is recycled into the tetrahydroanthraquinone extraction step [15]. The anthraquinone can be purified further by vacuum distillation [6].

Phthalic acid is isolated from the aqueous phthalic acid solution by crystallization and filtration. It is heated to convert it to phthalic acid anhydride. The phthalic acid anhydride can be further purified by distillation.

Kawasaki Kasei Chemicals operates a plant based on the naphthalene process that produces 3000 t/a. The plant has been in operation since 1980.

In Bayer's variation of the naphthalene process, naphthalene is oxidized with a mixture of air and recycled reaction gas on a vanadium pentoxide catalyst to give naphthoquinone, phthalic acid anhydride and unreacted naphthalene [16]–[18]. The products of the oxidation along with unreacted naphthalene are quenched and condensed into a liquid reaction mixture, which is reacted without purification with butadiene [19], [20]. The Diels-Alder product mixture that contains tetrahydroanthraquinone is reacted with air. The tetrahydroanthraquinone is oxydehydrogenated to give anthraquinone. The naphthalene is removed by distillation [20]–[22]. The two end products, anthraquinone and phthalic acid anhydride, are separated by fractional distillation [23].

## 3.4. Synthesis from Phthalic Anhydride and Benzene

In the first step o-benzoylbenzoic acid (**6**) is prepared from phthalic anhydride and benzene by a Friedel-Crafts reaction. In the second step the *o*-benzoylbenzoic acid is cyclized to anthraquinone by heating with concentrated sulfuric acid. The primary product of the Friedel-Crafts reaction is an aluminum chloride complex of the *o*-benzoylbenzoic acid, which can harden easily to form a compact mass in the reaction vessel. Several methods have been proposed to overcome this problem. Based on a process described by G. HELLER already in 1906, condensation is carried out in excess benzene to give yields of more than 95%. Patents issued to Klipstein & Sons and the I. G. Farbenindustrie from 1923 to 1927 describe a solvent-free process carried out in ball mills. The problem posed by the reaction mixture's baking can be reduced by adding ethylene glycol or 1,4-butanediol, which causes a reduction in the necessary excess benzene or makes a solvent-free process possible. For both processes batch processing is used even today.

Phthalic anhydride, benzene, and aluminum chloride in a molar ratio of $1:1:2$ are allowed to react below 45 °C in iron ball mills fitted with hollow axles for the addition of benzene and the removal of the hydrogen chloride formed during the reaction, or in a vessel equipped with a strong horizontal agitator, possibly propelled from the bottom. The reaction mass, which is heterogeneous at first, liquefies, then gradually becomes viscous as hydrogen chloride is continuously lost, and eventually forms a powder. During the reaction there is extensive foaming, producing several times the original volume, a fact that must be considered when the charge is measured out. The reaction is complete after 1 mol of hydrogen chloride per mol phthalic anhydride has been evolved. The reaction product is placed into dilute acid. The precipitated *o*-benzoylbenzoic acid is separated, washed, and dried, providing a yield of more than 95 %.

Pure *o*-benzoylbenzoic acid [*85-52-9*] crystallizes from water with 1 mol of water of hydration in triclinic crystals, *mp* 94 °C. The anhydrous acid crystallizes from xylene, *mp* 127 °C.

In the second step it is also possible to start with the aluminum chloride complex formed in the first step and cleave 1 mol of hydrogen chloride by heating to 200 °C. A good yield of anthraquinone is obtained. In practice, however, the condensation of the precipitated dried *o*-benzoylbenzoic acid with sulfuric acid is preferred.

The acid is dissolved in three to four times its weight in 95–98 % sulfuric acid and heated to 115–140 °C. The duration of heating depends on the quantity and concentration of the sulfuric acid and on the temperature. After completion of the reaction the anthraquinone is precipitated with water, filtered, and dried. A portion of the product also can be obtained in coarsely crystallized form by cooling the sulfuric acid solution directly. The yield is almost quantitative.

The condensation solution can be used directly for the manufacture of benzanthrone or anthraquinonesulfonic acids.

## 3.5. Styrene Process

This process was developed in pilot by BASF.

Styrene dimerizes in the presence of acid catalysts, such as sulfuric acid [24], [25], phosphoric acid, or boric acid [26], yielding primarily 1,3-diphenyl-2-butene, which cyclizes further on the same catalyst to 1-methyl-3-phenylindane [6416-39-3]. The yields obtained in this step are 85 – 90 %.

1-Methyl-3-phenylindane is converted directly to anthraquinone by oxidation with air in the vapor phase [27].

Basically, the same catalysts are used as those employed in the gas-phase oxidation of anthracene with air, i.e., vanadium compounds, primarily vanadium pentoxide in combination with other oxides, such as thallium oxide and antimony oxide [28].

Yields of $\approx$ 77 % are obtained in the oxidation stage. Byproducts are still attached to the precipitated reaction product, and further purification is necessary.

## 3.6. Environmental Considerations

Chromium salts should not get into the waste water from the chromic acid process. The chromium(III) sulfate lye is disposed of by processing it to tanning agents. The anthraquinone must be carefully washed free of chromium with small amounts of wash water.

In the gas-phase oxidation of anthracene with air and the naphthalene process, on the other hand, the emphasis is placed on purifying the large quantity of waste air.

The production of anthraquinone via the Friedel-Crafts reaction from phthalic anhydride and benzene is accompanied by a release of waste water containing 1.4 t of aluminum chloride and 4 – 6 t of sulfuric acid per ton of anthraquinone. Moreover, carcinogenic benzene must be eliminated from the waste air by thermal posttreatment.

## 4. Purification, Quality Requirements, and Analysis

Anthraquinone is most often purified by recrystallization, preferably from nitrobenzene. Other purification methods include sublimation or vacuum destillation.

At present, a degree of purity of at least 99% is required for the major uses of anthraquinone, i.e., production of nitroanthraquinones and anthraquinonesulfonic acids. A content of 98–99% is sufficient for use as catalyst in wood pulping. The admission of anthraquinone by the U.S. Food & Drug Administration (FDA) as catalyst in the manufacture of food packing paper requires a content of at least 98% [29].

Today the main method of analyzing the content is gas chromatography: The other components and solvents, which are present in ppm quantities, are detected directly. High-pressure liquid chromatography is another suitable method of analysis.

According to the classic analytical processes, still used today, the content is determined (1) by reduction to soluble anthrahydroquinone, filtration, back-oxidation, and weighing out the filtered anthraquinone or (2) by separating the anthraquinone by column chromatography and analyzing the eluate by UV spectroscopy.

## 5. Uses

Anthraquinone serves as the basis for the production of a large number of acid and base dyes, vat dyes, disperse dyes, and reactive dyes.

Anthraquinone-1-sulfonic acid, anthraquinone-2-sulfonic acid, anthraquinone-1,5-disulfonic acid, and anthraquinone-1,8-disulfonic acid as well as 1-nitroanthraquinone, 1,5-dinitroanthraquinone, 1,8-dinitroanthraquinone, anthrone, and benzanthrone are prepared in the first processing step. For environmental reasons the production of anthraquinone-2-sulfonic acid and anthraquinone-1,5- and anthraquinone-1,8-disulfonic acids is being replaced increasingly by the production of 1-nitroanthraquinone and 1,5- and 1,8-dinitroanthraquinone.

In the second stage of processing, the sulfonic or nitro groups generally are replaced by ammonia, amine, chlorine, alcoholate, phenolate, or other nucleophilic agents or are reduced to an amino group. Some of the anthraquinone dyes, however, are prepared from anthraquinone derivatives. These are synthesized not from anthraquinone, but from smaller molecules.

On a small scale, anthraquinone is used in the manufacture of denaturants to protect seed from crows, in the modification of stand oils, to brighten up colophoniy, and in the manufacture of hydrogen peroxide.

In recent years, the paper and pulp industry has become another rapidly expanding market for anthraquinone [30]. Anthraquinone is a redox catalyst in the production of pulp; the intermediate anthrahydroquinone catalyzes the alkaline hydrolysis of the

**Table 1.** Anthraquinone production capacities

| | |
|---|---|
| **Western Europe** | |
| Bayer | 11000 t/a |
| ICI | 2500 t/a |
| Ciba-Geigy | 2200 t/a |
| ACNA | 1000 t/a |
| Ugine Kuhlmann | 800 t/a |
| L.B. Holliday | 200 t/a |
| Yorkshire Chem. | 200 t/a |
| | 17900 t/a |
| **United States** | |
| Toms River Chem. | 1500 t/a |
| American Cyanamid | 400 t/a |
| | 1900 t/a |
| **Japan** | |
| Kawasaki Kasei | 3000 t/a |
| Nihon Joryn Kogyo | 2000 t/a |
| | 5000 t/a |
| **Eastern Europe** | |
| USSR | 2600 t/a |
| Poland | 1800 t/a |
| Czechoslovakia | 1200 t/a |
| | 5600 t/a |
| **Other Countries** | |
| Indian Dyestuff | 3000 t/a |
| Amar Dye (India) | 250 t/a |
| China | 400 t/a |
| | 3650 t/a |
| **Worldwide** | 34050 t/a |

polymeric lignin. When 0.03 – 0.05 % anthraquinone is added, the yield of pulp is increased 2.5 – 4 % and the cooking time is reduced up to 30 % [31].

The worldwide production capacity is adequate to cover the actual demand for anthraquinone [32].

# 6. Economic Aspects

The production capacity for anthraquinone worldwide is at present 34 000 t/a. About 50 % of this capacity is in Western Europe.

The most important manufacturers of anthraquinone and their production capacities are listed in Table 1.

# 7. Toxicology

Based on present knowledge, anthraquinone, unlike the "emodins" (hydroxyanthraquinone glycosides) from anthraquinone drugs [33], is biologically completely "inert," i.e., inactive, presumably as a consequence of its insolubility in water and lipids.

There is no toxicologic information about anthraquinone. Even in the *Toxic Substances List* [34] no $LD_{50}$ is found for anthraquinone. In contrast to benzoquinone, which causes severe local irritation and is included in the list of occupational hazards because of its damaging effect on the cornea, a fine dust of anthraquinone to which emulgators were added had no greater effect on the eyes of rabbits than a talcum suspension, as was shown in a BASF study.

An MAK value of 10 mg/m$^3$ was recommended [35]; therefore, anthraquinone dust is considered the same as inert dust.

# 8. References

General References

    *Beilstein*, **VII**, 781, **VII (1)**, 407, **VII (2)**, 709, **VII (3)**, 4059, **VII (4)**, 2556.
    E. de Barry Barnett: *Anthracene and Anthraquinone*, Baillière, Tindall & Co., London 1921.
    J. Houben: *Das Anthracen und die Anthrachinone*, Thieme Verlag, Leipzig 1929.

Specific References

[1]   Fiat Final Rep. no. 1313, vol. **2**, p. 19.
[2]   Fiat Final Rep. no. 1313, vol. **1**, p. 332.
[3]   Ciba, DE 1016694, 1954.
[4]   Bios Misc. Rep. no. 104.
[5]   Ciba, DE 1020617, 1954.
[6]   Kawasaki Kasei Chemicals Ltd., Research and Development in Japan awarded the Okochi Memorial Price, 1980, Okochi Memorial Foundation.
[7]   Kawasaki, JP 5108256, 1974.
[8]   Kawasaki, JP 5108257, 1974.
[9]   Kawasaki, JP 5322559, 1978.
[10]   Kawasaki, DE 3033341, 1980.
[11]   Kawasaki, JP 5251356, 1975.
[12]   Kawasaki, GB 2039897, 1978.
[13]   Kawasaki, JP 5422246, 1979.
[14]   Kawasaki, US 4412954, 1981.
[15]   Kawasaki, JP 5652434, 1981.
[16]   Bayer, DE 2532422, 1975.
[17]   Bayer, DE 2453232, 1974.
[18]   Bayer, DE 2532365, 1975.
[19]   Bayer, DE 2532388, 1975.

[20] Bayer, DE 2218316, 1972.
[21] Bayer, DE 2245555, 1972.
[22] Bayer, US 4284576, 1975.
[23] Bayer, DE 2532450, 1975.
[24] J. Risi, D. Gauvin, *Can. J. Res. Sect. B.* **14** (1936) 255.
[25] P. E. Spoerri, M. J. Rosen, *J. Am. Chem. Soc.* **72** (1950) 4918.
[26] BASF, DE 2064099, 1970.
[27] BASF, DE 1934063, 1969.
[28] BASF, DE 2135421, 1971.
[29] A. Budzinski, *Chem. Ind. (Düsseldorf)* **33** (1981) 332.
[30] H. H. Holton, *Pulp Pap. Can.* **78** (1977) 10.
[31] G. Gellerstedt, *Kem. Tidskr.* **58** (1979) no. 2.
[32] *Eur. Chem.* 1984, no. 27, 429.
[33] H. Auterhoff, *Arzneim. Forsch.* **3** (1953) 23.
[34] US-Department of Health; *Toxic Substances List,* Nat. Instit. for Occup. Safety and Health, Rockville, Maryland 1972.
[35] V. H. Volodishenko, *Biol. Abstr.* **53** (1972) no. 63983.

# Aziridines

DAVID N. ROARK (currently with Ethyl Corp.), Cordova Chemical Co., North Muskegon, Michigan 49445, United States (Chaps. 2–7)

BLAINE C. MCKUSICK, (E.I. Du Pont de Nemours & Co), Wilmington, Delaware 19803, United States (Chap. 8)

| | | | | | | |
|---|---|---|---|---|---|---|
| 1. | Introduction | 663 | 4.2. | The Dow Process | 667 |
| 2. | Physical Properties | 664 | 5. | Environmental Protection | 668 |
| 3. | Chemical Properties | 664 | 6. | Uses | 668 |
| 3.1. | Homopolymerization | 664 | 7. | Economic Aspects | 669 |
| 3.2. | Aziridine-Modified Polymers | 665 | 8. | Toxicology and Occupational Health | 669 |
| 3.3. | Polyfunctional Aziridines | 665 | 8.1. | Ethylenimine | 670 |
| 3.4. | Miscellaneous | 666 | 8.2. | Propylenimine | 670 |
| 4. | Production | 666 | 8.3. | Aziridine Derivatives | 670 |
| 4.1. | The Wenker Process | 667 | 9. | References | 671 |

# 1. Introduction

Aziridines are three-membered ring compounds containing a single nitrogen in the ring. All commercially available aziridines or aziridine derivatives are made from either ethylenimine (**1**) or propylenimine (**2**). Ethylenimine is also called aziridine [151-56-4] and propylenimine is called 2-methylaziridine [75-55-8].

Ethylenimine was first prepared in 1888 by GABRIEL [3], who mistakenly called it vinylamine. He prepared the ethylenimine by reacting 2-bromoethylamine hydrobromide [2576-47-8] with silver oxide [1301-96-8] or potassium hydroxide [1310-58-3]. Propylenimine was first prepared in 1890 by HIRSCH [4] from 2-bromopropylamine [20888-43-1] in a similar manner. Commercial production of ethylenimine began in Germany in 1938 by I.G. Farbenindustrie.

**Table 1.** Physical properties

|  | Ethylenimine | Propylenimine |
|---|---|---|
| $M_r$ | 43.07 | 57.10 |
| Density at 25 °C, g/mL | 0.832 | 0.802 |
| bp, °C | 57 | 66 |
| mp, °C | −74 | −65 |
| Vapor pressure, 25 °C, kPa | 28.5 | 18.7 |
| Refractive index $n_D^{25}$ | 1.4123 | 1.4084 |
| Viscosity, 25 °C, mPa · s | 0.418 | 0.491 |
| Flash point (closed cup), °C | −11 | −4 |

## 2. Physical Properties

Ethylenimine and propylenimine are clear, colorless liquids with amine-like odors. They are miscible with water and most organic liquids. Pure aziridines are generally stored over solid sodium hydroxide [1310-73-2] and when stored in this manner are stable indefinitely. Other properties are shown in Table 1.

## 3. Chemical Properties

Ethylenimine and propylenimine are highly reactive, versatile materials. These materials participate in two general types of reaction: (1) the ring-preserving reactions, involving alkylation or acylation of the ring nitrogen, and (2) the ring-opening reactions, involving protonation of the ring nitrogen followed by nucleophilic attack at one of the carbons. Use of aziridines frequently is the most economical and efficient method of incorporating an ethylamine group into a polymer or into a complex organic molecule.

### 3.1. Homopolymerization

Polyethylenimine [9002-98-6] is produced by the acid-catalyzed homopolymerization of ethylenimine [5]. The polymerization is usually carried out at 90 – 110 °C either in water or in a variety of organic solvents. The average molecular mass of the polyethylenimine prepared as described above is 10 000 – 20 000. Higher molecular mass polymers are prepared by addition of a difunctional alkylating agent, such as chloromethyloxirane [106-89-8] or 1,2-dichloroethane [107-06-2]. Likewise, polymers of lower molecular mass can be obtained by inclusion of a low molecular mass amine, such as 1,2-ethanediamine [107-15-3], during polymerization [6]. By using these techniques a range of molecular masses from 300 to 1 000 000 can be obtained.

All of these polyethylenimines are highly branched polymers having roughly a spherical shape. The amine distribution is approximately 30% primary, 40% secondary, and 30% tertiary. This distribution can be readily determined by $^{13}C$ nuclear magnetic resonance spectroscopy.

*Trade names:* Corcat (Cordova Chemical), Polymin (BASF), Epomin (Nippon Shokubai Kagaku Kogyo).

## 3.2. Aziridine-Modified Polymers

Ethylenimine and propylenimine graft onto polymers either having carboxyl groups attached to the polymer backbone [7] or having amines in their backbone [8]. This reaction improves the adhesion of the polymer to anionic surfaces or surfaces containing hydroxyl groups.

## 3.3. Polyfunctional Aziridines

Ethylenimine and propylenimine react with trifunctional acrylates, such as 2-ethyl-2-(hydroxymethyl)–1,3-propanediol triacrylate [*15625-89-5*], to produce trifunctional aziridines.

This reaction is normally carried out in the absence of solvent and does not require a catalyst. The resulting products are clear, slightly viscous liquids.

*Trade names:* Xama (Cordova Chemical), NeoCryl CX-100 (Polyvinyl Chemical Industries), Azcat (AZS Chemical), Pfaz (Ionac Chemical).

## 3.4. Miscellaneous

Ethylenimine reacts with thiols to give aminoethylsulfides. This reaction is used commercially [9] to produce 2-aminoethanethiol (**3**) [60-23-1] and 2,2'-thiobisethylamine (**4**) [871-76-1].

$$H_2S + \underset{CH_2}{\overset{CH_2}{\triangle}}NH \longrightarrow HSCH_2CH_2NH_2$$
**3**

$$H_2S + 2\ \underset{CH_2}{\overset{CH_2}{\triangle}}NH \longrightarrow H_2NCH_2CH_2SCH_2CH_2NH_2$$
**4**

Ethylenimine and propylenimine react with oxiranes, opening the oxirane ring while leaving the aziridine ring intact. For example, 1-aziridine ethanol [1072-52-2], 1-(2-hydroxyethyl)ethyleneimine, is produced by the reaction between ethylenimine and oxirane [75-21-8].

Another commercial application of ethylenimine is the reaction with isocyanates to give iminoureas. For example, ethylenimine reacts with 1,6-diisocyanatohexane [822-06-0]:

$$O=C=N(CH_2)_6N=C=O + 2\ \underset{CH_2}{\overset{CH_2}{\triangle}}NH \longrightarrow$$

$$\underset{CH_2}{\overset{CH_2}{\triangle}}N-\overset{O}{\underset{\|}{C}}NH(CH_2)_6NH\overset{O}{\underset{\|}{C}}-N\underset{CH_2}{\overset{CH_2}{\triangle}}$$

# 4. Production

Although many schemes for synthesizing ethylenimine have been described in the literature, only two are currently considered commercially viable: the Wenker process and the Dow process. A third, involving the reaction of 2-chloroethylamine hydrochloride [870-24-6] and sodium hydroxide, was used commercially to produce ethylenimine in Germany in the late 1930s:

$$ClCH_2CH_2NH_2 \cdot HCl + 2\ NaOH \longrightarrow \underset{CH_2}{\overset{CH_2}{\triangle}}NH + 2\ NaCl$$

However, this process is no longer considered commercially viable.

There has been recent interest in the direct conversion of 2-aminoethanol [141-43-5] into ethylenimine, but no commercial system has been developed to date [10].

$$\text{H}_2\text{NCH}_2\text{CH}_2\text{OH} \xrightarrow[\text{Nb cat.}]{380°\text{C}} \underset{\text{CH}_2}{\overset{\text{CH}_2}{\triangleright}}\!\text{NH} + \text{H}_2\text{O}$$

## 4.1. The Wenker Process

In 1935 HENRY WENKER [11] developed a process for making ethylenimine that is still the basis for all commercial production today. It is a two-step process, initially involving the reaction of 2-aminoethanol with sulfuric acid [7664-93-9] to give 2-aminoethyl hydrogen sulfate [926-39-6] [12]:

$$\text{H}_2\text{NCH}_2\text{CH}_2\text{OH} + \text{H}_2\text{SO}_4 \rightarrow \text{H}_2\text{NCH}_2\text{CH}_2\text{OSO}_3\text{H} + \text{H}_2\text{O}$$

The product is a water-soluble solid with properties similar to an amino acid. The 2-aminoethyl hydrogen sulfate reacts with aqueous sodium hydroxide to produce ethylenimine:

$$\text{H}_2\text{NCH}_2\text{CH}_2\text{OSO}_3\text{H} + 2\text{ NaOH} \rightarrow \underset{\text{CH}_2}{\overset{\text{CH}_2}{\triangleright}}\!\text{NH} + \text{Na}_2\text{SO}_4$$

Originally yields of only 26 % were obtained, but recent literature reports yields of 85 – 90 % [13]. The advantage of this process is the production of pure ethylenimine with no significant waste-disposal problems. The disadvantage is the relatively high raw material cost.

## 4.2. The Dow Process

In 1963 Dow Chemical began production of ethylenimine by a newly developed process that offered the possibility of significantly lower raw material costs [14]. This process is based on the reaction of 1,2-dichloroethane with excess ammonia [15]. Dow stopped production in 1978.

$$\text{ClCH}_2\text{CH}_2\text{Cl} + 3\text{ NH}_3 \rightarrow \underset{\text{CH}_2}{\overset{\text{CH}_2}{\triangleright}}\!\text{NH} + 2\text{ NH}_4\text{Cl}$$

The advantage of this process is its low raw material cost. However, there are difficulties caused by impurities in the ethylenimine, high corrosion rates, and waste stream disposal.

# 5. Environmental Protection

The primary environmental concern is the prevention of aziridine vapors from entering the environment (see Chap. 8). Efficient scrubbers should be used on any vent line or hood exhaust where aziridine vapor may be present. An effective scrubbing medium is an aqueous solution of diammonium thiosulfate [7783-18-8].

The only waste stream from the Wenker process is an aqueous one, primarily containing sodium sulfate [15124-09-1] along with small amounts of various organic nitrogen compounds. This stream can be used by a kraft pulp mill as a source of sulfate, making the Wenker process essentially a non-waste-producing system. No waste streams are produced in the preparations of the principal aziridine derivatives, i.e., homopolymer, aziridine-modified polymers, and polyfunctional aziridines.

# 6. Uses

Aziridines are sold primarily to manufacture products for the coatings, paper, water treatment, and petroleum industries.

In the *coatings* industry, ethylenimine and propylenimine are added to water-borne paints to increase wet adhesion. Polyfunctional aziridines are used to cross-link polymers with pendant carboxyl groups. These cross-linked polymers are used in a variety of performance coatings, such as those on exterior wood panels. Polyethylenimines are used as tie-coat adhesives, e.g., in the production of polypropylene laminated films [16], as adhesion promoters, and as pigment dispersants.

In the manufacture of *paper*, polyethylenimine and aziridine-modified polymers are used as retention and drainage aids [17], as wet-strength additives, and as pigment dispersants.

Polyethylenimines are used in *water treatment* as clarifying aids [18] and have been shown to be effective silica antiscalants [19]. They are also used in the manufacture of *reverse osmosis membranes* [20].

Polyethylenimines have several applications in the drilling and completion of *oil* and *gas* wells and in the subsequent production of oil. Polyethylenimines are used in conjunction with a sulfonated polymer, such as lignosulfonate or a condensed naphthalenesulfonate, to inhibit fluid loss from well cement [21]. They have also been used commercially as selective flocculants in drilling fluid and as demulsifiers for some crude oil emulsions.

An important emerging technology is the use of polyethylenimine to *immobilize enzymes* [22].

# 7. Economic Aspects

The cost of producing ethylenimine depends on the price of 2-aminoethanol and on the high capital requirements needed to build a safe and efficient ethylenimine plant. The cost of a plant is high because of both the highly toxic nature of the ethylenimine and the lack of gasket material that can withstand prolonged contact with ethylenimine. This means that a modern ethylen-imine production facility will be almost entirely of welded construction, using gasketless pumps, valves, and flow meters.

Because ethylenimine is significantly more expensive to produce than monomers used in most other water-soluble polymers, the price per kilogram of polyethylenimine is relatively high. Ethylenimine sold for $ 11 – 17/kg in 1982 and polyethylenimine sold for $ 11 – 15/kg.

Ethylenimine is currently produced commercially by BASF (FRG), Cordova Chemical Co. (USA), and Nippon Shokubai Kagaku Kogyo Co. (Japan). Propylenimine is produced commercially by Aceto Chemical Co. (USA).

# 8. Toxicology and Occupational Health [23] – [26]

Like many strong alkylating agents, aziridines are highly toxic. Their toxicity probably results from alteration of nucleic acids, proteins, and other important biochemicals by alkylation.

At one time substances with two or more aziridine rings, such as triethylenemelamine (5) [51-18-3], were considered promising antineoplastic drugs.

**5**

Many were prepared, a half dozen or more went to clinical trial, and some were used to treat cancer. However, although capable of inhibiting tumors and killing cancer cells, they were found to have serious side effects. These included severe damage to lymphatic cells and bone marrow and diminished fertility. Moreover, they often proved carcinogenic in animal tests. Because of these properties, they are rarely used as antineoplastic drugs today. However, the extensive testing of these agents, added to the testing of simple, commercially important aziridines, has provided more knowledge of the toxicological properties of aziridines than is available for most chemical classes. For example, the International Agency for Research on Cancer (IARC) has reviewed carcinogenic data of eleven aziridines [24]. For nine the data were sufficient to classify the substances as carcinogens. Clearly, any aziridines that have not been tested should be considered potential carcinogens.

## 8.1. Ethylenimine

In animal studies ethylenimine is highly toxic by ingestion, inhalation, skin or eye application, and injection. For example $LD_{50}$ (oral, rat) is 14 mg/kg; $LD_{50}$ (skin, rabbit) is 13 mg/kg; $LC_{50}$ (10-min inhalation, mouse) is 2236 ppm. In inhalation studies, lethal air concentrations typically cause irritation of nose and eyes during exposure, with edema and other signs of lung injury becoming evident a few hours afterwards, and death ensuing a few days later.

To humans ethylenimine is a skin sensitizer and a severe irritant to eyes, nose, and throat. Transient CNS, liver, and kidney damage has been observed. Two deaths have been reported, one from inhalation and one from skin contact.

Ethylenimine is a mutagen in microbial and fruit fly tests. It proved carcinogenic by the oral and subcutaneous route in four mouse tests. However, an epidemiological study of 144 aziridine workers found no evidence of carcinogenicity.

Ethylenimine is classified by OSHA as a "cancer suspect" agent and detailed standards and work practices are prescribed to prevent any significant exposure by inhalation, ingestion, or skin or eye contact. The ACGIH recommends a TLV of 0.5 ppm, with skin contact to be avoided [25]. The DFG classifies ethylenimine as an A2 carcinogen, requiring special protective and surveillance measures [26].

## 8.2. Propylenimine

The toxic properties of propylenimine are similar to those of ethylenimine. Limited studies in rats suggest it is one-fourth to one-eighth as toxic by inhalation, and this is part of the basis for the TLV of 2 ppm set by the ACGIH. In oral rat studies it produced a variety of malignant tumors. It is classified an A2 suspected carcinogen by ACGIH, an A2 proved animal carcinogen by DFG. It should be handled with the same care as ethylenimine.

## 8.3. Aziridine Derivatives

Substances derived from aziridines but no longer containing the aziridine ring are normal-ly of low toxicity. Examples are polymerized aziridines and polymers that have been modified by reaction with aziridines.

Table 2. Acute toxicity, doses in mg/kg

|  | EI | PI | Xama-7 | Corcat P-600 |
|---|---|---|---|---|
| Acute oral ($LD_{50}$) | 14 | 19 | 2000 | 7500 |
| Acute dermal ($LD_{50}$) | 13 | 43 | >3000 | >3000 |
| Eye irritation | severe | severe | moderate, transitory | irritant |

Polyethylenimine is used in personal care products and in food packaging materials. Table 2 shows toxicity data for ethylenimine (EI) and for propylenimine (PI) as well as for a typical polyfunctional aziridine (Xama-7) and a typical polyethylenimine (Corcat P-600).

# 9. References

General References

[1] O. C. Dermer, G. E. Ham: *Ethylenimine and Other Aziridines*, Academic Press, New York and London 1969.
[2] D. S. Zhuk, P. A. Gembitskii, V. A. Kargin, *Russian Chemical Reviews (Engl. Transl.)* **34** (1965) 515.

Specific References

[3] S. Gabriel, *Ber. Dtsch. Chem. Ges.* **21** (1888) 1049.
[4] P. Hirsch, *Ber. Dtsch. Chem. Ges.* **23** (1890) 964.
[5] I.G. Farbenindustrie, US 2223930, 1940 (R. Griessbach, E. Meier, H. Wassenegger).
[6] Dow Chemical, US 3519687, 1970 (J. G. Schneider, C. R. Dick, G. E. Ham).
[7] National Starch & Chemical, US 3372149, 1968 (J. Fertig, E. D. Mazzarella, M. Skoultchi).
[8] BASF, US 3642572, 1972 (H. Endres, R. Fikentscher, W. Maurer, E. Scharf, U. Soenksen).
[9] Dow Chemical, US 3463816, 1969 (G. F. Button, D. L. Childress).
[10] Dow Chemical, US 4337175, 1982 (E. G. Ramirez).
[11] H. Wenker, *J. Am. Chem. Soc.* **57** (1935) 2328.
[12] Shell Oil Co., US 3153079, 1964 (E. R. A. Forshaw).
[13] Chemirad, US 3326879, 1967 (R. G. Dunning).
[14] J. H. Olin, *Ind. Eng. Chem.* **55** (1963) no. 9, 10.
[15] Dow Chemical, US 3326895, 1967 (W. P. Coker).
[16] A. M. DeRoo in I. Skeist (ed.): *Handbook of Adhesives*, 2nd ed., Van Nostrand Reinhold Co., New York 1977, pp. 592–596.
[17] K. W. Britt, J. E. Unbehend, *Tappi* **63** (1980) 67.
[18] H. Yeh, M. M. Ghosh, *J. Am. Water Works Assoc.* **73** (1981) 211.
[19] J. E. Harrar, F. E. Locke, C. H. Otto, Jr., L. E. Lorensen, S. B. Monaco, W. P. Frey, *Soc. Pet. Eng. J.* **22** (1982) 17.
[20] H. K. Lonsdale, *J. Membrane Sci.* **10** (1982) 81.
[21] L. F. McKenzie, J. V. Eikerts, P. M. McElfresh, *Oil Gas J.* (1982) 146.
[22] Miles Laboratories, US 4355105, 1982 (O. J. Lantero, Jr.).

[23] C. F. Reinhardt, M. R. Brittelli: "Heterocyclic and Miscellaneous Nitrogen Compounds," in G. D. Clayton, F. E. Clayton (eds.):*Patty's Industrial Hygiene and Toxicology*, 3rd ed., vol. **2 A**, Wiley-Interscience, New York 1981, pp. 2652–2680.

[24] IARC: *Evaluation of Carcinogenic Risk of Chemicals to Man*, vol. **9**, International Agency for Research on Cancer, Lyon 1975, pp. 31–109.

[25] ACGIH (ed.): *Documentation of the Threshold Limit Values*, Cincinnati, Ohio 1983, pp. 186, 351.

[26] DFG: *Maximum Concentrations at the Workplace and Biological Tolerance Values for Working Materials*, Verlag Chemie, Weinheim 1984, Report 20.

# Benzaldehyde

FRIEDRICH BRÜHNE, Bayer AG, Krefeld-Uerdingen, Federal Republic of Germany (Chaps. 1–9)
ELAINE WRIGHT, General Motors Research Laboratories, Warren, Michigan 48090, United States (Chap. 10)

| | | | | | |
|---|---|---|---|---|---|
| 1. | Introduction | 673 | 7. | Uses | 683 |
| 2. | Physical Properties | 674 | 8. | Derivatives | 684 |
| 3. | Chemical Properties | 674 | 8.1. | Chlorobenzaldehydes | 684 |
| 4. | Production | 677 | 8.2. | Nitrobenzaldehydes | 685 |
| 4.1. | Hydrolysis of Benzal Chloride | 677 | 8.3. | Hydroxybenzaldehydes | 686 |
| 4.2. | Oxidation of Toluene | 679 | 8.4. | Aminobenzaldehydes | 687 |
| 4.3. | Other Production Processes | 681 | 8.5. | Methylbenzaldehydes | 688 |
| 5. | Quality Specifications and Test Methods | 682 | 9. | Economic Aspects | 689 |
| 6. | Storage, Transportation, and Safety Regulations | 682 | 10. | Toxicology | 689 |
| | | | 11. | References | 689 |

## 1. Introduction

Benzaldehyde is the simplest and industrially the most important aromatic aldehyde. It exists in nature, occurring in combined and uncombined forms in many plants. The best known natural source of benzaldehyde is amygdalin, in which it exists in a combined form as a glycoside and which is present in bitter almonds. The odor of bitter almonds arises from small amounts of free benzaldehyde formed by hydrolysis of the amygdalin. Owing to its occurrence in bitter almonds, the aldehyde was formerly referred to as "bitter almond oil." Benzaldehyde is also the main constituent of the essential oils obtained by pressing the kernels of peaches, cherries, apricots, and other fruits.

Benzaldehyde
[100-52-7]

In 1818–1819 VOGEL and MATRÈS reported for the first time, and independently of one another, that, in addition to hydrocyanic acid, a volatile oil could be obtained from bitter almonds. Later, in 1832, this oil was investigated by WÖHLER and LIEBIG; they determined its composition and recognized its relationship to benzoic acid and benzoyl chloride.

**Table 1.** Vapor pressure vs. temperature

| $t$, °C | 40 | 60 | 80 | 100 | 120 | 140 | 160 | 180 |
|---|---|---|---|---|---|---|---|---|
| $p$, kPa | 0.3 | 1.1 | 3.3 | 8.2 | 17.8 | 34.7 | 62.3 | 104.3 |

**Table 2.** Azeotropic mixtures of benzaldehyde

| Second component | $bp$ at 101.3 kPa, °C | Benzaldehyde, wt% |
|---|---|---|
| 3-Bromotoluene | 179.0 | 92.0 |
| Ethyl phenyl ether | 169.8 | 12.0 |
| Benzyl ethyl ether | 177.5 | 92.0 |
| 2-Octanol | 174.0 | 25.0 |
| α-Terpinene | 170.0 | 38.0 |
| Diisoamyl ether | 168.6 | 37.5 |

## 2. Physical Properties

**Benzaldehyde,** $C_7H_6O$, $M_r$ 106.13, is a colorless and highly refractive liquid with an odor of bitter almonds. Like many other aromatic aldehydes, it is volatile in steam, $bp$ 179 °C at 101.3 kPa, $mp$ −56 °C, refractive index $n_D^{20}$ 1.5450, density $\varrho$ at 0 °C 1.063 g/cm$^3$, at 20 °C 1.046 g/cm$^3$, at 50 °C 1.018 g/cm$^3$, specific heat capacity $c_p$ 1.676 J g$^{-1}$ K$^{-1}$ at 25 °C, heat of evaporation 371.0 J/g at 179 °C, standard heat of combustion 33.19 kJ/g, flash point 64.5 °C, autoignition temperature 190 °C, lower explosive limit 1.4 vol%. Vapor pressure versus temperature data for benzaldehyde are given in Table 1.

Benzaldehyde is miscible with many organic solvents; at 25 °C it is miscible with concentrated sulfuric acid, liquid carbon dioxide, liquid ammonia, methylamine, and diethylamine. The compositions of several binary azeotropic mixtures are given in Table 2. For further azeotropes, see [2]. Vapor–liquid equilibria data for binary mixtures of benzaldehyde with water or organic solvents will be found in [1]. One hundred milliliters of water dissolves 0.40 mL of benzaldehyde at 25 °C or 0.96 mL at 60 °C. Dynamic viscosity $\eta$ $1.40 \times 10^{-3}$ Pa · s at 25 °C, $1.11 \times 10^{-3}$ Pa · s at 40 °C, surface tension $\sigma$ $40.04 \times 10^{-3}$ N/m at 20 °C, dipole moment $\mu$ (measured in liquid benzene) 2.92 D ($9.74 \times 10^{-30}$ C m), dielectric constant $\varepsilon_r$ 17.7 at 25 °C.

## 3. Chemical Properties

The chemical behavior of benzaldehyde and of its nuclear substitution derivatives corresponds largely to that of the aliphatic aldehydes (→ Aldehydes, Aliphatic and Araliphatic). But the reactivity of the carbonyl group of the aromatic aldehydes is somewhat less because the π electrons of this group are included in the resonance system of the aromatic ring. Thus benzaldehyde forms Schiff bases with amines, an oxime with hydroxylamine, a hydrazone with phenylhydrazine, and acetals with alcohols. It adds hydrogen cyanide, sodium bisulfite, and Grignard compounds. If benzaldehyde is reacted with hydrogen cyanide in the presence of ammonia and the aminonitrile formed as an intermediate is then saponified, DL-2-phenylglycine is formed (Strecker synthesis).

Benzaldehyde undergoes autoxidation in air. This process, the final product of which is benzoic acid, is accelerated by light and by the addition of peroxides or heavy-metal salts. It is retarded by antioxidants, such as phenolic compounds and diphenylamine. Nitric acid, chromium(VI) oxide, and other oxidizing agents also oxidize benzaldehyde to benzoic acid.

The reduction or hydrogenation of benzaldehyde leads to benzyl alcohol, dibenzyl ether, benzoin, 1,2-diphenylethane-1,2-diol, stilbene, toluene, methylcyclohexane, and other products, depending on the reaction conditions. Catalytic hydrogenation of benzaldehyde is used industrially to produce benzyl alcohol. Reduction with aluminum alcoholates (Meerwein-Ponndorf-Verley reduction) also gives benzyl alcohol. This reaction also permits the reduction of unsaturated aldehydes, such as cinnamaldehyde, to the corresponding alcohols, in which the olefinic double bonds are retained.

The reaction of benzaldehyde with ammonia and hydrogen in the presence of hydrogenation catalysts leads to benzylamine and is an important industrial process. The side-chain chlorination of benzaldehyde gives benzoyl chloride [3].

Like the aliphatic aldehydes, benzaldehyde participates in condensation reactions with a number of organic compounds possessing active hydrogen atoms. Some of these reactions are used for industrial purposes. The Claisen-Schmidt condensation of benzaldehyde with acetaldehyde in the presence of aqueous alkali leads to cinnamaldehyde.

$$\text{PhCHO} + \text{CH}_3\text{CHO} \longrightarrow \text{PhCH=CH-CHO}$$

The Perkin condensation of benzaldehyde with acetic anhydride in the presence of sodium acetate or potassium acetate as a condensing agent is carried out industrially in the production of cinnamic acid (→ Acetic Anhydride).

$$\text{PhCHO} + (\text{CH}_3\text{CO})_2\text{O} \longrightarrow \text{PhCH=CH-COOH}$$

Cinnamic acid is also produced by Knoevenagel condensation of benzaldehyde with malonic acid in the presence of weakly basic catalysts, such as ammonia and amines.

$$\text{PhCHO} + \text{H}_2\text{C(COOH)}_2 \longrightarrow \text{PhCH=CH-COOH}$$

In the Reformatsky reaction, ethyl cinnamate is obtained from benzaldehyde and ethyl bromoacetate in the presence of activated zinc.

$$\text{C}_6\text{H}_5\text{CHO} + \text{BrCH}_2\text{COOC}_2\text{H}_5 \xrightarrow{\text{Zn}} \text{C}_6\text{H}_5\text{CH=CH-COOC}_2\text{H}_5$$

The Cannizzaro reaction is typical of benzaldehyde and other aldehydes that have no α-hydrogen atoms. If concentrated sodium or potassium hydroxide is present, this reaction results in disproportionation to benzoic acid and benzyl alcohol.

$$2\,\text{C}_6\text{H}_5\text{CHO} \xrightarrow{\text{NaOH}} \text{C}_6\text{H}_5\text{CH}_2\text{OH} + \text{C}_6\text{H}_5\text{COOH}$$

Sodium benzylate and aluminum benzylate catalyze the Claisen-Tishchenko condensation of benzaldehyde, which leads to benzyl benzoate.

$$2\,\text{C}_6\text{H}_5\text{CHO} \rightarrow \text{C}_6\text{H}_5\text{COOCH}_2\text{C}_6\text{H}_5$$

In comparison with their aliphatic analogues, benzaldehyde and other aromatic aldehydes have certain peculiar aspects in their reaction behavior. In the benzoin condensation, two molecules of benzaldehyde condense in the presence of cyanide to form the hydroxyketone benzoin. Thiazolium salts can be used as catalysts in place of cyanide [4].

$$2\,\text{C}_6\text{H}_5\text{CHO} \rightarrow \text{C}_6\text{H}_5\text{CH(OH)-CO-C}_6\text{H}_5$$

The reactivity of benzaldehyde towards ammonia also differs from that of aliphatic aldehydes. Instead of stopping at the aldehyde–ammonia adduct, the reaction continues until 1-phenyl-$N,N'$-bis(phenylmethylene)-methanediamine (hydrobenzamide) is obtained.

$$3\,\text{C}_6\text{H}_5\text{CHO} + 2\,\text{NH}_3 \rightarrow \text{C}_6\text{H}_5\text{CH(N=CH-C}_6\text{H}_5)_2$$

Benzyl alcohol and benzoic acid, but no copper(I) oxide, are formed when benzaldehyde is treated with Fehling's solution. Aromatic aldehydes do not show the polymerization or formation of cyclic compounds that is observed with aliphatic aldehydes. The condensation of benzaldehyde with phenols, aromatic amines, and benzene leads to triphenylmethane derivatives. This reaction is exploited industrially in the production of Malachite Green dyes.

Electrophilic substitution of the aromatic nucleus (chlorination, nitration, sulfonation) occurs preferentially in the meta position.

# 4. Production

Benzaldehyde is produced principally by the hydrolysis of benzal chloride or the partial oxidation of toluene. There are various other manufacturing processes, but at present they have no industrial importance.

## 4.1. Hydrolysis of Benzal Chloride

The hydrolysis of benzal chloride, which is readily obtainable by side-chain chlorination of toluene, is among the oldest industrial processes for the production of benzaldehyde. It can be carried out either in an alkaline or in an acidic medium.

$$\text{C}_6\text{H}_5\text{CHCl}_2 + \text{H}_2\text{O} \longrightarrow \text{C}_6\text{H}_5\text{CHO} + 2\,\text{HCl}$$

Hydrolysis under basic conditions can be carried out with calcium hydroxide, calcium carbonate, sodium hydrogencarbonate, or sodium carbonate. Excessive alkalinity increases the probability of side reactions. The preferred alkaline saponifying agent is sodium carbonate.

According to a fairly old process [5], benzal chloride is saponified with a small excess of 15% sodium carbonate solution at 138 °C. The reactor is a carbon steel agitator vessel lined internally with Heresite (a phenolic resin) and having an agitator made of a copper–silicon alloy. The cooler, separator, and pumps are made of Karbate (a graphitic material). On completion of the saponification, the benzaldehyde is isolated from the reaction mixture by steam distillation and distilled in a stainless steel column at 2–3 kPa (20–30 mbar). The chlorine content of the distilled product is less than 0.01%.

Instead of pure benzal chloride, a mixture of benzal chloride and benzotrichloride, as obtained industrially in the side-chain chlorination of toluene, can be hydrolyzed. Under the reaction conditions, benzotrichloride is saponified to sodium benzoate, which remains in the aqueous alkaline solution. It is converted separately to benzoic acid.

In a new continuous process [6], benzal chloride and the alkaline hydrolyzing agent are reacted as countercurrents in a flow reactor in an unreactive organic solvent (e.g., toluene, xylene). A temperature of 125–145 °C and a pressure of 1200–1800 kPa (12–18 bar) are maintained. In the extraction zone the dissolved aldehyde is extracted from the aqueous alkaline phase with an unreactive organic solvent flowing in the opposite direction. Water, likewise as a countercurrent, washes the solution of the crude benzaldehyde in the washing zone. The flow reactor, extraction zone, and washing zone are integrated in a special apparatus. The wash water and extract are returned to the hydrolysis zone in order to recover bezaldehyde (Fig. 1).

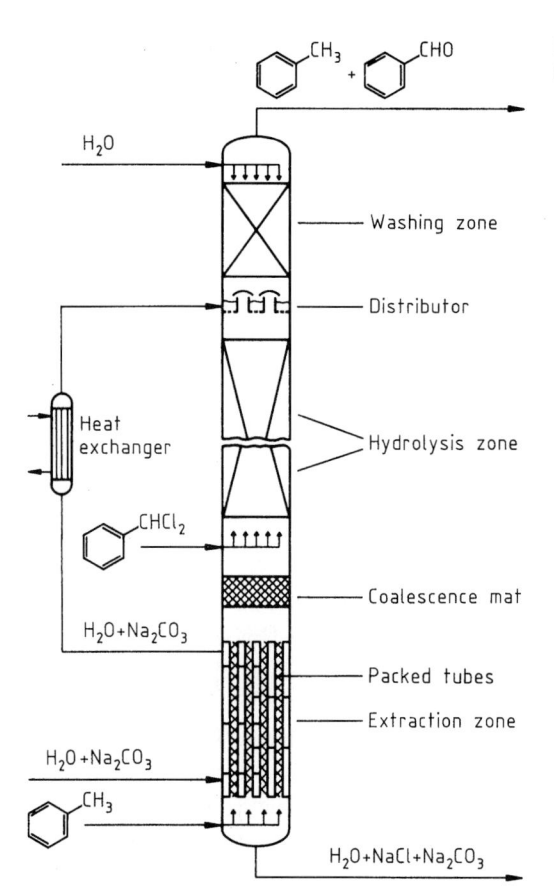

**Figure 1.** Hydrolysis of benzal chloride to benzaldehyde in a flow-through reactor [6]

Benzal chloride can also be converted into benzaldehyde by boiling with aqueous solutions of hexamethylenetetramine [7]. Because benzyl chloride also reacts with hexamethylenetetramine to give benzaldehyde (Sommelet reaction), industrial mixtures of benzyl chloride and benzal chloride can be used to form benzaldehyde in this way.

The acid hydrolysis of benzal chloride is carried out in the presence of acids and with metal salts as catalysts. It gives very high yields of benzaldehyde (more than 90%). Hydrogen chloride is formed simultaneously and is usually passed through an absorption unit to be recovered as concentrated hydrochloric acid.

The hydrolysis of benzal chloride was at one time commonly carried out in the presence of concentrated sulfuric acid. This process has the disadvantage that large amounts of dilute sulfuric acid are formed as a waste product. It has little current importance, except for the hydrolysis of certain substituted benzaldehydes that are difficult to saponify in other ways [8], [9]. Benzal chloride is also saponified in the presence of phosphoric acid or sulfonic acids [10], hydrochloric acid [11], and formic acid [12]. The reaction of benzal chloride with organic acids, such as acetic acid, in the presence of a tin(II) or tin(IV) chloride catalyst [13], gives benzaldehyde and the corresponding acid chloride.

$$\text{C}_6\text{H}_5\text{CHCl}_2 + \text{R-COOH} \longrightarrow \text{C}_6\text{H}_5\text{CHO} + \text{R-COCl}$$

R = alkyl, aryl, or cycloalkyl.

The hydrolysis of benzal chloride is catalyzed by metal salts, preferably those of iron or zinc. Water should not accumulate in the reaction mixture during the reaction because it would reduce the activity of the catalyst [14], [15].

Two parts by weight of anhydrous zinc chloride are added to 700 parts by weight of distilled benzal chloride, stirred, and heated to 105–110 °C. The mixture is maintained at 100–120 °C, while 85 parts by weight of water are added at the rate at which water is consumed. After all the water has been added, the reaction temperature is maintained for one hour to ensure removal of the liberated hydrogen chloride [15]. The yield of crude benzaldehyde is 95%, and the purity of the undistilled product is greater than 96.5%.

The catalyst is not deactivated by a momentary excess of water if insoluble zinc catalysts, such as zinc phosphate or zinc laurate, are used [16]. Tin chlorides [17] and copper(II) chloride [18] are also catalytic. The hydrolysis of benzal chloride can be carried out batchwise or continuously, e.g., in a cascade consisting of several reactors.

A new continuous process [19] hydrolyzes benzal chloride in the vapor phase at temperatures of 100–300 °C. The reaction is catalyzed by activated carbon which has been treated with an acid, such as sulfuric acid, or impregnated with a metal chloride, such as iron(III) chloride, or with a sulfate, such as copper(II) sulfate. The benzaldehyde yield is 97%. This process is particularly suitable for the hydrolysis of trifluoromethyl-substituted benzal chlorides, which are otherwise difficult to convert into the corresponding benzaldehydes. In a similar process, benzal chloride in the vapor phase at 300 °C is hydrolyzed to benzaldehyde using a silicon dioxide or aluminum oxide catalyst [20].

## 4.2. Oxidation of Toluene

The partial oxidation of toluene with oxygen to give benzaldehyde can be carried out in either the gas phase or the liquid phase. Benzaldehyde itself is easily further oxidized to benzoic acid and other products. Conditions must therefore be carefully chosen to favor only partial oxidation.

$$\text{C}_6\text{H}_5\text{CH}_3 + \text{O}_2 \longrightarrow \text{C}_6\text{H}_5\text{CHO} + \text{H}_2\text{O}$$

In the gas phase, the oxidation is carried out by passing toluene vapor, together with oxygen in a gaseous mixture such as air, through a catalyst bed in a tube-bundle or fluidized-bed reactor at a temperature of 250–650 °C. The reaction is highly exothermic, and effective cooling is necessary. It is advisable to dilute the mixture of toluene vapor and oxygen-containing gas with an unreactive gas, such as water vapor, nitrogen, or carbon dioxide. The benzaldehyde yield is favored by a low conversion rate (10–20%) per pass, a short residence time (0.1–1.0 s), and a precise adjustment of the amount of oxygen. Even then, it is only 40–60% of the theoretical yield based on toluene. The oxidation of toluene in the gas phase produces maleic anhydride, citraconic anhydride, phthalic anhydride, anthraquinone, cresol, acetic acid, and other compounds, in addition to substantial quantities of benzoic acid, carbon monoxide, and carbon dioxide [21]. The complete combustion of the toluene is retarded by applying high pressure [22] or by adding potassium sulfate or sodium fluoride to the catalyst [23].

Oxides of the elements of Groups V and VI of the periodic table are frequently used as catalysts, sometimes with other oxides. These oxide catalysts contain molybdenum and at least one additional element from among iron, nickel, cobalt, antimony, bismuth, vanadium, phosphorus, samarium, tantalum, tin, and chromium [24]. Some have palladium and phosphoric acid on activated carbon [25]. Others are mixed oxides of silver and transition metals, such as cobalt nickel, vanadium, platinum, or palladium [26], and one is a mixture of silver vanadate and iron vanadate [27].

A recently published process uses a mixed-oxide catalyst that has silicon dioxide as a carrier and contains uranium, copper, iron, phosphorus, tellurium, and lead, in addition to molybdenum [28]. At a toluene consumption of 35–50% per pass, the selectivity of the benzaldehyde formation at reaction temperatures of 475–550 °C is 40–70%. Catalysts based only on mixed oxides of molybdenum and uranium [29] require reaction temperatures of around 600 °C.

More important than gas-phase oxidation for the production of benzaldehyde is the oxidation of toluene in the liquid phase by oxygen in the form of air or other gaseous mixtures. This is carried out at 80–250 °C, preferably in the presence of cobalt, nickel, manganese, iron, or chromium compounds (alone or in combination) as catalysts. Lead compounds [30], ruthenium compounds [31], and thallium salts of organic acids [32] are also catalytic. A pressure at which the reaction medium remains liquid is chosen. Alkali, alkaline earth, and tertiary alkyl hypochlorites have been used as oxidation promoters [33] in place of the usual bromides. These promoters may cause corrosion, however.

Toluene can also be oxidized in an alcohol, e.g., methanol [34], an aliphatic aldehyde, e.g., acetaldehyde [35], or an acid, e.g., benzoic acid [36] or acetic acid [37]. Water can also be added to the reaction [38], but a recent process improves the benzaldehyde selectivity by continuously removing the water formed in the oxidation reaction [39]. When a cobalt catalyst is used and conversion rates of up to about 25% are achieved, this process gives aldehyde selectivities of 40–80%.

In another process, the selectivity of the benzaldehyde formation is improved by liquid-phase oxidation of the toluene with oxygen in the presence of a phosphoric acid–palladium catalyst that also contains certain phosphorus, sulfur, or nitrogen modifiers [40]. For an oxygen consumption of 63%, the yield of benzaldehyde is 41%.

In the large-scale production of phenol by the Dow process and of ε-caprolactam by the Snia Viscosa process via the toluene–benzoic acid synthesis route (→ Benzoic Acid), considerable amounts of benzaldehyde are formed as a by-product. Because a large proportion of this byproduct is worked up into pure benzaldehyde, oxidation processes intended exclusively to produce benzaldehyde are seldom used industrially at the present time.

Even when catalytic liquid-phase oxidation of toluene is carried out for the express purpose of manufacturing benzaldehyde, substantial quantities of benzoic acid are formed. Depending on the reaction conditions, carbon dioxide, carbon monoxide, formaldehyde, formic acid, benzyl alcohol, benzyl hydroperoxide, biphenyl, methylbiphenyls, and other compounds are formed also as byproducts. The crude benzaldehyde is normally refined by distillation at reduced pressure in a stainless steel column.

Benzaldehyde produced by oxidation of toluene often contains small amounts of impurities that are difficult to remove by distillation and which discolor the product. To remove these impurities the benzaldehyde can be treated with aqueous alkali [41] or with water and a base metal such as zinc powder [42]. A color-stable product can be obtained by treatment with oxidizing agents, e.g., hydrogen peroxide, followed by distillation [43].

Those processes in which toluene is oxidized with other agents, such as manganese dioxide in sulfuric acid, sodium persulfate, chromium(VI) oxide in acetic anhydride, or chromyl chloride (Etard reaction), cause waste water disposal problems and are without industrial importance.

## 4.3. Other Production Processes

Other processes for the manufacture of benzaldehyde include the reaction of benzene with carbon monoxide [44], the oxidation or dehydrogenation of benzyl alcohol [45], and the ruthenium-catalyzed oxidation of styrene with periodate or hypochlorite [46]. Benzaldehyde can also be made by hydrolysis of mixtures of benzyl chloride and benzal chloride in dilute nitric acid, using vanadium pentoxide as catalyst [47], or in the presence of an aqueous solution of hexamethylenetetramine [7]. The reduction of benzoyl chloride [48] or methyl benzoate [49] can also be used. These processes have no industrial importance in the production of benzaldehyde and are more likely to be used to make certain nuclear-substituted derivatives.

**Table 3.** Purity specifications of benzaldehyde grades

| Property | Industrial grade | |
|---|---|---|
| | Pure | Double-distilled |
| Assay, wt% | > 99.0 | > 99.5 |
| Benzoic acid, wt% | < 0.3 | < 0.1 |
| Water, wt% | < 0.3 | < 0.1 |
| Chlorine, wt% | < 0.02 | < 0.007 |
| Color (APHA) | < 20 | < 20 |

# 5. Quality Specifications and Test Methods

Benzaldehyde is commercially available in two grades (Table 3). *Pure* benzaldehyde, which is suitable for most uses, accounts for more than 95% of the amount sold. *Double-distilled* benzaldehyde is used mainly in the pharmaceutical, perfume, and flavor industries and must therefore be particularly pure and free from foreign odors.

Traces of chlorine compounds may be detectable in benzaldehyde obtained by the hydrolysis of benzal chloride, whereas benzaldehyde obtained by oxidation may contain byproducts formed in the oxidation of toluene. Benzaldehyde can now be produced, by hydrolysis of benzal chloride and by partial oxidation of toluene, to such a high standard of purity that it is suitable for all known applications.

A suitable quantitative determination method consists of reacting benzaldehyde with hydroxylamine hydrochloride and titrating the hydrochloric acid liberated. Gas chromatography in packed or capillary columns is becoming increasingly important, particularly in the determination of impurities. The pharmacopeias of the various countries should be consulted regarding the testing of benzaldehyde that is to be used for pharmaceutical purposes. United States Pharmacopeia XX includes tests for halogen compounds, hydrocyanic acid, and nitrobenzene.

# 6. Storage, Transportation, and Safety Regulations

Benzaldehyde is readily oxidized to benzoic acid in the presence of air and should therefore be stored under nitrogen. Tanks for benzaldehyde should be made of aluminum or stainless steel. Benzaldehyde can be transported in drums with baked enamel finishes or in tanks made of aluminum or stainless steel. The air content of containers used for shipping benzaldehyde must be minimized.

The transportation of benzaldehyde by rail and road is regulated by the provisions of Class 3, Figure 32c, of GGVE/RID and GGVS/ADR. The transportation of benzaldehyde by sea is not restricted by the IMDG Code. The labeling of benzaldehyde must comply with national regulations for dangerous substances or with Supplement I of the EEC Directive on the Packaging, Classification, and Labelling of Dangerous Substances, 6th Amendment (see Journal of the European Community no. L 360 of December 30, 1976).

Benzaldehyde belongs to Temperature Class T 4 (autoignition temperature 190 °C) and has a flash point of 64.5 °C. Electrical installations must be labeled accordingly. If it is finely divided in air, e.g., on cleaning cloths, sawdust, kieselguhr, or other materials with large surface areas, benzaldehyde shows a very strong self-heating tendency. Because of the relatively low autoignition temperature, there is a risk of spontaneous combustion under such circumstances.

# 7. Uses

Benzaldehyde is an important starting material for the manufacture of odorants and flavors. The compound is responsible for the odor of natural bitter almond oil and is incorporated directly in perfumes, soaps, foods, drinks, and other products. Substantial amounts are used in the production of derivatives that are also employed in the perfume and flavor industries. Examples of these derivatives are cinnamaldehyde, cinnamaldehyde dimethyl acetal, $\alpha$-amylcinnamaldehyde, $\alpha$-hexylcinnamaldehyde, dihydrocinnamaldehyde, cinnamyl alcohol, cinnamic acid, 4-phenyl-2-butanone (benzylacetone), $\beta$-bromostyrene, 2,2,2-trichloro-1-phenylethyl acetate, and benzyl benzoate.

Another application of benzaldehyde is the production of triphenylmethane dyes. For example, the leuco base of Malachite Green is obtained by condensation of benzaldehyde with dimethylaniline. The acridine dye benzoflavin is obtained from benzaldehyde and $m$-toluenediamine. In the pharmaceutical industry benzaldehyde is used as an intermediate in the manufacture of chloramphenicol, ephedrin, ampicillin, diphenylhydantoin, and other products.

Other important chemical intermediates obtained from benzaldehyde are benzoin, benzylamine, benzyl alcohol, mandelic acid, and 4-phenyl-3-buten-2-one (benzylideneacetone). Benzaldehyde is used in photochemistry, as a corrosion inhibitor and dyeing auxiliary, in the electroplating industry, and in the production of agricultural chemicals. The use of benzaldehyde in the extractive separation of isomeric amines has been proposed [50].

# 8. Derivatives

## 8.1. Chlorobenzaldehydes

2-Chlorobenzaldehyde [*89-98-5*], $C_7H_5ClO$, $M_r$ 140.57, is a colorless to yellowish liquid with a penetrating odor and low solubility in water, *mp* 11.6 °C, *bp* 212 °C at 101.3 kPa, *bp* 91–92 °C at 2.0 kPa, density $\varrho$ 1.248 g/cm$^3$ at 20 °C, refractive index $n_D^{20}$ 1.5660, flash point 90 °C.

2-Chlorobenzaldehyde is produced mainly by chlorination of 2-chlorotoluene to form 2-chlorobenzal chloride, which is then subjected to acid hydrolysis. Metal salts, such as iron(III) chloride, are used as catalysts [51]. The hydrolysis can also be accomplished using formic acid without a catalyst [12]. 2-Chlorobenzaldehyde can also be produced by oxidation of 2-chlorobenzyl chloride with N-oxides of tertiary amines [52] or with dilute nitric acid [47]. 2-Chlorobenzaldehyde is considerably more resistant to oxidation than benzaldehyde. When it is heated with sodium sulfite solution under pressure, benzaldehyde-2-sulfonic acid forms.

**3-Chlorobenzaldehyde** [*587-04-2*], $C_7H_5ClO$, $M_r$ 140.57, *mp* 17 °C, *bp* 213–214 °C at 101.3 kPa, *bp* 93–96 °C at 2.0 kPa, density $\varrho$ 1.236 g/cm$^3$ at 20 °C, refractive index $n_D^{20}$ 1.5641.

3-Chlorobenzaldehyde is formed in the nuclear chlorination of the benzaldehyde–aluminum chloride complex compound in the presence of chlorinated hydrocarbon solvents [53]. It can also be obtained from the diazonium salt of 3-aminobenzaldehyde by the Sandmeyer reaction.

**4-Chlorobenzaldehyde** [*104-88-1*], $C_7H_5ClO$ $M_r$ 140.57, is a colorless to yellowish crystalline mass with a penetrating odor and low solubility in water, *mp* 48–49 °C, *bp* 214.5–216.5 °C at 101.3 kPa, *bp* 98 °C at 1.87 kPa, flash point 88 °C.

4-Chlorobenzaldehyde is produced by the same methods as 2-chlorobenzaldehyde. According to a recently published method [54], 4-chlorobenzonitrile can be hydrogenated to form 4-chlorobenzaldehyde in the presence of an acid and a nickel catalyst that has been treated with a copper salt.

**2,4-Dichlorobenzaldehyde** [*874-42-0*], $C_7H_4Cl_2O$, $M_r$ 175.02, *mp* 72 °C, *bp* 233 °C at 101.3 kPa, *bp* 105–106 °C at 2.0 kPa, flash point 135 °C.

2,4-Dichlorobenzaldehyde is produced by hydrolysis of 2,4-dichlorobenzal chloride with sulfuric acid.

**2,6-Dichlorobenzaldehyde** [*83-38-5*], $C_7H_4Cl_2O$, $M_r$ 175.02, *mp* 71–72.5 °C.

2,6-Dichlorobenzaldehyde is formed by hydrolysis of 2,6-dichlorobenzal chloride with concentrated sulfuric acid. In aromatic sulfonic acids, hydrolysis can be catalyzed by zinc chloride, iron(III) chloride, or aluminum chloride [55]. 2,6-Dichlorobenzaldehyde has also been obtained by the oxidation of 2,6-dichlorotoluene with manganese dioxide and sulfuric acid. To separate the 2,6-dichlorobenzaldehyde from isomeric chlorine-containing benzaldehydes, the solution of the aldehyde mixture is extracted in an organic solvent with aqueous sodium bisulfite solution [56]. The 2,6-dichlorobenzaldehyde accumulates in the organic phase.

The chlorobenzaldehydes are important intermediates in the production of dyes, optical brighteners, agricultural chemicals, and pharmaceuticals.

## 8.2. Nitrobenzaldehydes

**2-Nitrobenzaldehyde** [552-89-6], $C_7H_5NO_3$, $M_r$ 151.12, mp 42 – 44 °C, bp 153 °C at 3.07 kPa, highly volatile in steam.

The classical method for the preparation of 2-nitrobenzaldehyde begins with cinnamic acid, which is nitrated to form a mixture of 2-nitro- and 4-nitrocinnamic acid. The two isomers are separated by fractional crystallization after which the 2-nitrocinnamic acid is oxidized with potassium permanganate to form 2-nitrobenzaldehyde. According to recent publications, 2-nitrobenzaldehyde can also be prepared by the oxidation of 2-nitrostyrene with ozone [57] or molecular oxygen [58]. Other processes begin with 2-nitrobenzyl bromide, which is oxidized to the aldehyde with dimethyl sulfoxide [59] or N-oxides of tertiary amines [52]. According to a similar process [60], 2-nitrobenzaldehyde can be obtained by the alkaline hydrolysis of 2-nitrobenzyl bromide followed by the oxidation of the intermediate 2-nitrobenzyl alcohol using nitric acid.

**3-Nitrobenzaldehyde** [99-61-6], $C_7H_5NO_3$, $M_r$ 151.12, mp 56 – 58 °C, bp 164 °C at 3.07 kPa.

3-Nitrobenzaldehyde is obtained by the nitration of benzaldehyde with a mixture of concentrated sulfuric acid and nitric acid at 5 – 10 °C. In this process, up to 20 % of the 2-isomer is formed.

**Warning:** *A risk of explosion exists while the crude product is being distilled under reduced pressure* [61]. The two isomers can be separated by exploiting the difference in their reactions with bisulfite.

**4-Nitrobenzaldehyde** [555-16-8], $C_7H_5NO_3$, $M_r$ 151.12, mp 107 °C, relatively involatile in steam.

4-Nitrobenzaldehyde can be obtained by the oxidation of 4-nitrotoluene with chromium(VI) oxide in glacial acetic acid. It can also be obtained by hydrolyzing 4-nitrobenzal bromide in 93 % sulfuric acid [62].

The nitrobenzaldehydes serve as intermediates in the production of dyes and pharmaceuticals.

## 8.3. Hydroxybenzaldehydes

**2-Hydroxybenzaldehyde** [*90-02-8*], salicylaldehyde, $C_7H_6O_2$, $M_r$ 122.12, is a yellowish, oily liquid with an odor resembling that of bitter almonds; it is volatile in steam. It occurs naturally in cassia oil and in the essential oils of several plants of the genus *Spirea*, mp 1.6 °C, bp 196 °C at 101.3 kPa, bp 92 °C at 2.93 kPa, density $\varrho$ 1.167 g/cm$^3$ at 20 °C, refractive index $n_D^{20}$ 1.5718, dynamic viscosity $\eta$ $2.50 \times 10^{-3}$ Pa · s at 25 °C, surface tension $\sigma$ $42 \times 10^{-3}$ N/m at 25 °C, dipole moment $\mu$ (measured in liquid benzene) $9.54 \times 10^{-30}$ C m (2.86 D) heat of combustion 27.29 kJ/g at 20 °C and constant pressure.

2-Hydroxybenzaldehyde is soluble in many organic solvents, in 20 % aqueous sodium carbonate, and in 10 % aqueous sodium hydroxide. At 86 °C, 1.7 g can be dissolved in 100 g of water. With iron(III) chloride, 2-hydroxybenzaldehyde gives an intense violet color. It forms yellow alkali salts through inner complexing.

The two most important processes for the manufacture of 2-hydroxybenzaldehyde begin with phenol. The best known process is based on the Reimer-Tiemann reaction, in which phenol is reacted with chloroform in the presence of an aqueous alkali. Moderate yields of 2-hydroxy- and 4-hydroxybenzaldehyde are obtained in a ratio of about 85:15. The use of aqueous methanol as a reaction medium [63] improves the yield, especially that of 4-hydroxybenzaldehyde. According to a modern variation [64] of the Reimer-Tiemann reaction, the process is carried out in the absence of water. An alkali-metal phenolate reacts with chloroform and a suspension of an alkali-metal hydroxide in an inert organic solvent.

One industrial process starts with phenol and forms 2-hydroxybenzyl alcohol (saligenin) by reacting triphenyl metaborate with formaldehyde [65]. The alcohol is catalytically oxidized in air to give 2-hydroxybenzaldehyde [66]. It is more economical to react phenol with formaldehyde in the presence of alkaline catalysts and in the absence of boric acid [67]. The yield is approximately 85 %. The ratio of 2-hydroxybenzyl alcohol to 4-hydroxybenzyl alcohol can be altered by varying the catalyst. The oxidation of the hydroxybenzyl alcohols is carried out in an aqueous solution with a platinum–lead–carbon catalyst and gives the corresponding hydroxybenzaldehydes in yields of more than 98 %.

Several other processes are based on side-chain chlorination of 2-cresol and saponification of the resulting dichloromethyl group to form the aldehyde group. The phenolic hydroxyl group is protected prior to chlorination by esterification with an inorganic or organic acid chloride, such as phosphorus oxychloride [68], phosgene [69], tetrachlorosilane [70], or acetyl chloride [71]. Hydrolysis splits the ester. 2-Hydroxybenzaldehyde can also be produced by electrolytic reduction of salicylic acid on a rotating amalgam cathode [72] and by catalytic reduction of a salicylic acid halide [73].

2-Hydroxybenzaldehyde is one of the most important derivatives of benzaldehyde. Estimated annual production worldwide is 4–6 kt. Because of its pleasant aromatic odor, 2- hydroxybenzaldehyde is used in perfumes and, on a large scale, as a starting material in the production of cumarin.

Its condensation products with amines have complex-forming properties and are used as additives, e.g., in petrochemistry. 2-Hydroxybenzaldehyde is also an intermediate for dyes, pharmaceuticals, plastics, photographic chemicals, agricultural chemicals, and electroplating chemicals.

**3-Hydroxybenzaldehyde** [*100-83-4*], $C_7H_6O_2$, $M_r$ 122.12, *mp* 108 °C, *bp* 240 °C at 101.3 kPa, heat of combustion 27.04 kJ/g at 20 °C and constant pressure, involatile in steam.

3-Hydroxybenzaldehyde can be produced by heating 3-chlorobenzaldehyde with aqueous potassium hydroxide under pressure at 250 °C [74] or by the reduction of 3-nitrobenzaldehyde to 3-aminobenzaldehyde and hydrolysis of the diazonium salt. It is used as an intermediate in producing dyes.

**4-Hydroxybenzaldehyde** [*123-08-0*], $C_7H_6O_2$, $M_r$ 122.12, *mp* 117 °C, sublimes at 110–120 °C and 4 Pa, heat of combustion 27.18 kJ/g at 20 °C and constant pressure involatile in steam. At 18 °C, 0.81 g dissolves in 100 g of water.

4-Hydroxybenzaldehyde can be produced similarly to 2-hydroxy- and 3-hydroxybenzaldehyde. It can also be produced by treating benzaldehydes or acetalized benzaldehydes that have a benzyloxy residue in the 4-position with aqueous hydrogen halides [75]. The benzaldehyde derivatives used as starting materials are obtained by catalytic or electrochemical oxidation of *p*-benzyloxytoluene.

4-Hydroxybenzaldehyde is used in electroplating and is an important intermediate for making dyes, pharmaceuticals, textile auxiliaries, odorants, and agricultural chemicals.

## 8.4. Aminobenzaldehydes

**2-Aminobenzaldehyde** [*529-23-7*], $C_7H_7NO$, $M_r$ 121.14, *mp* 39–40 °C, *bp* 80–85 °C at 270 Pa, low solubility in water.

2-Aminobenzaldehyde is obtained by the reduction of 2-nitrobenzaldehyde with iron(II) sulfate and ammonia or by catalytic hydrogenation. It is also formed in the hydrolysis of 2-dichloromethylphenyl isocyanate [76].

**4-Aminobenzaldehyde** [*556-18-3*], $C_7H_7NO$, $M_r$ 121.14, *mp* 71 °C, soluble in water.

The preferred method for the production of 4-aminobenzaldehyde is the conversion of 4-nitrotoluene with sodium polysulfide. The reaction is carried out in an aprotic polar solvent, such as dimethylformamide [77].

The aminobenzaldehydes have a strong tendency to undergo self-condensation, in which they expel water. The aldehydes should therefore be isolated in the form of their stable N-acetyl compounds. They are used as intermediates in the production of dyes. 2-Aminobenzaldehyde is a starting material in the manufacture of quinoline derivatives [76].

## 8.5. Methylbenzaldehydes

**2-Methylbenzaldehyde** [*529-20-4*], $C_8H_8O$, $M_r$ 120.15, liquid with the odor of bitter almonds, *bp* 199–200 °C at 101.3 kPa, *bp* 102–104 °C at 4.5 kPa, density $\varrho$ 1.0328 at 20 °C, refractive index $n_D^{20}$ 1.5462.

2-Methylbenzaldehyde can be produced by partial oxidation of *o*-xylene, e.g., in acetic acid with cobalt salts as catalysts and bromine compounds as promoters [78]. It can also be obtained from 2-methylbenzyl chloride by gas-phase oxidation at 350–450 °C in the presence of $Al_2O_3$–$V_2O_5$ catalysts [79]. In the Sommelet reaction, 2-methylbenzyl chloride reacts with hexamethylenetetramine to form 2-methylbenzaldehyde. This reaction can also use industrial chlorination mixtures containing small amounts of highly chlorinated products in addition to 2-methylbenzyl chloride and unreacted *o*-xylene [80].

**3-Methylbenzaldehyde** [*620-23-5*], $C_8H_8O$, $M_r$ 120.15, liquid with the odor of bitter almonds, *bp* 200 °C at 101.3 kPa, *bp* 83 °C at 2.13 kPa, density $\varrho$ 1.019 at 20 °C, refractive index $n_D^{20}$ 1.5410.

3-Methylbenzaldehyde can be prepared similarly to 2-methylbenzaldehyde.

**4-Methylbenzaldehyde** [*104-87-0*], $C_8H_8O$, $M_r$ 120.15, liquid with a pepper-like odor, *bp* 204 °C at 101.3 kPa, *bp* 85–86 °C at 2.0 kPa, density $\varrho$ 1.019 at 20 °C, refractive index $n_D^{20}$ 1.5454, dipole moment $\mu$ (in liquid benzene) $10.81 \times 10^{-30}$ C m (3.24 D).

4-Methylbenzaldehyde is produced industrially by a modified Gattermann-Koch reaction. Toluene is reacted with carbon monoxide in the presence of hydrogen fluoride and tantalum(V) fluoride [81] or hydrogen fluoride and boron trifluoride [82]. It can also be produced similarly to 2-methylbenzaldehyde.

The modified Gattermann-Koch synthesis for the production of 4-methylbenzaldehyde [82] is the first of two stages in a process for the production of terephthalic acid from the inexpensive raw materials toluene and carbon monoxide [83].

The methylbenzaldehydes serve as intermediates (e.g., in the production of terephthalic acid) and are also used in the flavors industry.

# 9. Economic Aspects

The benzaldehyde production capacities of Western Europe, the United States, and Japan in 1983 were approximately 14 kt, 7 kt, and 3.5 kt, respectively. The corresponding production figures were approximately 12 kt, 4.5 kt, and 2 kt.

# 10. Toxicology

Benzaldehyde is used as a flavoring and fragrance in food, cosmetics, pharmaceuticals, and soap and is "generally regarded as safe" (GRAS) by the U.S. Food and Drug Administration. Benzaldehyde has industrial and agricultural applications, but no environmental or workplace exposure limits have been proposed.

Acute toxicity of benzaldehyde is moderate, with oral $LD_{50}$ values reported to be 1.3 g/kg in rats and 1 g/kg in guinea pigs [84]. The probable lethal dose by mouth for a 70-kg human is estimated to be 50 mL. Subchronic oral administration to rodents produced no effects at daily doses of 400 mg/kg in rats and 300–600 mg/kg in mice. Higher doses caused necrotic and degenerative lesions in brain, kidney, and forestomach [85]. Tests for mutagenicity were negative in the Ames TA 100 *Salmonella* strain at concentrations of $0.1-10^3$ μg per plate [86]. No studies of carcinogenic, teratogenic, or reproductive effects have been reported.

Effects of toxic doses in acute and subchronic studies included depression, inactivity, tremors, seizures, and coma. Death may result from respiratory depression. Benzaldehyde has a weak local anesthetic effect and is mildly irritating to the eye and upper respiratory tract. Skin irritation is moderate, and allergic sensitivity may occur in some individuals [87].

Contact with eyes and skin should be avoided by use of gloves and protective clothing. Self-contained breathing apparatus should be used in poorly ventilated areas to prevent inhalation exposure.

# 11. References

General References

*Beilstein*, **7**, 174; **7 (1)**, 113; **7 (2)**, 145; **7 (3)**, 805; **7 (4)**, 505. *Houben-Weyl*, vol. **E 3** (1983).
*Ullmann*, 4th ed., vol. **8**, pp. 343–351.T. Laird in:
*Comprehensive Organic Chemistry*, vol. **1**, Pergamon Press, Oxford 1979, pp. 1105–1160.

Specific References

[1] J. Gmehling, U. Onken, W. Arlt: "Vapor-Liquid Equilibrium Data Collection," *Dechema Chemistry Data Series,* vol. **I,** part 2 d (1982) 474–475, 568–571, 606, 620; part 3 + 4 (1979) 62–63.
[2] L. H. Horsley: "Azeotropic Data-III," *Adv. Chem. Ser.* **116** (1973) 360–362.
[3] Stamicarbon B.V., DE-OS 2421268, 1973 (J. H. Gregoire).
[4] H. Stetter, R. Y. Rämsch, H. Kuhlmann, *Synthesis* 1976, 733.
[5] W. H. Shearon, H. E. Hall, J. E. Stevens, *Ind. Eng. Chem.* **41** (1949) 1812–1820.
[6] EKA AB, EP-A 64486, 1981 (R. K. Rantala, G. L. F. Hag).
[7] B. Ya. Libman et al., *Zh. Prikl. Khim. (Leningrad)* **39** (1966) 1669–1670; *Chem. Abstr.* **65** (1966)18518 f.
[8] Nippon Kayaku Co., JP-Kokai 7589337, 1973 (T. Ishikura); *Chem. Abstr.* **83** (1975) 205940 s.
[9] Diamond Shamrock Corp., US 3499934, 1967 (W. J. Pyne).
[10] Albright & Wildon, DE-OS 2261616, 1971 (H. Coates, W. E. Billingham).
[11] Bayer, DE-OS 2752612, 1977 (F. Brühne, K. A. Lipper).
[12] Bayer, EP-A 41672, 1980 (H. U. Blank, E. Wolters).
[13] Argus Chem. Corp., US 3691217, 1970 (T. J. McCann).
[14] Ciba-Geigy, DE-AS 2044832, 1969 (P. Liechti, F. Blattner).
[15] General Aniline & Film Corp., US 3087967, 1960 (D. E. Graham, W. C. Craig).
[16] Tenneco Chem. Inc., US 3524885, 1967 (A. J. Deinet).
[17] Instytut Przemyslu Organicznego, PL 93511, 1975 (D. Narkiewicz, J. Legocki); *Chem. Abstr.* **90** (1979) 87043 w.
[18] Hodogaya Chem. Co., JP 7107927, 1967 (Y. Murakami, T. Koizumi); *Chem. Abstr.* **75** (1971) 19945 g.
[19] Central Glass Co., DE-OS 3226490, 1981 (T. Kondow, K. Okazaki, Y. Katsuhara, K. Matsuoka).
[20] Ihara Chem. Ind. Co., JP 8202699, 1971; *Chem. Abstr.* **97** (1982) 5980 g.
[21] H. Pichler, F. Obenaus, *Brennst. Chem.* **45** (1964) 97–103.
[22] A. Guyer, DE-AS 1236493, 1963 (A. Guyer, P. Guyer, G. Gut).
[23] Mitsubishi Chem. Ind. Co., JP-Kokai 79151937, 1978 (T. Onoda, K. Wada, S. Fujii); *Chem. Abstr.* **92** (1980) 180826 e.
[24] BASF, DE-OS 2136779, 1971 (R. Krabetz, Ch. Dudeck, W. Fuchs).
[25] National Distillers & Chem. Corp., US 3946067, 1972 (J. Kwiatek, J. H. Murib, C. K. Brush).
[26] Standard Oil Co. (Indiana), US 4005049, 1975 (E. K. Fields).
[27] Stamicarbon B.V., DE-OS 2730761, 1976 (P. C. van Geem, A. J. Teunissen).
[28] Ashland Oil Inc., US 4390728, 1981 (C. Daniel).
[29] Institut Français du Pétrole, des Carburants et Lubrifiants, FR 1568763, 1967 (B. Delmon).
[30] Mitsubishi Chem. Ind. Co., JP 7528946, 1970 (Y. Suzuki, T. Maki, H. Nakajima, K. Mineda); *Chem. Abstr.* **84** (1976) 89823 a.
[31] Universal Oil Products Co., US 3775472, 1970 (S. N. Massie).
[32] Universal Oil Products Co., US 3723517, 1970 (S. N. Massie).
[33] Universal Oil Products Co., US 3790624, 1972 (S. N. Massie, H. S. Bloch).
[34] Mitsubishi Gas Chem. Co., Inc., JP-Kokai 8062029, 1978; *Chem. Abstr.* **93** (1980) 149992 c.
[35] Union Carbide Corp., US 3931330, 1972 (N. S. Aprahamian).
[36] Mitsubishi Chem. Ind. Co., JP-Kokai 7805132, 1976 (Y. Murao, T. Nakanome, T. Yamaura); *Chem. Abstr.* **88** (1978) 169778 n.
[37] Phillips Petroleum Co., US 4088823, 1975 (H. D. Holtz, L. E. Gardner).

[38] Mitsubishi Chem. Ind. Co., JP-Kokai 75108231, 1974 (A. Wada, N. Nakajima, T. Hironaka); *Chem. Abstr.* **84** (1976) 43580 r.
[39] Hoechst, EP-A 71166, 1981 (H. Kuckertz, G. Schaeffer).
[40] National Distillers & Chem. Corp., DE-OS 2745511, 1976 (J. A. Scheben).
[41] Mitsubishi Chem. Ind. Co., JP 7424467, 1969 (K. Okuno, K. Itagaki, T. Hironaka); *Chem. Abstr.* **82** (1975) 72638 e.
[42] Stamicarbon B.V., EP-A 15616, 1979 (C. Jongsma).
[43] Stamicarbon B.V., EP-A 1660, 1977 (J. Elmendorp).
[44] Mobil Oil Corp., US 3369048, 1963 (L. A. Hamilton, P. S. Landis).
[45] Stamicarbon B.V., EP-A 37149, 1980 (T. F. De Graaf, H. J. Delahaye).
[46] Mitsubishi Petrochemical Co., JP-Kokai 8087739, 1978; *Chem. Abstr.* **94** (1981) 103012 a.
[47] Mitsubishi Gas Chem. Co., JP-Kokai 8218644, 1980; *Chem. Abstr.* **97** (1982) 72084 f.
[48] Fuso Kagaku Kogyo Co., JP-Kokai 7823913, 1976 (K. Kobayashi, N. Ishino, Y. Ota, S. Wakita); *Chem. Abstr.* **89** (1978) 23964 u.
[49] Dow Chem. Co., US 4328373, 1980 (E. J. Strojny).
[50] UOP Inc., US 4174351, 1978 (J. P. Shoffner).
[51] Bayer, DE-OS 2026817, 1970 (J. Schneider).
[52] Dynamit Nobel, DE-OS 2948058, 1979 (G. Bernhardt, E. N. Petersen, G. Daum).
[53] Fujisawa Pharmaceutical Co., JP-Kokai 7777021, 1975 (K. Kariyone, H. Yagi); *Chem. Abstr.* **87** (1977) 201095 e.
[54] Sumitomo Chem. Co., EP-A 87298, 1982 (H. Yamachika, H. Nakanishi).
[55] S. Kobayashi, JP-Kokai 7966638, 1977; *Chem. Abstr.* **91** (1979) 175009 r.
[56] Hooker Chem. & Plastics Corp., US 4136122, 1978 (B. R. Cotter).
[57] Ciba-Geigy, DE-OS 2829346, 1977 (J. Gosteli).
[58] Bayer, DE-OS 2805402, 1978 (F. Hagedorn, L. Imre, K. Wedemeyer).
[59] Bayer, DE-OS 2808930, 1978 (W. Ertel).
[60] VEB Arzneimittelwerk Dresden, DE-OS 2708115, 1976 (W. Sauer, H. Goldner, H. J. Heidrich, G. Faust, W. Fiedler, E. Carstens, G. Heine).
[61] J. Lange, T. Urbański, *Chem. Ind. (London)* 1967, 1424.
[62] Nippon Kayaku Co., JP-Kokai 7589337, 1973 (T. Ishikura); *Chem. Abstr.* **83** (1975) 205940 s.
[63] Dow Chem. Co., US 3365500, 1964 (D. F. Pontz).
[64] Sumitomo Chem. Co., EP-A 68725, 1981 (K. Hamada, G. Suzukamo).
[65] Rhône-Poulenc S. A., FR 1328945, 1962 (P. Marchand, J. B. Grenet).
[66] Rhône-Poulenc S. A., FR 2305420, 1975 (J. Le Ludec).
[67] H. Fiege, K. Wedemeyer, K. A. Bauer, A. Krempel, R. G. Mölleken: "Fragrance and Flavor Substances," in R. Croteau (ed.): *Fragrance Flavor Subst. Proc. Int. Haarman & Reimer Symp. 2nd* **1979**, 63–73; *Chem. Abstr.* **95** (1981) 150058 a.
[68] Tenneco Chem., Inc., US 3641158, 1969 (A. J. Deinet, D. X. Klein).
[69] BASF, DE-OS 1925195, 1969 (H. Hoffmann, J. Datow, G. Wenner).
[70] Tokuyama Soda Co., JP-Kokai 83124729, 1982; *Chem. Abstr.* **99** (1983) 158016 p.
[71] Yoshitomi Pharmaceutical Ind., JP 7303831, 1968 (K. Saruwatari, T. Gono, K. Tsubone); *Chem. Abstr.* **79** (1973) 18387 x.
[72] K. S. Udupa, G. S. Subramanian, H. V. K. Udupa, *Ind. Chem.* **39** (1963) 238–241; *Chem. Abstr.* **59** (1963) 10986 b.
[73] Seimi Kagakagu Co., JP 6813204, 1965 (S. Abe, K. Sato, T. Asami, T. Amakasu, T. Itakura); *Chem. Abstr.* **70** (1969) 28646 j.
[74] Bayer, DE 942808, 1953 (W. Müller, W. Möllering, W. Schommer).

[75] BASF, DE-OS 2904315, 1979 (M. Barl, D. Degner, H. Siegel).
[76] Bayer, DE-OS 2730061, 1977 (V. Ehrig, H. S. Bien, E. Klauke, D. I. Schütze).
[77] K. Kato, M. Kawamura, T. Nishi, H. Hata, *Nippon Kagaku Kaishi* 1981, no. 2, 255–258; *Chem. Abstr.* **95** (1981) 6705 s.
[78] Agency of Industrial Sciences and Technology; Sanko Chemical Co., JP-Kokai 79157534, 1978 (J. Imamura, K. Kizawa); *Chem. Abstr.* **92** (1980) 215066 m.
[79] Rhône-Poulenc S. A., DE-OS 2062522, 1969 (Y. Colleuille, R. Perron).
[80] BASF, DE-OS 2942894, 1979 (W. Schoch, M. Kröner, R. Widder).
[81] Texaco Development Corp., US 4218403, 1979 (S. H. Vanderpool).
[82] Mitsubishi Gas Chemical Co., Inc., DE-OS 2460673, 1973 (S. Fujiyama, T. Takahashi, S. Kozao, T. Kasahara).
[83] Mitsubishi Gas Chemical Co., Inc., GB 2025956, 1978 (M. Komatsu, T. Ohta, T. Tanaka, R. Oda, Y. Takamizawa). S. Fujiyama, T. Kasahara, *Hydrocarbon Process.* **57** (1978) no. 11, 147–149.
[84] P. M. Jenner, E. C. Hagan, J. M. Taylor, E. L. Cook, O. G. Fitzhugh, *Food Cosmet. Toxicol.* **2** (1964) 327–343.
[85] W. M. Kluwe, C. A. Montgomery, *Food Cosmet. Toxicol.* **21** (1983) 245–250.
[86] W. H. Rapson, M. A. Nazar, V. V. Butsky, *Bull. Environ. Contam. Toxicol.* **24** (1980) 590–596.
[87] D. L. J. Opdyke, *Food Cosmet. Toxicol.* **14** Suppl. (1976) 693–698.

# Benzene

HILLIS O. FOLKINS, (Union Oil Company of California), Claremont, California 91711, United States

| | | | | | | |
|---|---|---|---|---|---|---|
| 1. | Introduction | 693 | 6. | Separation and Purification | 714 |
| 2. | Physical Properties | 694 | 6.1. | Removal of Dienes, Olefins, and Sulfur Compounds | 715 |
| 3. | Chemical Properties | 694 | | | |
| 4. | Raw Materials | 698 | 6.2. | Close Fractionation | 716 |
| 5. | Production | 700 | 6.2.1. | Azeotropic Distillation | 716 |
| 5.1. | Production from Coal | 701 | 6.2.2. | Extractive Distillation | 717 |
| 5.2. | Production from Petroleum | 704 | 6.2.3. | Liquid–Liquid Extraction | 717 |
| 5.2.1. | Benzene from Reformate | 704 | 6.3. | Processes for Aromatic Separation | 718 |
| 5.2.2. | Aromatics from Pyrolysis Gasoline | 706 | 7. | Quality Specifications | 726 |
| 5.3. | Methods of Increasing Production | 707 | 8. | Storage and Transportation | 728 |
| | | | 9. | Economic Aspects | 728 |
| 5.3.1. | Hydrodealkylation | 708 | 10. | Uses | 732 |
| 5.3.2. | Disproportionation | 710 | 11. | Toxicology and Occupational Health | 733 |
| 5.3.3. | Combination Dealkylation Processes | 711 | | | |
| 5.3.4. | Benzene from Other Sources | 714 | 12. | References | 737 |

# 1. Introduction

Benzene [71-43-2] is a single-ring, aromatic compound, $C_6H_6$, $M_r$ 78.11; colorless, flammable liquid, bp 80.1 °C, fp 5.5 °C. The term benzene denotes the pure compound; benzol is still used to a small degree in some countries to represent the compound or a material having benzene as its main component. Benzine, on the other hand, is a low-boiling hydrocarbon mixture or naphtha, often nonaromatic in composition.

Benzene

Benzene is thermally stable but is chemically reactive and thus serves as a source for the production of many petrochemicals and hydrocarbon materials. Other chemicals derived from benzene are styrene, phenol, and cyclohexane; manufactured products include plastics, resins, and other intermediates used in insecticides, drugs, dyes, and detergents.

Benzene is an excellent solvent, but because of its high toxicity its use has decreased greatly, with less toxic materials taking its place. Benzene has a high octane rating and is an important component of gasolines.

Benzene was first isolated by M. FARADAY in 1825 [1]; he isolated "bicarburet of hydrogen" from experiments on the pyrolysis of whale oil and other materials. A. W. HOFMANN and C. MANSFIELD of the Royal College of Chemistry also were working on liquids obtained from pyrolysis of coal. They developed commercial processes for obtaining benzene and other aromatics from coal tar between 1840 and 1850.

Prior to World War II, benzene was obtained mostly from coal. Catalytic processes developed in the petroleum industry have resulted in petroleum currently being the main source for benzene and related aromatics.

## 2. Physical Properties

The structure of benzene was postulated by A. KEKULÉ in 1865 as a planar, six-carbon ring with alternating single and double bonds and with one hydrogen atom attached to each carbon atom. Thermodynamic and spectral studies indicate that this simple picture cannot explain the true nature of benzene or related molecules. For instance, when benzene is formed from cyclohexane, less energy (151 kJ, 36 kcal) is required than is needed for the formation of three double bonds. The NMR spectrum of benzene shows the protons at lower field than would be expected for simple vinylogous ones. Spacing between the neighboring carbon atoms is constant (0.139 nm), which is less than single-bond spacing (0.154 nm), and more than, but close to, double-bond spacing (0.134 nm); C–H bond spacing is 0.108 nm. Benzene is thermally stable. It does not react as a compound having typical double bonds; however, it is more reactive than simple hydrocarbons.

Tables 1, 2, 3, 4, 5, and 6 give important physical data for benzene and for related aromatic compounds.

## 3. Chemical Properties

Benzene is the basic unit of the aromatic class of compounds. It is one of the largest volume organic chemicals, with the United States being the largest producer. Benzene is the source of a variety of organic chemicals, many of which are intermediates for the production of a host of commercial products. Benzene is thermally stable and its formation is kinetically and thermodynamically favored at temperatures $\geq 500$ °C.

**Table 1.** Physical data of $C_6$–$C_8$ aromatics

| | $M_r$ | Density $d_{20}$ | mp, °C | bp, °C | $n_D^{20}$ | Critical density, g/cm³ | Critical pressure, bar | Critical temp., °C | Flash point, °C |
|---|---|---|---|---|---|---|---|---|---|
| Benzene [71-43-2] | 78.11 | 0.87901 | + 5.533 | 80.099 | 1.50112 | 0.309 | 48.9 | 288.9 | −11 |
| Toluene [108-88-3] | 92.13 | 0.86694 | − 94.991 | 110.626 | 1.49693 | 0.291 | 40.7 | 319.9 | 4.0 |
| Ethylbenzene [100-41-4] | 106.16 | 0.8669 | − 94.975 | 136.186 | 1.49588 | 0.286 | 37.4 | 346.4 | 15 |
| o-Xylene [56004-61-6] | 106.16 | 0.88020 | − 25.182 | 144.411 | 1.50545 | 0.288 | 36.6 | 357.9 | 17 |
| m-Xylene [108-38-3] | 106.16 | 0.86417 | − 47.872 | 139.104 | 1.49722 | 0.286 | 37.0 | 353 | 23.2 |
| p-Xylene [41051-88-1] | 106.16 | 0.86105 | + 13.263 | 138.351 | 1.49582 | 0.286 | 36.5 | 343 | 25 |

| | Ignition temp., °C | Explosion limits in air, vol% | | Heat of fusion, J/g | Heat of combustion, kJ/mol | Gross heating value, J/g | Specific electrical conductivity, cm⁻¹ Ω⁻¹ |
|---|---|---|---|---|---|---|---|
| | | lower | upper | | | | |
| Benzene | 595 | 1.4 | 6.7 | 125.9 | − 3275.3 | 41932 | $3.8 \times 10^{-14}$ |
| Toluene | 552 | 1.4 | 6.7 | 71.8 | − 3911.3 | 42454 | $2 - 7 \times 10^{-13}$ |
| Ethylbenzene | 460 | 1.0 | – | 86.3 | − 4568.6 | 43060 | – |
| o-Xylene | 502 | 1.0 | 6.0 | 128.2 | − 4570.7 | 43079 | $< 10^{-15}$ |
| m-Xylene | 562 | 1.1 | 7.0 | 109.0 | − 4556.9 | 42949 | – |
| p-Xylene | 500 | 1.1 | 7.0 | 161.2 | − 4559.8 | 42977 | – |

Therefore elevated temperatures are required for its thermal decomposition or for condensation or dehydrogenation reactions to occur. As an example, at ca. 650 °C in contact with iron, lead, or over other catalytic materials, such as vanadium, *condensation* reactions take place to form diphenyl and other polyaromatic compounds.

Benzene is quite stable to *oxidation,* but under severe conditions it is oxidized to water and carbon dioxide. With a deficiency of air or oxygen under oxidizing conditions, partial decomposition and deposition of soot occur. Oxidation with air or oxygen in the vapor phase at 350 – 450 °C over a V–Mo catalyst produces maleic anhydride in yields of 65 – 70 % [2]. The use of pure oxygen offers no advantage over air [3]. Phenol can be obtained in low yield from the high-temperature oxidation of benzene with air [4].

*Substitution* reactions of benzene are of primary importance. Depending on reaction conditions one or more of the hydrogen atoms in the benzene ring may be exchanged for nitro or sulfonic acid radicals, for amine or hydroxyl groups, or for atoms, such as chlorine or bromine. Products include phenol, nitrobenzene, chlorobenzene, benzenesulfonic acid, and others. For disubstituted benzene three isomers are possible:

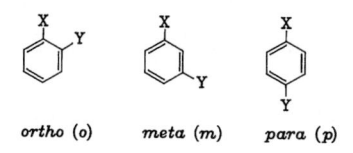

*ortho (o)*   *meta (m)*   *para (p)*

Other important reactions of benzene involve *addition* and include alkylation and hydrogenation. These reactions generally take place at elevated temperature and pressure, sometimes requiring active catalysts. Ethylbenzene results from the alkylation of

**Table 2.** Physical properties of benzene at different temperatures

| Temp., | Vapor pressure, | Specific heat, liquid, | Specific heat, vapor, | Heat of vaporization, | Thermal conductivity, | Viscosity, | Dielectric constant | Miscibility of water and benzene | |
|---|---|---|---|---|---|---|---|---|---|
| | | | | | | | | % water in benzene | % benzene in water |
| K | kPa | J/g | J/g | J/g | W m$^{-1}$K$^{-1}$ | mPa · s | | | |
| 223 | 0.103 | 1.097 | | | | | | | |
| 243 | 0.479 | 1.223 | | | | | | | |
| 263 | 1.950 | 1.491 | | | | | | | |
| 273 | 3.466 | 1.612 | 1.114 | 448.0 | | 0.906 | | | |
| 283 | 5.965 | 1.683 | | | | | | 0.034 | |
| 292 | | | | | | | 2.26 | | |
| 293 | 9.97 | 1.708 | | 435.0 | 14.58 | 0.654 | | | |
| 298 | | | | | | | 2.282 | 0.066 | 0.113 |
| 313 | 24.19 | 1.763 | | 421.6 | 14.06 | | | 0.095 | |
| 323 | 35.82 | | | | | 0.499 | | | |
| 328 | | | | | | | | 0.113 | |
| 330 | | | | | | | | | 0.248 |
| 333 | 38.4 | | | 408.2 | 13.54 | 0.398 | | | |
| 343 | 72.2 | 1.900 | | | | | | | |
| 353 | 99.7 | | | 394.4 | 12.02 | 0.336 | 2.17 | | |
| 363 | 135 | 1.943 | | | | | | | |
| 373 | 178 | | 1.260 | 379.3 | | | | | |
| 380 | | | | | | | | | 0.507 |
| 413 | 469.2 | | | 345.8 | | | | | |
| 433 | 765 | | | 329.1 | | | | | |
| 453 | 1015.3 | | | 310.2 | | | | | |
| 473 | 1419.6 | | 1.394 | 286.8 | | | | | |
| 493 | 1935.6 | | | | | | | | |
| 533 | 2579.4 | | | | | | | | |
| 553 | 4369.8 | | | | | | | | |
| 573 | | | 1.528 | | | | | | |
| 673 | | | 1.617 | | | | | | |

benzene with ethylene in the presence of a catalyst, such as aluminum chloride [5] (→ Ethylbenzene). The reaction is carried out at 40–100 °C, gauge pressure < 0.7 MPa. With less active catalysts higher temperature and pressure are employed [6].

The production of cumene by the vapor-phase catalytic alkylation of benzene with propene is another important primary addition reaction of benzene. The reaction takes place at 200–250 °C and gauge pressure of 2.7–4.2 MPa over active catalysts, such as phosphoric acid or kieselguhr, with yields of ca. 95 % [7].

Hydrogenation is also an addition reaction. One of the most important is the hydrogenation of benzene at elevated temperature and pressure to produce cyclohexane. The reaction may be carried out in the liquid phase [8] or in the vapor phase at higher temperatures.

The dehydrogenation of ethylbenzene to form styrene is an important commercial reaction. High temperature is necessary both kinetically and thermodynamically for this endothermic reaction to proceed.

**Table 3.** Binary azeotrope mixtures of benzene

| Component B | Benzene, wt% | Azeotrope bp, °C |
|---|---|---|
| Water | 91.17 | 69.25 |
| Formic acid | 31.0 | 71.05 |
| Nitromethane | 85.7 | 79.2 |
| Methanol | 60.4 | 58.34 |
| Acetonitrile | 60.0 | 73.7 |
| Acetic acid | 98.0 | 80.5 |
| Ethyl nitrate | 85.0 | 80.08 |
| Ethyl alcohol | 67.6 | 68.24 |
| Acrylonitrile | 53.0 | 73.3 |
| Allyl alcohol | 82.6 | 76.75 |
| 1,3-Dioxolane | 15.0 | 74.0 |
| Dimethyl carbonate | 99.0 | 80.17 |
| Isopropyl alcohol | 66.7 | 71.92 |
| Propyl alcohol | 83.1 | 77.12 |
| Diacetyl | 45.0 | 79.3 |
| 2-Butanone | 37.5 | 78.4 |
| Dioxane | 88.0 | 82.4 |
| Ethyl acetate | 6.0 | 76.95 |
| Methyl propionate | 48.0 | 79.45 |
| Propyl formate | 53.0 | 78.5 |
| Isobutyl alcohol | 90.7 | 79.84 |
| sec-Butyl alcohol | 84.0 | 78.8 |
| tert-Butyl alcohol | 63.4 | 73.95 |
| tert-Amyl alcohol | 85.0 | 80.0 |
| Cyclohexene | 85.0 | 79.45 |
| Cyclohexane | 55.0 | 77.5 |
| Methylcyclopentane | 10.0 | 71.4 |
| n-Hexane | 5.0 | 68.87 |
| 2,4-Dimethylpentane | 48.4 | 75.2 |
| 2,3-Dimethylpentane | 79.5 | 79.2 |
| n-Heptane | 99.3 | 80.1 |
| 2,2-Dimethylpentane | 46.3 | 75.85 |
| Trimethylbutane | 50.5 | 76.6 |
| 2,2,4-Trimethylpentane | 97.9 | 80.01 |

**Table 4.** Ternary azeotrope mixtures of benzeneA

| Component B | C | Benzene, wt% | B, wt% | Azeotrope bp, °C |
|---|---|---|---|---|
| Water | ethyl alcohol | 74.1 | 7.4 | 64.86 |
| Water | allyl alcohol | 82.1 | 8.6 | 68.3 |
| Water | n-propyl alcohol | 82.4 | 8.6 | 68.48 |
| Water | isopropyl alcohol | 73.8 | 7.5 | 66.51 |
| Water | 2-butanone | 73.6 | 8.9 | 68.9 |
| Water | sec-butyl alcohol | 85.55 | 8.63 | 69.0 |
| Water | tert-butyl alcohol | 70.5 | 8.1 | 67.3 |
| Chloroform | 1,2-dichloroethane | 66.4 | 5.0 | 79.2 |
| Water | acetonitrile | 68.5 | 8.2 | 66.0 |

**Table 5.** Volume (or cubic) expansion coefficient of benzene

| Temperature, K | $\gamma \times 10^6$ |
|---|---|
| 90–203 | 350 |
| 293 | 1237 |
| 273–303 | 1229 |
| 278–279 | 1160 |
| 279–283 | 1140 |
| 283–293 | 1180 |
| 203–313 | 1280 |
| 313–333 | 1380 |
| 333–353 | 1460 |
| 288–303 | 1060 |

**Table 6.** Boiling points and octane ratings of benzene and related compounds

| Compound | bp, °C | Octane rating, RON* |
|---|---|---|
| Benzene | 80.1 | 98 |
| Cyclohexane | 81.4 | 83 |
| Toluene | 110.6 | 120 |
| Ethylbenzene | 136.1 | 107 |
| o-Xylene | 144.4 | 120 |
| m-Xylene | 139.1 | 115 |
| p-Xylene | 138.4 | 116 |
| Cumene | 152.4 | 113 |

* Research octane number

Most of the reactions discussed in this chapter are primary reactions of benzene. A myriad of reactions and products rely on benzene, either directly or indirectly, as their source. Many of these are shown schematically in Figure 1.

# 4. Raw Materials

Benzene and its homologs, such as toluene and xylenes, are found in crude oils but in such small amounts that physical separation and recovery generally has not been economically feasible. Table 7 shows the aromatic, naphthenic, and paraffinic contents of the petroleum and naphtha fractions of Saudi Arabian crudes and those of a similar fraction of a United Kingdom crude oil [9], [10]. Because the naphtha fractions represent < 15 % of the total respective crudes, the percentages of benzene, toluene, and xylene (BTX) are quite low. The benzene contents in these fractions are ≦ 50 % of the total aromatic contents shown. However, the naphthenes, which are generally potential benzene precursors, are usually present in larger amounts than the aromatics.

With heavier crudes the aromatic contents of the light naphtha fractions may be somewhat greater. In synthetic crudes, such as those derived from tar sands, an increased concentration of aromatics in the light naphtha fraction can be expected. Such may not be the case, however. Table 8 contains an analysis of the $C_5$–150 °C

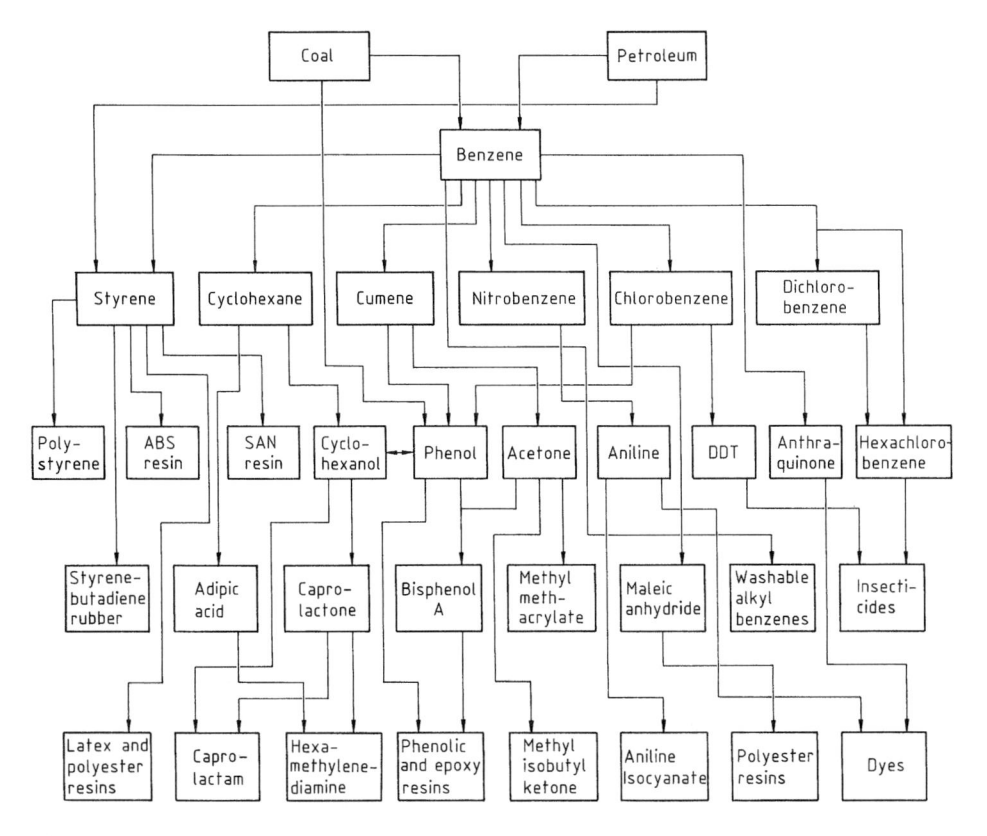

**Figure 1.** Products from benzene

naphtha fraction from a conventional crude and that of a synthetic naphtha derived from Athabasca tar sands [11]. The naphtha from the tar sands has lower contents of aromatics and naphthenes than does the fraction from the conventional crude. However, when the higher naphtha fractions are compared the synthetic crude has a considerably higher aromatic content (32 versus 19%), although the naphthene content is lower in the synthetic crude. The higher aromatic content of the heavier fraction can be explained on the basis of dealkylation or of ring opening of the higher molecular mass aromatics in the bitumen.

The commercial sources of benzene, as well as toluene and xylenes in primary products, are coal and petroleum. Historically benzene production has been a byproduct of the carbonization of coal to produce coke for the steel industry. More recently petroleum has become the primary source as catalytic and thermal methods have been developed for producing aromatic or related compounds. These materials are used to improve gasoline quality or as feedstocks to produce petrochemicals. For example, catalytic reformates, pyrolysis gasoline, and alkylated aromatics, such as toluene and xylenes, provide sources for the commercial production of benzene. Production methods are discussed in Chapter 5.

**Table 7.** Analysis of crude oil fractions

| Property | Saudi Arabian mixed | Thistle, U.K. | Saudi Arabian light |
|---|---|---|---|
| Crude gravity, °API | 34–34.9 | 37.4 | 34.1 |
| Sulfur, wt% | 1.63 | 0.31 | 1.72 |
| $C_5$ and lighter, vol% | 1.6 | 3.3 | 1.5 |
| Light naphtha, yield, vol% (debutanized) | 13.7 | 13.23 | 9.3 |
| T.b.p.* range, °C | $C_5$–126** | 31–105 | $C_5$–93** |
| End point, °C | 132 | | |
| Gravity, °API | 72.5 | 71.3 | 75.0 |
| Sulfur, wt% | 0.02 | 0.04 | 0.03 |
| Aromatics, vol% | 5 | 3.1 | 2.4 |
| Naphthenes, vol% | 5 | 30.5 | 12.3 |
| Paraffins, vol% | 90 | 66.4 | 85.3 |

\* T.b.p. is true boiling point distillation.
\*\* $C_5$–$t$ refers to $C_5$ plus content of the light naphtha.

# 5. Production

Until World War II, the coal industry supplied most of the commercial benzene in the United States and other countries. The pyrolysis of coal to yield metallurgical coke produced byproduct gas and aromatic liquid from which benzene and related aromatics were recovered by extraction and distillation. The small concentrations of benzene, toluene, xylene, etc., in crude oil fractions made this potential source noneconomic.

New processes discovered and commercialized in the petroleum industry during the 1930s and 1940s and the greatly increased demand for aromatics over this period changed the situation. Petroleum became more important as a raw material for the production of benzene and its homologs in the United States. Today, petroleum is the main source of the BTX aromatics – benzene, toluene, and xylenes.

During the 1930s catalytic cracking made its debut, first as fixed-bed processing and later as moving-bed processes. Cracked light naphthas contained a considerable aromatic content in contrast to the low concentrations found in virgin light naphthas. During this period catalytic reforming of virgin naphthas also became a reality. Aromatics, such as benzene, were made by dehydrogenation of cyclohexane or by isomerization and dehydrogenation of methylcyclopentane. *n*-Hexane was converted to benzene in smaller yields and under modified reaction conditions.

**Table 8.** Properties of naphthas from synthetic and conventional crudes

|  | Synthetic | Conventional |
|---|---|---|
| Boiling range | $C_5$ – 150 °C | |
| Density, g/cm³ | 0.697 | 0.718 |
| Sulfur, ppm | 1 | 30 |
| Hydrocarbons, vol%* | | |
|   Aliphatics | 69 | 43 |
|   Naphthenes | 25 | 46 |
|   Aromatics | 6 | 11 |
|   Benzene | 1.1 | 1.4 |
|   Toluene | 2.7 | 4.3 |
|   Xylenes | 2.0 | 0.5 |

* By GC analysis

## 5.1. Production from Coal

Light oils from the carbonization or coking of coal have long been sources of BTX. The carbonization process also produces a $C_5$ and lighter fraction and a heavier coal tar fraction. In the coking of coal, the yields and composition of the light oils depend on the type of coal being processed and on the carbonization temperature. *Low-temperature coking* was formerly employed to produce gas for heating and lighting. *Low-temperature (< 700 °C) carbonization* was important, especially in industrialized Europe before the advent of natural gas, petroleum, and other forms of energy. The benzene content of the BTX is lower in this case than when coking is carried out at a higher temperature. Low carbonization temperatures produce a rather low-density, clean burning coke.

*High-temperature carbonization* was developed to produce a hard coke for use in blast furnaces of the steel industry. *High-temperature coking* is done at $\geq$ 900 °C. A typical composition of light oils produced in coke-oven operation is (vol%): benzene (63), toluene (13), xylenes (8), higher aromatics (10), nonaromatics (6) [8].

Typical yields of liquid products obtained from slot-type coke-oven operations are 11 – 15 L of light oil and 30 – 38 L of coal tar pitch per ton of coal charged [12]. About 1% of ammonium sulfate and around 2800 L of coke-oven gas per ton of coke are produced. Coke and coke breeze yields are about 75%.

Whereas the benzene content of light oils derived from coke-oven operation is $\geq$ 65 wt%, the concentration is much lower (ca. 45%) in light oils derived from vertical retort operations. The typical, crude BTX material from coke oven operation mentioned previously contains considerable amounts of higher boiling aromatic and nonaromatic materials. These include around 5% of material boiling below benzene, which may be designated $C_5$ and lighter. These compounds may include aliphatics, simple olefins, and unsaturates, such as cyclopentadiene. Carbon disulfide, another low-boiling impurity, may be present at ca. 0.5%. It as well as thiophenes must be removed (see Chap. 6).

The overhead gas and the liquids from the coke-oven operation pass through a separation system. The gas and BTX fractions are separated from the coal tars and pass overhead to a recovery system where the $C_5$ and lighter fraction compounds are separated overhead from the BTX. The BTX fraction is recovered from the $C_5$ and lighter gas by countercurrent absorption using a high-boiling (generally 300–400 °C) petroleum fraction. Subsequent recovery of the BTX from the absorbent oil is accomplished by steam stripping. The BTX overhead products are condensed and separated from the water. One version of the process is covered in a U.S. patent issued to Bethlehem Steel [13]. A simplified flow sheet common to several processes is shown in Figure 2.

**Separation and Refining of BTX Fractions.** The crude BTX as obtained from the coke-oven recovery system must be refined to remove impurities so that the different BTX components meet purity specifications. Distillation removes $C_5$ and lighter hydrocarbons, carbon disulfide, and cyclopentadiene. Treatment with sulfuric acid removes thiophene and other sulfur compounds. Further purification by hydrotreating removes unsaturates, sulfur compounds, and nitrogen impurities. Hydrodealkylation reactions may accompany the hydrotreating process. Finally, recovery of the desired aromatics may be accomplished by solvent extraction and close fractionation to separate the individual aromatic components.

Individual processes for these different operations are being used. These include several combination processes that have been developed where the different chemical reactions are included in the overall process. The patent literature contains many references to combination processes designed to process coke-oven light oils into high-purity benzene or into high-quality BTX components. Modifications of these processes are used for the recovery of naphthalene from the crude BTX [14], [15].

One combination process, the *Houdry-Litol process,* has been in commercial operation since 1964. In 1981 there were six commercial installations operating in various countries [16], [17]. In this process a prefractionation removes the $C_5$ and lighter fraction, and the bottoms containing $C_9$ and heavier hydrocarbons. The BTX cut is passed to a pretreat catalytic reactor. Reactive unsaturates, e.g., cyclopentadienes and styrene, are selectively hydrogenated over a catalyst, such as cobalt molybdate on alumina. Hydrogenation of some sulfur compounds also occurs. Temperatures of ca. 250–350 °C and pressures of 5–7.5 MPa are used. More active catalysts allow lower reaction temperatures.

A second reaction system operates at $\geq$ 600 °C using a chromia–alumina catalyst. The reactions that take place include:

1) Desulfurization

$$CS_2 + 4\,H_2 \longrightarrow CH_4 + 2\,H_2S$$

2) Hydrocracking

$$C_6H_{12} + 3\,H_2 \longrightarrow 3\,C_2H_6$$

3) Aromatization

$$C_6H_{12} \longrightarrow C_6H_6 + 3\,H_2$$

or

$$C_6H_{14} \longrightarrow C_6H_6 + 4\,H_2$$

4) Hydrodealkylation

$$C_6H_5CH_3 + H_2 \longrightarrow C_6H_6 + CH_4$$

5) Dehydrogenation of naphthenes

$$C_6H_{12} \longrightarrow C_6H_6 + 3\,H_2$$

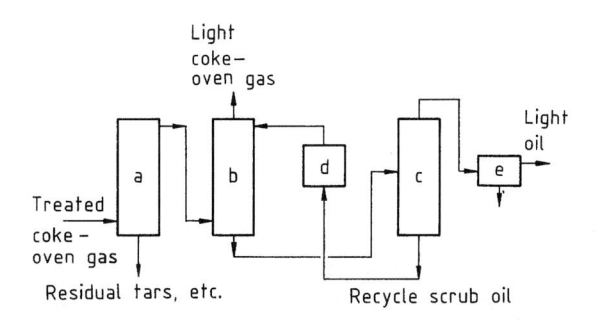

**Figure 2.** Recovery of light oil from coke-oven gas
a) Cooler and removal of residual tars and impurities; b) Light oil scrubber; c) Light oil stripper; d) Separation and cooling; e) Separator

A modified Litol process is more adaptable for complicated coke-oven aromatic materials [17]. In the Litol and related processes the readily polymerizable unsaturates should not exceed 2–3 % and nonaromatics should be limited to a similar amount. In the hydrodealkylation step about 50 % of the toluene and 75 % of the xylenes are dealkylated to benzene. After treatment of the Litol reactor effluent with clay, benzene is recovered and the remaining toluene separated by fractionation.

A small amount of benzene is available from the distillation of coke oven tars or tars from other coal carbonization processes. The low-boiling fractions containing BTX materials are treated with base to remove tar acids. Purification by hydrodealkylation produces benzene. Generally, however, the light oil fraction is 1.0 % or less of the coal tar [18]. Hence, this source is not of major importance in the production of benzene. Coal tars are more important as sources of pitches and of naphthalene, anthracene, or other condensed aromatic materials.

## 5.2. Production from Petroleum

Catalytic cracking and catalytic reforming of petroleum, adventing during the late 1930s and the 1940s, became a new source for the production of benzene, toluene, xylenes, and higher aromatics. Petroleum rather than coal began to emerge as the source of BTX materials, especially in the United States. In the United Kingdom and Europe the transition from coal to petroleum came around 1960.

Toluene was in great demand during World War II for the production of the explosive, trinitrotoluene (TNT). However, production of toluene from virgin naphthas and later from catalytically cracked and catalytically reformed naphthas was expensive. One of the early catalytic cracking processes was developed by Houdry Corp. [19], [20]. The silica–alumina catalyst used in this process was deposited in several fixed catalyst beds. The catalyst was regenerated frequently with air to burn deposited coke. Another process developed by Mobil Oil used a moving-bed silica–alumina catalyst [21]. The third type of process, developed by Exxon [22] and others, is known as the fluid process (FCC), and is now used exclusively.

Although the catalytic cracking processes were developed primarily for increasing the yield and octane rating of the motor fuels produced, the aromatic content of the gasolines was quite high [23] and so these processes provided a source for petrochemical benzene.

### 5.2.1. Benzene from Reformate

Catalytic reforming provided another petroleum source for BTX aromatics. Installation of these processes took place in the 1940s and early 1950s, at a time when the demand for benzene was expanding rapidly because of development of processes requiring it to manufacture resins, detergents, synthetic fibers, and other chemicals. Chemical reactions involved in reforming include dehydrogenation of naphthenes to aromatics, or isomerization of alkylnaphthenes followed by dehydrogenation. Likewise, paraffins may be dehydrocyclized to aromatics. The reaction is quite slow and process conditions must be adjusted to make it economic. A small amount of dealkylation of alkylated aromatics may occur.

Early catalytic reforming units used base-metal catalysts, generally molybdena on activated alumina. These processes operated on a short process cycle with frequent catalyst regeneration. The hydroforming process [24] operated at elevated temperatures of about 520 °C and at pressures of 2.5–7 MPa with hydrogen recycle. This process converted naphtha fractions into high-octane gasolines. The 10% molybdena on alumina catalyst required regeneration every 8–16 h. The process consisted of two parallel systems of two reactors in series. One pair of reactors could be in use while the catalyst in the other pair was being regenerated. Later catalytic reforming processes using base-metal catalysts employed either moving catalyst beds with the catalyst in granular form or used fluidized beds with a finely divided catalyst of small particle size.

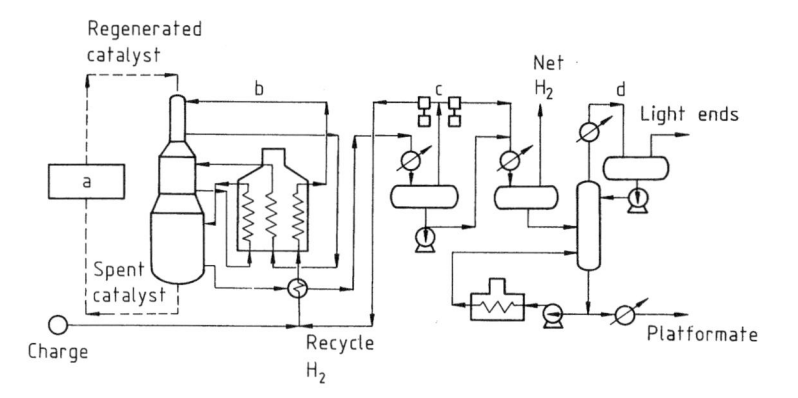

**Figure 3.** Continuous platformingprocess
a) Catalyst regeneration; b) Reactors; c) Product separation; d) Stabilizer
Reproduced with permission from [26]

Various catalytic processes using platinum on alumina and platinum with modifiers on alumina have been developed [25]. These methods may have process cycles as long as six months or much shorter cycles, depending on the feedstock, the process conditions, and the severity of operation. Pretreatment of the naphtha feedstock by catalytic hydrotreating is required to remove sulfur compounds. Commercial catalytic reforming processes include: catalytic reforming (Institut Français du Pétrole), magnaforming (Engelhard Industries), platforming (UOP), powerforming (Exxon Research & Engineering), rheniforming (Chevron Research Co.), and ultraforming (Standard Oil of Indiana).

Operating conditions cover a considerable range: temperatures vary from 425 to 525 °C and gauge pressures from 0.7 to 3.5 MPa. Hydrogen recycle rates depend on whether long process cycles or whether frequent or continuous catalyst regeneration is employed. Naphtha feedstocks vary but generally are in the gasoline boiling range. Narrower boiling range stocks may be used if aromatics are the desired end products. Regenerative-type products generally allow lower pressures, less recycle hydrogen, and higher severity operation. Naphthene conversions approach 100 %. Cyclization of paraffins is much lower. For example, for benzene production from $C_6$ and higher stocks, methylcyclopentane is isomerized to cyclohexane and dehydrogenated to benzene. Conversion of $n$-hexane to benzene is limited.

Catalytic reformer units may be operated at relatively high pressures with a series of fixed-bed reactors, the catalyst being regenerated at defined intervals. Recent units designed for higher conversion operate at lower pressure with more frequent catalyst regeneration. Shorter cycles, swing reactors, or continuous regeneration are employed. The version of platforming [26] shown in Figure 3 employs continuous catalyst regeneration. Another design developed by Standard Oil of Indiana uses a swing reactor in which the fixed-bed catalyst in one reactor is regenerated and brought back onstream, with the catalysts in each of the remaining reactors being regenerated in rotation. Gauge pressures of ca. 0.7–2.4 MPa are employed in the different processes. For gasoline production, naphthas used have boiling ranges of around 80–210 °C. If reforming is

**Table 9.** Reformer cuts and reformate composition (amounts in wt %)

|  | RON * clear | Benzene | Toluene | Xylene | $C_9$ aromatics | Non-aromatics |
|---|---|---|---|---|---|---|
| Total reformate | 95 | 5.5 | 19.5 | 27.1 | 13.1 | 34.8 |
| $C_6$ |  |  |  |  |  |  |
| Aromatic fraction |  | 8.4 | 29.9 | 41.6 | 20.1 |  |
| BTX fraction |  | 10.6 | 37.4 | 52.0 |  |  |
| Total reformate | 98 | 6.6 | 22.1 | 29.5 | 15.3 | 26.5 |
| $C_6$ |  |  |  |  |  |  |
| Aromatic fraction |  | 9.0 | 30.0 | 40.2 | 20.8 |  |
| BTX fraction |  | 11.3 | 38.0 | 50.7 |  |  |
| Total reformate | 101 | 7.8 | 25.9 | 32.0 | 16.3 | 18.0 |
| $C_6$ |  |  |  |  |  |  |
| Aromatic fraction |  | 9.5 | 31.6 | 39.0 | 19.9 |  |
| BTX fraction |  | 11.8 | 39.4 | 48.8 |  |  |

* Research octane number: indicates relative degree of reforming severity

carried out for the production of aromatic charge stocks, the composition and boiling range of the naphtha may be altered. The reformate may have RON clear octane rating as high as 100 and may contain up to 70 vol% aromatics. Typical compositions of reformate at different levels of reforming severity are shown in Table 9.

Over the years, the process has been made more effective. Reforming catalysts have been modified. Water in small amounts was found to enhance benzene yields with certain catalysts when producing BTX [27]. Higher xylene yields have resulted from operating at increased space velocities.

### 5.2.2. Aromatics from Pyrolysis Gasoline

Pyrolysis gasoline (dripolene), produced as a byproduct from the high-temperature, short-residence time cracking of paraffin gases, naphthas, gas oils, or other hydrocarbons used to produce ethylene, is another excellent source for BTX. The quantity of byproduct pyrolysis gasoline produced is a function of feedstock and operating conditions [28], [29]. Although the amount may be small with ethane and other gaseous paraffins as the ethylene charge stocks, it increases when heavier charge stocks are used. On a once-through basis, amounts may be 20% in the high-severity cracking of a medium range naphtha. As the boiling point of the feed increases, a larger yield of toluene and xylene, relative to that of benzene, is produced. Table 10 shows a typical composition of pyrolysis gasoline [30]. The BTX content of the pyrolysis gasoline is > 60%. Because of the high temperature necessary for ethylene production, the pyrolysis gasoline contains other unsaturates and diolefins. Diolefin content is typically $\geq$ 5% as is the total content of cyclic olefins and cyclic diolefins [31].

Table 11 gives an analysis of pyrolysis gasolines by carbon groups [30], [32]. Pyrolysis gasoline is quite unstable because of the appreciable amount of diolefinic material.

**Table 10.** Cracking severity and product composition from pyrolysis gasoline (mass fractions in %)

| Composition based on | Moderate severity | | | High severity | | |
|---|---|---|---|---|---|---|
| | Pyrolysis gasoline | Aromatics* | BTX* | Pyrolysis gasoline | Aromatics* | BTX* |
| Nonaromatics | 35.3 | | | 28.2 | | |
| Benzene | 23.7 | 36.6 | 39.4 | 43.2 | 60.2 | 62.2 |
| Toluene | 20.2 | 31.2 | 33.6 | 17.4 | 24.2 | 25.0 |
| Xylenes including ethylbenzene | 12.6 | | | 3.5 | | |
| Styrene | 3.6 | | | 5.4 | | |
| Total $C_8$ aromatics | 16.2 | 25.1 | 27.0 | 8.9 | 12.4 | 12.8 |
| Higher aromatics | 4.6 | 7.1 | | 2.3 | 3.2 | |

* Including styrene

**Table 11.** Composition of pyrolysis gasoline according to carbon groups

| Substances by type (mass fractions in %) | Groups | | | | | |
|---|---|---|---|---|---|---|
| | $C_5$ | $C_6$ | $C_7$ | $C_8$ | $C_{9+}$ | Total |
| Diolefin | 8 | 4 | 1 | 1 | 0 | 14 |
| Monoolefin | 2 | 2 | 1 | 3 | 0 | 8 |
| Saturates | 1 | 1 | 1 | 2 | 0 | 5 |
| Aromatics | 0 | 34 | 12 | 12 | 12 | 73 |
| Aromatics by type | 0 | 81.9 | 83.3 | 66.7 | 100 | 73 |

Lower boiling diolefins may be removed by distillation although their tendencies to polymerize and depolymerize may cause difficulties. Therefore a two-stage hydrotreating is generally employed. In the first stage a mild selective hydrotreating converts diolefins to olefins. If this gasoline is to be used as a fuel, further treatment may not be necessary. However, if the gasoline is a source of aromatics, a second stage is employed to saturate olefins and to remove residual sulfur.

## 5.3. Methods of Increasing Production

Depending on demand, benzene production from other aromatic compounds can be increased, particularly from the BTX fractions obtained from catalytic reformate, pyrolysis gasoline, or light oils from coal carbonization. Among the methods used are hydrodealkylation, disproportionation, or combination processes.

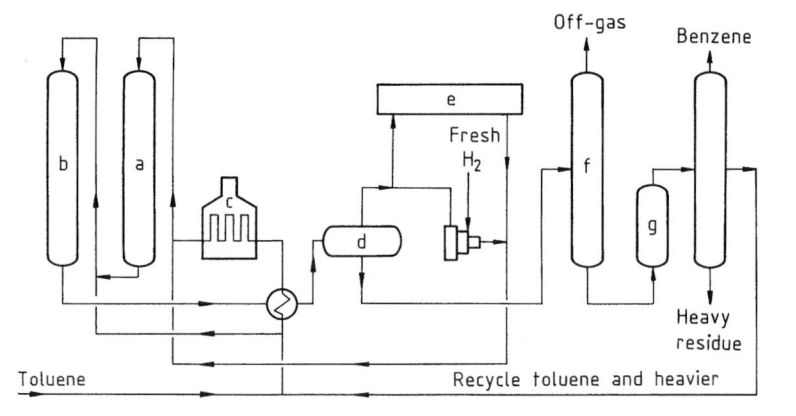

**Figure 4.** Detol process
a) Reactor I; b) Reactor II; c) Furnace; d) Separator; e) Hydrogen purification; f) Stabilizer; g) Clay treater

## 5.3.1. Hydrodealkylation

Hydrodealkylation of alkyl aromatics is a source of benzene. Toluene is the usual charge stock, although higher alkylated aromatics may be used:

$$C_6H_5CH_3 + H_2 \xrightarrow{Heat} C_6H_6 + CH_4$$

With more highly alkylated benzenes, the reaction proceeds stepwise:

$$C_6H_4(CH_3)_2 + H_2 \xrightarrow{-CH_4} C_6H_5CH_3 \xrightarrow{H_2} C_6H_6 + CH_4$$

Other alkylbenzenes, such as ethyl- and propylbenzene, dealkylate in a single step and form the corresponding alkanes. Both catalytic and thermal processes are used. The catalytic processes, e.g., Detol (Air Products & Chemicals) [33] and Hydeal (UOP) [34], operate at 575–650 °C and 2.5–6 MPa. The temperature of this exothermic reaction is controlled by recycling cold hydrogen.

Thermal dealkylation processes, such as HDA (Arco and Hydrocarbon Research) and THD (Gulf Oil), operate at higher temperatures than the catalytic, but yields and reaction systems are similar in both types. For a comparison of the different processes see [35].

The amount of toluene in BTX from catalytic reformate is greater than that of benzene. Whether or not hydrodealkylation is used is related directly to the demand and price of benzene relative to toluene. If the relative demand for the two products changes, the dealkylation units may be placed on standby.

**Detol Process** [33], [36]. This catalytic dealkylation process is illustrated in Figure 4. Benzene is produced from a feed consisting of toluene or mixtures of toluene and other alkylated benzenes. The alkylated aromatics and hydrogen pass at elevated temperature and pressure over a fixed-bed dealkylation catalyst, generally dispersed in more than one reactor in series. Heat exchangers cool the

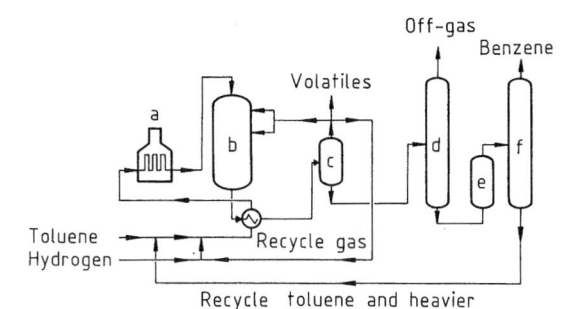

**Figure 5.** HDA process
a) Heater; b) Reactor; c) Separator; d) Stabilizer; e) Clay treater; f) Distillation

reactor products and condense benzene, unreacted toluene, and the heavier unreacted alkylated benzenes. A high-pressure flash drum separates recycle hydrogen and product gas. The gas is split into streams for fuel gas, for hydrogen purification if needed, and for recycle hydrogen.

The product gas from the flash drum is condensed and the liquid pumped to a stabilizer, where remaining gas and low-boiling hydrocarbons are driven off and utilized as fuel gas. The bottoms from the stabilizer are clay treated and passed to a column where the benzene is distilled overhead. Unreacted toluene and heavier aromatics are recycled. The benzene produced in yields of around 99 mol% is highly pure, typically 99.95%, with a freezing point of 5.45 °C. Minor side reactions, such as hydrocracking and hydrogenation, occur and produce small amounts of light hydrocarbons and also eliminate sulfur compounds (e.g., thiophene). Catalyst cycle life in these dealkylation processes is long, as it is in other high-pressure reactions carried out in the presence of hydrogen. As of 1981 twelve commercial Detol plants had been licensed.

**Hydeal Process** [34], [37]. This process was developed jointly by Ashland Oil and UOP. The process is similar to the Detol one described previously. Toluene alone or in mixtures containing xylenes and other alkylated benzenes is charged to the catalytic reactor, which contains a chromia–alumina catalyst. Hydrogen is added along with the alkyl aromatics. Nitration-grade benzene is obtained in yields of ca. 98 mol%. Unreacted aromatic charge material from the once-through operation is recycled to obtain practically complete conversion. The process is also used to hydrodealkylate alkylnaphthalenes.

**HDA Process** [38], [39]. This process, developed by Arco and Hydrocarbon Research, operates at 600–660 °C and at gauge pressures of 3.45–6.9 MPa (see Fig. 5). Operation is similar to that of the catalytic processes, with benzene yields of 99 mol%. The small amounts of diphenyls formed are recycled and a low-equilibrium concentration is maintained, permitting high benzene yields.

**THD Process** [40]. This Gulf process is similar to the other thermal processes.

**MHC Process** [41], [42]. Mitsubishi Petrochemical Co. has described a thermal dealkylation process known as MHC. Methods are emphasized for controlling condensation reactions and coke formation. The following advantages are claimed for the MHC process. Feedstocks containing up to 30% of nonaromatics can be handled without resorting to aromatic extraction and fractional distillation steps. The process will operate on low-purity hydrogen, thus reducing the amount of makeup hydrogen required. Benzene of 99.95% purity is readily achieved. Figure 6 is a flow sheet of the MHC process.

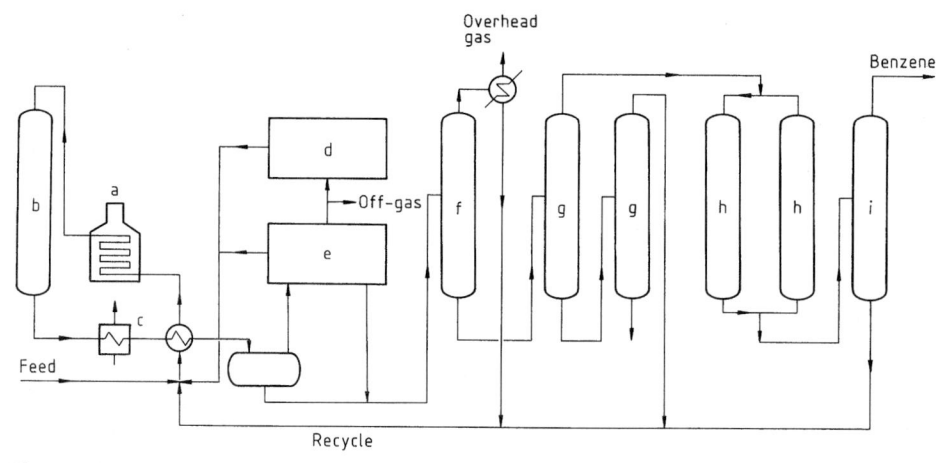

**Figure 6.** MHC process
a) Heater; b) Reactor; c) Waste-heat boiler; d) Hydrogen regeneration; e) Aromatic recovery; f) Fractionation; g) Vacuum columns; h) Clay treaters; i) Benzene column

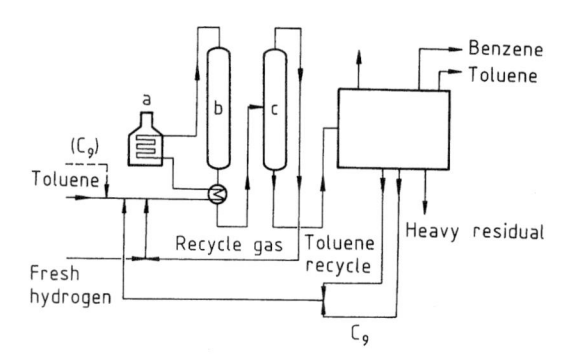

**Figure 7.** Tatoray process for combination toluene disproportionation and $C_9$-aromatic transalkylation
a) Preheater; b) Reactor; c) Separator

## 5.3.2. Disproportionation

Processes have been developed to disproportionate or transalkylate alkylated aromatics to produce benzene and alkylated benzenes. For example, the Tatoray process (UOP and Toray Industries, see Fig. 7) [43]–[45] produces benzene and equilibrium xylenes from toluene or from a mixture of toluene and $C_9$ aromatics:

$$2\ C_6H_5CH_3 \longrightarrow C_6H_6 + C_6H_4(CH_3)_2$$
$$C_6H_5CH_3 + C_6H_3(CH_3)_3 \longrightarrow 2\ C_6H_4(CH_3)_2$$

The reaction proceeds with a high $H_2$ recycle rate over a noble-metal or rare-earth catalyst. Operating conditions are 350–525 °C and 1–5 MPa.

Benzene of high purity ($fp$ 5.45 °C) and xylenes containing $< 10^{-9}$ parts of saturated hydrocarbons are produced. When the feedstock is limited to $C_9$ and $C_{10}$ aromatics, a mixture of benzene, toluene, and xylenes results. Typical yields with toluene as the

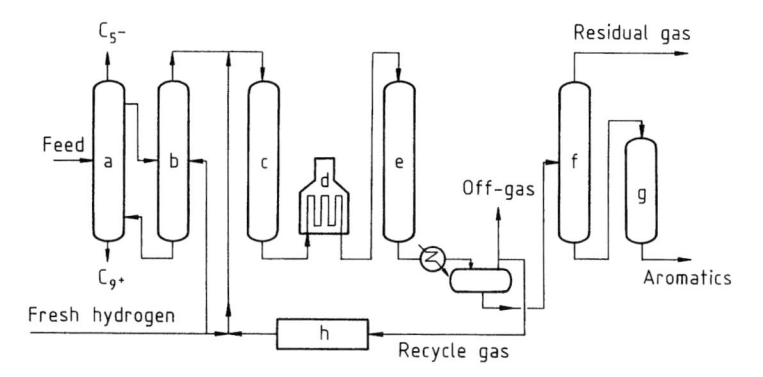

**Figure 8.** Pyrotol process
a) Distillation; b) Vaporizer; c) Prereactor; d) Heater; e) Pyrotol reactor; f) Stabilizer; g) Clay treater; h) Hydrogen purification

**Table 12.** Tatoray process yields, wt%

| Products* | Aromatic Feed | |
|---|---|---|
| | 70% Toluene 30% C$_9$+ | Pure toluene |
| Benzene | 29.6 | 41.2 |
| Toluene | | |
| Ethylbenzene | 2.3 | |
| Xylenes | 61.9 | 55.9 |
| C$_{10}$+ | | 1.0 |

*Balance of products are low-boiling hydrocarbons and C$_{10}$+ aromatics

charge are about 97%. Yields for both a pure toluene feed and a mixed feed of toluene and C$_9$ + aromatics are shown in Table 12.

Four commercial units were in operation in 1981 and others were being designed or built.

## 5.3.3. Combination Dealkylation Processes

Several combination dealkylation processes have been developed for producing benzene from impure BTX feedstocks. The Houdry Division of Air Products developed the *Pyrotol process* [46], [47] by a combination of hydrogenation and dealkylation (Fig. 8).

Essentially the process involves fractionation of the feed to remove C$_5$ minus and C$_9$ plus material from the crude BTX. The C$_6$–C$_8$ is vaporized and passes to a catalytic pretreat reactor for selective hydrogenation of diolefins, cyclic diolefins, and styrene. The effluent from the pretreat reactor is subsequently charged to the Pyrotol reactors, where aromatics are dealkylated to benzene. Other

**Table 13.** Pyrotol yield data and material balance, wt%

| Component | Dripolene raw charge | Fresh feed to pyrotol reactors | Benzene product |
|---|---|---|---|
| $C_5$s | 7.1 | 0.2 | |
| Benzene | 48.3 | 54.4 | 99.92 |
| Cyclohexane | 2.4 | 2.7 | |
| Toluene | 16.2 | 18.3 | 0.08 |
| Other $C_7$s | 8.7 | 9.8 | |
| Ethylbenzene | 1.8 | 2.0 | |
| Xylenes | 9.2 | 10.2 | |
| Other $C_8$s and heavier | 6.3 | 2.4 | |
| Total | 100.0 | 100.0 | 100.0 |
| wt% of charge | 100.0 | 88.63 | 68.88 |

reactions include desulfurization and hydrocracking of nonaromatics. Unreacted toluene and heavier aromatics from the first pass are recycled. The benzene product contains less than 0.5 ppm of thiophene and has a freezing point of at least 5.47 °C. Table 13 contains typical yield data. As of 1982 thirteen plants had been licensed.

A somewhat similar process was developed by the Houdry Division of Air Products and by Bethlehem Steel. The *Litol process* has been in operation in the United States and other countries. It produces pure benzene from aromatic light oil obtained from coal carbonization [48]. This process, as the Pyrotol process, utilizes two catalytic reaction stages. After prefractionation of the raw light oil to remove $C_5$ and lighter fractions overhead and $C_9$ and heavier as bottoms, the $C_6-C_8$ fraction is vaporized.

At the first stage diolefins and styrene are hydrogenated. The effluent then contacts the Litol chromia–alumina catalyst at 500–600 °C, where desulfurization, hydrocracking, and dealkylation occur. Benzene purity of 99.97% with freezing point of 5.5 °C and a thiophene content of less than 0.5 ppm are typical.

In 1968 a modified Litol process was developed [49]. This modified process is designed to include a heavier fraction of the light oil, mostly $C_9$s and heavier, along with the secondary light oil ($C_6-C_8$) normally used in the process. The lighter fraction typically contains around 70% benzene, 20% toluene, and 6% xylenes and ethylbenzene. The primary, or heavier fraction of the light oil, has a typical boiling range of 180–260 °C and is composed mainly of $C_9$ and heavier aromatics. Pilot plant operation indicates that a 50–50 mixture of the lower and higher boiling range light oils can be processed.

Mitsubishi Petrochemical Co. described a process [50] for producing benzene from pyrolysis gasoline by a combination method similar to those discussed previously. After a two-stage hydrogenation a solvent was used to extract the aromatics. Alternatively, the hydrogenated material was reacted in Mitsubishi's MHC process [42], [51], [52]. Advantages of the process are low-temperature liquid-phase operation of the first stage with no need for hydrogenated product recycling, and a reactor design incorporating an improved method for controlling temperature of the exothermic reactions.

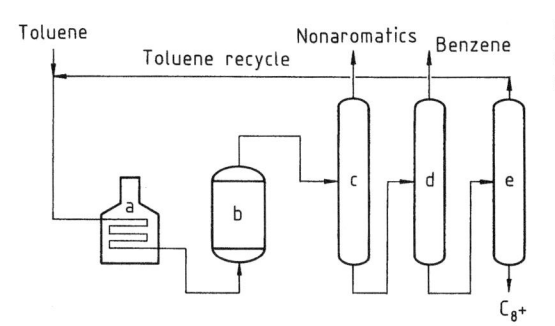

**Figure 9.** Mobil LTD process
a) Furnace; b) Reactor, c) Nonaromatics column; d) Benzene column; e) Toluene column

**Table 14.** LTD fresh feed and product composition, wt%

| Component | Feed | Product |
|---|---|---|
| Nonaromatics |  | 0.2 |
| Benzene |  | 43.8 |
| Toluene | 99.7 | 0.1 |
| Ethylbenzene | 0.1 | 0.2 |
| $p$-Xylene | 0.1 | 12.2 |
| $m$-Xylene | 0.1 | 27.3 |
| $o$-Xylene |  | 11.5 |
| Trimethylbenzenes |  | 4.0 |
| Other $C_9$ + aromatics |  | 0.7 |
| Total | 100.0 | 100.0 |

Mobil Chemical Co. has developed an aromatics processing catalyst and a process for the liquid-phase disproportionation of toluene to form benzene and xylenes (Fig. 9) [53], [54]. This *low-temperature–disproportionation process* (LTD) employs a zeolite-based catalyst of high activity which permits the reaction to proceed at temperatures as low as 260 °C; gauge pressure is 4.5 MPa. No hydrogen recycle is required in this liquid-phase process. An hourly mass-space velocity is standard. A catalyst life of 1.5 years is claimed with infrequent regenerations. With fresh or regenerated catalyst the initial reactor temperature is 260 °C. As the catalyst slowly deactivates, the temperature is raised in increments to 315 °C; then the catalyst is regenerated by burning accumulated coke.

After cooling the reactor effluent passes to three distillation columns arranged in series. In the first column non-aromatics plus a small amount of benzene are taken overhead. This benzene may be recovered by extraction if deemed necessary. Benzene is taken overhead as product in the second column and toluene is distilled overhead for recycle in the third column. Xylenes and heavier aromatics are withdrawn as bottoms from this last column. The $C_9$ + aromatics may be recycled. Toluene conversion on a once-through basis is around 45 % and the balance is recycled to extinction. Table 14 contains a typical wt% composition of the toluene fresh feed and the end product.

## 5.3.4. Benzene from Other Sources

The development of energy from solid fuels, such as coal, oil shale, and tar sands, may provide other potential sources for the production of benzene [55]. For example, if coal is developed as a source of liquid fuels, it has potential for benzene production [56]–[59]. Some of the processes being investigated include noncatalytic liquefaction, SRC-I and the related SRC-II. Others are COED, Exxon Donar Solvent, indirect liquefaction via the Fischer-Tropsch method, and the LCFFC catalytic process by Lummus. Coal gasification also may yield naphthas that can serve as charge material for benzene production.

Production of liquid fuels from coal has been known since 1911 when BERGIUS hydrogenated coal. Hydrogenation and Fischer-Tropsch plants were a source of liquid fuels in Germany during the 1940s. Two commercial processes for the production of liquid fuels from coal are the Sasol plants in South Africa [55] and the Mobil methanol-to-gasoline plant now under construction in New Zealand [60]–[62]. A third commercial process, using tar sands as raw material for the production of liquid fuels, has been described [55]. Other potential sources of liquid fuels include shale oil developments, biomass conversion, and the production of hydrocarbon liquid fuels from ethanol.

The yields of chemical feedstocks and gasolines from coal conversion depend on the process or combination of processes employed. For example, a BTX yield of 2.5%, based on coal charged, was projected using a combination of processing including the Fischer-Tropsch route [63]. Hydroliquefaction followed by processing of the naphtha produced can increase BTX. Another study indicates a liquid yield of ca. 50%. Distillate yields from other liquefaction processes, such as SRC-II and H-coal, are ca. 50%, with an aromatics content of ca. 25%. Fischer-Tropsch gasolines generally have a lower aromatic content than those from liquefaction processes. Unless secondary aromatization is carried out, the benzene content of distillates from Fischer Tropsch processes is generally lower than that found in petroleum derived gasolines.

Thus carbon-containing materials, such as coal, tar sands, and oil shale, are potential raw materials for benzene production. However, their commercial use is not imminent as long as petroleum is in ample supply at a reasonable price. Exceptions are light oils from coal carbonization and possibly coal conversion processes, such as Sasol and others in operation or under construction.

# 6. Separation and Purification

The separation and purification of benzene from crude BTX fractions or the separation of the different BTX components is accomplished by different methods and processes, such as those outlined in this chapter. The BTX fractions from catalytic reforming are the easiest to handle in that most of the objectionable impurities have been removed during reforming. Light oils from coal carbonization and pyrolysis gasoline from high-temperature manufacture of ethylene need additional treatment to remove impurities, such as dienes, olefins, and sulfur compounds, as well as small amounts of nitrogen and oxygen before proceeding with the recovery and separation of the aromatic components.

## 6.1. Removal of Dienes, Olefins, and Sulfur Compounds

Pyrolysis gasolines need pretreatment because of appreciable diene content. A two-stage hydrotreating process is generally employed. The dienes are hydrogenated to olefins in a mild first stage. In a second stage olefins are saturated and sulfur compounds are removed under more severe conditions. However, conditions are mild enough to avoid saturation of aromatics. Both base-metal and noble-metal catalysts are employed in diolefin removal. Nickel molybdate and cobalt molybdate generally are used in olefin and sulfur removal. Temperatures of 30–175 °C and pressures of 2–6 MPa are typical for diolefin operation. Noble-metal catalysts generally require lower temperatures than base-metal catalysts. Space velocities range from 2 to 10 vol of feed per volume of catalyst per hour, with higher values permitted with noble-metal catalysts.

Shell Development Co. has described a process for diene removal or, alternately, a two-stage process for diene removal followed by hydrogenation of sulfur and olefin compounds [64]. Diolefin hydrogenation is carried out in the first-stage reactor in a liquid flow with a special nickel-containing catalyst. Temperature control of the exothermic reaction is maintained by liquid product recycle. No desulfurization takes place in the first stage. If desired, olefins and sulfur may be removed in a second-stage hydrogenation in the presence of a nickel molybdate catalyst. Liquid product recycle may be used if necessary. Desulfurization and olefin saturation are nearly complete. Typical operating conditions when processing a $C_5$–200 °C pyrolysis gasoline charge are shown in Table 15.

A somewhat similar process has been described by the Institut Français du Pétrole. Reference [65] gives data for the hydrotreating of a $C_5$–205 °C pyrolysis gasoline. This is a one-stage process for the selective hydrogenation of diolefins or can be done as a two step process if complete hydrogenation of diolefins and olefins plus desulfurization is desired. Table 16 shows a partial listing of results.

In 1982, 65 units were in operation and others were under construction [65]. Several other processes of this type have been developed, including ones by British Petroleum [66] and Bayer.

Clay treating was formerly used for removal of diolefinic material and has been described in several publications. The process is outmoded, however, and is rarely used because of poor efficiency and disposal problems.

**Table 15.** Hydrogenation of pyrolysis gasoline (Shell Development)

| Conditions | First stage | Second stage |
|---|---|---|
| Temperature, °C | 80–130 | 230–380 |
| Pressure, MPa | 6 | 4.5–6.5 |
| Recycle ratio | 3–5 | 0.2 |
| Space velocity, t fresh feed per $m^3$ cat./h | 0.8–1.5 | 1.3 |
| $H_2$ makeup, $m^3$/t (STP) fresh feed | 50–100 | 50–125 |
| Recycle gas, $m^3$/t (STP) fresh feed | 200–500 | 450–1200 |

| Properties | Feed | Product | Feed | Product |
|---|---|---|---|---|
| Maleic anhydride number mg/g | 100–250 | 5–15 | 5–15 | 0 |
| Bromine number g/100 g | 60–100 | 25–80 | 25–80 | 1 |
| Sulfur, ppm | 100–400 | 100–400 | 100–400 | 0.5 |

**Table 16.** Hydrogenation of $C_5$–205°C pyrolysis gasoline (Institut Français du Pétrole)

| | Feedstock | After first stage | After second stage |
|---|---|---|---|
| Specific gravity | 0.815 | 0.813 | 0.860 |
| Diene value | 30 | 1 | ≈ 0 |
| Bromine number g/100 g | 75 | 56 | 0.1 |
| Sulfur, ppm | 400 | 400 | 0.5 |

## 6.2. Close Fractionation

Precise fractionation may be used in some cases to separate nonaromatics from benzene or from other aromatics if the concentration of nonaromatics is not too great. However, some of the nonaromatics, such as cyclohexane, form azeotropes with benzene and thus cannot be completely separated by close fractionation at atmospheric pressure. In some cases distillation under elevated pressure will change the respective vapor–liquid equilibria so that separation may be improved.

### 6.2.1. Azeotropic Distillation

Naphthenes and paraffins are separated from aromatics by azeotropic distillation. Olefins in the charge mixture are generally detrimental to efficient separation. Usually the light ends from the crude BTX material are removed, followed by distillation of the benzene fraction from the toluene-xylene mixture. Examples of azeotrope formers include acetone and methanol with added water. Acetone is particularly suitable for the benzene fraction.

The azeotropic mixtures of nonaromatics and promoting agent are distilled. The nonaromatics are recovered by breaking the azeotropic mixture with water. In the benzene application, benzene is removed as a bottom product, washed, and distilled.

Methanol is effective as an azeotrope forming agent in the treatment of the toluene–xylene fractions. By the azeotropic method, benzene with *fp* 5.5 °C is recovered in 98% yield. Pure toluene is separated in 95% yield. Generally the process is limited to materials containing < 12% nonaromatics. Azeotropic distillation has found limited application but is not as effective as other methods, such as extractive distillation and liquid-liquid extraction, in the isolation of high-purity aromatic compounds.

## 6.2.2. Extractive Distillation

Extraction and extractive distillation are commonly used to separate aromatics from crude BTX feeds. Extractive distillation is distinguished from normal distillation in that it is carried out in the presence of a high-boiling solvent having great solubility for aromatics and little or no solubility for paraffins and naphthenes. The solvent enters the distillation column near the top and above the point at which the BTX feed is introduced. The nonaromatics are distilled from the top of the column and the aromatic-rich solvent is withdrawn as a bottoms product. The aromatics are then distilled or stripped from the solvent in a second column and the lean solvent is recycled. Separation of the different aromatic components is carried out in subsequent distillation steps. Solvent to BTX feed ratios are generally in the order of 2:1 to 6:1 and conditions vary according to the nature of the solvent, column efficiency, and properties of the feedstock.

Extraction and extractive distillation methods as well as liquid–liquid extraction methods depend upon the use of very selective polar substances. Most of these compounds or solvents are high-boiling but some, such as $SO_2$, are not.

## 6.2.3. Liquid–Liquid Extraction

Liquid–liquid extraction in itself or in conjunction with other separation processes is the most widely used method for the isolation of aromatic fractions. All these processes involve the extraction of aromatics from nonaromatic material by the use of a polar solvent having a high-selective affinity for the former compounds.

The solvent, which is generally a high-boiling polar liquid, enters the top of the extractor vessel. Various designs of extractors have been used, including packed columns, sieve tray types, and others. A rotating disk contactor (RDC) has been used in some processes. The crude BTX or other aromatic-containing material enters near the middle of the extractors and thus contacts the solvent counter-currently in the column, or intimate contact is effected by other means. The nonaromatic raffinate leaves at the top of the extractor, whereas the aromatic-rich solvent is withdrawn at the bottom and passes to a distillation column, where the aromatics are taken overhead. The denuded solvent is then recycled to the extractor. The overhead aromatic stream is distilled to separate benzene and other aromatic components. The fundamentals of liquid–liquid extraction, extractive distillation, and various methods of aromatic recovery by extraction are discussed in [67], [68].

## 6.3. Processes for Aromatic Separation

Numerous methods and processes have been proposed and developed for the separation of pure aromatics from BTX feedstocks [69], [70]. The following are discussed in this article:

| Process | Solvent |
| --- | --- |
| Edeleanu | sulfur dioxide |
| Udex | diethylene glycol and others |
| Sulfolane | sulfolane |
| Arosolvan | 1-methyl-2-pyrrolidone |
| IFP | dimethyl sulfoxide |
| Formex | N-formylmorpholine |
| Morphylane | N-formylmorpholine |

**Edeleanu Process.** This process uses as solvent liquid sulfur dioxide [7446-09-5], $SO_2$, $M_r$ 68.06. It was developed in 1908 primarily for the removal of aromatics from kerosine lamp oil and is of historical interest only. Liquid sulfur dioxide is effective in extracting of aromatics but is now used only for special purposes. This process is not competitive for aromatic removal in stocks with a high-aromatic content.

**Udex Process (UOP-Dow).** One of the early liquid – liquid extraction processes for the recovery of BTX aromatics from catalytic reformate was the Udex process [71], [72]. The process is also effective for recovering aromatics from coal tar, light oil, and pyrolysis gasoline. The first commercial installation was in conjunction with the UOP platforming unit at Cosden Petroleum Corp. [73].

The Udex process (see Fig. 10) uses diethylene glycol [111-46-6], 2-hydroxyethyl ether, $(HOCH_2\text{-}CH_2)_2O$, $M_r$ 106.12. The glycol contains about 8% water to increase aromatic selectivity. This combination is effective as a solvent for separating aromatics from paraffins in the same boiling range. For aromatics production, a reformate fraction of 65 – 150 °C is charged to the liquid – liquid extractor where a multistage, countercurrent extraction takes place. The solvent enters at the top of the extractor column and passes countercurrent to the hydrocarbon feed, which is charged at an intermediate point. The raffinate passes overhead from the extractor. The aromatic-rich solvent is withdrawn from the bottom and passes to the solvent stripper, where the BTX material passes overhead and the denuded solvent is recycled to the extractor. The BTX mixture is washed with water and treated with clay before being separated into the benzene, toluene, and xylene by precise fractionation. Compounds of nitration grade specifications are obtained with only a small amount of xylene from the xylene distillation being lost from the bottoms. Typical results in the processing of catalytic reformate are shown in Table 17.

Improvements have been made in the Udex process in terms of solvent and of operating conditions. Aromatics recovery was improved and capacity was increased by using a solvent mixture of triethylene glycol and diethylene glycol or tetraethylene glycol. Pilot-plant studies carried out at Union Carbide showed that, where a solvent to feed ratio > 10 was required for high-aromatics recovery with diethylene glycol as the

**Table 17.** Results of Udex extraction (U-2)

| Properties of products | Udex charge | Benzene | Toluene | Xylene mixture |
|---|---|---|---|---|
| Distillation | | | | |
|   initial *bp*, °C | 66 | 79.7 | 110 | |
|   50% | 99 | 80.1 | 110.8 | |
|   end point | 150 | 80.3 | 111.0 | |
| *fp*, °C | | 5.49 | | |
| % Aromatics | 51.3 | 99.9 | 99.8 | >99.8 |
| Yields | Vol% of charge | Vol% of extract | % Recovery in extract | |
| Benzene | 7.6 | 15.3 | >99 | |
| Toluene | 21.2 | 42.5 | 95 | |
| Xylenes and ethylbenzene | 20.0 | 40.0 | 94 | |
| $C_9$ + Aromatics | 0.9 | 1.8 | 75 | |

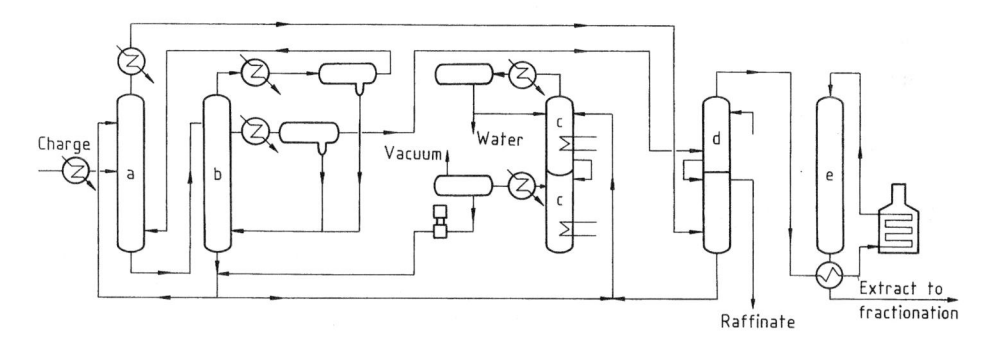

**Figure 10.** Udex extraction process
a) Extractor; b) Stripper; c) Water and solvent distillation; d) Water; e) Clay treater

solvent, the ratio could be reduced to around 6 with the ethylene glycol–propylene glycol mixture. With tetraethylene glycol as the solvent, the ratio could be reduced to as low as 4 [74]. Operation at Sun Oil's Corpus Christi, Texas, plant [75] was improved when tetraethylene glycol was substituted for propylene glycol–ethylene glycol. Solvent/feed ratio was decreased with a resulting increase in capacity. Although the original process was based on the use of diethylene glycol as the solvent, most plants now use other more effective solvents, such as a tetraethylene glycol–water mixture.

**Sulfolane Process.** Shell Development Co. and Royal Dutch Shell have developed an aromatic extraction process based on the use of sulfolane as a selective solvent [76]–[79].

Sulfolane [*126-33-0*], tetramethylene sulfone, $C_4H_8O_2S$, $M_r$ 120.17, is made by reacting sulfur dioxide and butadiene to form sulfolene, which is then hydrogenated to give sulfolane:

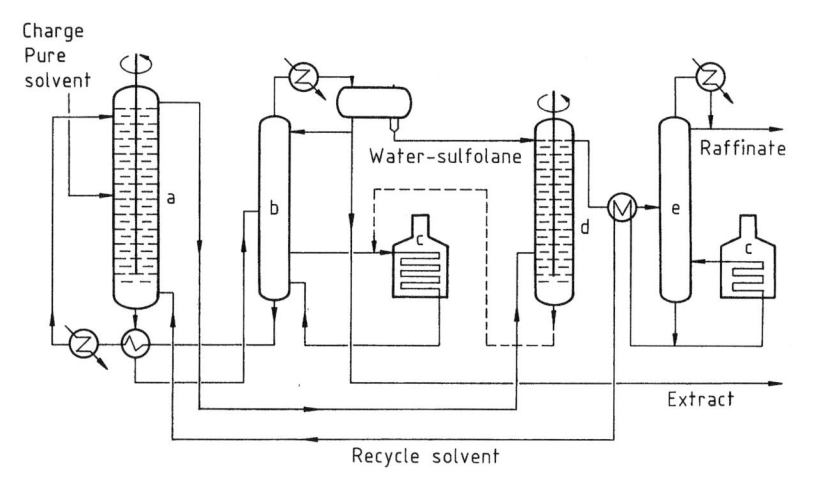

**Figure 11.** Sulfolane extraction process
a) Extractor; b) Extract stripper; c) Preheater; d) RDC washer; e) Rectifying column

Some of the advantages of the sulfolane process are high selectivities and high yields. Also sulfolane has suitable physical properties, such as high density, low-heat capacity, and a boiling point high enough to permit easy separation for recycling. One disadvantage is thermal instability at the boiling point. However, this shortcoming is met by adding small amounts of water, which depresses the boiling point.

The sulfolane process (see Fig. 11) is similar to other aromatic extraction processes but differs in some operational procedures. The BTX containing feed is charged to a rotating disk contactor or extractor. Aromatics are extracted with a countercurrent flow of the sulfolane solvent. The aromatic-rich solvent from the bottom of the extractor passes to an extractive stripper. The overhead from this stripper is returned to the extractor as a backwash stream, for recovery of any residual aromatics. The aromatic–solvent mixture is charged to a recovery column where the aromatic product passes overhead. The lean solvent from the bottom of the column is recycled to the extractor.

A comparison of sulfolane with diethylene glycol [*111-46-6*] in the treating of light catalytic reformate to separate and produce nitration grades of benzene, toluene, and xylenes is summarized in Table 18.

Yields and amounts of nonaromatic impurities (in parentheses) are: benzene, 99.9 % (0.01 %); toluene, 99.5 % (0.02 %); xylenes, 98.0 % (0.10 %).

As of 1980, over 55 sulfolane units were in operation with capacities ranging from a few hundred to over 45000 barrels/d [78].

**Table 18.** Comparison of sulfolane and diethylene glycol extraction

| Operating conditions | DEG* | Sulfolane |
|---|---|---|
| Solvent-to-feed, wt ratio | 20 | 6.8 |
| Backwash-to-feed, wt ratio | 1.1 | 0.5 |
| Stripping steam-to-feed, wt ratio | 0.6 | 0.13 |
| Stripper bottom temp., °C | 143 | 191 |
| Extractor top temp., °C | 143 | 100 |
| Feed temp., °C | 116 | 116 |
| Extractor gauge pressure, MPa | 0.76 | 0.10 |
| Water content of lean solvent | 7.5 | 1.3 |

* Diethylene glycol

**Table 19.** Comparison of properties of different aromatic solvents

| Property | 1-Methyl-2-pyrrolidone | Sulfolane | Diethylene glycol |
|---|---|---|---|
| bp, °C | 206 | 287 | 245 |
| fp, °C | −24 | 27.9 | −8 |
| Refractive index | | | |
| $n_D^{20}$ | 1.4703 | | 1.447 |
| $n_D^{30}$ | | 1.481 | |
| Density, g/mL | | | |
| 20 °C | 1.033 | | |
| 30 °C | 1.024 | 1.261 | 1.11 |
| 200 °C | | 1.116 | |
| Viscosity (mPa s) | | | |
| 30 °C | 1.017 | 10.2 | 2.7 |
| 100 °C | 0.97 | 2.5 | 2.5 |
| 200 °C | | 0.97 | |
| Interfacial surface tension, mN/m | | | |
| 30 °C | 8.2 | | |
| 120 °C | | 2.0 | 8.5 |

**Arosolvan Process.** The Arosolvan process [81]–[83], licensed by Lurgi, uses an aqueous 1-methyl-2-pyrrolidone solvent. The water content ranges from 10 to 20%, but is generally 12 to 14%. Aromatics are extracted at 30–35 °C and at about atmospheric pressure. A comparison of properties of 1-methyl-2-pyrrolidone [*872-50-4*] and other solvents is shown in Table 19. Advantages of the Arosolvan process include operation of the extraction tower at atmospheric pressure and at room temperature, and low consumption of solvent (< 0.09 mg/L) [82]. Feedstocks may contain considerable unsaturates, and depentanization of the feedstock is not required because of the low-temperature operation of the extractor. Other advantages are clay treating of recovered aromatics is not necessary and waste disposal is not a problem because wash water is recycled.

Figure 12 is a simplified flow sheet of the process. The BTX feedstock is fed to the middle of the extractor where the aromatics are removed by countercurrent extraction with the solvent. The raffinate, containing pentane, other nonaromatics, and a trace of solvent, passes overhead from the

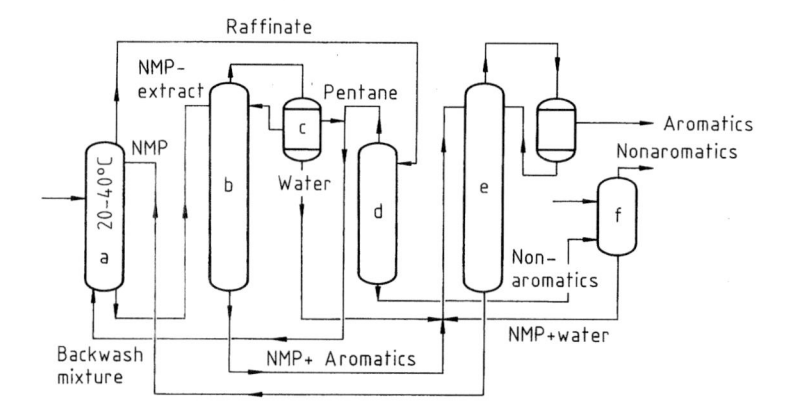

**Figure 12.** Arosolvan process
a) Extraction column; b) Stripper I; c) Separator; d) Pentane column; e) Stripper II; f) Recovery of 1-methyl-2-pyrrolidone (NMP)

extractor. The aromatic-rich solvent is passed to the first stripper, where pentane and some benzene passes overhead and, after joining with the overhead from the depentanizer, serves as a recycle stream to the bottom of the extractor. In the second stripper, high-purity aromatics and water are taken overhead while aromatics-free solvent is withdrawn from the bottom for recycle. The raffinate from the bottom of the depentanizing tower is separated from the solvent by water washing, and is removed from the process for fuel blending or other uses. The aromatic fraction from the overhead of the second stripper is separated by precise distillation.

Table 20 shows data obtained in the operation of two early commercial plants using the Arosolvan process. The Japanese plant operation involved the recovery of benzene, toluene, and xylene fractions from a feedstock composed of hydrogenated coal-gas and oil-gas benzene. The German operation used a hydrogenated pyrolysis gasoline as the feedstock.

**DMSO Process.** In 1965 the Institut Français du Pétrole described a process for the extraction of aromatics using dimethyl sulfoxide (DMSO) [67-68-5], $(CH_3)_2SO$, $M_r$ 78.13, admixed with small amounts of water as the solvent [84] – [86].

The DMSO process is shown in Figure 13. In the first extractor the BTX or other aromatic mixture is contacted at ambient temperature and atmospheric pressure in countercurrent fashion with the dimethyl sulfoxide solvent (containing < 9% water). To maximize aromatic purity a mixture of aromatics and paraffins is fed to the bottom of the first extractor, countercurrent to the solvent flow, so as to combine reflux and backwash effects. The aromatic – solvent mixture passes from the first contactor to the second extractor where the DMSO solvent is recovered by re-extraction of the aromatics with a paraffin solvent. Extraction and re-extraction are carried out at room temperature and atmospheric pressure in rotating disk contactors of 10 – 12 theoretical stages. The paraffin – aromatic mixture from the second stage is separated by distillation and the paraffin is recycled to the second contactor. The small amount of DMSO that remains in the paraffin phase may be recovered by water washing. The DMSO is separated from the wash water by vacuum distillation.

**Table 20.** Plant data from Arosolvan process

|  | Plant A Japan | | | Plant B Germany | |
|---|---|---|---|---|---|
| Feedstock | hydrogenated coal-gas and oil-gas benzene | | | hyrogenated pyrolysis gasoline | |
| Feedstock composition, wt% | | | | | |
| benzene | 56 | | | 35 | |
| toluene | 16 | | | 25 | |
| $C_8$ aromatics | 8 | | | 0 | |
| nonaromatics | 20 | | | 40 | |
| Yield, wt% in each fraction | | | | | |
| benzene | 99.9 | | | 99.9 | |
| toluene | 99.7 | | | 99.8 | |
| xylene + ethylbenzene | 96.8 | | | – | |
| Properties of fractions | B | T | X | B | T |
| fp, °C | 5.50 | – | – | 5.50 | |
| density, g/mL | 0.884 | 0.871 | 0.870 | 0.884 | 0.872 |
| benzene, wt% | 99.99 | <0.1 | – | 99.99 | <0.01 |
| toluene, wt% | 0.0 | >99.9 | 0.1 | 0.01 | 99.99 |
| $C_8$ aromatics, wt% | – | <0.1 | 99.8 | – | <0.01 |
| nonaromatics, wt% | 0.01 | 0.01 | <0.1 | 0.01 | <0.01 |

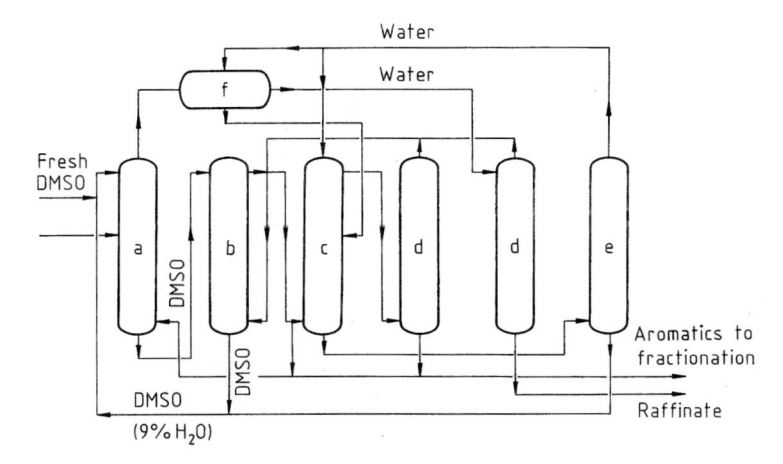

**Figure 13.** DMSO process
a) Extractor I; b) Extractor II; c) Washer for DMSO residue; d) Column for removal of pentane; e) Vacuum column for dehydration of DMSO; f) Raffinate container

Some of the advantages of this process are: DMSO has high selectivity for aromatic extraction; this selectivity is increased by the addition of small amounts of water; extraction and re-extraction are carried out at room temperature and atmospheric pressure; and DMSO is noncorrosive and is chemically stable under process conditions.

**Table 21.** DMSO process for aromatic extraction (yield data for different feedstocks)

|  | Catalytic reformate | | Hydrotreated pyrolysis gasoline | |
| --- | --- | --- | --- | --- |
|  | Feed components, wt% | Aromatic recovery, wt% | Feed components, wt% | Aromatic recovery, wt% |
| Benzene | 15 | >99.9 | 53.5 | 99.7 |
| Toluene | 40 | 99 | 17.5 | 99.7 |
| Xylenes | 10 | 75 | 6.0 | 90.0 |
| $C_9$ + aromatics | 2.5 | 20 | | |
| Total aromatics | 67.5 | 93 | 77.0 | 99.0 |
| Paraffins | 32.5 | | 23.0 | |

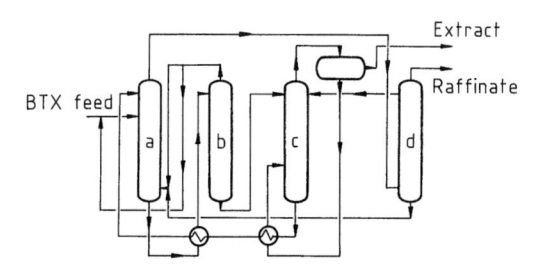

**Figure 14.** Formex process, BTX extraction
a) Extractor; b) Stripper; c) Extraction column; d) Raffinate washer

Yield data for benzene, toluene, and xylene in the treatment of a catalytic reformate and a pyrolysis gasoline, after hydrogenation of olefins and diolefins, are presented in Table 21.

**Formex Process.** The Formex aromatic extraction process (SNAM Progetti) uses $N$-formylmorpholine (FM) in admixture with water as the extraction solvent for aromatics [87], [88]. The process is designed for the extraction of benzene, toluene, and xylenes from BTX feedstocks, such as catalytic reformates or hydrogenated pyrolysis gasoline. The process can be used also for the recovery of $C_9$ and $C_{10}$ aromatics or for the removal of small amounts of aromatics from paraffinic hydrocarbon mixtures.

$N$-Formylmorpholine [4394-85-8], 4-formylmorpholine, 4-morpholinecarboxaldehyde, $C_5H_9NO_2$, $M_r$ 115.13, has the following structure (→ Amines, Aliphatic):

The Formex process applied to the recovery of BTX components is shown in Figure 14 [83]. The BTX feedstock is charged at the middle of the extractor column.

Solvent enters at the top. Extractor conditions (atmospheric pressure, ca. 40 °C) are controlled so that the raffinate, containing a minimum amount of aromatics, passes overhead. The aromatic-rich solvent from the extractor is fed to the extractive stripper, where a stream of residual saturated hydrocarbons is taken overhead and returned to the extractor as reflux. The bottoms from this extractive stripper pass to the extract recovery column. Here the aromatics are taken overhead and the lean solvent is recycled to the extractor. Typical yields and quality of products for a BTX charge containing 5–40% benzene, 15–30% toluene, and 15–30% xylenes are [89]:

|  | Benzene | Toluene | Xylenes |
|---|---|---|---|
| Yield, % | 100 | 99 | 95–97 |
| Nonaromatics, ppm | – | < 200 | < 200 |
| $fp$, °C | 5.5 | | |

The basic process may be modified. For benzene-rich feedstocks the liquid–liquid extraction column may be omitted and the feed charged in the vapor phase to a column where the raffinate, containing a minor amount of benzene, passes overhead and the aromatic-rich solvent from the extractor bottom is handled in the conventional manner. Other modifications include removal of minor amounts of aromatics from primarily saturated hydrocarbon stocks.

Commercial operations include a $500 \times 10^3$ t/a BTX extraction unit and a $24 \times 10^3$ t/a unit producing aromatic-free aliphatic solvents from a $C_6 - C_7$ feedstock.

**Morphylane Process.** Krupp-Koppers has described processes for separation and recovery of BTX components from aromatic feedstocks, such as catalytic reformate and hydrogenated pyrolysis gasoline [90]–[92]. The Morphylane process is essentially an extractive distillation. The solvents used are derivatives of morpholine, principally N-formylmorpholine.

Generally, extractive distillation processes are used for the recovery of one aromatic component. The addition of N-formylmorpholine to the BTX mixture makes nonaromatics that boil above benzene more volatile than benzene. Therefore, these nonaromatics can be extractively distilled from a binary mixture of aromatics (e.g., benzene and toluene).

The N-formyl-morpholine, without added components, such as water, is fed at ambient temperatures to the top of an extractive distillation column (Fig. 15). It contacts a countercurrent flow of the preheated BTX feedstock, which is charged at a middle point in the column. The nonaromatics are distilled and leave at the top of the column. The aromatic-rich solvent is withdrawn from the bottom and passes to the stripper, where the aromatics are discharged overhead. The cooled, lean solvent is recycled to the top of the extractive distillation column. The recovery of benzene from a hydrogenated fraction of pyrolysis gasoline is 97%, purity is 99.95 wt%, and $fp \geq 5.5$ °C. A toluene–xylene fraction from a feed consisting of catalytic reformate and hydrogenated pyrolysis gasoline, with morphylane processing under high-yield conditions, gives toluene in 99 wt% yield, purity > 99.9%; and xylenes in 99.3 wt% yield, purity of 98% [93].

A modified process for the combined and simultaneous recovery of BTX components from aromatic feedstocks, such as catalytic reformate or hydrogenated pyrolysis gasoline, has been described by

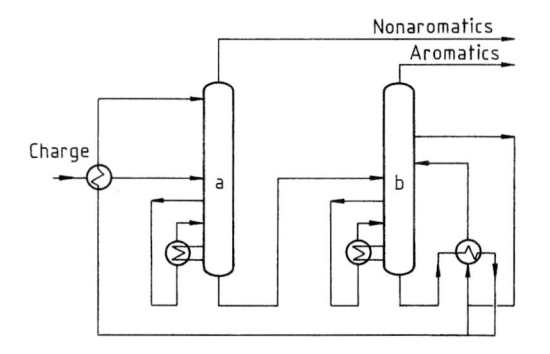

**Figure 15.** Morphylane extractive distillation process
a) Extractive distillation; b) Stripper

Krupp-Koppers [94]. This process, called Aromex or Morphylex, combines liquid – liquid extraction with extractive distillation. In the liquid – liquid extraction stage, high-boiling nonaromatics are sep arated. In the extractive distillation stage, lower boiling nonaromatics are separated by extractive distillation. The overhead, after water removal, is fed back to the liquid – liquid extractor. The bottoms from the extractive distillation are passed to a stripper where the aromatics are distilled overhead and the solvent is recycled to the two extraction columns.

Typical data for the processing of a BTX – containing feedstock are:

|  | Benzene | Toluene | Xylenes |
|---|---|---|---|
| Yield, wt % | > 99.9 | > 99.5 | > 97.0 |
| Purity after fractionation | 99.99 | 99.98 | 99.90 |

In 1982 six commercial units of the Morphylane process and one Morphylex unit were operating.

# 7. Quality Specifications

The ASTM has specified standards and test methods for benzene and related aromatic compounds. Likewise, specifications have been developed and used in other countries. As examples, the American National Standards have been used. The DIN Standards are used in Germany, and those of the Institute of Petroleum in England. The Technical Committee of ISO has standardized many of the specifications and test methods. Many of the ASTM methods have an Institute of Petroleum designation also.

The ASTM specifications for different grades of benzene are shown in Table 22. Test methods developed by ASTM are used in determining the specific properties of benzene:

1) Sampling: method D-270 for petroleum and petroleum products
2) ASTM D-891: specific gravity
3) ASTM D-1209: color
4) ASTM D-850: distillation
5) ASTM D-852: freezing point

6) ASTM D-848: wash color
7) ASTM D-847: acidity
8) ASTM D-849: copper corrosion
9) ASTM D-853: sulfur compounds
10) ASTM D-1685: thiophene
11) ASTM D-2360: nonaromatics

**Table 22.** Specifications for different grades of benzene

| Specification | Industrial grade (ASTM D836–71) | Refined benzene-535 (ASTM D2359–69) | Refined benzene-485 (nitration grade) (ASTM D835–71) |
|---|---|---|---|
| Appearance | | clear, water white | |
| Specific gravity (15.56/15.56 °C) | 0.875–0.886 | 0.882–0.886 | 0.882–0.886 |
| Color | not darker than 20 max. | 20 max. | 20 max. |
| Distillation range at 101.3 kPa | not more than 2.0 °C | not more than 1.0 °C | not more than 1.0 °C |
| Solidification point | | 5.35 °C min. (dry basis) | 4.85 °C min. (dry basis) |
| Acid wash color | not darker than no. 3 | no. 1 max. | no. 2 max. |
| Acidity | no free acid | nil | no evidence of acidity |
| $H_2S$, $SO_2$ | | free | |
| Thiophene | | 1 ppm | |
| Sulfur compounds | free of $H_2S$ and $SO_2$ | | free of $H_2S$ and $SO_2$ |
| Copper corrosion | copper strip must not show irridescence nor a gray or black deposit or discoloration | | |
| Nonaromatics | | 0.15% max. | |
| Carbon disulfide | | if needed use ASTM D2324 | |

Similar specifications and tests have been developed for related aromatic compounds as well as for reagent grades of other materials; e.g., ASTM methods D-362–75 and D-841–71 describe standard specifications for industrial and for nitration-grade toluene.

The concentration of nonaromatics permitted in different grades of benzene is limited and in most cases must not exceed 0.15%. For the 5.5 °C freezing point grade the maximum allowable is 0.01%. The amount of paraffinic and naphthenic materials in the finished benzene is reflected in the boiling range and freezing point of the benzene.

The tests outlined by the ASTM methods and similar tests in different countries characterize benzene purity and whether or not the desired specifications are met. Gas chromatography is an important and efficient method for determining the presence of impurities in benzene and also the amount of benzene in different potential charge stocks available for the production of benzene.

**Table 23.** Benzene production, United States

| Year | Amount, 1000 t | Reference | Year | Amount, 1000 t | Reference |
|---|---|---|---|---|---|
| 1953 | 911 | [95] | 1974 | 4970 | [97] |
| 1955 | 1031 | [95] | 1975 | 3420 | [97] |
| 1960 | 1539 | [95] | 1976 | 4760 | [97] |
| 1965 | 2770 | [95] | 1977 | 4866 | [97] |
| 1968 | 3340 | [96] | 1978 | 4970 | [97] |
| 1970 | 3788 | [96] | 1979 | 5588 | [97] |
| 1970 | 3812 | [95] | 1980 | 5294 | [97] |
| 1971 | 3594 | [96] | 1981 | 4472 | [97] |
| 1972 | 4182 | [97] | 1982 | 3583 | [97] |
| 1973 | 4853 | [97] | 1983 | 4175* | [98] |

* Projected

# 8. Storage and Transportation

Benzene is a flammable, volatile, and toxic material, and its storage and transportation is closely regulated. Labeling is particularly important for the shipment of benzene. United States agencies involved in these regulations include OSHA, EPA, DOT, NIOSH, and others. The regulations dealing with handling and shipping are updated and published each year in the CFR. Other countries have regulations and safety practices quite similar to those in the United States. Although specific regulations must be met for manufacture, laboratory testing, storage, handling, and transporting of benzene, the safety rules are similar to those for other flammable and toxic liquids or gases.

Benzene is stored and transported in steel drums or tanks. Adequate ventilation is necessary. Workers must be protected from skin contact or inhalation of fumes. Periodic physical examinations are recommended for those who work in areas where there is probable exposure to benzene. Recently, the handling of gasoline has caused concern because of its toxic components, which include benzene. Either $CO_2$ or dry chemical extinguishers are recommended for combatting benzene fires.

# 9. Economic Aspects

The demand for benzene as a raw material for the manufacture of petrochemicals has increased dramatically over the last 30 years. Before World War II, light oil from coal carbonization was the main source for the production of benzene and the other BTX compounds. The development of new processes in the petroleum industry, such as catalytic reforming, provided a new raw material source for benzene. The rapid growth in the demand and production of benzene during 1950–1970, using United States production figures as an example, is shown in Table 23. The same increase in benzene production over the period before 1970 occurred in many other countries, especially those of Western Europe. For example, 1965 benzene production in the Federal

**Table 24.** Benzene production, 1000 t

| Country | Year | | | | | | | References |
|---|---|---|---|---|---|---|---|---|
| | 1977 | 1978 | 1979 | 1980 | 1981 | 1982 | 1983 | |
| United States | 5102 | 5048 | 5769 | 4980 | 4494 | 3585 | 4175 | [95], [96], [98], [100] |
| United Kingdom | 1020 | 883 | 1144 | 897 | 816 | 625 | – | |
| Federal Republic of Germany | 824 | 892 | 1004 | 917 | 919 | 1024 | 1333 | [94], [98], [100] |
| Japan | 1951 | 2015 | 2178 | 2060 | 1899 | 1814 | 1750 | [94], [98], [100] |
| Canada | 383 | 555 | 493 | 604 | 571 | 504 | 580 | [94], [96], [97], [100] |
| Mexico | – | 79 | 71 | 79 | 77 | 96 | 139 | [97], [100] |
| Italy | – | – | 528 | 449 | 423 | 392 | 518 | [95], [100] |
| Brazil | 145 | 172 | 260 | 309 | 336 * | 336 * | – | [98], [99] |
| France | – | – | 631 | 546 | 499 | 514 | 623 | [97], [100] |

\* Estimates based on six-month values

Republic of Germany was ca. $330 \times 10^3$ t, but this had increased to ca. $820 \times 10^3$ t by 1970. Since the early 1970s the rate of growth in the quantities of benzene produced has leveled off somewhat, with downtrends apparent during periods of industrial recession. Because of decreased use of tetraethyllead in gasolines over the past several years, the demand for BTX aromatics in gasoline has increased. This may affect the overall demand and hence price level for benzene. Generally about 50% of the benzene produced by petroleum refiners is retained for use as a gasoline component.

Benzene production by country over the past several years is shown in Table 24. Benzene production in 1980 for the non-communist world was over $13 \times 10^6$ t, with the United States producing about 40% of the total [101]. An estimated annual growth of 2.0 – 2.5% is predicted over the next years. The amount of benzene available from light oils produced in coal carbonization has not increased for some years, especially in the United States, for several reasons. Depression of the steel industry and the adoption of more modern steel processing methods have resulted in decreased coke production and hence decreased amounts of light oil. In addition, much of the light oil formed is not segregated but is burned as fuel in the steel industry. Petroleum refiners, however, purchase and refine this material because of its high content of benzene and appreciable amounts of the other BTX components. Table 25 shows the decline in coal-derived benzene in the United States over the last 30 years. In 1953, 77% of the benzene produced was derived from coal, but in 1981 this had decreased to 7%. A further decline is projected. As mentioned earlier the change from coal-derived benzene to that produced by petroleum refiners came at a later date in Europe because of the greater dependency on coal. Thus in the Federal Republic of Germany in 1970, 38% of the benzene produced was coal derived, as compared to about 10% in the United States. About 50% of the light oils produced in the United States is processed into BTX materials for use as petrochemical charge stocks or as a motor fuel component.

Table 26 gives benzene production and sales by coke-oven operators and by petroleum refiners. The petroleum refinery figures include benzene derived from light oils purchased from coke-oven operators and processed by the petroleum refiners. The

**Table 25.** United States benzene production from petroleum and from coal, 1000 t

| Year | Production from | | Total | % From coal | Reference |
|---|---|---|---|---|---|
| | petroleum | coal | | | |
| 1953 | 211 | 702 | 913 | 76.9 | [95] |
| 1955 | 332 | 699 | 1031 | 67.8 | [95] |
| 1960 | 1034 | 495 | 1529 | 38.4 | [95] |
| 1965 | 2335 | 435 | 2770 | 15.7 | [95] |
| 1970 | 3415 | 402 | 3817 | 10.5 | [95] |
| 1978 | 4573 | 398 | 4971 | 8 | [101] |
| 1981 | 4170 | 314 | 4484 | 7 | [101] |
| 1986 * | 4930 | 315 | 5245 * | 6 | [101] |

* Projected

**Table 26.** Benzene production and sales of synthetic organic chemicals (USITC)

| Year | Production, 1000 t | | Sales, 1000 t | |
|---|---|---|---|---|
| | Petroleum refiners | Coke-oven operators | Petroleum refiners | Coke-oven operators |
| 1977 | 4580 | 216 | 1983 | 217 |
| 1978 | 4793 | 179 | 2347 | 181 |
| 1979 | 5383 | 204 | 2817 | 200 |
| 1980 | 6535 | 170 | 3661 | 169 |
| 1981 | 4368 | 105 | 2190 | 107 |

differences between the production and sales figures are indicative of the benzene retained by the petroleum refiners for use as fuel components.

Except for short intervals, production of benzene approximates demand. After a rapid growth in production facilities, the United States and other countries have production capacity far greater than demand. For the years shown, the ratio of benzene production to existing capacity was approximately [101]:

| | | |
|---|---|---|
| United States | 1980 | 0.62 |
| | 1981 | 0.56 |
| Western Europe | 1980 | 0.59 |
| Japan | 1980 | 0.71 |

Demand is expected to increase moderately during the 1980s and the ratio may improve. On the other hand, with projected increase in demand, production capabilities may increase, especially in industrial developing countries and in countries with excess hydrocarbon supplies. Illustrative of this projected increase in capacity are the benzene projects planned or under construction in late 1981 [102]. The survey showed that 32 projects were being considered or under construction. In North America there were five units and in Western Europe four units. Eastern Europe with eight units and the Asian area with seven units indicate a rather sizable expansion. Eight units were planned for other areas.

**Table 27.** Toluene and xylene production in different countries, 1000 t

| Country | 1976 | 1978 | 1979 | 1980 | 1981 | 1982 |
|---|---|---|---|---|---|---|
| *Toluene* | | | | | | |
| United States | 3279 | 4195 | 5379 | 5111 | 4680 | 3478 |
| United Kingdom | | | 293 | 190 | 179 | 169 |
| Federal Republic of Germany | 233 | 225 | 298 | 260 | 262 | 268 |
| Japan | 858 | 861 | 925 | 889 | 821 | 803 |
| Canada | | | 493 | 560 | 569 | 504 |
| Mexico | | 124 | 108 | 125 | 132 | 138 |
| Italy | | | 366 | 286 | 278 | 268 |
| Brazil | | 72 | 82 | 104 | 60 | 69 |
| France | | | 86 | 64 | 56 | 31 |
| *Xylenes* | | | | | | |
| United States | 2363 | 2794 | 3125 | 2902 | 2916 | 2399 |
| Federal Republic of Germany | 477 | 414 | 492 | 355 | 486 | 459 |
| Japan | 1169 | 1248 | 1317 | 1195 | 1202 | 1225 |
| Mexico | | 93 | 82 | 93 | 103 | 115 |
| Italy | | | 323 | 283 | 256 | 268 |
| Brazil | | | 117 | 134 | | |

**Table 28.** Aromatics prices, US $/L

| Year | Benzene* | Toluene* | Xylenes* | Reference |
|---|---|---|---|---|
| 1977 | 0.20 | 0.15 | 0.13 | [103] |
| 1979 | 0.08 | 0.28 | 0.28 | [103] |
| 1980 | 0.25 | 0.30 | 0.25 | [103] |
| 1981 | 0.45 | 0.33 | 0.39 | [103] |
| 1982 | 0.41 | 0.39 | – | [101] |
| Aug. 1983 | ≈ 0.41 | ≈ 0.29 | ≈ 0.30 | [104] |

* All grades, average

Because toluene and xylenes are related components of BTX feedstocks and because the alkylated benzenes may serve as feedstocks for benzene production, the amounts of toluene and xylenes produced are of direct significance. Table 27 contains production figures for toluene and xylenes.

The prices of the BTX chemicals in the United States are documented annually by the United States International Trade Commission (USITC) [103]. Prices have fluctuated but in general have increased over the years (see Table 28). The values given in this table are for all grades and include values assigned to the aromatics retained and used as components in fuel blends. Some small variations in price are evident depending on the source and quality of the aromatic. For benzene in 1981, the all-grades average price was $ 0.45/L. Benzene from coke-oven sources commanded $ 0.44/L and 1 °C and 2 °C benzene from petroleum refiners averaged $ 0.45/L. Spot prices for benzene, toluene, and xylenes are published regularly in technical and trade journals. In the August 17, 1983 issue of Chemical Week [104], the price of benzene was $ 0.41/L or less, with toluene and xylenes endeavoring to command $ 0.29/L and $ 0.30/L, respectively. A comparison of benzene prices in July, 1982 for United States, Japan, and the Federal

**Table 29.** Benzene consumption in various countries (1980), 1000 t

|  | United States | Western Europe | Canada | Mexico | Japan |
|---|---|---|---|---|---|
| Production | 5293 | 4059 | 604 | 79 | 2046 |
| Capacity | 8578 | 6880 | – | – | 2882 |
| Demand | 5166 | 4247 | – | – | 1825 |
| End use, % | | | | | |
|   Ethylbenzene | 51.0 | 48.6 | 67 | 22 | 50.4 |
|   Cumene | 20.7 | 19.3 | 7 | – | 12.1 |
|   Cyclohexane | 13.8 | 13.4 | 23 | 37 | 25.6 |
|   Nitrobenzene/aniline | 5.3 | 6.7 | – | 2 | – |
|   Detergent alkylate | 3.1 | 2.0 | – | 22 | – |
|   Chlorobenzenes | 2.6 | 5.2 | – | – | 3.7 |
|   Maleic anhydride | 2.8 | 3.3 | 1 | 6 | 2.5 |
|   Other | 0.8 | 1.5 | 2 | 11 | 5.7 |

**Table 30.** Benzene consumption in the United States

| Year | 1975 | 1977 | 1980 | 1986* |
|---|---|---|---|---|
| Demand, 1000 t | 3596 | 4886 | 5170 | 5610 |
| Consumption, wt% | | | | |
|   Ethylbenzene | 52 | 55 | 48 | 51 | 51 |
|   Cumene | 17 | 17 | 20 | 21 | 24 |
|   Cyclohexane | 17 | 16 | 17 | 14 | 14 |
|   Nitrobenzene/aniline | 4.7 | 5 | 5 | 5 | 5 |
|   Detergent alkylate | 3.4 | 2.8 | – | 3 | 2.7 |
|   Chlorobenzenes | 4.5 | 2.9 | – | 2.6 | 2.1 |
|   Maleic anhydride | 3.2 | 2.9 | 4 | 2.8 | – |
|   Other | – | – | 6 | 0.8 | 0.8 |
| Reference | [101] | [101] | [2] | [101] | [101] |

* Projected

Republic of Germany as published by SRI International [101] showed roughly equivalent prices, with the United States price range of $ 0.40 – 0.41/L; Japan, $ 0.40 – 0.42/L; and the Federal Republic of Germany $ 0.38 – 0.40/L.

# 10. Uses

With the exception of its use as a motor fuel component and with the restrictions on its use as a solvent, the three main applications for benzene are production of ethylbenzene, cumene, and cyclohexane. These three products account for 75 – 80 % of the benzene consumed as a chemical feedstock. Table 29 gives 1980 benzene consumption by product [101]. A distribution of products and end uses of benzene in the United States is shown in Table 30.

Most of the primary products of benzene may be classified as intermediates. For example, benzene is alkylated with ethylene to form ethylbenzene, which in turn is dehydrogenated to styrene (→ Ethylbenzene). Styrene is the monomer employed for the production of several important polymeric or copolymeric products. These include acrylic–butadiene–styrene resins, styrene–butadiene rubber, and others [105].

Cumene, another large-volume product made from benzene, is used as a charge stock for the production of phenol and acetone. Cumene is made by the vapor-phase alkylation of benzene with propene. Most of the phenol produced in the United States and non-communist countries is derived from the hydroperoxidation of cumene. Phenol is used mainly for the production of phenolic resins.

Cyclohexane is the third largest volume intermediate product derived from benzene. It is produced by the direct hydrogenation of benzene [106]. Cyclohexane is oxidized to form intermediates, such as adipic acid, and thus serves as the starting point for the production of nylons.

About 5 % of the benzene produced is nitrated to form nitrobenzene, which in turn is hydrogenated to form aniline.

The oxidation of benzene is an established process for the production of maleic anhydride but it is receiving stiff competition from other processes, notably those involving the oxidation of butane and butenes. Maleic anhydride serves as a precursor for polyester resins.

Other products of benzene include the halogenated benzenes and the linear alkylbenzenes. The latter are a source for the production of detergents.

# 11. Toxicology and Occupational Health

Benzene has long been recognized as a toxic compound with both acute and chronic effects. Poisoning occurs through inhalation, ingestion, and by rapid absorption through the skin. Manifestations of acute benzene poisoning include headaches, confusion, loss of muscular control, and irritation of the respiratory and gastrointestinal tract. Greater concentrations may result in unconsciousness and even in death. Exposure to long-term chronic poisoning or shorter term exposure at higher concentrations may have drastic aftereffects, such as anemia and probably leukemia. Brain damage and damage to the urinary tract, mucous membranes, and other body parts were found in postmortem studies of acute and fatal poisoning [107].

Inhalation of benzene at different concentrations in air is the most common source of benzene poisoning because benzene is rapidly absorbed by the lungs. Ingestion by mouth causes irritation of the mouth, esophagus, and stomach. Absorption of benzene by the blood may follow, with resultant chronic or acute poisoning. Absorption of benzene through the skin is not a major source of benzene poisoning [108].

The potential toxicity of benzene has been the subject of numerous studies [109], [110]. Benzene is selectively absorbed by the lungs [111]. About 80 % of the benzene inhaled by breathing air containing 3000–5000 ppm of benzene is retained. The benzene is absorbed by the blood and can become fixed in bone marrow, fatty tissues, and the liver. It is slowly metabolized to phenols and related compounds before being eliminated. Removal of benzene from the body by metabolism is determined by the presence of phenols in the urine. The phenols may act as toxic intermediates. The relationship between metabolism and benzene toxicity has been the subject of many studies [112], [113]. If benzene metabolism is inhibited by the addition of toluene, the destruction of red cell production is reduced [114]. Some inhaled benzene is slowly removed by exhalation. People breathing air containing 100 ppm benzene exhaled benzene 24 h after they ceased inhaling the benzene-contaminated air [115].

Susceptibility to benzene poisoning varies with the individual. Women, especially pregnant women, reportedly are more susceptible than men, but this has not been substantiated in other studies [116]. Individual susceptibility probably results from the natural resistance of the person or from the state of health at the time of exposure. Cases have been cited where workers, generally unconscious as a result of exposure to benzene fumes in enclosed areas, have recovered, whereas members of the rescue teams have died [117].

The literature is replete with references concerning the potential toxicity of benzene or that of its metabolized products and the various effects on the body. NTIS Report PB-289189 [118] reviews research findings of the effects of benzene poisoning at environmental exposure levels. Potential effects considered include leukemia, especially acute myelogenous leukemia, and pancytopenia (including aplastic anemia and chromosomal effects) and may result from long-term exposure. Again indications suggest that benzene toxicity and damage occur because of a toxic metabolite of benzene.

A group of chemical workers from seven industrial plants who had been exposed for 30 years to benzene for extended periods of time, but not necessarily continuously, were studied [119]. The incidence of various forms of cancer and the mortality rate of these workers was higher than, but not significantly above, those of the national norm.

Although conflicting reports have been issued over the years, it is generally believed that inhalation of benzene in occupational environments is potentially a cause of leukemia. An early indication of this relation was contained in a 1928 report [120]. In 1939, several cases of leukemia were reported in patients who had been chronically exposed to benzene [121]. In 1973, chronic benzene poisoning was thought to be responsible for four cases of acute leukemia. These patients had worked in an atmosphere containing 150–200 ppm benzene. According to other studies the connection between benzene poisoning and leukemia and other malignant conditions is much less certain. A 1974 survey included 38000 petroleum workers from eight refineries in Europe [122]. The purpose of the study was to determine if the incidence of leukemia could be related to workers chronically exposed to benzene at low-concentration levels. The conclusion was that the incidence of leukemia in the chronically exposed workers was not significantly different from that of the general population as a whole. However, this work has been criticized because of the different variables and assumptions made in the study. The results of other investigations indicate that benzene is suspected as a potential cause of leukemia.

Benzene toxicity has been related also to pancytopenia or destruction of elements in the blood. A related effect of benzene exposure is aplastic anemia in which bone marrow is damaged or destroyed. One early investigation studied workers in the printing industry who had been exposed for a long period to benzene vapors [123]. More recently the long-term effect of exposure to benzene has been studied. Partial recovery of blood and bone marrow damage may result on cessation of exposure if the damage has not been too great.

An EPA report [118] discusses the health effects caused by benzene at relatively low levels of concentration in terms of leukemia and blood disorders. Studies were made on humans and on animals.

Some investigations suggest that chromosome rearrangement may be caused by chronic exposure to benzene [110], [124]. Chromosomal damage of both stable and unstable nature can occur after benzene exposure. The stable effect was found to remain for several years. Alkylated benzenes did not cause this effect. These chromosome aberrations might result in the development of leukemia.

Although benzene has been universally recognized as a very toxic chemical compound with chronic and acute poisoning effects, questions still remain regarding its carcinogenicity. For instance, much research continues to determine whether benzene exposure results in leukemia. Studies have been done on animals and on groups of workers exposed to benzene vapors under varying operational conditions. Two examples of studies relating to leukemia indicate that much additional work is needed to sort out the complexities associated with the problem and to determine the quantitative level of exposure [125], [126].

Exposure to benzene emissions may occur from many sources. A major source of evaporative benzene emissions is from motor gasoline, both in exhaust gases from automobiles and from handling of gasoline during transport to service stations and from the filling of automobile tanks. Gasolines generally contain from 3–5 vol% benzene. Depending on control of emissions, a concentration of up to 3.2 ppm of benzene in the air may exist in the vicinity of the operation [127].

Another source of benzene emissions is in the manufacture of benzene and gasolines. Efficient process control in refineries, however, maintains evaporative losses at a minimum. Historically, a major source of benzene emissions was in the use of benzene or benzene-containing solvents in the formulation of paints, thinners, and adhesives. Because of the high-toxic characteristics of benzene its use in these applications has been practically eliminated. Regulations state that substitute solvents should be used wherever possible. At one time alkylated mononuclear aromatics, such as toluene and xylene, were considered to have toxic effects equal to that of benzene. However, benzene is unique in most of its toxic effects. Earlier reports on the toxic effects of toluene and xylene may have overlooked the fact that benzene was present in these compounds [128]. The lower relative volatility of these alkylated aromatics is another factor in their favor.

Coal carbonization, for the production of coal coke for steel mill use, produces large amounts of aromatic compounds with benzene present in high concentrations.

Treatment for acute benzene poisoning is similar to that used for many cases of poisoning from toxic volatile chemicals. These steps include removing the person from the source, artificial respiration, if necessary, and other standard first-aid procedures.

Good laboratory and plant practices are paramount in preventing benzene poisoning. These involve minimum concentrations of benzene, either in liquid or vapor form, in work areas. Nonpermeable gloves have been suggested and tested as a means of preventing skin contact. Viton rubber or poly(vinyl alcohol) materials were found to be the best of those tested. However, these gloves are expensive and gloves are not always a guarantee of protection [129], [130].

The TLVs for benzene adopted by the ACGIH for 1983–1984 are: TWA 10 ppm, 30 mg/m$^3$; and STEL 25 ppm, 75 mg/m$^3$. Benzene is listed as a class A 2 industrial substance suspected of carcinogenic potential for humans (10 ppm) [131].

The DFG does not give a MAK for benzene but lists it in Section III A 1 under working materials which have been unequivocably proved carcinogenic. It is a compound capable of inducing malignant tumors as shown by experience with humans [132].

With the spotlight on the toxicity of benzene, tighter restrictions will probably be imposed on the use, the handling, and the exposure limits of benzene. The new regulations may be in the 1-ppm range. Recently the EPA has issued standards mostly aimed at benzene, projected to reduce emissions of volatile organic compounds by 70 %. This would apply to petroleum refineries and chemical plants [133].

Indicative of the activity and studies related to the toxicity of benzene and other poisonous materials, the following may be cited. A report on different carcinogens and methods for defining their risk factors has been issued by the Interagency Regulatory Liaison Group (IRLG) listing 26 substances, including benzene, and relating these to occupations that show high cancer rates [134].

The International Register of Potentially Toxic Chemicals (IRPTC) represents an international agency that is studying and registering the various effects of toxic chemicals under different conditions for each material. These conditions include factors related to production sampling, carcinogenicity, treatment of poisoning, and others [135].

The IRPTC was initiated in 1971 and is planning to publish its findings. The initial list of chemicals and products number 300 and more are to be added each year. The register is hoped to bring an international uniformity of information and control.

# 12. References

[1] M. Faraday, *Trans. R. Soc. (London)* **115** (1825) 440.
[2] L. H. Hatch, S. Matar, *Hydrocarbon Process.* **57** (1978) no. 11, 291–301.
[3] R. Maux, *Hydrocarbon Process.* **55** (1976) no. 3, 90.
[4] *Hydrocarbon Process.* **60** (1981) no. 11, 196.
[5] A. C. MacFarlane, *Oil Gas J.* **74** (1976) no. 6, 99–102.
[6] *Hydrocarbon Process.* **56** (1977) no. 11, 144.
[7] P. R. Pujado, *Hydrocarbon Process.* 55 (1976) no. 3, 91–96.
[8] F. Treeny, *Erdöl Kohle* **20** (1967) no. 9, 629–634.
[9] *Oil Gas J.* **61** (1963) no. 15, 125.
[10] L. Aalund, *Oil Gas J.* **74** (1976) no. 27, 98–108.
[11] D. E. Steeve, W. A. MacDonald, D. H. Stone, *Hydrocarbon Process.* **61** (1981) no. 9, 263–268.
[12] M. Sittig: *Aromatic Hydrocarbons, Manufacture and Technology,* Noyes Data Corp., London 1976, p. 73.
[13] Bethlehem Steel Co., US 3312749, 1967 (C. J. Hess, W. M. Perry).
[14] Air Products, US 3564067, 1971 (W. C. Brenner, L. C. Doelph).
[15] Bethlehem Steel Co., US 3623973, 1971 (M. O. Tarhan).
[16] M. O. Tarhan, L. H. Windsor, *Chem. Eng.* 1966 (March 28) 96–98.
[17] J. M. Duffalo, D. C. Spence, W. A. Schwartz, *Chem. Eng. Prog.,* **77** (January 2, 1981) 56–62.
[18] D. McNeil: *Coal Carbonization Products,* Pergamon Press, 1966, p. 54.
[19] E. J. Houdry, W. F. Burt, A. E. Pew, W. A. Peters, *Pet. Refiner* **17** (November 1938) 574–582.
[20] Houdry Chemical, US 2078945, 1937 (E. J. Houdry).
[21] Socony, US 2384942, 1945 (M. M. Marisic).
[22] E. V. Murphree et al., *Pet. Refiner* **24** (1945) 97–100.
[23] R. T. Sebulsky et al., *Oil Gas J.* **66** (1968) no. 21, 103–109.
[24] A. N. Sachanen: *Conversion of Petroleum,* Reinhold Publ. Co., New York 1948, pp. 345–350.
[25] *Hydrocarbon Process.* **61** (1982) no. 9, 164–169.
[26] E. A. Sutton, A. R. Greenwood, F. H. Adams, *Oil Gas J.* **70** (1972) no. 21, 52–56.
[27] *Oil Gas J.* **65** (1967) no. 8, 70–75.
[28] S. B. Zdonik, E. J. Green, L. P. Hallee, *Oil Gas J.* **64** (1966) no. 51, 75–80.
[29] J. G. Freiling, A. A. Simone, *Oil Gas J.* **71** (1973) no. 1, 25–31.
[30] W. Kronig, *7th World Petroleum Congress,* **4** (1967) 235.
[31] D. J. Griffiths, D. M. Luntz, J. L. James, *Oil Gas J.* **66** (1968) no. 8, 107–108.
[32] S. Field, *Am. Pet. Inst. Proc. Refining Division* **50** (1970) 340–365.
[33] A. H. Weiss, J. A. Marker, R. Newwirth, *Oil Gas J.* **60** (1962) no. 4, 64–71.
[34] G. F. Asselin, R. A. Erickson, *Chem. Eng. Prog.* **58** (1962) no. 4, 47–52.
[35] *Ind. Eng. Chem.* **54** (1962) no. 2, 28–33.
[36] *Hydrocarbon Process.* **60** (1981) no. 11, 138.
[37] *Hydrocarbon Process.* **46** (1967) no. 11, 184.
[38] [37], p. 187.
[39] S. Feigelman, C. B. O'Connor, *Hydrocarbon Process.* **45** (1966) no. 5, 140–144.
[40] *Hydrocarbon Process.* **44** (1965) no. 11, 277.
[41] S. Masamune et al., *Hydrocarbon Process.* **46** (1967) no. 2, 155–158.
[42] S. Masamune, T. Kawatani, *Hydrocarbon Process.* **47** (1968) no. 12, 111–117.
[43] *Hydrocarbon Process.* **61** (1981) no. 11, 139; **50** (1971) no. 11, 133.

[44] S. Otani, *Jpn. Chem. Q.* **4** (1968) no. 6, 16.
[45] E. G. Hancock (ed.): *Benzene and Its Industrial Derivatives*, J. Wiley & Sons, New York 1975, pp. 66, 67.
[46] R. G. Craig, C. E. Fowler, M. L. Raczynski, *ACS Meeting*, Minneapolis, April 13–18, 1969.
[47] *Hydrocarbon Process.* **61** (1982) no. 9, 153.
[48] M. O. Tarhan, L. H. Windsor, *Chem. Eng.* **73** (March, 28, 1966) 96–98.
[49] W. Lorz et al., *Erdöl Kohle* **21** (1968) 610.
[50] H. Kubo, S. Masamune, R. Sako, *Hydrocarbon Process.* **49** (1970) no. 7, 111–113.
[51] S. Komatsu, *Ind. Eng. Chem.* **60** (1968) no. 2, 36.
[52] Mitsubishi Petrochemical Co., US 3400168 1968 (F. Fukudaetal).
[53] P. Grandio: *162nd ACS Annual Meeting*, Washington, D.C., September 12–17, 1971.
[54] P. Grandio, F. H. Schneider, A. B. Schwartz, J. J. Wise, *Oil Gas J.* **69** (1971) no. 48, 62–69.
[55] T. Wett, *Oil Gas J.* **78** (June 16, 1980) 55–61.
[56] R. Serrurier, *Hydrocarbon Process.* **55** (1976) no. 9, 253–257.
[57] R. G. Nene, *Hydrocarbon Process.* **60** (1981) no. 11, 287–290.
[58] R. C. Robinson, D. H. Broderick, H. A. Frumkin, *Hydrocarbon Process.* **61** (1982) no. 11, 102–108.
[59] H. Hiller, O. L. Garkisch, *Hydrocarbon Process.* **59** (1980) no. 9, 238–242.
[60] J. J. Wise, A. J. Silvesti, *Oil Gas J.* **74** (1976) no. 47, 140–142.
[61] M. Steinberg, P. T. Fallon, *Hydrocarbon Process.* **61** (1982) no. 11, 92–96.
[62] *Oil Gas J.* **82** (May 14, 1984) 76–80.
[63] J. B. O'Hara et al., *Hydrocarbon Process.* **57** (1978) no. 11, 117–121.
[64] *Hydrocarbon Process.* **61** (1982) no. 9, 126.
[65] [64], p. 125.
[66] R. Lester, *Hydrocarbon Process.* **40** (1961) 175.
[67] K. H. Eisenlohr, *Erdöl Kohle* **16** (1963) no. 6, 523–533.
[68] J. A. Gerster, *Chem. Eng. Prog.* **65** (1969) no. 9, 43–46.
[69] M. Sittig: *Aromatic Hydrocarbons*, Manufacture and Technology, Noyes Data Corp. N.J., London 1976, pp. 104–141.
[70] E. G. Hancock: *Benzene and its Industrial Derivatives*, J. Wiley & Sons, 1975, Chaps. 2, 3.
[71] H. W. Grote, *Chem. Eng. Progress.* **54** (1958) no. 8, 43–48.
[72] *Hydrocarbon Process.* **49** (1970) no. 9, 248.
[73] D. P. Thornton, Jr., *Pet. Process.* **8** (March 1953) 384–387.
[74] G. S. Somekh, B. I. Friedlander, *Hydrocarbon Process.* **48** (1969) no. 12, 127–130.
[75] T. S. Hoover, *Hydrocarbon Process.* **48** (1969) no. 12, 131–132.
[76] H. G. Staaterman, R. C. Morris, R. M. Stager, G. J. Pierotti, *Chem. Eng. Prog.* **43** (1947) no. 4, 148–151.
[77] G. H. Deal, Jr., H. D. Evans, E. D. Oliver, M. N. Papaopoulos, *Pet. Refiner* **38** (1959) no. 9, 185–192.
[78] *Hydrocarbon Process.* **59** (1980) no. 9, 203.
[79] D. B. Broughton, G. F. Asselin: *7th World Petroleum Congress*, Mexico City, April 2–8, 1967, Section 4, pp. 65–73.
[80] *Ullmann*, 4th ed., **8**: 395, 398.
[81] [67], p. 52.
[82] *Oil Gas J.* **64** (1966) no. 29, 83–84.
[83] K. H. Eisenlohr, W. Grosshans, *Erdöl Kohle* **18** (1965) no. 8, 614–618.
[84] J. Lautier, J. Durandet, C. Raumbault, M. Viguier,*Rev. Inst. Fr. Pet.* **20** (1965) 181–190.

[85] B. Choffe, C. Raumbault, F. P. Navarre, M. Lucas, *Hydrocarbon Process.* **45** (1966) no. 5, 188–192.
[86] *Hydrocarbon Process.* **51** (1972) no. 9, 185.
[87] E. Cinelli, S. Noe, G. Paret, *Hydrocarbon Process.* 51 (1972) no. 4, 141–144.
[88] K. H. Eisenlohr: *6th World Petroleum Congress*, Frankfurt/Main, June 19–26, 1963, Section 4, Paper 8.
[89] *Hydrocarbon Process.* **61** (1982) no. 9, 182.
[90] J. Morrison, *Oil Gas J.* **68** (1970) no. 47, 49–64.
[91] F. Trefny, *Erdöl Kohle* **23** (1970) no. 6, 337–340.
[92] M. Stein, *Hydrocarbon Process.* **52** (1973) no. 4, 139–141.
[93] *Hydrocarbon Process.* **61** (1982) no. 9, 186.
[94] *Hydrocarbon Process.* **61** (1982) no. 9, 187.
[95] *Ullmann,* 4th ed., **8**: 384.
[96] "Facts and Figures," *Chem. Eng. News* **57** (June 11, 1979) 32.
[97] "Facts and Figures," *Chem. Eng. News* **61** (June 13, 1983) 26.
[98] *Chem. Eng. News* **60** (Dec. 20, 1982) 48.
[99] "Facts and Figures", *Chem. Eng. News* **60** (June 14, 1982) 81.
[100] "Facts and Figures", *Chem. Eng. News* **62** (June 11, 1984) 32.
[101] C. S. Hughes, A. B. Abshire: "Benzene," *Chemical Economics Handbook,* SRI International, February, 1983, 300.7000 I, J.
[102] T. Wett, *Oil Gas J.* **80** (1982) no. 13, 81–84.
[103] "Synthetic Organic Chemicals (U.S. Production and Sales)" U.S. International Trade Commission*Publications 920* (1977), *1099* (1979), *1183* (1980), *1292* (1981).
[104] *Chem. Week* **133** (1983) no. 7, 29.
[105] T. C. Ponder, *Hydrocarbon Process.* **56** (1977) no. 7, 135–138.
[106] *Hydrocarbon Process.* **60** (1981) no. 11, 147, 148.
[107] H. W. Gerarde: *Toxicology and Biochemistry of Aromatic Hydrocarbons,* Elsevier, New York 1960, pp. 97–98.
[108] *H. E. W. Publication,* NIOSH, Washington, D.C., 1974, pp. 74–137.
[109] "Assessment of Benzene as a Potential Air Pollution Problem," *Report P. B. 258-356* National Technical Information Service, vol. **4**, 1976.
[110] "Health Effects of Benzene," *Report P.B. 254-388,* National Technical Information Service, June, 1976.
[111] "Metabolism versus Toxicity of Benzene," *Report P.B. 289-789,* National Technical Information Service, Sept. 1978, p. 8.
[112] A. Stulmann, C. P. Stewart: *Toxicology of Volatile Liquids,* vol. **1,** Academic Press, New York 1960, pp. 61–63.
[113] P. C. Parke, R. T. Williams, *Biochem. J.* **54** (1954) 231–238.
[114] L. S. Andrews, E. W. Lee, C. M. Witmer, J. J. Kocsis, *Biochem. J.* 1977 243.
[115] C. G. Hunter, *Proc. R. Soc. Med.* **61** (1968) 913–915.
[116] F. T. Hunter, *J. Ind. Hyg. Toxicol.* **21** (1939) 331.
[117] G. L. Winek, W. D. Collom, *J. Occup. Med.* **13** (1971) 250–261.
[118] *Report PB 289-789,* National Technical Information Service, Sept., 1978, p. 47.
[119] O. Wong: "An Industry-Wide Mortality Study of Chemical Workers Occupationally Exposed to Benzene," *Report to Chemical Manufacturers Assoc.,* Environmental Health Assoc., Berkely, Calif., Dec. 8, 1983.
[120] P. Delore, Borganamo, *J. Med. Lyon* **9** (1928) 227–233.

[121] T. B. Mallory, E. A. Gall, Brickley, *J. Ind. Hyg. Toxicol.* **21** (1939) 355–393.
[122] J. J. Thorpe, *J. Occup. Med.* **16** (1974) 375–382.
[123] L. J. Goldwater, Tewksbury, *J. Ind. Hyg.* **26** (1941) 957-973.
[124] E. Vigiliani, A. Forni, *J. Occup. Med.* **11** (1969) 148–149.
[125] *Chem. Eng. News* **59** (September 7, 1981) 68.
[126] *Chem. Eng. News* **54** (April 26, 1976) 7.
[127] G. S. Parkinson, *Ann. Occup. Hyg.* **14** (1971) 145–153.
[128] J. Aryanpur: *Tenth World Petroleum Congress,* Bucharest, 1979, vol. **5,** p. 237.
[129] E. B. Sansone, V. B. Tewari, *Am. Ind. Hyg. Assoc. J.* **39** (1978) no. 2, 169–174.
[130] *Chem. Week* **125** (October 31, 1979) no. 18, 20.
[131] ACGIH (ed.): *Threshold Limit Values (TLV)1983–1984,* ACGIH Cincinnati, Ohio 1984.
[132] DFG (ed.): *Maximum Concentrations at the Workplace (MAK),* Verlag Chemie, Weinheim 1982.
[133] *Chem. Eng. News* **62** (June 4, 1984) 5.
[134] *Chem. Eng. News* **60** (Feb. 12, 1982) 4.
[135] *Chem. Eng. News* **60** (May 10, 1982) 25.

# Benzenesulfonic Acids and Their Derivatives

*For nitrobenzenesulfonic acids, see* → *Nitro Compounds, Aromatic; sulfobenzoic acids are treated under* → *Benzoic Acid*

OTTO LINDNER, Bayer AG, Leverkusen, Federal Republic of Germany

| | | | | | |
|---|---|---|---|---|---|
| 1. | General Aspects | 742 | 2.3. | Alkylbenzenesulfonic Acids | 761 |
| 1.1. | Physical Properties | 742 | 2.4. | Formylbenzenesulfonic Acids | 767 |
| 1.2. | Chemical Properties | 742 | 2.5. | Chlorobenzenesulfonic Acids | 768 |
| 1.3. | Production | 744 | 2.6. | Aminobenzenesulfonic Acids | 769 |
| 1.3.1. | Direct Sulfonation | 744 | 2.7. | Hydroxybenzenesulfonic Acids | 774 |
| 1.3.2. | Oxidation of Sulfur Compounds | 748 | 2.8. | Chloro-alkylbenzenesulfonic Acids | 777 |
| 1.3.3. | Diazo Reaction | 748 | | | |
| 1.3.4. | Sulfite Reaction | 748 | 2.9. | Amino-alkylbenzenesulfonic Acids | 778 |
| 1.4. | Uses | 749 | | | |
| 1.5. | Analysis | 749 | 2.10. | Amino- and Hydroxy-carboxybenzenesulfonic Acids | 779 |
| 1.6. | Environmental Aspects | 750 | | | |
| 1.7. | General Characteristics of Benzenesulfonic Acid Derivatives | 751 | 2.11. | Chloro-hydroxybenzenesulfonic Acids | 780 |
| 1.7.1. | Benzenesulfonyl Chlorides | 751 | 2.12. | Amino-hydroxybenzenesulfonic Acids and Amino-chlorohydroxybenzenesulfonic Acids | 781 |
| 1.7.2. | Benzenesulfonamides | 753 | | | |
| 1.7.3. | Benzenesulfonohydrazides | 754 | | | |
| 1.7.4. | Benzenesulfonic Acid Esters | 754 | 2.13. | Amino-chlorobenzenesulfonic Acids | 784 |
| 2. | Individual Benzenesulfonic Acids and Derivatives | 755 | 2.14. | Amino-chloro-alkylbenzenesulfonic Acids | 785 |
| 2.1. | Benzenesulfonic Acid | 755 | | | |
| 2.2. | 1,3-Benzenedisulfonic Acid | 759 | 3. | References | 786 |

# 1. General Aspects

## 1.1. Physical Properties

The benzenesulfonic acids, with the exception of the aminobenzenesulfonic acids, are strong acids, which are almost entirely dissociated in aqueous solution. Some of them, including benzenesulfonic acid, can be distilled in a high vacuum without decomposition. They have a tendency to form hydrates, which are very stable in some cases and can be isolated from aqueous hydrochloric acid and dried at temperatures up to 100 °C without the water being lost. However, azeotropic distillation with benzene or chlorobenzene (depending on the heat stability of the hydrate) gives the anhydrous sulfonic acid.

Aminobenzenesulfonic acids are weak acids; they form zwitterions, which are poorly soluble in water, but their alkali salts are watersoluble. Unlike the free acids, the metal salts of the benzenesulfonic acids, particularly the alkali salts, are stable at temperatures up to and exceeding 300 °C. Most of them can be precipitated by adding an excess of salt to their aqueous solution. The calcium salts are more freely soluble in water than calcium sulfate. This is exploited industrially in the "liming" of the sulfonation mixture (see p. 760). In contrast to these metal salts certain organic salts are poorly soluble in water; for example, the salts with arylamines and *S*-benzylisothiourea have definite melting points and are therefore suitable derivatives for the identification of benzenesulfonic acids. For the physical properties of individual compounds, see Section 1.7 and Chap. 2 of this article.

## 1.2. Chemical Properties

*Desulfonation.* The sulfonation of aromatic compounds to form arylsulfonic acids is a reversible reaction. It proceeds through various activated complexes and intermediates; Figure 1 shows a schematic plot of the enthalpy along the "reaction coordinate." The ease by which sulfonation and desulfonation occur (kinetic control) and the position of the equilibrium (thermodynamic control) are dependent on the relative height of these enthalpies [8]. They are largely affected by other substituents present in the benzene nucleus. This is reflected by the various temperatures at which acidic desulfonation occurs [13]:

| | |
|---|---|
| benzenesulfonic acid | 227 °C |
| 2-methylbenzenesulfonic acid | 188 °C |
| 4-methylbenzenesulfonic acid | 186 °C |
| 3,4-dimethylbenzenesulfonic acid | 176 °C |
| 3-methylbenzenesulfonic acid | 155 °C |
| 2,4-dimethylbenzenesulfonic acid | 137 °C |
| 2,4,6-trimethylbenzenesulfonic acid | 100 °C |

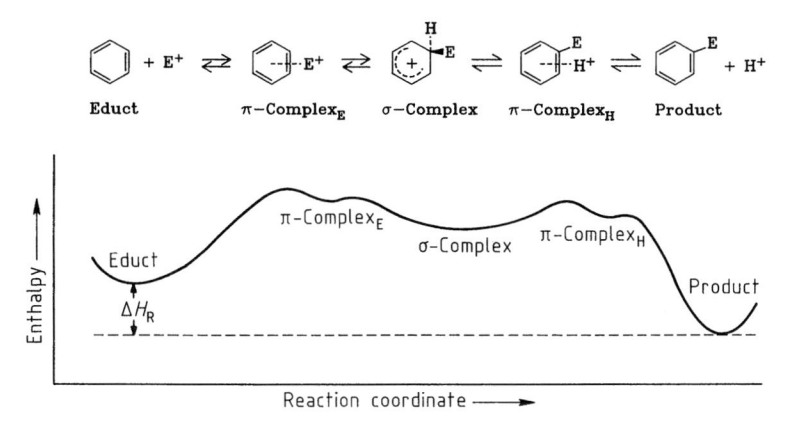

**Figure 1.** Sulfonation and desulfonation of benzene ($E^+ = SO_3$ or $HSO_3^+$), $\Delta H_R$ = reaction enthalpy

Figure 1 also explains the general rule according to which acidic desulfonation of benzenesulfonic acids occurs the more easily the easier it is to introduce the sulfonic acid group (higher sulfonation rate). This is often exploited industrially in the separation and purification of aromatic compounds as described in the next paragraph.

Pure p-xylene can be obtained from a mixture of o- and p-xylene by sulfonation and subsequent removal of water-soluble o-xylenesulfonic acid [14]. The same principle can be used to separate m-xylene from p-xylene [15], [16] and from ethylbenzene [17], or 1-ethyl-3,5-dimethylbenzene from other ethylxylenes [18], or 4-chloro-1,3-xylene from its isomers [19]. In the manufacture of DDT excess chlorobenzene is recovered by sulfonation, removal of the sulfonic acid, and subsequent desulfonation at 190 °C [20]. m-Dichlorobenzene can be separated as a sulfonic acid derivative from a mixture of isomeric dichlorobenzenes and obtained by subsequent desulfonation [21], [22]. Sulfonation and desulfonation presumably lead to the rearrangement of 1,3,4,5-tetraethylbenzene to 1,2,3,4-tetraethylbenzene (Jacobsen rearrangement) [23]. The sulfonic acid group also serves as a protective group and permits the selective introduction of further substituents, e.g., in the manufacture of o-chlorotoluene:

2,6-Dichlorophenol [24] and 6-halogeno-2-methylphenol can be produced analogously. In a similar manner 2,6- and 2,5-dichlorotoluene and 2,3,6-trichlorotoluene are obtained from toluenesulfonic anhydride by chlorination and subsequent desulfonation [25]. Sulfonation–desulfonation is also important in the production of 1,4-benzenesulfonic acid (Section 2.2).

*Alkali Fusion.* Alkaline cleavage of benzenesulfonic acids leads to phenols according to the following equation:

$ArSO_3Na + 2\ NaOH \longrightarrow ArONa + Na_2SO_3 + H_2O$

In former times this was the method of choice to produce phenol on an industrial scale; however, this process has mostly been replaced by the Hock process (→ Acetone).

*Sulfone Formation.* Benzenesulfonic acids react with aromatic compounds to form sulfones. This reaction is an undesired side reaction in industrial sulfonation (see Section 2.1); it is promoted by Friedel-Crafts reagents and inhibited by alkali ions [26].

*Reaction with Halogenides.* The reaction with phosphorus halogenides ($PCl_5$, $PBr_5$), chlorosulfuric acid, thionyl chloride, or phosgene leads to sulfonyl halogenides:

$$C_6H_5SO_2OH + PCl_5 \longrightarrow C_6H_5SO_2Cl + POCl_3 + HCl$$

If benzenesulfonic acid reacts with an excess of $PCl_5$, chlorobenzene is formed according to

$$C_6H_5SO_2Cl + PCl_5 \longrightarrow C_6H_5Cl + SOCl_2 + POCl_3$$

*Anhydride Formation.* Sulfonic anhydrides are made by the reaction of $P_2O_5$, $SOCl_2$, or $SO_3$ with benzenesulfonic acids [27]. They are also formed through side reactions in the preparation of benzenesulfonic acids or benzenesulfonylchlorides; in this case, however, they are frequently not observed because they hydrolyze fast in aqueous acid.

*Esterification.* Benzenesulfonic acid esters are formed in the reaction of the sulfonic acids or sulfonyl chlorides with alcohols or ethers (see Section 1.7.4).

## 1.3. Production

### 1.3.1. Direct Sulfonation

The direct introduction of the sulfonic acid group is one of the most important reactions in industrial organic chemistry. Together with nitration and chlorination, it belongs to the important group of electrophilic aromatic substitution reactions. It gives high yields under relatively mild conditions and usually results in well-defined benzene derivatives.

Aqueous sulfuric acid, at various concentrations from about 76% up to 100%, is frequently used as the sulfonation agent. In industry 100% sulfuric acid is often called *"monohydrate." Oleum* is a solution of sulfur trioxide in sulfuric acid; for sulfonation it is generally used at $SO_3$ concentrations of 20 or 65 wt% because the solidification points at these two concentrations are minimal (0 and 2 °C, respectively). Other concentrations are less favorable with regard to storage and transportation because the solidification points are higher (for example, 45% oleum solidifies already at 35 °C) [28].

The sulfonation with sulfuric acid is a reversible reaction, see also Figure 1:

$$ArH + H_2SO_4 \rightleftharpoons ArSO_3H + H_2O$$

It can be shifted optimally to the right if the water of the reaction is bound or removed by distillation, e.g., with benzene as an entrainer. Higher temperatures usually shift the equilibrium towards the right [29]. However, both methods increase sulfone formation.

The water of the reaction can also be bound by adding thionyl chloride:

$C_6H_6 + H_2SO_4 + SOCl_2 \longrightarrow C_6H_5SO_3H + SO_2 + 2\,HCl$

For example, the sulfonation of benzene with sulfuric acid and thionyl chloride gives a mixture consisting of 96.3% benzenesulfonic acid, 2.7% sulfuric acid, and 1.6% diphenyl sulfone [30]. Similar yields are obtained when sulfonating toluene or chlorobenzene in this way.

No reaction water is formed in sulfonations using chlorosulfuric acid or sulfur trioxide. However, increased sulfone formation occurs. As a sulfonation agent pure *sulfur trioxide* generally reacts too violently and leads to extensive side reactions including oxidation and sulfone formation. Hence it is used in complexed form, e.g., with pyridine, dioxane, trimethylamine, or dimethylformamide [31]. Many industrial processes use sulfur trioxide gas as the sulfonation agent, normally diluted with an inert gas [32]–[34]. In the liquid phase sulfur dioxide or dichloromethane are suitable as diluents. However, dichloromethane reacts with sulfur trioxide at temperatures above 0 °C and highly poisonous decomposition products are formed, such as bis(chloromethyl) ether. 1,1-Dichloroethane also reacts with sulfur trioxide at 80 °C; the decomposition products include phosgene.

In stoichiometric amounts, *chlorosulfuric acid* serves as a sulfonation agent according to

$ArH + ClSO_3H \longrightarrow ArSO_3H + HCl$

When using an excess of chlorosulfuric acid, sulfonyl chlorides are formed, see Section 1.7.1. The chlorosulfonation of toluene, for example, with chlorosulfuric acid at low temperature yields predominantly *o*-toluenesulfonyl chloride. This is the first step in the industrially important synthesis of saccharin.

**Reaction Mechanism.** The basic reaction mechanism of aromatic sulfonation is described in Figure 1. The nature of the electrophilic agent $E^+$, however, varies according to the sulfonation agent used and the solvent.

In monohydrate (100%) or only slightly dilute (92–99%) sulfuric acid, sulfur trioxide is the electrophilic agent [35], [36]. The rate-determining step in this reaction is the elimination of the proton.

In sulfonation with disulfuric acid and with chlorosulfuric acid the $SO_3H^+$ cation (sulfate cation) is assumed to be the sulfonation agent [37]:

$$H_2S_2O_7 + H^+ \xrightarrow{-H_2SO_4} {}^+\!\!\overset{\overset{O}{\|}}{\underset{\underset{O}{\|}}{S}}\!\!-OH \xleftarrow{-Cl^-} ClSO_3H$$

The $SO_3H^+$ cation is also formed as an intermediate when sulfur trioxide is used as the sulfonation agent [38]–[40]

$$ArSO_3H + SO_3 \longrightarrow ArSO_2-O-SO_3$$
$$\rightleftharpoons ArSO_2O^- + {}^+SO_3H \xrightarrow{+ArH} 2\,ArSO_3H$$

**Orientation Rules.** The sulfonation of benzene derivatives, $C_6H_5Y$, will be facilitated over that of benzene if the substituent Y supplies electron density to the aromatic ring. Examples are:

| Substituent Y | Electronic effect |
|---|---|
| Alkyl | + I |
| OR, $NR_2$ | + M > − I |
| $O^-$ | + M; + I |

where I = inductive effect and M = mesomeric effect

On the other hand, benzene derivatives will be sulfonated with greater difficulty compared with benzene if the substituent Y withdraws electron density from the aromatic ring. Examples are:

| Substituent Y | Electronic effect |
|---|---|
| COR, COOR, CN, $NO_2$, $N_2^+$, $SO_3H$ | − M; − I |
| Cl, Br | − I > + M |
| $NR_3^+$ | − I |

where I = inductive effect and M = mesomeric effect

The electron density-supplying substituents and the halogens direct the $SO_3H$ group predominantly into the *o*- and *p*-positions (*first-order substituents*). On the other hand, the electron density-withdrawing substituents (except for the halogens) generally direct the $SO_3H$ group into the *m*-position (*second-order substituents*). The seemingly exceptional role of the halogens is explained by the electronic stabilization of the various reaction intermediates ($\pi$complexes, $\sigma$-complex); this is discussed extensively in textbooks on organic chemistry.

This fundamental rule of aromatic substitution is valid particularly at low temperature. Raising the temperature may change the isomer distribution in the product mixture. For example, sulfonation of benzene with sulfuric acid at about 80 °C yields predominantly 1,3-benzenedisulfonic acid, see Section 2.2; however, by raising the temperature to 250 °C an isomer mixture consisting of 67 % 1,3-disulfonic acid and 33 % 1,4-disulfonic acid is formed. When sulfonating first-order-substituted benzene derivatives, the ratio of the ortho- and para-isomer formed is also being shifted in favor of the para-isomer if the temperature is raised.

**Technology of Sulfonation.** Sulfonation is carried out in cast-steel or enamelled steel vessels; in continuous processes, the vessels are arranged in cascades. In the batchwise mode the sulfonation agent is introduced into the vessel, the aromatic compound is subsequently added, and the mixture is then heated slowly to the reaction temperature. In continuously run cascades the sulfonation agent is often introduced into two or three vessels simultaneously.

On completion of the reaction the batch is forced into water, which causes unreacted aromatic compounds to be expelled. The dilute sulfonation mass is then cooled, and the free acid separated by filtration. If the free acid is too soluble, sodium sulfate or sodium chloride is added, after which the acid is separated as the sodium salt (salting out). Often it is also possible to neutralize a concentrated solution with sodium hydroxide and then remove the precipitated sodium sulfate at 70–90 °C. The filtrate then contains the sodium salt of the arylsulfonic acid almost free from inorganic salt. It can be dried in drum or spray dryers.

If isolation is not possible in this way, the excess sulfuric acid has to be neutralized by adding calcium carbonate. This leads to a large amount of gypsum, which is removed in the hot state ("liming" or "chalking"). The dissolved calcium arylsulfonate is then treated with soda and the precipitated calcium carbonate is removed by filtration. The filtrate contains the sodium arylsulfonate. In large-scale manufacture the economical removal of byproduct gypsum may become a problem. It should precipitate in well-formed crystals so that it can be separated easily from soluble organic impurities and is sufficiently dry after suction filtration to be dumped or used.

According to a new process the unreacted sulfuric acid is recovered by converting the sulfonic acid into its ammonium salt using a long-chain aliphatic amine. This salt is separated from the sulfuric acid as a liquid phase and then converted with sodium hydroxide solution into sodium sulfonate solution and the amine; the latter can be separated as a liquid phase and is re-used. Sulfonates practically free from inorganic salts are obtained in this way [41].

Special sulfonation and processing methods have been developed for compounds that withstand high temperatures and therefore allow the excess sulfuric acid to be distilled off in a vacuum (see Hooker process for the production of benzenedisulfonic acid, p. 760). This gives sulfonic acids almost free from sulfuric acid, which, with skillful mixing, can be neutralized and subjected to alkali fusion in a single step [42].

Aromatic amines may be sulfonated by a special solid-state process, called "baking": the ammonium sulfate that is first formed in the reaction between the amine and sulfuric acid is dewatered and simultaneously rearranged to aminobenzenesulfonic acid at 150–300 °C. The reaction product is solid, which requires the reaction to be carried out on trays in an oven.

The solids can also be suspended in an organic liquid such as diphenyl sulfone [43] or *o*-dichlorobenzene [44]; in this way the mass can be stirred during the entire reaction. This gives final products of higher quality because local overheating cannot occur (see also manufacture of *p*-sulfanilic acid, p. 771).

Many sulfonation methods using sulfur trioxide have been proposed for the purpose of enabling the sulfonation agent to be either diluted or evenly distributed so that the heat of reaction is dissipated quickly. This technology is substantially mature, for example, see [45]. When gaseous sulfur trioxide is used, it is diluted with a gas, such as air, nitrogen, or carbon dioxide [34].

## 1.3.2. Oxidation of Sulfur Compounds

Such sulfur compounds as thiophenols and diaryl disulfides can be oxidized with chlorine solution, permanganate, or nitric acid to form sulfonic acids. The process is industrially important wherever it is difficult or impossible to introduce the sulfonic acid group directly and the starting compounds are easily produced.

## 1.3.3. Diazo Reaction

Aryldiazonium halides are converted to aromatic sulfonic acids in glacial acetic acid containing $SO_2$ in the presence of copper(I) chloride [46]

$$HO_3S-Ar-N\equiv N^+Cl^- + 1/2\ Cu_2Cl_2$$
$$\rightarrow HO_3S-Ar-N\equiv N^+[CuCl_2]^- \rightarrow HO_3S-Ar\cdot + N_2 + CuCl_2$$
$$\xrightarrow{+SO_2} HO_3S-Ar-SO_2\cdot + CuCl_2$$
$$\xrightarrow{H_2O} HO_3S-Ar-SO_2OH + [CuCl_2]^- + H^+$$

This reaction, which was first observed by L. LANDSBERG, permits the production of, for example, 1,2-benzenedisulfonic acid from orthoanilic acid with a yield of 68 % [47].

## 1.3.4. Sulfite Reaction

Aromatic halogen compounds can react with sulfite to form sulfonic acids if the halogen substituent is activated by nitro groups. For example, 1-chloro-2,4-dinitrobenzene reacts with sulfite to give 1,3-dinitro-4-benzenesulfonic acid. The reaction is catalyzed by copper ions. 2-Formylbenzenesulfonic acid and 2-sulfobenzoic acid are obtained similarly from the corresponding chlorine compounds.

The addition of hydrogen sulfite to aromatic systems succeeds when their aromatic character is disturbed by the presence of certain substituents. The formation of 3-hydroxybenzenesulfonic acid from resorcinol [48] and the sulfitation of *m*-dinitrobenzene with simultaneous reduction of a nitro group [49] (Piria reaction) are examples:

Hydrogen sulfite addition, followed by the formation of a benzenesulfonic acid, gives satisfactory yields in special cases only and has therefore acquired little industrial importance so far.

## 1.4. Uses

Benzenesulfonic acids are used chiefly as intermediates. They are employed in the manufacture of sulfonic acid amides, hydrazides, and esters; of sulfinic acids, sulfones, phenols, and thiophenols; and of other compounds. Sulfonic acids that are substituted with OH and/or $NH_2$ groups serve as intermediates in the manufacture of finishing agents, optical brighteners, pickling agents, dyestuffs, tanning agents, water-soluble resins, insecticides, ion-exchange resins, wetting agents, pharmaceuticals, polymeric thickeners, plasticizers, etc. Benzenesulfonic acids are also used as such as acidic catalysts and standardizing agents in dyestuff manufacture.

## 1.5. Analysis

The degree of conversion in sulfonations can be checked by determining the residual amount of starting material. If this is an aromatic compound, gas chromatography is used. Paper chromatography or thin layer chromatography is used in the case of disulfonation. For example, the degree of completeness of benzene disulfonation can be estimated with sufficient accuracy after paper-chromatographic determination of the residual quantities of benzenemonosulfonic acid.

Aminobenzenesulfonic acids can be rendered distinctly visible on the paper chromatogram by diazotization and subsequent azo dye formation. Hydroxybenzenesulfonic acids can be stained well on the paper by coupling with a diazonium salt. Most other sulfonic acids can be recognized clearly from their fluorescence-extinguishing effects on chromatography paper previously sprayed with Pinakryptol Yellow and thus rendered fluorescent in UV light [50].

Benzenesulfonic acids can be determined quantitatively in the presence of sulfuric acid and sulfones by titrating the total acid with alkali and then determining the sulfuric acid content gravimetrically using barium chloride. The difference gives the content of sulfonic acid. Benzenemonosulfonic and benzenedisulfonic acid mixtures can be separated and analyzed quantitatively by GC analysis of their sulfofluorides [51]. Today they are separated by HPLC with high accuracy and efficiency [52]. By this method it is now possible, both in the laboratory and in industry, to analyze the exact composition of sulfonation mixtures, which frequently consist of many components, and to observe how the composition is affected by the reaction parameters.

## 1.6. Environmental Aspects

**Organic Compounds in Waste Water.** Benzenesulfonic acid is biologically degradable. However, the biological degradation of many substituted benzenesulfonic acids is difficult or impossible; these compounds are therefore not suitable for treatment in biological waste-water purification plants.

If the desired benzenesulfonic acid can be precipitated from sulfuric acid solution without salt being added, the spent acid can be upgraded in a sulfuric acid recovery plant. If the waste water contains salts and non-degradable sulfonic acids wet oxidation, the most costly method of treatment, will be the only possibility.

**Inorganic Compounds in Waste Water.** In the operation of large plants the treatment of such inorganic constituents as acids and salts is becoming an increasingly important cost factor in addition to the treatment of the organic constituents.

The following methods are available to decrease the acid or salt content of the waste water.

1) Azeotropic distillation of the reaction water with a solvent. This improves the utilization of the sulfuric acid.
2) Use of oleum, sulfur trioxide, or chlorosulfuric acid in place of sulfur acid. The resulting increase in sulfone formation can be counteracted, though not eliminated, with the aid of additives, see p. 751.
3) Precipitation of the final product from the sulfuric acid solution without adding salt, e.g., by cooling (see isolation of *p*-toluenesulfonic acid, see p. 763).
4) Distillative separation of the sulfuric acid (see manufacture of benzenedisulfonic acid, see p. 760).
5) Formation of a separate liquid phase consisting of an ammonium sulfonate. After phase separation, the sulfuric acid phase is returned to the sulfonation reactor.
6) Precipitation of the excess sulfuric acid as sodium sulfate if sodium sulfate is less soluble than the sodium sulfonate.
7) Precipitation of the excess sulfuric acid as calcium sulfate. The gypsum thus obtained is either dumped or used, depending on its quality.

8) Removal of the excess sulfuric acid by reacting it with additional aromatic compounds that are more reactive than the one used for the desired reaction. The resulting mixture could be subjected to alkali fusion to form phenols, which can be separated by distillation.

**Waste Gas.** The main organic constituents of waste air from benzenesulfonic acid production plants are unreacted portions of such starting products as benzene, toluene, or halogenated benzene derivatives. They are usually removed in a sulfuric acid scrubber. The waste sulfuric acid obtained in the scrubber can be recycled as a sulfonation agent. Because benzene is carcinogenic (group A 1 by MAK commission, group A 2 by ACGIH) the benzene emissions of newly built plants for the production of benzenesulfonic acid must be in accordance with the accepted lowest technically feasible levels.

Sulfur trioxide in waste gas forms aerosols that can only be removed effectively with specially designed scrubbers. In most sulfonations the oxidative effects of sulfuric acid or sulfur trioxide lead to the formation of sulfur dioxide, which can suitably be removed from the waste gas by alkali scrubbing.

In the production of benzenesulfonyl chlorides the main impurity of the waste gas is hydrogen chloride. It can be used to produce hydrochloric acid or chlorosulfuric acid, provided that organic matter has been removed from it by sulfuric acid scrubbing. Phosgene in the waste gas is hydrolyzed with water on activated carbon to form carbon dioxide and hydrogen chloride.

## 1.7. General Characteristics of Benzenesulfonic Acid Derivatives

### 1.7.1. Benzenesulfonyl Chlorides

**Properties.** The chlorides of benzenesulfonic acids are crystalline substances with definite melting points. They decompose, with formation of sulfur dioxide, at temperatures between 100 and 200 °C, depending on their structure. They undergo hydrolysis, which in some cases is very slow. Hydrolysis is accelerated by hydrogen ions; this explains why the decomposition of moist pastes may not occur until a considerable time has expired after their production, and why it may then take place exothermally and very rapidly, with the result that dangerous local overheating occurs. The chlorides of the benzenesulfonic acids react with amines to form sulfonamides and with alcohols to form esters.

**Production.** The most convenient method for the production of benzenesulfonyl chlorides is the *chlorosulfonation reaction* of benzene or substituted benzene with chlorosulfuric acid. Sulfonation and formation of the sulfonyl chloride take place in one reaction sequence:

$$Ar\text{–}H + Cl\text{–}SO_3H \longrightarrow ArSO_3H + HCl$$
$$ArSO_3H + Cl\text{–}SO_3H \rightleftharpoons ArSO_2Cl + H_2SO_4$$

Sulfone is formed in a side reaction of the chlorosulfonation (up to 7% of sulfone is formed when benzene is used as raw material and up to 35% when chlorobenzene is used). It is attributed to the reaction of the primarily formed sulfonic acid with unreacted aromatic compound:

$$ArSO_3H + ArH \longrightarrow ArSO_2Ar + H_2O$$
$$H_2O + ClSO_3H \longrightarrow H_2SO_4 + HCl$$

In this reaction the chlorosulfuric acid serves as a condensation agent.

The amount of sulfone formation can be reduced by diluting with a solvent, by using a large excess of chlorosulfuric acid, or by adding sulfone-inhibiting substances, e.g., alkali and ammonium salts, acetic acid, phosphoric acid, or dimethylformamide.

For *industrial chlorosulfonation* the chlorosulfuric acid is introduced into a cast-steel or enamelled steel vessel and 10–25 mol% of the aromatic compound is stirred in at 25–30 °C, whereupon sulfonation of the aromatic compound and HCl formation occur. The formation of sulfonyl chloride is initiated by heating the reactants to 50–80 °C. The reaction is exothermic. The temperature must be controlled accurately to ensure uniform release of the HCl gas. Restarting of the agitator after an interruption of the electricity supply is hazardous and may cause the contents of the vessel to foam over. The sulfonyl chloride is isolated by draining the reaction mass onto water and simultaneous cooling. Excess chlorosulfuric acid is decomposed, and the sulfonyl chloride either precipitates or separates as an organic liquid phase. The quality of the chlorosulfuric acid affects the yield.

In the case of aromatic compounds that easily take up two sulfochloride groups, e.g., anisol, monochlorosulfonation is carried out with only a little more than the calculated amount of chlorosulfuric acid at a low temperature (0 °C) and in the presence of a diluent such as chloroform or carbon tetrachloride. The reaction of benzenesulfonic acids with chlorosulfuric acid is an equilibrium reaction. Therefore, the yield can be increased by using an excess of chlorosulfuric acid [53].

*Other chlorinating agents,* such as phosgene, thionyl chloride, sulfuryl chloride, or phosphorus pentachloride, can be used instead of chlorosulfuric acid. When thionyl chloride is used the sulfonyl chloride is obtained from the sulfonic acid in high yield and without formation of sulfuric acid:

$$ArSO_3H + SOCl_2 \longrightarrow ArSO_2Cl + SO_2 + HCl$$

According to [54] the highest degree of conversion is obtained in the presence of at least 5% of a sulfonation agent, e.g., sulfur trioxide, or an excess of chlorosulfuric acid. If phosgene is used as a chlorinating agent, the reaction will be accelerated when adding dimethylformamide [54]. The crude sulfonyl chlorides can be purified either by fractional distillation or by crystallization from an anhydrous solvent.

In a *diazo reaction,* an $NH_2$ group can be ultimately replaced by the $SO_2Cl$ group; this reaction is suitable for the preparation of special sulfonyl chlorides whose $SO_2Cl$ group must be in a specific position [55]. A hydrochloric acid solution of the diazonium salt, whose concentration must be as high as possible, is allowed to flow into a $CuCl_2$-containing 30% solution of sulfur dioxide in glacial acetic acid [56]. Diazonium chlorides with substituents having a strong negative inductive effect ($NO_2$, $SO_3H$) give the best results. Poor yields can be improved by adding magnesium chloride or a water-immiscible solvent having a low dielectric constant (such as benzene or carbon tetrachloride). Disulfides can be converted into sulfonyl chlorides with chlorine in aqueous suspension. The reaction is also applicable to thiophenols.

$$(C_6H_5)_2S_2 + 5\ Cl_2 + 4\ H_2O \longrightarrow 2\ C_6H_5SO_2Cl + 8\ HCl$$

**Uses.** Benzenesulfonyl chlorides are important as intermediates. They are used to produce sulfonic acids, sulfonyl fluorides, sulfonamides, sulfonohydrazides, sulfonic acid esters, sulfinic acids, sulfones, and thiophenols.

## 1.7.2. Benzenesulfonamides

**Properties.** Benzenesulfonamides are readily crystallizing, colorless compounds with defined melting points and poor solubility in water. They are therefore suitable for the characterization of sulfonic acids (via the sulfonyl chlorides) and of primary and secondary amines and for the separation of amine mixtures (*Hinsberg method*) [57], [58].

Benzenesulfonamides are weak acids and form salts with bases. They are thermally stable and very difficult to hydrolyze with alkali; however, they are more easily hydrolyzed with mineral acids. In concentrated sulfuric acid sodium nitrite splits them into sulfonic acids and nitrogen. The hydrogen atoms bound to the nitrogen can be substituted.

*Dibenzenesulfonylamines* are acids similar in strength to the mineral acids [59]; they react with amines to form salts with definite melting points [60].

**Production.** Sulfonamides are produced by reacting the sulfochlorides with ammonia, primary, or secondary amines in water. Alternatively they are produced in the presence of a base such as sodium hydroxide solution, sodium hydrogen carbonate, calcium carbonate, sodium sulfite, a second mole of amine or pyridine, or sodium acetate in glacial acetic acid; in this case, the reaction proceeds in an inert solvent, e.g., benzene or acetone.

Dibenzenesulfonylamines are obtained analogously according to the following reaction [61]:

2 ArSO$_2$Cl + NH$_3$ $\longrightarrow$ ArSO$_2$NHSO$_2$Ar + 2 HCl

Unsymmetrical disulfonylamines are produced from sulfonamides and sulfochlorides in the absence of water [62] or in sodium hydroxide solution [63], [60].

**Uses.** Benzenesulfonamides serve as intermediates in the production of polysulfonamides, which are used as tanning agents and plastics. *N*-Alkylamides of the benzenesulfonic and toluenesulfonic acids can be used as plasticizers. Aminobenzenesulfonamides and diaryldisulfonylamines with amino groups serve as intermediates in the production of azo dyes.

Sulfonamides, especially those derived from *p*-sulfanilic acid, sulfonyl ureas and sulfonyl guanidines are important in medicine as antibiotics, antidiabetics, diuretics, and anthelmintics [64], [65].

### 1.7.3. Benzenesulfonohydrazides

**Properties.** Benzenesulfonohydrazides, unlike the benzenesulfonamides, are thermally unstable. They decompose, when heated in water in some cases, to form sulfides, thiosulfonates, and nitrogen; for this reason they are used as blowing agents in the production of foams.

**Production.** Benzenesulfonohydrazides are produced by reacting benzenesulfonyl chlorides with hydrazine in dioxane, alcohol, or water and in the presence of a second mole of hydrazine or of another base. They react with nitric acid to form benzenesulfonyl azides according to the following equation:

ArSO$_2$NHNH$_2$ + HNO$_2$ $\longrightarrow$ RSO$_2$N$_3$ + 2 H$_2$O

### 1.7.4. Benzenesulfonic Acid Esters

Esters of benzenesulfonic acid are liquids or low-melting crystalline substances. They have relatively high thermal stability and can be purified by vacuum distillation. In general, they can be hydrolyzed with alkali at elevated temperatures only. The esters of benzenesulfonic acid are produced by reacting benzenesulfonic acid halides with alcohols in the presence of a base [66], [67] or by reacting benzenesulfonic acids with ethers [68]. These compounds are used as alkylation agents and plasticizers.

## 2. Individual Benzenesulfonic Acids and Derivatives

### 2.1. Benzenesulfonic Acid

Benzenesulfonic acid [*98-11-3*], $C_6H_5SO_3H$, $M_r$ 158.17, was first obtained, together with diphenyl sulfone, by E. MITSCHERLICH in 1834 by heating benzene with fuming sulfuric acid. The industrially important reaction of benzenesulfonic acid with alkali hydroxide to form phenol (alkali fusion) was developed by A. WURTZ and A. KEKULÉ in 1867 and by P. O. DEGENER in 1878. Until the early 1960s benzenesulfonic acid was used chiefly in the manufacture of phenol. Other phenol syntheses are preferred now ($\rightarrow$ Phenol).

**Properties.** Benzenesulfonic acid crystallizes from aqueous solution as a hydrate with 1.5 $H_2O$ in the form of deliquescent needles; its melting point is 43–44 °C. The monohydrate melts at 45–46 °C. The anhydrous acid, which can be distilled without decomposition at 171–172 °C (0.13 mbar), melts at 65–66 °C. It is very easily soluble in water and alcohol, poorly soluble in benzene, and insoluble in ether and carbon disulfide. Its dielectric constant in aqueous solution is 0.2. For the density and refractive indices of aqueous solutions, see [69].

Sodium benzenesulfonate decomposes at about 450 °C. It is soluble in 1.75 parts of water at 30 °C and in 0.8 part of boiling water, and crystallizes from aqueous solutions with 1 mol of water of crystallization. The calcium and barium salts are also soluble in water.

Benzenesulfonic acid has the characteristic reactions of a strong aromatic sulfonic acid. Acid hydrolysis at 175 °C splits it up into benzene and sulfuric acid. Additional sulfonation with fuming sulfuric acid gives 1,3-benzenedisulfonic acid, which reacts further to 1,3,5-benzenetrisulfonic acid, and also diphenyl sulfone disulfonic acid (see Section 2.2).

Benzenesulfonic acid reacts with benzene to form diphenyl sulfone according to a Friedel-Crafts-type reaction:

$$C_6H_5SO_3H + C_6H_6 \xrightarrow{SO_3} C_6H_5SO_2C_6H_5 + H_2SO_4$$

Benzenesulfonic acid reacts with alkali hydroxide at 320–350 °C to form sodium phenolate according to

$$C_6H_5SO_3Na + 2\ NaOH \longrightarrow C_6H_5ONa + Na_2SO_3 + H_2O$$

This reaction was used in the first industrial synthesis of phenol.

**Production.** Benzenesulfonic acid is formed from benzene and sulfuric acid in an exothermic reaction according to

$$C_6H_6 + H_2SO_4 \longrightarrow C_6H_5SO_3H + H_2O$$

CROOKS and WHITE investigated the effects of temperature and of the composition of the sulfonation mixture on the reaction rate and developed a formula for calculating this rate [70]. The reaction stops at a particular sulfuric acid concentration between 74 and 78%, which depends on the temperature and the amount of water formed. A proportion of the sulfuric acid (45% in the classical process) is not utilized, but serves as a solvent and diluent and ensures that byproducts, particularly diphenyl sulfone, are formed in low yield only. In industry the excess acid is removed by adding calcium carbonate.

Any method that serves to reduce the excess of sulfuric acid increases the sulfone formation to a greater or lesser extent. The following methods have been used industrially:

1) Use of oleum or sulfur trioxide instead of sulfuric acid
2) Extraction of the benzenesulfonic acid from the reaction mixture with benzene
3) Azeotropic removal of the reaction water with benzene

*Classical Process.* The sulfonation by the classical process [71] is now of historical interest only, because it is coupled with the synthesis of phenol (use of sulfite from phenol production to neutralize the benzenesulfonic acid). Details of the process will be found in [72].

*Continuous Sulfonation with Oleum* (Monsanto process) [73]. Benzene and oleum are pumped simultaneously into the first of six sulfonation vessels forming a cascade. The vessels, which have propeller stirrers, are connected to one another by overflow pipes so that the reaction mixture can flow from one vessel to the next.

The first two vessels are cooled, whereas the others are heated. If optimal temperature conditions are provided, the reaction is complete when the sulfonation mixture leaves the last vessel. The excess sulfuric acid is removed by neutralizing the reaction mixture with sodium sulfite or sodium hydroxide solution. Sodium sulfate precipitates and is separated in a number of centrifuges. The solution of sodium benzenesulfonate is concentrated in an evaporator, which causes more sodium sulfate to precipitate. After centrifuging the benzenesulfonic acid can be reacted directly to phenol or it is dried. According to CARSWELL oleum containing 35.6% of sulfur trioxide is used and the temperatures of the first and last sulfonation vessels are held at 70–80 °C and 110 °C, respectively. About 1% of the benzene is converted to diphenyl sulfone.

*Continuous Extraction Process.* Under vigorous stirring excess benzene and sulfur trioxide are introduced into a vessel through separate pipes leading into the bottom of the vessel [74]. A benzene layer saturated with benzenesulfonic acid collects in the upper part of the vessel and passes through an overflow pipe into a second agitator vessel. In this vessel the benzenesulfonic acid is removed from the benzene extract by continuous washing with water or sodium hydroxide solution. The benzene is separated from the aqueous benzenesulfonic acid solution in a separator. After being dried it is returned to the reaction vessel. The sulfuric acid consumption is 1260 kg for 1000 kg of reacted benzene. Diphenyl sulfone formation is below 2% [75].

*Azeotropic Removal of Reaction Water* [76], [77]. Sulfuric acid with a concentration of, for example, 79% is heated to 170 °C in a cast-iron or enamelled steel sulfonation vessel and finely divided benzene vapor is introduced through a perforated plate situated in the lower part of the vessel. A portion of the benzene vapor, of which an excess is used, is sulfonated, whereas the unreacted portion continuously removes the water from the reaction mixture as it is formed. The mixture of benzene and water vapor leaving the vessel is condensed; after phase separation the benzene is returned to the evaporator.

Continuous operation of this process gives a final product containing 80.2% of benzenesulfonic acid and 14.3% of sulfuric acid; in batch operation the final product contains 93.1% of sulfonic acid and 4.8% of sulfuric acid. When adding sodium benzenesulfonate to the sulfonation mixture the sulfone formation can be kept below 2% [78].

*Analysis.* A sample is taken from the sulfonation reactor and is diluted with water. Unreacted benzene is determined by gas chromatography, benzenesulfonic acid and diphenyl sulfone by high performance liquid chromatography. The sulfate concentration is determined at pH 2–3 by photometric titration with lead nitrate solution and dithizone (1,5-diphenylthiocarbazone) indicator in aqueous acetone solution [79].

*Quality.* Benzenesulfonic acid generally is sold as sodium salt. This is almost colorless and without substantial impurities except for a small percentage of sodium sulfate.

*Uses.* Benzenesulfonic acid is used as an acid catalyst. The sodium salt is used to standardize dyes.

**Benzenesulfonyl Chloride** [*98-09-9*], $C_6H_5SO_2Cl$, $M_r$ 176.62, $bp_{1013\ mbar}$ 251.5 °C, $bp_{13.3\ mbar}$ 120 °C, *mp* 14.5 °C, $d_{15}^{15}$ 1.3842. Benzenesulfonyl chloride is made by reacting benzene with chlorosulfuric acid at 30–35 °C [80]:

$$C_6H_6 + 2\ ClSO_3H \longrightarrow C_6H_5SO_2Cl + HCl + H_2SO_4$$

The sulfuric acid formed simultaneously inhibits the reaction. Therefore the yield of sulfonyl chloride is about 75%, based on benzene. Benzenesulfonic acid and diphenyl sulfone are formed as byproducts in yields of approximately 18% and 5–7%, respectively.

The yield can be raised by partly replacing the chlorosulfuric acid with thionyl chloride. If this is done, the raw material and waste-water treatment costs are reduced, but the flue gas contains sulfur dioxide in addition to hydrogen chloride. Absorption of the gases in water recovers impure hydrochloric acid. The sulfur dioxide leaves the top of the hydrochloric acid recovery plant and can be used further in a sulfuric acid plant.

At the end of the reaction the products are cooled and simultaneously forced into a vessel containing water. The benzenesulfonyl chloride separates from the aqueous phase as a liquid organic phase. It is then washed twice with water and distilled; before that, the residual water is removed in a vacuum. The main bottom product of the distillation column is diphenyl sulfone. The reaction can be carried out in vessels of cast-iron or enamelled steel.

*Uses.* Benzenesulfonyl chloride is produced on an industrial scale in Europe, the United States, and Japan. It is used as an intermediate for dyes and in the manufacture of benzenesulfinic acid, sulfonamides, and sulfonic acid esters. Benzenesulfonohydrazide serves as a blowing agent in the production of foams (trade name: Porofor BSH, Bayer).

**Methyl Benzenesulfonate** [*80-18-2*], $C_7H_8O_3S$, $M_r$ 172.20, $bp_{20\,mbar}$ 150 °C, $d_4^{17}$ 1.2730, is produced by reacting benzenesulfonyl chloride with methanol in the presence of sodium carbonate [81]. Yield: 87% (All yields in this article are expressed as theoretical yields; moles of product divided by moles of starting materials). Used as an alkylation agent.

**Ethyl Benzenesulfonate** [*515-46-8*], $C_8H_{10}O_3$ S, $M_r$ 186.23, $bp_{20\,mbar}$ 156 °C, $d_4^{17}$ 1.2192. Its manufacture and use are analogous to those of the methyl ester.

**2-Chloroethyl Benzenesulfonate** [*27887-43-0*], $C_8H_9ClO_3S$, $M_r$ 220.67, $bp_{12\,mbar}$ 184 °C, $d_4^{15}$ 1.353, is produced by reacting 2-chloroethanol and benzenesulfonyl chloride at 12–15 °C (at most 25 °C) and adding sodium carbonate. The reaction product is diluted with water and rendered alkaline by adding sodium carbonate. The organic lower layer is removed and distilled in a vacuum [82]. Yield: 87.8%, calculated on benzenesulfonyl chloride.

***p*-Chlorophenyl Benzenesulfonate** and **2,4-Dichlorophenyl Benzenesulfonate** are important as acaricides.

**Benzenesulfonamide** [*98-10-2*], $C_6H_5SO_2NH_2$, $M_r$ 157.19, needles from water, flakes from alcohol, *mp* 156 °C. Obtainable from benzenesulfonyl chloride and ammonia. Used to manufacture chloramine B and dichloramine B (→ Chloroamines).

***N*-(Methyl)benzenesulfonamide** [*5183-78-8*], $C_7H_9NO_2S$, $M_r$ 171.22, $bp_{22.5\,mbar}$ 202 °C, *mp* 30 °C, freely soluble in dilute alkaline solutions. Produced in the same way as the amide. Used as a plasticizer.

***N*-(Ethyl)benzenesulfonamide** [*5339-67-3*], $C_8H_{11}NO_2S$, $M_r$ 185.25, *mp* 57–58 °C. Obtained as crystals from alcohol. Soluble in excess alkaline solution. Produced from benzenesulfonyl chloride with ethylamine and sodium carbonate. Yield: 98.7% [83].

***N*-(Butyl)benzenesulfonamide** [*3622-84-2*], $C_{10}H_{15}NO_2S$, $M_r$ 213.23, oil. Used as a plasticizer.

**Benzenesulfonohydrazide** [80-17-1], $C_6H_8N_2O_2S$, $M_r$ 172.21, mp 101–103 °C, white powder, $\varrho$ 1.48 g/cm$^3$, is used as a blowing agent in foam manufacture (Porofor BSH, Bayer).

**4-[N-(Phenylsulfonyl)amino]acetanilide** [565-20-8], $C_{14}H_{14}N_2O_3S$, $M_r$ 290.34, mp 157 °C, crystallized from alcohol. Manufactured by reacting 4-(amino)acetanilide with benzenesulfonyl chloride in water at 65 °C to which lime is added [84]. The compound is used to manufacture N-(4-aminophenyl)benzenesulfonamide.

$$\text{Ph-SO}_2\text{Cl} + \text{H}_2\text{N-C}_6\text{H}_4\text{-NHCOCH}_3 \longrightarrow \text{Ph-SO}_2\text{NH-C}_6\text{H}_4\text{-NHCOCH}_3$$

**N-(4-Aminophenyl)benzenesulfonamide** [5466-91-1], $C_{12}H_{12}N_2O_2S$, $M_r$ 248.3, mp 171–172 °C. Made from the 4'-acetylaminoanilide by splitting off the acetyl group in an alkaline medium. Used as an intermediate for dyes.

**N-(Phenylsulfonyl)benzenesulfonamide** [2618-96-4], $C_{12}H_{11}NO_4S_2$, $M_r$ 297.35, mp 157–158 °C, poorly soluble in water and alcohol, insoluble in ether, slightly soluble in acetone. Sodium salt: mp 314–316 °C; anilinium salt: mp 176 °C. Manufactured by reacting benzenesulfonamide in 5 % sodium hydroxide solution with benzenesulfonyl chloride at 50–55 °C (1–2 h). During the reaction the pH value must be kept at 7.2 by adding 5 % sodium hydroxide solution. At the end of the reaction the disulfonylamine is precipitated from the clear solution by adding hydrochloric acid. Yield: 70–98 % [85]. Additive for electroplating baths [86].

## 2.2. 1,3-Benzenedisulfonic Acid

1,3-Benzenedisulfonic acid [98-48-6] (**1**), $C_6H_6O_6S_2$, $M_r$ 238.24, is a hygroscopic substance with a solidification point of 137 °C.

$$\underset{\mathbf{1}}{\text{C}_6\text{H}_4(\text{SO}_3\text{H})_2}$$

It is formed as the main product in the sulfonation of benzene or benzenesulfonic acid with excess oleum at temperatures of 80–250 °C. In proportion to the oleum concentration and reaction temperature 1–29 % of diphenyl sulfone disulfonic acid is formed as an undesired byproduct [87]. 1,4-Benzenedisulfonic acid is formed also, especially at high temperature, through rearrangement; mercury accelerates this process. At temperatures of 240–250 °C an equilibrium exists at a ratio of the p-acid to the m-acid of 1:2.

The disodium salt of 1,3-benzenedisulfonic acid is soluble in water at 27.5 °C to the extent of 41%. With potassium or sodium hydroxide in aqueous solution under pressure it forms 3-hydroxybenzenesulfonic acid at 180–200 °C. At higher temperature resorcinol is formed.

**Industrial Production** [88]. Benzene is introduced into a cast-steel agitator vessel containing 100% sulfuric acid. The contents of the reactor are heated to 100 °C and kept at this temperature for one hour. The products of the reaction (monosulfonation) are transferred to an agitator vessel containing 65% oleum.

The temperature rises from 30 to 80 °C and is maintained at this level for another 1–2 h. The sulfonation mass is freed from sulfate ions in the normal way by adding calcium hydroxide or calcium carbonate and then removing the gypsum. It is then converted with sodium carbonate into the disodium salt, which is dried on a drum dryer. The yield is 93% and the product has a purity of 96%.

In a process patented by Farbwerke Hoechst [26] benzenesulfonic acid is mixed simultaneously with benzene and sulfur trioxide at temperatures above 120–140 °C. Sulfone formation is prevented by adding 3–15% of the disodium salt of benzenedisulfonic acid. The excess sulfur trioxide is about 0.2 mol per mole of benzene. The process can be operated continuously or discontinuously. After being neutralized with sodium hydroxide solution the final product can be melted to resorcinol directly. There is no necessity to remove sulfate ions. Excess sulfonation agent can be removed by subsequent reaction with benzene, toluene, or xylene – partly to the corresponding monosulfonic acids. A mixture of resorcinol and phenol, cresol, or xylenol, is then formed by alkali fusion; the components of this mixture can be separated by distillation [89]–[91].

If benzene is sulfonated with 30% oleum at 120 °C in the presence of sodium sulfate and the excess sulfuric acid is then converted into benzenemonosulfonic acid by introducing benzene vapor, a mixture of monosulfonic and disulfonic acid is obtained which can be used directly to produce phenol and resorcinol in the ratio 8:1 [89].

Particularly pure 1,3-benzenedisulfonic acid is obtained according to the process of Hooker Chemical Corp. [92]. Benzene is sulfonated with 96% sulfuric acid at a molar ratio of the reactants of 1:3.4. The temperature is initially 65 °C, but is then raised to 80 °C. The water of the reaction is then removed by evaporation at 33 mbar and 200 °C. Finally, during disulfonation at 1.3 mbar and 210–235 °C, sulfuric acid and reaction water are distilled. The reaction product has a purity of 98.7% and a solidification point of 136 °C. The process also can be operated continuously.

In a patent Koppers Co. [93] proposes that the disulfonation and simultaneous removal of water and excess sulfuric acid be carried out in a thin-film evaporator at a film thickness of 0.5–3 mm and a residence time of one minute. A vacuum of 13 mbar is claimed to be sufficient.

**Uses.** 1,3-Benzenedisulfonic acid is used to produce resorcinol and 3-hydroxybenzenesulfonic acid.

**1,3-Benzenedisulfonyl Chloride** [585-47-7], $C_6H_4Cl_2O_4S_2$, $M_r$ 275.13, prisms from ether, mp 63 °C, $bp_{1.3\,mbar}$ 145 °C, $bp_{27\,mbar}$ 210.7 °C. Manufactured from *m*-benzenedisulfonic acid and phosphorus pentachloride or from benzenesulfochloride and chlorosulfuric acid. The compound is used to produce 1,3-benzenedisulfonohydrazide.

**1,3-Benzenedisulfonamide** [3701-01-7], $C_6H_8N_2O_4S$, $M_r$ 236.27, obtained as needles from water, mp 229 °C. Manufactured from *m*-benzenedisulfochloride and aqueous ammonia.

**1,3-Benzenedisulfonohydrazide,** $C_6H_{10}N_4O_4S_2$, $M_r$ 266.31, white powder, mp 163 °C (decomp.), $\varrho$ 1.24 g/cm$^3$, is suitable as a blowing agent for foams (Porofor B 13 CP 50, Bayer).

## 2.3. Alkylbenzenesulfonic Acids

This group of compounds comprises methylbenzenesulfonic acids (toluenesulfonic acids), [25231-46-3], $C_7H_8O_3S$, $M_r$ 172.20, methylbenzenedisulfonic acids (toluenedisulfonic acids) $C_7H_8O_6S_2$, $M_r$ 252.26, dimethylbenzenesulfonic acids (xylenesulfonic acids) [58723-02-7], $C_8H_{10}O_3S$, $M_r$ 186.23, and their derivatives.

**Production.** The sulfonation of toluene is accomplished more easily than that of benzene and is accompanied by less sulfone formation. Low temperatures favor *o*-substitution; *m*-substitution occurs only to the extent of about 2–3%. At temperatures around 100 °C the ratio of *o*- to *p*-substitution is about 3:7. The sulfuric acid concentration has a strong influence on the isomer ratio [94]. Surprisingly, almost 100% *p*-substitution is obtained at −10 °C if sulfur dioxide is used as the solvent [95].

At a sulfonation temperature of 100 °C and a sulfuric acid concentration of 97.4%, 31% *m*-substitution is achieved [96]. Through rearrangement at 140–200 °C the concentration of *m*-toluenesulfonic acid rises to about 46% [97], [98]. If this is followed by hydrolysis with steam at 170–200 °C, the concentration of *m*-isomer rises to more than 90% [99].

**Isolation from the Mixture.** *p*-Toluenesulfonic acid can be crystallized from 66–71% sulfuric acid, or from concentrated hydrochloric acid at temperatures below 0 °C [100]. Then, according to the reaction conditions of the preceding sulfonation, the *o*- or *m*-toluenesulfonic acid can be obtained from the mother liquor. If the compounds are required to be very pure, the route via the toluidines (diazotization and reaction with sulfur dioxide) is preferred [46].

2,6-Toluenedisulfonic acid can be obtained by adding 65% oleum to the sulfonation reaction mixture while the temperature is below 100 °C and then raising the temperature to 125 °C. In 90% sulfuric acid at 190 °C, 64% of the 2,4-toluenedisulfonic acid rearranges to 3,5-toluenedisulfonic acid. The 3,5-isomer is isolated from hydrochloric acid solution at −15 °C [101].

**Uses.** Toluenesulfonic acids are used as isomer mixtures, or as mixtures with xylenesulfonic acids; they serve as solubilizers, as acidic catalysts, and as additives for detergents [102]. They are used as intermediates in the production of cresols and sulfobenzoic acids.

**2-Methylbenzenesulfonic Acid** [88-20-0] (**2**), $o$-toluenesulfonic acid, $C_7H_8O_3S$, $M_r$ 172.20, flaky crystals (dihydrate), mp 62.1 °C.

Produced together with 4-methylbenzenesulfonic acid by sulfonation of toluene with 96 % sulfuric acid at 40 °C, or preferably with 1 mol of chlorosulfuric acid at 0 to −10 °C. Used to produce $o$-sulfobenzoic acid.

**2-Methylbenzenesulfonyl Chloride** [133-59-5], $C_7H_7ClO_2S$, $M_r$ 190.65, $bp_{13\ mbar}$ 126 °C, mp 15.5 °C. Manufactured by reacting toluene with chlorosulfuric acid at temperatures not exceeding 3 °C, forcing the reaction mixture into ice, siphoning off the water, and centrifuging 4-methylbenzenesulfonyl chloride. Used in the production of toluenesulfonamide.

**2-Methylbenzenesulfonamide** [88-19-7], $C_7H_9NO_2S$, $M_r$ 171.22, mp 156.5 °C, is obtained by reacting 2-methylbenzenesulfonyl chloride with aqueous ammonia ($d$ 0.910). Purified by fractional precipitation: hydrochloric acid is added to a solution which has been rendered alkaline with sodium hydroxide; the impurities are precipitated first. Used in the manufacture of saccharin.

**3-Methylbenzenesulfonic Acid** [617-97-0] (**3**), $m$-toluenesulfonic acid, $C_7H_8O_3S$, $M_r$ 172.20. The pure acid can be obtained from $m$-toluidine. Isomer mixtures with $m$-isomer concentrations of more than 50 % are obtained from toluene sulfonation mixtures by bubbling in toluene vapor at 180–205 °C [103], see also [99]. Used to manufacture $m$-cresol (→ Cresols and Xylenols ).

**3-Methylbenzenesulfonyl Chloride** [1899-93-0], $C_7H_7ClO_2S$, $M_r$ 190.65, $bp_{29\ mbar}$ 146 °C, mp 11.7 °C. Obtained from $m$-toluidine according to [34]. Yield: 71.2 %.

**3-Methylbenzenesulfonamide** [1899-94-1], $C_7H_9NO_2S$, $M_r$ 171.22, mp 103 °C.

**4-Methylbenzenesulfonic Acid** [104-15-4] (**4**), *p*-toluenesulfonic acid, $C_7H_8O_3S$, $M_r$ 172.20, *mp* 38 °C, monohydrate *mp* 106 °C, trihydrate *mp* 93 °C, $bp_{26.3\,mbar}$ 140 °C.

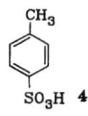

*Production.* According to the process of Allied Chemical and Dye Corp. [104], boiling toluene is sulfonated with 90–95% sulfuric acid. Evaporating toluene is condensed, separated from water, dried, and recycled. A mixture with a *p*-isomer content of 75–85%, an *o*-isomer content of 10–20%, a *m*-isomer content of 2–5%, and less than 1% of sulfuric acid is present at the end of the reaction. There is very little sulfone formation. The process can also be performed continuously in a vertical reactor [105].

For some purposes, for example, for use as an acidic catalyst or in cresol manufacture, the crude toluenesulfonic acid obtained according to the above processes does not have to be purified. Purification is possible by crystallization from 66% sulfuric acid or via the barium salt.

*Uses.* The applications of 4-methylbenzenesulfonic acid include the manufacture of 4-formylbenzenesulfonic acid, *p*-sulfobenzoic acid, 2-chlorotoluene-4-sulfonic acid, and 4-(chloromethyl)phenylmethanesulfonic acid.

**4-Methylbenzenesulfonyl Chloride** [98-59-9], tosyl chloride, $C_7H_7ClO_2S$, $M_r$ 190.65, $bp_{25\,mbar}$ 138–139 °C, *mp* 69 °C. Formed as a byproduct in the manufacture of 2-methylbenzenesulfonyl chloride, from which it can be separated by remelting several times under water.

This compound is used in the production of sulfonamides, arylides, nuclear-substituted toluenesulfonyl chlorides, 4-(chloromethyl)benzenesulfonyl chloride, and *p*-toluenesulfinic acid. The tosylation of hydroxyl groups, i.e., their esterification with *p*-toluenesulfonyl chloride, serves to protect these groups; this method is employed particularly often in the chemistry of natural substances [106].

**4-Methylbenzenesulfonamide** [70-55-3], $C_7H_9NO_2S$, $M_r$ 171.22, *mp* 137 °C, dihydrate *mp* 105 °C, weak acid, soluble in aqueous alkali. Produced from the sulfonyl chloride with aqueous ammonia. Used in the production of chloroamines, e.g., chloramine T and dichloramine T (→ Chloroamines).

**Methyl 4-Methylbenzenesulfonate** [80-48-8], $C_8H_{10}O_3S$, $M_r$ 186.23, $bp_{17\,mbar}$ 168–170 °C, *mp* 28 °C, is obtained from the chloride of the acid by reaction with methanol; important alkylation agent.

**Ethyl 4-Methylbenzenesulfonate** [80-40-0], $C_9H_{12}O_3S$, $M_r$ 200.26, $bp_{0.4\,mbar}$ 137–139 °C, $bp_{12\,mbar}$ 165–166 °C, *mp* 34 °C, is obtained from the sulfonyl chloride and ethyl alcohol at −5 to 0 °C after addition of 45–50% sodium hydroxide solution [107]. Used as an alkylation agent and as a plasticizer.

**4-(Chloromethyl)benzenesulfonic Acid** [*46062-27-5*], $C_7H_7ClO_3S$, $M_r$ 206.65, hygroscopic crystals, is obtained at a yield of 90% through the action of chlorine on sodium 4-methylbenzenesulfonate in the presence of dichlorobenzene [108]. Used in the manufacture of surfactants.

**4-(Chloromethyl)benzenesulfonyl Chloride** [*2389-73-3*], $C_7H_6Cl_2O_2S$, $M_r$ 225.09, mp 64–65 °C, $bp_{20\,mbar}$ 183–195 °C, is obtained by reacting 4-methylbenzenesulfonyl chloride with chlorine in the presence of $PCl_5$ at 120–140 °C.

**Methyl 4-(Chloromethyl)benzenesulfonate** [*89981-68-0*], $C_8H_9ClO_3S$, $M_r$ 219.67, is used as a quaternizing agent. Reactions with tertiary amines, e.g., trimethylamine, give sulfobetaines with fungistatic properties [109].

$$R_3N + ClCH_2-\phantom{x}\!\!\!\!\!\!\!\!\!\!\!\!\!\!-SO_2OCH_3 \longrightarrow R_3\overset{+}{N}-CH_2-\phantom{x}\!\!\!\!\!\!\!\!\!\!\!\!\!\!-SO_3^-$$

**4-(Acetylaminomethyl)benzenesulfonyl Chloride** [*39169-92-1*] (**5**), $C_9H_{10}ClNO_3S$, $M_r$ 247.70, mp 95–97 °C, is obtained from *N*-acetylbenzylamine and chlorosulfuric acid [110].

**4-(Acetylaminomethyl)benzenesulfonamide** [*2015-14-7*], $C_9H_{12}N_2O_3S$, $M_r$ 228.27, mp 177 °C (from water or aqueous alcohol), is produced from the sulfonyl chloride with aqueous ammonia at 15 and 70 °C; yield 80%. It is an important intermediate in the manufacture of 4-(aminomethyl)benzenesulfonamide.

**4-Methyl-1,3-benzenedisulfonic Acid** [*121-04-0*] (**6**), $C_7H_8O_6S_2$, $M_r$ 252.26, is a viscous oil. Obtained by additional sulfonation of the *o*- and *p*-toluenesulfonic acid mixture with 66% oleum at 125 °C. Used in the manufacture of 4-formyl-1,3-benzenedisulfonic acid, 4-carboxy-1,3-benzenedisulfonic acid, and 4-methylresorcinol.

**4-Methyl-1,3-benzenedisulfonyl Chloride** [*2767-77-3*], $C_7H_6Cl_2O_4S_2$, $M_r$ 289.16, mp 56 °C, is produced from toluene and chlorosulfuric acid at 120 °C [111]. Addition of phosphorus pentoxide raises the yield [112].

**4-Methyl-1,3-benzenedisulfonamide** [*717-44-2*], $C_7H_{10}N_2O_4S_2$, $M_r$ 250.30, mp 190–191 °C.

**2,4-Dimethylbenzenesulfonic Acid** [88-61-9] (**7**), *m*-xylenesulfonic acid, $C_8H_{10}O_3S$, $M_r$ 186.23, dihydrate: *mp* 57 °C, prisms or flakes. Produced by sulfonating *m*-xylene with concentrated sulfuric acid; the acid is precipitated by dilution with water. Of the three isomeric xylenes, *m*-xylene is the one most easily sulfonated. The ease of sulfonation decreases in the order $m > p > o$, whereas the ease of hydrolysis increases in the order $o < m < p$. *m*-Xylenesulfonic acid is hydrolyzed at 180 °C. Used in the production of nitroxylenesulfonic acid, chloroxylenes, and chloroxylenesulfonic acids.

**2,4-Dimethylbenzenesulfonyl Chloride** [609-60-9], $C_8H_9ClO_2S$, $M_r$ 204.68, *mp* 34 °C.

**2,4-Dimethylbenzenesulfonamide** [7467-12-1], $C_8H_{11}NO_2S$, $M_r$ 185.25, *mp* 138 °C.

**2,5-Dimethylbenzenesulfonic Acid** [609-54-1] (**8**), $C_8H_{10}O_3S$, $M_r$ 186.23, *mp* 48 °C, *bp* 149 °C (cathode vacuum), is soluble in chloroform. The dihydrate, *mp* 95 °C, is obtained by sulfonating *p*-xylene with 93% sulfuric acid and removing the water by distillation [113]. Used in the production of 2,5-xylenol.

**2,5-Dimethylbenzenesulfonyl Chloride** [19040-62-1], $C_8H_9ClO_2S$, $M_r$ 204.68, *mp* 25.5 °C, $bp_{29\,\text{mbar}}$ 152–153 °C.

**2,5-Dimethylbenzenesulfonamide** [6292-58-6], $C_8H_{11}NO_2S$, $M_r$ 185.25, *mp* 148 °C.

**4-Ethylbenzenesulfonic Acid** [98-69-1], $C_8H_{10}O_3S$, $M_r$ 186.21, is produced by sulfonating ethylbenzene with sulfuric acid. It is separated from the isomers via the aniline salt [114]. Used in the production of *p*-ethylphenol.

**4-(2-Bromoethyl)benzenesulfonic Acid** [54322-31-5] (**9**), $C_8H_9BrO_3S$, $M_r$ 265.13; *S*-benzylisothiuronium salt, *mp* 149–150 °C. The acid is obtained by sulfonating 1 mol of (bromoethyl)benzene with 1.1 mol of $SO_3$ in methylene chloride [115], see also [116], [117]. The potassium salt is obtained by reacting the sulfonyl chloride with potassium carbonate in water [118].

**4-(2-Bromoethyl)benzenesulfonyl Chloride** [64062-91-5], $C_8H_8BrClO_2S$, $M_r$ 283.58, is produced from 2-(bromoethyl)benzene and chlorosulfuric acid at temperatures not exceeding 25 °C [119]. Used in the production of 4-vinylbenzenesulfonic acid.

**4-(2-Bromoethyl)benzenesulfonamide** [5378-84-7], $C_8H_{10}BrNO_2S$, $M_r$ 264.15, *mp* 185.5–186 °C.

**4-Vinylbenzenesulfonic Acid** [98-70-4] (**10**), $C_8H_8O_3S$, $M_r$ 184.21; *p*-toluidine salt, *mp* 182–183 °C, is obtained in the form of the potassium salt from 4-(2-bromoethyl)-benzenesulfonic acid with methanolic KOH [120] or from 4-(2-bromoethyl)benzenesulfonyl chloride with alcoholic KOH [121]. The monomer can be stabilized by adding 0.5–5% of sodium nitrite [122].

4-Vinylbenzenesulfonic acid has been proposed for copolymerization with acrylonitrile to improve the dyeing properties of the fibers [123] and as a starting product in the manufacture of polymeric styrenesulfonic acid [120], [121].

**4-Vinylbenzenesulfonamide** [2633-64-9], $C_8H_9NO_2S$, $M_r$ 183.23, *mp* 138–139 °C [124].

**4-*tert*-Butyl-2,6-dimethylbenzenesulfonic Acid** [28188-48-9] (**11**), $C_{12}H_{18}O_3S$, $M_r$ 242.31, is prepared from 1-*tert*-butyl-3,5-dimethylbenzene with 5% oleum at 20–45 °C. Used in the manufacture of 4-*tert*-butyl-2,6-dimethylphenol and 2,6-xylenol [125].

**2,3-Dihydro-1*H*-indene-5-sulfonic Acid** [40117-41-7] (**12**), $C_9H_{10}O_3S$, $M_r$ 198.24, is obtained by precipitation as a sodium salt, after 2,3-dihydro-1*H*-indene and concentrated sulfuric acid have been reacted at 150 °C [126]. At low temperatures mainly 1*H*-indene-4-sulfonic acid is formed. 1*H*-Indene-5-sulfonic acid has been proposed as a starting material for the manufacture of sulfonylureas with hypoglycemic effects [127].

**2,3-Dihydro-1*H*-indene-5-sulfonyl Chloride** [52205-85-3], $C_9H_9ClO_2S$, $M_r$ 216.69, *mp* 49 °C, $bp_{21\,mbar}$ 180 °C, is obtained by reacting the sodium salt of 1*H*-indene-5-sulfonic acid with $PCl_5$, or by reacting 2,3-dihydro-1*H*-indene and chlorosulfuric acid at 0–15 °C, pouring onto ice, and extracting with chloroform [128].

**2,3-Dihydro-1*H*-indene-5-sulfonamide** [35203-93-1], $C_9H_{11}NO_2S$, $M_r$ 197.26, *mp* 135 °C.

## 2.4. Formylbenzenesulfonic Acids

**2-Formylbenzenesulfonic Acid** [*91-25-8*] (**13**), $C_7H_6O_4S$, $M_r$ 186.18, is obtained as a thick syrup by treatment of its barium salt with sulfuric acid. Acquires a deep reddish-violet color if Schiff's reagent is added. In contrast to the poorly soluble barium salt, the sodium salt is freely soluble in water; it is also freely soluble in hot, but not in cold, ethyl alcohol.

*Production.* The starting product is *o*-chlorobenzaldehyde, which is heated with sodium hydrogen sulfite solution in an autoclave to 190 – 200 °C, with the result that its chloro substituent is replaced by the sulfonic acid group. The batch is allowed to cool, sulfuric acid is added, and excess sulfur dioxide and unreacted chlorobenzaldehyde are driven off by boiling. The resulting solution is used directly in the manufacture of dyes. The acid can also be produced by oxidation of *o*-toluenesulfonic acid with oxygen in the presence of bromine and cobalt as catalysts [129].

*Uses.* 2-Formylbenzenesulfonic acid is an intermediate in the manufacture of optical brighteners [130], trade names including Uvitex NFW (for textiles) and Tinopal CBS (for detergents). Triphenylmethane dyes are produced by condensation of the acid with various *N,N*-dialkylanilines, *N,N*-dialkyl-*m*-aminophenols, and their sulfonic acids.

**4-Formylbenzenesulfonic Acid** [*5363-54-2*] (**14**), $C_7H_6O_4S$, $M_r$ 186.18. The thiosemicarbazone decomposes at 230 °C. 4-Formylbenzenesulfonic acid is produced by oxidation of *p*-toluenesulfonic acid with manganese dioxide in 25 % oleum or by oxidation of *p*-toluenesulfonyl chloride in glacial acetic acid – acetic anhydride with chromium trioxide and concentrated sulfuric acid at 0 °C and hydrolysis of the sulfonyl chloride of the aldehyde-hydrate diacetate with dilute hydrochloric acid [131].

**4-Formyl-1,3-benzenedisulfonic Acid** [*88-39-1*] (**15**), $C_7H_6O_7S_2$, $M_r$ 266.25, is freely soluble in water. The sodium salt crystallizes with 2 mol of $H_2O$ and is freely soluble in water and poorly soluble in ethyl alcohol. It is made by reaction of 2,4-dichlorobenzaldehyde with sodium hydrogen sulfite liquor or oxidation of 2,4-toluenedisulfonic acid with manganese dioxide in sulfuric acid. The acid is used in the production of triphenylmethane dyes and wetting agents.

## 2.5. Chlorobenzenesulfonic Acids

**4-Chlorobenzenesulfonic Acid** [*98-66-8*] (**16**), $C_6H_5ClO_3S$, $M_r$ 192.62, *mp* 68 °C, deliquescent needles, is soluble in water and ethanol. Prepared by sulfonation of chlorobenzene with sulfuric acid; the water formed in the reaction is removed continuously. This compound is also a waste product in the manufacture of DDT, and it can be hydrolyzed to chlorobenzene with steam at 190 °C [132]. In the manufacture of DDT, 4-chlorobenzenesulfonic acid can also be converted – without being isolated – with chlorosulfuric acid into the sulfochloride and then into the *p*-chlorophenolate [133]. The compound is used in the manufacture of 4-chloro-3-nitrobenzenesulfonic acid.

**4-Chlorobenzenesulfonyl Chloride** [*98-60-2*], $C_6H_4Cl_2O_2S$, $M_r$ 211.01, $bp_{16\ mbar}$ 140 °C, *mp* 55 °C, is manufactured by mixing 4-chlorobenzene and chlorosulfuric acid at 35 °C, heating for two hours at 80 °C, and discharging the contents of the reactor into ice water; yield 81 % of theory, together with a small amount of bis(4-chlorophenyl) sulfone [134], [135]. Used in the manufacture of chlorobenzenesulfinic acid and sulfimides; three-step reaction to the sulfone in a single reactor is possible without intermediate isolation [136].

**4-Chlorobenzenesulfonamide** [*98-64-6*], $C_6H_6ClNO_2S$, $M_r$ 191.63, *mp* 443 – 444 °C, is freely soluble in ethyl alcohol, ether, and hot water.

**4-Chloro-N-[(4-chlorophenyl)sulfonyl]benzenesulfonamide** [*2725-55-5*], $C_{12}H_9Cl_2NO_4S_2$, $M_r$ 366.24, *mp* 298 °C (from water), is prepared by reacting the corresponding sulfonyl chloride with 0.45 mol of ammonium chloride at 0 – 3 °C, with dropwise addition of sodium hydroxide solution at pH 8; yield 91 % [137]. Used as a reagent for identifying amines.

**2,5-Dichlorobenzenesulfonic Acid** [*88-42-6*] (**17**), $C_6H_4Cl_2O_3S$, $M_r$ 227.06, crystals, freely soluble in water. Obtained by sulfonating 1,4-dichlorobenzene with oleum.

**2,5-Dichlorobenzenesulfonyl Chloride** [*5402-73-3*], $C_6H_3Cl_3O_2S$, $M_r$ 245.51, *mp* 39 °C, needles (from ethyl alcohol), is made by reacting 1,4-dichlorobenzene with chlorosulfuric acid at 150 °C and stirring into ice water; yield 85 % [138].

**3,4-Dichlorobenzenesulfonic Acid** [*939-95-7*] (**18**), $C_6H_4Cl_2O_3S$, $M_r$ 227.06, is poorly soluble in water; the calcium salt is freely soluble, barium and lead salts are slightly soluble. Manufactured by sulfonating *o*-dichlorobenzene with oleum [139].

**3,4-Dichlorobenzenesulfonyl Chloride** [98-31-7], $C_6H_3Cl_3O_2S$, $M_r$ 245.51, fp 22.4 °C, is manufactured from o-dichlorobenzene and chlorosulfuric acid; used as an inexpensive acylating agent.

**3,4-Dichlorobenzenesulfonamide** [23815-28-3], $C_6H_5Cl_2NO_2S$, $M_r$ 226.08, mp 140 °C.

**17**   **18**   **19**

**2,4,5-Trichlorobenzenesulfonic Acid** [6378-25-2] (**19**), $C_6H_3Cl_3O_3S$, $M_r$ 261.51, is obtained in the form of the sodium salt by sulfonating 1,2,4-trichlorobenzene with 25 % oleum and pouring the clear sulfonation mixture into a solution of sodium chloride in water; yield 94 – 95 %.

**2,4,5-Trichlorobenzenesulfonyl Chloride** [15945-07-0], $C_6H_2Cl_4O_2S$, $M_r$ 279.96, mp 65 – 67 °C (from petroleum ether), $bp_{0.6\,mbar}$ 138 °C, is obtained from trichlorobenzene by chlorosulfonation with chlorosulfuric acid at 90 °C and pouring the reaction mass into water [140] – [142].

Esters and amides of several chlorobenzenesulfonic acids (particularly 4-chlorobenzenesulfonic acid, 2,5-dichlorobenzenesulfonic acid, and 2,4,5-trichlorobenzenesulfonic acid) have acquired some importance as crop protection products and agents for the control of fiber pests.

## 2.6. Aminobenzenesulfonic Acids

The rearrangement of the N-phenylsulfamic acid initially formed in the sulfonation of aniline is dependent on temperature. At elevated temperatures it gives 4-aminobenzenesulfonic acid almost quantitatively.

Even at low temperatures, the two other isomers are formed in only relatively small amounts. They are therefore manufactured industrially in other ways (see below).

For identification, aromatic aminosulfonic acids cannot be converted into amides via the sulfonyl chlorides; they must first be converted to the corresponding chlorobenzenesulfonic acids by Sandmeyer's reaction [143], and then derivatized. Today, identification is by HPLC.

**2-Aminobenzenesulfonic Acid** [88-21-1], $C_6H_7NO_3S$, $M_r$ 173.19, is obtained as a byproduct in the manufacture of 3-aminobenzenesulfonic acid by nitration of benzenesulfonic acid at 100 °C, separation of the resulting nitrobenzenesulfonic acids via the magnesium salts, and reduction with iron turnings.

In the reaction between amidosulfuric (sulfamic) acid and excess aniline at temperatures up to 160 °C, 2-aminobenzenesulfonic acid and 4- aminobenzenesulfonic acid are obtained in a ratio of 3:2 [144]. The two isomers are separated by means of their different solubilities in aqueous KOH [145].

2-Aminobenzenesulfonic acid is obtained from 2-nitrothiophenol by boiling in a mixture of dioxane and water (ratio 20:1) for seven hours; the yield is 86.7% of theory [146]. The acid can also be produced as follows: 2-chloronitrobenzene is reacted with sodium disulfide in aqueous alcoholic solution to form 2,2′-dinitrodiphenyldisulfide. After having been dissolved in a mixture of hydrochloric acid and nitric acid this is oxidized with chlorine to form 2-nitrobenzenesulfonyl chloride [147], which is then hydrolyzed with soda solution and finally reduced with iron turnings [148]. Alternatively 2,2′-dinitro-diphenyldisulfide can first be reduced to 2,2′-diamino-diphenyldisulfide and then oxidized with hydrogen peroxide in aqueous sulfuric acid to 2-aminobenzenesulfonic acid [149], [150].

2,2′-Diamino-diphenyldisulfide can also be arrived at by alkaline hydrolysis of benzothiazole or mercaptobenzothiazole, followed by oxidation with hydrogen peroxide. The additional oxidation leading to 2-aminobenzenesulfonic acid can be performed in alkaline solution. Thus the synthesis starting from benzothiazole can be carried out exclusively in alkaline solution in a single reactor [151]. Yield: 70%, based on benzothiazole.

**2-Aminobenzenesulfonamide** [*3306-62-5*], $C_6H_8N_2O_2S$, $M_r$ 172.21, *mp* 153 °C, crystals (from water), is freely soluble in water, glacial acetic acid, ethyl alcohol, acetone, and methanol; insoluble in benzene.

**2-Amino-N-methylbenzenesulfonamide** [*16288-77-0*], $C_7H_{10}N_2O_2S$, $M_r$ 186.23, is obtained by reacting 2-nitrobenzenesulfonyl chloride with methylamine and subsequent reduction. Used as a dye component [152].

The amides of 2-aminobenzenesulfonic acid can be reacted with aldehydes or orthoformic acid esters to 1,2,4-benzothiadiazine-1,1-dioxides or their dihydro compounds; these substances have been proposed as diuretics [153].

**2-Amino-N-(phenylsulfonyl)benzenesulfon-amide,** $C_{12}H_{12}N_2O_4S_2$, $M_r$ 312.37, *mp* 193–194 °C, is obtained by reacting 2-nitrobenzenesulfonyl chloride with benzenesulfonamide at 70 °C in the presence of aqueous sodium hydroxide solution and subsequent reduction of the nitro group [77]. Used as a component of azo dyes.

**Phenyl 2-Aminobenzenesulfonate** [*68227-69-0*], $C_{12}H_{11}NO_3S$, $M_r$ 249.29, *mp* 70 °C, is obtained by reacting 2-nitrobenzenesulfonyl chloride with sodium phenolate and subsequent reduction of the nitro group. Used as an azo dye component.

**3-Aminobenzenesulfonic Acid** (metanilic acid) [*121-47-1*], $C_6H_7NO_3S$, $M_r$ 173.19, can be obtained by sulfonation of aniline sulfate with oleum and boric acid at 20–50 °C. The yield is 98% [154].

*Production.* For industrial production, however, nitrobenzene is sulfonated to *m*-nitrobenzenesulfonic acid, which is then reduced either with iron [155], [156] or catalytically with hydrogen [157].

Crude 3-aminobenzenesulfonic acid can be used directly, either as a solution or dried, to obtain 3-aminophenol by alkali fusion. It is not sufficiently pure for use in dye manufacture, however. Therefore the crude solution is acidified; after filtration the product thus obtained is of satisfactory purity.

3-Aminobenzenesulfonic acid is used in the manufacture of dyes and optical brighteners [158]. It is also used to manufacture 3-hydroxybenzenesulfonic acid and 3-aminophenol. The latter is an important intermediate in the manufacture of 4-amino-2-hydroxybenzoic acid (*p*-aminosalicylic acid) used to fight tuberculosis.

**3-Aminobenzenesulfonamide** [*98-18-0*], $C_6H_8N_2O_2S$, $M_r$ 172.21, *mp* 142 °C, is poorly soluble in cold water.

**3-Amino-*N*-(phenylsulfonyl)benzenesulfonamide,** $C_{12}H_{12}N_2O_4S_2$, $M_r$ 312.37, is obtained by reacting 3-nitrobenzenesulfonyl chloride with benzenesulfonamide and subsequent reduction of the nitro group. It is used as a dye component [159]. The use of the *N*-acrylsulfonamide in the polymerization of acrylonitrile has been proposed to improve the dyeing properties of the fibers [160].

**4-Aminobenzesulfonic Acid** (sulfanilic acid) [*121-57-3*], $C_6H_7NO_3S$, $M_r$ 173.19, is made by the reaction of aniline with sulfuric acid at temperatures above 190 °C:

*Production.* Equimolar amounts of aniline and sulfuric acid are reacted in a lead-lined reactor. The reaction mass is spread on trays and baked in an oven at 190–220 °C, the water of the reaction being removed continuously. The process is complete if a sample dissolves clearly in a dilute alkaline solution and has practically no smell of aniline. The crude product is dissolved as a sodium salt in water that has been rendered alkaline with sodium carbonate solution; traces of aniline are removed by blowing with steam. The solution is clarified by filtration and can then be processed further immediately. The baking operation can be carried out in a vacuum, which facilitates the removal of the water of the reaction.

In industrial manufacture the aniline hydrogen sulfate is placed on trays in trolleys and these are passed at constant speed through a tunnel-shaped oven against an air current at 260–280 °C. The aniline hydrogen sulfate is thus brought to the baking temperature very gradually. It leaves the oven after 12.5 h [161]. According to a recent patent the reaction can be carried out in a fluidized bed consisting of sulfanilic acid [162].

The baking process can also be carried out in a multiphase reactor, with a resultant shortening of reaction time and improvement in quality of the end product [163]. The baking process can be simplified if the formation of the aniline salt and subsequent rearrangement are carried out successively or simultaneously in a solvent, e.g., diphenyl sulfone [57] or 1,2-dichlorobenzene [164]; this enables the reaction mass to be stirred throughout the production process [165]. Particularly high yield and end product purity can be achieved if the reactants are used in a molar ratio of 1:1; the rearrangement is carried out under a slight positive pressure (1–3 bar) and at 200–240 °C; water vapor is expelled from the reactor simultaneously [166]. The reaction time is three hours.

*Analysis.* The assay is carried out by diazotization. Impurities (aniline, 2,4-disulfanilic acid, 2- and 3-aminobenzenesulfonic acid) are determined by HPLC.

*Uses.* 4-Aminobenzenesulfonic acid is an important intermediate in the production of dyes, pesticides, pharmaceuticals, etc.

**4-Aminobenzenesulfonamide** [*63-74-1*], $C_6H_8N_2O_2S$, $M_r$ 172.21, forms the basis for chemotherapeutical sulfonamides.

**4-(Acetylamino)benzenesulfonic Acid** [*121-62-0*], $C_8H_9NO_4S$, $M_r$ 215.23, is produced by dissolving aminobenzenesulfonic acid in water, adding sodium carbonate, acetylating with acetic anhydride at 45–50 °C, and salting out with sodium chloride; the yield is 82 % [167].

**4-(Acetylamino)benzenesulfonyl Chloride** [*121-60-8*], $C_8H_8ClNO_3S$, $M_r$ 233.67, mp 149 °C (from benzene), is relatively stable to water; therefore it can be precipitated from alcoholic solutions with water. It is freely soluble in ether and ethyl acetate; obtainable as an anhydrous product from the aqueous paste if this is subjected to azeotropic distillation with a water-immiscible solvent at a low temperature.

*Production* (see also [168]). Acetanilide is introduced into stirred chlorosulfuric acid solution at 20 °C. The solution is then heated to 55 °C and kept at this temperature for 1.5 h. The solution is cooled to 20 °C and introduced as a thin jet into water, the temperature of which is kept at 0–5 °C by external cooling. The precipitated sulfochloride is filtered and washed with cold water until neutral to Congo Red. Yield is 80–82 %. The crude product contains about 60 % water; the dried product melts at 144–147 °C. About 1–2 % of bis(acetylaminophenyl) sulfone is obtained as a byproduct.

If thionyl chloride is added, less chlorosulfuric acid is needed because it can be replaced partly by oleum. The product is stable for several days if kept in a cool place. It is generally used directly in the moist state. 4-(Acetylamino)benzenesulfonyl chloride is an important intermediate in the manufacture of sulfonamides.

**4-(Acetylamino)benzenesulfonamide** [*121–61-9*], $C_8H_{10}N_2O_3S$, $M_r$ 214.24, *mp* 219 °C, is obtained from 4-acetylaminobenzenesulfonyl chloride and aqueous ammonia at temperatures not exceeding 30 °C. Uses include the manufacture of sulfanilamide.

**5-Amino-1,3-benzenedisulfonic Acid** [*5294-05-3*] (**20**), $C_6H_7NO_6S_2$, $M_r$ 253.25, is obtained by reduction of 5-nitro-1,3-benzenedisulfonic acid.

**5-Amino-1,3-benzenedisulfonyl Chloride,** $C_6H_5Cl_2NO_4S_2$, $M_r$ 290.14, is obtained from the sodium salt or free acid with chlorosulfuric acid at 100 – 160 °C [169].

**5-Amino-N,N-bis(2-chloroethyl)–1,3-benzenedisulfonamide** [*22480-69-9*], $C_{10}H_{15}Cl_2N_3O_4S_2$, $M_r$ 376.28, *mp* 126 – 128 °C, is formed by reacting the dichloride with 2-chloroethylamine [170], [171].

**4-Amino-1,3-benzenedisulfonic Acid** [*137-51-9*] (**21**), $C_6H_7NO_6S_2$, $M_r$ 253.25, crystallizes with two moles of $H_2O$. The compound is freely soluble in water and ethyl alcohol. It is prepared by sulfonating 4-aminobenzenesulfonic acid with oleum at 20 °C and 130 °C (4 – 6.5 h), draining the sulfonation mixture into water under pressure, adding lime at 70 – 75 °C, separating the resulting gypsum, precipitating the calcium ions with soda, removing the calcium carbonate formed, and concentrating the sodium salt solution by evaporation. Yield 90 – 91 % [172].

**4-Amino-1,3-benzenedisulfonamide** [*40642–90–8*], $C_6H_9N_3O_4S_2$, $M_r$ 251.28, *mp* 235 °C, crystals (from water). The compound is used as a dye intermediate.

**2-Amino-1,4-benzenedisulfonic Acid** [*98-44-2*] (**22**), $C_6H_7NO_6S_2$, $M_r$ 253.25, is obtained by sulfonating metanilic acid in 100 % sulfuric acid with 50 % oleum at 160 °C (5 – 6 h); the sulfonation mixture is diluted with water, lime is added, and the resulting calcium sulfate is removed; the product is then precipitated with excess concentrated hydrochloric acid in hot solution; yield 90 – 91 % [173].

The trisulfonic and tetrasulfonic acids formed in this sulfonation process are unstable in hot aqueous solution. When water is added to the sulfonation mass they eliminate sulfonic acid groups, thus forming the end product; see also [174]. It is also possible to start from 4-chloro-3-nitrobenzenesulfonic acid, in which case sulfite is used to replace the chlorine by the sulfonic acid group and the resulting 2-nitro-1,4-benzenedisulfonic acid is then reduced [175].

Compound **22** is used as an intermediate for dyes; for example, see [176]. 2-Amino-1,4-benzenedisulfonic acid is converted to triazine derivatives, which are intermediates in the manufacture of dyes and optical brighteners [177], [178].

**2,4-Diaminobenzenesulfonic Acid** [*88-63-1*] (**23**), $C_6H_8N_2O_3S$, $M_r$ 188.2, monoclinic plates or triclinic prisms, poorly soluble in cold water, is obtained by sulfonating a solution of *m*-phenylenediamine in 100% sulfuric acid with oleum at 155 °C, adding water to the sulfonation mixture, desulfonating the resulting disulfonic acids at 140 °C, and isolating the free acid at 10 °C; yield 93% [179]. Another process starts from 1-chloro-2,4-dinitrobenzene, which, as an alcoholic solution, is reacted with sodium sulfite solution to form 2,4-dinitrobenzenesulfonic acid [*89-02-1*], from which the end product is obtained by reduction with iron according to BÉCHAMP [175, pp. 99, 100].

**4,6-Diamino-1,3-benzenedisulfonic Acid** [*137-50-8*] (**24**), $C_6H_8N_2O_6S_2$, $M_r$ 268.27, is obtained from *m*-phenylenediamine by sulfonation with excess oleum at 70–130 °C. Used as an intermediate for dyes [180], [181].

**2,5-Diamino-1,4-benzenedisulfonic Acid** [*7139-89-1*] (**25**), $C_6H_8N_2O_6S_2$, $M_r$ 268.27, is obtained from *p*-phenylenediamine and 25% oleum at 140 °C [182] or by heating *p*-phenylenediamine sulfate and 98% sodium hydrogen sulfate in a mixture of 1,2,4-trichlorobenzene and 1-chloronaphthalene to the boiling point of the solvent mixture (215 °C), removing the solid product, dissolving it in water, neutralizing it with sodium hydroxide solution to pH 7.5–8, and clarifying the solution with kieselguhr. The product contains 5% of monosulfonic acid, but only a small amount of the isomeric 2,6-disulfonic acid; yield 90% [183]. The compound is used as an intermediate for dyes.

## 2.7. Hydroxybenzenesulfonic Acids

The chemical properties of the phenolsulfonic acids are characterized by the reactive aromatic hydroxyl group. The reactions of these acids therefore include azo coupling reactions with diazonium salts to form azo dyes. Hydroxybenzenesulfonic acids with more than one hydroxyl group on the benzene nucleus form complexes with metal ions.

**2-Hydroxybenzenesulfonic Acid** [*609-46-1*] (**26**), $C_6H_6O_4S$, $M_r$ 174.17, *mp* 145 °C (monohydrate); aniline salt *mp* 165 °C; potassium salt (dihydrate) *mp* 235–240 °C. If phenol is sulfonated with an equal amount of 100 % sulfuric acid under mild conditions, a mixture consisting of 2-hydroxy- and 4-hydroxybenzenesulfonic acid in a ratio of roughly 2:3 is obtained. The 2-hydroxybenzenesulfonic acid can be isolated from this mixture via the monobarium salt [184]. The pure compound is obtained by diazotizing 2-aminobenzenesulfonic acid and boiling it under acid conditions. Because it is relatively costly to manufacture, this substance has little importance as an intermediate for dyes.

**3-Hydroxybenzenesulfonic Acid** [585-38-6] (**27**), $C_6H_6O_4S$, $M_r$ 174.17; potassium salt (monohydrate) *mp* 200–210 °C; sodium salt (from ethyl alcohol) *mp* 314 °C, is synthesized by reacting 1,3-benzenedisulfonic acid and sodium hydroxide in water in an autoclave for 30 h at 250 °C. The reaction mass is neutralized and evaporated to dryness. The sodium salt of the acid is extracted with 78 % ethyl alcohol; yield 78 % [185]. The crude solution obtained by acidifying the reaction mass with hydrochloric acid and freeing it from sulfur dioxide by boiling can be used in the manufacture of dyes [186].

**4-Hydroxybenzenesulfonic Acid** [98-67-9] (**28**), $C_6H_6O_4S$, $M_r$ 174.17, aniline salt *mp* 170 °C, *S*-benzylisothiuronium salt (monohydrate) *mp* 168.7 °C.

*Production:* phenol (25 kg) and 96 % sulfuric acid (28 kg) are heated to 50 °C; then 1.25 kg additional sulfuric acid is added and the reaction mixture is heated at 110 °C for 5–6 h. Reaction water, together with about 5 % of the phenol, is distilled off during this period. Yield 95 % [187].

Crude 4-hydroxybenzenesulfonic acid obtained by hot sulfonation, i.e., mixed with 2-hydroxybenzenesulfonic acid, is used as such in the manufacture of synthetic tanning agents. Condensation products of phenolsulfonic acid mixtures with formaldehyde improve the dyeing properties of polyamide fibers [188].

4-Hydroxybenzenesulfonic acid and 2-halogenomalonic esters form the corresponding ethers, which are used in the manufacture of polyesters with improved affinity for basic dyes [189].

If a mixture of hydroxybenzenesulfonic acids and bis(hydroxyphenyl) sulfone is condensed with formaldehyde, and hydrogen sulfite is subsequently added, substances with good dispersing effects on dyes are obtained. They also are used as thinning agents for cement [190], see also [191].

Most of the 4-hydroxybenzenesulfonic acid is used as an additive for electroplating baths. By comparison, the acid is relatively unimportant as a dye intermediate. With ethylene oxide it can be reacted to form poly(ethoxy)oxybenzenesulfonic acid, which can be used as a plasticizer and emulsifier.

**4-Methoxybenzenesulfonic Acid** [5857-42-1] (**29**), $C_7H_8O_4S$, $M_r$ 188.20, ammonium salt mp 285 °C. The acid is obtained by sulfonation of anisol. The ammonium salt of **29** can be obtained in high yield by reacting anisol with an equal mass of sulfamic acid at 140–150 °C [192]. The compound is used in the manufacture of 4-methoxybenzenesulfonylureas, which have hypoglycemic effects.

**4-Hydroxy-1,3-benzenedisulfonic Acid** [96-77-5] (**30**), $C_6H_6O_7S_2$, $M_r$ 254.24, hygroscopic crystals, is obtained by hydrolysis of the dichloride with ethyl alcohol and/or water. Nitration and reduction of this acid gives 5-amino-4-hydroxy-1,3-benzenedisulfonic acid. 4-Hydroxy-1,3-benzenedisulfonic acid accelerates the rearrangement of cumene hydroperoxide in the Hock-Lang phenol process [193].

**4-Hydroxy-1,3-benzenedisulfonyl Chloride** [1892-33-7], $C_6H_4Cl_2O_5S_2$, $M_r$ 291.13, mp 89 °C, needles (from gasoline), is obtained by reacting phenol with chlorosulfuric acid and introducing the sulfonation mixture into concentrated hydrochloric acid [194].

**2,4-Dihydroxybenzenesulfonic Acid** [6409-58-1] (**31**), $C_6H_6O_5S$, $M_r$ 190.17, ammonium salt (from ethyl alcohol) mp 190 °C [195]. The acid is prepared by sulfonating resorcinol with 96 % sulfuric acid in a molar ratio of 1:1. The compound is used as a catalyst in the curing of resins [196].

**3,4-Dihydroxybenzenesulfonic Acid** [7134-09-0] (**32**), $C_6H_6O_5S$, $M_r$ 190.17, deliquescent crystals; the ammonium salt (flakes, from aqueous ethanol) melts at 260 °C (decomp.). Synthesis: 0.3 mol of disodium 4-hydroxy-1,3-benzenedisulfonate is reacted with 1.2 mol of sodium hydroxide for 36 min at 340 °C; the reaction mass is diluted with water, acidified with sulfuric acid, and boiled to remove sulfur dioxide; the product is extracted with ether. Yield 80 %. Pyrocatechol is obtained by hydrolyzing the sulfonic acid group at 190 °C [197].

**4,5-Dihydroxy-1,3-benzenedisulfonic Acid** [149-46-2] (**33**), $C_6H_6O_8S_2$, $M_r$ 270.23, is produced by sulfonation of pyrocatechol with oleum [198] or by alkali fusion of 2-hydroxy-1,3,5-benzenetrisulfonic acid: 10 kg of the tetrasodium salt of phenoltrisulfonic acid is introduced into 16 kg of sodium hydroxide at about 20 °C. After adding some water, the temperature is gradually raised to 160 °C; at the end of the reaction the melt is diluted with water and acidified with sulfuric acid, and sulfur dioxide is driven off. After cooling, precipitated sodium sulfate is removed by filtration and excess sulfate ions are precipitated in fractions. Finally the barium salt, which has poor solubility in cold water, is precipitated.

The disodium salt of **33**, which is referred to in the literature as Tiron [199], serves to detect traces of iron in blood and is used as a reagent for colorimetric determination of $Ti^{4+}$ and $Mo^{4+}$. The ability to form water-soluble complex salts, e.g., with antimony, has been exploited in pharmacology [200]. Tiron is claimed to improve the storability of silver halogenide emulsions [201].

## 2.8. Chloro-alkylbenzenesulfonic Acids

**3-Chloro-4-methylbenzenesulfonic Acid** [98-34-0] (**34**), $C_7H_7ClO_3S$, $M_r$ 206.65, decomposes above 110 °C. Production is from toluene by sulfonation with anhydrous sulfuric acid, followed by chlorination with gaseous chlorine [202].

3-Chloro-4-methylbenzenesulfonic acid is an intermediate in the manufacture of 2-chlorotoluene, 5-chloro-4-methyl-2-nitrobenzenesulfonic acid, 2,4-dichlorobenzoic acid, and 2-chloro-4-sulfobenzoic acid.

**3-Chloro-4-methylbenzenesulfonyl Chloride** [42413-03-6], $C_7H_6Cl_2O_2S$, $M_r$ 225.09, $bp_{32\,mbar}$ 166 °C, $mp$ 38 °C, is obtained by chlorination of 4-methylbenzenesulfonyl chloride in the presence of iron trichloride or antimony trichloride.

A high yield of sulfochloride with only a small percentage of isomers is obtained if toluene is sulfonated with sulfur trioxide in 1,2-dichloroethane. The resulting sulfonic acid is chlorinated in the presence of iodine, and the resulting chlorobenzenesulfonic acid is reacted with chlorosulfuric acid. The crude product can be used immediately in the production of 2,4-dichloro-α,α,α-trifluorotoluene, which is a precursor of several herbicides [203].

**3-Chloro-4-methylbenzenesulfonamide** [51896-27-6], $C_7H_8ClNO_2S$, $M_r$ 205.66, melts at 137 °C.

**5-Chloro-2-methylbenzenesulfonic Acid** [133-73-3] (**35**), $C_7H_7ClO_3S$, $M_r$ 206.65; the sodium salt crystallizes from water with 1/2 mol of $H_2O$. The acid is formed as the main product in the sulfonation of 4-chlorotoluene with 100% sulfuric acid at 100 °C. It is purified by recrystallizing the barium salt. The compound is used in the production of 5-chloro-2-formylbenzenesulfonic acid.

**4-Chloro-3-(trifluoromethyl)benzenesulfonyl Chloride** [*32333-53-2*] (**36**), $C_7H_3Cl_2F_3O_2S$, $M_r$ 279.06, $bp_{16\,mbar}$ 128–131 °C, $n_D^{20}$ 1.5158, is obtained from 2-chloro-α,α,α-trifluorotoluene and chlorosulfuric acid in 65% oleum [204]. The compound is used as an intermediate in the production of pesticides and dyes [204].

**3,5-Dichloro-4-methylbenzenesulfonic Acid** [*2225-18-5*], $C_7H_6Cl_2O_3S$, $M_r$ 241.09, is obtained (in addition to tri- and tetrachlorotoluenesulfonic acids) from 4-toluenesulfonic acid by chlorination with chlorine or sulfuryl chloride in sulfuric acid [205], or by hydrolysis of the sulfonyl chloride (see next paragraph). The compound is used as an intermediate in the manufacture of 2,6-dichlorotoluene.

**3,5-Dichloro-4-methylbenzenesulfonyl Chloride** [*24653-79-0*], $C_7H_5Cl_3O_2S$, $M_r$ 259.54, is formed from 4-toluenesulfonyl chloride by chlorination with chlorine, antimony trichloride being added as a catalyst [206]. The compound is used as an intermediate in the manufacture of 2,6-dichlorotoluene [205].

## 2.9. Amino-alkylbenzenesulfonic Acids

**2-Amino-5-methylbenzenesulfonic Acid** [*88-44-8*] (**37**), $C_7H_9NO_3S$, $M_r$ 187.22, crystallizes with 1 mol of $H_2O$. It is produced by baking *p*-toluidine sulfate at 180 °C and 97–101 mbar, dissolving the product in 3% sodium hydroxide solution, and precipitating the free acid by adding concentrated hydrochloric acid [207], [209].

**2-Amino-3,5-dimethylbenzenesulfonic Acid** [*88-22-2*] (**38**), $C_8H_{11}NO_3S$, $M_r$ 201.24; the barium salt crystallizes from water with 2 mol of $H_2O$; the lead salt is very poorly soluble in water. Production is from 4-amino-1,3-dimethylbenzene (*m*-xylidine) and 96% sulfuric acid by baking the sulfate paste in a vacuum, subsequently purifying the product with boiling milk of lime, and finally precipitating the sodium salt with sodium chloride [210]. 2-Amino-3,5-dimethylbenzenesulfonic acid can also be produced by sulfonation of *m*-xylidine in 1,1,2,2-tetrachloroethane with chlorosulfuric acid and subsequent boiling until no more hydrogen chloride is formed.

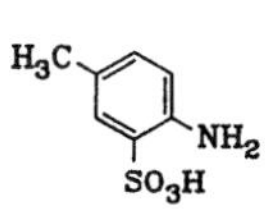

**3-Amino-4-methylbenzenesulfonic Acid** [618-03-1] (**39**), and **5-Amino-2-methyl-benzenesulfonic Acid** [118-88-7] (**40**), $C_7H_9NO_3S$, $M_r$ 187.22; both compounds are produced analogously to the 3-aminobenzenesulfonic acids from the corresponding nitrotoluenesulfonic acids by Béchamp reduction.

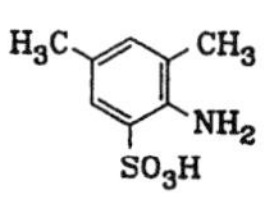

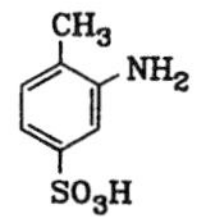

**4-Amino-3-methylbenzenesulfonic Acid** [98-33-9] (**41**), $C_7H_9NO_3S$, $M_r$ 187.22, crystallizes with 1 mol of $H_2O$. It is prepared from o-toluidine and 98% $H_2SO_4$ via the sulfate, which is then baked. The sodium salt can be isolated after liming and precipitation of calcium sulfate [211]. The solution can be used directly in the manufacture of dyes [212].

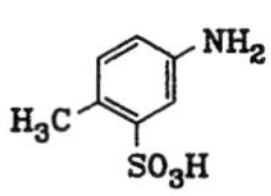

**4-Amino-2-methylbenzenesulfonic Acid** [133-78-8] (**42**), $C_7H_9NO_3S$, $M_r$ 187.22, crystallizes from water with 3 mol of $H_2O$. It is produced analogously to 4-amino-3-methylbenzenesulfonic acid by baking of m-toluidine sulfate [213].

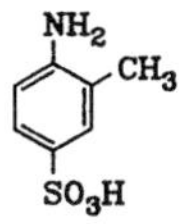

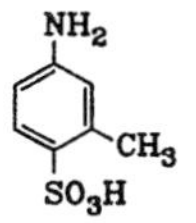

## 2.10. Amino- and Hydroxy-carboxy-benzenesulfonic Acids

**2-Amino-4-sulfobenzoic Acid** [98-43-1] (**43**), $C_7H_7NO_5S$, $M_r$ 217.2, crystallizes with 1 mol of $H_2O$; it is freely soluble in hot water, less soluble in cold water. The compound is obtained at a moderate yield by reacting 4-methyl-3-nitrobenzenesulfonic acid with sodium hydroxide solution:

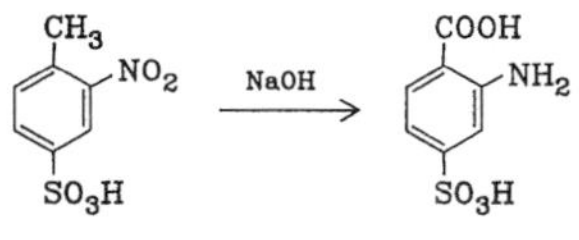

This acid is used mainly in the manufacture of diazoamino compounds of the R–N=N–NH–R′ type. As mixtures with naphthols these compounds have acquired importance as printing inks (Rapidogen dyes).

**2-Amino-5-sulfobenzoic Acid** [3577-63-7], (**44**) $C_7H_7NO_5S$, $M_r$ 217.2, is produced from anthranilic acid by sulfonation with chlorosulfuric acid in anhydrous sulfuric acid or nitrobenzene or by baking the sulfate of anthranilic acid at 200 °C [214], [215]. This compound is used, like its N-alkyl derivatives, as a stabilizer for diazo compounds (see next paragraph).

**2-Isobutylamino-5-sulfobenzoic Acid,** $C_{11}H_{15}NO_5S$, $M_r$ 273.31, is obtained in crystalline form by sulfonating N-isobutylanthranilic acid in 98% sulfuric acid with 65% oleum at 80 °C and pumping the sulfonation mixture into water; yield 93–94% [216].

**2-Hydroxy-5-sulfobenzoic Acid** [97-05-2] (**45**), $C_7H_6O_6S$, $M_r$ 218.18, needles, mp 120 °C, soluble in water, ethyl alcohol, and ether, is obtained by heating one part of salicylic acid with 5 parts of concentrated sulfuric acid [217]. It can also be prepared from the sulfochloride.

2-Hydroxy-5-sulfobenzoic acid forms a dihydrate and is used in analytical chemistry to detect albumins, iron, and mercury [218]. Applications described in patents include the use as a heat stabilizer for polyacrylonitrile [219], as an additive used in the anodizing of metals [220]–[222], and – as a tributyltin compound – as an agent for the antibacterial protection of organic materials [223]. The acid is also claimed to be suitable for a pregnancy test for horses [224].

**5-(Chlorosulfonyl)-2-hydroxybenzoic Acid** [17243-13-9], $C_7H_5ClO_5S$, $M_r$ 236.63, mp 171 °C (decomp.), is obtained from salicylic acid and chlorosulfuric acid [225], [226]. The compound is used to manufacture the corresponding sulfinic acid and sulfones.

## 2.11. Chloro-hydroxybenzenesulfonic Acids

**3-Chloro-4-hydroxybenzenesulfonic Acid** [46060-27-9] (**46**), $C_6H_5ClO_4S$, $M_r$ 208.62, is prepared by chlorination of 4-hydroxybenzenesulfonic acid or by sulfonation of 2-chlorophenol with 98% sulfuric acid [227]. The compound can be converted by nitration into 3-chloro-4-hydroxy-5-nitrobenzenesulfonic acid, from which 3-amino-5-chloro-4-hydroxybenzenesulfonic acid is obtained by reduction. The sulfonic acid group is eliminated easily at 180–190 °C.

**5-Chloro-2-hydroxybenzenesulfonic Acid** [2051-65-2] (**47**), $C_6H_5ClO_4S$, $M_r$ 208.62, is obtained from 2,5-dichlorobenzenesulfonic acid by partial hydrolysis with sodium hydroxide solution in water at 170–190 °C and precipitation with concentrated hydrochloric acid. Used to prepare 5-chloro-2-hydroxy-3-nitrobenzenesulfonic acid, which is then reduced to 3-amino-5-chloro-2-hydroxybenzenesulfonic acid.

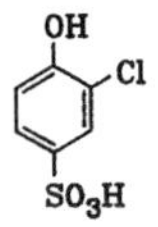

**3,5-Dichloro-2-hydroxybenzenesulfonic Acid** [26281-43-6] (**48**), $C_6H_4Cl_2O_4S$, $M_r$ 243, forms a poorly water-soluble barium salt.

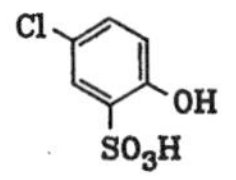

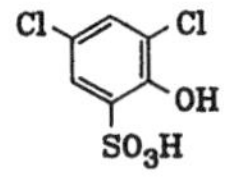

**3,5-Dichloro-2-hydroxybenzenesulfonamide** [35337-99-6], $C_6H_5Cl_2NO_3S$, $M_r$ 242.08, melts at 230 °C.

## 2.12. Amino-hydroxybenzenesulfonic Acids and Amino-chlorohydroxy-benzenesulfonic Acids

All the compounds described in this section are used in the manufacture of dyes.

**3-Amino-4-hydroxybenzenesulfonic Acid** [98-37-3] (**49**), $C_6H_7NO_4S$, $M_r$ 189.19, is freely soluble in hot water, slightly soluble in cold, dilute hydrochloric acid. Obtained by boiling 4-chloro-3-nitrobenzenesulfonic acid with sodium hydroxide solution, followed by Béchamp reduction [228]:

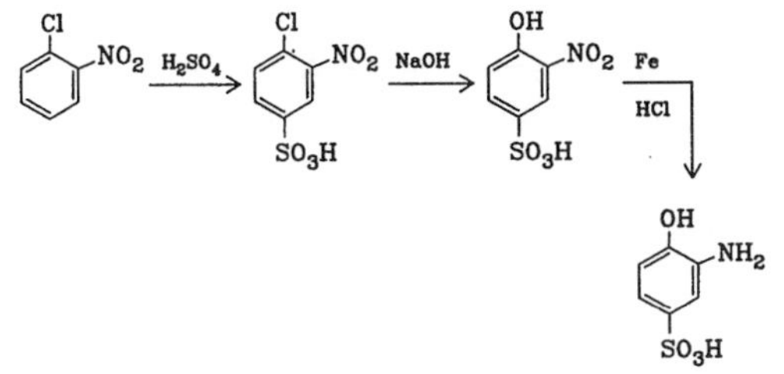

3-Amino-4-hydroxybenzenesulfonic acid is one of the most important intermediates in the production of dyes and is used to manufacture Diamond Black PV and 3-amino-4-hydroxy-5-nitrobenzenesulfonic acid.

**3-Amino-4-hydroxybenzenesulfonamide** [98-32-8], $C_6H_8N_2O_3S$, $M_r$ 188.2. Manufacture: 2-nitrochlorobenzene is sulfochlorinated with chlorosulfuric acid; the resulting sulfochloride is reacted with ammonia to 4-chloro-3-nitrobenzenesulfonamide; chlorine is replaced by hydroxyl by heating the product at 100 °C in aqueous sodium hydroxide solution. Finally, the hydroxybenzenesulfonamide is reduced with iron turnings [229].

**3-Amino-4-methoxybenzenesulfonic Acid** [98-42-0] (**50**), $C_7H_9NO_4S$, $M_r$ 203.22, needles, is freely soluble in water. The compound is obtained from 2-anisidine and 98% sulfuric acid at 50–55 °C [230].

**3-Acetylamino-4-methoxybenzenesulfonyl Chloride** [3746-67-6] (**51**), $C_9H_{10}ClNO_4S$, $M_r$ 263.7, mp 152–163 °C, is obtained from 2-acetylaminoanisol and chlorosulfuric acid at 35–40 °C [231]–[233]. The compound is used to produce the corresponding sulfinic acid and Fast Red ITR base.

**3-Acetylamino-4-methoxybenzenesulfonamide** [85605-29-4], $C_9H_{12}N_2O_4S$, $M_r$ 244.27, melts at 226 °C.

**3-Acetylamino-4-methoxy-N-methylbenzene-sulfonamide,** $C_{10}H_{14}N_2O_4S$, $M_r$ 258.3, is obtained by reacting the chloride with 2 mol of ethylamine at 25 °C. Other N-alkylamides of this sulfonic acid are produced in the same way [234].

**5-Amino-4-hydroxy-1,3-benzenedisulfonic Acid** [120-98-9] (**52**), $C_6H_7NO_7S_2$, $M_r$ 269.25, is obtained by nitration of 4-hydroxy-1,3-benzenedisulfonic acid and reduction of the hydroxynitrobenzenedisulfonic acid [74, p. 148].

**5-Acetylamino-2-hydroxybenzenesulfonic Acid** [55034-25-8] (**53**), $C_8H_9NO_5S$, $M_r$ 231.23, needles, freely soluble in water, decomposition on melting. Manufactured by sulfonating N-(4-hydroxyphenyl)acetamide with 98% sulfuric acid at 80 °C [235] or by acetylating 5-amino-2-hydroxybenzenesulfonic acid with acetic anhydride and sodium acetate in water.

**3-Amino-6-ethoxybenzenesulfonic Acid** [6375-02-6] (**54**), $C_8H_{11}NO_4S$, $M_r$ 217.24, needles, poorly soluble in water and ethyl alcohol. Manufactured by reacting o-dichlorobenzene with 96% sulfuric acid and baking the resulting sulfate.

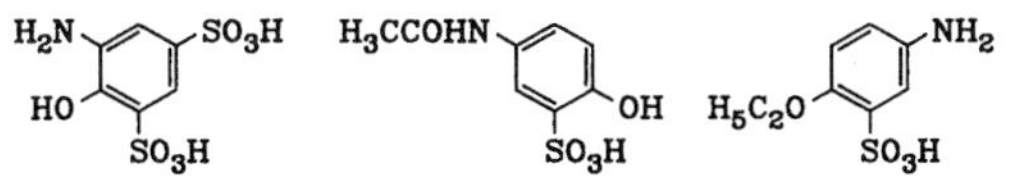

**4-Amino-6-methoxy-1,3-benzenedisulfonic Acid** (**55**), $C_7H_9NO_7S_2$, $M_r$ 283.28.

**4-Amino-6-methoxy-1,3-benzenedisulfonyl Chloride** [*670-02-0*], $C_7H_7Cl_2NO_5S_2$, $M_r$ 320.17, mp 129–130 °C, is obtained by adding sodium chloride to *m*-anisidine and chlorosulfuric acid while cooling with ice and then heating [236].

**3-Amino-5-chloro-2-hydroxybenzenesulfonic Acid** [*88-23-3*] (**56**), $C_6H_6ClNO_4S$, $M_r$ 223.63, colorless crystals, freely soluble in hot water, less soluble in cold water, very easily soluble in sodium hydroxide solution. The industrial product has a weak brownish-pink color and is produced by sulfonating 2-amino-4-chlorophenol or from *p*-chlorophenol:

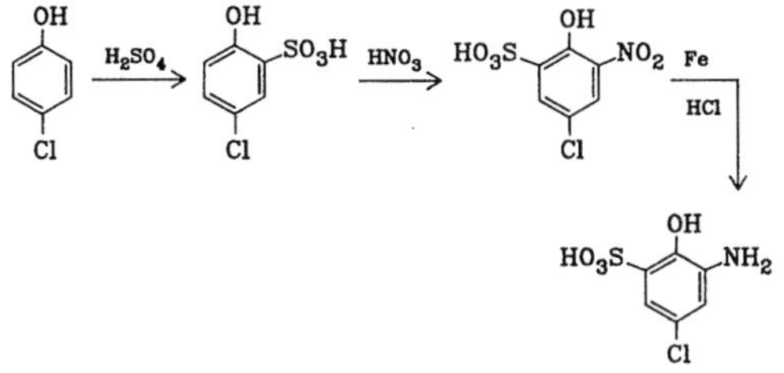

The 5-chloro-2-hydroxy-3-nitrobenzenesulfonic acid obtained from *p*-chlorophenol by sulfonation and nitration is reduced with iron and hydrochloric acid without being precipitated. After filtering from the iron sludge, the highly alkaline reduction liquor is concentrated by evaporation and decolorized with activated carbon. Upon acidification 3-amino-5-chloro-2-hydroxybenzenesulfonic acid precipitates. Yield 72%, based on *p*-chlorophenol. The acid is an important intermediate for dyes. It is used mainly in the manufacture of Diamond Black P 2 B and of acid chrome blue dyes.

**3-Amino-5-chloro-4-hydroxybenzenesulfonic Acid** (**57**) is similar in properties and manufacture to the isomeric acid described in the previous paragraph but it is less important industrially [237].

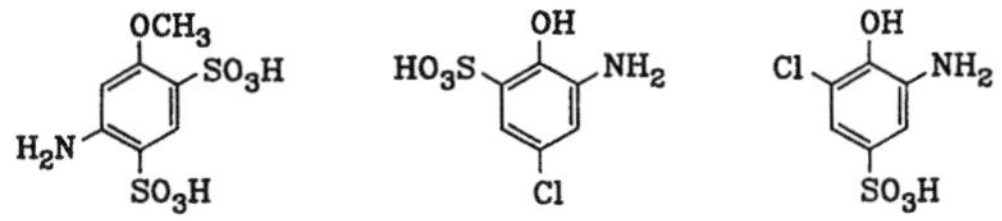

## 2.13. Amino-chlorobenzenesulfonic Acids

**2-Amino-5-chlorobenzenesulfonic Acid** [133-74-4] (**58**), $C_6H_6ClNO_3S$, $M_r$ 207.64, decomposes at 280 °C and crystallizes from water with 1 mol of $H_2O$. Manufactured by baking 4-chloroaniline hydrogen sulfate in a vacuum; after the mass has been pumped into water, the acid is neutralized with calcium oxide; the solution is suitable for use in the manufacture of azo dyes [238].

**3-Amino-4-chlorobenzenesulfonic Acid** [98-36-2] (**59**), $C_6H_6ClNO_3S$, $M_r$ 207.64, needles (from water), crystallizes with 1 mol of $H_2O$. It is obtained from 4-chloro-3-nitrobenzenesulfonic acid by reduction; the compound is used for the production of azo dyes.

**2-Amino-4,5-dichlorobenzenesulfonic Acid** [6331-96-0] (**60**), $C_6H_5Cl_2NO_3S$, $M_r$ 242.08, crystallizes as needles from water. Manufactured by baking 3,4-dichloroaniline sulfate. The compound is used as a component of azo dyes.

**2-Amino-4,5-dichlorobenzenesulfonyl Chloride** [36110-12-0], $C_6H_4Cl_3NO_2S$, $M_r$ 260.53, is obtained by reacting 3,4-dichloroaniline with chlorosulfuric acid in the presence of sodium chloride.

**2-Amino-4,5-dichlorobenzenesulfonamide** [16948-63-3], $C_6H_6Cl_2N_2O_2S$, $M_r$ 241.1, mp 183–185 °C, crystals from 50% alcohol, is obtained by reacting the chloride with aqueous ammonia [239].

**4-Amino-2,5-dichlorobenzenesulfonic Acid** [88-50-6] (**61**), $C_6H_5Cl_2NO_3S$, $M_r$ 242.08, is obtained by sulfonating 2,5-dichloroaniline with 65% oleum in anhydrous sulfuric acid; the compound is purified via the sodium salt [240].

**4-Amino-6-chloro-1,3-benzenedisulfonic Acid** (**62**), $C_6H_6ClNO_6S_2$, $M_r$ 287.7.

**4-Amino-6-chloro-1,3-benzenedisulfonyl Chloride** [671-89-6], $C_6H_4Cl_3NO_4S_2$, $M_r$ 324.59, mp 142–144 °C, is obtained by chlorosulfonation of 3-chloroaniline with

11 mol of chlorosulfuric acid at 130 °C [241]–[243]; after the chlorosulfonation mixture is treated with 4 mol of thionyl chloride at 80 °C, the yield is 90% [244].

**4-Amino-6-chloro-1,3-benzenedisulfonamide** [*121-30-2*], $C_6H_8ClN_3O_4S_2$, $M_r$ 285.73, *mp* 258–260 °C, is obtained by reacting the chloride with ammonia in *tert*-butanol and concentrating the solution by evaporation; the yield is 80% [244]. Condensation with formic acid gives 6-chloro-7-sulfamoyl-1,2,4-benzothiadiazine 1,1-dioxide, an important diuretic [245], [246]. The 1-monoamide **63** is an intermediate for the production of azo dyes for polyamide fibers. For manufacture, see [247].

**4-Amino-6-chloro-$N^1$,N-dimethyl-1,3-benzene-disulfonamide** [*60385-32-2*], $C_8H_{12}ClN_3O_4S_2$, $M_r$ 313.78, *mp* 183 °C (crystals from aqueous alcohol), is prepared by boiling the chloride with aqueous methylamine in dioxane at reflux temperature for 3 h and precipitating with water [241].

## 2.14. Amino-chloro-alkylbenzenesulfonic Acids

**2-Amino-5-chloro-4-methylbenzenesulfonic Acid** [*88-53-9*] (**64**), $C_7H_8ClNO_3S$, $M_r$ 221.66, is a colorless to slightly reddish crystalline powder. The compound is obtained by chlorination of *p*-toluenesulfonic acid to form 3-chloro-4-methylbenzenesulfonic acid, subsequent nitration of this acid to an isomeric nitration mixture, precipitation of the 2-chloro-4-methyl-6-nitrobenzenesulfonic acid with sodium chloride solution, and, finally, Béchamp reduction of the nitro group with iron and acetic acid [248].

A different synthesis route in which no isomers are formed is as follows [249]:

The compound is used as Lake Red C and CLT Acid as a diazo component for azo dyes and azo pigments.

**3-Amino-5-chloro-4-methylbenzenesulfonic Acid** [*6387-27-5*] (**65**), $C_7H_8ClNO_3S$, $M_r$ 221.66, is a moist, slightly reddish paste in raw form. It is manufactured by sulfonation of 1-chloro-2-methyl-3-nitrobenzene [*83-42-1*] with oleum, followed by Béchamp reduction [250]. The compound is used as a diazo component for azo dyes in the dyeing of leather.

**4-Amino-5-chloro-3-methylbenzenesulfonic Acid** [*6387-14-0*] (**66**), $C_7H_8ClNO_3S$, $M_r$ 221.66, needles from water, splits off the sulfonic acid group in 75% sulfuric acid at 150–160 °C. It is made by sulfonation of 2-chloro-6-methylaniline in *o*-dichlorobenzene with anhydrous sulfuric acid in a vacuum, during which the water of the reaction distills off [251]. Alternative: chlorination of 4-acetylamino-3-methylbenzenesulfonic acid in water and deacetylation. The compound is used as a dye component.

**4-Amino-5-chloro-2-methylbenzenesulfonic Acid** (**67**), $C_7H_8ClNO_3S$, $M_r$ 227.66, is obtained from *m*-toluidine sulfate by chlorination, followed by sulfonation in an inert solvent [252].

**2-Amino-4-chloro-5-methylbenzenesulfonic Acid** [*88-51-7*] (**68**), $C_7H_8ClNO_3S$, $M_r$ 221.66, is insoluble in cold water. The compound is made by baking 3-chloro-4-methylbenzeneamine hydrogen sulfate in a vacuum at 195 °C [253], and it is used for dyes; see, for example, [254].

**6-Amino-2,4-dichloro-3-methylbenzenesulfonic Acid** (**69**), $C_7H_7Cl_2NO_3S$, $M_r$ 256.11, is obtained by sulfonation of 3,5-dichloro-4-methylaniline in *o*-dichlorobenzene with chlorosulfuric acid [255].

# 3. References

General References

[1] B. T. Brooks, C. E. Brooks, S. S. Kurtz, L. Schmerling: "Sulfonation of Aromatic Hydrocarbons," in: *The Chemistry of Petroleum Hydrocarbons*, vol. **III**, Reinhold, New York 1955.

[2] H. Cerfontain: *Mechanistic Aspects in Aromatic Sulfonation and Desulfonation*, J. Wiley & Sons, New York 1968.

[3] E. E. Gilbert: "The Reaction of Sulfur Trioxide and of its Adducts with Organic Compounds," *Chem. Rev.* **62** (1962) 549–589.

[4] C. W. Suter: *The Organic Chemistry of Sulfur. Tetravalent Sulfur Compounds*, J. Wiley & Sons, New York 1945.

[5] C. M. Suter, A. W. Weston: "Direct Sulfonation of Aromatic Hydrocarbons and their Halogen Derivatives," in: *Organic Reactions*, vol. **III**, J. Wiley & Sons, New York 1946, pp. 141–197.

[6] N. N. Woroshzow: *Grundlagen der Synthese von Zwischenprodukten und Farbstoffen*, Akademie-Verlag, Berlin 1966.
[7] E. E. Gilbert: *Sulfonation and Related Reactions*, Interscience Publ., New York 1964.
[8] K. LeRoi Nelson in G. A. Olah (ed.): *Friedel-Crafts and Related Reactions*, vol. **3**, Interscience Publ., New York 1964, pp. 1355–1392.
[9] F. Muth: "Methoden zur Herstellung und Umwandlung aromatischer Sulfonsäuren," in: *Houben-Weyl*, vol. **9**, 1955, pp. 435–556.
[10] W. J. Hickingbottom, in *Rodd's Chemistry of Carbon Compounds*, vol. **III A**, Elsevier, Amsterdam 1954, pp. 230–248.
[11] C. Grundmann:"Desulfonation," in *Houben-Weyl*, vol. **5/2 b**, 1981, pp. 354–357.
[12] H. Cerfontain et al.: "Aromatic Sulfonation" (a series of articles), *J. Chem. Soc. Perkin Trans. 2* 1978,719; *ibid.* **1979**, 224; *J. Am. Chem. Soc.* **100** (1978) 8244; *J. Chem. Soc. Perkin Trans 2* **1979**, 673–679; *Tetrahedron Lett.* **1978**, 3263; *J. Chem. Soc. Perkin Trans 2* **1979**, 844–850; *ibid.* **1979**, 851–852; *ibid.* **1980**, 13–38; *ibid.* **1980**, 358–362; *ibid.* **1980**, 358–62; *ibid.* **1980**, 904–914.

Specific References

[13] V. Vesely, T. Stojanova, *Collect. Czech. Chem. Commun.* **9** (1937) 465.
[14] G. Bourjol, *Chem. Ind.* **78** (1957) 214.
[15] Texaco, US 3311670, 1963.
[16] California Res. Corp., US 2511711, 1946.
[17] Rütgerswerke, DE 1468913, 1965.
[18] California Res. Corp., US 2541959, 1947.
[19] Hoechst, DE 950464, 1954.
[20] I. T. Rodeanu, RO 60761, 1972.
[21] Y. and G. Erykalov, V. G. Chirtulov, A. A. Spryskov, *Izv. Vyssh. Uchebn. Zaved. Khim. Khim. Tekhnol.* **14** (1971) no. 1, 79–82.
[22] Mitsui Toatsu Chem., JP 53044528, 1976.
[23] O. Jacobsen, *Ber. Dtsch. Chem. Ges.* **19** (1886) 1209.
[24] V. D. Simonov, M. A. Ikrina, N. F. Popova, L. A. Kozlova, *Zh. Prikl. Khim. (Leningrad)* **45** (1972) no. 12, 2765–2766.
[25] Nitto Chem. Ind., JP-KK 20541/65, 1962.
[26] Hoechst, DE 1063151, 1957.
[27] L. Field, P. H. Settlage, *J. Am. Chem. Soc.* **76** (1954) 1222.
[28] C. M. Gable, H. F. Betz, S. H. Maron, *J. Am. Chem. Soc.* **72** (1950) 1445–1448.
[29] Allied Chem. & Dye Corp., US 2697117, 1951.
[30] J. A. Bradley, H. H. Harkins, *Abstr. Pap. Am. Chem. Soc.* **127th**, 1955.
[31] E. E. Gilbert, *Chem. Rev.* **62** (1962) no. 6, 549.
[32] L. Leiserson et al. *Ind. Eng. Chem.* **40** (1948) 508.
[33] Monsanto, GB 679827, 1949, US-Prior. 1948.
[34] Matsuyama Sekiyuka, JP 52133945, 1976.
[35] V. Gold, D. P. N. Satchell, *J. Chem. Soc.* 1956, 1635.
[36] M. Kilpatrick et al., *J. Phys. Chem.* **64** (1960) 1433.
[37] J. C. D. Braml, *J. Chem. Soc.* 1950, 1004, 1952, 3922, 3927.
[38] J. E. Woodbridge, *J. Am. Oil Chem. Soc.* **35** (1958) 528.
[39] Vulcan Chem., GB 747659, 1952.
[40] A. W. Kaandorp, H. Cerfontain, F. L. J. Soxma, *Recl. Trav. Chim. Pays-Bas* **81** (1962) 969–992.

[41] BASF, EP 41134, 1981 (F. Brunnmueller, W. Boehm, V. Weberndoerfer).
[42] Hoechst, DE 1493663, 1964.
[43] Z. Skrowaczewska, *Pr. Wroclaw, Tow. Nauk. Ser. B.* **61** (1953) 5; *Chem. Abstr.* **48** (1954) 7568.
[44] Bayer, EP 63271, 1981 (H. Emde, H. U. Blank, P. Schnegg).
[45] *Chem. Eng.* **69** (1962) no. 5, 70–72.
[46] H. Meerwein et al., *Ber. Dtsch. Chem. Ges.* **90** (1957) 846.
[47] H. Meerwein et al., *Ber. Dtsch. Chem. Ges.* **90** (1957) 851.
[48] Z. J. Allan, J. Podstata, *Collect. Czech. Chem. Commun.* **31** (1966) 3573.
[49] W. H. Hunter, M. M. Sprung, *J. Am. Chem. Soc.* **53** (1931) 1432.
[50] J. Borecky, *J. Chromatogr.* **2** (1959) 612.
[51] V. V. Kharitonov, A. A. Spryskov, V. P. Leshchev, *Izv. Vyssh. Uchebn. Zaved. Khim. Khim. Tekhnol.* **14** (1971) no. 2, 238.
[52] E. Toinlinson, T. M. Jefferies, C. M. Riley, *J. Chromatogr.* **159** (1978) 315–358.
[53] L. I. Levina, S. N. Patrakova, L. A. Patrusev, *Zh. Obshch. Khim.* **28** (1958) no. 9, 2427.
[54] Bayer, EP 1276, 1978 (H. U. Blank, T. Pfister).
[55] H. Meerwein et al., *Ber. Dtsch. Chem. Ges.* **90** (1957) 841.
[56] Bayer, DE 859461, 1942.
[57] C. R. Gambill, T. D. Roberts, H. Shechter, *J. Chem. Educ.* **49** (1972) no. 4, 287.
[58] O. Hinsberg, J. Kessler, *Ber. Dtsch. Chem. Ges.* **38** (1905) 906.
[59] F. Runge et al., *Fresenius Z. Anal. Chem.* **158** (1957) 266.
[60] F. Runge et al., *Ber. Dtsch. Chem. Ges.* **88** (1955) 533.
[61] M. L. Crossley et al., *J. Am. Chem. Soc.* **60** (1938) 2222.
[62] VEB Fahlberg-List, DD 9132, 1942.
[63] Bayer, DE 694974, 1955.
[64] E. Schlittler, G. de Stevens, L. Werner, *Angew. Chem.* **74** (1962) no. 9, 317–326.
[65] J. A. Bogan, *Med. Actual.* 1977, 393–395.
[66] Chem. Fabrik Griesheim, DE 840240, 1943.
[67] H. R. Slagh, E. C. Britton, *J. Am. Chem. Soc.* **72** (1950) 2808.
[68] Esso, DE 1227009, 1962.
[69] H. Kohner, M. L. Gressmann, *Z. Phys. Chem. Abt. A* **144** (1929) 144.
[70] R. C. Crooks, R. R. White, *Chem. Eng. Progr.* **46** (1950) 249.
[71] BIOS Final Report no. 664, 8; no. 1149, 128.
[72] *Ullmann*, 3rd ed., vol. **4**, p. 305.
[73] R. L. Kenyon, N. Boehmer, *Ind. Eng. Chem.* **42** (1950) 1446.
[74] A. N. Planowski, S. Z. Kagan, *Chem. Zentralbl.* 1941, 1075.
[75] G. F. Lisk, *Ind. Eng. Chem.* **40** (1948) 1678.
[76] A. Guyot, *Chim. Ind. Paris* **2** (1919) 879–891, 1167.
[77] D. H. Killeffer, *Ind. Eng. Chem.* **16** (1924) 1066.
[78] E. Sobczak, M. Badzynski, S. Chojnocky, K. Blanowicz, *Pr. Wydz. Nauk Tech., Bydgoskie Tow. Nauk., Ser. A* **13** (1979) 109–112; *Chem. Abstr.* **91**, 211020 y.
[79] H. Soep, P. Demoen, *Microchem. J.* **4** (1960) 82–87.
[80] BIOS 986, I, 56.
[81] BIOS 986, II, 415.
[82] FIAT 1313, I, 101.
[83] BIOS 986, II, 414.
[84] BIOS 1153, 150.
[85] N. N. Dykhanov, *J. Gen. Chem. (USSR Engl. Transl.)* **29** (1959) 3563.

[86] Harshaw Chem., US 2757133, 1952.
[87] A. A. Spryskov, S. P. Starkov, *Zh. Obshch. Khim.* **27** (1957) 2780.
[88] BIOS 986, II, 385.
[89] Ube Ind., JP 71/41529, 1971.
[90] Koppers Co. Inc., US 4302403, 1976.
[91] Taoka Chemical, JP-KK 80034776, 1969.
[92] Hooker Chem., DE 1468040, 1964.
[93] Koppers, BE 698821, 1967.
[94] H. Cerfontain, F. L. J. Sixma, L. Vollbracht, *Recl. Trav. Chim. Pays-Bas* **82** (1963) no. 7, 659–670.
[95] Tennessee Corp., US 2841612, 1956.
[96] V. S. Patwardhan, R. E. Eckert, *Ind. Eng. Chem. Process Des. Dev.* **20** (1981) no. 1, 82–85.
[97] A. A. Spryskov, V. A. Kozlov, *Khim. Khim. Tekhnol. Alma Ata 1962* **12** (1969) no. 2, 166–169 (1969).
[98] A. A. Spryskov, V. A. Kozlov, *Khim. Khim. Tekhnol. Alma Ata 1962* **12** (1969) no. 7, 900–902 (1969).
[99] Koppers, BE 844860, 1975 (US Appl. 602768, 1975).
[100] A. A. Spryskov, *Khim. Khim. Tekhnol.* **4** (1961) no. 6, 981–984.
[101] T. I. Potapova, A. A. Spryskov, *Khim. Khim. Tekhnol.* **10** (1967) no. 8, 885–887.
[102] Witco Chem., DE 1812708, 1968.
[103] Taoka Dye MFG., JP 68/22861, 1964.
[104] Allied Chem. & Dye Corp., US 2362612, 1942.
[105] Standard Oil Develop. Co., US 2540519, 1946.
[106] *Org. Synth.*, Coll. Vol. **III** (1955) 366.G. Storck, R. Borch, *J. Am. Chem. Soc.* **86** (1964) 937.
[107] M. J. Morgan, L. H. Cretcher, *J. Am. Chem. Soc.* **70** (1948) 375.
[108] Phillips Petroleum Co., US 2678947, 1949.
[109] BASF, DE-AS 1157629, 1962.
[110] F. H. Gereim, W. Braker, *J. Am. Chem. Soc.* **66** (1944) 1459.
[111] W. Herzog, *Angew. Chem.* **39** (1926) 728.
[112] Sankyo Kasei Co., JP 67/13937, 1965.
[113] H. Meyer, *Justus Liebigs Ann. Chem.* **433** (1923) 333.
[114] Sugai Chemical, JP-KK 52136144, 1976.
[115] Dow, US 2821549, 1954.
[116] Toyo Soda MFG, JP-KK 55031059, 1973.
[117] Toyo Soda MFG, JP-KK 55064565, 1978.
[118] Du Pont, US 2837500, 1953.
[119] G. E. Inskeep, *J. Am. Chem. Soc.* **69** (1947) 2237.
[120] R. H. Wiley, *J. Am. Chem. Soc.* **76** (1954) 720.
[121] R. H. Wiley, S. F. Reed, Jr., *J. Am. Chem. Soc.* **78** (1956) 2171.
[122] Du Pont, US 2822385/6, 1955.
[123] Dow., US 2913438, 1955.
[124] R. H. Wiley, C. C. Ketterer, *J. Am. Chem. Soc.* **75** (1953) 4520.
[125] Mitsubishi Petro-Chem., JP 70/36498, 1967.
[126] W. Borsche, M. Pommer, *Ber. Dtsch. Chem. Ges.* **54** (1921) 105.
[127] Chem. Fabrik von Heyden, DE-AS 1159937, 1960.
[128] P. Cagniant, *Bull. Soc. Chim. Fr.* 1950, 29.
[129] Agency of Ind. Sci. Techn., JP 54100343, 1978.

[130] Ciba-Geigy, DE-OS 2209223, 1971; DE-OS 2241304, 1972.
[131] T. P. Sycheva, M. N. Shchukina, *Sb. Statei Obshch. Khim.* 1953, 527.
[132] J. N. Tay, M. L. Dey, *J. Chem. Soc.* **117** (1920) 1407.
[133] Y. and E. Briskin et al., SU 192192, 1966.
[134] FIAT 949, 23.
[135] Weiler – Ter Meer,DE 385049, 1921; *Friedländer* **14** (1926) 386.
[136] Union Carbide, DE-OS 2704972, 1977 (U. A. Steiner; US Appl. 657209, 1976).
[137] F. Runge, F. Pfeiffer, *Ber. Dtsch. Chem. Ges.* **90** (1957) 1757.
[138] J. Stewart, *J. Chem. Soc.* **121** (1922) 2557.
[139] G. M. Kraay, *Recl. Trav. Chim. Pays-Bas* **49** (1930) 1083.
[140] Phillips, FR 1119189, 1954.
[141] W. V. Farrar, *J. Chem. Soc.* 1960, 3063.
[142] SU 166678, 1962.(E. V. Sergeev, V. I. Zetkin).
[143] C. F. H. Allen, G. F. Frame, *J. Org. Chem.* **7** (1942) 15.
[144] Sugai Chem. Ind., JP 70/19892, 1967.
[145] M. Saito, M. Dehara, O. Manabe, *Nippon Kagaku Kaishi* 1972, 2, 380 – 382.
[146] Imperial Chemical Industries, EP 25274, 1979.
[147] H. E. Fierz et al., *Helv. Chim. Acta* **12** (1929) 665.
[148] E. Wertheim, *Org. Synth.* **15** (1935) 55 – 58.
[149] H. E. Fierz et al., *Helv. Chim. Acta* **12** (1929) 663.
[150] Amer. Cyanamid Co., US 3038932, 1958.
[151] Bayer, DE-OS 3224155, 1982 (M. Michna, H. Henk).
[152] Eastmann Kodak, US 2529924, 1948.
[153] J. H. Freemann, E. C. Wagner, *J. Org. Chem.* **16** (1951) 815.
[154] Chelevin RN, SU 740758, 1978.
[155] FIAT Final Report no. 1313, vol. **1** (1948) 187 – 191.
[156] H. E. Fierz-David: *Grundlegende Operationen der Farbenchemie,* Springer Verlag, Wien 1952, p. 117.
[157] BASF, FR 1336648, 1962.
[158] Bayer, US 3532692, 1967.
[159] BASF, DE 1115384, 1957.
[160] Bayer, DE 1089548, 1959.
[161] FIAT Final Report no. 1313, vol. **1** (1948) 255 – 256.
[162] BASF, DE-OS 2049639, 1970.
[163] Hoechst, DE-OS 2439297, 1974.
[164] J. Casper, W. Petzold, PB Rept. 73911, FIAT Microfilm no. 87, 4648-60.
[165] SU 667550, 1977 (B. I. Kissin, A. V. Tarasevich).
[166] Bayer, EP 63271, 1981 (H. Emde, H. U. Blank, P. Schnegg).
[167] BIOS 1149, 125.
[168] *Org. Synth.*, Coll. Vol. 1 (1932) 8.
[169] O. Lustig, E. Katscher, *Monatsh. Chem.* **48** (1927) 94.
[170] I.G. Farbenindustrie, DE 743766, 1940.
[171] Bayer, DE 899536, 1951.
[172] BIOS 1153, 171.
[173] FIAT 1313, I, 54.
[174] P. K. Maarsen, R. Bregman, H. Cerfontain, *J. Chem. Soc. Perkin Trans 2* 1977, 14, 1863 – 1868.

[175] H. E. Fierz-David: *Grundlegende Operationen der Farbenchemie,* Springer Verlag, Wien 1952, 102.
[176] Imperial Chemical Industries, DE 2921309, 1978.
[177] Imperial Chemical Industries, DE 2045086, 1969.
[178] Geigy, US 3589921, 1969.
[179] Sumitomo Chemical, JP-KK 57048961, 1980.
[180] Imperial Chemical Industries, GB 1205017, 1966.
[181] Imperial Chemical Industries, JP 4706532, 1970.
[182] Ciba-Geigy, CH 523865, 1965.
[183] Hoechst, DE 2162963, 1971.
[184] J. Obermiller, *Ber. Dtsch. Chem. Ges.* **40** (1907) 3637.
[185] F. Willson, K. H. Meyer, *Ber. Dtsch. Chem. Ges.* **47** (1914) 3162.
[186] Bayer, DE-OS 1921046, 1969.
[187] M. Hazard-Flammand, DE 141751, 1897; *Friedländer* **6,** 1295.
[188] E. N. Anischuk, SU 678109, 1977.
[189] FMC Corp., DE 2064944, 1971.
[190] C. Stopanski, DE 3004543, 1979.
[191] Bayer, DE 2934980, 1979.
[192] K. A. Hofmann, E. Biesalski, *Ber. Dtsch. Chem. Ges.* **45** (1912) 1396.
[193] V. I. Burmiskov et al., SU 213892, 1969.
[194] J. Pollak, *Monatsh. Chem.* **46** (1925) 395.
[195] A. Quilico, *Gazz. Chim. Ital.* **57** (1927) 799.
[196] Bakelite, FR 1012550, 1949.
[197] Ube Ind., JP 7021501, 1966.
[198] Eastman Kodak, US 3772379, 1971.
[199] *Römpp,* 7th ed., p. 3604.
[200] V. Fischl, H. Schlossberger: *Handbuch der Chemotherapie,* Thieme Verlag, Leipzig 1934, p. 583.
[201] Eastman Kodak, DE 1302776, 1971.
[202] FIAT 1313, I, 124.
[203] Dow, US 4131619, 1978.
[204] Bayer, BE 758231, 1969.
[205] Ishihara Sangyo Kaisha, JP 66/10810, 1962; GB 1017976, 1963.
[206] Shell, BE 625993, 1962; GB-Prior. 1961.
[207] BIOS 1153, 176.
[208] FIAT 1313, I, 51.
[209] D. W. Hein, E. S. Pierce, *J. Am. Chem. Soc.* **76** (1954) 2729.
[210] BIOS 1153, 177.
[211] *Org. Synth* **27** (1947) 88–91
[212] BIOS 1153, 173; 986, 441.
[213] BIOS 986, 399.
[214] Ciba, US 2353351, 1943.
[215] W. F. Harris, T. R. Sweet, *J. Am. Chem. Soc.* **77** (1955) 2893.
[216] FIAT 1313, I, 184.
[217] BIOS 1153, II, 255.
[218] SU 251911, 1967 (L. N. Lapin, I. V. Reis).
[219] SU 318595, 1969 (Z. I. Mironova,L. N. Smirnov, A. G. Temnova).
[220] Kaiser Aluminium and Chem. Corp., DE 1446002, 1959.

[221] Sumitomo Chem., DE 2243178, 1971.
[222] G. Scholze, DD 96261, 1972.
[223] P. B. Hutson, US 3534077, 1968.
[224] Teikoku Hormone MFG., JP 70/24372, 1966.
[225] FIAT 1313, I, 77.
[226] BIOS 1153, 255.
[227] BIOS 1153, 223.
[228] BIOS 1153, 214.
[229] BIOS 1153, 217.
[230] BIOS 1153, 179.
[231] BIOS 1149, 24.
[232] BIOS 1153, 113.
[233] FIAT 1313, I, 18.
[234] BIOS 1153, 114, 117.
[235] BIOS 1153, 234.
[236] Merck, DE 1153761, 1957.
[237] BIOS 1153, 222.
[238] BIOS 1153, 182.
[239] A/S Ferrosan, DE 1157628, 1960.
[240] BIOS 1153, 185.
[241] W. Logemann et al., *Justus Liebigs Ann. Chem.* **623** (1959) 162.
[242] F. C. Novello et al., *J. Org. Chem.* **25** (1960) 965.
[243] Merck, DE 1088504, 1958; DE 1153761, 1957.
[244] Merck, DE 1161910, 1958.
[245] F. C. Novello, J. M. Sprague, *J. Am. Chem. Soc.* **79** (1957) 2028.
[246] Merck, US 2809194, 1956; DE 1119282, 1957.
[247] Ciba-Geigy, EP 39306, 1980.
[248] BIOS 986, 130–134.
[249] Hodogaya Chem. Ind., JP-KK 4151945, 1978.
[250] BIOS 986, 129–130.
[251] BIOS 986, 128.
[252] Dainichisaika Color Chem., JP 4125636, 1978.
[253] BIOS 1153, 188.
[254] Du Pont, US 2744027, 1952.
[255] Amer. Cyanamid Co., US 2754294, 1953.

# Benzidine and Benzidine Derivatives

HANS SCHWENECKE, Hoechst Aktiengesellschaft, Frankfurt/Main, Federal Republic of Germany (Chaps. 2–7)

DIETER MAYER, Hoechst Aktiengesellschaft, Frankfurt/Main, Federal Republic of Germany (Chap. 8)

| | | | |
|---|---|---|---|
| 1. | Introduction ............. 793 | 6.3. | Halogenated Benzidines..... 807 |
| 2. | Production ............. 794 | 6.4. | Alkoxybenzidines.......... 808 |
| 2.1. | Reduction .............. 794 | 6.5. | Benzidinecarboxylic Acids ... 810 |
| 2.2. | Benzidine Rearrangement ... 797 | 6.6. | Benzidinesulfonic Acids and |
| 2.3. | Isolation ............... 798 | | Benzidinesulfones ......... 810 |
| 3. | Storage and Transportation .. 799 | 7. | Economic Aspects ......... 812 |
| 4. | Occupational Health ....... 799 | 8. | Toxicology............... 812 |
| 4.1. | General Safety Precautions... 799 | 8.1. | Benzidine ............... 812 |
| 4.2. | Special Safety Regulations ... 800 | 8.2. | 3,3'-Dichlorobenzidine...... 813 |
| 5. | Environmental Protection ... 802 | 8.3. | o-Tolidine (3,3'-Dimethylben- |
| 6. | Characteristics of the | | zidine) ................. 814 |
| | Diphenyl Bases ........... 803 | 8.4. | o-Dianisidine (3,3'-Dimethoxy- |
| 6.1. | Benzidines............... 803 | | benzidine) .............. 815 |
| 6.2. | Tolidines................ 805 | 9. | References............... 815 |

# 1. Introduction

Benzidine [92-87-5], 4,4'-biphenyl-diyldiamine, and its substitution products represent the group called the diphenyl bases. They are used mainly as intermediates in the production of azo dyes and azo pigments. Symmetrically or asymmetrically coupled products can be produced by simultaneous or successive diazotization (coupling), respectively.

$$H_2N-\underset{5'\ 6'}{\overset{3'\ 2'}{\underset{4'}{\bigcirc}}}-\underset{6\ 5}{\overset{2\ 3}{\underset{1\ 4}{\bigcirc}}}-NH_2$$

The diphenyl bases Benzidine have recently aroused interest as cross-linking agents, e.g., in polyurethane plastics [1], in which they can noticeably increase the temperature stability [2]. The diphenyl radical has a chain-stiffening effect in polyamides [3]. The ability of the diphenyl bases to react with numerous cations, anions, and organic substances, such as oxidizing agents and blood, is utilized for analytical and diagnostic purposes.

The economic importance of the diphenyl bases is apparent from the fact that world production in 1983 was about 15 kt. The major producing countries were the United States, Japan, and the Federal Republic of Germany.

The most important of the diphenyl bases are $o$-tolidine, 3,3′-dichlorobenzidine, and $o$-dianisidine. The physical and chemical properties of benzidine and its derivatives are treated in detail in Chapter 6. The following sections describe the general methods for the production and handling of benzidine and benzidine derivatives.

## 2. Production

Benzidine and the other diphenyl bases are produced in three separate processing stages:

1) Reduction of nitro groups to form hydrazo compounds
2) Benzidine rearrangement
3) Isolation of the bases

## 2.1. Reduction

The alkaline reduction of nitrobenzene to hydrazobenzene can be represented by the simple equation

$$2\ C_6H_5NO_2 + 10\ H \longrightarrow C_6H_5NH–NHC_6H_5 + 4\ H_2O$$

It passes, according to the Haber process, through several stages, including the intermediate condensation of a nitroso with a hydroxylamine compound to form the azoxy compound.

$$C_6H_5NO_2 \longrightarrow C_6H_5NO \longrightarrow C_6H_5NHOH \longrightarrow C_6H_5NH_2$$

$$C_6H_5N{=}NC_6H_5\!\!\downarrow\!\!O \quad (-H_2O,\ -2\ H_2O)$$

$$C_6H_5N{=}NC_6H_5 \longrightarrow C_6H_5NH{-}NHC_6H_5$$

This method always forms symmetrical hydrazo compounds.

The main byproduct is the monocyclic primary amine (e.g., aniline) corresponding to the original nitro compound. This is formed not only by further reduction of the phenylhydroxyamine, but also by disproportionation of the unstable hydrazo compound.

Various reduction methods are used in industry.

**Reduction with Zinc Dust.** The reduction of aromatic nitro compounds with zinc dust in alkaline medium takes place according to the equation:

$$2\ C_6H_5NO_2 + 5\ Zn + 10\ NaOH \longrightarrow C_6H_5NH\text{–}NHC_5H_5 + 5\ Na_2ZnO_2 + 4\ H_2O$$

In industry, the nitro compound is reduced in batches of 500 kg or more. The nitro compound is dissolved in a high-boiling solvent, e.g., $o$-dichlorobenzene or solvent naphtha, and alcohols (methanol, ethanol) are often added to keep the resultant hydrazo compound in solution. Zinc dust is then suspended in this solution and sodium hydroxide solution is emulsified with it. A quantity of zinc dust as much as 50% above the theoretical requirement is usually needed. The necessary alkali is introduced in the form of a concentrated (40–50%) sodium hydroxide solution. Only a small percentage (ca. 2–10%) of the theoretical quantity of sodium hydroxide is required because the sodium zincate quickly regenerates alkali by hydrolysis.

$$Na_2ZnO_2 + H_2O \rightleftharpoons 2\ NaOH + ZnO$$

The reduction is highly exothermic. It can be regulated by controlling the addition of one or more of reactants. The optimum reduction temperature is between 80 and 120 °C, depending on the aromatic compound. Higher temperatures can be used with the unsubstituted nitrobenzene without causing overreduction than are possible with $o$-alkyl and $o$-alkoxy nitrobenzenes. In many cases the addition of small amounts of zinc dust and water is necessary to complete reduction and convert zinc oxide into zinc hydroxide.

The reduction is terminated when the red color of the azo stage has disappeared. The hydrazobenzene solution is then separated, either directly or after dilution with hot water, from the zinc hydroxide, which may still contain unconverted zinc dust. The zinc hydroxide is extracted with fresh solvent to increase the yield. The hydrazobenzene solutions are usually subjected to a subsequent benzidine rearrangement without intermediate isolation of the hydrazo compound. The zinc hydroxide is freed of solvent in a vacuum or with steam and either further processed into zinc salts or dried and returned to the zinc works [4], [5]. The reduction process using zinc and alkali is technically the most important and can be used for all hydrazo compounds.

However, because of the high price of zinc dust there have been numerous attempts to replace it with cheaper reducing agents.

**Reduction with Iron.** Reduction with iron and sodium hydroxide solution, only of minor importance in preparative chemistry, has been investigated extensively for commercial production ever since 1900.

$$2\ C_6H_5NO_2 + 4\ Fe + 6\ H_2O \longrightarrow C_6H_5NH\text{–}NHC_6H_5 + 2\ Fe(OH)_2 + 2\ Fe(OH)_3$$

Finely ground cast iron turnings, iron powder obtained by decomposition of iron carbonyl, carbide-containing iron, or ferrosilicon are used. The iron turnings are first slightly corroded with 60% sodium hydroxide solution. To prevent overreduction the concentration of sodium hydroxide is maintained at a high level and sometimes even increased during reduction. Reduction is carried out using nitrobenzene, iron, and sodium hydroxide in the molar ratio 1:2:4 at a temperature of 100–120 °C.

The process can be used for the reduction of nitrobenzene to hydrazobenzene but less often for its substitution products, especially because it is difficult to separate the hydrazo compound from the ferruginous mud. A yield of 88–92% has been reported for the reduction of nitrobenzene to hydrazobenzene using ferrosilicon (15% Si) [5].

**Reduction with Sodium Amalgam.** Sodium amalgam at 70–100 °C in an upright, cylindrical nickel reactor is brought into contact with nitrobenzene emulsified in water or sodium hydroxide. The process is carried out in batches to produce azobenzene. The reduction conditions can be varied by adding solvent (solvent naphtha), changing the concentration of the sodium hydroxide solution or the amalgam, or by adding decomposition catalysts (graphite, activated charcoal).

Nevertheless, in the further reduction from the azo to the hydrazo stage, some overreduction to the monocyclic amine cannot be avoided. The product formed contains up to 90% azobenzene in addition to hydrazobenzene, aniline, and nitrobenzene. It is therefore isolated and further reduced to the hydrazo stage with zinc dust and sodium hydroxide solution. The amalgam reduction is used industrially only with unsubstituted nitrobenzene.

**Electrolytic Reduction.** The electrochemical reduction of nitro to hydrazo compounds is also carried out on a commercial scale.

$$2\ C_6H_5NO_2 + 10\ H^+ + 10\ e^- \longrightarrow C_6H_5NH-NHC_6H_5 + 4\ H_2O$$

Electrolytic cells, with cathodes, diaphragms, and anodes positioned either side by side or above each other, are used.

The satisfactory functioning of electrolytic reduction depends largely on the condition of the surface of the cathode. Sheet iron cathodes with a lead sponge are used. Overreduction is less likely to occur with a lead sponge than with shiny lead surfaces. Even lead cathodes with spongy zinc have been used, but the spongy zinc was partially used up by chemical reduction. Nickel cathodes have also been described [5], [6].

The catholyte consists of a vigorously stirred emulsion of the nitro compound in 8–10% sodium hydroxide solution. Ethanol and toluene can be used as diluents. Depending on the geometry of the cell, solvents with high specific gravities (chlorinated benzene, chlorinated toluene, or chlorinated naphthalene) are used to keep the cathode liquid away from the anode compartment above it. Cement or special ceramic compounds are used to make the diaphragms. The anolyte, consisting of 10% sodium hydroxide solution, contains the lead or iron anode, which has a $PbO_2$ coating or, for preparative purposes, may also be nickel-plated.

Electrochemical reduction is carried out at temperatures of 65–90 °C, current densities of about 20 A/dm$^2$ and voltages of 4–10 V. Electrochemical reduction is technically satisfactory and may also be cheaper than zinc dust reduction if it is interrupted when only about 50 % of the starting material has reached the hydrazo stage. After rearrangement and separation of the diphenyl base salts, the azo components and the solvent are recovered for the next batch. Nitrobenzene, 2-chloronitrobenzene, 2,5-dichloronitrobenzene, 2-nitrotoluene, and 2-nitroanisol can be electrochemically reduced to hydrazo compounds in yields of 70–90 % [5], [6].

**Catalytic Reduction.** Nitrobenzene can be reduced to hydrazobenzene with hydrogen in the presence of a palladium–carbon catalyst. Industrially, reduction is carried out in dilute alcohol (ethyl or isopropyl alcohol) in the presence of a base, such as 50 % sodium hydroxide solution. The reaction takes place in an autoclave at 80–85 °C and with a slightly elevated pressure. Overreduction is suppressed by the high concentration of alkali. The yield can also be influenced by buffering the reaction medium, e.g., by the addition of acetates, borates, or alkanolamines. This reduction method also applies to 2-nitrotoluene and 2-nitroanisol.

For preparative purposes, Raney nickel can be used at elevated temperatures and with mixed catalysts. Instead of hydrogen, hydrazine hydrate or sodium borohydride can be used with palladium–calcium carbonate, Raney nickel, ruthenium–carbon, or copper ions as the catalyst [5].

**Other Reduction Methods.** In preparative chemistry, numerous other reducing agents are used to produce hydrazo compounds. Lithium, sodium, potassium, magnesium, and aluminum are used for reduction, either as metals or as alcoholates. Silicon, lithium, and aluminum are used as amalgams. Of the nonmetal reducing agents, special mention should be made of hydrogen sulfide and its salts (sulfide reduction). Methanol, formaldehyde, and glucose, sometimes combined with quinonoid catalysts, are organic reducing agents.

## 2.2. Benzidine Rearrangement

Aromatic hydrazo compounds, formed by the alkaline reduction of aromatic nitro compounds, are easily rearranged by mineral acids into diaminodiphenyl compounds (**1**)–(**3**) and aminodiphenylamine compounds (**4**), (**5**). This is the second stage in the production of diphenyl bases.

Ph—NH—NH—Ph → H$_2$N—C$_6$H$_4$—C$_6$H$_4$—NH$_2$

Hydrazobenzene     Benzidine (**1**)

Traces of *o*-benzidine (2,2′-diaminodiphenyl) (**2**) are formed as a byproduct by ortho rearrangement, and diphenyline (2,4′-diaminodiphenyl) (**3**) is also formed in appreciable quantities by ortho-para rearrangement.

Partial rearrangements yield *o*-semidine (2- aminodiphenylamine) (**4**) and *p*-semidine (4-aminodiphenylamine) (**5**).

The type and quantity of the rearrangement products are largely dependent on the chemical constitution of the reactant and can only be influenced to a limited extent by the reaction conditions. The byproducts, mainly diphenyline (**3**), are formed in quantities of up to 15 % and have no commercial value.

The benzidine rearrangement is a true intramolecular reaction because no mixed benzidines are formed from mixtures of different hydrazo compounds and because only the corresponding unsymmetrical benzidine is obtained from unsymmetrically substituted hydrazobenzenes [7].

The industrial benzidine rearrangement usually starts with the hot solution that is obtained from the reduction of the nitro compound to the hydrazo compound. Intermediate isolation of the hydrazo compound is not necessary, but it is advisable in some cases, such as amalgam reductions. Because hydrazo compounds disproportionate to form azo compounds and the corresponding monocyclic amines at higher temperatures, some reducing agent, such as sodium hydrosulfite (sodium dithionite) or zinc dust, can be added to reduce the azo content of the hydrazo compound.

$$2\ C_6H_5NH-NHC_6H_5 \longrightarrow C_6H_5N=NC_6H_5 + 2\ C_6H_5NH_2$$

The hydrazo solution is mixed with a mineral acid in an acid-resistant mixer. Depending on the substituents, suitable mineral acids are 10 – 30 % hydrochloric acid, 20 – 80 % sulfuric acid, or a mixture of these two. The optimum rearrangement temperature varies from 0 – 5 °C for dianisidine to 100 °C for the less heat-sensitive diphenyl bases. Temperatures higher than necessary diminish the yield, which should be 70 – 95 %.

## 2.3. Isolation

The most important stage of the benzidine rearrangement is the formation of a hydrochloric or sulfuric acid salt of the diphenyl base. This is either isolated directly, e.g., by salting out with sodium chloride or sodium sulfate, or first converted into the free base. For this a dilute alkali, such as sodium hydroxide solution or ammonia solution, should be used. The undesirable byproducts, especially the monocyclic amine and diphenyline, can be separated out relatively easily because of the greater solubility of these bases and their salts. Finally, only the azo compound remains in the inert solvent after acid extraction. It is returned to the reduction process along with the solvent.

## 3. Storage and Transportation

The diphenyl bases and their salts are usually introduced onto the market in the form of moist solids to prevent dust from forming when they are handled. They are packed in corrugated-iron drums with plastic liners; in addition, plastic-lined fiber drums, which can be burnt after they have been emptied and rinsed out, are often used as disposable packs. Wooden drums are avoided. The packs should be kept closed if possible so that dust does not form as a result of drying out of the diphenyl bases.

Benzidine is often isolated by collecting it on suction filters as a slurry and then transporting it in this form. The earlier procedure of drying, distilling, and grinding diphenyl bases is no longer used.

If the diphenyl bases are to be further processed by the producer in an integrated system, they are transported through pipe systems as aqueous suspensions.

## 4. Occupational Health

### 4.1. General Safety Precautions

Benzidine has properties which are dangerous to health and the health effects of the substituted diphenyl bases are unknown. The diphenyl bases should therefore be produced and processed in special, separated areas. Walls and floors must not be capable of absorbing chemicals. The floors should have adequate drainage and wood should not be used either as a building material or for equipment. Good ventilation must be ensured and provisions must be made for air circulation. Ventilation of the workrooms must not be restricted. The exhaust air must be effectively cleansed using filters or exhaust-gas scrubbers. Wherever possible, equipment should operate as a closed system. Filter presses and open suction filters must be avoided. A vacuum in the system is a precaution against leakage or the uncontrolled escape of vapors. When evacuated machines are opened, attention must be paid to the formation of sublimates. Tasks that involve an increased risk of human contact include sampling during manufacturing and processing, isolation, packing, and filling. If possible, the work is carried out with automated machines or in separately ventilated areas. Before undertaking repairs, special care must be taken to ensure that the equipment has been decontaminated.

Scalding out should only be allowed if the vapors can be carefully condensed and do not enter the atmosphere. If necessary and if permitted by the construction material, purification can be carried out by the addition of chemicals. Reaction with hypochlorite or even diazotization is suitable for the chemical destruction of amines. When using dilute hypochlorite solution, care must be taken to ensure that the solution is always kept considerably basic by the appropriate addition of alkali. This avoids formation of explosive nitrogen trichloride. Access to contaminated containers should only be permitted if complete protective clothing is worn and breathing apparatus is used. Diphenyl bases penetrate the surface of protective articles made of rubber or plastic. The workplaces must be monitored for emissions of benzidine and its derivatives [8].

## 4.2. Special Safety Regulations

In the United States and most other industrialized countries, rules and regulations have been issued for benzidine, which is regarded as carcinogenic in humans, and for 3,3′-dichlorobenzidine, which has proved to be carcinogenic in animals [9]. The aim is to protect the worker from contamination. Mixtures containing 0.1 wt% or more of benzidine are subjected to detailed handling regulations. The production of benzidine and its salts is prohibited in Japan and has been stopped in the Soviet Union.

Nowadays the most commonly produced and used diphenyl base is 3,3′-dichlorobenzidine (as the free base or one of its salts). The appropriate U.S. regulations are therefore presented. They apply to substances that contain at least 1 wt% 3,3′-dichlorobenzidine or its salts [10].

1) Areas in which these substances are produced, processed, used, repackaged, released, handled, or stored are designated *controlled areas.*
2) The number of people entering and leaving these areas must be restricted and monitored. In these areas a distinction is made between
    a) isolated systems, e.g., glove boxes
    b) closed systems, e.g., sealed containers or piping systems
    Special permission must be obtained for access to closed systems; employees shall be required to wash themselves upon each exit from the regulated areas.
    c) open vessel system operations are prohibited.
3) Upon completion of their work in isolated or closed systems, employees must observe certain minimum regulations in terms of personal hygiene.
4) When opening a system employees shall be required to wear clean, full-body protective clothing, shoe covers, and gloves. Employees engaged in 3,3′-dichlorobenzidine (or its salts) handling operations shall be required to use a half-face filter-type respirator for dusts, mists, and fumes.
5) For the cleanup of leaks or spills and maintenance or repair operations on contaminated systems, impervious garments, gloves, boots, and a hood with a continuous supply of air should be worn.
6) Prior to each exit from a regulated area, employees shall be required to remove and leave behind protective clothing and to place used clothing and equipment in impervious containers for purposes of decontamination or disposal.

7) Employees shall be required to wash on each exit from the regulated area and to shower after the last exit of the day.
8) A daily roster of employees entering regulated areas shall be established and maintained. Before taking up employment they must be informed of the carcinogenic hazard of 3,3'-dichlorobenzidine, of the possibilities of contact with the product in their work, and of the medical surveillance program. They are subjected to an annual medical examination.
9) Signs bearing the following inscription should be posted at the entry points of regulated areas:
10) CANCER-SUSPECT AGENT
    AUTHORIZED PERSONNEL ONLY
11) More detailed warnings apply to contaminated areas. Additional signs should inform employees of the procedures that must be followed upon entering or leaving a regulated area.
12) The storage, consumption, and use of foodstuffs, drinks, tobacco, cosmetics, and the like are prohibited in regulated areas.
13) Decontamination procedures shall be established. Dry sweeping and dry mopping are prohibited.
14) The contents of the containers must be identified by suitable labels. The words *Cancer-Suspect Agent* must be added, the minimum size print being stipulated.
15) Any incidents must be reported to the relevant authorities.
16) Deposits on machinery or parts of the building are detected by a swipe test. Sodium hypochlorite solution must be used for decontamination.

In Great Britain the use (for research purposes) of *Prohibited Substances*, such as benzidine and its salts, is only possible if special permission is obtained for work in completely closed systems or if the benzidine, in the form of a hydrochloride, is always kept moist with at least one part water to two parts benzidine hydrochloride.

Dichlorobenzidine, *o*-tolidine, *o*-dianisidine, and their salts are designated *Controlled substances*, and employees should be protected from contamination by these substances. Workers are medically monitored and records kept [11].

In the Federal Republic of Germany the important diphenyl bases are included among the regulated dangerous working substances.

1) Benzidine and its salts are regarded as carcinogenic hazards in concentrations above 0.01% by weight in the working material.
2) 3,3'-Dichlorobenzidine is regarded as a carcinogenic hazard in concentrations above 0.1% by weight in the working material.
3) *o*-Tolidine is considered harmful to health.
4) *o*-Dianisidine and its salts are considered toxic.

When working with these substances, all the necessary measures relating to safety, industrial medicine, and occupational hygiene should be taken to protect employees from the effects of these substances and to keep their health under constant medical check. The air at the workplace should be examined to monitor the benzidine and 3,3'-dichlorobenzidine contents. If benzidine or 3,3'-dichlorobenzidine is marketed, the package must bear the words *May cause cancer* [12].

Equipment in which benzidine or its salts may be present must be operated in such a way that breakdowns are prevented and the harmful effects are kept to a minimum. All the necessary precautions are summarized in [13].

In accordance with the technical regulations for dangerous working substances, the presence of an effect must always be assumed when handling benzidine. Here "effect" means the risk of damage to workers' health. If the concentration of 3,3′-dichlorobenzidine in the air is below 0.1 mg/m$^3$ at the workplace, it can be assumed to have no effect on the worker [14].

In Belgium [15], Japan [16], and numerous other countries there are similar regulations for handling benzidine and its derivatives [17]. Before starting a job in which diphenyl bases can be formed or are used, it is high advisable to obtain exact details of the local regulations.

# 5. Environmental Protection

The acute fish toxicity LC$_{50}$ is about 5 – 50 ppm, depending on the type of fish and the diphenyl base. In terms of acute toxicity, diphenyl bases can be regarded as only slightly toxic to mammals. The oral LD$_{50}$ for rats is 3820 mg/kg for 3,3′-dichlorobenzidine, 404 mg/kg for $o$-tolidine, and 1920 mg/kg for $o$-dianisidine.

Benzidine achieves only a slight, and 3,3′- dichlorobenzidine a moderate bioaccumulation in bluegill sunfish [18]. Thus benzidine is unlikely reach a significant level in the food chain [19].

Because of the very low vapor pressures of the diphenyl bases, emissions into the ambient air from production or processing works are only possible in the form of dust. Such emissions can certainly be prevented by filtering or by using wet separators. Diphenyl base concentrations in the ambient air are monitored using TLC or HPLC [20].

In aqueous solution 3,3′-dichlorobenzidine base is very quickly degraded by photolysis, with monochlorobenzidine and benzidine being formed as intermediates [21], [22]. 3,3′-Dichlorobenzidine, $o$-tolidine, and $o$-dianisidine can be slowly biodegraded. More than 90 % of these two compounds can be eliminated in biological waste water treatment plants. Elimination takes place mainly by adsorption on sediments followed by slow biodegradation. Only a portion can be extracted again. This part declines as the adsorption process continues, because the 3,3′-dichlorobenzidine reacts with the sediments [21].

3,3′-Dichlorobenzidine cannot be removed quantitatively from waste water by hypochlorite solution because higher chlorinated benzidines are formed. Tetrazotization and conversion into 3,3′-dichloro-4,4′-dihydroxybiphenyl are more suitable procedures; they reduce the 3,3′-dichlorobenzidine content to less than 0.1 mg/kg [23]. Other appropriate waste water purification processes are oxidation with peroxide, adsorption on suspended adsorption agents, precipitation with iron(III) hydroxide, and combinations of these methods.

For decontamination of workrooms, buildings, and machines, a hot solution of 5 wt % tetrapotassium pyrophosphate and 10 wt % sodium ethyl hexyl sulfate is sprayed on with a hydraulic jet cleaner in the ratio 1:10 with water under pressure, so that 90–99 % of the adherent 3,3′-dichlorobenzidine is removed [23].

Wastes, such as production residues, contaminated disposable protective equipment, and empty packaging, which may contain diphenyl bases, are best incinerated. Suitable incinerators must have a secondary incineration compartment in which a dwell time of at least 0.3 s and a minimum temperature of 900 °C are maintained [24].

# 6. Characteristics of the Diphenyl Bases

## 6.1. Benzidines

**Benzidine** [*92-87-5*], 4,4′-biphenyl-diyldiamine, $C_{12}H_{12}N_2$, $M_r$ 184.24, *mp* (coarse rods) 128 °C, crystallizes out of the melt usually in a metastable modification with a melting point of 122–125 °C, *bp* 400–401 °C at 98.7 kPa.

$$H_2N-\phantom{x}\!\!\!\!\bigcirc\!\!\!\!-\!\!\!\!\bigcirc\!\!\!\!-NH_2$$

Benzidine crystallizes out of water above 80 °C in the anhydrous form; below 60 °C, it retains one molecule of water of crystallization; the monohydrate melts at 104–105 °C. Pure benzidine is colorless. One part by weight of benzidine dissolves in 2447 parts of water at 12 °C, in 106.5 parts of water at 100 °C, and in 45 parts of ether or 13 parts of absolute ethyl alcohol at 20 °C. Benzidine is diacidic; at 30 °C, $K_1 = 9.3 \times 10^{-10}$ and $K_2 = 5.6 \times 10^{-11}$. The heat of neutralization is 106.5 kJ/mol.

Benzidine slowly becomes discolored on exposure to air. It is chemically resistant to water. Blue, green, or red colorations and precipitates, many of which are based on quinonoid structures, form depending on the nature of the oxidizing agent and the reaction conditions used. These color reactions can be used to detect numerous oxidizing agents.

When chlorine is passed into a suspension of benzidine hydrochloride in concentrated hydrochloric acid, 3,3′,5,5′-tetrachlorobenzidine [*41687-08-5*] is formed. Bromine forms the corresponding tetrabromobenzidine [*62477-23-0*]. 2-Nitrobenzidine [*2243-78-9*], 2,2′-dinitrobenzidine, and a litte 2,3′-dinitrobenzene are formed from benzidine sulfate in concentrated sulfuric acid upon the addition of potassium nitrate [25]. Sulfonic acids or sulfones of benzidine are formed under sulfonating conditions. The N-acetylation products, which also occur as metabolites in animal digestion, are formed with acetic anhydride. They are *N*-acetyl-benzidine [*3366-61-8*], $C_{14}H_{14}N_2O$, and *N,N'*-diacetylbenzidine [*613-35-4*], $C_{16}H_{16}N_2O_2$.

The most important commercial reaction is the diazotization of the two amino groups. Reaction with nitrous acid converts benzidine into the tetrazonium compound [4], in which the first diazonium group is coupled very vigorously whereas the second reacts more slowly. As a result it is possible to produce asymmetrical diazo dyes. Gradual diazotization is also possible.

In the presence of acids, benzidine forms salts that are difficult to dissolve. Among these are benzidine monohydrochloride [*14414-68-7*], $C_{12}H_{12}N_2 \cdot HCl$, needles, sparingly soluble in water, readily soluble in dilute hydrochloric acid; benzidine dihydrochloride [*531-85-1*], $C_{12}H_{12}N_2 \cdot 2\ HCl$, flakes, readily soluble in water and in alcohol. The dihydrochloride hydrolyzes to form the monohydrochloride and hydrochloric acid.

Benzidine sulfate [*531-86-2*], $C_{12}H_{12}N_2 \cdot H_2SO_4$, forms microscopic flakes and is virtually insoluble in water and alcohol. Because of its poor solubility, benzidine sulfate can be used for the quantitative determination of sulfuric acid. Benzidine disulfate, $C_{12}H_{12}N_2 \cdot 2\ H_2SO_4$, is formed from benzidine sulfate and sulfuric acid.

Benzidine has been produced from nitrobenzene on an industrial scale since about 1880. Commercial production methods include alkaline iron reduction [5], amalgam reduction, and electrochemical reduction. The resultant hydrazobenzene is rearranged with hydrochloric acid or sulfuric acid during cooling. The base is then isolated in the form of benzidine hydrochloride or benzidine sulfate. The conversion of these salts to the free base is avoided as much as possible because of the chronic toxicity of benzidine.

An aqueous benzidine solution gives a blue coloration in the presence of aqueous potassium ferricyanide. The sensitivity is about 10 mg/kg.

In very dilute colorless bromine water, benzidine yields a blue coloration. The solution turns green when more reagent is added and finally becomes colorless again upon the precipitation of reddish flakes. Quantitative determination is by titration with 0.1 N sodium nitrite solution in hydrochloric acid, using potassium iodide and starch paper as the indicator. Quantitative determination of benzidine in dyestuffs involves petroleum ether extraction, column chromatographic separation, and colorimetric determination [26]. Benzidine can be detected in urine or air by extraction with *m*-cresol methyl ether and reaction with potassium 1,2-naphthoquinone-4-sulfonate [*5908-27-0*]; the detection limit is 0.33 mg/kg [27], [28]. A simple and rapid method for the detection of benzidine in the urine is based on its the color-forming reaction with cyanogen bromide; the detection limit is 0.05 mg/kg [29]. Benzidine in the urine can be separated from its metabolites by means of paper chromatography [30].

Benzidine is used as the reagent base for the production of a large number of dyes, particularly azo dyes for wool, cotton, and leather. However, because benzidine has been found to be carcinogenic in humans, there has been a marked decline in the use of the benzidine dyes.

Benzidine is used for the quantitative determination of sulfuric acid and for the detection and determination of numerous anions and metal ions. The reaction of benzidine with pyridine in the presence of elemental chlorine is suitable for detecting traces of free chlorine or pyridine in drinking water [29]. The green to blue coloration which occurs when benzidine reacts with hydrogen peroxide in the presence of peroxidases can be used to detect blood [31]. Benzidine still plays a role in many chemical syntheses.

**3,3',4,4'-Tetraaminodiphenyl** [*91-95-2*], diphenyl tetramine, 3,3',4,4'-biphenyl-tetrayltetramine, $C_{12}H_{14}N_4$, $M_r$ 214.27, *mp* 178 °C, forms fused blocks and flakes which turn a dark color upon exposure to air; it is soluble in hot water.

The hydrochloride [*7411-49-6*], $C_{12}H_{14}N_4 \cdot 4\,HCl \cdot 2\,H_2O$, forms needles and is readily soluble in water. The sulfate, $C_{12}H_{14}N_4 \cdot H_2SO_4$, forms small needles, is sparingly soluble in cold water, alcohol, and ether, and is soluble in hot water.

3,3',4,4'-Tetraaminodiphenyl can be produced from benzidine by acetylation followed by nitration, saponification, and reduction with iron powder and hydrochloric acid [32].

The reduction of 3,3'-dinitrobenzidine can be carried out with hydrogen using a catalyst [33], with sodium dithionite in aqueous methanol, or (for preparative purposes) with tin(II) chloride and hydrochloric acid. The reaction of 3,3'-dichlorobenzidine with ammonia at high temperature and under pressure can also be used [34]. The crude product can be purified by recrystallization from aqueous sulfuric acid [35].

3,3',4,4'-Tetraaminodiphenyl is suitable for the production of fibers based on polybenzimidazole. These fibers are resistant to high temperature and to chemicals. Fabrics made from them do not burn, do not melt in contact with a flame, and do not become hard when carbonized. They are suitable for the production of protective clothing and flue–gas filters and as substitutes for asbestos [36].

## 6.2. Tolidines

*o*-**Tolidine** [*119-93-7*], 3,3'-dimethyl-4,4'-biphenyl-diyldiamine, $C_{14}H_{16}N_2$, $M_r$ 212.30, *mp* 130 °C, colorless (in technical grades also brownish flakes), sparingly soluble in water, readily soluble in alcohol and ether. *o*-Tolidine is diacidic; at 30 °C, $K_1 = 6.2 \times 10^{-10}$, and $K_2 = 2.9 \times 10^{-10}$.

*o*-Tolidine barely becomes discolored upon exposure to air. It is chemically resistant to water. Numerous color reactions occur as a result of oxidation, e.g., an intense blue coloration forms on exposure to bromine or chlorine vapors. As with benzidine, these color reactions form the basis of a number of methods of detection and determination, such as those used for chlorine in air or water, for nitrates, and for hydrogen peroxide.

Mono- and disulfonic acids and *o*-tolidine-6,6-sulfone can be formed from *o*-tolidine sulfate.

*o*-Tolidine forms acid salts, such as *o*-tolidine hydrochloride, $C_{14}H_{16}N_2 \cdot HCl$, flakes. At 12 °C, one part dissolves in 112.4 parts of water. *o*-Tolidine dihydrochloride [*612-82-8*], $C_{14}H_{16}N_2 \cdot 2\,HCl$, forms crystals. At 12 °C, one part dissolves in 17.3 parts of water. *o*-Tolidine sulfate, $C_{14}H_{16}N_2 \cdot H_2SO_4$, forms needles or flakes. Water at 20 °C dissolves 0.12 g per 100 mL and more in the presence of hydrochloric acid. *o*-Tolidine disulfate [*64969-36-4*], $C_{14}H_{16}N_2 \cdot 2\,H_2SO_4$ forms crystals and is sparingly soluble in water.

*o*-Nitrotoluene undergoes alkaline reduction with zinc dust [4], electrolytic reduction [37], and catalytic reduction to form 2,2′-dimethylhydrazobenzene. This is rearranged using dilute hydrochloric acid or 20% sulfuric acid at 5–50 °C. The free base or the dihydrochloride can be isolated.

The color reactions previously described can also be used for the detection of *o*-tolidine. Quantitatively, it is determined by titration with nitrite, as was described for benzidine.

*o*-Tolidine can be detected in the urine by extraction with 2,4-dichloroisopropylbenzene and conversion with potassium 1,2-naphthoquinone-4-sulfonate [38]. A rapid and simple procedure is based on the color reaction with cyanogen bromide; the detection limit in both cases is 0.05 mg/kg [29].

*o*-Tolidine is marketed as the free base and as the dihydrochloride. Both are supplied as moist products with a water content of 10–30%.

*o*-Tolidine is a starting material in the production of a large number of azo dyes and pigments. *o*-Tolidine is used in the determination of oxygen and chlorine present in water, and for the colorimetric determination of cations of gold, cerium, and manganese.

An important derivative of *o*-tolidine is its diacetoacetyl compound, 4,4′-bisacetoacetylamino-3,3′-dimethyl diphenyl, which is marketed under the trade name Naphtol AS-G. It is a coupling agent which is frequently used and in combination with chloroanilines gives yellow shades. *o*-Tolidinediisocyanate is employed as a crosslinking agent to make polymers.

**m-Tolidine** [*84-67-3*], 2,2′-dimethyl-4,4′-bi-phenyl-diyl-diamine, $C_{14}H_{16}N_2$, $M_r$ 212.30, mp 109 °C, forms prisms from water, soluble in cold water, alcohol, and ether.

*m*-Tolidine gives no color in the presence of iron(III) chloride. It is tetrazotized with nitrite and acid and couples to form dyes.

*m*-Tolidine sulfate, $C_{14}H_{16}N_2 \cdot H_2SO_4$, is insoluble in cold alcohol, sparingly soluble in cold water, and soluble in hot water.

*m*-Tolidine is made by the reduction of *m*-nitrotoluene with zinc dust and sodium hydroxide solution [4]. The subsequent rearrangement is carried out with hydrochloric acid or sulfuric acid and gives rise to *m*-tolidine dihydrochloride or *m*-tolidine sulfate, which are isolated as salts. The free base is difficult to isolate and is therefore not a commercial product.

*m*-Tolidine was used in the past for the production of azo dyes.

## 6.3. Halogenated Benzidines

**2,2'-Dichlorobenzidine** [*84-68-4*], 2,2'-dichloro-4,4'-biphenyl-diyldiamine, $C_{12}H_{10}Cl_2N_2$, $M_r$ 253.13, *mp* 166.8 °C, forms prisms from alcohol. It is almost insoluble in water but readily soluble in alcohol.

2,2'-Dichlorobenzidine dihydrochloride, $C_{12}H_{10}Cl_2N_2 \cdot 2$ HCl, crystallizes in the form of flakes from water and is relatively soluble in water.

2,2'-Dichlorobenzidine is made from *m*-nitrochlorobenzene, preferably using zinc dust and sodium hydroxide solution. It is subsequently rearranged and isolated in the form of a dihydrochloride.

**3,3'-Dichlorobenzidine** [*91-94-1*], 3,3'-dichloro-4,4'-biphenyl-diyldiamine, $C_{12}H_{10}Cl_2N_2$, $M_r$ 253.13, *mp* 132–133 °C, forms needles from alcohol or benzene. The commercial product is light brown to violet in color. It is virtually insoluble in water but readily soluble in alcohol, benzene, and glacial acetic acid.

3,3'-Dichlorobenzidine is barely discolored on exposure to air and is chemically resistant to water. Oxidizing agents, such as bromine water, potassium dichromate, and iron(III) chloride, cause the formation of a green color. With gold, the green color is still detectable at a dilution of 1 : 5 000 000.

*N*-Acetyl-3,3'-dichlorobenzidine and *N,N'*diacetyl-3,3'-dichlorobenzidine form when 3,3'- dichlorobenzidine is treated with acetic anhydride in dilute alcohol. The tetrazotizing behavior is similar to that of benzidine and *o*-tolidine.

3,3'-Dichlorobenzidine dihydrochloride [*612-83-9*], $C_{12}H_{10}Cl_2N_2 \cdot 2$ HCl, is sparingly soluble in water and readily soluble in alcohol. It is marketed in the form of a colorless to pale grey crystalline powder.

3,3'-Dichlorobenzidine is made from *o*-nitrochlorobenzene by reduction with zinc dust and sodium hydroxide solution and subsequent rearrangement with dilute hydrochloric acid or sulfuric acid. It is assayed by titration with sodium nitrite solution in dilute hydrochloric acid and detected in human urine by colorimetry using chloramine. The detection limit is 0.1 mg/kg [29].

Like *o*-tolidine, 3,3′-dichlorobenzidine is marketed as a free base and as the dihydrochloride. Both are supplied as moist products with water contents of 5 – 30%. The so-called *urethane quality* is taken to mean an anhydrous base, free of hydrochloric acid.

3,3′-Dichlorobenzidine was introduced about 1932 and is now the most important diphenyl base. It is used as the starting material for pigments with yellow and red shades. These are used for coloring printing inks, paints, plastics, and rubbers. The important diarylide yellow pigments, which are incorrectly known as benzidine yellows, are formed by the combination of 3,3′-dichlorobenzidine with acetic acid arylides. 3,3′-Dichlorobenzidine is also used in the production of polyurethane rubbers.

**2,2′,5,5′-Tetrachlorobenzidine** [*15721-02-5*], 2,2′,5,5′-tetrachloro-4,4′-biphenyl-diyldiamine, $C_{12}H_8Cl_4N_2$, $M_r$ 322.02, mp 134 – 137 °C.

2,2′,5,5′-Tetrachlorobenzidine is made by the reduction of 2,5-dichloronitrobenzene with zinc dust and sodium hydroxide solution. The subsequent rearrangement is carried out using sulfuric acid or hydrochloric acid. The dihydrochloride melts at 230 °C. Tetrachlorobenzidine is marketed as the free base and is used to produce yellow pigments.

## 6.4. Alkoxybenzidines

*o*-**Dianisidine** [*119-90-4*], 3,3′-dimethoxy-4,4′-biphenyl-diyldiamine, $C_{14}H_{16}N_2O_2$, $M_r$ 244.30, crystallizes dimorphically, rarely in needles, with *mp* 133 °C, often in flakes with *mp* 137 – 138 °C. It forms colorless crystals, but commercial products have a tinge of violet. It is sparingly soluble in water but soluble in alcohol, ether, and benzene. One gram of ethyl acetate dissolves 0.285 g of *o*-dianisidine at 73 °C.

The pure compound is stable upon exposure to air, but commercial products turn violet. *o*-Dianisidine is resistant to water but sensitive to oxidizing agents.

*o*-Dianisidine dihydrochloride [*20325-40-0*], $C_{14}H_{16}N_2O_2 \cdot 2$ HCl, forms prisms with *mp* 272 °C (decomp.). It is readily soluble in hot water and sparingly soluble in cold water and alcohol.

The hydrazo compound (2,2′-dimethoxyhydrazobenzene [787-77-9]) is obtained by the reduction of o-nitroanisole [91-23-6] with zinc dust and sodium hydroxide solution, often in a medium containing ethyl alcohol.

The reduction can also be performed electrochemically or with a palladium catalyst in an aqueous solution of isopropyl alcohol and sodium hydroxide. Rearrangement is carried out at a maximum of 20 °C with 20% sulfuric acid [4], [5].

Like benzidine and o-tolidine, o-dianisidine forms colors with numerous oxidizing agents, e.g., copper, cobalt, and gold ions. o-Dianisidine is quantitatively determined by titration with nitrite using potassium iodide and starch paper as the indicator. o-Dianisidine can be detected in the urine using potassium 1,2-naphthoquinone- 4-sulfonate after extraction [38]. A rapid and easy method is based on the color formed with cyanogen bromide. Its detection limit is 0.05 mg/kg [29].

The free o-dianisidine base and the dihydrochloride are marketed in moist forms with a 10–30% water content.

o-Dianisidine is a starting material for the production of disazo dyes and pigments.

**o-Diphenetidine** [6264-77-3], 3,3′-diethoxy-4,4′-biphenyl-diyldiamine, $C_{16}H_{20}N_2O_2$, $M_r$ 272.35, mp 119 °C, forms needles or flakes. It assumes a grey or violet shade in commercial products. It is insoluble in cold water, slightly soluble in boiling water, and very soluble in alcohol, ether, and chloroform.

The aqueous solution turns red on contact with oxidizing agents, such as bromine, iron(III) chloride, or potassium permanganate. The dihydrochloride melts at 254 °C with decomposition.

o-Diphenetidine is made from 2-nitro-1-ethoxybenzene [610-67-3] by reduction using zinc dust, followed by rearrangement using hydrogen chloride gas in ethanol or dilute sulfuric acid.

**Dichlorodianisidine** [5855-70-9], 2,2′-dichloro-5,5′-dimethoxy-4,4′-biphenyl-diyldiamine, $C_{14}H_{14}Cl_2N_2O_2$, $M_r$ 313.19, mp 175 °C, is a violet crystalline powder.

Dichlorodianisidine is made from 4-chloro-2- nitro-1-methoxybenzene [89-21-4] by zinc dust reduction. Dichlorodianisidine is used in special pigments.

## 6.5. Benzidinecarboxylic Acids

**Benzidine-2,2'-dicarboxylic acid** [*17557-76-5*], 4,4'-diaminobiphenyl-2,2'-dicarboxylic acid, $C_{14}H_{12}N_2O_4$, $M_r$ 272.26, mp 265 °C, crystallizes from water in needles. It is sparingly soluble in water, alcohol, and ether. The dihydrochloride, $C_{14}H_{12}N_2O_4 \cdot 2\,HCl$, crystallizes in columns or needles.

Benzidine-2,2'-dicarboxylic acid is made from 3-nitrobenzoic acid [*121-92-6*] by reduction with zinc dust and sodium hydroxide solution to form the hydrazo compound. This is followed by rearrangement using hydrochloric acid to form the dihydrochloride.

**Benzidine-3,3'-dicarboxylic acid** [*2130-56-5*], 4,4'-diaminobiphenyl-3,3'-dicarboxylic acid, $C_{14}H_{12}N_2O_4$, $M_r$ 272.26, crystallizes into needles which are converted into benzidine by decarboxylation at temperatures above 250 °C. It is slightly soluble in alcohol and ether.

Benzidine-3,3'-dicarboxylic acid is made by reducing 2-nitrobenzoic acid [*552-16-9*] with zinc dust and sodium hydroxide solution. The resultant hydrazobenzene-2,2'-dicarboxylic acid [*612-44-2*] is rearranged by heating with hydrochloric acid. The free benzidine-3,3'-dicarboxylic acid is precipitated from the hydrochloric acid solution by adding ammonia or sodium acetate [4]. Electrolytic reduction followed by rearrangement with dilute sulfuric acid is also possible.

Benzidine-3,3'-dicarboxylic acid is used for the manufacture of azo dyes containing copper.

## 6.6. Benzidinesulfonic Acids and Benzidinesulfones

**Benzidine-2,2'-disulfonic acid** [*117-61-3*], 4,4'-diaminobiphenyl-2,2'-disulfonic acid, $C_{12}H_{12}N_2O_6S_2$, $M_r$ 344.37, crystallizes in monoclinic prisms, chars without melting. One hundred grams of the aqueous solution at 22 °C contains 0.0791 g of the anhydrous acid. The compound is virtually insoluble in alcohol and ether.

$$\text{H}_2\text{N}-\underset{\text{HO}_3\text{S}}{\bigcirc}-\underset{\text{SO}_3\text{H}}{\bigcirc}-\text{NH}_2$$

Benzidine-2,2′-disulfonic acid is made from the potassium or sodium salt of 3-nitrobenzenesulfonic acid by a gradual reduction using zinc dust and sodium hydroxide solution in a dilute aqueous solution. The product is subsequently rearranged using hydrochloric acid or dilute sulfuric acid [4]. Electrochemical or catalytic reductions are also possible [39].

The sulfonic acid must be diazotized indirectly because of its poor solubility. This is achieved by dissolving it in the necessary amount of sodium hydroxide solution, mixing the neutral solution with sodium nitrite and allowing the mixture to trickle into hydrochloric or sulfuric acid in a thin stream [4].

Benzidine-2,2′-disulfonic acid is used in the production of yellow and red azo dyes.

**Benzidine-3,3′-disulfonic acid** [3365-90-0], 4,4′-diaminobiphenyl-3,3′-disulfonic acid, $C_{12}H_{12}N_2O_6S_2$, $M_r$ 344.37, forms tetrahedral flakes, virtually insoluble in alcohol and ether, very slightly soluble in boiling water.

$$\text{H}_2\text{N}-\underset{}{\bigcirc}\overset{\text{SO}_3\text{H}}{|}-\overset{\text{SO}_3\text{H}}{|}\underset{}{\bigcirc}-\text{NH}_2$$

Benzidine-3,3′-disulfonic acid is made by reacting one part of benzidine sulfate with two parts of sulfuric acid monohydrate at 210 °C.

**o-Tolidine-6,6′-disulfonic acid** [83-83-0], 4,4′-diamino-5,5′-dimethylbiphenyl-2,2′-disulfonic acid, $C_{14}H_{16}N_2O_6S_2$, $M_r$ 372.42, crystallizes from water in needles. One hundred grams of the aqueous solution at 18 °C contains 0.226 g of the anhydrous acid. The substance is insoluble in alcohol, ether, and glacial acetic acid.

$$\text{H}_2\text{N}-\underset{\text{HO}_3\text{S}}{\overset{\text{CH}_3}{\bigcirc}}-\underset{\text{SO}_3\text{H}}{\overset{\text{CH}_3}{\bigcirc}}-\text{NH}_2$$

o-Tolidine-6,6′-disulfonic acid is produced from 2-nitrotoluene-4-sulfonic acid [97-06-3] by reduction with zinc and sodium hydroxide solution and subsequent rearrangement.

o-Tolidine-6,6′-disulfonic acid is used to make yellow dyes for wool.

**Benzidinesulfone** [6259-19-4], 3,7-diaminodi-benzothiophene-5,5-dioxide, $C_{12}H_{10}N_2O_2S$, $M_r$ 246.29, forms yellow needles or flakes, *mp* 330–332 °C, is virtually insoluble in boiling water, boiling alcohol, ether, and benzene.

$$\text{H}_2\text{N}-\underset{}{\bigcirc}\underset{\overset{|}{\text{O}_2}}{\text{S}}\underset{}{\bigcirc}-\text{NH}_2$$

The dihydrochloride, $C_{12}H_{10}N_2O_2S \cdot 2\,HCl$, crystallizes as flakes from dilute hydrochloric acid and is decomposed by water. The sulfate, $C_{12}H_{10}N_2O_2S \cdot H_2SO_4 \cdot 1.5\,H_2O$, crystallizes as needles or flakes, is slightly soluble in water containing sulfuric acid, and is decomposed by water.

Benzidinesulfone is produced by heating benzidine sulfate with an excess of 20 – 40 % oleum in a water bath.

**Benzidinesulfone-disulfonic Acid,** 3,7-diaminodibenzothiophene-2,8-disulfonic acid 5,5-dioxide, $C_{12}H_{10}N_2O_8S_3$, $M_r$ 406.41, crystallizes in pale yellow needles, soluble in hot water, slightly soluble in alcohol, virtually insoluble in cold hydrochloric acid and in dilute sulfuric acid.

Benzidinesulfone-disulfonic acid is made by heating benzidine or benzidine sulfate with 40 % oleum at 150 °C.

# 7. Economic Aspects

Because of the carcinogenic properties of benzidine, a number of older benzidine plants have been replaced by new ones built to meet the industrial hygiene requirements. At the same time, every effort is being made to replace benzidine with less dangerous bases. Thus the demand for benzidine, which once amounted to several thousand tons per year, has been falling for some time.

There are no statistics available on the world production of diphenyl bases, but in 1983 about 15 kt was produced. Dichlorobenzidine and its salts accounted for about 7 – 10 kt, followed by *o*-tolidine with 2 – 3 kt. In contrast, *o*-dianisidine production was only about 1 – 2 kt and the production of tetrachlorobenzidine hardly more than 100 t.

# 8. Toxicology

## 8.1. Benzidine

The *acute* oral $LD_{50}$ of benzidine administered to male rats is 1.57 g/kg [40]. *Subacute* dietary exposure of mice to concentrations of 100 – 800 ppm resulted in cloudy swelling of the liver, vacuolar degeneration of the renal tubules, and hyperplasia of myeloid elements of the bone marrow and lymphoid cells in the thymus and spleen [41]. Little is known about the *dermal* or *pulmonary absorption,* but the toxic systemic manifestations following these modes of exposure indicate that proportional amounts are absorbed [42].

If benzidine is injected intravenously, its half-life in blood exhibits several phases. The fourth and final phase has a half-life of 65 h in rats and 88 h in dogs [43]. The routes of elimination differed in the various animal species investigated. In rats, approximately 80% of an i.v. dose was excreted via feces. In dogs and monkeys 67% and in monkeys 50% was eliminated via the urine [43].

Benzidine is metabolized to *N*-acetyl-benzidine and *N,N'*-diacetylbenzidine, which is converted by an NADPH-dependent reaction to *N*-hydroxy-*N,N'*-diacetylbenzidine and 3-hydroxy-*N,N'*-diacetylbenzidine; the latter binds to nucleic acids [44].

Benzidine is positive in the Ames test following metabolic activation with liver homogenate from rats [45], [46] or humans [47]. The Ames test is also positive for the urine of rats that have been exposed to benzidine by oral administration [48]. *N*-Acetylbenzidine [48] and *N*-hydroxy-*N,N'*-diacetylbenzidine [44] also induce point mutations in the Ames system. Benzidine induces unscheduled DNA synthesis in HeLa cells [49] and in rat hepatocytes [50]. After metabolic activation, DNA strand breaks [51], [52] and cell transformation [53], [54] occur in V-79 cells.

Benzidine hydrochloride produces hepatocellular carcinoma in mice at a dose level of 150 mg/kg [55]–[58]. Evidence of the carcinogenic properties of benzidine has been found in rats [59] and hamsters [60].

The first information about the carcinogenic properties of benzidine in humans was published by Oppenheimer in 1927 [61]. He observed tumors of the bladder among workers in the dyestuffs industry. This observation has been confirmed by many other researchers [62]–[71]. The incidence of bladder cancer in workers decreased after reduction of the industrial exposure [71].

In summary, there is sufficient evidence that benzidine is carcinogenic in mice, rats, and hamsters. Ample evidence exists that benzidine also is carcinogenic in humans.

The MAK commission has classified benzidine in group A 1 (clearly carcinogenic). The American Conference of Governmental Industrial Hygienists (ACGIH) lists benzidine in group A 1 b: "no exposure or contact by any route – respiratory, skin, or oral, as detected by the most sensitive methods – shall be permitted."

## 8.2. 3,3'-Dichlorobenzidine

The oral $LD_{50}$ (rat) of 3,3'-dichlorobenzidine is 7.07 g/kg [72].

In chronic studies 100 mg was administered orally to beagle dogs three times weekly for six weeks and five times weekly thereafter for seven years. Serum-glutamic-oxalacetic transaminase levels were increased [73], a sign of a slightly toxic effect upon the liver. Carcinoma of the bladder, hepatocellular carcinoma, and mammary gland tumors were also observed.

3,3'-Dichlorobenzidine is rapidly metabolized. After an oral dose of one gram, dogs excrete only 2% unchanged in the feces and urine [74]. Rats and dogs treated with 3,3'-dichlorobenzidine showed multiphasic blood clearances. The half-life of the final phase was 68 h in rats and 86 h in dogs. The majority of the administered dose was recovered from the bile, thereby demonstrating the importance of the hepato-biliary excretion [43].

In the Ames test, 3,3'-dichlorobenzidine was positive with and without adding a metabolic activation system [46], [75], [76]. Positive results were also obtained in the unscheduled DNA synthesis test in HeLa cells [49] and in the cell transformation assay in baby hamster kidney cells [77].

When mice were treated with 3,3'-dichlorobenzidine in a dietary concentration of 1000 ppm for one year, all of the animals developed hepatomas [78]. In another experiment, rats developed tumors in various organs after an exposure for one year followed by an unlimited observation period [79]. The results are difficult to evaluate because of the absence of an adequate control group. When rats were treated with 1000 ppm of 3,3'-dichlorobenzidine in the diet, significant statistical increases were found for granulocytic leukemias, mammary adenocarcinomas, and zymbal gland carcinomas [80]. Hamsters did not develop tumors at the same dose level, but following doses of 3000 ppm, tumors of the bladder and the liver were observed [60], [81].

In epidemiological studies of workers exposed to 3,3'-dichlorobenzidine, no bladder tumors were found [72], [82], [83]. In summary, 3,3'-dichlorobenzidine is carcinogenic in laboratory animals, but proof of its oncogenic properties in humans does not exist. Both commissions (MAK and ACGIH) have classified this compound in group A II. Neither MAK nor TLV values have been established. 3,3'-Dichlorobenzidine is absorbed by the skin.

## 8.3. o-Tolidine (3,3'-Dimethylbenzidine)

$o$-Tolidine can enter the body by percutaneous absorption, ingestion, or inhalation [84]. It is metabolized in exposed workers to diacetyl-$o$-tolidine and to a hydroxyamino metabolite (probably 5-hydroxy-$o$-tolidine), which are excreted via the urine [85]. Dogs are unable to acetylate aromatic amines such as $o$-tolidine [86].

In the Ames test, $o$-tolidine is a weak mutagen [76], [87]. $o$-Tolidine acts as a systemic carcinogen in Sherman rats when administered subcutaneously at a dose level of 60 mg per rat per week. It induces mainly tumors of the zymbal glands [88]. No carcinogenic potential could be observed in hamsters after feeding them a diet containing 1000 ppm of $o$-tolidine [60], [81]. No epidemiological studies in humans are available.

Exposure limits have been established. The MAK value is 5 mg/kg (22 mg/m$^3$, group III B), and the TLV is 2 mg/kg (9 mg/m$^3$, group A 2).

## 8.4. o-Dianisidine (3,3'-Dimethoxybenzidine)

Information about the biotransformation of *o*-dianisidine is rather limited. In dogs, 3,3'-dihydroxybenzidine is formed as a major metabolite [74].

In the presence of rat liver homogenate, *o*-dianisidine is positive in the Ames test [76]. It is capable of inducing unscheduled DNA synthesis in HeLa cells [49] and in primary rat hepatocytes [89], and it provokes cell transformation in baby hamster kidney cells in vitro [77]. *o*-Dianisidine is carcinogenic in rats. An oral dose of 30 mg administered three times per week for 13 months produced tumors in the zymbal glands [89], [90] and in other tissues [91], [92].

No conclusive epidemiological studies have been reported on the carcinogenicity of *o*-dianisidine in humans. No MAK value has been established. *o*-Dianisidine is classified in group III B.

# 9. References

General References

*Beilstein,* **13,** 214, 227, 234, 255, 256, 340, 807, 808; **14,** 567, 568, 770, 794, 795, 796; **18,** 591, 636; **13 (1),** 58, 66, 67, 79, 331, 332; **14 (1),** 647, 737, 743; **18 (1),** 590; **13 (2),** 90, 102, 106, 116, 152, 502, 504; **14 (2),** 344, 345, 479; **18 (2),** 490; **13 (3),** 425, 437, 447, 484, 2310; **14 (3),** 2264, 2265; **18 (3/4),** 7289, 8286.

*Houben-Weyl,* vol. **X/2,** pp. 693–743; vol. **XI/1,** pp. 839–848.

T. S. Scott: *Carcinogenic and Chronic Toxic Hazards of Aromatic Amines,* Elsevier, Amsterdam–New York 1962.

R. Powell, M. Murray, C. Chen, A. Lee: *Survey of the manufacture, import and uses for benzidine, related substances and related dyes and pigments,* Report 1979 EPA/560/13–79/005; Order No. PB-296544. Avail. NTIS. *Gov. Rep. Announce. Index (U.S.),* **79** (1979) no. 20, 120; *Chem. Abstr.* **92** (1980) 7831 e.

T. J. Haley: "Review of the literature and problems associated with the use of benzidine and its congeners," *Clin. Toxicol.* **8** (1975) no. 1, 13–42 (106 refs.).

W. Seidenfaden: *Künstliche organische Farbstoffe und ihre Anwendungen,* Enke Verlag, Stuttgart 1957.

K. Venkataraman: *The Chemistry of Synthetic Dyes,* vol. **I.** (1952); vol. **III** (1970); vol. **V** (1971), Academic Press, New York–London.

W. Herbst, K. Hunger: "Azo-Pigmente, Eigenschaften, Anforderungen, Entwicklungstendenzen," *Progr. Org. Coat.* **6** (1978) 105–210.

*Ullmann,* 4th ed., vol. **18,** pp. 661–695.

Specific References

[1] W. A. Rye, P. F. Woolrich, R. P. Zanes: "Facts and Myths Concerning Aromatic Diamine Curing Agents," *JOM J. Occup. Med.* **12** (1970) no. 6, 211–215.

[2] R. Merten: "Die Synthese heterocyclischer Ringsysteme für wärmebeständige Kunststoffe aus Polyisocyanaten," *Angew. Chem.* **83** (1971) 339–347;*Angew. Chem. Int. Ed. Engl.* **10** (1974) 294.

[3] A. Horvath, B. Vollmert: "Über den Einfluß von Triäthylamin auf die Viskosität von Polyamidcarbonsäuren aus Pyromellitsäuredianhydrid und Benzidin in Dimethylacetamid," *Angew. Chem.* **83** (1971) 375; *Angew. Chem. Int. Ed. Engl.* **10** (1971) 348.

[4] H. E. Fierz-David, L. Blangey: *Grundlegende Operationen der Farbenchemie*, 8th ed., Springer Verlag, Wien 1982.

[5] P. H. Groggins: *Unit Processes in Organic Synthesis*, 5th ed., McGraw-Hill, New York–Toronto–London 1958.

[6] *Houben-Weyl*, vol. **IV/2**, p. 494.

[7] D. V. Banthorpe et al., *J. Chem. Soc. Perkin Trans. 2* 1973,551–56H. J. Shine et al., *J. Am. Chem. Soc.* **99** (1977) 3719–23 Z. J. Allan, *Justus Liebigs Ann. Chem.* 1978, no. 5, 705–09.

[8] International Labour Office: "Occupational Safety and Health Series No. 39," *Occupational Cancer-Prevention and Control*, Geneva 1977.M. R. Zavon, U. Hoegg, E. Bingham: "Benzidine exposure as a cause of bladder tumors," *Arch. Environ. Health* **27** (1973) no. 1, 1–7.

[9] *IARC Monogr.* **29** (1982) 149–181 (Benzidine and its Sulfate, Hydrochloride and Dihydrochloride); 239–256 (3,3'-Dichlorobenzidine and its Dihydrochloride).

[10] Department of Labor. Occupational Safety and Health Administration: *Carcinogens*, Occupational Health and Safety Standards, Federal Register, vol. **39**, no. 20, Part III, Washington, D. C., 1974, pp. 3757, 3771–3773.

[11] Statutory Instruments 1967, no. 879: Factories. The Carcinogenic Substances Regulations 1967. Publ. by Her Majesty's Stationery Office 1967.

[12] Verordnung über gefährliche Arbeitstoffe vom 11. 2. 1982, Bundesgesetzblatt 1982, part I, p. 144; Berufsgenossenschaft der chemischen Industrie, Unfallverhütungsvorschrift Nr. 47; Schutzmaßnahmen bei Umgang mit krebserzeugenden Arbeitsstoffen vom 1. 10. 1982.

[13] Zwölfte Verordnung zur Durchführung des Bundes-Immissionsschutzgesetzes – Störfall-Verordnung – vom 27. 6. 1980. Bundesgesetzblatt 1980, part I, p. 772.

[14] Hauptverband der gewerblichen Berufsgenossenschaften, Spezifische Einwirkungsdefinitionen, ZH 1/600.16, proposal of 10.1979, Carl Heymanns Verlag, Köln.

[15] Arrêté Royal du 3. 10. 1973, Moniteur Belge 23. Nov. 1973.

[16] Labour Safety and Hygienic laws 2. 6. 1972.

[17] R. Montesano, L. Tomatis: "Legislation Concerning Chemical Carcinogens in Several Industrialized Countries," *Cancer Res.* **37** (1977) 310–315.

[18] H. T. Appleton, H. C. Sikka: "Accumulation, Elimination and Metabolism of Dichlorobenzidine in the Bluegill Sunfish," *Environ. Sci. Technol.* **14** (1980) no. 1, 50–54.

[19] P. Y. Lu, R. L. Metcalf, N. Pummer, D. Mandel: "The environmental fate of three carcinogens," *Arch. Environ. Contam. Toxicol.* **6** (1977) no. 2–3, 129–42; *Chem. Abstr.* **88** (1978) 27196.

[20] R. Morales, S. M. Rappaport, R. E. Hermes: "Air sampling and analytical procedures for benzidine, 3,3'-dichlorobenzidine and their salts," *Am. Ind. Hyg. Assoc. J.* **40** (1979) 970.

[21] H. C. Sikka, H. T. Appleton, S. Banerjee; *Fate of 3,3'-Dichlorobenzidine in Aquatic Environments*, U.S.-E.P.A.-600/3-78-068, National Technical Information Service, Springfield, Virginia 22161, USA, 1978.

[22] S. Banerjee, H. C. Sikka, R. Gray, C. M. Kelly: "Photodegradation of 3,3'-Dichlorobenzidine," *Environ. Sci. Technol.* **12** (1978) no. 13, 1425–27.

[23] R. J. Hackman, T. Rust: "Removal and decontamination of residual 3,3'-Dichlorobenzidine," *Am. Ind. Hyg. Assoc. J.* **42** (1981) 341–47.

[24] Erste Allgemeine Verwaltungsvorschrift zum Bundes-Immissionsschutzgesetz (Technische Anleitung zur Reinhaltung der Luft) vom 28. 8. 1974. Gemeinsames Ministerialblatt Ausgabe A, **25**, 4. 9. 1974, p. 442.

[25] *Ullmann,* 4th ed., vol. **8,** p. 359.

[26] F. Jones, D. Patterson, D. Srinivasan: "The determination of free benzidine in dyes," *J. Soc. Dyers Color.* **96** (1980) no. 12, 628–31.

[27] P. Engelbertz, E. Babel: "Nachweis von Benzidin und seinen Umwandlungsprodukten im Harn und in Organteilen," *Zentralbl. Arbeitsmed. Arbeitsschutz* **3** (1953) 161–68. P. Engelbertz, E. Babel: "Über die Anwendung der Benzidin-Testmethode mit 1,2-naphthochinon-4-sulfonsaurem Kalium zur Bestimmung der gesundheitlichen Gefährdung in Benzidinbetrieben," *Zentralbl. Arbeitsmed. Arbeitsschutz* **4** (1954) 40–42.

[28] P. Engelbertz, E. Babel: "Nachweis und Bestimmung von 3,3′-Dichlorbenzidin und 3-Oxybenzidin," *Zentralbl. Arbeitsmed. Arbeitsschutz* **6** (1956) 58–60.

[29] O. Winkler: "Beitrag zum Nachweis von Di- phenylbasen im Harn," *Zentralbl. Arbeitsmed. Arbeitsschutz* **9** (1959) 140–42.

[30] S. Laham, J. P. Farant, M. Potvin, *Ind. Med. Surg.* **39** (1970) 142–47.

[31] H. Steinberg: "The hazard of benzidine to criminal justice personnel," *NBS Spec. Publ. U.S.* 1977, 480–421;*Chem. Abstr.* **87** (1977) 178486.

[32] W. Zönnchen, DD 31935, 1963.

[33] Bayer, DE 2232095, 1972.

[34] Celanese Corp., US 3865876, 1969.

[35] Celanese Corp., US 3943175, 1969.

[36] D. R. Coffin, G. A. Serad, H. L. Hikcs, R. T. Montgomery: "Properties and Applications of Celanese PBI-Polybenzimidazole Fiber,"*Text. Res. J.* **52** (1982) no. 7, 466–72.

[37] Borma BV, DE 3020846, 1979.

[38] P. Engelbertz, E. Babel: "Nachweis von o-Tolidin und o-Dianisidin sowie ihrer Umwandlungsprodukte im Harn und in Organteilen," *Zentralbl. Arbeitsmed. Arbeitsschutz* **4** (1954) 179–83.

[39] Clayton Aniline Co., EP 32784, 1980.

[40] J. Marhold, M. Matrka, M. Hub, F. Ruffer, *Neoplasma* **15** (1968) 3–10.

[41] K. V. N. Rao, J. H. Rust, N. Mihailovich, S. D. Vesselinovitch, J. M. Rice, *Fed. Proc. Fed. Am. Soc. Exp. Biol.* **30** (1971) 444.

[42] G. Ghetti, *Med. Lav.* **51** (1960) 102–114.

[43] H.-M. Kellner, O. E. Christ, K. Lötzsch, *Arch. Toxicol.* **31** (1973) 61–79.

[44] K. C. Morton, F. A. Belland, F. E. Felland, N. F. Fulland, F. F. Kandlubarr, *Cancer Res.* **40** (1980) 751–757.

[45] B. N. Ames, W. E. Durston, E. Yamasaki, F. D. Lee, *Proc. Natl. Acad. Sci. U.S.A* **70** (1973) 2281–2285.

[46] D. Anderson, J. A. Styles, *Br. J. Cancer* **37** (1978) 924–930.

[47] S. Haworth, T. Lawlor, in: 9th Annual Meeting of the Environmental Mutation Society, San Francisco 1978, Abstract No. 1 a–8.

[48] K. J. Tanaka, S. Marui, T. Mii, *Mutat. Res.* **79** (1980) 173–176.

[49] C. N. Martin, A. C. McDermid, R. C. Garner, *Cancer Res.* **38** (1978) 2621–2627.

[50] G. M. Williams, *Cancer Let.* **4** (1978) 69–75.

[51] J. A. Swenberg, G. L. Petzold, P. R. Harbach, *Biochem. Biophys. Res. Commun.* **72** (1976) 732–738.

[52] G. L. Petzold, J. A. Swenberg, *Cancer Res.* **38** (1978) 1589–1594.

[53] J. Ashby, J. A. Styles, D. Paton, *Br. J. Cancer* **38** (1978) 34–50.

[54] R. J. Pienta, in F. J. de Serres, E. Hollaender (eds.): *Chemical Mutagens. Principles and Methods for their Detection*, vol. **6**, Plenum, New York 1980, pp. 175–202.
[55] S. D. Vesselinovitch, K. V. N. Rao, N. Mihailovich, *Cancer Res.* **35** (1975) 2814–2819.
[56] S. D. Vesselinovitch, K. V. N. Rao, N. Mihailovich, *Natl. Cancer Inst. Monogr.* **51** (1979) 239–250.
[57] C. H. Frith, K. P. Baetcke, C. J. Nelson, G. Schieferstein, *Toxicol. Lett.* **4** (1979) 507–518.
[58] C. H. Frith, K. P. Baetke, C. J. Nelson, G. Schieferstein, *Eur. J. Cancer* **16** (1980) 1205–1216.
[59] E. Boyland, J. Harris, E. S. Horning, *Br. J. Cancer* **8** (1954) 647–654.
[60] U. Saffiotti, F. Cefis, R. Montesano, A. R. Sellakumar, in W. B. Deichmann, K. F. Lampe (eds.): Bladder Cancer: A Symposium, Birmingham, Aesculapius Publ. Co., Birmingham 1967, pp. 129–135.
[61] R. Oppenheimer, *Z. Urol. Chir. (Gynaekol.)* **21** (1927) 336–350.
[62] A. Muller, *Z. Urol. Chir. (Gynaekol.)* **36** (1933) 202–219.
[63] G. Di Maio, *Arch. Ital. Urol.* **14** (1937) 283–385.
[64] M. W. Goldblatt, *Br. J. Ind. Med.* **6** (1949) 65–81.
[65] M. Barsotti, E. C. Vigliani, *Med. Lav.* **40** (1949) 129–138.
[66] T. S. Scott, *Br. J. Ind. Med.* **9** (1952) 127–132.
[67] P. Aboulker, G. Smagghe, *Arch. Mal. Prof.* **14** (1953) 380–386.
[68] F. Uebelin, A. Pletscher, *Schweiz. Med. Wochenschr.* **84** (1954) 917–920.
[69] M. Douilett, J. Bourret, A. Convert, *Arch. Mal. Prof.* **20** (1959) 713–733.
[70] J. L. Billard-Duchesne, *Acta Unio Int. Cancrum* **16** (1960) 284–288.
[71] K. M. Ferber, W. J. Hill, D. A. Cobb, *Am. Ind. Hyg. Assoc. J.* **37** (1976) 61–68.
[72] H. W. Gerarde, D. F. Gerarde, *J. Occup. Med.* **16** (1974) 322–344.
[73] E. F. Stula, H. Sherman, W. R. Clayton, *Toxicol. Appl. Pharmacol.* **31** (1975) 159–176.
[74] L. J. Sciarmy, J. W. Meigs, *Arch. Environ. Health* **2** (1961) 584–588.
[75] R. C. Garner, A. L. Walpole, F. L. Rose, *Cancer Lett.* **1** (1975) 39–42.
[76] E. J. Lazear, S. C. Louie, *Cancer Lett.* **4** (1977) 21–25.
[77] I. A. Styles, *Br. J. Cancer* **37** (1978) 931–936.
[78] H. Osanai, *J. Sci. Labour* **52** (1976) 179–201.
[79] G. B. Pliss, *Vopr. Onkol.* **5** (1959) 524–533.
[80] E. F. Stula, J. R. Barnes, H. Sherman, C. F. Reinhardt, J. A. Zapp, *J. Environ. Pathol. Toxicol.* **1** (1978) 475–490.
[81] A. R. Sellakumar, R. Montesano, U. Saffiotti, *Proc. Am. Assoc. Cancer Res.* **10** (1969) Abstr. 309.
[82] T. Gadian, *Chem. Ind.* 1975, 4th Oct., 821–831.
[83] I. McIntyre, *J. Occup. Med.* **17** (1975) 23–26.
[84] J. W. Meigs, L. J. Sciarini, W. A. Van Sandt, *Arch. Ind. Hyg.* **9** (1954) 122.
[85] H. M. L. Dieteren, *Arch. Environ. Health* **12** (1966) 30.
[86] IARC, *Monographs on the evaluation of carcinogenic risk of chemicals to man*, vol. **I**, International Agency for Research on Cancer, Lyon 1972, pp. 88–91.
[87] J. I. Ferretti, W. Lu, M. B. Liu, *Am. J. Clin. Pathol.* **67** (1977) 526–527.
[88] G. B. Pliss, M. A. Zabezhinsky, *J. Natl. Cancer Inst.* **45** (1970) 283.
[89] G. B. Pliss, *Gig. Tr. Prof. Zabol.* **9** (1965) 18.
[90] G. B. Pliss, *Acta Unio Int. Cancer* **19** (1963) 499.
[91] ACGIH, Documentation of the Threshold Limit Values, 3rd ed., American Conference of Governmental Industrial Hygienists, Cincinatti, Ohio, 1971.
[92] Z. Hadidian, T. N. Fredrickson, E. K. Weisburger, J. H. Weisburger, R. M. Glass, N. Mantel, *J. Natl. Cancer Inst.* **41** (1968) 985.

# Benzoic Acid and Derivatives

Takao Maki, Dia Research Martech Inc., Yokohama, Japan (Chaps. 1–8)
Kazuo Takeda, Mitsubishi Chemical Safety Institute Ltd., Yokohama, Japan (Chap. 9)

| | | | | | |
|---|---|---|---|---|---|
| 1. | Introduction | 819 | 8.3. | Benzoyl Chloride | 827 |
| 2. | Physical Properties | 820 | 8.4. | Benzonitrile | 829 |
| 3. | Chemical Properties | 821 | 8.5. | Alkyl and Acyl Analogues | 831 |
| 4. | Production | 821 | 8.6. | Chlorobenzoic Acids | 832 |
| 5. | Quality Specifications | 823 | 8.7. | Aminobenzoic Acids | 833 |
| 6. | Storage, Transportation and Legal Aspects | 823 | 8.8. | Nitrobenzoic Acids | 835 |
| 7. | Uses and Economic Aspects | 825 | 8.9. | 3-Sulfobenzoic Acid | 836 |
| 8. | Derivatives of Benzoic Acid | 825 | 8.10. | Hexahydrobenzoic Acid | 836 |
| 8.1. | Salts of Benzoic Acid | 825 | 9. | Toxicology | 836 |
| 8.2. | Esters of Benzoic Acid | 826 | 10. | References | 838 |

# 1. Introduction

**Benzoic acid** [65-85-0], $C_7H_6O_2$. The name benzoic acid originates from gum benzoin, a balsamic resin obtained from a South Asian plant called styrax. The extraction of benzoic acid was carried out by Scheele in 1775. Its structure was determined by Liebig and Wöhler in 1832. The initial production methods were developed in the late 1800s. They were based on the hydrolysis of benzotrichloride or the decarboxylation of phthalic anhydride. Today, benzoic acid is produced by oxidation of toluene with air, which has displaced dichromate and nitric acid oxidation processes.

**Occurrence.** Benzoic acid and its derivatives are widely distributed in nature. Gum benzoin contains from 12–18% benzoic acid in free and esterified forms. Other natural products containing benzoic acid are the bark, foliage, fruits, and seeds of various plants, including cherries and prunes. Hippuric acid, found in the urine of herbivores, is a glycine derivative of benzoic acid.

## 2. Physical Properties

Benzoic acid, benzenecarboxlic acid, $M_r$ 122.12, *mp* 122 °C, *bp* 250 °C, forms white monoclinic crystals that begin to sublime at about 100 °C. It is volatile in steam and forms azeotropes with biphenyl, butyl benzoate, catechol, diphenyl ether, diphenylmethane, and naphthalene. At 89.7 °C, a mixture of excess benzoic acid and water forms two stable liquid phases. The water content of the benzoic acid phase is 26.5 wt%. The two phases become homogeneous at 117.2 °C. The mixture then contains 32.34% benzoic acid and 67.66% water.

Some physical properties of benzoic acid are listed in the following:

| | |
|---|---|
| Solubility in water, g/100 g | |
|   at 0 °C | 0.17 |
|   at 20 °C | 0.29 |
|   at 40 °C | 0.60 |
|   at 60 °C | 1.20 |
|   at 80 °C | 2.75 |
|   at 95 °C | 6.85 |
| Solubility in organic solvents at 25 °C, g/100 g | |
|   Acetone | 55.60 |
|   Benzene | 12.17 |
|   Tetrachloromethane | 4.14 |
|   Ethanol (absolute) | 58.40 |
|   Hexane | 0.94 (at 17 °C) |
|   Methanol | 71.50 (at 23 °C) |
|   Toluene | 10.60 |
| Vapor pressure, kPa | |
|   at 96 °C | 0.133 |
|   at 132 °C | 1.33 |
|   at 186 °C | 13.3 |
| Vapor density relative to air | 4.2 |
| Refractive index | |
|   $n_D^{15}$ | 1.53974 |
|   $n_D^{131.9}$ | 1.504 |
| Relative density | |
|   $d_4^{20}$ | 1.321 |
|   $d_4^{122.375}$ | 1.0819 |
|   $d_4^{180}$ | 1.02942 |
| Specific heat | |
|   solid (20–122 °C) | 1.204 J g$^{-1}$ K$^{-1}$ |
|   liquid (122–322 °C) | 1.775 J g$^{-1}$ K$^{-1}$ |
| Heat of fusion | 18 kJ/mol |
| Heat of sublimation at 110 °C | 86.2 kJ/mol |
| Heat of vaporization at 150 °C | 68.2 kJ/mol |
| Heat of formation | – 384.9 kJ/mol |
| Heat of combustion at 25 °C | 3227 kJ/mol |
| Free energy | |
|   solid | – 251 kJ/mol |
|   liquid | – 247 kJ/mol |
|   aqueous solution | – 242 kJ/mol |
| Flash point | 121.1 °C |
| Ignition temperature | |
|   in air | 573 °C |

| | |
|---|---|
| in oxygen | 556 °C |
| Dissociation constant at 25 °C, $K_a$ | $6.335 \times 10^{-5}$ |
| pH of the saturated solution at 25 °C | 2.8 |
| Change in volume upon freezing | $-0.138$ cm$^3$/g |
| log POW* | 1.870 |

\* Bioaccumulation factor (octanol/water-partition factor)

# 3. Chemical Properties

Benzoic acid is stable towards common oxidizing agents. Air, permanganate, chromic acid, hypochlorite, and dilute nitric acid do not affect it. However, above 220 °C it reacts with copper(II) salts to form phenol and its derivatives [1]. This reaction, followed by regeneration of the copper(II) with molecular oxygen, is the basis for the commercial production of phenol from benzoic acid [2] (→ Phenol). Benzoic acid reacts with ammonia under similar conditions to form aniline [3]. Upon heating to 150 °C, some dehydration takes place to form benzoic anhydride. Decarboxylation occurs when benzoic acid is heated above 370 °C or as low as 245 °C in the presence of catalysts [4]. Benzene and a small amount of phenol are formed. Copper and cadmium powder accelerate the decarboxylation. When potassium benzoate is heated in carbon dioxide, it disproportionates to terephthalate and benzene. Cadmium and zinc salts catalyze this reaction. Benzoic acid is converted into hydroxybenzoic acids by fused potassium hydroxide [5].

Benzoic acid can be directly hydrogenated with molecular hydrogen to afford benzaldehyde in quantitative yield. Zirconium and other metal oxides catalyze the reaction [31]. Hydrogenation on noble metal catalysts gives cyclohexanecarboxylic acid (hexahydrobenzoic acid). Chlorination yields chiefly 3-chlorobenzoic acid [6]. Nitration forms 3-nitrobenzoic acid [7], and sulfonation forms 3-sulfobenzoic acid.

# 4. Production

**Liquid-Phase Oxidation of Toluene with Molecular Oxygen.** This process was developed by I.G. Farbenindustrie in Germany during World War II and was operated until the end of the war [8]. The plant consisted of a bubble-column oxidation reactor made of aluminum-lined steel and equiped with an air inlet tube, a separator, and an absorber to recover toluene. The reactor contained heat-exchange coils to remove the great amount of heat generated in the reaction. The purity of the toluene was critical because sulfur compounds, nitrogen compounds, phenols, and olefins inhibit the oxidation reaction. The reaction was carried out at 140 °C and ca. 0.2 MPa with a cobalt naphthenate catalyst (0.1%) for 30 h. About 50% conversion of toluene was achieved, 80% of which formed benzoic acid. Other oil-soluble cobalt salts were also used as catalysts. The crude oxidation product was neutralized with sodium carbonate to produce sodium benzoate. Unreacted toluene was recovered from the spent oil phase by distillation.

The elemental mechanism of this oxidation is a free-radical chain process [9]. The cobalt catalyst shortens the induction period of the reaction and retards the accumulation of inhibitors. Peroxides are reaction intermediates. Manganese behaves similarly to cobalt as a catalyst, but its performance is weaker. Copper adversely affects the oxidation. High pressures retard the reaction, especially at lower catalyst concentrations. Above 130 °C, the reaction rate is determined by the diffusion of oxygen. Phenolic compounds gradually accumulate during the reaction and eventually inhibit it.

In a typical modern process the oxidation is conducted at 165 °C and 0.9 MPa. The reaction heat is removed by external circulation of the contents of the reactor. The pressure of the liquid discharged form the reactor is reduced to atmospheric, and unreacted toluene is recovered. Benzoic acid is purified by rectification. The bottom residue is extracted to recover the cobalt catalyst. The exhaust gas is cooled to recover most of the toluene, purified, and vented. The oxidation reactor can be a bubble column or a stirred tank. Stainless steel is generally used as the construction material. Heat removal is still a serious problem, and air distribution within the reactor is critical to prevent a runaway reaction. The oxygen content of the exhaust gas should be strictly monitored to avoid an explosion.

The oxidation produces several byproducts. Benzaldehyde is formed in fairly large amounts, especially at lower conversions. It can be recovered by distillation. A limited amount of benzyl alcohol is formed, but its recovery is impractical because of the unavoidable esterification with benzoic acid. Benzyl benzoate can be recovered from the bottoms of the benzoic acid column [10]. Other esters, including benzyl formate and benzyl acetate, are also present. Biphenyl and methylbiphenyls are formed in smaller amounts. Because their vapor pressures are close to that of benzoic acid, they play a role in the rectification of benzoic acid. The product is contaminated by small amounts of phthalic acid. Carbon monoxide and carbon dioxide are detected in the exhaust gas. The yield of benzoic acid is about 90%, but it varies with toluene conversion. At lower conversions, the selectivity for benzaldehyde and benzyl alcohol exceeds 10%.

Conventional technical-grade benzoic acid is usually of as-rectified quality. Further processing is necessary for higher grade products. Sublimation, recrystallization, and neutralization processes have been proposed. Thermal treatment in an inert atmosphere is attempted [11]. To remove phthalic acid, whose presence is not allowed for food or pharmaceutical uses, treatment with amines [32] and rinsing [33] methods have been reported. Methylbiphenyls are responsible for the odor of the product. For their removal, extraction [34] and treatment in an inert gas stream [35] have been proposed.

The oxidation reactions of alkylaromatic hydrocarbons are dramatically enhanced by adding bromine compounds to the cobalt catalyst in acetic acid [12]. This method is now widely used in the production of terephthalic acid and is also applicable to the benzoic acid process. Under certain conditions, benzoic acid itself can be used as a solvent. Manganese has synergistic effect on the cobalt catalyst. This process can realize high reaction rates and better toluene conversion, as well as eliminating reaction inhibition. However, to avoid corrosion, the equipment must be constructed of expensive titanium.

Another type of cobalt catalyst activator is a readily oxidizable carbonyl compound such as methyl ethyl ketone or acetaldehyde. These additives permit a considerable reduction in reaction temperature, but because of the high consumption of these activators, this technology does not seem to be practical.

The catalytic oxidation of toluene in aqueous sodium hydroxide directly affords sodium benzoate. This process has been tried since 1940s, but satisfactory yields have not been achieved.

**Vapor-phase oxidation of toluene** at higher temperatures is another potential process. Various catalysts (vanadium, tungsten, uranium, molybdenum, silver, etc.) have been investigated, but yields were not comparable to those of the liquid-phase process. However, a recent patent claims a significant improvement in yield by using complex vanadia catalysts [36].

# 5. Quality Specifications

Quality specifications for benzoic acid are listed in Tables 1 and 2. An EU guideline for the use of benzoic acid and its derivatives has been issued [37]

# 6. Storage, Transportation and Legal Aspects

Benzoic acid is commercially handled as flakes packed in 25 kg paper bags or in 500 kg flexible containers. Bulk quantities are transported as molten liquid in stainless steel vessels. For transport, benzoic acid is not regulated as a hazardous material (IMO, IATA, ADR/SDR, and RID/SDR). There is no IMDG code. The reportable quantity in USA is 2270 kg (40 CFR 355.40). The European hazard-labeling symbol is Xn (harmful). The MSDN is available on the Internet [38], [39].

Benzoic acid can be stored without deterioration, but the low-grade product tends to solidify on storage. Precautions should be taken against dust during handling since it irritates the eyes and skin. Dust/air mixtures can be explosive above the flash point (121°C). The explosion limit of the vapor/air mixture is 0.085-0.99g/m$^3$.

Benzoic acid has the following registry number for the legal control: TSCA 65-85-0, EINECS (European Inventory of Existing Chemical substances) 200-618-2, ECNS (Japan) (3)-1397, RTECS DG0875000.

**Table 1.** Quality specifications for technical-grade benzoic acid

| Items | Specifications Japan (JIS K 4127, 1995) | Specifications for commercial products | | | |
|---|---|---|---|---|---|
| | | Japan Commercial grade | USA Industrial grade | USA Technical grade | Europe Commercial grade |
| $mp$, °C | 120–123 | 121–123 | 121–123 | 121.5–123 | 122–122.5 |
| Assay, % min. | 99.0 (by glc) | 99.6 | 97.5 | 99.0 | 99.5 |
| Moisture, % max. | 0.3 (by titration) | | | | |
| Low-boiling substances, % max. | 0.5 (by glc) | | | | |
| High-boiling substances, % max. | 0.5 (by glc) | | | | |
| Heavy metals, ppm max. | | | | 10 | 10 |
| Phthalic acid, % max. | 0.5 (by glc) | | | | |
| Ignition residue, % max. | 0.1 | 0.1 | 0.1 | 0.1 | 0.1 |
| Appearance | white | off-white | brownish | white | white |

**Table 2.** Quality specifications of pharmaceutical- and food-grade benzoic acid

| | Pharmaceutical | | | Food additive | |
|---|---|---|---|---|---|
| | Japan (JP12 1993) | USA (USP23 1995) | Germany | Japan (Official) | USA (FCC 1996) |
| Identification | pass test | pass test | pass test | | |
| $mp$, °C | 121–124 | 121–123 (fp) | 121–124 | 121–123 | 121–123 (fp) |
| Assay, % min. (by titration) | 99.5 | 99.5–100.5 | 99.0–100.5 | 99.5 | 99.5–100.5 |
| Moisture, % max. | 0.5 (by weight loss) | 0.7 | | 0.5 | 0.7 (by titration) |
| Heavy metals, ppm max. | 20 | 10 | 10 | 10 | 10 |
| Arsenic, ppm max. | | 3 | | 4 | |
| Readily oxidizable substances (by permanganate) | pass test | pass test | pass test | | pass test |
| Readily carbonizable substances (by sulfuric acid) | pass test | pass test | pass test | | pass test |
| Phthalic acid (by fluorescence) | pass test | | | pass test | |
| Halogen compounds | pass test | | pass test | 0.014 % max. | |
| Ignition residue, % max. | 0.05 | 0.05 | 0.1 | | 0.05 |
| Solubility | | | pass test | | |

**Table 3.** Benzoic acid consumption (in %)

| Item | USA 1991 | Japan 1991 |
|---|---|---|
| Plasticizers | 50 | 0 |
| Sodium benzoate | 25 | 55 |
| Benzoyl chloride | 10 | 15 |
| Alkyd resins | 5 | 10 |
| Others | 10 | 20 |

**Table 4.** Major benzoic acid producers

| Company | Capacity, $10^3$ t/a |
|---|---|
| Kalama Chemical (USA) | 95 |
| Velsicol Chemical (USA) | 30 |
| DSM (NL) | 120 |
| Chimica del Fruita (IT) | 30 |
| Velsicol Eesti (Estonia) | 30 |
| Liquid Quimica (E) | 5 |
| Mitsubishi Chemical (JP) | 10 |
| Nippon Steel Chemical (JP) | captive |

# 7. Uses and Economic Aspects

The majority of the benzoic acid is utilized captively to produce phenol. The rest comes onto the market. The benzoic acid market in the USA and Japan is summarized in Table 3. Major producers are listed in Table 4.

Glycol esters of benzoic acid are used as plasticizers for vinyl resins. Demand has increased recently in the USA as a replacement for butyl benzyl phthalate in poly(vinyl acetate) emulsion adhesives [40]. Another major market is sodium benzoate for preservatives and anticorrosives. Benzoic acid is used in modified alkyd resins for automotive refinishing enamels to improve surface properties. Benzoic acid is also used as an intermediate for perfumes, pharmaceuticals, and cosmetics. The market in USA has experienced growth rate of 4 % per annum [42].

# 8. Derivatives of Benzoic Acid

## 8.1. Salts of Benzoic Acid

**Sodium Benzoate** [532-32-1], $C_7H_5O_2Na$, $M_r$ 144.11, forms white granules, flakes, or a crystalline powder with a sweetish taste. One gram dissolves in 2 mL of water. Between 0 °C and 50 °C the solubility in water is nearly constant, but at higher temperature it increases sharply. At −13.5 °C, the eutectic mixture with water consists of 44.9 g of the salt and 100 g of water. The aqueous solution is slightly alkaline (pH ca. 8). Sodium benzoate is soluble to the extent of 0.8 g in 100 mL of ethanol, 8.2 g in

**Table 5.** Specifications for food-grade sodium benzoate

|  | Japan | USA (FCC, 1996) |
|---|---|---|
| Appearance | pass test | pass test |
| Solubility | pass test |  |
| Assay, % min. (by titration) | 99.0 | 99.0 – 100.5 |
| Moisture, % max. (by weight loss) | 1.5 | 1.5 |
| Free alkali, % max. (as NaOH) | pass test | 0.04 |
| Free acid | pass test |  |
| Heavy metals, ppm max. (as Pb) | 10 | 10 |
| Arsenic, ppm max. | 4 |  |
| Readily oxidizable substances (by permanganate) | pass test |  |
| Phthalic acid (by fluorescence) | pass test |  |
| Halogen compounds, % max. (by Ag) | 0.014 |  |
| Sulfate, % max. | 0.3 |  |

100 mL of methanol, and over 20 g in 100 mL of ethylene glycol. Aqueous and alcoholic solutions of sodium benzoate passivate the surfaces of metals and alloys.

Dry sodium benzoate is electrically charged by friction and forms an explosive mixture when its dust is dispersed in air. It absorbs moisture from air to some extent. Upon heating in an inert atmosphere, sodium benzoate carbonizes and forms sodium carbonate. In the presence of air, it burns to give carbon dioxide and sodium carbonate. Sodium benzoate reacts with benzoyl chloride to form benzoic anhydride [93-97-0]. Other salts of benzoic acid can be prepared by metathesis reactions with the salts of particular cations. Sodium benzoate is produced by the neutralization of benzoic acid with sodium hydroxide. Direct oxidation of toluene in sodium hydroxide solution is not currently in use.

The principal use of sodium benzoate is an anticorrosive. The expansion of the motor industry has increased demand. It is added mainly to antifreeze coolants. It is also used as a preservative for foods and cosmetics. Specifications for food-grade sodium benzoate are listed in Table 5. The CFR GRAS level is 0.1 % max., and its use below pH 4.5 is recommended [41].

**Potassium Benzoate** [582-25-2], $C_7H_5O_2K$, $M_r$ 160.22. At 17.5 °C, a saturated aqueous solution contains 41.1 wt% of the salt. It has similar properties to the sodium salt. It is used in the food industries where the presence of sodium is undesirable.

## 8.2. Esters of Benzoic Acid

**Methyl Benzoate** [93-58-3], $C_8H_8O_2$, $M_r$ 136.15, mp −12.21 °C, bp 199.5 °C, colorless liquid with a pleasant odor, $n_D^{20}$ 1.516, $d_4^{15}$ 1.09334, solubility in water at 20 °C 0.15 wt%, miscible with common organic solvents, vapor pressure 0.133 kPa at 39 °C, 13.3 kPa at 130.8 °C, flash point 83 °C, ignition point 505 °C, heat of combustion

3950 kJ/mol. Methyl benzoate forms an azeotrope (20.8 wt% ester) with water, *bp* 99.08 °C. Stable against oxidation and easily saponified by strong base.

Methyl benzoate is prepared by reacting methanol with benzoic acid. It is used as a perfume and as an intermediate in the production of other benzoic esters.

**n-Butyl Benzoate** [136-60-7], $C_{11}H_{14}O_2$, $M_r$ 178.23, *mp* −22.4 °C, *bp* 250.3 °C, $d_4^{20}$ 1.0061, flash point 111 °C; autoignition temperature 440 ± 5 °C, heat of combustion 5828 kJ/mol. Butyl benzoate forms an azeotrope (6 wt% ester) with water, *bp* 99.8 °C. Butyl benzoate is widely used as a dye carrier in polyester fibers.

**Glycol Benzoates.** *Diethylene glycol dibenzoate* [120-55-8], $C_{18}H_{18}O_5$, $M_r$ 314.34, *mp* −28 °C, *bp* 240 °C at 0.7 kPa, $d_4^{20}$ 1.178, $n_D^{25}$ 1.5424, flash point 232 °C.

*Propylene glycol dibenzoate* [19224-26-1], $C_{17}H_{16}O_4$, $M_r$ 284.31, *fp* −3 °C, *bp* 232 °C at 1.6 kPa.

*Dipropylene glycol dibenzoate* [27138-31-4], $C_{20}H_{22}O_5$, $M_r$ 342.40, *mp* −40 °C, *bp* 232 °C at 0.7 kPa, $d_4^{20}$ 1.129, $n_D^{25}$ 1.5282, flash point 212 °C.

*Glycerine tribenzoate* [614-33-5].

The glycol benzoates are used as plasticizers in poly(vinyl chloride), poly(vinyl acetate), and polacrylate resins.

**Phenyl Benzoate** [93-99-2], $C_{13}H_{10}O_2$, $M_r$ 198.21; monoclinic crystals, *mp* 70 °C; *bp* 314 °C, $d_4^{20}$ 1.235, slightly soluble in water, soluble in hot ethanol. Phenyl benzoate is an intermediate in the production of phenol from benzoic acid.

**Benzyl Benzoate** [120-51-4], $C_{14}H_{12}O_2$, $M_r$ 212.25, leaflets or oily liquid, aromatic odor, *mp* 21 °C, *bp* 323–324 °C, *bp* 189–191 °C at 2.13 kPa, *bp* 156 °C at 0.6 kPa. It decomposes above 200 °C and is slightly volatile in steam and insoluble in water.

Benzyl benzoate is produced by esterifying dry sodium benzoate with benzyl chloride [13] or by the dimerization of benzaldehyde [14]. It is also obtained from the bottoms of benzoic acid rectification plants(see p. 821).

Benzyl benzoate is used as a dye carrier. Other uses include solvents for cellulose derivatives, plasticizers, and fixatives in the perfume industry.

## 8.3. Benzoyl Chloride

**Physical Properties.** Benzoyl chloride [98-88-4], $C_7H_5ClO$, $M_r$ 140.57, *mp* −0.6 °C, *bp* 198.3 °C, vapor pressure 0.133 kPa at 32.1 °C, 1.33 kPa at 73.8 °C, 13.3 kPa at 128 °C, colorless liquid with a penetrating odor, lachrymatory, fumes on exposure to air, $n_D^{20}$ 1.55369, $d_4^{20}$ 1.2113, decomposes in water and alcohols, miscible with benzene, flash point 102 °C, autoignition temperature 600 °C, heat of combustion 3282 kJ/mol.

**Chemical Properties.** Benzoyl chloride reacts violently with water to give benzoic acid and hydrochloric acid. If the amount of water is limited, benzoic anhydride is formed. Reaction with ammonia forms benzamide [55-21-0]. Benzoyl chloride is an excellent benzoylation reagent for compounds with active hydrogen, such as alcohols, phenols, or amines, under Friedel–Crafts and Schotten–Baumann reaction conditions.

**Production.** Benzoyl chloride is produced commercially from benzotrichloride [98-07-7], which is available by the chlorination of toluene. The chlorination reaction proceeds stepwise via benzyl chloride [100-44-7] and benzal chloride [98-87-3].

$$C_6H_5CH_3 + Cl_2 \longrightarrow C_6H_5CH_2Cl + HCl$$
$$C_6H_5CH_2Cl + Cl_2 \longrightarrow C_6H_5CHCl_2 + HCl$$
$$C_6H_5CHCl_2 + Cl_2 \longrightarrow C_6H_5CCl_3 + HCl$$

The benzotrichloride is then partially hydrolyzed with a limited amount of water.

$$C_6H_5CCl_3 + H_2O \rightarrow C_6H_5COCl + 2\ HCl$$

This series of reactions consumes a large quantity of chlorine. An alternative process uses benzoic acid.

$$C_6H_5CCl_3 + C_6H_5COOH \longrightarrow 2\ C_6H_5COCl + HCl$$

Phosphorous chlorides or iron chlorides are used as catalysts at the solvolytic stage of both processes. Benzoyl chloride can also be made by the reaction of benzoic acid with phosgene, thionyl chloride, or phosphorous pentachloride [15]; chlorination of benzaldehyde, benzyl alcohol, or benzyl benzoate [16]; and direct reaction of benzene with phosgene or the combination of carbon monoxide and chlorine [17].

The benzoyl chloride prepared by the conventional processes may contain trace amounts of chlorobenzoyl chloride, but the product obtained by the reaction of molten phthalic anhydride with hydrogen chloride at 200 °C is free of such impurities.

**Uses.** Benzoyl chloride is used in making organic peroxides. Benzoyl peroxide [94-36-0] and *tert*-butyl peroxybenzoate [614-45-9] are widely used as polymerization initiators and as curing agents in the polymer industries. Other uses include intermediates for rubber additives, UV absorbers, dyes, insecticides, and pharmaceuticals.

**Specifications.** Assay 99.5 % min., *sp* above −0.5 °C, benzal chloride 0.01 % max., trichlorobenzal chloride 0.01 % max., color (APHA) 10 max.

## 8.4. Benzonitrile

**Physical Properties.** Benzonitrile [100-47-0], C$_7$H$_5$N, $M_r$ 103.12, mp −13.8 °C, bp 191.1 °C, vapor pressure 1.37 kPa at 70 °C, 2.27 kPa at 79 °C, 4.53 kPa at 94 °C, colorless liquid, almond oil odor, $n_D^{15}$ 1.53056, $d_4^{20}$ 1.0052, solubility in water 1 wt% at 100 °C, miscible with common organic solvents, heat capacity $c_p$ 1.840 J g$^{-1}$ K$^{-1}$, heat of vaporization 46 kJ/mol, heat of formation −155 kJ/mol, heat of combustion 3063 kJ/mol, flash point 70 °C.

**Chemical Properties.** Benzonitrile is stable in air and light. In the presence of a strong base or acid, benzonitrile is hydrolyzed to benzoic acid and ammonia (via benzamide). Benzylamine [100-46-9] and dibenzylamine can be prepared from benzonitrile by hydrogenation.

Benzonitrile is a powerful solvent. At room temperature it dissolves poly(vinyl chloride), poly(vinyl acetate) resins, polystyrene, polymethacrylate, and nitrocellulose. It does not dissolve polyethylene, polyamide, poly(vinyl alcohol), or fluoropolymers.

**Production.** Benzonitrile is produced by the high-temperature vapor-phase oxidation of toluene in the presence of ammonia. The traditional catalysts are vanadium and molybdenum, but they suffers from low selectivity and serious decomposition of ammonia [18]. A tungsten–manganese complex catalyst shows better performance [19]. With the latter catalyst, the reaction is carried out in a fixed-bed reactor at 450 °C. The concentration of toluene in the feed stream is 1.5 vol%. The ratio of ammonia to toluene is 4 : 1 with a contact time of 2.4 s. The reaction converts 97% of the toluene and 30% of the ammonia. The selectivity is 87.4%. The crude product, which contains ammonia, hydrogen cyanide, toluene, and high-boilers, is purified by distillation.

Benzonitrile has also been produced from benzoic acid and ammonia in the vapor phase at 400–410 °C over alumina [20] or in the liquid phase at 225–245 °C [21]. Other methods include liquid-phase ammoxidation of toluene in the presence of a cobalt or manganese bromide catalyst [22], dehydrogenative condensation of benzyl alcohol or benzaldehyde with ammonia, the high-temperature reaction of toluene with nitrous oxide, and cyanation of benzene with cyanogen chloride or dicyanogen.

**Uses.** The major derivative of benzonitrile is benzoguanamine [91-76-9], which is made by the reaction of benzonitrile with dicyanodiamide in the presence of a strong base [23]. Benzoguanamine reacts with formaldehyde and/or alcohols to afford a heat-resistant thermosetting resin.

**Table 6.** Physical properties of the methylbenzoic acids

| | mp, °C | bp, °C | Density, g/cm³ | Solubility in water 25 °C, wt% | 100 °C, wt% | in chlorobenzene, 31.8 °C, mol% | in corresponding xylene, 14 °C, wt% | Heat capacity, J g⁻¹ K⁻¹ | Heat of fusion, kJ/mol | Heat of combustion, kJ/mol | Heat of formation, kJ/mol | Flash point, °C | Ignition point, °C |
|---|---|---|---|---|---|---|---|---|---|---|---|---|---|
| 2-Methylbenzoic acid | 105 | 259 | 1.062 at 115 °C | 0.118 | | 12.73 | 6.64 | 1.285 | 20.2 | 3887 | −416.5 | 148±3 | 495±5 |
| 3-Methylbenzoic acid | 110 | 263 | 1.0543 at 112 °C | 0.098 | | 14.07 | 7.89 | 1.219 | 15.7 | 3886 | −426.1 | 146 | |
| 4-Methylbenzoic acid | 180 | 274 | 1.23 at 20 °C | 0.034 | 1.16 | 1.76 | 1.45 | 1.204 | 22.7 | 3880 | −429.2 | 181±1 | 570±5 |

## 8.5. Alkyl and Acyl Analogues

**Methylbenzoic Acids,** $C_8H_8O_2$, $M_r$ 136.15.

The physical properties of the methylbenzoic acids are presented in Table 6.

The chemical properties of the methylbenzoic acids are similar to those of benzoic acid. They are better preservatives than benzoic acid, but their higher toxicity limits their application. Their alkali metal salts have better anticorrosion effects than benzoates.

**2-Methylbenzoic Acid** [118-90-1], *o*-toluic acid, can be produced by the oxidation of *o*-xylene. It is an intermediate for agrochemicals, such as the fungicide mepronil and the herbicide londax.

**3-Methylbenzoic Acid** [99-04-7], *m*-toluic acid, is predominantly used for the production of an insect repellant, N,N'-diethyl-*m*-toluamide [134-62-3]. Other uses are as an intermediate for poly(vinyl chloride) stabilizers and agrochemicals.

**4-Methylbenzoic Acid** [99-94-5], *p*-toluic acid, is an intermediate in the production of terephthalic acid by oxidation of *p*-xylene. The methyl group of 4-methylbenzoic acid, unlike that of toluene or *p*-xylene, is resistant to oxidation by molecular oxygen with a conventional cobalt catalyst. The addition of bromide overcomes this difficulty. 4-Methylbenzoic acid can be prepared from *p*-xylene by vapor-phase oxidation or nitric acid oxidation. 4-Methylbenzoic acid is used for producing intermediates for antibiotic pharmaceuticals, and organic pigments.

**4-*tert*-Butylbenzoic Acid** [98-73-7], $C_{11}H_{14}O_2$, $M_r$ 178.23, crystals, *mp* 168.5–169 °C, dissociation constant $6.99 \times 10^{-7}$ at 25 °C.

4-*tert*-Butylbenzoic acid is obtained by the liquid-phase air oxidation of 4-*tert*-butyltoluene (prepared from toluene and isobutylene) [24]. It is a modifier for alkyd resins and a polymerization regulator for polyesters. It is also used as an additive in cutting oils and as a corrosion inhibitor. It is hydrogenated to afford the fragrance 4-*tert*-butylbenzaldehyde [939-97-9].

**2-Benzoylbenzoic Acid** [85-52-9], benzophenone-*o*-carboxylic acid, $C_{14}H_{10}O_3$, $M_r$ 226.23, *mp* 127 °C.

The Friedel–Crafts acylation of benzene with phthalic anhydride is generally used to produce 2-benzoylbenzoic acid. 2-Benzoylbenzoic acid is principally used for the production of photosensitizers.

Table 7. Physical properties of the monochlorobenzoic acids

|  | 2-Chlorobenzoic acid | 3-Chlorobenzoic acid | 4-Chlorobenzoic acid |
|---|---|---|---|
| mp, °C | 141 | 154 | 241.5 |
| bp, °C | 279 | 275 | 276 |
| Density, g/cm$^3$ | 1.544 at 20 °C | 1.517 at 25 °C | 1.571 at 24 °C |
| Solubility, wt% | | | |
| in water at 25 °C | 0.21 | 0.045 | 0.0077 |
| at 100 °C | 4.02 | 0.5 | 0.11 |
| in benzene at 14–16 °C | 0.92 | 0.66 | 0.017 |
| Heat capacity, J g$^{-1}$K$^{-1}$ | 1.042 | 1.042 | 1.073 |
| Heat of fusion, kJ/mol | 25.68 | 23.85 | 32.25 |
| Heat of combustion, kJ/mol | 3084 | 3069 | 3066 |
| Dissociation constant at 25 °C, $K_a$ | $1.2 \times 10^{-3}$ | $1.5 \times 10^{-4}$ | $1.05 \times 10^{-4}$ |
| Flash point, °C | 173±6 | 150 | 238±1 |

## 8.6. Chlorobenzoic Acids

The physical properties of the monochlorobenzoic acids are listed in Table 7.

**2-Chlorobenzoic Acid** [*118-91-2*], $C_7H_5ClO_2$, $M_r$ 156.57.

2-Chlorobenzoic acid is the strongest acid among the three isomers. Its chlorine atom is easily exchangable. It is decarboxylated in aqueous solution at 300 °C. With sodium sulfite it gives 2-sulfobenzoic acid. 2-Aminobenzoic acid is formed with ammonia. 2-Chlorobenzoic acid is produced by oxidation of 2-chlorotoluene [*95-49-8*] with molecular oxygen [35] or by the hydrolysis of α,α,α,2-tetrachlorotoluene [*2136-89-2*]. 2-Chlorobenzoic acid is used for the manufacture of sweeteners, anti-inflammatory pharmaceuticals, and fungicides.

**3-Chlorobenzoic Acid** [*535-80-8*], $C_7H_5ClO_2$. The reactivity of the chlorine atom of 3-chlorobenzoic acid is sluggish compared with the other isomers. Decarboxylation at higher temperatures is insignificant, and it resists oxidation. 3-Chlorobenzoic acid is made by the oxidation of 3-chlorotoluene [*108-41-8*] or by the hydrolysis of 3-chlorobenzoyl chloride [*618-46-2*], which is derived from benzoyl chloride. 3-Chlorobenzoic acid is used for making diphenyl ether type herbicides.

**4-Chlorobenzoic Acid** [*74-11-3*], $C_7H_5ClO_2$ is produced by the oxidation of 4-chlorotoluene [*106-43-4*].

**2,4-Dichlorobenzoic Acid** [*50-84-0*], $C_7H_4Cl_2O_2$, $M_r$ 191.01, mp 164 °C, needles, slightly soluble in water.

2,4-Dichlorobenzoic acid is an important pharmaceutical chemical used in diuretics, antirheumatics, and antimalarials. It is produced by the oxidation of 2,4-dichlorotoluene [*95-73-8*], which is obtained by the direct chlorination of toluene.

**2,5-Dichlorobenzoic Acid** [*50-79-3*], $C_7H_4Cl_2O_2$, mp 154.4 °C, bp 301 °C.

Table 8. Physical properties of the aminobenzoic acids

|  | 2-Aminobenzoic acid | 3-Aminobenzoic acid | 4-Aminobenzoic acid |
|---|---|---|---|
| $mp$, °C | 146.1 | 177.9 | 188.0 |
| Crystal structure | rhombic | monocline | monoclinic |
| Density at 20 °C, g/cm$^3$ | 1.367 | 1.5105 | |
| Solubility, wt % | | | |
| in water | 0.35 at 14 °C | 0.59 at 15 °C | 0.59 at 25 °C, 1.1 at 100 °C |
| in 90 % ethanol | 10.7 at 9.6 °C | 2.2 at 9.6 °C | 11.3 at 9.6 °C |
| in benzene | 0.18 at 11.4 °C | | 0.06 at 11 °C |
| Dissociation constant at 25 °C | | | |
| $K_{acid}$ | $1.07 \times 10^{-5}$ | $1.67 \times 10^{-5}$ | $1.2 \times 10^{-5}$ |
| $K_{base}$ | $1.1 \times 10^{-12}$ | $7.9 \times 10^{-12}$ | $1.7 \times 10^{-12}$ |

2,5-Dichlorobenzoic acid is made from benzoyl chloride by chlorination and hydrolysis. It is used as a starting material in the production of herbicides.

## 8.7. Aminobenzoic Acids

Aminobenzoic acids, $C_7H_7NO_2$, $M_r$ 137.14.
The physical properties of aminobenzoic acids are presented in Table 8.

**2-Aminobenzoic Acid** [118-92-3], anthranilic acid, was first obtained from indigo by alkali fusion. The methyl derivative is found in essential oils. It is an off-white crystalline powder of sweetish taste, sublimable, and freely soluble in hot water. It has an amphoteric character as a weak acid and a weak base. Therefore, it is soluble in strong acids and bases. It is easily decarboxylated to form aniline.

2-Aminobenzoic acid is usually made from phthalic anhydride via sodium phthalamate.

Oxidative decarboxylation (the Hofmann reaction) is carried out by the addition of sodium hypochlorite solution at 60–100 °C followed by acidification.

The product is purified with activated carbon and bisulfite. The overall yield is 88 %. Other synthetic routes include the amination of 2-chlorobenzoic acid and the reduction of 2-nitrobenzoic acid.

2-Aminobenzoic acid is called vitamin $L_1$; it enhances the milk production of cows. The major use of 2-aminobenzoic acid is as an intermediate for dyes. As its occurrence suggests, it is an intermediate for indigo synthesis. Mordant Brown 40 and Vat Violet 13 are other derivatives. 4-Hydroxy-1-methyl-carbostyril [1677-46-9] is an important derivative used in making dyes and pigments. Its methyl and ethyl esters are used as fragrances for toiletries. In the pharmaceutical industry, it is used as an intermediate for tranquilizers and antiphlogistics.

**3-Aminobenzoic Acid** [99-05-8] is prepared by the reduction of 3-nitrobenzoic acid and used in dye syntheses. Its derivative N,N-dimethyl-3-aminobenzoic acid [99-64-9] is also used in the production of dyes.

**4-Aminobenzoic Acid** [150-13-0] is widely found in nature as a factor in vitamin B complex. Baker's yeast contains 5 or 6 mg/kg of 4-aminobenzoic acid, and brewer's yeast 10–100 mg/kg. It enhances the growth of various microorganisms, and it is essential to the anaerobic metabolism of some bacteria. It is known as bacterial vitamin H and is antagonistic to sulfonamide drugs. It darkens slightly on exposure to air or light. It is slightly soluble in cold water. The pH of a 0.5 wt% aqueous solution is 3.5. It is soluble in ethyl acetate and acetic acid but practically insoluble in petroleum ether.

The conventional method of producing 4-aminobenzoic acid is the reduction of 4-nitrobenzoic acid. The Hofmann degradation reaction of terephthalic acid is another source of 4-aminobenzoic acid. Dimethyl terephthalate reacts with one equivalent of potassium hydroxide in methanol. The resultant salt is then amidated with liquid ammonia in dimethylformamide at 130–140 °C to give potassium terephthalamate. This then reacts with sodium hypochlorite solution at 50–92 °C. The overall yield is approximately 70% [26].

A two-stage Hofmann reaction has been developed. The yield is 93% [27].

4-Aminobenzoic acid is used principally in the pharmaceuticals industry. Other uses are as a cross-linking agent for polyurethane resins, dyes, and feedstock additives. Ethyl 4-aminobenzoate [94-09-7], anesthesin, mp 90–91 °C is a local anesthetic of low toxicity, irritation, and persistency. It is used especially by dentists.

Folic acid [59-30-3], another derivative of 4-aminobenzoic acid, is a hematopoietic vitamin.

**Table 9.** Physical properties of the mononitrobenzoic acids

|  | 2-Nitrobenzoic acid | 3-Nitrobenzoic acid | 4-Nitrobenzoic acid |
|---|---|---|---|
| mp, °C | 148 | 142 | 240 |
| Density, g/cm$^3$ | 1.575 | 1.494 | 1.610 |
| Solubility, g per 100 mL of solvent at 10 °C |  |  |  |
| in water | 0.75 (at 25 °C) | 0.24 (at 15 °C) | 0.02 (at 15 °C) |
| in methanol | 42.72 | 47.34 | 9.6 |
| in benzene | 0.294 | 0.795 | 0.017 (at 12.5 °C) |
| Heat capacity, J g$^{-1}$K$^{-1}$ | 1.149 | 1.035 | 1.077 |
| Heat of fusion, kJ/mol | 28.02 | 29.33 | 37.02 |
| Heat of combustion, kJ/mol | 3078 | 3053 | 3881 |
| Dissociation constant, $K_a$ | $6.1 \times 10^{-3}$ at 18 °C | $3.48 \times 10^{-4}$ at 25 °C | $3.93 \times 10^{-4}$ at 25 °C |
| Decarboxylation temperature, °C | 180 | 238 | > 240 |

## 8.8. Nitrobenzoic Acids

The physical properties of the mononitrobenzoic acids are given in Table 9.

**2-Nitrobenzoic Acid** [27178-83-2], $C_7H_5NO_4$, $M_r$ 167.12, can be prepared by the oxidation of 2-nitrotoluene [88-72-2] with nitric acid.

**3-Nitrobenzoic Acid** [121-92-6] is prepared by nitrating benzoic acid at low temperature. Approximately 20 % of the 2-nitro isomer and 1.5 % of the 4-nitro isomer are co-produced. The 3-Nitrobenzoic acid can be purified by recrystallization of the sodium salt. The oxidation of 3-nitrobenzaldehyde [99-61-6], prepared by the controlled oxidation of benzaldehyde, gives a higher yield. 3-Nitrobenzoic acid is an intermediate in the preparation of 3-aminobenzoic acid and azo dyes. Its derivative 4-chloro-3-nitrobenzoic acid [96-99-1] is an intermediate in the production of dyes.

**4-Nitrobenzoic Acid** [62-23-7] is produced commercially by the oxidation of 4-nitrotoluene [99-99-0] with molecular oxygen. Oxidation with 15 % nitric acid at 175 °C produces the acid in 88.5 % yield. An interesting method involves the nitration and subsequent oxidation of polystyrene [28]. This method uses the steric hindrance of the polymer chain to improve the *para* to *ortho* ratio of the product.

4-Nitrobenzoic acid is an intermediate for 4-aminobenzoic acid. 4-Nitrobenzoyl chloride [122-04-3], mp 75 °C, is a starting material for procaine hydrochloride and folic acid.

## 8.9. 3-Sulfobenzoic Acid

3-Sulfobenzoic acid [121-53-9], mp 141 °C, mp of dihydrate 98 °C, hygroscopic, soluble in water and ethanol. It is produced by the direct sulfonation of benzoic acid with oleum in ca. 85 % yield. 3-Sulfobenzoic acid is chiefly used for the production of 3-hydroxybenzoic acid [99-06-9] by alkali fusion.

## 8.10. Hexahydrobenzoic Acid

Hexahydrobenzoic acid [98-89-5], cyclohexanecarboxylic acid, $C_7H_{12}O_2$, $M_r$ 128.17, mp 30 – 31 °C, bp 232 – 234 °C, bp 150.7 °C at 6.53 kPa, bp 120 – 121 °C at 1.87 kPa, $d_4^{20}$ 1.0274, slightly soluble in water, soluble in ethanol and petroleum ether.

Hexahydrobenzoic acid reacts with nitrosylsulfuric acid to produce caprolactam. This is the basic reaction of the toluene-based caprolactam process. Hexahydrobenzoic acid can be oxidized by molecular oxygen at 220 °C in the presence of a copper catalyst [29]. The main products are cyclohexene and cyclohexanone. It can be dehydrocarbonylated in the vapor phase over a heteropoly acid catalyst to yield cyclohexene [30]. Hexahydrobenzoic acid is produced in quantitative yield by the hydrogenation of benzoic acid over noble metal catalysts at 150 – 170 °C, at elevated pressure.

# 9. Toxicology

Acute toxicity data for benzoic acid and its derivatives are given in Table 10.

Benzoic acid is not accumulated in the human body. After administration it reacts with glycine to form hippuric acid, which is excreted in the urine. The FAO/WHO guideline for the upper limit of daily intake of benzoic acid is 5 – 10 mg/kg. The reproductive toxicity test for rats showed no abnormality.

Benzoic acid has an irritating effect on human mucous membranes. When dust formation is anticipated, protections should be arranged.

Benzoyl chloride is significantly toxic, like other acid chlorides. It irritates mucous membranes strongly.

Benzonitrile is less toxic than aliphatic nitriles. It is, nevertheless, absorbed through the skin and may cause convulsions of tissue and paralysis of nerves.

The esters of 4-aminobenzoic acid have a local anesthetic effect. After absorption they stimulate the central nervous system. At the workplace, some of these derivatives may have an allergenic and skin hypersensitizing effect.

4-Nitrobenzoic acid has a significant bactericidal action against *Staphylococci* and *Streptococci*. It is reported to be slightly mutagenic.

**Table 10.** Acute toxicity data for benzoic acid and its derivatives (RTECS)

| Compound | Species | Route | Acute toxicity |
|---|---|---|---|
| Benzoic acid | rat | oral | $LD_{50}$ 1700 mg/kg |
| [65-85-0] | mouse | oral | $LD_{50}$ 1940 mg/kg |
| | rat | inhalation | $LC_{50} > 26$ mg/m$^3$ |
| | rat | i.p. | $LD_{50}$ 1600 mg/kg |
| | mouse | i.p. | $LD_{50}$ 1460 mg/kg |
| | human | skin | $TDL_0$ 6 mg/kg |
| | rabbit | s.c. | $LDL_0$ 2 g/kg |
| | rabbit | irritation (skin) | mild |
| | rabbit | irritation (eye) | severe |
| Sodium benzoate | rat | oral | $LD_{50}$ 4070 mg/kg |
| [532-32-1] | mouse | oral | $LD_{50}$ 1600 mg/kg |
| | dog | oral | $LD_{50}$ 2 g/kg |
| | mouse | i.v. | $LD_{50}$ 1440 mg/kg |
| | rat | s.c. | $LD_{50}$ 2 g/kg |
| | mouse | i.m. | $LD_{50}$ 2306 mg/kg |
| Methyl benzoate | rat | oral | $LD_{50}$ 1177 mg/kg |
| [93-58-3] | mouse | oral | $LD_{50}$ 3330 mg/kg |
| | rabbit | irritation (skin) | mild |
| | rabbit | irritation (eye) | mild |
| n-Butyl benzoate | rat | oral | $LD_{50}$ 735 mg/kg |
| [136-60-7] | rabbit | skin | $LD_{50}$ 4 g/kg |
| | rabbit | irritation (skin) | severe |
| | rabbit | irritation (eye) | mild |
| Diethylene glycol dibenzoate | rat | oral | $LD_{50}$ 2830 mg/kg |
| [120-55-8] | rabbit | skin | $LD_{50}$ 20 g/kg |
| | rabbit | irritation (skin) | mild |
| | rabbit | irritation (eye) | mild |
| Phenyl benzoate | mouse | oral | $LD_{50}$ 1225 mg/kg |
| [93-99-2] | | | |
| Benzyl benzoate | rat | oral | $LD_{50}$ 1700 µL/kg |
| [120-51-4] | mouse | oral | $LD_{50}$ 1400 µL/kg |
| | dog | oral | $LD_{50} > 22.44$ g/kg |
| | mouse | i.p. | $LD_{50} > 500$ mg/kg |
| | rabbit | skin | $LD_{50}$ 4 g/kg |
| Benzoyl chloride | rat | oral | $LD_{50}$ 1900 mg/kg |
| [98-88-4] | rat | inhalation | $LC_{50}$ 1870 mg/m$^3$ |
| Benzonitrile | rat | oral | $LDL_0$ 720 mg/kg |
| [100-47-0] | mouse | oral | $LD_{50}$ 971 mg/kg |
| | rat | inhalation | $LCL_0$ 950 ppm |
| | mouse | inhalation | $LC_{50}$ 1800 mg/m$^3$ |
| | mouse | i.p. | $LD_{50}$ 400 mg/kg |
| | rabbit | s.c. | $LDL_0$ 200 mg/kg |
| | rabbit | skin | $LD_{50}$ 1250 mg/kg |
| | rabbit | irritation (skin) | moderate |
| Benzaldehyde | rat | oral | $LD_{50}$ 1300 mg/kg |
| [100-52-7] | mouse | oral | $LD_{50}$ 28 mg/kg |
| | rat | inhalation | $LC_{50} > 500$ mg/m$^3$ |
| | rat | s.c. | $LDL_0$ 5 g/kg |
| | mouse | i.p. | $LD_{50}$ 9 mg/kg |
| | rabbit | irritation (skin) | moderate |
| 2-Methylbenzoic acid | mouse | i.p. | $LD_{50}$ 422 mg/kg |
| [118-90-1] | | | |
| 3-Methylbenzoic acid | rat | oral | $LD_{50} > 5$ g/kg |
| [99-04-7] | mouse | oral | $LD_{50}$ 1630 mg/kg |

**Table 10.** (continued)

| Compound | Species | Route | Acute toxicity |
|---|---|---|---|
| | mouse | i.p. | $LD_{50}$ 562 mg/kg |
| 4-Methylbenzoic acid | mouse | oral | $LD_{50}$ 2340 mg/kg |
| [99-94-5] | rat | i.p. | $LD_{50}$ 874 mg/kg |
| | mouse | i.p. | $LD_{50}$ 916 mg/kg |
| 4-tert-butylbenzoic acid | rat | oral | $LD_{50}$ 700 mg/kg |
| [98-73-7] | rat | inhalation | $LC_{50} > 1900$ mg/m$^3$ |
| 2-Chlorobenzoic acid | rat | oral | $LD_{50} > 500$ mg/m$^3$ |
| [118-91-2] | rat | i.p. | $LD_{50}$ 2350 mg/kg |
| | rabbit | irritation (skin) | mild |
| | rabbit | irritation (eye) | moderate |
| 3-Chlorobenzoic acid [535-80-8] | rat | i.p. | $LD_{50}$ 750 mg/kg |
| 4-Chlorobenzoic acid | rat | oral | $LD_{50}$ 1170 mg/kg |
| [74-11-3] | mouse | oral | $LD_{50}$ 1170 mg/kg |
| | rat | i.p. | $LD_{50}$ 1 g/kg |
| 2,4-Dichlorbenzoic acid | mouse | oral | $LD_{50}$ 830 mg/kg |
| [50-84-0] | mouse | s.c. | $LD_{50}$ 1200 mg/kg |
| 2,5-Dichlorobenzoic acid | mouse | i.p. | $LD_{50}$ 237 mg/kg |
| [50-79-3] | mouse | s.c. | $LD_{50}$ 1200 mg/kg |
| 2-Aminobenzoic acid | rat | oral | $LD_{50}$ 5410 mg/kg |
| [118-92-3] | mouse | oral | $LD_{50}$ 1400 mg/kg |
| | mouse | i.v. | $LD_{50} > 500$ mg/kg |
| | mouse | i.p. | $LD_{50}$ 2500 mg/kg |
| 3-Aminobenzoic acid | mouse | oral | $LD_{50}$ 6300 mg/kg |
| [99-05-8] | mouse | i.p. | $LD_{50}$ 500 mg/kg |
| 4-Aminobenzoic acid | rat | oral | $LD_{50} > 6$ g/kg |
| [150-13-0] | mouse | oral | $LD_{50}$ 2850 mg/kg |
| | dog | oral | $LD_{50}$ 1 g/kg |
| | rabbit | i.v. | $LD_{50}$ 2 g/kg |
| | rat | i.p. | $LD_{50} > 3450$ mg/kg |
| 3-Nitrobenzoic acid | mouse | i.v. | $LD_{50}$ 640 mg/kg |
| [121-92-6] | mouse | i.p. | $LD_{50}$ 610 mg/kg |
| 4-Nitrobenzoic acid | rat | oral | $LD_{50}$ 1960 mg/kg |
| [62-23-7] | mouse | i.v. | $LD_{50}$ 770 mg/kg |
| | mouse | i.p. | $LD_{50}$ 880 mg/kg |
| | rabbit | irritation (eye) | moderate |
| Hexahydrobenzoic acid [98-89-5] | rat | oral | $LD_{50}$ 3265 mg/kg |

# 10. References

[1] W. W. Kaeding, *J. Org. Chem.* **26** (1961) 3144.
[2] W. W. Kaeding, R. O. Lindblom, R. G. Temple, *Ind. Eng. Chem.* **53** (1961) no. 10, 805.
[3] American Cyanamid, DE 2 258 227, 1973.
[4] C. R. Kinney, D. P. Langlois, *J. Am. Chem. Soc.* **53** (1931) 2189.
[5] Mitsubishi Chem. Ind., JP 47-44215, 1972.
[6] J. T. Bornwater, A. F. Holleman, *Recl. Trav. Chim. Pays-Bas* **31** (1922) 221.
[7] W. J. LeNoble, G. W. Wheland, *J. Am. Chem. Soc.* **80** (1958) 5397.
[8] BIOS 1786; BIOS (Misc.) 112.

[9] N. Ohta, T. Tezuka, *Kogyo Kagaku Zasshi* **59** (1956) 71.
[10] Mitsubishi Chem. Ind., JP-Kokai 56-39045, 1981.
[11] Bayer, JP-Kokai 51-88934, 1976.
[12] Y. Kamiya, *Adv. Chem. Ser.* **76** (1968) 193.
[13] I. P. Tharp, H. A. Nottorf, *Ind. Eng. Chem.* **39** (1947) 1300.
[14] *Org. Synth.* **2** (1922) 5.
[15] SU 56 693, 1936 (Uwarov,Stepanov).
[16] Chemische Werke Witten, DE 1 070 616, 1956.
[17] ICI, GB 987 516, 1965.
[18] Allied Chemical, US 2 486 934, 1949; US 2 499 055, 1950; US 2 828 325, 1958.
[19] T. Ohara, Z. Iwao, M. Ninomiya, Y. Nakagawa, *Yuki Gosei Kagaku Kyokaishi* **26** (1968) 213.
[20] BIOS **986**$^2$, no. 209, 417.
[21] Rohm & Haas, US 2 770 641, 1954.
[22] G. D. Shik et al., *Neftekhimiya* **21** (1981) 133.
[23] *Org. Synth.* **33** (1953) 13.
[24] G. W. Hearne et al., *Ind. Eng. Chem.* **47** (1955) 2311.
[25] Toray Industries, JP-Kokai 57-175143, 1982.
[26] C. S. Rondestvedt Jr., *J. Org. Chem.* **42** (1977) 3118.
[27] Akzo, DE 2313580, 1973.
[28] C. S. Rondestvedt, Jr., J. R. Jeffrey, J. E. Miller, *Ind. Eng. Chem. Prod. Res. Dev.* **16** (1977) no. 4, 309.
[29] J. A. Bigot, P. L. Kerkhoffs, *Recl. Trav. Chim. Pays-Bas* **82** (1963) 677.
[30] Mitsubishi Chem. Ind., JP 55-37532, 1980; JP 55-50928, 1980.
[31] T. Yokoyama, T. Setoyama, N. Fujita, M. Nakajima, T. Maki, K. Fujii, *Appl. Catal. A* **88** (1992) 149.
[32] Bayer, DE 3 420 111, 1984(W. Schulte-Huermann). Stamicarbon, EP-A 453 022, 1990 (U. F. Kragten, M. K. Frohn-Schloesser).
[33] Mitsubishi Chem. Ind., JP-Kokai 4-69357, 1990 (K. Fujii, M. Nakamura).
[34] Stamicarbon, EP-A 183 318, 183 319, 1984 (J. J. P. M. Goorden, A. J. F. Simons, L. A. L. Kleintjens).
[35] Stamicarbon, EP-A 188 298, 1985 (S. M. P. Mutsers, M. H. Willems, W. P. Wolvers).
[36] Nippon Kokan, JP-Kokai 5-255 181, 1991 (I. Osada, A. Imai, A. Miki).
[37] European Parliament Council, Official Journal, no. L061 (1995) 0001–0040.
[38] http://www1.iastate.edu/~chem163/benzoic.html.
[39] http://checfs1.ucsd.edu/Courses/CoursePages/Uglabs/MSDS/benzoic.acid.-fisher.html.
[40] C. Barron: "Benzoic acid" in *SRI International Chemical Economics Handbook*, 1995, **618**, 8000A.
[41] 21 CFR 184. 1733 (Report no. 88, US Dept. of Agriculture).
[42] *Chemical Market Reporter* Sept. 29, 1997.

# Benzoquinone

K. THOMAS FINLEY, State University of New York, Brockport, New York 14420, United States

| | | | | | |
|---|---|---|---|---|---|
| 1. | Introduction | 841 | 5. | Uses of the Quinones | 845 |
| 2. | Physical Properties | 842 | | | |
| 3. | Chemical Properties | 842 | 6. | Toxicology of Benzoquinones | 846 |
| 4. | Production of 1,4-Benzoquinone | 845 | 7. | References | 846 |

## 1. Introduction

Benzoquinones, six-carbon cyclic dienediones, play significant and varied roles in metabolic processes and in synthetic reactions [1]–[4].

The benzoquinones exist in two isomeric modifications: 1,2-benzoquinone (**1**) [583-63-1] and 1,4-benzoquinone (**2**) [106-51-4]. Both of these structures can be substituted with a wide variety of alkyl, aryl, halo, oxygen, nitrogen, and sulfur groups.

The quinones are most often named as derivatives of **1** or **2**, e.g., 2-methyl-1,4-benzoquinone (**3**) [553-97-9]. The CAS nomenclature of the quinones, e.g., 2,5-cyclohexadiene-1,4-dione for **2** is seldom used. Trivial names like quinone, *p*-benzoquinone, and *p*-tolyquinone (**3**) exist and are used occasionally. Two important exceptions are the oxidizing agents *p*-chloranil (**4**) [118-75-2], and DDQ (**5**) [84-58-2].

These two compounds are widely used and are frequently referred to by their trivial names.

These "rules" are well illustrated by the *1984 – 1985 Aldrich Catalog/Handbook of Fine Chemicals* where one finds: "DDQ, see 2,3- dichloro-5,6-dicyano-1,4-benzoquinone", but "2,5-dichloro-3,6-dihydroxy-1,4-benzoquinone, see chloranilic acid" (**6**), [*87-88-7*] [5].

$$\underset{\textbf{6}}{\text{Cl, OH, HO, Cl structure}}$$

## 2. Physical Properties

1,4-Benzoquinone, (**2**), $C_6H_4O_2$, $M_r$ 108.10, *mp* 113 °C, $d_4^{20}$ 1.318, $p_v^{20}$ 12 Pa, is soluble in most oxygenated organic solvents (ether, alcohol), and hot ligroin, slightly soluble in petroleum ether, and insoluble in water. Crystallization from alcohol or sublimation produces yellow monoclinic prisms.

1,2-Benzoquinone (**1**), $C_6H_4O_2$, $M_r$ 108.10, *mp* 60 – 70 °C (decomp.), crystallizes in red plates or prisms. It is much less stable than **2**, e.g., it decomposes in water. Compound **1** is soluble in ether, acetone, and benzene, but insoluble in petroleum ether.

2,3,5,6-Tetrachloro-1,4-benzoquinone, (**4**), *p*-chloranil, $C_6Cl_4O_2$, $M_r$ 245.88, *mp* 290 °C (sealed tube); and 2,3-dichloro-5,6-dicyano-1,4-benzoquinone (**5**), DDQ, $C_8Cl_2N_2O_2$, $M_r$ 227.01, *mp* 213 – 216 °C, are yellow to bright yellow in color and form monoclinic prisms and plates on crystallization.

The lower molecular mass benzoquinones have high-vapor pressures and pungent, irritating odors.

## 3. Chemical Properties

The quinones are important *oxidants* in synthetic [6] and metabolic processes [7]. The presence of an extensive array of conjugated systems, especially the $\alpha,\beta$-unsaturated ketone arrangement, allows the quinones to participate in a variety of reactions. The acid-catalyzed Michael addition is most characteristic of these reactions, but Diels-Alder, electrophilic, and radical additions, as well as substitutions are well documented [8]. There is also a growing literature related to the role of quinones in photochemical processes [9].

In an effort to understand the question of valence in organic molecules, A. MICHAEL carried out one of the earliest quinone addition reactions [10]:

The subsequent chemistry, as shown in Equation 1, involves cross-oxidation to give **7**, 2-phenylthio-1,4-benzoquinone [*18232-03-6*]. Further addition and oxidation yields the bis(phenylthio) derivative (**8**) [*17058-53-6*]. The resulting molecular change is generally due to the electronic influence of the substituent that is introduced. However, steric effects can also play an important role [11]

Quinones with high-oxidation potentials have made practical a great many conversions involving delicate oxidations. 2,3-Dichloro-5,6-dicyano-1,4-benzoquinone (**5**) is often chosen as the oxidizing reagent because of its favorable combination of high-oxidation potential (ca. 1000 mV) and stability [12]:

Chloranil (**4**) often finds application in the selective dehydrogenation of steroid ketones [13]:

The useful change in product structure with DDQ (**5**) has been studied in detail [14].

Quinone chemistry in general presents problems involving instability, sequential addition, and poor regiospecificity. One method that minimizes these difficulties is in situ generation of the quinone in the presence of the nucleophile [15]. This technique has been particularly useful with the less stable 1,2-benzoquinones [16]:

It also gives excellent results with Diels-Alder reactions of sensitive quinones [17]. Products, such as 4α-acetyl-4α,5,8,8α-tetrahydro-6-methyl-5-(2-(phenylmethoxy)ethyl-1,4-naphthalene-dione [80866-98-4], are useful in the synthesis of antitumor compounds:

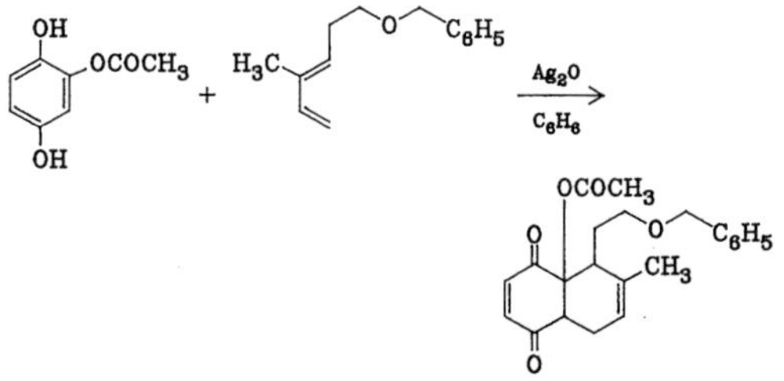

The quinone monoacetals, e.g., 3,4,4-trimethoxycyclohexa-2,5-dienone (**9**) [64701-03-7], are useful in reactions requiring bases, to which the quinones are especially sensitive [18]:

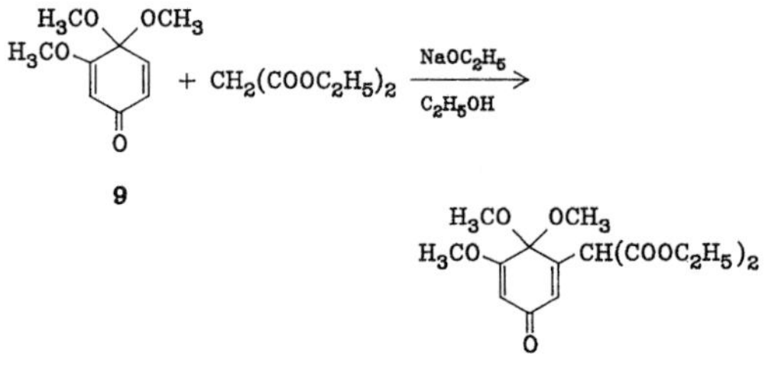

Quinones possessing unsaturated side chains are important in respiratory, blood clotting, and oxidative phosphorylation processes and, as a result, have stimulated the search for good synthetic methods. The conversion of 2,3-dimethoxy-5-methyl-1,4-benzoquinone (**10**) [605-94-7] to 2,3-dimethoxy-5-methyl-6-(3-methyl-2-butenyl)-1,4-benzoquinone or coenzyme $Q_1$ (**11**) [727-81-1] using an allyltin reagent has been described recently [19]:

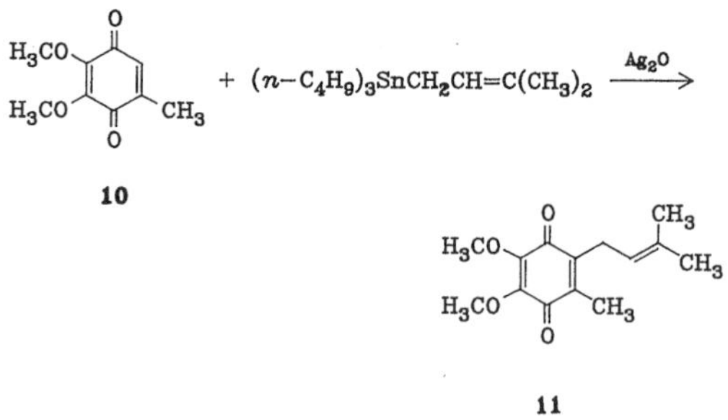

The synthetic potential of benzoquinone radical chemistry awaits further research. As Equation 2 suggests, past work has dealt primarily with the naphthoquinones [20]. The studies of L. Fieser and his students with compounds related to 2-methyl-1,4-naphthoquinone (**12**) [58-27-5] produced alkylated products, such as ethyl 9-(2-methyl-1,4-naphthoquinone-3)pelargonate (**13**) [80632-67-3]:

$$\text{12} + [C_2H_5OOC(CH_2)_8COO]_2O_2 \longrightarrow$$

$$\text{13} \quad (2)$$

## 4. Production of 1,4-Benzoquinone

With the exception of 1,4-benzoquinone, the simple quinones do not have a substantial market. Many syntheses have been examined, but industrial scale preparation of this quinone is still carried out by the oxidation of aniline or phenol (Eq. 3). The product is steam distilled, chilled, and obtained in high yield and purity [21].

$$2\, C_6H_5NH_2 + 4\, MnO_2 + 5\, H_2SO_4 \xrightarrow{cold} 2\, C_6H_4O_2 + 4\, MnSO_4 + (NH_4)_2SO_4 + 4\, H_2O \quad (3)$$

Recent prices for this compound in industrial quantities are about $ 10/kg. A substantial number of quinones are also available as fine organics in small quantities. They are generally prepared by oxidative methods.

## 5. Uses of the Quinones

Aside from their utility in laboratory syntheses, the benzoquinones find application as oxidants, bactericides, and chemical intermediates. 1,4-Benzoquinone is also used as an intermediate in the preparation of hydroquinone, which is important in the photographic and dye industries, and is used in the tanning of leather. The closely related naphthoquinones are used as vitamin K substitutes, antihemorrhagic agents, and fungicides ($\rightarrow$ Naphthoquinones).

# 6. Toxicology of Benzoquinones

The high-vapor pressures of the lower molecular mass quinones, coupled with their penetrating odor, present a health hazard [22]–[25]. Quinone vapor, and the quinone resulting from hydroquinone dust in moist air, produce eye injuries that can lead to appreciable loss of vision. In concentrations of 0.1–0.15 ppm, the odor of 1,4-benzoquinone becomes definite, and at 0.5 ppm becomes irritating. Both the MAK and TLV (TWA) are 0.1 ppm and 0.4 mg/m$^3$; STEL are 0.3 ppm and 7 mg/m$^3$ [26], [27]. However, notably Russia has fixed the threshold limit in air at 0.01 ppm [28]. Both solid quinones and their solutions must be handled with care because they can cause severe local damage to the skin and mucous membranes.

Limited studies have been conducted with mice to determine the carcinogenicity of quinones. Results for *p*-chloranil are conclusive, while those for 1,4-benzoquinone are only suggestive [29].

# 7. References

[1] S. Patai (ed.): *The Chemistry of the Quinonoid Compounds*, Wiley-Interscience, New York 1974, 1274 pp.

[2] S. M. Bruce:"Benzoquinones and Related Compounds," in S. Coffey (ed.): *Rodd's Chemistry of Carbon Compounds*, 2nd ed.,vol. **33**,Elsevier, Amsterdam , 1974,Chap. 8.

[3] "A Literature Review of Hydroquinone and *p*-Benzoquinone," Eastmann Kodak Co., Kingsport Tenn., 1977, 64 pp.

[4] T. Laird: "Quinones," in J. F. Stoddart (ed.): *Comprehensive Organic Chemistry, the Synthesis and Reactions of Organic Compounds*, vol. **1**, Pergamon Press, Oxford 1979, Chap. 5.5.

[5] *Aldrich Catalog/Handbook of Fine Chemicals*, Aldrich Chemical Co., Milwaukee, Wisc., 1984–1985.

[6] H.-D. Becker: "Quinones as Oxidants and Dehydrogenating Agents," in S. Patai (ed.): *The Chemistry of the Quinonoid Compounds*, Part **1**, Chap. 7, Wiley-Interscience, New York 1974.

[7] R. Bentley, I. M. Campbell: "Biological Reactions of Quinones," in S. Patai (ed.): *The Chemistry of the Quinonoid Compounds*, Part **2**, Chap. 13, Wiley-Interscience, New York 1974.

[8] K. T. Finley: "The Addition and Substitution Chemistry of Quinones," in S. Patai (ed.): *The Chemistry of the Quinonoid Compounds*, Part **2**, Chap. 17, Wiley-Interscience, New York 1974.

[9] J. M. Bruce: "Photochemistry of Quinones," in S. Patai (ed.): *The Chemistry of the Quinonoid Compounds*, Part **1**, Chap. 9, Wiley-Interscience, New York 1974.

[10] A. Michael, *J. Prak. Chem.* **79** (1909) 418.

[11] E. R. Brown, K. T. Finley, R. L. Reeves, *J. Org. Chem.* **36** (1971) 2849.

[12] H.-D. Becker, A. Björk, E. Adler, *J. Org. Chem.* **45** (1980) 1596.

[13] E. J. Agnello, G. D. Laubach, *J. Am. Chem. Soc.* **82** (1960) 4293.

[14] D. Burn, D. N. Kirk, V. Petrov, *Proc. Chem. Soc.* 1960 14.

[15] W.-H. Wanzlick: *Newer Methods of Preparative Organic Chemistry*, vol. **4**, Academic Press, New York 1968, pp. 139–154.

[16] J. Daneke, U. Jahnke, B. Pankow, W.-H. Wanzlick, *Tetrahedron Lett.* 1970 1271.

[17] G. A. Kraus, M. J. Taschner, *J. Org. Chem.* **45** (1980) 1174.
[18] K. A. Parker, S.-K. Kang, *J. Org. Chem.* **45** (1980) 1218.
[19] K. Maruyama, Y. Naruta, *J. Org. Chem.* **43** (1978) 3796.
[20] L. F. Fieser, R. B. Turner, *J. Am. Chem. Soc.* **69** (1947) 2338.
[21] W. H. Shearon, L. G. Davy, H. Von Bramer, *Ind. Eng. Chem.* **44** (1952) 1730.
[22] L. T. Fairhall: *Industrial Toxicology,* 2nd ed., Hafner, New York 1969, pp. 333, 334.
[23] R. E. Gosselin, R. P. Smith, H. C. Hodge: *Clinical Toxicology of Commercial Products,* 5th ed., Williams & Wilkins, Baltimore 1984, pp. II-191, II-1183.
[24] N. I. Sax: *Dangerous Properties of Industrial Materials,* 6th ed., Van Nostrand-Reinhold, New York 1984, p. 388.
[25] G. Weiss: *Hazardous Chemicals Data Book,* Noyes Data Corp., Park Ridge, N.J., 1980, pp. 993, 1150.
[26] DFG (ed.): *Maximum Concentrations at the Workplace (MAK),* Verlag Chemie, Weinheim 1982.
[27] ACGIH (ed.): *Threshold Limit Values (TLV),* ACGIH, Cincinnati, Ohio, 1983.
[28] K. Verschueren: *Handbook of Environmental Data on Organic Chemicals,* 2nd ed., Van Nostrand, New York 1977, pp. 280, 350.
[29] N. I. Sax: *Cancer Causing Chemicals,* Van Nostrand-Reinhold, New York 1981, pp. 321, 322.

# Benzyl Alcohol

FRIEDRICH BRÜHNE, Bayer AG, Krefeld-Uerdingen, Federal Republic of Germany (Chaps. 1–9)
ELAINE WRIGHT, General Motors Research Laboratories, Warren, Michigan 48090, United States (Chap. 10)

| | | |
|---|---|---|
| 1. | Introduction | 849 |
| 2. | Physical Properties | 850 |
| 3. | Chemical Properties | 851 |
| 4. | Production | 852 |
| 4.1. | Hydrolysis of Benzyl Chloride | 852 |
| 4.2. | Hydrogenation and Reduction of Benzaldehyde | 853 |
| 4.3. | Oxidation of Toluene | 855 |
| 4.4. | Other Manufacturing Processes | 855 |
| 5. | Quality Standards and Test Methods | 856 |
| 6. | Storage, Transportation, and Safety Regulations | 857 |
| 7. | Uses | 857 |
| 8. | Derivatives | 857 |
| 8.1. | Dibenzyl Ether | 858 |
| 8.2. | Benzyl Acetate | 858 |
| 8.3. | Benzyl Benzoate | 859 |
| 8.4. | Benzyl Salicylate | 859 |
| 9. | Economic Aspects | 860 |
| 10. | Toxicology | 860 |
| 11. | References | 861 |

## 1. Introduction

Benzyl alcohol [100-51-6] is the simplest and also the industrially most important aromatic alcohol. LIEBIG and WÖHLER first prepared benzyl alcohol from bitter almond oil (benzaldehyde) in 1832 [1]. The structure of benzyl alcohol was determined in 1853 by CANNIZZARO [2]. CANNIZZARO used the reaction named after him, in which benzaldehyde is disproportionated into benzoic acid and benzyl alcohol through the action of an alkali.

Benzyl alcohol occurs in nature both free and in combined forms. In the latter case it exists as esters of acetic, benzoic, salicylic, and cinnamic acids, among others. It is known to occur in the balsams of Peru and Tolu, in the flower oils of hyacinths and wallflowers, in ylang-ylang oil, and in other essential oils. It also occurs as a glucoside in maize.

**Table 1.** Vapor pressure of benzyl alcohol vs. temperature

| Temperature, °C | Vapor pressure, kPa |
|---|---|
| 60 | 0.18 |
| 80 | 0.66 |
| 100 | 2.02 |
| 120 | 5.28 |
| 140 | 12.20 |
| 160 | 25.47 |
| 180 | 48.88 |
| 200 | 87.43 |

**Table 2.** Azeotropes of benzyl alcohol

| Component | $bp$ (101.3 kPa), °C | Benzyl alcohol, wt % |
|---|---|---|
| Water | 99.9 | 9.0 |
| Hexachloroethane | 182.0 | 12.0 |
| Ethylene glycol | 193.4 | 46.5 |
| Nitrobenzene | 204.2 | 62.0 |
| Dimethylaniline | 193.9 | 6.5 |
| Diethylaniline | 204.2 | 72.0 |
| Naphthalene | 204.1 | 60.0 |
| $p$-Cresol | 206.8 | 62.0 |
| $p$-Bromotoluene | 184.5 | 8.0 |

## 2. Physical Properties

Benzyl alcohol, $C_7H_8O$, is a colorless liquid with a faint aromatic odor and a mildly irritating effect on the mucous membranes.

$M_r$ 108.14, $bp$ 205.4 °C at 101.3 kPa, $mp$ −15.4 °C, refractive index $n_D^{20}$ 1.5400, density $\varrho$ 1.061 g/cm$^3$ at 0 °C, 1.045 g/cm$^3$ at 20 °C, 1.030 g/cm$^3$ at 40 °C, 1.015 g/cm$^3$ at 60 °C.

Specific heat 1972 J kg$^{-1}$ K$^{-1}$ at 20 °C, 2135 J kg$^{-1}$ K$^{-1}$ at 50 °C, heat of fusion 82.9 J/g, heat of evaporation 467.0 J/g at 205.4 °C, standard combustion enthalpy 34.58 kJ/g.

Flash point 101 °C, autoignition temperature 435 °C, lower explosive limit (at 170 °C and 101.3 kPa) 1.3 vol %, upper explosive limit 13.0 vol %.

Table 1 shows the vapor pressure in relation to temperature.

Benzyl alcohol is miscible with many organic solvents. At 20 °C, 4.0 g of benzyl alcohol is soluble in 100 g of water and 5.1 g of water is soluble in 100 g of benzyl alcohol. The compositions of several binary azeotropic mixtures are given in Table 2. For further azeotropes, see [3]. Liquid–vapor equilibrium data for binary mixtures of benzyl alcohol with water or organic solvents are given in [4].

Dynamic viscosity $\eta$ 5.584 × 10$^{-3}$ Pa · s at 20 °C, surface tension $\sigma$ 39.96 × 10$^{-3}$ N/m at 20 °C. Dipole moment $\mu$ (measured in liquid benzene) 5.571 × 10$^{-30}$ C · m (1.67 D), relative dielectric constant $\varepsilon_r$ 11.92 at 30 °C, 9.81 at 60 °C.

# 3. Chemical Properties

The chemical properties of benzyl alcohol are determined mainly by the hydroxyl group. This group shows the typical reactions of aliphatic alcohols (→ Alcohols, Aliphatic). The reactivity of the hydroxyl group is increased by the proximity of the aromatic nucleus. Benzyl alcohol is less acidic than the isomeric cresols and therefore not completely soluble in aqueous alkalis.

If benzyl alcohol is heated in the presence of dehydrating compounds (e.g., aluminum oxide), dibenzyl ether, toluene, and benzaldehyde are formed. Dibenzyl ether is also formed when benzyl alcohol is heated with strong acids or bases. Mixed ethers can be prepared simply through the action of alkyl halides on benzyl alcohol in the presence of bis(acetylacetonato)nickel as a catalyst [5].

Oxidation results in benzaldehyde or benzoic acid, depending on the nature of the oxidizing agent and the reaction conditions. Thus nitric acid as an oxidizing agent gives benzaldehyde [6], whereas solid sodium permanganate monohydrate gives benzoic acid [7]. Under the conditions of the Oppenauer oxidation, benzyl alcohol is oxidized to benzaldehyde in the presence of furfural as a hydrogen acceptor [8]. Benzaldehyde is also formed slowly if benzyl alcohol is exposed to air. Benzyl alcohol can be dehydrogenated in the gas phase using catalysts containing copper or noble metals. The main product of the dehydrogenation is benzaldehyde.

Depending on the reaction conditions, the hydrogenation of benzyl alcohol gives toluene, benzene, methylcyclohexane, cyclohexane, and hydroxymethylcyclohexane.

The reactions between benzyl alcohol and hydrogen halides give the corresponding benzyl halides [9]. Under the conditions for chlorinating the side chains of aromatic compounds, benzyl alcohol forms benzoyl chloride [10].

With ammonia or amines, benzyl alcohol can be converted into $N$-alkyl-, $N,N$-dialkyl-, $N,N,N$-trialkylamines. In the presence of Friedel-Crafts catalysts, aromatic hydrocarbons can be alkylated on the aromatic nucleus. Thus benzyl alcohol and benzene form diphenylmethane, whereas benzyl alcohol and phenol give a mixture of 2- and 4-benzylphenols.

With organic and inorganic acids, acid halides, or acid anhydrides, benzyl alcohol reacts in the normal way to give the corresponding esters. The acid-catalyzed reactions of benzyl alcohol with aldehydes give acetals. The addition of carbon monoxide to benzyl alcohol in the presence of carbonyl catalysts, e.g., rhodium trichloride with benzyl iodide, leads to phenylacetic acid [11].

The treatment of benzyl alcohol with Friedel-Crafts catalysts (e.g., anhydrous aluminum chloride, zinc chloride, concentrated sulfuric acid, or perchloric acid) results in the elimination of water and the formation of resinous products having the composition $H(C_7H_6)_n OH$ and corresponding to the polybenzyl made from benzyl chloride. At temperatures above 150 °C, as little as 1% hydrogen bromide and 0.04% iron (II) can cause the exothermic polycondensation of benzyl alcohol to proceed very rapidly. In this reaction the temperature rises to 240 °C, and the sudden pressure increase may be sufficient to break a closed vessel [12]. Thus, for safety reasons, benzyl alcohol should not be heated to temperatures above 100 °C unless it is free from acidic matter and dissolved iron.

## 4. Production

Benzyl alcohol can be produced in many ways. At present there are only two processes of substantial industrial importance.

1) The hydrolysis of benzyl chloride

$$C_6H_5-CH_2Cl + H_2O \longrightarrow C_6H_5-CH_2OH + HCl$$

2) The hydrogenation of benzaldehyde

$$C_6H_5-CHO + H_2 \longrightarrow C_6H_5-CH_2OH$$

## 4.1. Hydrolysis of Benzyl Chloride

The hydrolysis of benzyl chloride is a reversible reaction which leads to the almost quantitative formation of benzyl alcohol only in the presence of alkaline saponifying agents that combine with the hydrogen chloride as it is formed. It is therefore carried out by heating benzyl chloride with stoichiometric excesses of aqueous solutions of oxides, hydroxides, or carbonates of the alkali or alkaline earth metals. As in the following example, soda is the preferred saponifying agent.

To 610 parts of boiling 10% soda solution 126.5 parts of benzyl chloride is added with stirring. The reaction mixture is refluxed and stirred until carbon dioxide no longer escapes; this takes an average of five to six hours. An alkali-resistant steel reactor with brick walls is used. The stirrer and heating coil are made of an alloy with a high nickel content. After the reaction mixture has cooled, the upper layer, consisting of crude benzyl alcohol, is removed. The sodium chloride solution below still contains some soda. Dissolved benzyl alcohol can be obtained from it by adding salt or by extracting with benzene or toluene. The crude benzyl alcohol is carefully fractionated at a reduced pressure in an efficient column. The yield is 85 parts of benzyl alcohol and 10 parts of dibenzyl ether.

In the discontinuous hydrolysis of benzyl chloride the reaction time can be reduced by hydrolyzing the final residue of benzyl chloride with sodium hydroxide solution after the saponification with soda solution has been in progress for three hours [13]. The total saponification time required is then only four hours. Normally 8–12% of dibenzyl ether is formed as an unavoidable byproduct of the alkaline hydrolysis of benzyl chloride. Pure dibenzyl ether is recovered by distillation from the residue in the distillation column. The use of alkali hydroxides instead of carbonates reduces the reaction time but favors the formation of dibenzyl ether, which may then exceed 20% in the crude benzyl alcohol. The percentage of dibenzyl ether can be reduced if the hydrolysis is performed in the presence of an inert solvent, such as benzene, toluene, or xylene.

Benzyl alcohol can be produced continuously if benzyl chloride and the aqueous alkaline solution of saponifying agent react in a flow reactor at an elevated temperature (150 – 350 °C) and sufficient pressure to maintain the reactants in the liquid phase. Thorough mixing of the reactants is important; it can be achieved by the use of suitable baffles in the flow reactor [14] or by maintaining an adequate flow rate [15]. The residence time need then be no more than a few minutes. The mixture emerging from the hydrolysis zone is processed in the normal way. Only a small amount of dibenzyl ether is formed and this can be further reduced by using an inert solvent.

In a new continuous process, benzyl chloride, flowing in one direction, and the alkaline saponifying agent, flowing in the other, are reacted as counter-currents in an inert organic solvent (e.g., toluene, xylene) in a flow reactor. In the extraction zone, the alcohol dissolved in the aqueous alkaline phase is extracted by an inert organic solvent flowing in the opposite direction. Water, likewise as a counter-current, washes the solution of the crude benzyl alcohol in a washing zone. The flow reactor, extracting zone, and washing zone are integrated in a special apparatus (Fig. 1). The wash water and extract are returned to the hydrolysis zone. This process can be operated at low reaction temperatures (120 – 150 °C) and with a small stoichiometric excess of saponifying agent (5 – 25%); a small reactor can be used [16].

In a two-step process, benzyl chloride reacts with sodium acetate to form benzyl acetate, which is then saponified to form benzyl alcohol [17]. In a similar process, benzyl chloride reacts with an alkali or alkaline earth formate in the presence of a catalyst to form benzyl formate. The benzyl alcohol is liberated by subsequent transesterification using an alcohol and a catalyst [18]. Neither of these processes has acquired much importance in the manufacture of benzyl alcohol. However, they could have advantages in the hydrolysis of certain substituted benzyl chlorides that react with soda solution to give dibenzyl ethers and not the desired alcohol.

## 4.2. Hydrogenation and Reduction of Benzaldehyde

The industrial production of benzyl alcohol by the hydrogenation of benzaldehyde gained importance because substantial quantities of benzaldehyde were available as a byproduct of the production of phenol (Dow process) and ε-caprolactam (Snia Viscosa process) by the oxidation of toluene (→ Benzaldehyde).

Depending on the reaction conditions, benzaldehyde can be hydrogenated to form benzyl alcohol, toluene, hydroxymethylcyclohexane, methylcyclohexane, and other products. However, very high yields of benzyl alcohol are obtainable if suitable reaction conditions and catalysts are chosen. Raney nickel doped with transition metals [19],

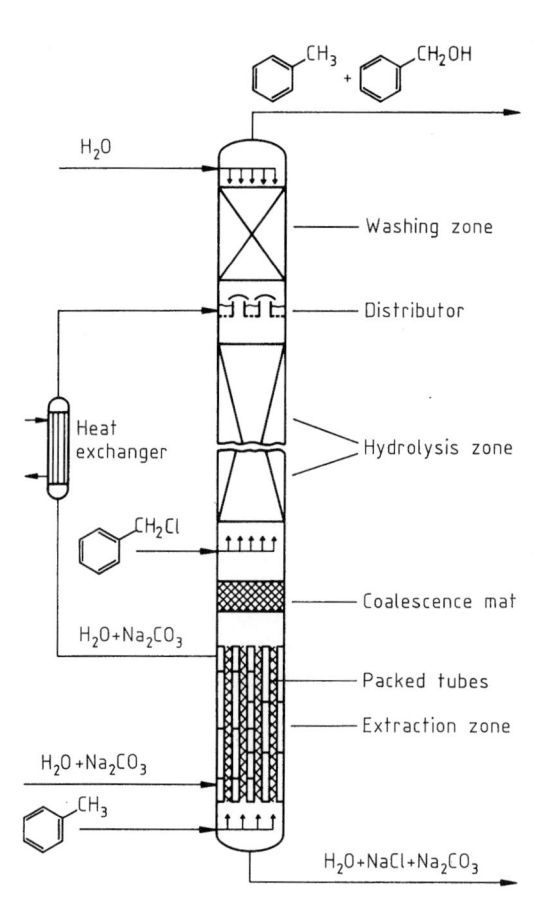

**Figure 1.** Hydrolysis of benzyl chloride to form benzyl alcohol in a continuous-flow reactor [16]

nickel or platinum metals with phosphines or phosphine oxides [20], palladium combined with an organic nitrogen [21] or alkali base [22], water [23], or another transition metal [24] can serve as the catalyst. If benzaldehyde is hydrogenated at temperatures of 70–200 °C and a hydrogen pressure of 1–4 MPa, high yields of benzyl alcohol are obtained in short reaction times. In a well-known continuous process [25], benzaldehyde is hydrogenated to form benzyl alcohol with a high degree of efficiency and a high selectivity in the presence of a platinum–aluminum oxide–lithium oxide catalyst. Stannane [26], sodium hydride [27], zinc [28], and microorganisms [29] can also be used to reduce benzaldehyde to benzyl alcohol, but these processes have no industrial importance. Benzyl alcohol is no longer produced industrially using the Cannizzaro reaction.

## 4.3. Oxidation of Toluene

The catalytic oxidation of toluene [30] gives only low yields of benzyl alcohol because the reaction conditions support further oxidation to benzaldehyde and benzoic acid. Therefore intermediates are isolated in most benzyl alcohol production processes that start with toluene. The liquid-phase oxidation of toluene with oxygen-depleted air at 170 – 220 °C and pressure sufficient to keep the reaction medium liquid is carried out in the presence of stabilizers for organic hydroperoxides (e.g., sodium pyrophosphate or sodium fluoride). It leads mainly to benzyl hydroperoxide [31] if not more than 10 % of the toluene is permitted to react. The subsequent decomposition of the oxidation product at 165 °C in the presence of soluble cobalt salts yields benzyl alcohol. Benzaldehyde and benzoic acid are formed in fairly large quantities as byproducts. Benzyl hydroperoxide can also be reduced to benzyl alcohol using alkali metal sulfites [32].

The oxidation of toluene as a liquid phase in air is conducted in the presence of acids, acid chlorides, or acid anhydrides; it forms the benzyl esters of these acids. These are then saponified to form benzyl alcohol. For example, the oxidation of toluene in the presence of acetic anhydride at temperatures of 140 – 240 °C and pressures of 1 – 3 MPa yields benzyl acetate [33]. The oxidation is discontinued after 10 % of the toluene has reacted. After the saponification of the oxidation product, 350 g of benzyl alcohol, 55 g of benzaldehyde, and 67 g of benzoic acid are obtained from 500 g of reacted toluene. Phenolic impurities, e.g., cresols, are unavoidably present in the benzyl alcohol obtained by the oxidation of toluene. These impurities can be removed by washing the benzyl alcohol vapor with a countercurrent of alkali benzylate solution in a plate column or a packed column [34].

## 4.4. Other Manufacturing Processes

The hydrogenation of benzoic acid esters to form benzyl alcohol is of particular interest in cases where benzoic acid esters are formed in substantial amounts. For example, methyl benzoate is formed in the production of dimethyl terephthalate by the Witten process. Catalysts containing copper are preferred. Benzoic acid esters can be hydrogenated to form benzyl alcohol with the aid of a copper catalyst and a carrier consisting of alkaline earth oxides or carbonates [35]. The selectivity can be increased if the catalyst also contains chromium. Temperatures of 100 – 300 °C and pressures above 6 MPa are used.

The hydrogenation of methyl benzoate in the presence of a copper–chromium catalyst is a feature of a three-step process for the production of benzyl alcohol from toluene [36]. The first step is the oxidation of toluene to form benzoic acid. The acid is esterified with methanol in the second step. The hydrogenation of methyl benzoate to form benzyl alcohol is performed in the final step. This process produces benzyl alcohol at an exceptionally low cost.

**Table 3.** Specifications for certain grades of benzyl alcohol (wt%)

| Compound | Grades | | | |
|---|---|---|---|---|
| | Technical | Pure | Double distilled | Photographic |
| Benzyl alcohol | > 97.0 | > 99.0 | > 99.8 | > 99.5 |
| Benzaldehyde | < 0.5 | < 0.2 | < 0.04 | < 0.03 |
| Chlorine | < 0.2 | < 0.02 | < 0.01 | < 0.005 |
| Dibenzyl ether | < 2.0 | < 0.05 | < 0.02 | < 0.02 |

Ruthenium, rhodium, platinum, and palladium, activated by alkali metal arenes, ketyls, or alkoxides, are catalysts for the hydrogenation of the esters [37]. Using these catalysts, benzoic acid esters can be hydrogenated to form benzyl alcohol under mild conditions and with high selectivity.

Other processes for the production of benzyl alcohol include the hydrogenation of benzoic acid [38], the electrochemical reduction of benzoic acid [39], the hydrolysis of benzylsulfonic acid [40], and the decarboxylation of benzyl formate [41]. These processes have no importance in the industrial production of benzyl alcohol, but they may be used to produce derivatives substituted on the aromatic nucleus. Benzyl alcohol can also be obtained from the benzyl benzoate which is formed in the manufacture of benzoic acid [42].

# 5. Quality Standards and Test Methods

Several grades of benzyl alcohol are commercially available (Table 3). *Technical* grade is obtained as the first and last fractions in the distillation of the crude alcohol, it has little importance. *Pure* benzyl alcohol is the most important grade in terms of volume and is required for most applications. *Double-distilled* benzyl alcohol is used mainly in the perfume and pharmaceutical industries and must therefore be exceptionally pure and free of foreign odors. At 25 °C it forms a clear 4% aqueous solution. In recent years the *photographic* grade has become increasingly important. It normally contains 0.01 – 0.02% of hydroquinone monomethyl ether to inhibit oxidation.

Gas chromatography using packed or capillary columns is suitable for the quantitative determination of benzyl alcohol. It is more accurate than the method involving acetylation of the hydroxyl group with pyridine and acetic anhydride. Traces of chlorine compounds are unavoidably present in the benzyl alcohol produced by hydrolysis of benzyl chloride. Byproducts, e.g., cresols, may be present in the benzyl alcohol obtained by the oxidation method or via benzaldehyde. Chlorine is quantitatively determined by Wickbold combustion [43], and cresols can be detected by color reactions with aminoantipyrine [34]. Benzaldehyde can be quantitatively determined using hydroxylamine hydrochloride.

The pharmacopoeias of the various countries should be consulted regarding the testing of benzyl alcohol which is to be used in the pharmaceutical industry.

# 6. Storage, Transportation, and Safety Regulations

Benzyl alcohol is stored in stainless steel tanks. Because benzyl alcohol oxidizes readily, it is advisable to cover the surface of the liquid with nitrogen. Benzyl alcohol can be transported in drums protected by stoving finishes and in tank wagons of aluminum or stainless steel.

Benzyl alcohol belongs to Temperature Class T 2 (autoignition temperature 435 °C). The transportation of benzyl alcohol by rail, road, or water is not subject to special regulations. The labeling of benzyl alcohol must comply with the national and international regulations on dangerous substances.

# 7. Uses

By virtue of its good solvent properties, benzyl alcohol is an important solvent for surface-coating materials and resins. It dissolves cellulose esters and ethers, alkyd resins, acrylic resins, and fats; it is also used as an ingredient in the inks for ball-point pens. It is added in small amounts to surface-coating materials to improve their flow and gloss. In the textile industry, it is used as an auxiliary in the dyeing of wool, polyamides, and polyesters. Because it has only a relatively faint odor, it is used as a solvent and diluting agent in the manufacture of perfumes and flavors. In color photography it is increasingly important as a development accelerator. In pharmacy it is used as a local anesthetic and, because of its microbicidal effect, as an ingredient of ointments and other preparations.

Benzyl alcohol is a starting material for the preparation of numerous benzyl esters that are used as odorants, flavors, stabilizers for volatile perfumes, and plasticizers. Benzyl alcohol is also used in the extractive distillation of *m*- and *p*-xylenes and *m*- and *p*-cresols.

# 8. Derivatives

The most important derivatives of benzyl alcohol are dibenzyl ether, benzyl acetate, benzyl benzoate, and benzyl salicylate.

## 8.1. Dibenzyl Ether

Dibenzyl ether [103-50-4], $C_{14}H_{14}O$, $M_r$ 198.27, is a colorless liquid with a faint odor; mp 3.6 °C, bp 296–298 °C at 101.3 kPa, bp 157–158 °C at 1.2 kPa, density $\varrho$ 1.043 g/cm³ at 20 °C, refractive index $n_D^{20}$ 1.5622, flash point 135 °C. The compound is very slightly soluble in water.

⟨⟩–CH₂–O–CH₂–⟨⟩

**Production.** Dibenzyl ether is formed when benzyl alcohol is heated with strong acids or bases. It is produced almost exclusively by the alkaline hydrolysis of benzyl chloride. Heat treatment decomposes dibenzyl ether into benzaldehyde and toluene.

**Use.** Dibenzyl ether is used as a plasticizer in the surface coatings industry, for special purposes in the rubber and textile industries, and as a solvent for musk and other odorants.

Excess dibenzyl ether can be converted into a mixture of benzotrichloride and benzoyl chloride under the conditions of side chain chlorination with chlorine [44]. Treatment of dibenzyl ether with hydrochloric acid in the presence of trialkyl- or tetraalkylammonium salts gives benzyl chloride [45].

## 8.2. Benzyl Acetate

Benzyl acetate [140-11-4], $C_9H_{10}O_2$, $M_r$ 150.18, is a colorless liquid with an odor of pears and jasmine. It is a constituent of jasmine, gardenia, and other flower oils; mp −51.5 °C, bp 215 °C at 101.3 kPa, density $\varrho$ 1.056 g/cm³ at 20 °C, refractive index $n_D^{20}$ 1.5020, flash point 102 °C, very slightly soluble in water.

⟨⟩–CH₂–O–CO–CH₃

**Production and Use.** Benzyl acetate is produced by reacting benzyl chloride with an alkali acetate or by esterifying benzyl alcohol with acetic anhydride. It is also formed in the oxidation of toluene in the presence of acetic acid or acetic anhydride. Benzyl acetate is one of the most used odorants. It is also used as a flavor and, to a small extent, as a high-boiling solvent.

## 8.3. Benzyl Benzoate

Benzyl benzoate [*120-51-4*], $C_{14}H_{12}O_2$, $M_r$ 212.25, is a colorless, viscous liquid with a pleasant odor. It is the main constituent of Peru balsam, Tolu balsam, and certain flower oils; mp 19–21 °C, bp 323–324 °C at 101.3 kPa, density $\varrho$ 1.112 g/cm³ at 25 °C, refractive index $n_D^{24}$ 1.5672, flash point 148 °C, insoluble in water and glycerol.

$$\text{C}_6\text{H}_5\text{-CH}_2\text{-O-CO-C}_6\text{H}_5$$

**Production.** Benzyl benzoate is produced from benzyl alcohol and sodium benzoate in the presence of triethylamine or by transesterification of methyl benzoate with benzyl alcohol in the presence of an alkali benzyl oxide. In another manufacturing process benzaldehyde is condensed to form benzyl benzoate in the presence of sodium (Claisen-Tishchenko condensation). The presence of a small amount of an aliphatic ether improves this reaction [46]. Benzyl benzoate is a byproduct in the manufacture of benzoic acid by the oxidation of toluene; it is present in the benzoic acid distillation residue.

**Use.** Benzyl benzoate is used in perfume manufacture, mainly to stabilize volatile odorants. It has vasodilating and spasmolytic effects and is present in many asthma and whooping cough drugs. It is used as an insect repellent and to treat scabies. Benzyl benzoate is used as a plasticizer in the surface coatings and plastics industries.

## 8.4. Benzyl Salicylate

Benzyl salicylate [*118-58-1*], $C_{14}H_{12}O_3$, $M_r$ 228.25, is a colorless, viscous liquid with a pleasant odor. It occurs naturally in several essential oils; mp 24 °C, bp 211 °C at 2.67 kPa, density $\varrho$ 1.175 g/cm³ at 25 °C, refractive index $n_D^{25}$ 1.5787, slightly soluble in water.

$$\text{C}_6\text{H}_5\text{-CH}_2\text{-O-CO-C}_6\text{H}_4\text{-OH}$$

**Production and Use.** Benzyl salicylate can be manufactured by reacting benzyl chloride with an alkali salicylate or by the transesterification of methyl salicylate with benzyl alcohol. It is used as a perfume stabilizer and, because it absorbs ultraviolet light, as an ingredient in sunscreening preparations.

## 9. Economic Aspects

The benzyl alcohol production capacities of Western Europe and the United States in 1983 were approximately 10 kt and 6 kt, respectively. About 7 kt of benzyl alcohol was actually produced in Western Europe in 1983. Approxi-mately 3.4 kt of benzyl alcohol was produced in the United States in 1979.

## 10. Toxicology

Benzyl alcohol is used in pharmaceuticals, soaps, detergents, and cosmetics as a fragrance, preservative, and antimicrobial, and is "generally regarded as safe" (GRAS) for use in food by the U.S. Food and Drug Administration. It has industrial applications as a solvent and dye assistant, but no environmental or workplace exposure limits have been established.

Benzyl alcohol is moderately toxic with acute oral $LD_{50}$ values of 1.2 g/kg in rats and 1.6 g/kg in mice [47]. The $LD_{50}$ for inhalation exposure is approximately 1.0 g/kg (eight hours) in rats [48]. Benzyl alcohol can be absorbed through the skin in sufficient quantities to cause toxicity; the dermal $LD_{50}$ in guinea pigs is $< 5$ mL/kg [49]. In vitro screening tests for chemical carcinogens in a repair-deficient bacterial strain (P 3478 *E. coli*) were negative [50]. Benzyl alcohol is not considered to be a carcinogen and no data are available regarding teratogenic or reproductive effects.

Benzyl alcohol is widely used as a preservative in solutions of drugs for intravenous administration. The minimum lethal dose in dogs (i.v.) is between 88 and 113 mL/kg of a 0.9 vol% solution of benzyl alcohol in water or in physiological saline solution [51]. Similar doses have been fatal for newborn humans receiving benzyl alcohol in the course of intravenous drug treatment [52]. Benzyl alcohol is estimated to be more than twenty times more toxic than ethyl alcohol (i.v.) [51].

Benzyl alcohol is severely toxic and highly irritating to the eye. Contact with benzyl alcohol causes mild to moderate skin irritation [53]. Recent reports suggest that hypersensitivity reactions may occur in some individuals with both topical and parenteral administration [54].

Signs of intoxication include vomiting and diarrhea, collapse, central nervous system depression, excitability, muscle paralysis, convulsions, dyspnea, and respiratory arrest [53]. Exposure should be avoided by the use of adequate ventilation systems, self-contained respiration equipment, protective goggles, gloves, and clothing.

# 11. References

General References

*Beilstein* **6**, 428; **6 (1)**, 217; **6 (2)**, 403; **6 (3)**, 1445; **6 (4)** 2222.

M. Watanabe, K. Namikawa, *CEER Chem. Econ. Eng. Rev.* **2** (1970) 50–52.

Specific References

[1]  F. Wöhler, *Ann. Pharm. (Lemgo, Ger.)* **3** (1832) 249, 254.
[2]  S. Cannizzaro, *Justus Liebigs Ann. Chem.* **88** (1853) 129.
[3]  L. H. Horsley: "Azeotropic Data-III," *Adv. Chem. Ser.* 1973, no. 116, 374–376.
[4]  J. Gmehling, U. Onken, W. Arlt: "Vapor-Liquid Equilibrium Data Collection," *Dechema Chem. Data Ser.*, vol. **I**, part 1a, 447–450, 1981, part 2b, 434, 1978, part 2d, 563–577, 1982.
[5]  M. Yamashita, Y. Takegami, *Synthesis* 1977, 803.
[6]  A. McKillop, M. E. Ford, *Synth. Commun.* **2** (1972) 307–313.
[7]  F. M. Menger, C. Lee, *Tetrahedron Lett.* **22** (1981) 1655–1656.
[8]  SCM Corp., DE 2556161, 1975 (W. J. Ehmann, W. E. Johnson Jr.).
[9]  J. F. Norris, *Am. Chem. J.* **38** (1907) 631.
[10] ICI, GB 310909, 1928 (B. W. Henderson).
[11] A. Masuda, H. Mitani, K. Oku, Y. Yamazaki, *Nippon Kagaku Kaishi* 1982, 249–256; *Chem. Abstr.* **96** (1982) 141989 y.
[12] E. Iselin, F. Barfuss, *Chem. Ing. Tech.* **44** (1972) 462.
[13] National Oil Products Co., US 2221882, 1938 (L. T. Rosenberg).
[14] All-Union Scientific Research Institute of Synthetic and Natural Perfumes, SU 225183, 1967 (V. V. Kashnikov, N. I. Gel'perin, S. A. Voitkevich, O. N. Zhuchkova, T. P. Khrenkova); *Chem. Abstr.* **70** (1969) 28614 x.
[15] Velsicol Chem. Corp., US 3557222, 1968 (H. W. Withers, J. L. Rose).
[16] EKA, EP 64486, 1981 (R. K. Rantala, G. L. F. Hag).
[17] Chemische Werke Witten, DE 1108677, 1959 (G. Renckhoff, H. L. Hülsmann).
[18] Dynamit Nobel, DE 2825364, 1978 (G. Bernhardt, G. Daum).
[19] Kazakh Chemical-Technological Institute, SU 235 002, 1966 (F. B. Bizhanov, D. V. Sokol'skii, N. I. Popov, A. M. Khisametdinov, N. Ya. Malkina); *Chem. Abstr.* **70** (1969) 114813 t.
[20] Hoechst, DE 2837022, 1978 (K. Kümmerle, H. Heise).
[21] Mitsui Petrochem. Industries, JP-Kokai 77 27738, 1975 (K. Teranishi, T. Shimizu, K. Koga); *Chem. Abstr.* **87** (1977) 134502 v.
[22] Mitsui Petrochem. Industries, JP-Kokai 77 27737, 1975 (K. Teranishi, T. Shimizu, K. Koga); *Chem. Abstr.* **87** (1977) 134501 u.
[23] Sumitomo Chem. Co., JP-Kokai 82 26634, 1980; *Chem. Abstr.* **96** (1982) 199279 h.
[24] Mitsui Petrochem. Industries, JP-Kokai 76 86432, 1975 (K. Teranishi, T. Ito); *Chem. Abstr.* **85** (1976) 192342 n.
[25] Universal Oil Products Co., US 3663626, 1969 (J. T. Arrigo, N. J. Christensen).
[26] M & T Chemicals Inc., US 3708549, 1970 (G. H. Reifenberg, W. J. Considine).
[27] Sagami Chem. Res. Center, JP-Kokai 76 39609, 1974 (T. Fujisawa, K. Sugimoto, H. Ota); *Chem. Abstr.* **85** (1976) 122719 s.
[28] Sumitomo Chem. Co., JP-Kokai 74 135933, 1973 (T. Mizutani, Y. Ume, T. Matsuo); *Chem. Abstr.* **82** (1975) 155745 p.

[29] Politechnika Wroclawska, PL 114065, 1979 (J. Wisniewski, A. Poranek, T. Winnicki); *Chem. Abstr.* **97** (1982) 214282 v.
[30] Universal Oil Products Co., US 3775472, 1970 (S. N. Massie).
[31] Rhône-Poulenc, FR 1366078, 1963 (J. Bonnard, G. Poilane).
[32] Nippon Oils & Fats Co., JP-Kokai 78 21133, 1976 (S. Suyama, H. Ishigaki); *Chem. Abstr.* **88** (1978) 190371 j.
[33] Institute of Petrochemical Synthesis, DE 1912078, 1969 (A. N. Baskirov, M. M. Grozhan, V. V. Kamzolkin, Yu. A. Lapitskii).
[34] Rhône-Poulenc, DE 1668646, 1967 (J. Bonnart, P. Rey).
[35] Nippon Mining Co., DE 1768262, 1967 (M. Watanabe, K. Mamikawa, K. Yasuda, M. Takagi).
[36] K. Kato, K. Namikawa, M. Watanabe, *Bull. Jpn. Pet. Inst.* **14** (1972) 206–216; *Chem. Abstr.* **78** (1973) 60190 g.
[37] Allied Chem. Corp., EP 36939, 1980 (R. A. Grey, G. P. Pez).
[38] Sumitomo Chem. Co., EP 28422, 1979 (T. Takano, G. Suzukamo, M. Ishino, K. Ikimi).
[39] Union Rheinische Braunkohlen Kraftstoff AG, DE 2237612, 1972 (G. Schwarzlose).
[40] Universal Oil Products Co., US 3641165, 1967 (L. Schmerling, R. A. Dombro).
[41] Dynamit Nobel, DE 2825362, 1978 (G. Bernhardt).
[42] Stamicarbon B. V., EP 19341, 1979 (C. Jongsma, L. H. B. Frijns, P. A. M. Raven-Donners).
[43] F. Ehrenberger, S. Gorbach: *Methoden der organischen Elementar- und Spurenanalyse*, Verlag Chemie, Weinheim 1973, p. 238 ff.
[44] Bayer, DE 1909523, 1969 (W. Böckmann).
[45] Odessa State University, SU 872525, 1980 (G. L. Kamalov, D. G. Chikhichin, L. Ya. Glinskaya, G. P. Fedorchenko, F. P. Filek); *Chem. Abstr.* **96** (1982) 85217 f.
[46] Dow Chem. Co., US 3387020, 1964 (C. E. Handlovits, J. B. Louch).
[47] P. M. Jenner, E. C. Hagan, J. M. Taylor, E. L. Cook, O. G. Fitzhugh, *Food Cosmet. Toxicol.* **2** (1964) 327–343.
[48] H. F. Smyth, Jr., C. P. Carpenter, C. S. Weil, *AMA Arch. Ind. Hyg. Occup. Med.* **4** (1951) 120.
[49] D. L. J. Opdyke, *Food Cosmet. Toxicol.* **11** (1973) 1011–1081.
[50] E. R. Fluck, L. A. Poirier, H. W. Ruelius, *Chem. Biol. Interact.* **15** (1976) 219–231.
[51] E. T. Kimura, T. D. Darby, R. A. Krause, H. D. Brondyk, *Toxicol. Appl. Pharmacol.* **18** (1971) 60–68.
[52] J. Gershanik, B. Boecler, H. Ensley, S. McCloskey, W. George, *N. Engl. J. Med.* **307** (1982) 1384–1388.
[53] V. K. Rowe, S. B. McCollister in G. D. Clayton, F. E. Clayton (eds.): *Patty's Industrial Hygiene and Toxicology*, vol. **2 C**, John Wiley & Sons, New York 1982, pp. 4636–4703.
[54] J. A. Grant, O. A. Bilodeau, B. G. Guernsey, F. H. Gardner, *N. Engl. J. Med.* **307** (1982) 108.

# Benzylamine

WOLFGANG KIEL, Bayer AG, Leverkusen, Federal Republic of Germany

| | | | | | |
|---|---|---|---|---|---|
| 1. | Physical Properties | 863 | 5. | Uses | 865 |
| 2. | Chemical Properties | 863 | 6. | Storage and Transportation | 865 |
| 3. | Production | 864 | 7. | Ecology | 865 |
| 4. | Quality Specifications and Analysis | 864 | 8. | Toxicology | 866 |
| | | | 9. | References | 866 |

# 1. Physical Properties

Benzylamine [*100-46-9*], $C_7H_9N$, $M_r$ 107.16, is a colorless liquid with a smell similar to that of ammonia.

Benzylamine is miscible with water, alcohol, and ether; bp 184.5 °C at 101.3 kPa, bp 90 °C at 1.6 kPa.

Examples of azeotropic mixtures are given in Table 1.

Heat of vaporization 49 kJ/mol, heat of combustion 4058.7 kJ/mol at 101.3 kPa and 20 °C, refractive index $n_D^{20}$ 1.5401, density $d_4^{20}$ 0.9813, $d_4^{86.6}$ 0.9272, dielectric constant 4.6 at 20.6 °C, electrolytic dissociation constant in water $2.35 \times 10^{-5}$ at 20 °C, viscosity 1.78 mPa · s at 21.2 °C, 0.295 mPa · s at 178.2 °C, surface tension $38.82 \times 10^{-5}$ N/cm at 21.1 °C, $31.70 \times 10^{-5}$ N/cm at 88 °C, autoignition temperature 390 °C, flash point 65 °C.

# 2. Chemical Properties

Benzylamine is a stronger base than its toluidine isomers. It forms addition compounds with, e.g., phenol (1:3, *mp* 15.3 °C), *p*-cresol (1:1, *mp* −6 °C), and formic acid (1:1, *mp* 81 °C). Examples of the salts of benzylamine are the hydrochloride, $C_6H_5CH_2NH_3^+Cl^-$, *mp* 260 °C (decomp.), and the picrate, *mp* 194 °C. Benzylamine

**Table 1.** Azeotropic mixtures containing benzylamine

| Other component, A | bp of A, °C | bp of azeotropic mixture, °C | Component A in the azeotropic mixture, wt% |
|---|---|---|---|
| Aniline | 184.35 | 186.2 | 42 |
| Phenol | 182.2 | 196.8 | 45 |
| m-Cresol | 202.2 | 207.2 | 56 |

absorbs carbon dioxide from the air to form the solid carbamic acid salt. If benzylamine is boiled with glacial acetic acid, N-acetylbenzylamine is formed. Benzylamine reacts with isocyanates to give substituted ureas, $C_6H_5CH_2NHCONHR$. Benzylamine is nitrated on the aromatic nucleus (8% ortho-, 49% meta-, 43% para-nitrobenzylamine) by nitric acid. Catalytic hydrogenation gives hexahydrobenzylamine, $C_6H_{11}CH_2NH_2$.

# 3. Production

Benzylamine is produced by the reaction of benzyl chloride with ammonia [1], the catalytic hydrogenation of benzonitrile [2]–[4], or the catalytic hydrogenation of benzaldehyde in the presence of ammonia [5], [6].

In the benzaldehyde process, 8 kg of Raney nickel that has been kept under water (corresponding to about 4 kg of 100% Ni), 10 g of glacial acetic acid, and 110 kg of ammonia are added to a mixture of 250 kg of methanol and 500 kg of benzaldehyde. The resulting batch is hydrogenated in a 1200-L steel autoclave at 100 °C and 15 MPa. The process takes 3–5 h. The hydrogen consumed in the reaction is made up at intervals of about 10 min. The contents of the reaction vessel are maintained at the same temperature and pressure for 30 min after hydrogen consumption ends. The catalyst is then separated from the reaction products on a pressure filter [5].The filtrate contains 470 kg of benzylamine, about 1 kg of benzyl alcohol, less than 1 kg of dibenzylamine, and about 2 kg of Schiff base (N-benzalbenzylamine), in addition to methanol, reaction water, and ammonia. The yield is 93%. Vacuum distillation yields benzylamine of the required technical purity.

# 4. Quality Specifications and Analysis

Commercial benzylamine has an assay of about 99% and a density of 0.98 g/cm$^3$. In this form benzylamine can be kept indefinitely in closed containers.

The most suitable method for the quantitative determination of benzylamine is gas chromatography. For example, the reaction mixture obtained from the benzaldehyde process can be fractionated in a column packed with 82.7% Chromosorb W, 4.1%

potassium hydroxide, 3.3% Bayer Silicone Fluid PM 200, and 9.9% of an alkane sulfonate (Bayer K 30).

# 5. Uses

Benzylamine is used as a corrosion inhibitor [7]. In the formulation of insecticides it is used as a cell-active penetration aid [8]. Benzylamine is a valuable key compound in the production of pharmaceutically active substances [9].

# 6. Storage and Transportation

Pure benzylamine keeps indefinitely in closed containers. Steel drums are suitable as transportation containers.

Benzylamine (EC No. 612-047-00-x) is included in the list of substances contained in EC Guideline 67/548/EEC and is therefore to be considered a dangerous industrial material. The containers must be marked as follows:

C        = corrosive
R 34     = corrosive
S 26     = if splashed into eyes, rinse thoroughly with water and consult a doctor.

Symbols required by the Gefahrgutverordnung Transport (Dangerous Goods Transport Order) of the Federal Republic of Germany:

Railways (international) = RID Category 8, Figure 35
Road (international) = ADR Category 8, Figure 35
Inland vessels = ADN-R Category 8, Figure 35
Ocean-going vessels = IMDG Code Category 8, UN No. 1719.

# 7. Ecology

Laboratory investigations to determine the ecological behavior of benzylamine gave the following BOD values: At a COD value of 2802 mg/L the $BOD_5$ value is 625 mg/L, the $BOD_{10}$ value 1700 mg/L, and the $BOD_{20}$ value 2130 mg/L. At aqueous concentrations of up to 500 mg/L benzylamine causes no harm to *Bacterium coli* or *Pseudomonas fluorescens* [10]. In the golden orfe test, aqueous benzylamine at 100 mg/L killed all the fish within 1.5 h, whereas an aqueous solution of 20 mg/L was tolerated for 48 h without harm.

## 8. Toxicology

Skin compatibility: corrosive (rabbit, 4 h exposure). Acute oral toxicity: $LD_{50}$ 1.15 mg/kg (rat). Acute percutaneous toxicity: $LD_{50}$ 1.34 g/kg (rat) [11].

## 9. References

General References

*Beilstein* **12**, 1013; **12 (1)**, 445; **12 (2)**, 540; **12 (3)**, 2194.
*Houben-Weyl*, **11/1**, 1957.

Specific References

[1]   California Research Corp., US 2987548, 1961.
[2]   Miles Laboratories, Inc., US 541494, 1975.
[3]   BASF, FR 2165515, 1939.
[4]   M. Grunfeld, US 2449036, 1941.
[5]   C. F. Winans, *J. Am. Chem. Soc.* **61** (1939) 3566.
[6]   BASF, FR 146354, 1966.
[7]   B. Wanderott, *Z. Metallkd.* **56** (1965) no. 1, 63–64.
[8]   *Chem. Abstr.* **47** (1950) 7149 h.
[9]   H. Ippen: *Index Pharmacorum,* Thieme Verlag, Stuttgart 1968.
[10]  *Deutsche Einheitsverfahren zur Wasser-, Abwasser- und Schlamm-Untersuchung,* 3rd. ed., Verlag Chemie, Weinheim 1960–1971.
[11]  Bayer AG, Toxicological Institute, Elberfeld 1977.

# Bromine Compounds, Organic

MICHAEL J. DAGANI, Ethyl Corp., Baton Rouge, Louisiana 70821, United States (Sections 1 – 1.4.10, 1.5)

HENRY J. BARDA, Ethyl Corp., Baton Rouge, Louisiana 70821, United States (Section 1.4.11 – 1.4.11.4)

THEODORE J. BENYA, Ethyl Corp., Baton Rouge, Louisiana 70821, United States (Section 1.6)

| | | |
|---|---|---|
| 1. | Introduction | 867 |
| 1.1. | Physical Properties | 868 |
| 1.2. | Chemical Properties | 868 |
| 1.2.1. | Nucleophilic Displacement of Bromine | 869 |
| 1.2.2. | Displacement of Bromine by Metals | 870 |
| 1.3. | Production | 871 |
| 1.3.1. | Addition Reactions | 871 |
| 1.3.2. | Substitution Reactions | 872 |
| 1.4. | Commercial Products | 873 |
| 1.4.1. | 1,2-Dibromoethane | 873 |
| 1.4.2. | 1-Bromododecane | 874 |
| 1.4.3. | 1-Bromo-3-chloropropane | 875 |
| 1.4.4. | 1-Bromo-2-phenylethane | 876 |
| 1.4.5. | 1-(Bromomethyl)-3-phenoxybenzene | 877 |
| 1.4.6. | Bromoacetic Acid | 877 |
| 1.4.7. | Bromomethane | 878 |
| 1.4.8. | 3-Bromo-1-propene | 879 |
| 1.4.9. | Bromochloromethane and Dibromomethane | 880 |
| 1.4.10. | Trifluorobromomethane and Difluorobromochloromethane | 881 |
| 1.4.11. | Flame Retardants | 882 |
| 1.4.11.1. | Tetrabromobisphenol-A | 882 |
| 1.4.11.2. | Decabromobiphenyl Oxide | 885 |
| 1.4.11.3. | Tetrabromophthalic Anhydride | 885 |
| 1.4.11.4. | Hexabromocyclododecane | 886 |
| 1.5. | Manufacturers | 887 |
| 1.6. | Toxicology and Occupational Health | 887 |
| 2. | References | 892 |

# 1. Introduction

In almost all of its properties, both physical and chemical, bromine is intermediate between the other active halogens, chlorine and iodine. These same intermediate characteristics are also found in organic compounds in which halogen is covalently bonded to carbon. In general the organic compounds of bromine tend to be less stable and more reactive than the corresponding chlorine compounds, but more stable and less reactive than their iodine counterparts. Because of their unique physical, chemical, and biological properties, organic bromides have found wide use in numerous commercial products and industrial applications.

Annual production of organic bromine compounds is also intermediate between that of the organic compounds of chlorine and iodine, mainly because of the cost of the elemental halogens. On a mass basis the relative costs of chlorine, bromine, and iodine are about $1:4:78$, but on a molar basis these relative costs rise to about $1:9:280$, respectively. In spite of apparent economic disadvantages, the unique properties and greater reactivity of organic bromine compounds compared to their chlorine analogs are sufficient in many cases to justify the higher cost of bromine. Furthermore, recycle of byproduct inorganic bromides to bromine can almost eliminate the premium cost of bromine over chlorine.

In terms of bromine consumption, organic compounds accounted for about 75–80% of elemental bromine production, and inorganic compounds accounted for the remainder in the early 1980s. Ethylene dibromide, largely because of its use as an additive for leaded gasoline, is still the largest volume organic bromine compound produced. Organic bromides are used in many other applications, such as flame retardants, chemicals for agriculture and water treatment, pharmaceuticals, disinfectants, flame extinguishing agents, and dyes. Bromine and its organic compounds are also widely employed as chemical intermediates for the production of other commercial organic compounds.

The scope of this article is limited to those organic bromides having industrial significance. General reference works should be consulted for the numerous other small-volume applications in industry and in basic and applied chemistry [1]–[6].

## 1.1. Physical Properties

Except for the gaseous lower aliphatic bromides, organic bromides are either liquids or solids at ambient temperature and pressure. In general the boiling points fall between those of the corresponding chlorine and iodine compounds, as do the melting points, relative densities, and refractive indexes. Simple organic compounds containing bromine are generally soluble in most common aliphatic and aromatic solvents, such that some bromine compounds are classified as solvents. On the other hand their water solubility is usually low and is roughly inversely proportional to their relative molecular mass [7].

Because many organic bromine compounds have been known for years, their physical properties have been documented [8]. Their spectro-scopic and electronic properties, in conjunction with their chemical analysis, are discussed in general texts [9], [10]. Representative lists of aliphatic and aromatic bromine compounds and their physical properties are given in Tables 1 and 2, respectively.

## 1.2. Chemical Properties

The chemistry of organic bromine compounds is so extensive and varied that only a few of the reactions used in the chemical and related industries are briefly discussed. In general, organic bromides undergo these displacement and other chemical reactions with greater facility and usually in higher yield than the less expensive organic chlor-

**Table 1.** Physical properties of aliphatic bromine compounds

| Compound | CAS registry no. | Formula | $M_r$ | mp, °C | bp, °C (101.3 kPa) |
|---|---|---|---|---|---|
| Bromomethanes | | | | | |
| Bromomethane | [74-83-9] | $CH_3Br$ | 95.0 | −93.6 | 3.6 |
| Dibromomethane | [74-95-3] | $CH_2Br_2$ | 173.8 | −52.6 | 97.0 |
| Bromochloromethane | [74-97-5] | $CH_2BrCl$ | 129.4 | −86.5 | 68.1 |
| Tribromomethane | [75-25-2] | $CHBr_3$ | 252.8 | 8.3 | 149.5 |
| Tetrabromomethane | [558-13-4] | $CBr_4$ | 331.6 | 90.5 | 189.5 |
| Trichlorobromomethane | [75-62-7] | $CCl_3Br$ | 198.3 | −5.6 | 104.7 |
| Trifluorobromomethane | [75-63-8] | $CF_3Br$ | 148.9 | −168.0 | −57.8 |
| Difluorobromochloromethane | [353-59-3] | $CF_2BrCl$ | 165.4 | −159.5 | −3.3 |
| Bromoalkenes | | | | | |
| Bromoethylene | [593-60-2] | $CH_2=CHBr$ | 107.0 | −137.8 | 15.8 |
| 1,1-Dibromoethene | [593-92-0] | $CH_2=CBr_2$ | 185.9 | – | 91.7 |
| cis-1,2-Dibromoethene | [540-49-8] | $BrCH=CHBr$ | 185.9 | −53.0 | 112.5 |
| 3-Bromo-1-propene | [106-95-6] | $CH_2=CHCH_2Br$ | 121.0 | −119.4 | 71.3 |
| 2,3-Dibromo-1-propene | [513-31-5] | $CH_2=CBrCH_2Br$ | 199.9 | – | 141.5 |
| trans-1,4-Dibromo-2-butene | [821-06-7] | $BrCH_2CH=CHCH_2Br$ | 213.9 | 53.5 | 205.0 |
| Tribromoethene | [598-16-3] | $BrCH=CBr_2$ | 264.8 | −52.9 | 164.0 |
| Tetrabromoethene | [79-28-7] | $Br_2C=CBr_2$ | 343.7 | 56.5 | 226.0 |
| Bromoalkanes | | | | | |
| Bromoethane | [74-96-4] | $CH_3CH_2Br$ | 109.0 | −118.6 | 38.4 |
| 1,1-Dibromoethane | [557-91-5] | $CH_3CHBr_2$ | 187.9 | −63.0 | 108.0 |
| 1,2-Dibromoethane | [106-93-4] | $BrCH_2CH_2Br$ | 187.9 | 9.8 | 131.4 |
| 1-Bromopropane | [106-94-5] | $CH_3CH_2CH_2Br$ | 123.0 | −110.0 | 71.0 |
| 2-Bromopropane | [75-26-3] | $CH_3CHBrCH_3$ | 123.0 | −89.0 | 59.4 |
| 1,2-Dibromopropane | [78-75-1] | $CH_3CHBrCH_2Br$ | 201.9 | −55.2 | 140.0 |
| 1,3-Dibromopropane | [109-64-8] | $BrCH_2CH_2CH_2Br$ | 201.9 | −34.2 | 167.3 |
| 1-Bromobutane | [109-65-9] | $CH_3(CH_2)_3Br$ | 137.0 | −112.4 | 101.6 |
| 2-Bromobutane | [78-76-2] | $CH_3CH_2CHBrCH_3$ | 137.0 | −111.9 | 91.2 |
| 1-Bromopentane | [110-53-2] | $CH_3(CH_2)_4Br$ | 151.0 | −87.9 | 129.6 |
| 1-Bromohexane | [111-25-1] | $CH_3(CH_2)_5Br$ | 165.1 | −84.7 | 155.3 |
| 1-Bromooctane | [111-83-1] | $CH_3(CH_2)_7Br$ | 193.1 | −55.0 | 200.8 |
| 1-Bromodecane | [112-29-8] | $CH_3(CH_2)_9Br$ | 221.2 | −29.2 | 240.6 |
| 1-Bromododecane | [143-15-7] | $CH_3(CH_2)_{11}Br$ | 249.2 | −9.5 | 275.9 |
| 1-Bromotetradecane | [112-71-0] | $CH_3(CH_2)_{13}Br$ | 277.3 | 5.6 | 307 |
| 1-Bromohexadecane | [112-82-3] | $CH_3(CH_2)_{15}Br$ | 305.4 | 17 – 19 | 336 |
| 1-Bromooctadecane | [112-89-0] | $CH_3(CH_2)_{17}Br$ | 333.4 | 28.2 | 362 |

ides. This comparative advantage, coupled with their more efficient preparation, accounts for much of their industrial importance.

## 1.2.1. Nucleophilic Displacement of Bromine

Bromine bonded to tetravalent carbon is a good leaving group and is readily displaced by neutral or anionic nucleophiles to form a wide variety of other useful organic derivatives. Aliphatic bromides undergo these substitution reactions by either unimolecular ($S_N1$) or bimolecular ($S_N2$) reaction mechanisms that may be accompa-

**Table 2.** Physical properties of aromatic bromine compounds

| Compound | CAS registry no. | Formula | $M_r$ | mp, °C | bp, °C (101.3 kPa) |
|---|---|---|---|---|---|
| Bromobenzene | [108-86-1] | $C_6H_5Br$ | 157.0 | −30.8 | 156.0 |
| 1,2-Dibromobenzene | [583-53-9] | $C_6H_4Br_2$ | 235.9 | 7.1 | 225 |
| 1,3-Dibromobenzene | [108-36-1] | $C_6H_4Br_2$ | 235.9 | − 7 | 218 |
| 1,4-Dibromobenzene | [106-37-6] | $C_6H_4Br_2$ | 235.9 | 87.3 | 218.5 |
| 1-Bromo-4-chlorobenzene | [106-39-8] | $ClC_6H_4Br$ | 191.5 | 68.0 | 196.0 |
| 1-Bromo-4-fluorobenzene | [460-00-4] | $FC_6H_4Br$ | 175.0 | − 8 | 152 |
| 1-Bromo-4-iodobenzene | [589-87-7] | $IC_6H_4Br$ | 282.9 | 92.0 | 252 |
| 4-Bromophenol | [106-41-2] | $BrC_6H_4OH$ | 173.0 | 66.4 | 238 |
| 2-Bromotoluene | [95-46-5] | $BrC_6H_4CH_3$ | 171.0 | −27.7 | 181.7 |
| 4-Bromotoluene | [106-38-7] | $BrC_6H_4CH_3$ | 171.0 | 28.5 | 184.4 |
| 4-Bromoanisole | [104-92-7] | $BrC_6H_4OCH_3$ | 187.0 | 9–10 | 223 |
| 3,5-Dibromosalicylic acid | [3147-55-5] | $Br_2C_6H_2(OH)CO_2H$ | 295.9 | 223 | – |
| 1,3,5-Tribromobenzene | [626-39-1] | $C_6H_3Br_3$ | 314.8 | 121–124 | 271 |
| 2,4,6-Tribromoaniline | [147-82-0] | $Br_3C_6H_2NH_2$ | 329.8 | 122 | 300 |
| 2,4,6-Tribromophenol | [118-79-6] | $Br_3C_6H_2OH$ | 330.8 | 95–96 | 282–290 |
| 4,4'-Dibromo-1,1'-biphenyl | [92-86-4] | $BrC_6H_4–C_6H_4Br$ | 312.0 | 164 | 355–360 |
| 1-Bromonaphthalene | [90-11-9] | $C_{10}H_7Br$ | 207.1 | − 6.2 | 281.0 |

nied by competing elimination reactions (E 1 or E 2). The rates of substitution or elimination are influenced by a number of factors, such as the substrate structure, the presence of neighboring groups, the nature of the solvent, and the concentration of reactants, to cite a few. Appropriate reaction conditions can usually be chosen so that either substitution or elimination reactions will predominate:

$$RCH_2CH_2Br + Nuc^- \begin{cases} \rightarrow RCH_2CH_2Nuc + Br^- \quad (S_N1 \text{ or } S_N2) \\ \rightarrow RCH=CH_2 + NucH + Br^- \\ \quad (E1 \text{ or } E2) \end{cases}$$

where $Nuc^-$ is any nucleophile. In general, aryl and vinyl bromides undergo these reactions only with difficulty, although there are exceptions. Some examples of the many nucleophiles used to displace bromine in commercial processes are water, secondary and tertiary amines, cyanide and phenoxide ions, the carbanions of active methylene compounds, and trisubstituted phosphines and phosphites.

## 1.2.2. Displacement of Bromine by Metals

Both aliphatic and aromatic bromides can react with magnesium to form alkyl- and arylmagnesium bromides. These *Grignard reagents* are very useful reaction intermediates because they are capable of reacting with a wide variety of organic functional groups under mild reaction conditions. Organic bromides can also react with lithium to produce the more reactive but more expensive organolithium compounds, which are more convenient to use than Grignard reagents in some cases. Because of economic considerations these organometallic compounds are predominantly used in the synthesis of high-value products, such as pharmaceuticals or perfumes.

## 1.3. Production

Organic bromine compounds can be produced by a number of different chemical reactions [1]–[6]; however, addition and substitution reactions are the methods most commonly employed in industrial processes.

### 1.3.1. Addition Reactions

The uncatalyzed addition of bromine to olefins is rapid and quantitative, and can be used to analytically determine the amount of unsaturation in a molecule (bromine number). In the case of unsymmetrical reagents, such as hydrogen bromide (HBr), the addition to unsaturated compounds can proceed by ionic or free-radical reaction mechanisms. Under ionic reaction conditions and in the absence of peroxides, the product is the one that is formed from the most stable carbocation intermediate generated by the addition of a proton to the double bond (normal or Markownikoff addition). Because carbocation stability follows the order tertiary > secondary > primary, a 2-bromoalkane is the expected product when HBr is added to a 1-alkene:

$$RCH=CH_2 \xrightarrow{H^+} R\overset{+}{C}HCH_3 \xrightarrow{Br^-} RCHBrCH_3$$

Under free-radical reaction conditions the reverse order of addition to olefins occurs (abnormal or anti-Markownikoff addition). Thus the reaction of HBr with a 1-alkene in the presence of a peroxide or other free-radical initiator produces a 1-bromoalkane. The reaction proceeds by radical chain-initiating and chain-propagating steps, and the direction is based on the most stable free radical (tertiary > secondary > primary) generated by the addition of a bromine atom to the olefin:

$$RCH=CH_2 \xrightarrow{Br\cdot} R\dot{C}HCH_2Br \xrightarrow{HBr} RCH_2CH_2Br + Br\cdot$$

The free-radical addition of HBr is unique because the other hydrogen halides undergo addition to olefins only by the ionic pathway.

## 1.3.2. Substitution Reactions

Bromine and other reagents containing bromine can react with various organic compounds by substitution reactions that involve the replacement of hydrogen by bromine. In the case of saturated hydrocarbon substrates, such as alkanes and alkyl aromatic compounds, the bromination reaction occurs by a free-radical chain reaction and requires thermal, photolytic, or other forms of initiation:

$$Br_2 \longrightarrow 2\ Br\cdot\ \text{(initiation)}$$
$$RH + Br\cdot \longrightarrow R\cdot + HBr\ \text{(propagation)}$$
$$R\cdot + Br_2 \longrightarrow RBr + Br\cdot\ \text{(propagation)}$$
$$R\cdot + Br\cdot \longrightarrow RBr\ \text{(termination)}$$

where R is alkyl or alkylaryl.

The reaction with most linear aliphatic hydrocarbons is not generally useful because hydrogen abstraction by bromine atoms occurs at random in the hydrocarbon, and the radical chains are short and not self-sustaining. Reaction selectivity is low because the carbon–hydrogen bond dissociation energy differences between the many secondary hydrogens in a long-chain alkane are negligible. Reaction selectivities are much greater with branched alkanes because of the greater energy differences between the different carbon–hydrogen bonds, such that tertiary alkyl bromides can be the predominant substitution product.

In the case of alkyl aromatic compounds, such as ethylbenzene, the substitution reaction is very selective because of the much greater energy differences between the radicals generated in the hydrogen abstraction step. Because the benzylic radical ($C_6H_5\dot{C}CH_3$) is much more stable than the primary radical ($C_6H_5CH_2CH_2\cdot$), 1-bromo-1-phenylethane is essentially the only product formed.

Aromatic compounds (ArH) undergo nuclear substitution reactions with bromine by a ratelimiting electrophilic addition to form an arenium ion intermediate followed by the rapid loss of a proton:

$$ArH + Br_2 \longrightarrow \left[Ar{<}^H_{Br}\right]^+ Br^- \longrightarrow ArBr + HBr$$

Replacement of hydrogen by bromine usually requires a Lewis acid catalyst when the aromatic compound contains deactivating (nitro) or weakly activating (alkyl) substituents. No catalyst is required for the bromination of more activated aromatics, such as phenol or aniline. The amount and orientation of bromine substitution is dependent on a number of factors, such as stoichiometry, catalyst activity, reaction conditions, and substituent effects.

An equimolar mixture of bromine and chlorine (bromine chloride) can also be used for these substitution reactions. In free-radical substitutions the overall reaction is faster than with bromine because the more energetic chlorine atoms lower the energy of activation in the hydrogen abstraction step. Although minor amounts of chlorinated products are usually produced, bromine utilization is improved because HCl rather than HBr is produced as the byproduct. This reagent is often used in place of bromine for both aliphatic and aromatic substitution reactions when there are not adequate facilities to recycle or use HBr.

Other substitution reactions that are often used in commercial chemical processes include: the replacement of allylic and benzylic hydrogen atoms by bromine using *N*-bromo compounds, such as *N*-bromosuccinimide and 1,3-dibromo-5,5-dimethylhydantoin; the substitution of hydroxyl groups of alcohols by bromine using phosphorus compounds, such as phosphorus tribromide or triphenylphosphonium dibromide; and the replacement of chlorine atoms of many chlorinated hydrocarbons by bromine with various Lewis acids and phase-transfer catalysts. Various reagents can also be used to substitute the hydrogen atoms adjacent to carbonyl groups to produce the α-bromo derivatives of aldehydes, ketones, carboxylic acids, and active methylene compounds.

## 1.4. Commercial Products

Of the numerous organic bromine compounds that are commercially produced, the individual compounds discussed in this chapter were selected because of their industrial importance and because they illustrate the most common patented production processes. The order of treatment is by chemical reaction classification and begins with addition reactions, followed by substitution reactions of bromine for hydrogen, hydroxyl, and halogen. Because brominated flame retardants consume a significant amount of bromine, the production of this growing and important class of compounds is treated separately.

Most of these commercial products have different product quality specifications because they are generally produced by several manufacturers. Additional physical and chemical properties, first aid and personal protection information, toxicity and safety data, and proper precautions for handling, storage, and transportation are available from the manufacturers.

### 1.4.1. 1,2-Dibromoethane

**Production.** 1,2-Dibromoethane [*106-93-4*], $BrCH_2CH_2Br$ (ethylene dibromide), is manufactured by the uncatalyzed addition of bromine to ethylene.

The commercial process is carried out in a glass-column reactor consisting of a lower packed section and an unpacked upper section containing a number of superimposed, high-capacity, coil heat exchangers. Liquid bromine is continuously added above the packed section while a slight excess of ethylene is continuously fed countercurrently from the bottom of the packed section. The exothermic reaction between ethylene and bromine occurs in the liquid phase on the surfaces of the cooling coils, and heat is removed at a rate sufficient to maintain a maximum temperature of 100 °C in this section of the column. Some reaction also occurs in the gas phase above the bromine feed where product is condensed and separated from the vent gas (ethylene, HBr, and inert material). As the crude liquid product passes downward through the packed section, it provides a contacting surface for rising ethylene to convert any residual dissolved bromine. Ethylene dibromide is continuously withdrawn from the reactor into a hold-up tank where it is irradiated with ultraviolet light to eliminate minor amounts of unconverted starting materials. The technical-grade product is obtained in 99.5% purity as a clear, colorless liquid in close to quantitative yield [11].

A continuous process that claims lower corrosivity and high output of ethylene dibromide involves the countercurrent addition of bromine and ethylene plus the continuous addition of a cold concentrated aqueous solution of inorganic bromide, which functions as the heat exchange medium [12]. A process that uses ethylene dibromide as the reaction medium is also reported to allow good control of the exothermic reaction between bromine and ethylene [13].

**Uses.** The largest market for ethylene dibromide is in gasoline. It acts as a scavenger for lead alkyls and prevents buildup of lead oxides in automobile engines. This use, which amounted to about 35–40% of United States bromine consumption in 1981, has been declining since 1974 as a result of the lead phasedown mandated by the U.S. Environmental Protection Agency (EPA). Government agencies in many other countries are following this course of action to decrease the amount of lead in the atmosphere. The continued production of nonlead tolerant automobiles and conservation policies will more than likely result in further declines.

The other major application for ethylene dibromide is for soil and space fumigation, and the postharvest fumigation of fruit. It was used as a substitute for DBCP (1,2-dibromo-3-chloropropane), $BrCH_2CHBrCH_2Cl$ [96-12-8], which was banned by the EPA in the late 1970s because workplace exposure led to azoospermia; it was applied to control pests, such as soil nematodes, rodents, and most species of insects. In late 1983 the EPA took emergency action to cancel the pesticide registrations for ethylene dibromide formulations used in most soil fumigation applications in the United States because of groundwater contamination. Some minor fumigation uses were allowed to continue, but the predominant volume of ethylene dibromide used in soil fumigation will cease.

Smaller amounts of ethylene dibromide are used as a reaction intermediate in the manufacture of other chemicals, such as the reactive flame retardant monomer, vinyl bromide [593-60-2]. Minor amounts are used as a highdensity, nonflammable solvent in a variety of applications.

### 1.4.2. 1-Bromododecane

**Production.** 1-Bromododecane [143-15-7], $n$-$C_{12}H_{25}Br$ (lauryl or $n$-dodecyl bromide), is produced by the free-radical addition of gaseous HBr to 1-dodecene. Almost all industrial processes to manufacture the higher 1-bromoalkanes containing 6–18 carbon atoms employ this addition reaction. The alternative manufacturing process, the reaction of the corresponding alcohols with HBr, is not as economical because of the higher cost and lower reactivity of the alcohols compared to the olefins.

The continuous process is conducted in the liquid phase by contacting 1-dodecene containing a free-radical initiator with anhydrous HBr in a glass-lined reactor. Teflon (polytetrafluoroethylene) or Kynar (poly(vinylidene fluoride)) reactor linings are suitable alternatives. The reaction is usually carried out under moderate pressure to provide good mass transfer. Reaction temperatures of less than 20 °C are usually necessary to obtain high regioselectivity to 1-bromododecane. Various continuous reactor designs that have been used to produce high ratios of primary to secondary alkyl bromides at higher temperatures (20–30 °C) include a two-stage reactor [14], and a thin turbulent rising film reactor [15]. The latter reactor operates by forming a thin film of olefin in a vertical tubular reactor, which is contacted with an upward stream of gaseous HBr to maintain turbulence and intimate mixing. Because residence times are short and the exothermic reaction is easily controlled by cooling the large surface area of the reactor tube, a 22:1 molar ratio of 1-bromo- to 2-bromododecane is obtained at 27 °C. When HBr is added to the olefin under batch reaction conditions, the rate of the competing ionic addition reaction increases and the molar ratio of primary to secondary bromides falls to 8:1 at comparable temperatures (32 °C).

The crude reaction product can be either flashed or contacted with an aqueous solution of sodium hydroxide to remove excess HBr. The conversion of 1-dodecene and the yield of 1- and 2-bromododecane are essentially quantitative.

Free-radical initiators that have been employed for this reaction are preformed olefin ozonides [16], olefin peroxides generated by preblowing air through the olefin [17], oxygen [18] or phosgene [19] cofed with HBr, organic peroxides alone [20] or in the presence of phase-transfer catalysts [21], molecular sieves [22], and ultraviolet light [23].

**Uses.** *n*-Alkyl bromides have assumed considerable importance as starting materials for the production of many types of organic derivatives. The major use for *n*-alkyl bromides with 12–16 carbon atoms is in the production of alkyldimethylamines by the nucleophilic displacement of bromide ion with dimethylamine. These amines are used as intermediates in the manufacture of fatty amine oxides and alkyl dimethyl benzyl quaternary ammonium salts, which are used in a wide range of products because of their biological (germicidal) or physical (cationic) properties. Minor amounts of *n*-alkyl bromides are used to produce insect pheromones, phasetransfer catalysts, and other useful specialty chemicals.

## 1.4.3. 1-Bromo-3-chloropropane

**Production.** The versatile synthetic intermediate 1-bromo-3-chloropropane [*109-70-6*], $Br(CH_2)_3Cl$ (trimethylene chlorobromide), is virtually always produced by the free-radical addition of anhydrous HBr to allyl chloride.

Continuous or batch-type processes in glass-lined or impervious polymer-coated equipment can be employed. Gaseous HBr is added under slight positive pressure to a solution of allyl chloride dissolved in an inert, low-dielectric solvent using a free-radical initiator, such as oxygen [24], molecular sieves

[25], organic peroxides [26], or ultraviolet irradiation [27]. Because of the inductive effect of the neighboring chloride group, the addition reaction is not as rapid or exothermic as the corresponding hydrobromination of 1-propene. The competing ionic addition to form 2-bromo-1-chloropropane [*3017-95-6*] is retarded to a much greater extent than the free-radical reaction so that low reaction temperatures are not necessary to achieve a high ratio of primary to secondary bromide. Metal ions that accumulate during allyl chloride storage in metal tanks or drums are usually removed because they can promote ionic addition.

The yield of crude product, which typically consists of 95% primary and 5% secondary bromochloropropanes, is about 95% based on converted allyl chloride. The minor isomer can easily be separated by distillation if product specifications call for higher purity.

**Uses.** Trimethylene chlorobromide is a useful reaction intermediate to prepare γ-chloropropyl derivatives because of the greater reactivity of the bromine atom compared to the chlorine atom in nucleophilic displacement reactions. For example, the cyanide displacement of trimeth-ylene chlorobromide forms γ-chlorobutyronitrile, an intermediate used for the production of herbicides, pesticides, pharmaceuticals, and other commercial products.

### 1.4.4. 1-Bromo-2-phenylethane

**Production.** The useful intermediate, 1-bromo-2-phenylethane [*103-63-9*], $C_6H_5CH_2CH_2Br$ (β-phenethyl bromide), is usually manufactured by the free-radical addition of HBr to styrene.

The typical batch process is run in a glass-lined reactor by sparging gaseous HBr through a solution of styrene in an inert solvent containing a free-radical initiator. The reaction, which is rapid and moderately exothermic, can be controlled by introducing HBr at about the rate it is consumed. When styrene is completely converted, any excess HBr that is present can be recovered by various techniques prior to recovery of solvent and product by distillation. Styrene conversions are usually quantitative and β-phenethyl bromide yields are typically 95%. Although most patents describe batch reactions, the process can also be carried out continuously. The essential features necessary to obtain high selectivities to the desired β-isomer and negligible amounts of the undesired α-isomer, 1-bromo-1-phenylethane [*585-71-7*], are the use of dilute solutions of styrene (< 30 wt%) in an inert, nonpolar solvent, such as heptane; reaction temperatures in the range of 40 – 95 °C; and the absence of water and free-radical inhibitors. Organic peroxides [28], oxygen or air [29], azo compounds [30], and irradiation by ultraviolet light or by $^{60}$Co [31] are examples of free-radical initiators that have been employed for this reaction.

Although β-phenethyl bromide can be prepared in comparable yields by reaction of the more expensive β-phenethyl alcohol with phosphorus tribromide [32] or with other reagents, this alternative method is seldom used because of economic considerations.

**Uses.** β-Phenethyl bromide is an important starting material for the production of various β-phenethyl derivatives, pharmaceuticals, fragrances, and other fine chemicals.

## 1.4.5. 1-(Bromomethyl)-3-phenoxybenzene

**Production.** The synthetic pyrethroid starting material, 1-(bromomethyl)-3-phenoxybenzene [*51632-16-7*], 3-$C_6H_5OC_6H_4CH_2Br$ (*m*-phenoxybenzyl bromide), is produced by the thermal side chain bromination of *m*-phenoxytoluene. The usual byproduct from this benzylic bromination is 1-(dibromomethyl)-3-phenoxybenzene [*53874-67-2*], 3-$C_6H_5OC_6H_4CHBr_2$ (*m*-phenoxybenzal bromide).

The batch reaction is carried out in a glass reactor equipped with an appropriate HBr recovery system by feeding 20 % molar excess bromine vapor to neat *m*-phenoxytoluene at 265 °C. The rate of addition should be slow enough to avoid high local concentrations of bromine so that only negligible amounts of undesired ring-brominated byproducts are formed. Metal contamination must be avoided for the same reason and also to prevent Friedel-Crafts alkylation reactions. This uncatalyzed thermal bromination process produces the benzyl and benzal bromides in a 25:1 molar ratio in about 65 % yield based on 90 % converted *m*-phenoxytoluene. The low amount of *m*-phenoxybenzal bromide produced is partly due to its thermal instability, which results in the formation of some oligomeric byproducts at these high reaction temperatures [33]. Other thermal processes for brominating *m*-phenoxytoluene employ a catalyst, such as phosphorus trichloride [34], or activation with ultraviolet radiation [35]; however, appreciable amounts of ring-brominated byproducts are formed in these batch reactions. Continuous bromination of *m*-phenoxytoluene with ultraviolet light at 225 °C is claimed to eliminate formation of ring-brominated byproducts, but the product distribution of the benzyl and benzal bromides was not reported [35].

Additional *m*-phenoxytoluene brominating agents that have been used are $BrCCl_2CCl_2Br$ [36] and *N*-bromosuccinimide [37]. The former reagent is reported to produce mostly the benzyl bromide; the latter claims the predominant formation of the benzal bromide.

The competitive and analogous free-radical chlorination of *m*-phenoxytoluene to produce the corresponding benzyl chloride has been reported, but the reaction is not as selective as bromination and more ring-halogenated by-products are formed.

**Uses.** Both *m*-phenoxybenzyl and *m*-phenoxy- benzal bromides are important starting materials for the production of synthetic pyrethroid reaction intermediates. They can be converted by various reaction sequences to the alcohol, *m*-phenoxybenzyl alcohol, or to the aldehyde, *m*-phenoxybenzaldehyde, the key intermediates common to these synthetic pyrethroids. Both the alcohol and the aldehyde must contain less than 0.5 % ring-brominated byproduct for use in this application, hence the importance of minimizing the aromatic substitution side reaction.

## 1.4.6. Bromoacetic Acid

**Production.** Bromoacetic acid [*79-08-3*], $BrCH_2CO_2H$, and other α-bromocarboxylic acids are usually prepared by the well-known Hell-Volhard-Zelinski reaction in which the α-hydrogens of the carboxylic acid are replaced by bromine. A catalyst, such as a phosphorus trihalide, is necessary to convert the acid to the acid halide, the actual reaction intermediate. If the acid chloride or bromide is available as a starting material, no catalyst is needed and only one equivalent of bromine is required versus the two equivalents necessary when the carboxylic acid is the starting material. The resulting bromoacetyl halide can either be hydrolyzed or esterified to produce bromoacetic acid or its esters.

High yields of acid bromides can be obtained by adding one equivalent of bromine to the carboxylic acid containing phosphorus tribromide at 25 °C [38]. In practice a portion of the acid is converted to the acid bromide at 25–30 °C and then the remainder of the two equivalents of bromine is fed over time at higher temperatures to complete the reaction. The intermediate, bromoacetyl bromide [598-21-0], is hydrolyzed by the controlled addition of water and the resulting bromoacetic acid is separated from the major impurity, dibromoacetic acid [631-64-1], by distillation. The hydrogen bromide byproduct that is evolved during the reaction and the hydrolysis is usually recovered as a concentrated aqueous solution for subsequent use in other processes or for recycle to bromine. Bromine utilizations can be improved by using bromine chloride as the brominating agent to produce bromoacetyl chloride [22118-09-8], which is subsequently hydrolyzed to bromoacetic acid and HCl [39]. If the corresponding α-chlorocarboxylic acid is available, it can be converted to the α-bromo-carboxylic acid by reaction with gaseous HBr catalyzed by aluminum chloride [40] or with concentrated hydrobromic acid and an azeotropic solvent to remove water and hydrochloric acid [41]. The nucleophilic displacement reaction of α-chlorocarboxylic acids with inorganic bromides, such as KBr or NaBr, is an alternative method [42]. The reaction of chloroacetic esters with hydrobromic acid to produce bromoacetic acid and an alkyl bromide from the ester portion of the molecule is another variation of the reaction [43].

**Uses.** Bromoacetic acid as well as other α-bromocarboxylic acids and their esters are used in many chemical processes as starting materials or as reaction intermediates.

## 1.4.7. Bromomethane

**Production.** Bromomethane [74-83-9], $CH_3Br$ (methyl bromide), is produced from methanol and hydrogen bromide, which is generated in situ from bromine and a reducing agent like sulfur or hydrogen sulfide.

The commercial process is conducted by continuously feeding streams of liquid bromine, elemental sulfur, and methanol to a reactor maintained at 75–85 °C under a slight positive pressure. Methyl bromide vapor (99% purity) is continuously withdrawn from the reactor overhead system and scrubbed free of dimethyl ether by contacting the gas with a sulfuric acid scrubber. Excess methanol and sulfur are used because some methanol is entrained in coproduced sulfuric acid and because some sulfur is consumed in formation of a thio acid byproduct. Sulfuric acid containing thio acid, sulfur, and methanol is continuously purged from the reactor and either processed for resale or treated for waste disposal. By operating at a higher temperature and pressure than described in other similar processes, this process allows for good control of the exothermic reaction, retards the formation of dimethyl ether, and produces methyl bromide in high yield and purity [44]. Some methyl bromide is produced by the decomposition of a brominated ester of the thio acid, thus the stoichiometry deviates somewhat from that shown in Equation 1; therefore, this process produces a 6:1 molar ratio of methyl bromide to sulfur, another process using hydrogen sulfide as the reducing agent [45] has the advantage of producing an 8:1 ratio (Eq. 2):

$$6\ CH_3OH + 3\ Br_2 + S \longrightarrow 6\ CH_3Br + H_2SO_4 + 2\ H_2O \quad (1)$$
$$8\ CH_3OH + 4\ Br_2 + H_2S \longrightarrow 8\ CH_3Br + H_2SO_4 + 4\ H_2O \quad (2)$$

Methyl bromide can also be coproduced with other useful organic bromine compounds by the reaction of methanol solvent with byproduct hydrogen bromide. Thus, when a solution of phenol in methanol is reacted with bromine, aromatic substitution produces a 96% yield of the flame retardant intermediate, 2,4,6-tribromophenol [118-79-6], and hydroxyl substitution of the solvent produces methyl bromide in 34% yield [46]. The flame retardant tetrabromo-bisphenol-A [79-94-7] and methyl bromide are obtained in 95% and 61% yields, respectively, by the similar bromination of bisphenol-A in methanol [47]. Even though methyl bromide yields are much less than theory because of the equilibrium, $CH_3OH + HBr \rightleftharpoons CH_3Br + H_2O$, favorable economics are achieved because less methanol and HBr need to be recovered, and a saleable coproduct is produced.

**Uses.** Methyl bromide is a colorless, odorless, nonflammable, poisonous gas at ambient temperature and pressure. It is primarily used as a space and soil fumigant to control a wide range of animal and plant pests and disease organ-isms. In some of these fumigation applications methyl bromide containing the odorant, chloropicrin ($Cl_3CNO_2$), competes with ethylene dibromide.

A smaller but important use of methyl bromide is in the synthesis of many pharmaceuticals and fine chemicals, either as a methylating agent or as the Grignard reagent, methylmagnesium bromide [75-16-1].

## 1.4.8. 3-Bromo-1-propene

**Production.** The useful intermediate, 3-bromo-1-propene [106-95-6], $CH_2=CHCH_2Br$ (allyl bromide), is usually produced commercially from allyl alcohol and various brominating agents. The laboratory method of reacting allyl alcohol with hydrobromic acid and sulfuric acid is probably the most used process even though other reactants, such as red phosphorus and bromine [48], can be employed.

The batch reaction of allyl alcohol with a 35% molar excess of 60% hydrobromic acid is carried out in a stirred, glass-lined reactor at reflux (68 °C) for 3 h. The aqueous phase containing excess hydrobromic acid is separated and the organic phase is washed successively with water and aqueous sodium carbonate. Flash distillation produces allyl bromide (98–99% purity) in about 75% yield. An additional 6% yield of allyl bromide can be recovered from the aqueous phases by extraction. The major impurity, 1,2-dibromopropane [78-75-1] formed by addition of HBr to the product, is readily separated by the distillation. Although the yield is lower than that from the similar process employing sulfuric acid (92% yield), the process using 60% hydrobromic acid has the advantage of higher reactor loading and does not require recycle of sulfuric acid [49]. Allyl bromide can also be prepared in 50% yield by the halogen exchange reaction of allyl chloride with excess hydrobromic acid in the presence of a copper bromide catalyst [50]. The allylic bromination of propene by N-bromo derivatives or by bromine in the gas phase are alternative methods; however, they are not widely used commercially.

**Uses.** Allyl bromide is a very reactive intermediate capable of undergoing facile displacement reactions with many nucleophiles to produce synthetic perfumes, pharmaceuticals, and other compounds containing the allyl functionality.

## 1.4.9. Bromochloromethane and Dibromomethane

**Production.** Bromochloromethane [74-97-5], $CH_2BrCl$ (methylene chlorobromide), can be produced from dichloromethane by a halogen exchange reaction using either liquid bromine or gaseous HBr:

$$6\ CH_2Cl_2 + 3\ Br_2 + 2\ Al \longrightarrow 6\ CH_2BrCl + 2\ AlCl_3$$

$$CH_2Cl_2 + HBr \xrightarrow{AlCl_3} CH_2BrCl + HCl$$

The unisolated bromochloromethane can be further brominated to produce dibromomethane [74-95-3], $CH_2Br_2$ (methylene bromide), by adjusting the reactant stoichiometry and some process variables:

$$6\ CH_2BrCl + 3\ Br_2 + 2\ Al \longrightarrow 6\ CH_2Br_2 + 2\ AlCl_3$$

$$CH_2BrCl + HBr \xrightarrow{AlCl_3} CH_2Br_2 + HCl$$

In the process using bromine to produce bromochloromethane, a stirred reactor is charged with aluminum filings and one-sixth of the total dichloromethane charge, followed by a small amount of bromine to initiate the reaction. When the reaction temperature approaches 35 °C the remaining dichloromethane, which is used in excess, is added, followed by the slower addition of bromine at a rate sufficient to maintain a mild reflux. After completion of the reaction, the crude reaction mixture is pumped to a hydrolyzer vessel from which the halogenated methanes are steam distilled. The crude product, which is separated from water, consists of 90% bromochloromethane and 10% dibromomethane in addition to unconverted starting material. Fractional distillation is employed to separate all three components. The overall yield is about 85% based on the bromine that was used [51]. In the process employing HBr to produce dibromomethane, three reactors are connected for passage of a gas in series and each is charged with dichloromethane and aluminum chloride catalyst (5 mol%). Gaseous HBr is fed into the liquid phase of the first vessel and the evolved gas from each vessel is fed into the next vessel in sequence. The gas exiting from the last vessel is principally hydrogen chloride. When the contents of the first vessel consist of 99% $CH_2Br_2$ and 1% $CH_2BrCl$, the composition in the second vessel is 46% $CH_2Br_2$, 47% $CH_2BrCl$, and 7% $CH_2Cl_2$, and the composition in the third vessel is 18% $CH_2Br_2$, 55% $CH_2BrCl$, and 27% $CH_2Cl_2$. This overall product composition corresponds to a 91% utilization of the HBr fed. At this point the first reactor is discharged for recovery of product, recharged with starting materials, and reconnected to function as the last vessel in a new reaction sequence. The HBr feed line is switched from the first reactor to the second reactor, which becomes the first vessel in the next cycle. Thus, the third reactor in the original sequence operates as the second vessel in the new sequence [40].

Other patented processes to produce these materials include the gas-phase bromination of methyl chloride with a mixture of HBr and chlorine [52] or a mixture of bromine and chlorine [53], and the liquid-phase nucleophilic displacement reaction of dichloromethane with inorgan-ic bromides in aprotic [54] and protic [55] solvents.

**Uses.** The major use for bromochloromethane is as a fire-extinguishing fluid in portable systems. Dibromomethane is primarily used as a solvent and as a gauge fluid. Both compounds have been employed as reaction intermediates for the production of several herbicides and pesticides.

## 1.4.10. Trifluorobromomethane and Difluorobromochloromethane

**Production.** Both trifluorobromomethane [75-63-8], $CF_3Br$, and difluorobromochloromethane [353-59-3], $CF_2BrCl$, are usually produced by substitution reactions involving the replacement of either hydrogen or chlorine atoms of fluorocarbons by bromine.

These reactions are carried out in the gas phase at elevated temperatures (450–650 °C) by continuously feeding the reactants through a tube reactor. To obtain high reaction selectivities to the desired halomethane, the most important process variables are contact times and the proper ratios of reactants, which in turn are dependent on the temperature and whether or not a catalyst is employed. In most cases starting material conversions range from about 40 to 90% with product selectivities of 90% or higher. Trifluorobromomethane can be produced along with byproduct HBr by the gas-phase substitution reaction of trifluoromethane with bromine in a tube reactor packed with catalysts, such as alumina [56], or aluminum oxide–fluoride containing nickel [57]. Alternatively the reaction can be carried out by cofeeding trifluoromethane and bromine containing minor amounts of either sulfur trioxide [58] or chlorine [59] at elevated temperatures.

Other patented processes to produce $CF_3Br$ include the disproportionation reaction of difluorodibromomethane ($CF_2Br_2$) catalyzed by aluminum chloride [60], and the zinc bromide–carbon catalyzed chloride exchange reaction of trifluorochloromethane ($CF_3Cl$) with hydrogen bromide [61].

Difluorobromochloromethane is produced by similar halogen exchange and disproportionation reactions. The gas-phase reaction of $CF_2ClH$ with bromine in a tube lined with alumina (560 °C) produces a 96% yield of a mixture of $CF_2BrCl$ and $CF_2Br_2$ (15:1 molar ratio) based on a 91% conversion of starting material [62]. The byproduct, $CF_2Br_2$, can be separated and converted to $CF_2BrCl$ by the gas-phase reaction with chlorine [63].

**Uses.** Both halogenated hydrocarbons are employed as fire-extinguishing agents, especially in high-value applications, such as electronics and computer installations. They are just as effective as bromochloromethane but are much less toxic. Trifluorobromomethane is mainly used in automatic fire-extinguishing systems, and $CF_2BrCl$ is primarily used in portable fire extinguishers.

## 1.4.11. Flame Retardants

Fires in industrialized countries inflict a heavy toll in terms of economic loss, human suffering, and death. In response to this problem, the chemical industry has developed and commercialized flame retardants to reduce the flammability and to improve the fire safety of polymeric materials.

Flame retardants are incorporated into polymers either by physical blending techniques, *additive type*, or by chemical methods in which part of a conventional monomer is replaced with a flame retardant monomer in the polymerization process, *reactive type*. Although reactive flame retardants have the potential for greater performance in the polymer (migration and leaching are traditional problems with polymer additives in general), economic considerations have limited their acceptance mostly to thermoset polymers, while the additive types are widely used in both thermoplastic and thermoset polymers.

In the late 1950s pentabromobiphenyl oxide was the first aromatic bromine flame retardant to gain commercial significance. Since then the development of new flame retardants has focused on products with increased molecular mass, such as decabromobiphenyl oxide and poly(tribromostyrene). The benefits of higher mass are lower volatilization during processing and reduced surface migration during polymer use. Other significant parameters that have been improved over the years with the introduction of new flame retardants are enhanced thermal, light, and chemical stability, reduced toxicity, and improved physical and engineering properties of the resulting polymer formulations. The major markets for flame retardants are the electrical, electronic, appliance, automotive, textile, and furniture industries.

A number of industrially important brominated flame retardants used in polymers are listed in Table 3. Although manufacturing procedures for these materials are not generally published, those described in this section are from the patent literature and illustrate the different types of processes that can be employed.

### 1.4.11.1. Tetrabromobisphenol-A

**Production.** Tetrabromobisphenol-A (TBBPA), 4,4'-(1-methylethylidene)bis-(2,6-dibromophenol) [79-94-7], the largest selling brominated flame retardant, is produced by the bromination of bisphenol-A. The reaction is an example of substitution on an activated aromatic system and does not require catalysis.

The bromination reaction is generally conducted in solvents, such as a halocarbon [64], a halocarbon and water [65]–[67], a halocarbon and 50% hydrobromic acid [68], or aqueous alkyl monoethers [69]. Aqueous acetic acid is also a satisfactory medium [70], and acetic acid with added sodium acetate is reported to improve product color [71].

The HBr byproduct can either be recovered and recycled to recover bromine values or utilized in situ by means of one of two techniques. The first method involves the

**Table 3.** Physical properties of brominated flame retardants Tetrabromo-*seepsee*-xylene

| Compound | CAS registry no. | Structure | $M_r$ | Bromine content, % | mp, °C |
|---|---|---|---|---|---|
| Dibromoneopentyl glycol | [3296-90-0] | HOCH$_2$C(CH$_2$Br)$_2$CH$_2$OH | 262.0 | 61.0 | 109–110 |
| Tetrabromo-*p*-xylene | [23488-38-2] | H$_3$C–C$_6$Br$_4$–CH$_3$ | 421.8 | 75.8 | 254–255 |
| 1-Dibromoethyl-3,4-dibromo-cyclohexane | [3322-93-8] | Br,Br-cyclohexyl–CHBrCH$_2$Br | 427.8 | 74.7 | 70 (65–68*) |
| Tetrabromophthalic anhydride | [632-79-1] | Br$_4$-phthalic anhydride | 463.8 | 68.9 | 275–280 (270 |
| Tetrabromobisphenol-A | [79-94-7] | (HO–C$_6$H$_2$Br$_2$)$_2$C(CH$_3$)$_2$ | 543.9 | 58.8 | 179 (179–181 |
| Pentabromoethylbenzene | [85-22-3] | Br$_5$C$_6$–CH$_2$CH$_3$ | 500.6 | 79.8 | 138 (136–138 |
| Pentabromochlorocyclohexane | [87-84-3] | Br$_5$C$_6$H–Cl | 513.1 | 77.9 | 203 |
| Pentabromobiphenyl oxide | [32534-81-9] | (C$_6$H$_5$)$_2$O, Br$_5$ | 564.8 | 70.7 | liquid |
| Hexabromocyclododecane | [3194-55-6] | cyclododecane-Br$_6$ | 641.8 | 74.7 | 170–180 (185 |
| Hexabromobenzene | [87-82-1] | C$_6$Br$_6$ | 551.6 | 86.9 | 327 |
| 1,2-Bis(2,4,6-tribromophenoxy)ethane | [37853-59-1] | (Br$_3$C$_6$H$_2$–OCH$_2$)$_2$ | 687.7 | 69.7 | 223–225 |

**Table 3.** contiued

| Compound | CAS registry no. | Structure | $M_r$ | Bromine content, % | mp, °C |
|---|---|---|---|---|---|
| Octabromobiphenyl oxide | [32536-52-0] | | 801.5 | 79.8 | 70–150 |
| 1,2-Bis(tetrabromo-phthalimido)ethane | [32588-76-4] | | 951.6 | 67.2 | 446–450 |
| Decabromobiphenyl oxide | [1163-19-5] | | 959.3 | 83.3 | 304 (300–315*) |
| 1,2-Bis(pentabromo-phenoxy)ethane | [61262-53-1] | | 1003.3 | 79.6 | 322 |
| Poly(tribromostyrene) | [57137-10-7] | | (340.9)$_n$ | 70.3 | – |

addition of an oxidizing agent to the reaction medium to convert evolved HBr to bromine, which is subsequently consumed in the reaction. Suitable oxidizing agents are hydrogen peroxide [72], [73], sodium bromate [74], sodium chlorate [75], or chlorine [76], [77]. In the case of chlorine, the amount and rate of addition must be carefully controlled to limit chlorination.

The other approach is the use of an alcohol as the solvent to coproduce an economically useful alkyl bromide by the reaction of byproduct HBr with the alcohol [78]. Methanol is the solvent of choice because methyl bromide is widely marketed as a fumigant.

In the typical batch reaction, bromine in slight excess is added to a stirred bisphenol-A solution in methanol in a suitable reactor at 25–30 °C. The exothermic reaction is controlled by the rate of bromine addition, which is dependent on the refrigeration capacity of the reactor cooling system. After bromine addition, the contents of the reactor are heated to reflux (65 °C) for a short period to allow the slower reaction between methanol and dissolved HBr to occur. During this postreaction period, methyl bromide and some HBr are vented through the overhead condenser system into a series of scrubbers that neutralize and dry methyl bromide, which is produced in about 60% yield based on byproduct HBr. Water is then added to precip-itate TBBPA and the resulting slurry is cooled, filtered, washed, and dried. The yield of TBBPA is generally 94–97% based on bisphenol-A [47].

**Uses.** Commercial TBBPA is available in two grades, an epoxy resin grade and a higher quality polycarbonate grade. TBBPA is a reactive flame retardant; both hydroxyl groups can be reacted with epichlorohydrin under basic conditions to form the diglycidyl ether, which is widely used in epoxy resin formulations. TBBPA is also used in polycarbonate and ether polyester resins requiring low color and good clarity.

### 1.4.11.2. Decabromobiphenyl Oxide

**Production.** Decabromobiphenyl oxide (DBBPO), decabromodiphenyl ether, 1,1'-oxybis(2,3,4,5,6-pentabromobenzene) [1163-19-5], is manufactured by the exhaustive bromination of phenyl ether. As the bromination reaction progresses, partial substitution of hydrogen by bromine deactivates the aromatic ring toward further substitution, thus necessitating more vigorous reaction conditions than those used in the preparation of TBBPA. Lewis acid catalysis and an excess of bromine are used to obtain high conversions of the less reactive, under-brominated intermediates to DBBPO.

Decabromobiphenyl oxide can be prepared at atmospheric pressure by reacting bromine with phenyl ether in ethylene dibromide solvent and in the presence of aluminum bromide catalyst [79]. The use of bromine in an organic solvent requires long reaction times, results in low productivity per reactor volume, and necessitates recycling the solvent. These limitations have led to the use of bromine as both the reactant and the solvent [80] – [82].

A typical batch process to produce DBBPO [82] involves the addition of phenyl ether over one hour to an agitated reactor containing 150% excess of the stoichiometric amount of bromine and catalytic amounts of aluminum bromide. The reaction temperature is maintained at 44 – 55 °C throughout the addition. After a 2-h postaddition reaction at reflux (59 °C), the product is recovered by first steam distilling the excess bromine from the reactor and then filtering the aqueous slurry. The resulting solid cake is washed with water and oven dried at 110 °C. The crude DBBPO, which consists of about 96% deca-, 3% nona-, and 1% octabromobiphenyl oxides, is obtained in high yield and contains small amounts of occluded bromine and hydrogen bromide. These occluded impurities can be substantially removed by first grinding and subsequently heating the crude DBBPO [83].

**Uses.** Decabromobiphenyl oxide is an additive-type flame retardant that has excellent thermal stability under conditions necessary to process high-impact polystyrene, the polymer system in which it is most widely used. It is also suitable for applications in ABS, rubbers, and some polyolefins and epoxy resins.

### 1.4.11.3. Tetrabromophthalic Anhydride

**Production.** Tetrabromophthalic anhydride (TBPA), 4,5,6,7-tetrabromo-1,3-isobenzofurandione [632-79-1], is prepared by phthalic anhydride bromination. Phthalic anhydride is a deactivated aromatic compound and is unreactive toward bromine under the usual bromination reaction conditions. TBPA can be prepared by reacting phthalic anhydride and bromine in a mixture of concentrated sulfuric acid and 70% hydrogen peroxide in the presence of catalytic amounts of iodine [84], or in chlorosulfonic acid containing a small amount of sulfur [85].

The most common reaction medium is fuming sulfuric acid containing 45–65% sulfur trioxide (oleum) and catalytic amounts of iron powder and iodine. If a mixture of bromine and chlorine is employed as the brominating agent, an excess of bromine must be present to limit the incorporation of chlorine in the product [86]. Use of bromine as the halogenating agent requires an excess of sulfur trioxide because it oxidizes byproduct HBr to bromine during the course of the reaction.

The batch reaction is typically carried out by adding 90% of the bromine (2.1 mol/mol of phthalic anhydride) to the anhydride in excess oleum containing the catalyst at 90 °C. The reaction is completed by heating the reaction mixture to 105 °C followed by the addition of the remaining bromine. Excess sulfur trioxide and bromine are then distilled from the reactor and crude product is recovered from the cooled distillation residue by filtration. The product is thoroughly washed with water to minimize residual sulfuric acid, dried, and obtained in 94% yield based on phthalic anhydride [87].

**Uses.** Tetrabromophthalic anhydride is primarily employed as a reactive flame retardant in unsaturated polyester and epoxy resin formulations. It is also used as an intermediate in the manufacture of other flame retardants, such as diallyl tetrabromophthalate and tetrabromophthalate polyol esters.

## 1.4.11.4. Hexabromocyclododecane

**Production.** The aliphatic flame retardant, hexabromocyclododecane (HBCD), is produced by the addition reaction of bromine to *trans,trans,cis*-1,5,9-cyclododecatriene, the cyclic trimer of butadiene. The HBCD that is obtained from this reaction consists of a mixture of stereoisomers of 1,2,5,6,9,10-hexabromocyclododecane [*3194-55-6*] ranging from oils to an isomer with a 205–208 °C melting point. Commercially available HBCD contains a mixture of solid isomers and has a melting point range of 170–180 °C (185–195 °C).

Addition of bromine to the cyclic triene in nonpolar solvents results in substantial amounts of free-radical substitution reactions which lead to decreased yields and lower quality product. Use of polar solvents, such as ethanol, reduces the substitution reaction but can give rise to vicinal bromoethoxy byproducts. Mixed solvent systems, such as benzene–*tert*-butyl alcohol [88] or perchloroethylene–ethanol [89], can be employed to improve product purities and yields.

Another technique to reduce the formation of undesired byproducts is to add the bromine and cyclododecatriene simultaneously and separately to ethanol [90]. This process is carried out by separately feeding bromine and cyclododecatriene in a molar ratio of 3:1 to 96% ethanol in a stirred reactor. The reaction temperature is kept at 20–30 °C and a 1–5% excess of bromine is maintained during the 2.5–3.0 h addition. The precipitated product is recovered by filtration or centrifugation, and the mother liquor is recovered and reused in subsequent reactions after adding makeup ethanol. The total yield of dried HBCD from nine separate batches using recycled mother liquor is 92 % [90].

**Uses.** The main use for HBCD is as an additive-type flame retardant for extruded and expanded polystyrene foam. Other applications include crystal and high-impact polystyrene, SAN resins, adhesives, and coatings.

## 1.5. Manufacturers

Organic bromine compounds that are produced in annual volumes of 100 t or more are generally manufactured by companies that also produce elemental bromine. Manufacturers with organic bromine production facilities near their bromine-producing plants have lower production costs than nonbromine producers because bromine shipping and handling costs are minimized and byproduct inorganic bromides can be readily processed to recover bromine values.

The major integrated producers are Dow Chemical, Ethyl Corp., Great Lakes Chemical Corp., and their various subsidiary companies in the United States; the Dead Sea Bromine Group, a member of Israel Chemicals Ltd. (Israel); Chemische Fabrik Kalk GmbH, a subsidiary of Kali and Salz AG (Federal Republic of Germany); ATOCHEM, a subsidiary of Elf Aquitaine (France); and Asahi Glass Company Ltd. and Toyo Soda Manufacturing Company Ltd. (Japan). Associated Octel Company Ltd. (United Kingdom), which manufactures large volumes of ethylene dibromide, can also be included in this group of producers.

## 1.6. Toxicology and Occupational Health

Organic bromine compounds having adopted TLVs and MAKs are given in Table 4.

**1,2-Dibromoethane.** Ethylene dibromide (EDB) is severely irritating to skin, eyes, and lungs; acutely toxic; and a profound CNS depressant [93]–[99]; sensitizer [93], [94], [97]; mutagen [100]–[105]; and carcinogen [95], [106]–[111].

Reproductive studies in mammals have demonstrated a no-effect or reversible effect at nontoxic dose levels. Reduced egg weights of hens have been reported [112], [113]. Teratospermia in roosters could not be induced [113]; reproduction in rats could not be affected [114]–[116]. Ethylene dibromide has been reported to be nonteratogenic in rats [116]. Reversible spermicidal effects have been demonstrated in rats, rams, and bulls [117]–[122]. A dominant lethal effect in mice or rats, respectively, could not be demonstrated [123], [124].

**Table 4.** Maximum concentrations in the work environment

| | TWA | | STEL | | MAK | |
|---|---|---|---|---|---|---|
| | ppm | mg/m$^3$ | ppm | mg/m$^3$ | ppm | mg/m$^3$ |
| Bromomethane [74-83-9], skin | 5 | 20 | 15 | 60 | 5 | 20 |
| Tribromomethane [75-25-2], skin | 0.5 | 5 | – | – | – | – |
| Tetrabromomethane [558-13-4] | 0.1 | 1.4 | 0.3 | 4 | – | – |
| Bromochloromethane [74-97-5] | 200 | 1050 | 250 | 1300 | 200 | 1050 |
| Trifluorobromomethane [75-63-8] | 1000 | 6100 | 1200 | 7300 | 1000 | 6100 |
| Difluorodibromomethane [75-61-6] | 100 | 860 | 150 | 1290 | 100 | 860 |
| Bromoethane [74-96-4] | 200 | 890 | 250 | 1110 | 200 | 890 |
| 1,2-Dibromoethane [106-93-4], skin | * | * | – | – | ** | ** |
| 1,1,2,2-Tetrabromoethane [79-27-6] | 1 | 15 | 1.5 | 20 | 1 | 14 |
| 2-Bromo-2-chloro-1,1,1-trifluoroethane [151-67-7] | – | – | – | – | 5 | 40 |
| 1,2-Dibromo-3-chloropropane [96-12-8] | – | – | – | – | ** | ** |
| Bromoethylene [593-60-2] | 5 * | 20 * | – | – | – | – |

\* Suspect of carcinogenic potential for humans.
\*\* Carcinogenic in animal experiments

Two metabolic pathways have been postulated. One involves a glutathione transferase mediated conjugation of EDB; the other is based on oxidative dehalogenation by microsomal enzymes yielding a free-radical intermediate [125].

Disulfiram (tetraethylthiuram disulfide) has been reported to potentiate EDB toxicity in mice and rats and carcinogenicity in rats [126]–[128]. Disulfiram, an inhibitor of aldehyde dehydrogenase, blocks detoxification of the EDB intermediate, 2-bromoacetaldehyde, which results in mutagenicity and DNA binding [129].

EDB has now been prohibited for use as a soil fumigant by EPA [130], and OSHA has proposed a reduction in permissible industrial exposure limits from 20 ppm to 0.1 ppm [131].

**I-Bromododecane.** No toxicology records exist in computer literature data bases (N.L.M., S.D.C., Dialog) from 1965 to 1984.

**I-Bromo-3-chloropropane (TMCB).** TMCB vapors cause severe eye and skin irritation, typical haloalkane cardiovascular effects [132], and CNS depression [133]. $LC_{50}$ (fish): 75 mg/L/24 h [134]; BOD: 0.02 g/g, [135]; it is nonmutagenic (Ames); nondominant lethal; spermatogenesis was not impaired in mice treated at 300, 600, or 1200 mg/kg/d for 5 d [133].

**1-Bromo-2-phenylethane.** Phenylethylbromide has been reported to be metabolized by rats and rabbits to metabolites that are common to styrene and styrene oxide, and that are ultimately excreted in the urine [136], [137].

**1-(Bromomethyl)-3-phenoxybenzene.** 3-Phenoxybenzyl bromide is a skin sensitizer [133], [138]; $LD_{50}$ (rats, oral): 5 g/kg male, and 2.3 g/kg female [133].

**Bromoacetic Acid.** Bromoacetic acid has an $LD_{50}$ (oral, mice) of 100 mg/kg [139]; is irritating and corrosive to skin and mucous membranes [140], [141]. Biological in vivo or in vitro studies are based upon bromoacetate.

Bromoacetate administered as sodium bromoacetate to rabbits causes retinal degeneration, [142]–[144] and inhibits human erythrocyte carbonic anhydrase B [145].

Neuromuscular blockage by bromoacetate has been demonstrated [146]. Inhibition of neuroblastoma in A/J mice following intratumor injection has been reported. A mechanism involving irreversible binding of bromoacetate to cholinergic receptors on neuroblastoma cell membrane and/or to cholinesterase was proposed [147]. The antitumor action may be based on alkylation of DNA, not inhibition of cholinergic receptors [148].

The carcinogenic potential of chloroacetic acid has been studied in a 2-year (1982–1984) chronic gavage bioassay [149].

**Bromomethane.** Acute inhalation exposure causes pulmonary edema, convulsions, hyperthermia, and coma; chronic exposure can lead to neuropathies and behavioral disturbances. Significant changes occurred in rat brain monoamines following inhalation exposure to less than 15 ppm [150].

In addition to skin irritation and blistering, dermal exposure has been associated with systemic effects typical of inhalation exposure [151].

Methyl bromide has been reported to be mutagenic in a number of systems, but nonteratogenic [151]. It is currently under evaluation for carcinogenic potential by inhalation [149].

For further background, an excellent review article is recommended [151]. The present TLVs and MAKs are shown in Table 4. Ethyl Corp. has set an internal standard of 1 ppm, 8-h TWA [133].

**3-Bromo-1-propene.** Allyl bromide is extremely irritating and painful to the eyes, lungs, mucosal membranes, and skin [152].

Direct mutagenicity in the absence of activat-ing enzymes has been demonstrated; the direct mutagenicity (allyl iodide > allyl bromide > allyl chloride) either decreased or was totally lost following addition of microsomal enzyme S-9 supernatant [153], [154]. Also in vitro alkylation of DNA by allyl bromide was shown [155], [156].

**Dibromomethane.** The toxicity of dibromomethane has been summarized as follows [157]: more toxic than either methylene chloride or methylene chlorobromide, but less toxic than tribromomethane; can produce significant liver and kidney damage in animals; may cause cardiac arrhythmias; $LD_{50}$ (rats, oral): > 1000 mg/kg; slightly irritating to the eyes and skin of rabbits and does not appear to be absorbed significantly even when applied repeatedly; vapors are anesthetic; disorder occurs in the protein-prothrombin and glycogenesis functions of the liver and in filtration capacity of the kidneys in animals; metabolized to carbon monoxide. Mutagenicity (Ames) has been demonstrated [158]; it has been reported as the most mutagenic among dihalomethanes [159].

Rats were observed to have elevated blood levels of carboxyhemoglobin following [$^{13}$C] dichloromethane administration. It was later demonstrated in vitro that dihalomethanes were metabolized to carbon monoxide by a microsomal mixed function oxidase system [160], [161]. The reported rate and amount of carbon monoxide produced in vivo was as follows: dibromide > dichloride > chloride – bromide > diiodide [162]. The rate of metabolism by isolated rat hepatocytes was as follows: diiodide > dibromide > chloride – bromine = dichloride [163]. Further studies have shown that trihalomethanes are also metabolized to carbon monoxide [164]. The P-450 pathway yields a reactive formyl halide intermediate which may either acylate tissue nucleophiles or decompose to carbon monoxide, hydrogen ion, and inorganic halide [164] – [167].

Dihalomethanes are also metabolized to formaldehyde, formic acid, and inorganic halide. Dihalomethanes were metabolized by rat liver homogenates to formaldehyde and organic halide [168]. An enzymatic pathway yielding a highly reactive intermediate capable of either alkylating tissue nucleophiles or undergoing metabolism to inorganic halide, formaldehyde, and formic acid was proposed [164], [166], [169]. Elevated urinary formic acid levels have been reported in workers exposed to high levels of dichloromethane [166]. The mutagenic activity of dibromo- and diiodomethane, involving an electrophilic intermediate, is mediated by microsomal enzymes and glutathione-S transferase [170].

**Bromochloromethane.** The most significant toxicological effects of bromochloromethane, as well as haloalkanes in general, are central nervous system depression [171] and impaired cardiac performance [172], [173], arrhythmias due to sensitization to epinephrine [174], [175], and decreased arterial blood pressure [176]. Two distinct activation in vivo pathways have been observed: a high-affinity, saturable oxidation reaction involving microsomal enzymes, and a much lower affinity pathway involving glutathione-S transferase [177]. The rate and amount of carbon monoxide metabolized is less than that following exposure to dibromomethane, but greater than dichloromethane [160]. Studies have also demonstrated adverse effects upon the liver, kidney, and adrenal cortex [178] – [180]; it is irritating to the eyes and skin [181]; $LD_{50}$ (rats, oral), 5000 mg/kg [182]. Inhalation toxicity was reported to be greater than either dichloromethane or carbon tetrachloride [183]. The TLVs and MAKs are given in Table 4.

**Trifluorobromomethane.** Trifluorobromomethane causes central nervous system and respiratory depression [184], [185]; cardiac arrhythmias due to sensitization to epinephrine [174], [175]; reversible hypotension due to ganglionic blockage [176]; and impaired cardiac performance [172], [173].

The compound has been observed to inhibit acetylcholinesterase activity [186], [187]. For TLVs and MAKs, see Table 4.

**Difluorobromomethane** causes central nervous system depression [171], [188], [189], cardiac sensitization to epinephrine induced arrythymias, hypotension, and reduced myocardial performance [172], [173], [188], [189], and liver damage [190], [191].

**Tetrabromobisphenol-A (TBBPA).** Tetrabromobisphenol-A has a low order of acute toxicity [$LD_{50}$ (rat, oral) > 5 g/kg, $LD_{50}$ (rabbit, dermal) > 2 g/kg]; it is nonirritating to the eyes and skin (Draize scores of 0/24 h – 7 d and 0/24 h – 72 h, respectively); nonsensitizer; nonmutagenic (Ames test); and nonbromacnegenic [133].

The uncoupling of oxidative phosphorylation by TBBPA in isolated rat liver mitochondria has been demonstrated; oxygen uptake decrease was also observed with the same concentration of biphenyl or biphenyl ether [192]. Adverse effects upon mitochondrial energy transfer reactions by polychlorinated biphenyls have been observed [193]. Also, TBBPA induced liver microsomal enzyme systems [194]. Ethyl Corp. has adopted a TLV of 2 mg/m$^3$, 8-h TWA [133].

**Octabromobiphenyl Oxide (OBBPO).** Preliminary studies have indicated that OBBPO causes developmental toxicity when administered as a corn oil suspension via gavage to pregnant rats on days 5 through 16 of gestation.

Administration of 26 mg/kg per day of the mixture, ranging from 6 bromine atoms to 9 bromine atoms per molecule, resulted in siginificant effects on the conceptuses including reduced average fetal body weights, increased embryo and fetal death (resorption), fetal malformation, and delayed skeletal ossification. The embryo and fetal no-effect level was defined at 2.5 mg/kg per day.

**Decabromobiphenyl Oxide (DBBPO).** Decabromobiphenyl oxide has also been shown to have a low order of acute toxicity [$LD_{50}$ (rat, oral) and (rabbit, dermal) > 5 g/kg] [195]; nonirritating to the eyes [195], [196], but may cause transient irritation to the conjunctival membrane [196]; nonsensitizer to humans [197]; nonbromacnegenic [196]; nonteratogenic to rats [195], [196], [198]; no adverse reproductive effects in rats [198]; neither mutagenic [195] nor cytotoxic [198]; readily photodegraded by ultraviolet light [196]; and nonbioconcentrating in fish [196].

The principal target organs which may be adversely affected by DBBPO include the thyroid, liver, and kidney. Thyroid hyperplasia has been observed in rats following a 30-d feeding study at 80 and 800 mg/kg per day [196], [199]; no evidence of thyroid alterations was observed following a 2-year rat feeding study at 0.01 – 1.0 mg/kg per day [200].

Bromine concentration in the liver and adipose tissue of rats increased during a one-year feeding study at 0.1 mg/kg per day [198]; liver and kidney lesions were observed following a 30-d feeding study at dose levels > 80 mg/kg per day [196], [199]; liver alterations were observed following a rat inhalation study at 0.05 mg/L of air [195].

Liver enlargement in animals has been reported to occur with all brominated diphenyl ethers that are capable of inducing xenobiotic metabolism [201]. Neither DBBPO nor DBB (decabromobiphenyl) showed any porphyrinogenic potential [197]. There is evidence that porphyrinogenic effects of halogenated aromatic hydrocarbons are mediated through reactive intermediates, and apparently DBBPO is very resistant to metabolic degradation [202].

DBBPO was neither toxic nor carcinogenic in rats fed 0.01 – 1.0 mg/kg per day for 2 years [200]. DBBPO carcinogenicity potential is presently under evaluation [149].

**Tetrabromophthalic Anhydride (TBPA)** has a low order of acute toxicity [$LD_{50}$ (rat, oral) and $LD_{50}$ (rabbit, dermal) > 10 g/kg]; may cause moderate eye irritation (Draize score of 12/4/24 h); it is nonirritating to the skin, but is a potential sensitizer [133].

Neither TBPA [133] nor its hydrolysis product, tetrabromophthalic acid [203], are mutagenic (Ames).

**Hexabromocyclododecane (HBCD)** has a low order of acute toxicity: $LD_{50}$ (rat, oral) > 10 g/kg, $LD_{50}$ (rabbit, dermal) >8 g/kg; $LC_{50}$ (rat, 1 h): >200 mg/L; nonirritating to eyes or skin (Draize scores of 0); nonmutagenic (Ames) [131].

# 2. References

[1] Z. E. Jolles (ed.): *Bromine and its Compounds,* Academic Press, New York 1966.

[2] S. Patai (ed.): *The Chemistry of the Carbon – Halogen Bond,* J. Wiley & Sons, New York 1973, Part 1 & 2.

[3] *Houben-Weyl* **5/4**.

[4] W. K. R. Musgrave, vol. 1 A, 2nd ed., p. 478; W. J. Feast, W. K. R. Musgrave, vol. **3 A**, p. 241, both in S. Coffey (ed.): *Rodds Chemistry of Carbon Compounds,* Elsevier, Amsterdam 1964, 1971, respectively.

[5] W. J. Feast, vol. 1 AB, 2nd ed. supplement, p. 31; vol. 3A, 2nd ed. supplement, p. 141, in M. F. Ansell (ed.): *Rodds Chemistry of Carbon Compounds,* Elsevier, Amsterdam 1975, 1983, respectively.

[6] R. D. Chambers, S. R. James in D. Barton, W. D. Ollis (eds.): *Comprehensive Organic Chemistry*, vol. **1**, Pergamon, Oxford 1979, p. 493.
[7] A. L. Horvath: *Halogenated Hydrocarbons: Solubility-Miscibility with Water*, Dekker, New York 1982.
[8] R. R. Dreisbach: *Physical Properties of Chemical Compounds*, vol. **1–3**, Advances in Chemistry Series, No. 15, 22, and 29, Amer. Chem. Soc., Washington, D.C., 1955, 1959, 1961.
[9] E. F. Mooney, M. Goldstein, in [1] p. 802.
[10] J. Zabicky, S. Ehrlich-Rogozinski, A. G. Loudon, [2] p. 63, p. 223.
[11] Associated Ethyl Co., GB 804995, 1958 (W. J. Read, D. B. Clapp, D. R. Stephens, J. E. Russell).
[12] Dead Sea Bromine Co. Ltd., US 2921967, 1960 (F. Yaron).
[13] Dow Chemical, US 2746999, 1956 (A. A. Gunkler, D. E. Lake, B. C. Potts).
[14] Procter & Gamble Co., US 3336403, 1967 (A. Kessler).
[15] Procter & Gamble Co., US 3396204, 1968 (C. B. McCarty, K. W. Thiele).
[16] Procter & Gamble Co., US 3321538, 1967 (K. W. Thiele, G. R. Wyness, C. B. McCarty).
[17] Gulf Research & Development Co., US 3699179, 1972 (J. P. Boyle, C. R. Murphy, W. W. Walsh).
[18] Distillers Co., Ltd., US 3108141, 1963 (E. J. Gasson, D. J. Badley).
[19] Great Lakes Chemical Co., US 3655788, 1972 (R. M. Thomas, F. R. Gerns).
[20] Distillers Co., Ltd., GB 843234, 1960 (E. J. Gasson).
[21] Continental Oil Co., US 3812212, 1974 (R. D. Gordon).
[22] Shell Oil Co., CA 749982, 1967 (L. C. Fetterly, K. F. Koetitz).
[23] Chevron Research Co., US 3519695, 1970 (S. Suzuki).
[24] Thomas A. Edison, Inc., US 2412882, 1946 (C. B. Gardenier).
[25] Shell International Research, NL-A 6410368, 1965.
[26] Esso Research & Engineering Co., US 3847985, 1974 (J. Linder, L. L. Maravetz, G. M. Schmit).
[27] Société Nationale des Pétroles d'Aquitaine, DE-OS 2530675, 1976 (M. Gellato, J. L. Seris, J. Suberlucq).
[28] Dow Chemical, US 2935535, 1960 (A. A. Asadorian).
[29] Du Pont, US 3321539, 1967 (F. J. Plesmid).
[30] M.C.P. Chemical Processes Ltd., US 4228106, 1980 (M. Martan).
[31] Dow Chemical, DE 1240059, 1967 (C. E. Grabiel, J. L. Russell, H. Volk, B. W. Wilkinson et al.).
[32] Chemische Fabrik Kalk GmbH, GB 1103650, 1968.
[33] Ethyl Corp., US 4326089, 1982 (K. A. Keblys).
[34] Sumitomo, US 4104940, 1979 (Y. Ume, T. Matsuo, N. Itaya).
[35] Shell Oil Co., US 4010087, 1977 (D. A. Wood, R. F. Mason).
[36] Toyo Soda Manufacturing Co., Ltd., EP-A 19804, 1980 (K. Katsuragawa, H. Sakka, K. Kihara).
[37] Roussel UCLAF, DE-OS 2810305, 1978 (J. Warnant, J. Jolly).
[38] Chemische Fabrik Kalk GmbH, GB 1210475, 1970 (H. Jenkner, H. Nestler).
[39] Zaklady Chemiczne "Organica-Azot", PL 99987, 1975 (J. Swietoslawski, A. Ratajczak). *Chem. Abstr.* **91** (1979) 91182 d.
[40] Dow Chemical, US 2553518, 1951 (D. E. Lake, A. A. Asadorian).
[41] Katayama Kagaku Kogyo Kenkyusho Co., Ltd., JP-Kokai 78105419, 1978 (O. Umekawa, K. Takemori, S. Katayama).
[42] Katayama Kagaku Kogyo Kenkyusho Co., Ltd., DE-OS 2640943, 1977 (O. Umekawa, K. Takemori, S. Katayama).
[43] Chemische Fabrik Kalk GmbH, GB 1381919, 1975 (H. Jenkner, R. Karsten).
[44] Degussa, GB 768893, 1957.

[45] Degussa, US 2717911, 1955 (L. Huter, H. Veith).
[46] Nippon Kayaku Co., JP-Kokai 74108003, 1974 (K. Matsuda, M. Sugino, S. Kaji).
[47] Dow Chemical, US 3182088, 1965 (H. E. Hennis).
[48] Whiffen & Sons Ltd., GB 836653, 1960 (P. F. Pasco, P. R. E. Lewkowitsch).
[49] Ethyl Corp., unpublished.
[50] Chemische Fabrik Kalk GmbH, DE-OS 2204194, 1973 (H. Jenkner, W. Buettgens).
[51] I.G. Farbenind. AG, DE 727690, 1942 (O. Scherer, F. Dostal).
[52] Solvay et Cie., DE 1283214, 1968 (U. Giacopelli, M. Manca).
[53] Società Chimica Dell'Aniene SpA, GB 874062, 1959.
[54] Shell International Research, FR 1441233, 1966.
[55] Ethyl Corp., US 3923914, 1975 (P. Kobetz, K. L. Lindsay).
[56] Du Pont, US 2875254, 1959 (F. J. Gradishar).
[57] Hoechst, DE-OS 3046330, 1982 (H. Block, J. Mintzer, J. Wittman, J. Russow).
[58] Central Glass Co., Ltd., JP Kokai 7976504, 1979 (N. Kouketsu, F. Inoue, T. Komatsu, K. Matsuoka).
[59] Onoda Cement Co., Ltd., JP Kokai 7762208, 1977 (Y. Ikubo, K. Kunihiro).
[60] Kali-Chemie AG, DE-OS 1913405, 1970 (H. Paucksch).
[61] Kali-Chemie AG, BE 856233, 1977 (H. Boehn, K. H. Hellburg).
[62] Hoechst, BE 631933, 1963.
[63] Kali-Chemie AG, DE-OS 1946509, 1974 (H. Paucksch, J. Massonne).
[64] Bayer, DE 1151811, 1964 (J. Nentwig).
[65] Chemische Fabrik Kalk GmbH, FR 1412959, 1964.
[66] Dow Chemical, US 3546302, 1970 (A. A. Asadorian, R. G. Tigner).
[67] Asahi Glass Co., Ltd., JP-Kokai 7654538, 1976 (M. Ichimura, T. Nishiyama, K. Suzuki).
[68] Chemische Fabrik Kalk GmbH, DE 1768444, 1972 (H. Jenkner, O. Rabe).
[69] Dow Chemical, US 3363007, 1968 (T. E. Majewski, L. R. Collins).
[70] Ethyl Corp., US 4013728, 1977 (D. R. Brackenridge).
[71] Chemische Fabrik Kalk GmbH, DE-OS 2511981, 1976 (H. Jenkner).
[72] Degussa, DE-OS 2227439, 1974 (W. Weigert, H. Hein, H. Mechler, E. Meyer-Simon et al.).
[73] Produits Chimiques Ugine Kuhlmann, FR 2274586, 1976 (A. Isard, D. Pellet).
[74] Chemische Fabrik Kalk GmbH, DE-OS 2613969, 1977 (H. Jenkner, R. Strang).
[75] Zaklady Chemiczne "Organica-Azot", US 4112242, 1978 (J. Swietoslawski, A. Silowiecki, A. Ratajczak, B. Nocon et al.).
[76] Chemische Fabrik Kalk GmbH, DE 1129957, 1962 (H. Hahn).
[77] Società Italiana Resine SpA, DE-OS 2162859, 1972 (F. Montanari, B. Calcagno, L. Conti).
[78] Dow Chemical, US 3029291, 1962 (A. J. Dietzler).
[79] I.S.C. Chemical Ltd., GB 1472383, 1977 (L. J. Belf).
[80] Produits Chimiques Ugine Kuhlmann, DE OS 2400455, 1975 (D. Balda, D. Pitiot).
[81] Michigan Chemical Corp., US 3965197, 1976 (H. Stepniczka).
[82] Great Lakes Chemical Corp., US 4287373, 1981 (J. A. Garman, R. I. Mamuzic, R. B. McDonald, J. L. Sands et al.).
[83] Great Lakes Chemical Corp., US 4327227, 1982 (J. T. Ayres, D. L. McAllister, J. L. Sands).
[84] Degussa, DE-OS 2250550, 1974 (H. Hein, K. Janzon, W. Weigert, C. Liedtke et al.).
[85] Unassigned, DD 14750, 1958 (H. Ohle).
[86] Chemische Fabrik Kalk GmbH, US 3194817, 1965 (H. Hahn).
[87] Chemische Fabrik Kalk GmbH, US 3382254, 1968 (H. Jenkner, O. Rabe, R. Strang).
[88] Cities Service Co., US 3833675, 1974 (J. Newcombe, A. O. Dotson).

[89]   Toa Gosei Chemical Industry Co., Ltd., JP 7505187, 1975 (T. Kowaguchi, N. Hisanega, H. Naito).
[90]   Chemische Fabrik Kalk GmbH, US 3558727, 1971 (H. Jenkner, O. Konigstein).
[91]   ACGIH (ed.): *Threshold Limit Values (TLV)*, ACGIH, Cincinnati, Ohio, 1983.
[92]   DFG (ed.): Maximale Arbeitsplatzkonzentrationen (MAK), Verlag Chemie, Weinheim 1982.
[93]   *Clin. Toxicol.* **14** (1979) 473–478.
[94]   V. K. Rowe, H. C. Spencer, D. D. McCollister, R. L. Hollingsworth et al., *AMA. Arch. Ind. Hyg. Occup. Med.* **6** (1952) 158–173.
[95]   S. F. Stinson, G. Reznik, J. M. Ward, *Cancer Lett.* **12** (1981) 121–129.
[96]   G. Reznik, S. F. Stinson, J. M. Ward, *Arch. Toxicol.* **46** (1980) 233–240.
[97]   *Vet. Hum. Toxicol.* **22** (1980) 101–103.
[98]   I. Jakobson, J. E. Wahlberg, B. Holmberg, E. Johansson, *Toxicol. Appl. Pharmacol.* **63** (1982) 181–187.
[99]   V. K. Rowe, H. C. Spencer, D. D. McCollister, R. L. Hollingsworth et al., *AMA. Arch. Ind. Hyg. Occup. Med.* **6** (1952) 158–173.
[100]  D. Clive, K. O. Johnson, J. F. S. Spector, A. G. Batson, *Mutat. Res.* **59** (1979) 61–108.
[101]  H. Bren, A. B. Stein, H. S. Rosenkranz, *Cancer Res.* **34** (1974) 2576–2579.
[102]  E. Vogel, J. L. R. Chandler, *Experientia* **30** (1974) 621–623.
[103]  U. Ranning, *Mutat. Res.* **76** (1980) 269–295.
[104]  E-L. Tan, A. W. Hsie, *Mutat. Res.* **90** (1981) 183–191.
[105]  P. A. Brimer, E-L. Tan, A. W. Hsie, *Mutat. Res.* **95** (1982) 377–388.
[106]  National Cancer Institute DHHS Publication No. (NIOSH) 81-1766, 1982, 1–163.
[107]  B. L. VanDuuren, B. M. Goldschmidt, G. Loewengart, A. C. Smith et al., *J. Natl. Cancer Inst.* **63** (1979) 1433–1439.
[108]  W. A. Olson, R. T. Habermann, E. K. Weisburger, J. M. Ward et al., *J. Natl. Cancer Inst.* **51** (1973) 1993–1995.
[109]  M. B. Powers, R. W. Voelker, N. P. Page, K. Weisburger, *Toxicol. Appl. Pharmacol.* **33** (1975) 171.
[110]  E. Nachtomi, D. S. R. Sarma, *Biochem. Pharmacol.* **26** (1977) 1941–1945.
[111]  K. D. Nitschke, R. J. Kociba, D. G. Keyes, J. J. McKenna, *Fund. Appl. Toxicol.* **1** (1981) 437–442.
[112]  A. Bondi, E. Olomucki, M. Claderon, *J. Sci. Food Agric. Abstr.* **6** (1955) 600–602.
[113]  E. Alumot, E. Nachtomi, O. Kempenish-Pinto, E. Mandel et al., *Poult. Sci.* **41** (1968) 1079–1985.
[114]  D. Amir et al., *Anim. Repro. Sci.* **6** (1983) 35–50.
[115]  D. Amir, E. Ben-David, *Ann. Biol. Anim. Biochim. Biophys.* **13** (1973) 165–170.
[116]  R. D. Short, J. L. Minor, J. M. Winston, J. Seifter et al., *Toxicol. Appl. Pharmacol.* **46** (1978) 173–182.
[117]  D. Amir, R. Volcani, *Nature (London)* **206** (1965) 99–100.
[118]  D. Amir, C. Esnault, J. C. Nicolle, M. Courot, *J. Reprod. Fertil.* **51** (1977) 453–456.
[119]  K. Edwards, H. Jackson, A. R. Jones, *Biochem. Pharmacol.* **19** (1970) 1783–1789.
[120]  D. Amir, *J. Reprod. Fertil.* **35** (1973) 519–525.
[121]  A. H. El-Jack, F. Hrudka, *J. Ultrastruct. Res.* **67** (1979) 124–134.
[122]  D. Amir, B. L. Gledhill, D. L. Garner, J. C. Nicholle et al., *Animal Repro. Sci.* **6** (1983) 35–50.
[123]  S. S. Epstein, E. Arnold, J. Andrea, W. Bass et al., *Toxicol. Appl. Pharmacol.* **23** (1972) 228–235.

[124] R. D. Short, J. M. Winston, C. Hong, J. L. Minor et al., *Toxicol. Appl. Pharmacol.* **49** (1979) 97–105.

[125] A. Tomasi, E. Albano, M. U. Dianzani, T. F. Slater, *FEBS Lett.* **160** (1983) 191–194.

[126] L. C. K. Wong, J. M. Hong, C. B. Lee, C. C. Bhandari et al., *Pharmacologist* **20** (1978) 174.

[127] L. C. K. Wong, J. M. Winston, C. B. Hong, H. Plotnick, *Toxicol. Appl. Pharmacol.* **63** (1982) 155–165.

[128] A. M. El-Hawari, *Pharmacologist* **20** (1978) 213.

[129] T. W. Shih, D. L. Hill, *Res. Commun. Chem. Pathol. Pharmacol.* **33** (1981) 449–461.

[130] Federal Register 48, 1983, 49918–46248.

[131] Federal Register 48, 1983, 45956–46003.

[132] E. W. VanStee (ed.): *Target Organ Toxicology Series: Cardiovascular Toxicology,* Raven Press, New York 1982, pp. 281–326.

[133] Ethyl Corp., *Toxicology Reports,* 1979, 1984.

[134] A. L. Bridie, C. J. M. Wolff, M. Winter, *Water Res.* **13** (1979) 623–626.

[135] [134] pp. 627–630.

[136] S. P. James, D. A. White, *Biochem. J.* **104** (1967) 914–921.

[137] B. F. Seuher, L. P. C. Delbressine, F. L. M. Smeets, I. A. P. Wagenaars-Zegers, *Xenobiotica* **9** (1979) 311–316.

[138] K. S. Rao, J. E. Betso, K. J. Olson, *Drug Chem. Toxicol.* **4** (1981) 331–351.

[139] *J. Pharmacol. Exp. Ther.* **86** (1946) 336.

[140] J. Apffel, *Arch. Mal. Prof.* **32** (1971) 603–606.

[141] *Merck Index*, 9th ed., Merck & Co., Inc., Rahway, N.J., 1976.

[142] D. R. Lucas, J. P. Newhouse, J. B. Davery, *Br. J. Ophthalmol.* **41** (1957) 313–316.

[143] D. R. Lucas, J. P. Newhouse, *Br. J. Ophthalmol.* **43** (1959) 147–158.

[144] A. Sorsby, J. P. Newhouse, D. R. Lucas, *Br. J. Ophthalmol.* **41** (1957) 309–312.

[145] P. L. Whitney, *Eur. J. Biochem.* **16** (1970) 126–135.

[146] B. V. Rama Sastry, C. Y. Chiou, *Toxicol. Appl. Pharmacol.* **17** (1970) 303.

[147] C. Y. Chiou, *J. Pharm. Sci.* **67** (1978) 331–333.

[148] C. E. Stratton, W. E. Ross, S. Chapman, *Biochem. Pharmacol.* **30** (1981) 1497–1500.

[149] National Toxicology Program Management Status Report, Dec. 6, 1983.

[150] T. Honma, A. Sudo, M. Miyagawa, M. Sato et al., *Neurobehav. Toxicol. Teratol.* **4** (1982) 521–524.

[151] G. V. Alexeeff, W. W. Kilgore, *Residue Rev.* **88** (1983) 101–153.

[152] R. E. Goddelin, H. C. Hodge, R. P. Smith, M. N. Gleason (eds.): *Clinical Tox. of Commercial Products,* 4th ed., Williams & Wilkins Co., Baltimore, Md., 1976, p. 113.

[153] W. Liginsky, A. W. Andrews, *Teratogenesis Carcinog. Mutagen.* **1** (1980) 259–267.

[154] E. Eder, T. Neudecker, D. Lutz, D. Henschler, *Biochem. Pharmacol.* **29** (1980) 993–998.

[155] E. Eder, D. Lutz, D. Henschler, *Naunyn-Schmiedebergs Arch. Pharmacol.* **51** (1981) 316–supplement.

[156] E. Eder, T. Neudecker, D. Lutz, D. Henschler, *Chem. Biol. Interact.* **38** (1982) 303–315.

[157] T. R. Torkelson, V. K. Rowe in G. D. Clayton, F. E. Clayton (eds.): *Patty's Industrial Hygiene and Toxicology,* vol. **2 B,** 3rd ed., J. Wiley & Sons, New York 1981, pp. 3433–3601.

[158] M. Bignami, G. Cardamine, P. Comba et al., *Mutat. Res.* **46** (1977) 243–244.

[159] V. F. Simmer: "Structural Correlates of Carcino-genesis and Mutagenesis," *Second FDA Science Summer Symposium*, Washington, D. C., 1977, pp. 163–171.

[160] V. L. Kubic, M. W. Anders, R. R. Engel, C. H. Barlow et al., *Drug Metab. Dispos.* **2** (1974) 53–57.

[161] V. L. Kubic, M. W. Anders, *Drug Metab. Dispos.* **3** (1975) 104–112.
[162] F. L. Rodkey, H. A. Collison, *Toxicol. Appl. Pharmacol.* **40** (1977) 39–47.
[163] J. L. Stevens, J. H. Ratnayake, M. W. Anders, *Toxicol. Appl. Pharmacol.* **55** (1980) 484–489.
[164] A. E. Ahmed, V. L. Kubic, J. L. Stevens, M. W. Anders, *Fed. Proc.* **39** (1980) 3150–3155.
[165] V. L. Kubic, M. W. Anders, *Biochem. Pharmacol.* **27** (1978) 2349–2355.
[166] A. E. Ahmed, M. W. Anders, *Biochem. Pharmacol.* **27** (1978) 2021–2025.
[167] M. W. Anders, V. L. Kubic, A. E. Ahmed, *Xenobiotics* **440** (1978) 22–24.
[168] L. A. Heppel, V. T. Porterfield, *J. Biol. Chem.* **176** (1948) 763–769.
[169] A. E. Ahmed, M. W. Anders, *Drug Metab. Dispos.* **4** (1976) 357–361.
[170] P. J. VanBladeren, D. D. Breimer, G. M. T. Rotteveel-Smijs, G. R. Mohn, *Mutat. Res.* **74** (1980) 341–346.
[171] K. C. Back, E. W. VanStee, *Annu. Rev. Pharmacol. Toxicol.* **17** (1977) 83–95.
[172] E. W. VanStee, A. M. Harris, M. L. Horton, K. C. Back, *Toxicol. Appl. Pharmacol.* **34** (1975) 62–71.
[173] P. S. Beck, D. G. Clark, D. J. Tinston, *Toxicol. Appl. Pharmacol.* **24** (1973) 20–29.
[174] J. H. Wills, P. Bradley, H. Kao, H. Grace et al., *Toxicol. Appl. Pharmacol.* **22** (1972) 305–306.
[175] D. G. Smith, D. J. Harris, *Aerosp. Med.* **44** (1973) 198–201.
[176] E. W. VanStee, K. C. Back, *Toxicol. Appl. Pharmacol.* **23** (1972) 428–442.
[177] M. L. Gargas, M. E. Andersen, *Toxicol. Appl. Pharmacol.* **66** (1982) 55–68.
[178] C. C. Comstock, R. W. Fogleman, F. W. Oberst, *Arch. Ind. Hyg. Occup. Med.* **7** (1953) 526.
[179] T. R. Torkelson, F. Oyen, F. K. Rowe, *Am. Ind. Hyg. Assoc. J.* **21** (1960) 275.
[180] B. Highman, J. L. Svirbely, W. F. von Oettingen, W. C. Alford, L. J. Pecora, *Arch. Pathol.* **45** (1948) 299.
[181] *Toxic and Hazardous Ind. Chem. Safety Manual,* International Technical Information Institute, Tokyo, Japan, 1979, p. 80.
[182] W. B. Deichmann: *Toxicology of Drugs & Chemicals,* Academic Press, New York 1969, p. 390.
[183] J. L. Svirbely, B. Highman, W. C. Alford, W. F. von Oettinger, *J. Ind. Hyg.* **29** (1947) 382–389.
[184] D. C. Call, *Aerosp. Med.* **44** (1973) 202–204.
[185] R. A. Rhoder, K. L. Gabriel, *Toxicol. Appl. Pharmacol.* **21** (1972) 166–175.
[186] W. Young, J. A. Parker, *Combust. Toxicol.* **2** (1975) 286–297.
[187] V. A. Voronin, A. A. Denisenko, L. A. Linyucheva, N. M. Petushkov et al., *Gig. Tr. Prof. Zabol.* **2** (1982) 52–53.
[188] E. W. VanStee, K. C. Back, *Toxicol. Appl. Pharmacol.* **25** (1973) 469.
[189] P. S. Beck, D. G. Clark, D. J. Tinston, *Toxicol. Appl. Pharmacol.* **24** (1973) 20–29.
[190] R. S. Weinstein, B. U. Pauli, J. Alroy, R. Fresco, *U.S. National Technical Information Service Report No. ADA 046051,* 1977, p. 19.
[191] J. P. Murphy, E. W. VanStee, K. C. Back, *Toxicol. Appl. Pharmacol.* **41** (1977) 175.
[192] B. Inouye, Y. Katayama, T. Ishida, M. Ogata et al., *Toxicol. Appl. Pharmacol.* **48** (1979) 467–477.
[193] I. J. Stotz, Y. A. Greichus, *Bull. Environ. Contam. Toxicol.* **19** (1978) 319–325.
[194] P. Lundberg: "Drug Oxidation," *Int. Symp. Microsomes,* 4th ed., 1980, p. 853.
[195] R. Millischer, F. Girault, R. Heywood, G. Clark et al., *Toxicol. Eur. Res.* **2** (1979) 155–161.
[196] J. M. Norris, J. W. Ehrmantraut, C. L. Gibbons, R. J. Kociba et al., *Appl. Polym. Symp.* **22** (1973) 195–219.
[197] P. Koster, F. M. H. Debets, J. J. T. W. A. Strik, *Bull. Environ. Contam. Toxicol.* **25** (1980) 313–315.

[198] J. M. Norris, R. J. Kociba, B. A. Schwetz, J. Q. Rose et al., *Environ. Health Perspect.* **11** (1975) 153–161.

[199] J. M. Norris, J. W. Ehrmantraut, C. L. Gibbons, R. J. Kociba, *J. Fire Flammability Combust. Toxicol.* **1** (1974) 52–77.

[200] R. J. Kociba, L. O. Frauson, C. G. Humiston, J. M. Norris et al., *J. Combust. Toxicol.* **2** (1975) 267–285.

[201] G. P. Carlson, *Toxicol. Lett. (Netherlands)* **5** (1980) 19–25.

[202] M. Goto, M. Hattori, K. Sugiura, *Chemosphere* **3** (1975) 177.

[203] J. T. MacGregor, M. Friedman, *Mutat. Res.* **56** (1977) 81–84.

# Butadiene

Hans Joachim Müller, Erdölchemie GmbH, Köln, Federal Republic of Germany (Chaps. 2–7)
Eckhard Löser, Bayer AG, Wuppertal, Federal Republic of Germany (Chap. 8)

| | | | | | |
|---|---|---|---|---|---|
| 1. | Introduction | 899 | 6. | Stabilization, Storage, and Transportation | 914 |
| 2. | Physical Properties | 900 | 7. | Uses and Economic Importance | 916 |
| 3. | Chemical Properties | 900 | 8. | Toxicology | 917 |
| 4. | Production | 906 | 8.1. | Experimental Toxicity Data | 918 |
| 4.1. | Dehydrogenation of Butane and Butenes | 907 | 8.2. | Metabolism | 919 |
| 4.2. | Isolation of Butadiene from $C_4$ Steam Cracker Fractions | 911 | 8.3. | Human Data | 920 |
| 5. | Specifications of Butadiene | 914 | 9. | References | 921 |

# 1. Introduction

The name butadiene generally refers to 1,3-butadiene [106-99-0], $C_4H_6$, $M_r$ 54.092. The thermodynamically less stable 1,2-butadiene [590-19-2], which has two cumulated double bonds, has no technical importance. The 1,3-isomer, on the other hand, is economically the most important unsaturated $C_4$-hydrocarbon.

$CH_2=CH–CH=CH_2$ 1,3-Butadiene
$CH_2=C=CH–CH_3$ 1,2-Butadiene

E. Caventou was the first to isolate butadiene by means of the pyrolysis of amyl alcohol. M. Berthelot produced it by passing a mixture of acetylene and ethylene through a red hot iron tube. In 1885 the structure of butadiene was elucidated by G. Ciamician and P. Magnaghi. H. E. Armstrong and A. K. Miller, in 1886, discovered butadiene in the products obtained on cracking petroleum. Ever since S. Lebedew in 1910 discovered that butadiene forms rubber-like polymers, practical interest in this compound and its derivatives developed. Butadiene was produced on a large scale in Germany prior to World War II and in the USA during the war. Simultaneously, methods were developed for the manufacture of useful butadiene polymers [1]. After the war, the work of K. Ziegler and G. Natta, among others, on polymerization with organometallic catalytic agents led to a better quality of rubber.

## 2. Physical Properties

Butadiene is a colorless gas under normal conditions. The most important physical properties are summarized in Table 1. Table 2 presents the vapor pressure with reference to the temperature.

The technical data [2] important for reasons of safety are, above all, the flash point, – 85 °C, the ignition temperature, 415 °C, and the explosion limits when mixed with air and oxygen, see Table 3. Unstabilized or insufficiently stabilized butadiene forms explosive peroxides with oxygen present in the air.

Table 4 gives a list of the azeotropic mixtures [9] which are important for the distillation of butadiene-containing hydrocarbons.

Butadiene is sparingly soluble in water, see Table 5, soluble in methanol and ethanol, and very soluble in higher-boiling polar solvents, e.g., methylpyrrolidone.

## 3. Chemical Properties

Butadiene has two conjugated double bonds and, therefore, can take part in numerous reactions, which include 1,2- and 1,4-additions with itself (polymerization) and with other reagents, linear dimerization and trimerization, and ring formation.

*Polymerization* [10]. Polymerization by means of 1,2- and 1,4-addition is the most important butadiene reaction:

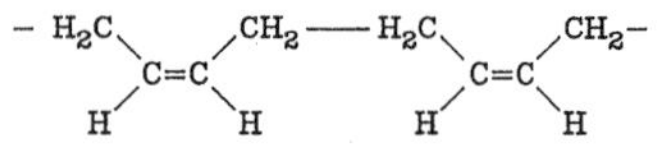

cis–1,4–Addition

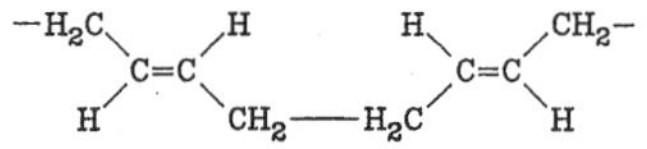

trans–1,4–Addition

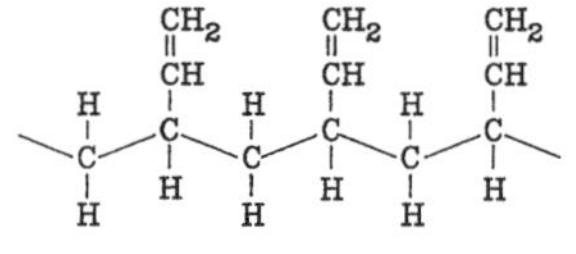

isotactic 1,2–Addition

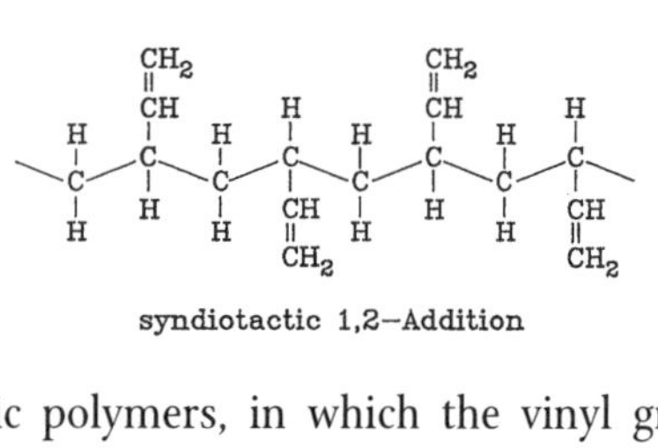

syndiotactic 1,2–Addition

On 1,2-addition, atactic polymers, in which the vinyl group has an arbitrary steric position, can also be formed.

**Table 1.** Physical properties of butadiene

| | |
|---|---|
| $M_r$ | 54.092 |
| bp at 0.1013 MPa | − 4.411 °C |
| mp at 0.1013 MPa | − 108.902 °C |
| Critical temperature | 152.0 °C |
| Critical pressure | 4.32 MPa |
| Critical density | 0.245 g/cm$^3$ |
| Density, liquid, at 0 °C | 0.6452 g/cm$^3$ |
| at 15 °C | 0.6274 g/cm$^3$ |
| at 20 °C | 0.6211 g/cm$^3$ |
| at 25 °C | 0.6194 g/cm$^3$ |
| Gas density (air = 1) | 1.9 |
| Viscosity, liquid, at 0 °C | 0.25 mPa · s |
| at 40 °C | 0.20 mPa · s |
| Enthalpy of vaporization at 25 °C | 20.88 kJ/mol |
| at − 4.41 °C | 21.98 kJ/mol |
| Enthalpy of formation, gaseous, at 298 K, 0.1013 MPa | 110.16 kJ/mol |
| Free enthalpy of formation, gaseous, at 298 K, 0.1013 MPa | 150.66 kJ/mol |
| Enthalpy of combustion, gaseous, at 298 K, 0.1013 MPa | 2541.74 kJ/mol |
| Enthalpy of hydrogenation to butane, gaseous, at 298 K, 0.1013 MPa | 236.31 kJ/mol |
| Entropy of formation, gaseous, at 298 K, 0.1013 MPa | 278.74 J mol$^{-1}$ K$^{-1}$ |
| Enthalpy of melting at 164.2 K, 0.1013 MPa | 7.988 kJ/mol |

**Table 2.** Vapor pressure of butadiene with respect to the temperature

| $t$, °C | − 4.413 | 0 | + 20 | + 40 | + 60 | + 80 | + 100 |
|---|---|---|---|---|---|---|---|
| $p$, MPa | 0.1013 | 0.1173 | 0.2351 | 0.4288 | 0.7247 | 1.1505 | 1.7342 |

**Table 3.** Explosion limits of butadiene in air (concentrations of butadiene stated)

| | At 0.1013 MPa, 20 °C | | At 0.4904 MPa, 30 °C | |
|---|---|---|---|---|
| | vol% | g/m$^3$ | vol% | g/m$^3$ |
| Lower limit | 1.4 | 31 | 1.4 | 150 |
| Upper limit | 16.3 | 365 | approx. 22 | approx. 2400 |

The nature of these polymers depends greatly on the way in which they are prepared and on the proper choice of the catalyst system employed.

*Further Addition Reactions.* Butadiene reacts with a series of reagents according to the 1,2- or 1,4-addition mechanism. The relative proportions of 1,4- to 1,2-addition products are dependent on the reaction conditions, e.g., temperature, the solvent, and the duration of the reaction. The addition products are important intermediates in the manufacture of chloroprene, adipic acid, anthraquinone etc. A typical example of an electrophilic addition to butadiene is the reaction with hydrogen chloride.

**Table 4.** Binary azeotropic mixtures of 1,3-butadiene and of other $C_4$-hydrocarbons

|  | bp, °C |  | Composition |  |
|---|---|---|---|---|
| Butane – butadiene | − 6.5 | (0.0933 MPa) | 20 | vol % butane |
| cis-2-Butene – 1-butyne | + 1.6 | (0.0933 MPa) | 20 | vol % 1-butyne |
| trans-2-Butene – 1-butyne | − 1.5 | (0.0933 MPa) | 9.5 | vol % 1-butyne |
| 1-Butene – vinylacetylene | − 9 | (0.0980 MPa) | ca. 0.7 | vol % vinylacetylene |
| cis-2-Butene – vinylacetylene | − 0.2 | (0.0933 MPa) | 33 | vol % vinylacetylene |
| trans-2-Butene – vinylacetylene | − 2.2 | (0.0933 MPa) | 25 | vol % vinylacetylene |
| Ammonia – butadiene | − 37 | (0.1013 MPa) | 45 | wt % butadiene |
| Methylamine – butadiene | − 9.5 | (0.1013 MPa) | 58.6 | wt % butadiene |
| Acetyldehyde – butadiene | + 5.53 | (0.1013 MPa) | 94.8 | wt % butadiene |

**Table 5.** Solubility $\alpha$ of butadiene in water at 0.1013 MPa (Bunsen's solubility coefficient) and solubility $L$ of water in liquid butadiene

| $t$, °C | $\alpha$, m$^3$/m$^3$ | $L$, g $H_2O$/kg butadiene |
|---|---|---|
| 10 | 0.29 | 0.53 |
| 20 | 0.23 | 0.66 |
| 30 | 0.19 | 0.82 |
| 40 | 0.16 |  |

The *manufacture of chloroprene* [126-99-8] requires the chlorination of butadiene followed by isomerization and alkaline dehydrochlorination:

$$CH_2=CH-CH=CH_2 + Cl_2 \longrightarrow \underset{Cl\ \ \ \ Cl}{CH_2-CH-CH=CH_2} + \ 60\%$$

$$\underset{40\%}{Cl-CH_2-CH=CH-CH_2-Cl} \xrightarrow{\text{Isomeri-zation}}$$

$$\underset{Cl\ \ \ Cl}{CH_2-CH-CH=CH_2} \xrightarrow{-HCl} \underset{Cl}{CH_2=C-CH=CH_2}$$

In the *production of adipic acid* according to a BASF procedure (→ Adipic Acid), butadiene reacts with carbon monoxide and methanol in two steps under different reaction conditions. At a higher temperature, approximately 185 °C, and at a lower pressure pentene acid ester reacts again with carbon monoxide and methanol to give adipic acid dimethyl ester. Hydrolysis then leads to the formation of adipic acid.

A crude $C_4$-fraction from an ethylene plant, containing approximately 44 % butadiene, in addition to butene, butane, 1,2-butadiene, and $C_4$-acetylenes, can be used for the first esterification step.

$$CH_2=CH-CH=CH_2 \xrightarrow{+ CO, + CH_3OH}$$
$$CH_3CH=CHCH_2-COOCH_3 \xrightarrow{+ CO, + CH_3OH}$$
$$H_3COOC-CH_2CH_2CH_2CH_2-COOCH_3$$

BASF plans to build a large-scale commercial plant producing 60 000 t/a, in Ludwigshafen.

The *hydroformylation of butadiene to give valeric aldehyde* is described in [11].

In the *production of hexamethylenediamine [124-09-4]* [12] (→ Hexamethylenediamine), hydrogen cyanide reacts with butadiene in two steps and the adiponitrile thus obtained is hydrogenated to give the diamine.

Several processes have been developed for the *production of 1,4-butanediol* from butadiene. In the three-step Mitsubishi process [13], [14], butadiene catalytically reacts with acetic acid to give 1,4-diacetoxy-2-butene, which in turn is hydrogenated to 1,4-diacetoxybutane and hydrolyzed to 1,4-butanediol.

$$CH_2=CH-CH=CH_2 + 2\ CH_3COOH + \tfrac{1}{2}O_2 \longrightarrow$$
$$H_3CCOO-CH_2-CH=CH-CH_2-OOCCH_3 + H_2O$$

A similar process has been published by BASF [15].

Toyo Soda's method [16] for the preparation of 1,4-butanediol involves the reaction of the chlorine addition products of butadiene, 1,4-dichloro-2-butene and 1,2-dichloro-3-butene, with sodium acetate to give, first, 1,4-diacetoxy-2-butene, which is then hydrogenated directly to 1,4-butanediol.

According to a Shell patent [17], butadiene can be treated with a peroxide to give diperoxybutene, which is converted to 1,4-butanediol on hydrogenation. 1,4-Butanediol is the starting product in the synthesis of *tetrahydrofuran [109-99-9]* (→ Tetrahydrofuran), which can also be obtained from 1,2-epoxy-3-butene, (Chevron) [18]. The epoxide in turn is made by, e.g., treating butadiene with a peroxide [19].

$$CH_2=CH-\overset{\overset{O}{\diagdown}}{CH}-CH_2 \longrightarrow \text{(epoxide)} \xrightarrow{H_2} \text{(THF)}$$

*Linear Dimerization and Trimerization.* Butadiene forms linear dimers or trimers in the presence of Ni, Co, Pd, or Fe catalysts. A good review is provided in [20].

Since the linear oligomers of butadiene differ from each other by a chain length of four carbon atoms, the separation of these oligomers is easier than that of ethylene oligomers. Dimerization results in the formation of 1,3,7-octatriene *[1002-35-3]* and of 1,3,6-octatriene [21], [22], which are, however, technically uninteresting. The aromatic hydrocarbons xylene and ethylbenzene [23] can be made by the dehydrocyclization of 1,3,7-octatriene.

If the dimerization is carried out under hydrogenating conditions, 1,7-octadiene with its terminal double bonds and also 1,6-octadiene are formed [24], [25].

$$CH_2=CH-CH_2-CH_2-CH_2-CH_2-CH=CH_2$$
1,7-Octadiene

$$CH_2=CH-CH_2-CH_2-CH_2-CH=CH-CH_3$$
1,6-Octadiene

1,7-Octadiene is converted into a $C_{10}$-diol on hydroformylation or into a $C_{10}$-diamine on treatment with hydrogen cyanide and hydrogen.

The manufacture of 1-octene, which is used as a copolymer in the synthesis of high-quality, linear low-density polyethylene, may get of some technical importance. Dimerization of butadiene and simultaneous reaction with carbon monoxide and alcohol leads to the synthesis of *pelargonic acid* [26]:

$$2\ CH_2=CH-CH=CH_2 + CO + ROH \longrightarrow$$
$$CH_2=CH-CH_2CH_2CH_2CH=CH-CH_2COOR \xrightarrow[\text{Saponification}]{H_2}$$
$$CH_3(CH_2)_7COOH$$

Pelargonic acid is, e.g., a starting material in the production of heat-resistant lubricants.

*Cyclization, Diels-Alder Reaction.* The Diels-Alder reaction is one of the best known reactions of butadiene [27]. Usually, a dienophile, i.e., an olefin with an activated double bond, reacts with butadiene forming a cyclohexane ring. This addition reaction, which is exclusively a 1,4-addition, can also take place with a second molecule of butadiene as the dienophile component, forming *4-vinylcyclohexene-1* [*100-40-3*] [28].

The reaction can take place with or without a catalyst [29]–[31]. Vinylcyclohexene, when subjected to dehydrogenation or oxidation, gives *styrene* [*100-42-5*] [32].

Two molecules of butadiene can dimerize to form 1,5-cyclooctadiene [*111-78-4*] and three molecules of butadiene can catalytically trimerize to 1,5,9-cyclododecatriene [*4904-61-4*] [33].

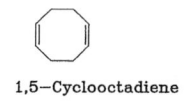

1,5–Cyclooctadiene

In the synthesis of *anthraquinone* (→ Anthraquinone), butadiene undergoes a Diels-Alder reaction with naphthoquinone to give tetrahydroanthraquinone, which in turn is oxidized to anthraquinone.

Butadiene readily undergoes a 1,4-addition with sulfur dioxide forming a cyclic sulfone, *2,5-dihydrothiophene-1,1-dioxide* [*1708-32-3*], [34], [35]. This compound is converted into sulfolan [*126-33-0*], a heat-stable and highly polar solvent, on catalytic hydrogenation.

2,5–Dihydrothiophene–1,1–dioxide     Sulfolan

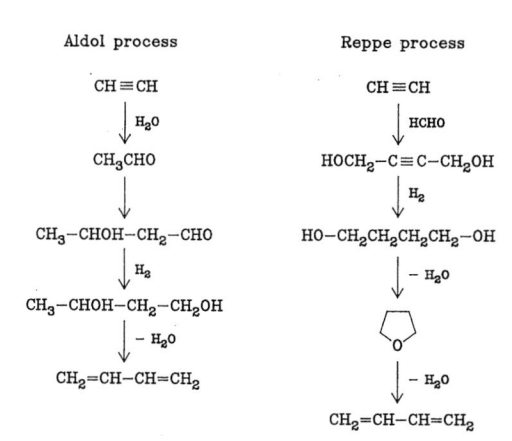

**Figure 1.** Production of butadiene from acetylene

**Aldol process**

$$CH \equiv CH$$
$$\downarrow H_2O$$
$$CH_3CHO$$
$$\downarrow$$
$$CH_3-CHOH-CH_2-CHO$$
$$\downarrow H_2$$
$$CH_3-CHOH-CH_2-CH_2OH$$
$$\downarrow -H_2O$$
$$CH_2=CH-CH=CH_2$$

**Reppe process**

$$CH \equiv CH$$
$$\downarrow HCHO$$
$$HOCH_2-C \equiv C-CH_2OH$$
$$\downarrow H_2$$
$$HO-CH_2CH_2CH_2CH_2-OH$$
$$\downarrow -H_2O$$
$$\text{(tetrahydrofuran ring)}$$
$$\downarrow -H_2O$$
$$CH_2=CH-CH=CH_2$$

*Formation of Complexes.* Butadiene reacts with numerous metal compounds to form complexes, e.g., with Cu(I) salts, which are used in the extraction of butadiene from $C_4$-hydrocarbon mixtures. However, this method has been replaced by modern extractive distillation techniques, see Section 4.2. Complexes with iron, nickel, cobalt, palladium, and platinum are also well known.

# 4. Production

A series of articles were published in 1942 describing all the methods used to produce butadiene [36].

*Production from Acetylene* [37], [38]. The large-scale industrial production of butadiene from acetylene has been carried out in Germany using two processes (see Fig. 1).

These methods are no longer used as the production of acetylene requires a large amount of energy and is very expensive. Only the first steps of the Reppe process leading to 1,4-butanediol and to tetrahydrofuran are commercially employed today.

*The production from ethanol* [39], [40] was first developed by W. IPATJEW and I. I. OSTROMISLENSKY and was modified by S. W. LEBEDEW into a commercial process, which has been used in the USSR ever since 1928 and was also employed in the United States during World War II.

$$2\ CH_3CH_2OH \longrightarrow CH_2=CH-CH=CH_2 + H_2 + 2\ H_2O$$

The raw material primarily used today is the $C_4$-fraction obtained by cracking naphtha and gas oil to ethylene. However, if the scarcity of this starting material persists, the production from ethanol could regain importance. It could also become the method of choice in countries which manufacture cheap ethanol from carbohydrates or in the European Economic Community with its enormous, subsidized agricultural surplusses (e.g., wine).

Nowadays butadiene is preferentially isolated from $C_4$-fractions obtained by cracking naphtha and gas oil to ethylene ($\rightarrow$ Ethylene). However, the supply of these raw materials is at the present time variable [41]. NGL, natural gas liquids (mixture of ethane, propane, *n*-butane and isobutane) and LPG, liquified petroleum gas (mixture of propane, butane) are increasingly being used as a starting material. The ethane component of gas, found in the North Sea or in the Middle East, is cracked to ethylene, leading to a decrease in the amount of $C_4$-fraction available. Hence, on a long-term basis, the dehydrogenation of butenes and butane to butadiene will gain new importance.

## 4.1. Dehydrogenation of Butane and Butenes

The dehydrogenation reactions are endothermic, the following values being valid at 430 °C:

| | | | | |
|---|---|---|---|---|
| Butane | → 1-Butene | + $H_2$ | $\Delta H$ = 131 kJ/mol |
| Butane | → *cis*-2-Butene | + $H_2$ | $\Delta H$ = 118 kJ/mol |
| *cis*-2-Butene | → Butadiene | + $H_2$ | $\Delta H$ = 126 kJ/mol |

According to Le Chatelier's principle, the yield is increased by decreasing the partial pressure of the reaction products. Practically, the dehydrogenation can be improved either by conducting the reaction under vacuum or by the addition of steam. The latter has the following advantages:

1) It reduces coking of the dehydrogenation catalyst.
2) It provides the required amount of heat for the endothermic dehydrogenation process.
3) Steam can be easily separated from the reaction products by means of condensation.

A further increase in yield is achieved by raising the reaction temperature.

Undesirable side reactions are cracking, isomerization, and polymerization. Compounds which tend to undergo these reactions are removed before the dehydrogenation process.

As the dehydrogenation process does not proceed to completion, the reaction products must be separated using suitable methods; the starting materials must be recovered and refed into the dehydrogenation process.

**Dehydrogenation of *n*-Butane.** The best known one-step dehydrogenation is the Houdry Catadiene process (Fig. 2), which has been in operation on a commercial scale since 1943 [42]–[45]. In this adiabatic process, several packed bed reactors, arranged parallel to each other, are operated alternatingly. Aluminum oxide mixed with approximately 20 % chromium oxide is the catalyst. *n*-Butane is subjected to dehydrogenation as such or in a mixture of *n*-butenes at 600–700 °C and 10–25 kPa. The use of high temperatures results in byproducts like $C_1$–$C_3$ hydrocarbons, hydrogen, and carbon

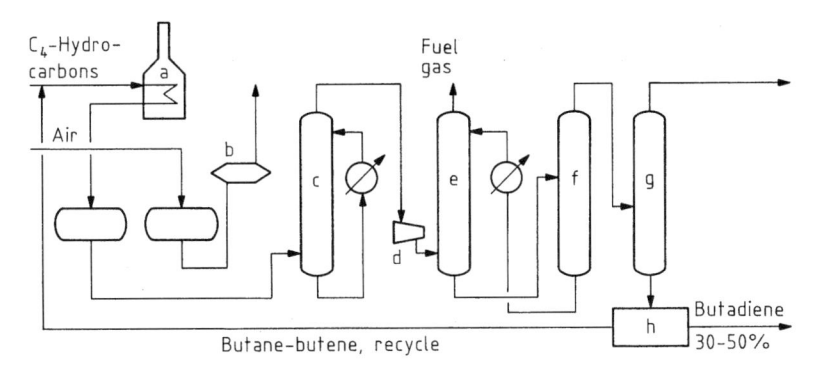

**Figure 2.** Houdry Catadiene process
a) Reactor; b) Waste-heat boiler; c) Quench column; d) Compressor; e) Absorber; f) Stripper; g) $C_3$ Column; h) $C_4$ Separation

deposits on the catalytic agent. After 5 – 15 min of running time, the reactor is switched to regeneration. The heat generated by burning the coke residue during the regeneration phase is stored in the catalyst and in the added inert material and is then reused in the next reaction phase.

The concentration of butadiene at the outlet of the reactor is 15 – 18 %. During the recovery process, which includes absorption of the $C_3$- and $C_4$-hydrocarbons, compression, stripping, and separation from unreacted n-butane and n-butenes, the concentration of butadiene is increased to 30 – 50 %. Approximately 550 t of butadiene are obtained from 1000 t of n-butane.

Phillips Petroleum has developed a two-step dehydrogenation process [46], [47]. It consists of the following:

1) n-Butane is catalytically dehydrogenated at 600 °C and 1 bar with $Cr_2O_3$–$Na_2O$–$Al_2O_3$ to butene.
2) The n-butenes are separated by means of extractive distillation, auxiliary substances being, e.g., acetone, acetonitrile, and furfural.
3) The n-butenes are dehydrogenated to butadiene in an isothermic tubular reactor, heated with flue gas to a temperature of 600 °C, at a pressure of 1 bar, and upon addition of over-heated steam on a $Fe_2O_3$–$K_2O$–$Al_2O_3$ catalyst.
4) Butadiene is recovered by means of extractive distillation using auxiliary substances, as mentioned above, and purified. The advantages of this method are the longer running times without catalyst regeneration (approx. 1 h) and higher butadiene yields (65 % of the starting butane).

**Dehydrogenation of n-Butenes.** n-Butenes are formed in the production of standard gasoline by catalytic cracking, in the production of olefins by thermal cracking, and in the dehydrogenation of LPG.

**Table 6.** Catalytic dehydrogenation of butenes

| Catalyst | Phillips 1490 | Shell 205 | Dow Type B |
|---|---|---|---|
| Composition | $Fe_2O_3$–Bauxite | $Fe_2O_3$–$Cr_2O_3$ | Ca/Na phosphate–$Cr_2O_3$ |
| Mole ratio $H_2O$/butene | 9–12/1 | 8/1 | 18–20/1 |
| Conversion per pass, % | 26–28 | 27–33 | ≤45 |
| Selectivity, % | 76 | 70 | 90 |
| Regeneration after | | 24 h | 30 min |
| Duration of regeneration | | 1 h | 15 min |

Normally, butenes are a part of a $C_4$-hydrocarbon mixture. The $C_4$-paraffins can be separated from the $C_4$-olefins by means of extractive distillation using auxiliary substances, as routinely employed in the recovery of butadiene [46], see above.

The boiling points of isobutene and $n$-butene are so close together that these isomers can only be separated using special methods. The gas phase method of BASF employing 40–45% $H_2SO_4$ selectively converts isobutene into *tert*-butanol [48]. Today, separation is carried out either by conversion to methyl *tert*-butyl ether or to isobutene oligomers [49]–[52].

Table 6 shows a list of the especially active and selective catalytic agents which have been developed for the dehydrogenation of $n$-butenes to butadiene.

**Oxidative Dehydrogenation of *n*-Butenes** [57]–[58]. The conversion and the selectivity of the dehydrogenation of $n$-butenes to butadiene can be significantly improved by removing the hydrogen from the equilibrium. The addition of oxygen causes the oxidation of hydrogen to water:

$$C_4H_8 + 1/2\ O_2 \longrightarrow C_4H_6 + H_2O$$

The exothermic oxidation of hydrogen partially covers the heat requirements of the endothermic dehydrogenation reaction and, in addition, the oxygen, together with steam added during the reaction, reduces the coke deposits on the catalyst. Especially productive methods are: the Oxo-D process of Petro-Tex [59] and the O-X-D process of Phillips [60], [61].

The *Oxo-D process* was first applied on a large scale in 1965. The advantages of this method are the low consumption of steam and heating energy, high conversion and selectivity per reactor cycle, longer life span of the catalyst, and no necessity for catalyst regeneration. Petro-Tex achieved a 65% conversion and a butadiene selectivity of 93% using a steam/butene ratio of 12 moles/1 mole.

Phillips has been running a plant in Borger (Texas) since 1970 based on the *O-X-D process* producing 125 000 t/a of butadiene. Phillips also obtains a high conversion and selectivity. An experimental plant claimed a 75–80% conversion and a 88–92% selectivity for butadiene.

The addition of oxygen to the dehydrogenation reaction of butane is not meaningful because at the high temperatures required, oxygen reacts with the reaction products, giving rise to undesired byproducts.

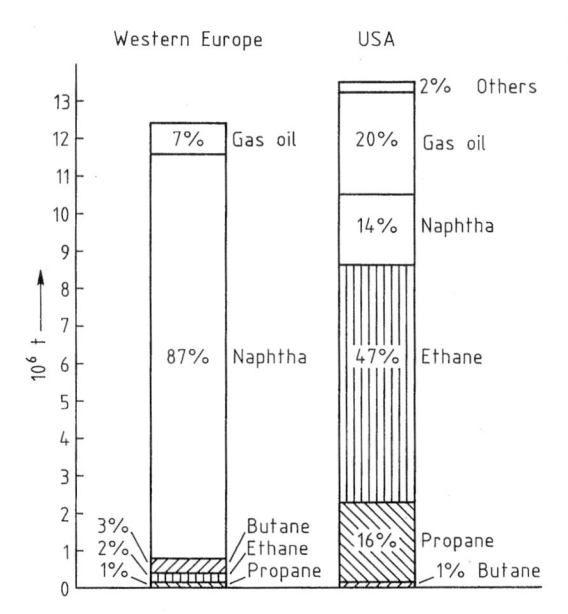

**Figure 3.** Raw materials for the production of ethylene in Western Europe and USA, 1979

**Oxidative Dehydrogenation of Butane with Halogen** [62], [63]. Shell has developed a one-step dehydrogenation of butane to butadiene using iodine as the hydrogen acceptor. The addition of iodine enables a high conversion and yield of butadiene, but has the disadvantage of causing serious corrosion problems in the plant.

## 4.2. Isolation of Butadiene from C$_4$ Steam Cracker Fractions

Today, the isolation of butadiene, worldwide, is based predominantly on the butadiene-containing C$_4$-fractions, which are produced during the steam-cracking of ethane, LPG, naphtha, gas oil, and other higher boiling hydrocarbon fractions to ethylene and homologous compounds. In Western Europe, naphtha and gas oil are primarily used as raw materials. In the USA, however, ethane and propane make up more than 60% of the starting materials, see Figure 3 [64].

Table 7 shows typical cracking yields and their relationship to the starting materials [65]–[67]. The influence of the cracking intensity, i.e., of the cracking temperature and duration, on the butadiene yield is minor compared to the influence exerted by the starting materials. On the other hand, as shown in Table 8, the composition of the C$_4$-fraction itself is greatly dependent on the cracking intensity.

Butadiene cannot be separated from this C$_4$-hydrocarbon mixture by means of simple distillation, because 1,3-butadiene and butane form an azeotrope (see Table 4) and the azeotrope of vinylacetylene with *cis*- and *trans*-2-butene has a boiling point only slightly above that of butadiene.

**Table 7.** Feedstocks and steam cracking yields (in wt%)

| | Light hydrocarbons | | | Naphthas and heavier | | |
|---|---|---|---|---|---|---|
| | Ethane | Propane | $n$-Butane | Medium range naphthas | Atm. gas oil | Light vacuum gas oil |
| **Feedstocks** | | | | | | |
| Methane | 0.6 | | | | | |
| Ethane | 95.2 | | | | | |
| Propane | 4.2 | 100 | | | | |
| $n$-Butane | | | 100 | | | |
| Paraffins | | | | 79.9 | | |
| Olefins | | | | – | | |
| Naphthenes | | | | 17.4 | | |
| Aromatics | | | | 2.7 | | |
| Total | 100 | 100 | 100 | 100 | | |
| TBP cut range, °C | | | | | 204–343 | 343–454 |
| ASTM boiling range, °C | | | | 47–148 | | |
| Rel. density | | | | 0.692 | 0.844 | 0.901 |
| **Cracking yields** | | | | | | |
| Hydrogen and methane | 15.1 | 29.7 | 23.8 | 17.7 | 12.1 | 9.5 |
| Ethylene | 77.7 | 42.0 | 40.0 | 34.0 | 25.9 | 20.5 |
| Propylene | 2.8 | 16.8 | 17.2 | 15.7 | 16.2 | 14.1 |
| Butenes and butanes | 0.8 | 1.3 | 6.7 | 4.3 | 4.8 | 6.3 |
| Butadiene | 1.9 | 3.0 | 3.5 | 4.7 | 4.6 | 5.4 |
| Pyrolysis gasoline | 1.7 | 6.6 | 7.1 | 18.8 | 18.4 | 19.3 |
| Pyrolysis fuel oil | – | 0.6 | 1.7 | 4.8 | 18.0 | 25.0 |
| Total | 100.0 | 100.0 | 100.0 | 100.0 | 100.0 | 100.0 |

Hence, butadiene is isolated using liquid–liquid extraction or extractive distillation, which is the method of choice today.

**CAA Liquid–Liquid Extraction** [46]. The CAA method (CAA=cuprous ammonium acetate), introduced by ESSO, is based on the ability of butadiene to form a complex with Cu(I) compounds. Butadiene is separated from the $C_4$-hydrocarbon mixture by means of countercurrent extraction, using Cu(I) acetate dissolved in an aqueous ammonia solution. This method is advantageous in the processing of $C_4$-fractions with a low butadiene and $C_4$-acetylene content, which is the case in dehydrogenation fractions [70]. A higher content of α-acetylenes, e.g., approximately 100 ppm of vinylacetylenes, strongly interferes with this process as a result of foam formation. Acetylenes, especially the unstable vinylacetylene, can be removed by means of selective hydrogenation before the isolation of butadiene [71]–[73]. Further advantages of the hydrogenation process are described in the following Section.

**Extractive Distillation.** The method of choice today for the isolation of butadiene from $C_4$ fractions is extractive distillation using selective organic solvents. The affinity of hydrocarbons to polar solvents depends directly upon their degree of unsaturation. A

**Table 8.** Typical analyses of $C_4$-fractions (in vol%)

| | Cracking intensity | | | | Dehydro-genates |
|---|---|---|---|---|---|
| | moderate | medium | high | very high | |
| $C_3$-Hydrocarbons | 0.3 | 0.3 | 0.3 | 0.16 | * |
| Butane | 4.2 | 5.2 | 2.8 | 0.54 | * |
| Isobutane | 2.1 | 1.3 | 0.6 | 0.53 | * |
| 1-Butene | 20.0 | 16.0 | 13.6 | 9.18 | * |
| cis-2-Butene | 7.3 | 5.3 | 4.8 | 1.61 | * |
| trans-2-Butene | 6.6 | 6.5 | 5.8 | 3.63 | * |
| Isobutene | 32.4 | 27.2 | 22.1 | 10.13 | * |
| 1,3-Butadiene | 26.1 | 37.0 | 47.4 | 70.10 | 15–45 |
| 1,2-Butadiene | 0.12 | 0.15 | 0.2 | 0.40 | ** |
| Methylacetylene | 0.06 | 0.07 | 0.08 | 0.10 | ** |
| Vinylacetylene | 0.15 | 0.3 | 1.6 | 2.99 | ** |
| Ethylacetylene | 0.04 | 0.1 | 0.2 | 0.53 | ** |
| $C_5$-Hydrocarbons | 0.5 | 0.5 | 0.5 | 0.1 | ** |

\* sum of fractions = 85–55 vol%
\*\* sum of fractions = 0.03–0.2 vol%

**Table 9.** Relative volatility of $C_3$- and $C_4$-hydrocarbons in comparison to butadiene (50°C, infinite dilution)

| | Without solvent | NMP* | DMF* | Acetonitrile | DMAC* | Furfural |
|---|---|---|---|---|---|---|
| Butane | 1.17 | 3.66 | 3.43 | 3.13 | 3.13 | 2.89 |
| 1-Butene | 1.08 | 2.38 | 2.17 | 1.92 | 2.07 | 1.78 |
| trans-2-Butene | 1.23 | 1.90 | 1.76 | 1.59 | 1.71 | 1.20 |
| cis-2-Butene | 1.37 | 1.63 | 1.56 | 1.45 | 1.52 | 1.26 |
| 1,3-Butadiene | 1.00 | 1.00 | 1.00 | 1.00 | 1.00 | 1.00 |
| 1,2-Butadiene | 1.79 | 0.74 | 0.72 | 0.73 | 0.71 | 0.65 |
| Methylacetylene | 2.16 | 0.81 | 0.72 | 1.00 | 0.73 | 1.04 |
| Ethylacetylene | 1.62 | 0.42 | 0.42 | 0.48 | 0.44 | 0.52 |
| Vinylacetylene | 1.44 | 0.21 | 0.23 | 0.39 | 0.23 | 0.41 |

\* NMP N-Methylpyrrolidone; DMF Dimethylformamide; DMAC Dimethylacetamide

highly unsaturated hydrocarbon is not only more soluble in a polar solvent, but the solvent is also more effective in decreasing the volatility of the hydrocarbon (see Table 9).

The first plants using furfural as solvent were built in the United States during World War II (the process was developed by Phillips Petr. Co.) [46], [75]. The method introduced by the Shell Dev. Co. in the early 1950s involves the use of acetonitrile as solvent. It has been developed from a one-step procedure involving separate removal of acetylene into a two-step process and is still applied today [46]. Union Carbide Corp. published a method in the mid 1960s using dimethylacetamide [76].

Modern plants employ either the BASF procedure [77]–[81] with N-methylpyrrolidone or the dimethylformamide method [82], [83] of Nippon Zeon.

The BASF procedure is described here as an example of extractive distillation, see Figure 4. N-Methylpyrrolidone (bp at 0.1013 MPa is 203 °C), containing 5–10% water to increase the selectivity, is used as solvent. In the main washer (a), the more

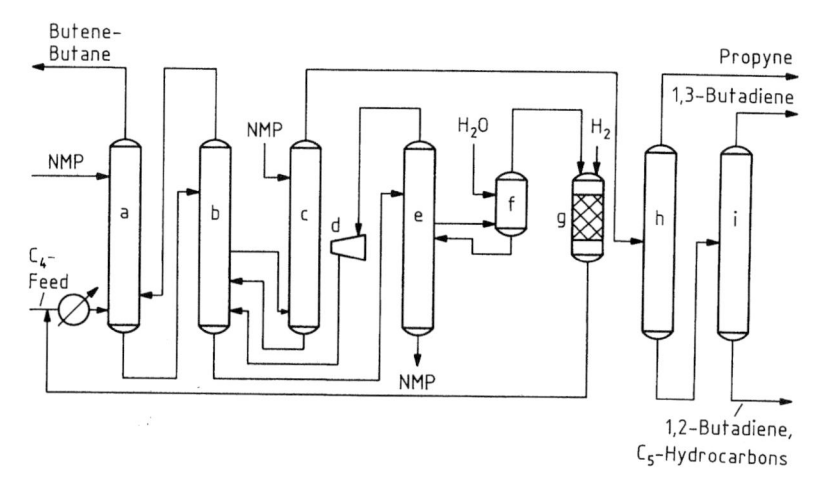

**Figure 4.** BASF Extractive distillation process with NMP
a) Main washer; b) Rectifier; c) After washer; d) Centrifugal compressor; e) Degassing tower; f) Water washer; g) Hydrogenation reactor; h) First distillation; i) Second distillation

unsaturated hydrocarbons, like acetylenes and diolefins, are dissolved in NMP and the butenes/butanes are removed from the top of the washer. In a second extractive distillation step, in tower (c) and in the lower part of tower (b), the $C_4$-acetylenes and partly 1,2-butadiene, which in turn are more soluble in NMP than butadiene, are separated in the sump. Crude butadiene, which, however, corresponds to the specification as far as the alkanes, alkenes, and $C_4$-acetylenes are concerned, is removed from the top of tower (c) and is separated from the more volatile propyne in tower (h) and from the less volatile 1,2-butadiene and $C_5$-hydrocarbons in tower (i). The loaded solvent is led from the sump of tower (b) into tower (e) for complete degassing. Here the $C_4$-acetylenes are removed as a side stream and, after selective hydrogenation, returned to tower (a). Recoveries of 99–100%, based on the butadiene content of the $C_4$-feed, are achieved on selective hydrogenation of vinylacetylenes to butadiene [84]. An additional advantage of the hydrogenation is the safe removal of the very unstable $C_4$-acetylenes. Improper handling of $C_4$-acetylenes led to a severe explosion and complete destruction of a U.S. plant working with dimethylacetamide [85], [86].

Other efficient solvents for extractive distillation of butadiene are given in [87]; e.g., methoxypropionic acid nitrile is used in some plants.

**Table 10.** Typical specifications of butadiene

| Compound | Specification | Test method |
|---|---|---|
| 1,3-Butadiene | $\geq$ 99.6% | ASTM-D 2593-78 |
| Propyne | max. 5 ppm | ASTM-D 2593-78 |
| Propadiene | max. 2 ppm | ASTM-D 2593-78 |
| Vinylacetylene | max. 5 ppm | ASTM-D 2593-78 |
| Ethylacetylene | max. 30 ppm | ASTM-D 2593-78 |
| $\Sigma$ Acetylenes (calculated as vinylacetylene) | max. 30 ppm | ASTM-D 1020-76 |
| Dimerics | max. 250 ppm * | ISO/DIS 7381 |
| 1,2-Butadiene | max. 30 ppm | ASTM-D 2593-78 |
| Carbonyls (calculated as $CH_3CHO$) | max. 25 ppm | ASTM-D 1089-78 |
| Sulfur | max. 1 ppm | ASTM-D 3246-76 |
| Peroxides (calculated as $H_2O_2$) | max. 1 ppm | ASTM-D 1022-76 |
| Methanol | max. 10 ppm | ISO/DIS 8174 |
| $H_2O$ | max. 40 ppm | ISO/DIS 6191 |
| Residue | max. 10 ppm | ASTM-D 1025-76 |
| Inhibitor TBC ** | 75 – 125 ppm | ASTM-D 1157-78 |
| Oxygen in the gas phase | max. 0.3 vol% | ASTM-D 2504-77 |

\* Maximum set by battery producers;
\*\* TBC=*tert*-butylpyrocatechol

# 5. Specifications of Butadiene

The type of butadiene specification necessary to meet the market requirements today is shown in Table 10. The purity requirements have been intensified because the organometallic catalysts (Ziegler-type catalysts) employed in the production of stereo-specific rubbers can be damaged by, e.g., allenes and acetylenes. The purity requirements are less stringent for butadiene used in emulsion polymerization, such as in the synthesis of styrene – butadiene rubber and latex [88]. However, a high degree of purity is generally required today for butadiene in world trade. In individual cases, gas chromatographic values of 20 ppm for the sum of the acetylenes and 15 ppm for 1,2-butadiene should not be exceeded.

# 6. Stabilization, Storage, and Transportation

Stabilization [83] involves protection against the action of oxygen and against polymerization. Butadiene forms dangerous polymeric peroxides [89], [90] with oxygen, which are viscous liquids sparingly soluble in liquid butadiene. They are deposited at the bottom of butadiene containers. Hence, the handling of butadiene requires complete exclusion of oxygen. In addition, inhibitors, like TBC (*p-tert*-butylpyrocatechol) or

TBK (2,6-di-*tert*-butyl-*p*-cresol), are added because they are especially effective at scavenging radicals. They also prevent a spontaneous polymerization of butadiene and can easily be removed by washing with aqueous sodium hydroxide. An aqueous sodium nitrite solution is also used as an antioxidant in the production of butadiene.

Butadiene can polymerize during its production, storage, and transportation in three different ways: First, it can dimerize to vinylcyclohexene [91]. This reaction is dependent on time and temperature and can be stopped by adding a potassium cyanide solution to butadiene [92]. Secondly, butadiene can polymerize under the influence of oxygen and high temperatures, especially during its manufacture, to give rubber-like polymers (fouling). Antioxidants, like TBC, exclusion of oxygen, low temperatures, and short exposures to higher temperatures reduce this polymerization [93]. Thirdly, butadiene can polymerize, initiated by oxygen, high temperatures, and rust, to give a glassy, granular, opalescent, and very hard polymer, the so-called Popcorn [93], [94]. It forms preferentially in off-stream tubes. Its growth is favored when Popcorn seeds are present in the plant. The formation of Popcorn in closed areas can lead to pressures of > 1000 bar, causing the bursting of off-stream tubes or containers. It can be prevented by the addition of TBC and by the careful elimination of all Popcorn seeds. Popcorn is spontaneously inflammable in air and, hence, must be kept moist during its removal. Liquified butadiene is stored at normal temperatures in liquid-gas containers or, today, for safety reasons at a temperature of 4 °C in almost pressureless containers. It is transported at normal temperatures and raised pressures in tankers, railroad tank wagons, and in ships. Especially in ships, temperatures of 4 °C and pressureless containers are required [95]–[97].

The following regulations must be followed when labeling and transporting butadiene:

By road: FRG: GGVS, Class 2, no. 3 C; Europe: ADR, USA: DOT Regulations.
By railroad: FRG: GGVE, Class 2, no. 3 C; Europe: RID; USA: DOT Regulations.
By inland shipping: FRG: GGVB, Class 2, no. 6 F; Europe: ADNR.
By sea freight: FRG: GGV Sea, Class 2; International: IMDG-Code.
By air freight: International: IATA, Class 2.

Butadiene is safely transported today in amounts greater than 100000 t/a from, e.g., Western Europe to Japan and to the United States.

# 7. Uses and Economic Importance

The largest part of the butadiene produced worldwide is used, as a monomer or comonomer, in the manufacture of synthetic rubbers, above all for styrene–butadiene rubber and latex (SBR), polybutadiene rubber (BR), acrylonitrile–butadiene rubber and latex (NBR), and for chloroprene rubber (CR). Important plastics containing butadiene as a monomeric component are shock-resistant polystyrene, a two-phase

**Table 11.** Butadiene production capacities in 1983 (in $10^3$ t/a)

|  | From steam cracker | Dehydrogenation | Total amount | Dehydrogenation, in % of the total amount |
|---|---|---|---|---|
| North America | 1811 | 282 | 2093 | 13.5 |
| South America | 187 | 95 | 282 | 33.7 |
| Western Europe | 2040 | – | 2040 | – |
| Eastern Europe | 788 | 950 | 1738 | 54.7 |
| Asia and Pacific | 883 | 45 | 928 | 4.8 |
| Total | 5709 | 1372 | 7081 | 19.4 |

Source: De Witt & Company Inc.

**Table 12.** Butadiene production and consumption in 1983 (in $10^3$ t/a)

|  | Production | Consumption | Import | Export |
|---|---|---|---|---|
| North America | 1200 | 1544 | 344 |  |
| South America | 206 | 195 |  | 11 |
| Western Europe | 1741 | 1228 |  | 513 |
| Eastern Europe | 1065 | 1061 |  | 4 |
| Asia and Pacific | 815 | 865 | 50 |  |
| Africa and others | – | 65 | 65 |  |
| Total | 5027 | 4958 | 459 | 528 |

Source: De Witt & Company Inc.

system consisting of polystyrene and polybutadiene; ABS polymers, consisting of acrylonitrile, butadiene, and styrene; and a copolymer of methyl methacrylate, butadiene, and styrene (MBS), which is used as a modifier for poly(vinyl chloride).

In addition, butadiene is an intermediate in the synthesis of several important chemicals, see Chapter 3.

The worldwide production capacity for butadiene is summarized in Table 11. Butadiene-containing $C_4$ steam cracker fractions are the starting material for 80% of the production. In Western Europe, naphtha and gas oil are the principal raw materials for steam cracking. High yields of butadiene are obtained by this method, which explains the absence of dehydrogenation capacities. In the United States, 87% of the butadiene production capacity is based on steam cracking. The available dehydrogenation plants are only used at times of butadiene shortage.

The production and consumption of butadiene are summarized in Table 12. The United States and Japan import butadiene, especially from Western European surplusses. In USA, Japan, and Western Europe, most of the butadiene is required by the automobile industry. The automobile still requires the largest amounts of rubber. Table 13 shows the worldwide butadiene usage arranged according to the different butadiene products [99].

In the 1950s and 1960s, new steam cracking plants were built worldwide, and the accumulation of butadiene, as a byproduct in the production of ethylene, resulted in the steady fall in the price of butadiene. The prices then dramatically rose after the oil crises

**Table 13.** Butadiene uses in 1981

|  | Butadiene, $10^3$ t/a | % |
|---|---|---|
| Styrene – butadiene rubber/latex | 2705 | 56 |
| Polybutadiene rubber | 1080 | 22 |
| Nitrile – butadiene rubber/latex | 200 | 4 |
| Polychloroprene rubber | 290 | 6 |
| Acrylonitrile – butadiene – styrene polymer | 200 | 4 |
| Hexamethylenediamine | 205 | 4 |
| Others | 180 | 4 |
| Total | 4860 | 100 |

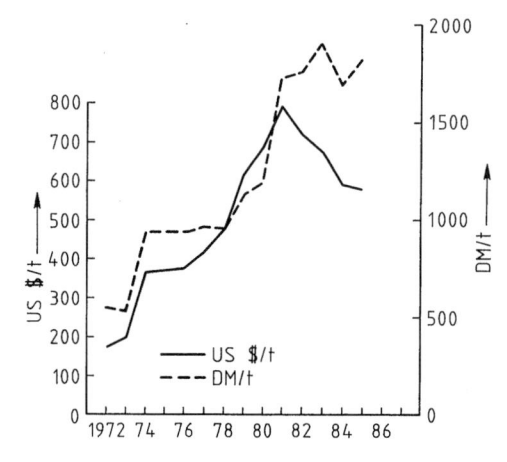

**Figure 5.** Development of the price of butadiene in Western Europe, time span 1972 till the first quarter of 1985

of 1973/74 and 1979/80. Figure 5 shows the effect on the butadiene prices in Western Europe.

The changing raw material situation for the production of ethylene could stimulate this price trend and simultaneously favor the manufacture of butadiene using the dehydrogenation and oxidative dehydrogenation methods. The dehydrogenation plants must be situated in areas with ready access to energy and raw materials so that the butadiene thus produced is able to compete with the butadiene from steam crackers [100].

In Western Europe, USA and Japan, the utilization of $C_4$ steam cracker fractions must be expanded in the future. It would then be worthwhile to dehydrogenate the butene – butane mixture left over after the isolation or conversion of butadiene and isobutene.

# 8. Toxicology

Workers usually are exposed to butadiene by inhalation. Contact of liquid butadiene with skin may exceptionally occur.

## 8.1. Experimental Toxicity Data

**Acute Toxicity.** Butadiene is of low acute toxicity. Lethal concentrations are: $LC_{50}$ 12.5 vol% (rat, inhal., 2 h) and 11.5–12 vol% (mouse, inhal., 2 h) [101], [102]. Like some other hydrocarbons butadiene causes narcotic effects after inhalation of high concentrations, sometimes preceded by excitation and hyperventilation [103].

**Repeated Short-Term Exposure.** Rats, guinea pigs, rabbits, and one dog were exposed to butadiene at concentrations of 0, 600, 2300, and 6700 ppm for 7.5 h per day, 6 days a week during 8 months with no significant progressive injury occurring; only the highest concentration slightly retarded the growth and caused light cloudy swelling of liver in some cases. The data on fertility, hematology, body weights, blood biochemistry, urinalysis, and histopathology did not reveal any adverse effect to the animals [103].

In a similar study on Sprague-Dawley rats no clinical sign of toxicity was observed with the exception of moderately increased salivation, particularly at higher concentrations (4000, 8000 ppm). No adverse effects on growth rate, food consumption, neuromuscular function, hematological and biochemical parameters were recorded in the second, sixth, or thirteenth week of the test. Histopathologic investigations of the rats also did not demonstrate any change [104].

Mice were exposed to butadiene at levels of 0, 625–8000 ppm during 15 days or 14 weeks with no effect on survival or pathology during 15 days. Only the body weight decreased slightly at the 8000 ppm level. Exposure for 14 weeks resulted in dose dependent lower body weights and markedly increased mortality. No other effect has been reported [105].

**Long-term Exposure.** Groups of Sprague-Dawley CD rats were exposed to butadiene in air (0, 1000, 8000 ppm) on 6 h/day, 5 days/week for up to 105 (females) or 111 (males) weeks. The effects were: transient nasal secretion, slight ataxia, early increased mortality (at 8000 ppm concentration), suggestive higher frequency of spontaneous nephropathy in males (8000 ppm), higher liver weight, and a slightly increased incidence of metaplasia in the lung (males at 8000 ppm). Various tumors were identified with higher incidence in the exposed groups, others appeared earlier than in controls [106].

The authors concluded that butadiene was a "weak oncogen" under the conditions of this experiment. It is, however, questionable whether the small increase in mostly benign tumors and tumors with high background incidence like mammary tumors are biologically significant at these extremely high exposure levels.

Groups of B6C3F$_1$ mice were exposed to butadiene in air (0, 625, 1250 ppm) on 6 h/day, 5 days/week. Following week 60 for males and week 61 for females the study was terminated because of excessive mortality. Toxic and proliferative lesions of the nasal cavity occurred at increased incidence in the 1250-ppm group, but no neoplastic lesions were observed in the nasal cavity. Butadiene was also associated with liver necrosis as well as ovarial and testicular atrophy.

There was an early onset and a high incidence of several tumors including malign ones in the exposed mice. The authors concluded that there was clear evidence of carcinogenicity for butadiene. Mice obviously are much more sensitive to butadiene than rats are with respect to toxicity and carcinogenicity [105].

**Mutagenicity.** Gaseous butadiene was mutagenic in the Ames test only in the presence of liver microsomal enzyme preparations. This suggests the formation of a mutagenic intermediate (see Section 8.2) [107], [108].

**Embryotoxicity and Teratogenicity.** Groups of rats were exposed from day 6 through day 15 of gestation to various concentrations of butadiene [109]. There was an effect on body weight of the dams. The fetuses showed reduction of fetal weight at 8000 ppm (highest concentration) and an increased incidence of normally occurring variations, especially in wavy ribs, indicating retarded development, most prominent in the high-level animals. Major malformations occurred in two pups of one litter of the 8000-ppm group only. The 1000-ppm group did not exhibit higher incidences of malformed fetuses [109].

## 8.2. Metabolism

**In vitro.** 1,2-Epoxy-3-butene is formed when butadiene is incubated with liver microsomes and NADPH. Further metabolism involving epoxide hydratase yields 3-butene-1,2-diol [110]–[114]. The diepoxybutane and epoxybutanediol as well as conjugation products are postulated as further intermediates:

$$H_2C=CH-CH=CH_2 \xrightarrow{MFO/NADPH} H_2\overset{O}{C-CH}-CH=CH_2$$

with subsequent conversion via MFO/epoxide hydratase to

$$H_2C-CH-CH=CH_2$$
with OH groups on the first two carbons,

and via MFO/NADPH to the diepoxide

$$H_2\overset{O}{C-CH}-\overset{O}{CH-CH_2}$$

MFO = monofunctional oxidase

The formation of 1,2-epoxy-3-butene is significantly more pronounced in mouse liver tissue than in liver cells from rats, rhesus monkeys, and humans [114]. The experiments using lung tissues of the same species demonstrated the very rapid formation of 1,2-epoxy-3-butene in mice. However, rat lung preparations were capable of producing only one tenth of the amount seen in mouse lung preparations, whereas the epoxide could not be detected using monkey or human lung tissue. The formation of diepoxybutane could not be detected in any experiment [114].

**In vivo.** When rats were exposed to butadiene in a closed system, 1,2-epoxy-3-butene also was detected in the exhaled air [113], [114]. Mice in a similar experiment exhaled more of the monoepoxide than rats.

## 8.3. Human Data

Adverse effects after occupational exposure to butadiene are not reported in the literature. Volunteers exposed to a very high level of 8000 ppm of butadiene in air complained of eye irritation, blurring of vision, coughing, nasal congestion, and drowsiness. Subsequent repeated exposures did not suggest any cumulative effect [103], [115].

A hematology survey of workers at a styrene–butadiene rubber plant showed no pronounced evidence of hematological abnormalities in the peripheral blood. Exposure to butadiene was up to an average of 20 ppm in the tank farm area; in other departments less than 2 ppm occurred [116].

A retrospective study was conducted at two styrene–butadiene rubber plants involving 2756 male workers with an average duration of employment of approx. 10 years. No significant excess of total or cause-specific mortality has been found [116], [117]. Another mortality study covered a period of 36 years and reviewed a population of approx. 14000 workers from styrene–butadiene rubber production facilities [119]. The study indicated that the overall mortality was lower as compared to general population (Healthy Worker Effect). There were no statistically significant differences in tumor mortality in total or for any specific cause of death. This study is currently being updated.

**Hygiene Standards.** As a result mainly from the long-term mice study, butadiene is listed as an Λ2 carcinogen with a TLV of 10 ppm by the ACGIH (1984/1985). The German MAK committee has withdrawn the MAK value, and butadiene is now listed in category A2 of the 1984 MAK list; a TRK value (lowest technically feasible level) will be established as a consequence.

The in vitro studies on the formation of critical intermediates in tissues from different species (mice, rat, monkey, man), however, suggest that the data from the long-term rat and especially from the mice study may not considered appropriate for direct use in risk evaluation for man.

# 9. References

[1] G. S. Whitby: *Synthetic Rubber*, John Wiley & Sons, New York, 1954, Chapter 2.
[2] Berufsgenossenschaft chem. Industrie, BR Deutschland: *1,3-Butadien*, Merkblatt M 049, Verlag Dr. Otto Pfeffer, Heidelberg 1985.
[3] H. Grosse-Wortmann, *Chem. Ing. Techn.* **46** (1974) 111.
[4] American Petr. Inst.: *Technical Data Book*, Petroleum Ref., New York 1966.
[5] N. V. Steere, *J. Chem. Educ.* **45** (1968) no. 3, A 199.
[6] R. N. Mac Callum, J. J. McKetta, *Hydrocarbon Process.* **42** (1963) no. 5, 191.
[7] L. D. Hood, C. L. Yaws, *Chem. Eng.* **83** (1976) no. 17, 81.
[8] R. W. Elliot, H. Watts, *Can. J. Chem.* **50** (1972) no. 1, 31.
[9] Am. Chem. Soc.: *Azeotropic Data III* (Advances in Chem. Ser. 116), Washington, D.C., 1973.
[10] J. P. Kennedy et al.: *Polymer Chemistry of Synthetic Elastomers*, Interscience, New York 1968.
[11] J. Falbe: *Synthesen mit Kohlenmonoxyd*, Springer Verlag, Berlin 1967.
[12] DuPont, US 3547972, 1970; US 3818067, 1974; US 3846474, 1974.
[13] Mitsubishi Chem., DE-OS 2345160, DE-OS 2424539, 1974; DE-OS 2504637, DE-OS 2510088, DE-OS 2510089, DE-OS 2505749, 1975.
[14] Y. Tanabe, *Hydrocarbon Process.* **60** (1981) no. 9, 187.
[15] BASF, DE-OS 2444004, DE-OS 2454768, 1976.
[16] Toyo Soda, US 3720704, 1973.
[17] Shell Oil, US 4384146, 1983.
[18] Chevron Research, US 3932468, 1976.
[19] Shell Research, GB 1303403, 1973.
[20] R. Baker, *Chem. Rev.* **73** (1973) 487.
[21] E. J. Smutny, *J. Am. Chem. Soc.* **89** (1967) 6793.
[22] A. D. Josey, *J. Org. Chem.* **39** (1974) no. 2, 139.
[23] T. Anstock et al., *Ind. Eng. Chem. Prod. Res. Dev.* **21** (1982) 415.
[24] P. Roffio et al., *J. Organomet. Chem.* **55** (1973) 405.
[25] C. U. Pittman, Jr., R. M. Hanes, *Ann. N.Y. Acad. Sci.* **239** (1974) 76.

[26] Celanese, US 4269781, 1981.
[27] J. Hamer: *1,4-Cycloaddition Reactions, The Diels-Alder-Reaction in Heterocyclic Synthesis,* Academic Press, New York 1967.
[28] J. Tkatchenko, *J. Organomet. Chem.* **124** (1977) no. 3, C 39.
[29] Esso Res. & Eng., US 3454665, 1969.
[30] Borg-Warner, DE-OS 2000974, 1970.
[31] Union Chimie Elf-Aquitaine, DE-OS 2350689, 2350690, 1974.
[32] Maruzen Oil, US 3502736, 1970.
[33] C. Wilke et al., *Angew. Chem.* **69** (1957) 397; **71** (1959) 574.
[34] H. Staudinger, B. Ritzenthaler, *Ber. Dtsch. Chem. Ges.* **68** (1935) 455.
[35] L. R. Drake et al., *J. Am. Chem. Soc.* **68** (1946) 2521.
[36] G. Egloff, G. Hulla, *Oil Gas. J.* **41** (1942) no. 26, 40, no. 27, 228, no. 28, 41, no. 29, 124, no. 30, 36, no. 31, 45, no. 32, 36.
[37] S. A. Miller: *Acetylenes,* vol. **1,** Academic Press, New York 1965.
[38] W. Reppe: *Chemie und Technik der Acetylen-Druckreaktionen,* Verlag Chemie, Weinheim/Bergstraße 1952.
[39] G. S. Whitby: *Synthetic Rubber,* J. Wiley & Sons, New York 1954, p. 86.
[40] A. Talalay, M. Magat: *Synthetic Rubber from Alcohol,* Interscience, New York 1945.
[41] G. Gale, *Chem. Bus.* 1982, Apr. 5, 33.
[42] R. G. Craig, J. M. Duffalo, *Chem. Eng. Prog.* **75** (1979) no. 2, 62.
[43] R. G. Craig, E. A. White, *Hydrocarbon Process.* **59** (1980) no. 12, 111.
[44] *Hydrocarbon Process.* **55** (1976) no. 9, 229.
[45] H. A. Foster, *Hydrocarbon Process.* **52** (1973) no. 9, 119.
[46] J. C. Reidel, *Oil Gas J.* **54** (1957) Nov. 11, 166, Dec. 2, 88, Dec. 9, 114, Dec. 16, 110, Dec. 23, 74.
[47] K. K. Kearby: "Catalytic Dehydrogenation," in: *Catalysis,* vol. **3,** Reinhold, New York 1955.
[48] H. Kröper et al., *Erdöl Kohle Erdgas Petrochem.* **22** (1969) no. 10, 605.
[49] V. Fattore et al., *Hydrocarbon Process.* **60** (1981) no. 8, 101.
[50] J. A. Convers et al., *Hydrocarbon Process.* **60** (1981) no. 3, 95.
[51] W. Krönig, G. Scharfe, *Erdöl Kohle Erdgas Petrochem.* **19** (1966) no. 7, 497.
[52] D. Stadermann, *Chem. Techn.* **35** (1983) no. G, 290.
[53] J. C. Reidel, *Oil Gas J.* **54** (1957) Dec. 9, 114.
[54] P. M. Reilly, *Chem. Can.* **5** (1953) no. 3, 25.
[55] T. H. Arnold, *Chem. Eng.* **68** (1961) Oct. 30, 90.
[56] R. J. Harbour, 5th World Petr. Congr., Sect. IV, Paper 10, 121, New York 1959.
[57] D. S. Alexander, 7th World Petr. Congr. Proc., vol. **5,** Elsevier, London 1967.
[58] P. A. Batist et al., *J. Catal.* **7** (1967) 33, 12 (1968) 45.
[59] L. M. Welch et al., *Hydrocarbon Process.* **57** (1978) no. 11, 131.
[60] T. Hudson, Jr. et al., *Hydrocarbon Process.* **55** (1974) no. 6, 133.
[61] P. C. Husen et al., *Oil Gas J.* **69** (1971) Aug. 2, 60.
[62] C. Wolf, *Chem. Week* 1966, May 28, 113.
[63] R. B. Stobaugh, *Hydrocarbon Process.* **46** (1967) no. 6, 141.
[64] H. Kuper, *Erdöl Kohle Erdgas Petrochem.* **35** (1982) no. 3, 119.
[65] J. G. Freiling, A. A. Simone, *Oil Gas J.* **71** (1973) Jan. 1, 25.
[66] H. Ritzer, *Erdöl Kohle Erdgas Petrochem.* **35** (1982) no. 3, 124.
[67] L. F. Hatch, S. Matar, *Hydrocarbon Process.* **57** (1978) no. 3, 129.
[68] B. Schleppinghoff, *Erdöl Kohle Erdgas Petrochem.* **27** (1974) 240.
[69] F. Asinger: *Die petrochemische Industrie,* Akademie-Verlag, Berlin 1971, pp. 418–422.

[70] H. K. Kroper et al., *Hydrocarbon Process.* **41** (1962) no. 11, 191.
[71] H. Lauer, *Erdöl Kohle Erdgas Petrochem.* **36** (1983) no. 6, 249.
[72] M. Derrien et al., *Hydrocarbon Process.* **58** (1979) no. 5, 175.
[73] G. Ritzert, W. Berthold, *Chem. Ind. Techn.* **45** (1973) no. 3, 131.
[74] W. A. Gorschkow et al., *Khim. Prom.* **11** (1971) 807.
[75] C. Buell, *Ind. Eng. Chem.* **39** (1947) no. 6, 695.
[76] W. W. Coogler, Jr., *Hydrocarbon Process.* **46**(1967) no. 5, 166.
[77] H. Kroeper et al., *Hydrocarbon Process.* **41** (1962) no. 11, 191.
[78] B. Hausdoerfer et al., *Chem. Ind. Techn.* **40** (1968) no. 23, 1147.
[79] H. Klein et al., *Hydrocarbon Process.* **47** (1968) no. 11, 135.
[80] U. Wagner, H. M. Weitz, *Ind. Eng. Chem.* **62** (1970) no. 4, 43.
[81] K. Volkamer et al., *Erdöl Kohle Erdgas Petrochem.* **34** (1981) no. 8, 343.
[82] Japanese Geon Co., US 3436436, 1969; US 3436438, 1969; GB 1177040, 1970.
[83] S. Takao, *Hydrocarbon Process.* **45** (1966) no. 11, 151.
[84] A. Lindner, *Chem. Ing. Techn.* **55** (1983) no. 1, 68.
[85] S. Griffith, R. G. Keister, *Hydrocarbon Process.* **49** (1970) no. 9, 323.
[86] J. H. Buehler et al., *Chem. Eng.* **77** (1970) Sept. 7, 77.
[87] H. Asatani, W. Hayduk, *Can. J. Chem. Eng.* **61** (1983) Apr., 227.
[88] A. Hahn et al., *Hydrocarbon Process.* **54** (1975) no. 2, 89.
[89] Mitteilungsblatt Berufsgenossenschaft Chem. Ind., Jedermann-Verlag, Heidelberg 1980, Jan.
[90] D. G. Hendry et al., *Ind. Eng. Chem. Prod. Res. Dev.* **7** (1950) 136.
[91] R. F. Robey et al., *Ind. Eng. Chem.* **36** (1944) no. 1, 3.
[92] Erdölchemie, DE-OS 2051548, 1972.
[93] M. S. Kharasch et al., *Ind. Eng. Chem.* **39** (1947) 830.
[94] G. Whitby: *Synthetic Rubber*, J. Wiley & Sons, New York 1954, p. 80.
[95] Pittsburgh – Des Moiness Steel Co., *Chem. Metallurg. Eng.* 1942, Nov., 117.
[96] J. H. Boyd, *Ind. Eng. Chem.* **40** (1948) no. 9, 1703.
[97] Petroquimica Argentina, Informative Bulletin and Specification, no. 1, "Butadiene".
[98] De Witt & Comp.: 1984 Butadiene Annual, Houston, Texas, USA, June 1984.
[99] J. Steel, BP Chemicals, ECMRA Oslo, 11th–13th Oct., 1983.
[100] J. R. Hodson, *Chemical Marketing Reporter* 1982, May 5.
[101] I. B. Batkina, *Gig. Sanit.* **31** (1966) 18–22.
[102] B. B. Shugaer, *Arch. Environ. Health* **18** (1969) 878–882.
[103] C. P. Carpenter, C. B. Shaffer, C. S. Weil, H. F. Smyth, Jr., *J. Ind. Hyg. Toxicol.* **26** (1944) 69–78.
[104] C. N. Crouch, D. H. Pullinger, I. F. Gaunt, *Am. Ind. Hyg. Assoc. J.* **40** (1979) 796–802.
[105] Toxicology and Carcinogenesis Studies of 1,3-Butadiene in B6C3F1 Mice (Inhalation Studies), National Toxicology Program, Technical Report Series no. 288 (1984).
[106] Hazleton Laboratories Europe, The Toxicity and Carcinogenicity of Butadiene Gas Administered to Rats by Inhalation for Approximately 24 Months, Final Report, vol. **1–4**, Report no. 2653-522/2, Hazleton Labs., Harrogate, England, 1981.
[107] C. De Meester, F. Poncelet, M. Roberfroid, M. Mercier, *Toxicol. Lett.* **6** (1980) 125–130.
[108] F. Poncelet, C. de Meester, M. Duverger-van Bogaert, M. Lambotte-Vandepaer, M. Roberfroid, M. Mercier, *Arch. Toxicol. Suppl.* **4** (1980) 63–66.
[109] Hazleton Laboratories Europe: "1,3-Butadiene: Inhalation Teratogenicity Study in the Rat," Final Report, Report no. 2788-522/3, Hazleton Labs., Harrogate, England, 1981.
[110] E. Malvoisin, G. Lhoest, F. Poncelet, M. Roberfroid, M. Mercier, *J. Chromatogr.* **178** (1979) 419–425.

[111] E. Malvoisin, M. Roberfroid, *Xenobiotica* **12** (1982) 137–144.
[112] E. Malvoisin, M. Mercier, M. Roberfroid, *Adv. Exp. Med. Biol.* **136 A** (1982) 437–444.
[113] H. M. Bolt, G. Schmiedel, J. G. Filser, H. P. Rolzhäuser, K. Lieser, D. Wistuba, V. Schurig, *J. Cancer Res. Clin. Oncol.* **106** (1983) 112–116.
[114] U. Schmidt, E. Löser, *Archives of Toxicology*, in press.
[115] R. H. Wilson, G. V. Hough, W. E. McCormick, *Ind. Med.* **17** (1948) 199–207.
[116] H. Checkoway, T. M. Williams, *Am. Ind. Hyg. Assoc. J.* **43** (1982) 164–169.
[117] T. J. Meinhardt, R. J. Young, R. W. Hartle, *Scand. J. Work Environ. Health* **4** (1978) 240–246.
[118] T. J. Meinhardt, R. A. Lemen, M. S. Crandall, R. J. Young, *Scand. J. Work Environ. Health* **8** (1982) 250–259.
[119] G. M. Matanoski, L. Schwartz, J. Sperrazza, J. Tonascia: *Mortality of Workers in the Styrene–Butadiene Rubber Polymer Manufacturing Industry*, Johns Hopkins University, School of Hygiene & Public Health, Baltimore MD, June 1982.
[120] TLVs, Threshold Limit Values for Chemical Substances in the Work Environment Adopted by ACGIH for 1984–85.
[121] DFG: *Maximale Arbeitsplatzkonzentrationen und Biologische Arbeitsstofftoleranzwerte*, Verlag Chemie, Weinheim 1984.
[122] *Gesundheitsschädliche Arbeitsstoffe, Toxikologisch-arbeitsmedizinische Begründung von MAK-Werten*, Verlag Chemie, Weinheim 1984.

# Butanals

HANSWILHELM BACH, Ruhrchemie AG, Oberhausen, Federal Republic of Germany
RODERICH GÄRTNER, Ruhrchemie AG, Oberhausen, Federal Republic of Germany
BOY CORNILS, Ruhrchemie AG, Oberhausen, Federal Republic of Germany

1. Introduction .............. 925
2. Physical Properties ........ 925
3. Production .............. 926
4. Quality Specifications and Testing ................. 929
5. Handling, Storage, and Shipment ................ 929
6. Chemical Reactions and Applications ............. 930
7. Toxicology ............... 934
8. References ............... 934

## 1. Introduction

Butanals are saturated aliphatic $C_4$ aldehydes, $C_4H_8O$, $M_r$ 72.11. There are two isomers: the straight-chain butyraldehyde [123-72-8] (butanal) and the branched isobutyraldehyde [78-84-2] (2-methylpropanal, dimethylacetaldehyde).

$CH_3CH_2CH_2CHO$  $\quad$ $(H_3C)_2CHCHO$
Butyraldehyde  $\qquad$ Isobutyraldehyde

Both aldehydes have been found in small amounts in various essential oils, in plants as well as in the dry distillation products of various natural substances [1], [2]. Both, butyraldehyde and isobutyraldehyde are highly reactive intermediate products for a large number of chemical syntheses, above all leading to the formation of compounds with four to eight carbon atoms.

The annual world production of butyraldehydes is several million tons; industrial-scale production is almost exclusively by hydroformylation[3].

## 2. Physical Properties

$n$-Butyraldehyde and isobutyraldehyde are colorless, flammable liquids with a pungent penetrating odor. Their vapors have a narcotic effect and irritate mucous membranes. Both isomers are miscible with organic solvents, such as alcohols, ethers, and benzene. The solubilities of the butyraldehydes in water are given in Table 1.

The most important physical data of the butyraldehydes are compiled in Table 2.

**Table 1.** Solubility of butyraldehydes in water and of water in butyraldehydes (mass fractions in %)

| Temperature, | Aldehyde in water | | Water in aldehyde | |
|---|---|---|---|---|
| | Butyraldehyde | Isobutyraldehyde | Butyraldehyde | Isobutyraldehyde |
| 0 | 9.8 | 9.3 | 3.2 | 2.1 |
| 10 | 8.6 | 7.8 | 2.8 | 2.0 |
| 20 | 7.6 | 6.7 | 2.6 | 1.9 |
| 30 | 6.8 | 5.8 | 2.4 | 1.9 |
| 40 | 6.1 | 5.0 | 2.4 | 2.0 |
| 50 | | | 2.4 | 2.1 |

**Table 2.** Physical properties of butyraldehydes

| | Butyraldehyde | Isobutyraldehyde |
|---|---|---|
| Melting point, °C | −97 | −66 |
| Boiling point, °C | 74.8 | 64 |
| Density at 20 °C, g/cm$^3$ | 0.803 | 0.788 |
| Refractive index $n_D^{20}$ | 1.3805 | 1.3730 |
| Viscosity at 20 °C, mPa s | 0.45 | 0.45 |
| Specific heat capacity at 25–30 °C, J g$^{-1}$ K$^{-1}$ | 2.123 | 2.544 |
| Heat of vaporization at the boiling point, J/g | 436 | 409 |
| Heat of combustion, kJ/mol | −2478.7 | −2510 |
| Crit. pressure, MPa | 4.06 | ca. 4.3 |
| Crit. density, g/cm$^3$ | 0.259 | |
| Crit. temperature, °C | 248 | ca. 267 |
| Surface tension at 20 °C, mN/m | 24.6 | 24 |
| Dielectric constant at 20 °C | 14.9 | 13.5 |
| Vapor pressure at 20 °C, kPa | 12 | 17 |
| Dipole moment (vapor), Debye | 2.72 | |
| Coefficient of expansion at 20–30 °C, K$^{-1}$ | 0.0013 | 0.0014 |

Table 3 gives the composition of some binary and ternary azeotropes. Further physical properties are given in [1], [2], [4].

# 3. Production

The most widely used process for the manufacture of butyraldehydes is the hydroformylation of propylene. The hydrogenation of crotonaldehyde prepared by the dimerization of acetaldehyde or the dehydrogenation of butanols play only a subordinate role.

**Hydroformylation of Propylene.** In the hydroformylation process discovered by OTTO ROELEN at Ruhrchemie AG in 1938, olefins are reacted with synthesis gas – an equimolar mixture of carbon monoxide and hydrogen – in the presence of catalysts to

**Table 3.** Azeotropic mixtures with butyraldehydes

| Compound A | Compound B | Compound C | Boiling point of a azeotrope, °C | Concentration of B (C) in the azeotrope, wt% |
|---|---|---|---|---|
| Butyraldehyde | ethanol | | 70.7 | 60.6 |
| Butyraldehyde | water | | 68 | 8.8 |
| Butyraldehyde | hexane | | 60 | 74 |
| Butyraldehyde | ethanol | water | 67.2 | 11 (9) |
| Isobutyraldehyde | water | | 60.5 | 6.0 |

form aldehydes with one more carbon atom. Hydrido transition-metal carbonyls having the general formula $HM(CO)_4$ (M = Co, Rh, Ru) serve as catalysts; they are normally used in the homogeneous phase. On an industrial scale hydroformylation is carried out at 90–180 °C and 2.5–35 MPa (25–350 bar) [3].

Basically, all olefins, functionally substituted olefins, diolefins, etc. can be used as the starting material. On an industrial scale olefins from $C_2$ (ethylene) to $C_{18}$ are most often used. Propylene has gained particular significance, being converted to a mixture of *n*- and isobutyraldehyde in an initial reaction stage.

$$H_3C-CH=CH_2 \xrightarrow[cat.]{CO, H_2} H_3C-CH_2-CH_2-CHO + \underset{H_3C}{\overset{H_3C}{>}}CH-CHO$$

Depending on the oxo catalyst, the reaction conditions, and the process variant, reaction mixtures with a ratio of *n*- to isobutyraldehyde between 1:1 and ≈ 20:1 are obtained. The choice of oxo process variant is determined essentially by the intended use of the butyraldehydes. In addition to the aldehydes that constitute 90–98% of the reaction mixture, minor amounts of byproducts and secondary products of the reactive butyraldehydes are formed: butyl formate; butanols; aldehyde dimers, trimers, and condensation products [3].

Throughout the world ≈ $4 \times 10^6$ t of *n*- and isobutyraldehyde are manufactured each year by the hydroformylation of propylene. About $3.2 \times 10^6$ t are *n*-butyraldehyde, and $0.8 \times 10^6$ t are isobutyraldehyde. Large-scale plants such as those in operation at BASF, CWH, Ruhrchemie, or UCC each have an annual output of 200–300 000 t.

**Hydrogenation of Crotonaldehyde.** Before the discovery and commercial exploitation of the oxo synthesis the reaction sequence: acetaldehyde → aldol condensation → crotonaldehyde → hydrogenation → *n*-butyraldehyde was the most important method of obtaining n-butyraldehyde.

$$2\ H_3C-CHO \xrightarrow[-H_2O]{cat.} \underset{\text{Crotonaldehyde}}{H_3C-CH=CH-CHO} \xrightarrow[cat.]{H_2} \underset{n\text{-Butyraldehyde}}{H_3C-CH_2-CH_2-CHO}$$

The acetaldehyde can be manufactured from acetylene (by addition of water), ethylene (via the Wacker process), or ethanol (dehydrogenation) resulting in alternative reaction routes to *n*-butyraldehyde with $C_2$ or $C_3$ compounds as starting materials (Fig. 1).

*n*-Butanol is a byproduct, arising in a secondary reaction, the hydrogenation of the carbonylic group, which takes place during the selective hydrogenation of the double bond of crotonaldehyde.

Selective hydrogenation can be conducted industrially in the gas phase or in the liquid phase (sump phase hydrogenation or trickle-bed hydrogenation). The usual catalysts are supported catalysts on the basis of Cu, Ni, Pd, Ru, Fe, Cr, or Mn [1], [6]–[11].

In the 1950s this production route was still the main method of manufacturing *n*-butyraldehyde [12], but the convenient, one-stage, economical process of propylene hydroformylation has replaced to a great extent crotonaldehyde hydrogenation in the last 25 years, the lower price of the $C_3$ starting material, propylene, being a contributory factor. Possibly, countries with a high agricultural ethanol production, such as Brazil, will rediscover the crotonaldehyde route for producing *n*-butyraldehyde [13].

**Dehydrogenation of Butanols.** Butanols from other production processes (byproduct of the Fischer-Tropsch synthesis [14], alcoholic fermentation of sugar with *Clostridium saccharobutylacetonium liquefaciens* or *Bacillus amylobacter* [15], Reppe carbonylation [16]) can be dehydrogenated to butyraldehydes, for example:

$$CH_3CH_2CH_2CH_2OH \rightleftharpoons CH_3CH_2CH_2CHO + H_2$$

The reaction is usually carried out in the gas phase with a heterogeneous catalyst [17]. The addition of hydrogen acceptors such as $O_2$, either pure or as air, increases the aldehyde yield but also the formation of byproducts. Zinc- and/or chromium-doped copper catalysts are particularly effective [18], [19].

**Other Processes.** The processes of isobutene oxidation [20], [21] or catalytic rearrangement of unsaturated alcohols such as methallyl or crotyl alcohol [22]–[24] have no industrial significance.

$$CH_3CH=CHCH_2OH \rightleftharpoons CH_3CH_2CH_2CHO$$

The Fischer-Tropsch synthesis produces small amounts of *n*-butyraldehyde, which does have importance on some local markets [14], [59].

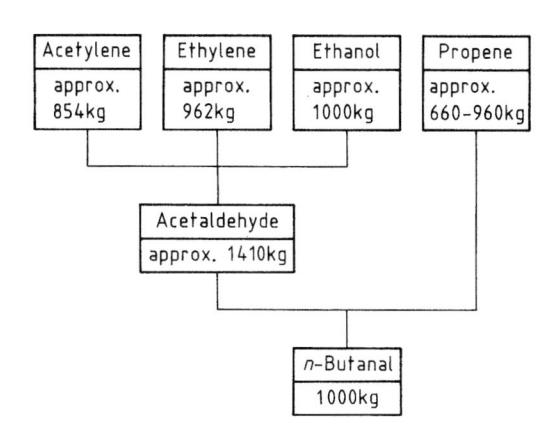

**Figure 1.** Yields of various routes to butyraldehyde [5]

```
[Acetylene approx. 854kg]  [Ethylene approx. 962kg]  [Ethanol approx. 1000kg]  [Propene approx. 660-960kg]
                                     |
                          [Acetaldehyde approx. 1410kg]
                                     |
                              [n-Butanal 1000kg]
```

## 4. Quality Specifications and Testing

The most important method of testing (Table 4) is gas chromatography with low injection temperature to avoid decomposition of e.g. aldols.

In the analysis of isobutyraldehyde it is especially important to detect the trimeric aldehyde, which forms easily if unstabilized product is stored. In conducting the Karl Fischer water determination special precautions must be observed to avoid acetal formation, which is accompanied by the release of water. This may be done through the use of a solvent mixture consisting of four parts pyridine and one part ethylene glycol. Density and refraction of the water-containing product are strongly affected by the equilibrium hydrate concentrations, which are established slowly. During determination of the acid number care must be taken to exclude atmospheric oxygen. The boiling range of these aldehydes is strongly affected by trace amounts of water.

Table 4 outlines the standard quality specifications for butyraldehydes. The admissible limits are largely determined by the intended use.

## 5. Handling, Storage, and Shipment

Both aldehydes are stored and transported in containers of corrosion-resistant steel, aluminum, glass, or ceramic as well as specially coated tanks of normal steel and drums with removable polyethylene inserts. Even with water contents < 0.1% and storage under inert gas, $n$-butyraldehyde in steel tanks takes up iron, causing discoloration of the product.

Table 4. Delivery specifications for butyraldehydes

|  | Method | Butyraldehyde | | Isobutyraldehyde | |
|---|---|---|---|---|---|
|  |  | moist | dry | moist | dry |
| Butyraldehyde, wt% | GLC | $\geq$ 96.0 | $\geq$ 99.0 | $\leq$ 0.5 | $\leq$ 0.2 |
| Isobutyraldehyde, wt% | GLC | $\leq$ 0.3 |  | $\geq$ 95.0 | $\geq$ 99 |
| Density at 20 °C, kg/L | DIN 51757, ASTM D 1298 | 0.807–0.812 | 0.801–0.803 | 0.790–0.794 | 0.786–0.790 |
| Refractive index $n_D^{20}$ | DIN 51423, ASTM D 1747 | 1.380–1.382 |  | 1.373–1.376 | 1.372–1.374 |
| Color, Pt-Co scale | DIN 53409, ASTM D 1209 | $\leq$ 15 | $\leq$ 10 | $\leq$ 15 | $\leq$ 15 |
| Acid number, mg KOH/g | DIN 53402, ASTM D 1613 | $\leq$ 3.0 | $\leq$ 2 | $\leq$ 3.0 | $\leq$ 2.0 |
| Water, wt% | Karl Fischer | $\leq$ 3.0 | $\leq$ 0.2 | $\leq$ 2.0 | $\leq$ 0.1 |
| Distillation range (95 vol%) at 101.3 kPa, °C | ASTM D 1078 | 67–80 | ca. 74–76 ($\triangle t \leq 2$ °C) | 57–70 | 63–67 |

On contact with air butyraldehydes oxidize readily to the carboxylic acids. Therefore, the entry of air must be prevented above all by storage under inert gas. Isobutyraldehyde shows a particular tendency to trimerize in the presence of even traces of strong acids. In general this can be prevented by the addition of a suitable amine (e.g., a few ppm of triethanolamine, [61]).

Owing to the low flash points and ignition temperatures (Table 5) as well as the high volatility of these aldehydes the safety regulations are to be strictly observed.

These aldehydes are shipped in tank cars or trucks, tankers, or drums. The classification and designations for the various modes of transport must be observed (Table 5).

# 6. Chemical Reactions and Applications

Butyraldehydes exhibit the typical reactions of saturated aliphatic aldehydes. Owing to their reactivity they are important starting materials for a wide variety of products (see, e.g., [25]). Figure 2 outlines the reactions used on an industrial scale.

*n*-Butyraldehyde is therefore the intermediate product for a large number of important industrial $C_4$ products, including *n*-butanol, *n*-butyric acid, and *n*-butylamines; for the important plasticizer alcohol, 2-ethylhexanol; and for $C_6$ products such as the 2-ethylbutyl compounds.

Isobutyraldehyde is processed mainly to isobutanol and neopentyl glycol. Via isobutyric acid it is a potential intermediate product for alternative manufacturing processes leading to methyl methacrylate.

**Table 5.** Safety information for butyraldehydes*

|  |  | Butyraldehyde | Isobutyraldehyde |
|---|---|---|---|
| Flash point, closed cup (DIN 51755) | °C | − 13 | < − 18 |
| Lower and upper explosive limit in air at 101.3 kPa | vol% | 1.9 – 12.5 | 1.6 – 10.6 |
| Autoignition temperature (DIN 51794, ASTM D 2155) | °C | 230 | 225 |
| Odor threshold |  | low | low |
| $LD_{50}$ (oral, rat) | g/kg | 2.5 – 5.9 | 2.8 – 3.7 |
| $LD_{50}$ (dermal, rabbit) | g/kg | 3.56 | 3.7 – 7.1 |
| $LC_{Lo}$ (inhal., rat) |  | 8000 ppm/4 h | 8000 ppm/4 h |
| $LD_{Lo}$ (subcut., rat) | g/kg | 10 |  |
| Marking according to Regulations of the European Community (Council directive 67/548/EEC, 5th Adaptation, 29th June, 1983) | Symbol<br>R-Phrases<br>S-Phrases | F<br>11<br>9-29-33 | F **<br>11 **<br>9-29-33 ** |
| Hazard classification for transport: |  |  |  |
| RID/ADR (Class, Item Number) |  | 3, 1 a<br>(3.3 b new) | 3, 1 a<br>(3, 3 b new) |
| IMDG-Code (Class, UN No., Packing Gp.) |  | 3.2, 1129, II | 3.1, 1245, II |
| IATA-DGR (ICAO Techn. Instr.) (Class, UN No., Packing Gp.) |  | 3, 1129, II | 3, 2045, II |
| Mail (FRG) |  | max. 125 $cm^3$ per receptacle and max. 500 $cm^3$ per consignment |  |
| DOT (USA) |  | flammable liquid |  |

\* Vapor pressure at 20 °C, solubility of butyraldehydes in water and vice versa: see Tables 1 and 2
\** No EEC regulation, marking corresponds to *n*-butyraldehyde.

**Hydrogenation.** The butyraldehydes are converted into the corresponding butanols by catalytic hydrogenation, whereby both liquid-phase and gas-phase processes are used. The mild, highly selective gas-phase hydrogenation is carried out on nickel- or copper-supported catalysts [19].

**Oxidation.** Pure oxygen or air converts butyraldehydes to carboxylic acids easily and in high yields, the reaction taking place in the presence or absence of a catalyst. Transition metals of the 5th to the 8th sub-group as well as Cu, Ag, Ce, the alkali metals, and alkaline-earth metals are suitable catalysts. If oxidation is carried out in the liquid phase, the metal naphthenates or carboxylates are used. In gas-phase processes generally the metal oxides, deposited on carriers such as silica gel, kieselguhr, silicates, quartz, or alumina, are used [1]. For further information, see [1], [26] – [28].

The use of special metal oxide catalysts permits the oxidation of isobutyraldehyde at high temperature to be controlled in such a manner that methacrolein or methacrylic acid, or a mixture of the two, is formed [1], [29] – [32]. Both products are gaining significance as starting materials for the formation of methyl methacrylate. By variation of the reaction conditions the oxidation of isobutyraldehyde can produce acetone and

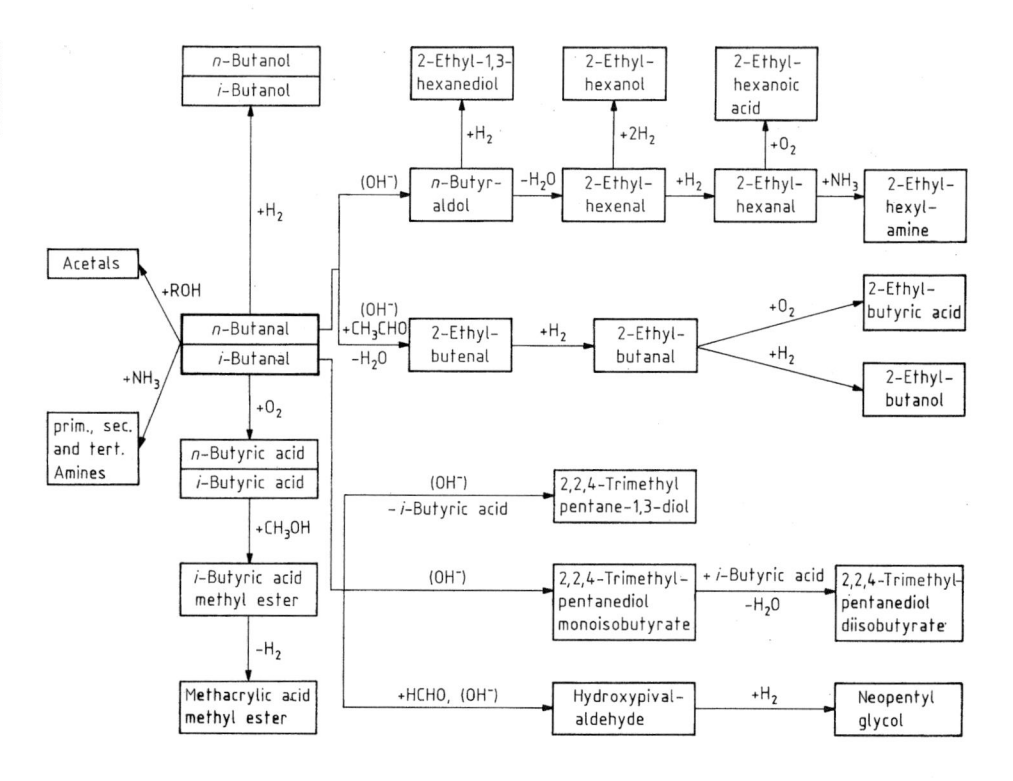

**Figure 2.** Reactions of butyraldehydes

isopropanol [1], [33]. *n*-Butyric acid anhydride can also be produced from *n*-butyraldehyde [1].

**Addition and Condensation Reactions.** In a large number of addition and condensation reactions butyraldehydes react with themselves, other carbonyl compounds, amines, alcohols, nitriles, etc. In industry, the most important synthesis of this kind is the self-condensation of *n*-butyraldehyde in the presence of basic catalysts. This leads primarily to 2-ethylhexenal, from which 2-ethylhexanol, the alcohol component of the most important PVC plasticizer, dioctyl phthalate (DOP), can be obtained by hydrogenation [1], [3], [5], [34]. Mild reaction conditions during aldolization permit isolation of "butyraldol" (2-ethyl-3-hydroxyhexanal), which can be hydrogenated to 2-ethyl-1,3-hexanediol.

The condensation product of *n*-butyraldehyde, 2-ethylhexenal, can be partially hydrogenated in the presence of precious metal catalysts, whereby 2-ethylhexanal is formed, opening the way to 2-ethylhexanoic acid and 2-ethylhexylamine [1], [3], [5].

Condensation of *n*-butyraldehyde with two moles of formaldehyde and subsequent hydrogenation yield trimethylol propane [1], [35], [36]. Phenol and *n*-butyraldehyde condense to yield oil-soluble paints, and condensation with urea produces alcohol-soluble paints [1].

Condensation of *n*-butyraldehyde with acetaldehyde gives 2-ethylbutyl products [37]. Their properties and applications are described in [38].

With poly(vinyl alcohol) *n*-butyraldehyde condenses to poly(vinyl butyral), which is used in the manufacture of laminated safety glass or as a bonding agent in paints [1].

Condensation products of *n*-butyraldehyde with aniline, butylamine, thiourea, diphenylguanidine, or methylthiocarbamate are used on a large scale as vulcanizing accelerators for rubber [1].

Reductive amination, i.e., the reaction with hydrogen and ammonia or amines, gives the primary, secondary and tertiary butylamines [1], [39]–[41].

The self-condensation of isobutyraldehyde leads via the intermediate isobutyraldoxane (4-hydroxy-5,5-dimethyl-2,6-diisopropyl-1,3-dioxane) to isobutyraldol or secondary products such as 2,2,4-trimethylpentane-1,3-diol, 2,2,4-trimethylpentane-1,3-diol mono-, and -di-isobutyrate, or 2,2,4-trimethylpentane-1-ol [1], [42]–[45].

The reaction of isobutyraldehyde with formaldehyde forms primary hydroxypivalaldehyde, a starting material for products with the neopentyl structure. Especially neopentyl glycol, from the hydrogenation of hydroxypivalaldehyde, has become important [46], [50]. The wide variety of its uses is described in [47].

The D,L-hydroxy-$\beta,\beta$-dimethyl-butyrolactone, known under the name pantolactone, also has the neopentyl structure. It can be obtained from isobutyraldehyde, formaldehyde, and sodium cyanide [1], [48], [49].

The reaction of isobutyraldehyde with urea produces fertilizers which are characterized by their long-term activity [1]. The amino acids valine and leucine can be obtained from isobutyraldehyde [1].

The Tishchenko reaction is used to obtain isobutyl isobutyrate, a speciality solvent, from isobutyraldehyde. Acetals of the butyraldehydes with various alcohols serve as solvents for cellulose, resins, and rubber [1], [50].

The crossed Tishchenko reaction of *n*-butyraldehyde with acetaldehyde provides commercial ester mixtures. Butyraldehydes and hydroxylamine yield the corresponding aldoximes or their salts [1], [50].

**Other Reactions.** Isobutyraldehyde can be catalytically isomerized to *n*-butyraldehyde and methyl ethyl ketone or converted to diisopropyl ketone [1], [50].

The chlorination of butyraldehydes produces fragrances and pharmaceutical intermediates [1], [50]–[53].

*n*-Butyraldehyde is used as a catalyst component for special polymerization reactions [54]–[57].

Isobutyraldehyde can be converted to isobutene on special molybdenum catalysts [58].

Additional reactions and uses can be found in [1], [50] and the literature cited there.

# 7. Toxicology

Butyraldehydes have low acute toxicity (Table 5). Exposure limits have not yet been established. Their vapors and liquids greatly irritate the skin, eyes, and respiratory organs, possibly due to rapid oxidation to the carboxylic acids on contact with air. The symptoms are coughing, watering eyes, and a burning sensation on the skin as well as in the nose and throat. On exposure to higher doses the central nervous system is impaired, symptomized by depression, dizziness, headaches, and a slight narcotic effect. The repugnant smell of the two aldehydes can occasionally cause nausea and vomiting. Subcutaneous injection of *n*-butyraldehyde causes heemolysis (→ Aldehydes, Aliphatic and Araliphatic).

# 8. References

[1] *Ullmann*, 4th ed., vol. **9**, p. 42.
[2] *Kirk-Othmer*, 3rd. ed., vol. **4**, p. 376.
[3] B. Cornils in J. Falbe (ed.): *New Syntheses with Carbon Monoxide*, Springer Verlag, Berlin-Heidelberg-New York 1980.
[4] J. McGarry, *Ind. Eng. Chem.* **22** (1983) 313.
[5] B. Cornils, A. Mullen, *Hydrocarbon Process.* 1980 (11) 93.
[6] V. N. Kulakov et al., SU 172744, 1964.
[7] Usines de Melle, DE 814444, 1949.
[8] BASF AG, DE-OS 2839474, 1980; DE-OS 3130805, 1983 (M. Horner et al.).
[9] V. Macho, CS 131903, 1966.
[10] L. S. Pakhomova et al., SU 335227, 1969.
[11] Chem. Werke Hüls, DE-OS 3151086, 1982 (L. Fischer et al.).
[12] O. Horn in *Ullmann*, 3rd ed., vol. **4**, p. 796.
[13] W. Swodenk, *Chem. Ing. Techn.* **55** (1983) 683.
[14] H. Schulz, J. H. Cronjéin *Ullmann*, 4th ed., p. 329.
[15] B. Cornils, E. Zilly in F. Korte (ed.): *Methodicum Chimicum*, vol. **5**, p. 88.
[16] see [15], p. 20.
[17] L. Bexten, J. Weber in F. Korte (ed.): *Methodicum Chimicum*, vol. **5**, p. 281.
[18] C. D. Frohning, G. Horn in J. Falbe, U. Hasserodt (eds.): *Katalysatoren, Tenside und Mineralöladditive*, G. Thieme Verlag, Stuttgart 1978, p. 63.
[19] Hoechst AG: *Katalysatoren Hoechst*, Frankfurt 1979, p. 44.
[20] Allied Chem. Corp., US 2808429, 1954.
[21] Universal Oil Prod. Co., US 2683174, 1949.
[22] V. M. Polyakov et al., *Khim. Promst. (Moscow)* 1972, no. 10, 48, 743; *Chem. Abstr.* **78** (1973)42812.
[23] W. Rupilius in F. Korte (ed.): *Methodicum Chimicum*, vol. **5**, p. 325.
[24] J. Simonik, L. Beranek, *J. Catal.* **24** (1972) no. 2, 348.
[25] C. E. Loeffler, L. Stautzenberger, I. D. Unruh, *Encycl. Chem. Process. Des.* 1977, no. 5, 358–405.

[26] Mitsubishi Chemical Industries Co. Ltd., DE-OS 2928002, 1980 (T. Masuko, S. Fukaya, N. Murai, J. Noma); JP 78105412, 1977 (T. Maki); JP 78105413, 1977 (T. Maki); JP 7909203, 1977 (T. Maki); JP 78108915, 1977 (T. Maki).
[27] Chisso Corp., JP 7378116, 1973 (M. Fukin, T. Hirai, T. Ohashi, S. Furukawa, I. Koga).
[28] A. Benning, L. Mußler, H. Tummes in F. Korte (ed.): *Methodicum Chimicum,* vol. **5,** p. 558.
[29] Mitsubishi Rayon Co., Ltd., JP 7882715, 1978 (H. Matsuzawa, M. Kobayashi, H. Ishii, M. Kato); JP 7214085, 1972 (T. Kita, C. Ishii); DE-OS 2633593, 1978 (T. Onoda, M. Otake).
[30] Union Carbide Corp., DE-OS 2836309, 1979 (D. W. McNeil, B. Phillips).
[31] Chisso Corp., JP 7946705, 1979 (M. Fukui, N. Otake, N. Nagata).
[32] Johnson, Matthey PLC, GB 2094782, 1982 (E. Shutt).
[33] Rhône-Progil, US 3987103, 1976 (G. Gobrou, C. Falize, H. Dufour).
[34] Ruhrchemie AG, DE-OS 2713434, 1978 (H. Tummes, H. Noeske, B. Cornils, W. Kascha); DE-OS 2437957, 1976 (G. Kessen).
[35] Bayer AG, DE-OS 2702582, 1978 (O. Immel, H. H. Schwarz, O. Weissel, H. Krimm).
[36] L. Cairati et al., *Chim. Ind. (Milan)* **63** (1981) no. 11, 723–725.
[37] Mitsubishi Chemical Industries Co., Ltd., JP 7424891, 1974 (A. Matsukuma, I. Takakishi, K. Yoshida); JP 7406887, 1974 (J. Takakishi, K. Yoshida).
[38] B. Cornils, H. Feichtinger, *Chem. Ztg.* **101** (1977) 107–117.
[39] Texaco Inc., US 4299985, 1981 (J. F. Knifton,P. H. Moss).
[40] BASF AG, DE-OS 2725669, 1978 (K. Merkel et al.).
[41] B. Cornils, H. Feichtinger in J. Falbe, U. Hasserodt (eds.): *Katalysatoren, Tenside und Mineralöladditive,* G. Thieme-Verlag, Stuttgart 1978, p. 45.
[42] BASF AG, DE-OS 820518, 1979 (F. Merger, H. J. Förster).
[43] Eastman Kodak Co., US 4225726, 1980 (D. L. Morris, A. W. McCollun).
[44] Chisso Corp., JP 7715582, 1977 (J. Tsuchiya et al.); JP 7323411, 1973 (R. Takakashi, J. Tsuchiya); JP 7334574, 1973 (S. Matsumoto, K. Gunji, T. Hamahata).
[45] Hoechst AG, DE-OS 3102826, 1982 (H. Baltes, E. J. Leupold).
[46] Eastman Kodak Co., DE-OS 2652224, 1977 (B. W. Palmer, H. N. Wright).
[47] B. Cornils, H. Feichtinger, *Chem. Ztg.* **100** (1976) 504–514.
[48] BASF AG, DE-OS 2758883, 1979 (H. Distler, W. Goetze).
[49] VEB-Jenapharm, GB 1345459, 1974 (J. Schmidt et al.); DE-OS 2228641, 1973 (J. Schmidt, H. Grunert, C. Weigelt).
[50] H. J. Hagemeyer, G. DeCroes: *The Chemistry of Isobutyraldehyde and its Derivatives,* Tennessee Eastman Comp., Kingsport, Tenn., 1953.
[51] Allied Chemical Corp., US 4096187, 1978 (J. H. Bonfield, A. Murthy, D. Pickens).
[52] Dow Chemical Co., US 3801645, 1974 (D. A. Dalman).
[53] Shell International Research Maatschappij, DE-OS 2641356, 1977 (A. Reinink, J. Grendelman).
[54] Daicel Ltd., JP 7411480, 1974 (A. Tanaka et al.).
[55] Thiokol Chemical Corp., DE 1545025, 1974 (R. M. Gobran, S. W. Osborn).
[56] Japan Synthetic Rubber Co. Ltd., DE-OS 2444681, 1975 (M. Ikeda et al.).
[57] Ube Industries Ltd., DE-OS 2445776, 1975 (H. Ueno et al.).
[58] Toa Nenryo Kogyo K. K., JP 7748603, 1977 (K. Kaneko, H. Furukawa).
[59] C. D. Frohning et al. in J. Falbe (ed.); *Chemierohstoffe aus Kohle,* G. Thieme Verlag, Stuttgart 1977, p. 289.
[60] Hoechst AG: *Delivery specifications,* Frankfurt 1984. Chemische Werke Hüls AG: *Data sheet for n-butyraldehyde, dry,* 1983.
[61] Ruhrchemie AG, DE 2905267, 1981 (J. Weber, V. Falk, C. Kniep).

# Butanediols, Butenediol, and Butynediol

HEINZ GRÄFJE, BASF Aktiengesellschaft, Ludwigshafen, Federal Republic of Germany (Chap. 1.1)
WOLFGANG KÖRNIG, BASF Aktiengesellschaft, Ludwigshafen, Federal Republic of Germany (Chap. 1.2)
HANS-MARTIN WEITZ, BASF Aktiengesellschaft, Ludwigshafen, Federal Republic of Germany (Chap. 1.3)
WOLFGANG REISS, BASF Aktiengesellschaft, Ludwigshafen, Federal Republic of Germany (Chap. 1.3)
GUIDO STEFFAN, Bayer AG, Leverkusen, Federal Republic of Germany (Chap. 2.1)
HERBERT DIEHL, Bayer AG, Leverkusen, Federal Republic of Germany (Chap. 2.1)
HORST BOSCHE, BASF Aktiengesellschaft, Ludwigshafen, Federal Republic of Germany (Chap. 2.3)
KURT SCHNEIDER, BASF Aktiengesellschaft, Ludwigshafen, Federal Republic of Germany (Chap. 2.3)
HEINZ KIECZKA, BASF Aktiengesellschaft, Ludwigshafen, Federal Republic of Germany (Chap. 3)

| | | | | | |
|---|---|---|---|---|---|
| 1. | 1,4-Diols | 937 | 2.1. | 2,3-Butanediol | 945 |
| 1.1. | 2-Butyne-1,4-diol | 937 | 2.2. | 1,3-Butanediol | 946 |
| 1.2. | 2-Butene-1,4-diol | 940 | 2.3. | 1,2-Butanediol | 947 |
| 1.3. | 1,4-Butanediol | 941 | 3. | Toxicology | 947 |
| 2. | Other Butanediols | 945 | 4. | References | 948 |

# 1. 1,4-Diols

## 1.1. 2-Butyne-1,4-diol

**Physical Properties.** 2-Butyne-1,4-diol [110-65-6], HOCH$_2$–C≡C–CH$_2$ OH, $M_r$ 86.09, $mp$ 58 °C, $bp$ 150 °C (at 1.8 kPa), is a colorless, hygroscopic, orthorhombic crystalline compound, which is readily soluble in water (374 g in 100 g at 25 °C) and in strongly polar solvents, e.g., alcohol, acetone, but only slightly soluble in ether and almost insoluble in hydrocarbons.

| | |
|---|---|
| Heat of vaporization | 50.28 kJ/mol, 584 kJ/kg (at 101.3 kPa) |
| Heat of combustion | 2204 kJ/mol, 25.5 kJ/g |
| Flash point (DIN 51758) | 152 °C |
| Ignition point | 335 °C |

**Chemical Properties.** 2-Butyne-1,4-diol decomposes relatively slowly between 160 and 200 °C, but at temperatures significantly above 200 °C violent decomposition can occur. Explosive decomposition can be caused or accelerated by alkali hydroxide, strong anhydrous acids, and certain heavy-metal salts, particularly those of mercury. 2-Butyne-1,4-diol is sensitive to oxidation.

Substitution of the hydroxyl groups is analogous to that of other diols. In the case of substitution by halogens very reactive compounds are formed. The reaction with bifunctional compounds leads to linear polymers only. The carbon–carbon triple bond undergoes the usual addition reactions of acetylene. Two moles of chlorine, but only one mole of bromine or iodine, are added per mole of diol. Chlorination in the presence of hydrochloric acid gives 2,3-dichloro-3-formylacrylic acid (mucochloric acid) [87-56-9]:

$$\underset{H}{\overset{O}{\succ}}C-CCl=CCl-COOH$$

The addition of water, catalyzed by mercury salts, leads to the rather unstable 2-oxobutanediol.

**Production.** The synthesis of 2-butyne-1,4-diol, based on the work of Reppe and his coworkers [1]–[3], is carried out by reaction under pressure of acetylene and an aqueous solution of formaldehyde, catalyzed by copper acetylide:

$$2\,CH_2O + HC\equiv CH \longrightarrow HOCH_2-C\equiv C-CH_2OH \qquad \Delta H = -100.5\ kJ/mol$$

This reversible reaction proceeds via propargyl alcohol, 2-propyn-1-ol [107-19-7], which mostly remains on the catalyst and thus only partially goes into solution. The reaction is first order in respect to formaldehyde and, in industrial processes, effectively zero order in respect to acetylene.

The prime objective in the synthesis is to minimize the formaldehyde concentration in the resulting butynediol solutions. In principle, it is possible to react rather concentrated formaldehyde solutions partially, distill off the unreacted formaldehyde under pressure, and reintroduce it into the reactor. However, this process is complex and detrimental to the catalyst. Usually, the production of the butynediol is carried out continuously in cascades of three to five reactors. In these reactors the catalyst is present either as a fixed bed of strands 3–5 mm in diameter and 10 mm in length (Fig. 1) or as a fluidized bed of particles of diameter 0.1–2 mm. In the first case the formaldehyde solution, together with gaseous acetylene, is fed from the top [4]. In the fluidized-bed process the required amount of acetylene is dissolved in the solution and the solution is then fed from the bottom, suspending the catalyst (Fig. 2) [5].

In either case the reactors are equipped with liquid-circulating systems. In the fixed-bed process the volume of the liquid is determined by external cooling. In the fluidized-bed process the volume of the circulating liquid is determined by the amount of acetylene which is introduced by absorbing columns or injectors.

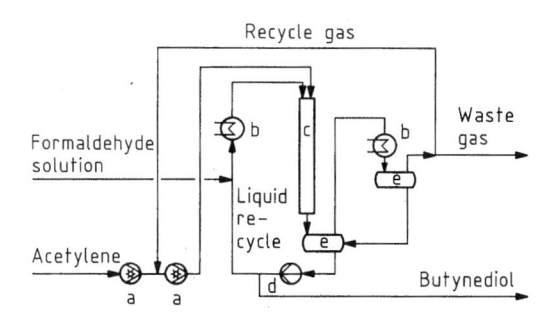

**Figure 1.** Synthesis of butynediol, fixed-bed reactor
a) Liquid seal pump; b) Cooler; c) Reactor; d) Recycle pump; e) Gas–liquid separator

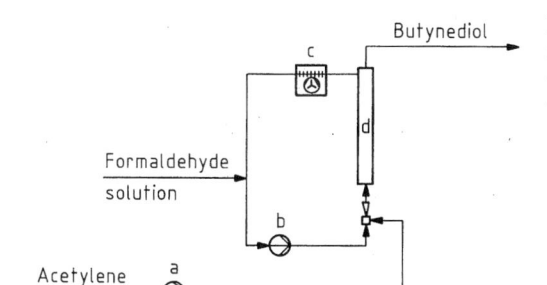

**Figure 2.** Synthesis of butynediol, fluidized-bed reactor
a) Liquid seal pump; b) Recycle pump; c) Air cooler; d) Reactor

The catalyst is 3–6% bismuth(III) oxide and 10–20% copper(II) oxide mostly supported on silica. Copper(II) oxide is converted by acetylene and formaldehyde at 60–90 °C to copper(I) acetylide, which complexes further acetylene [6] and thus forms the catalyst proper. Bismuth oxide inhibits the formation of water-insoluble polymers, the "cuprenes," from oligomeric acetylene complexes. Conditions of synthesis: 80–100 °C; 30–50% aqueous formaldehyde solution; partial pressure of acetylene 2–6 bar, pH 5–8. Dilute alkali is added to the cycle solution to adjust the pH.

The product solution contains 33–55% butynediol, 1–2% propargyl alcohol, 0.4–1% unreacted formaldehyde, and 1–2% byproducts. These byproducts are mainly high-boiling materials, along with some sodium formate. The butynediol can be recovered from this solution by vacuum distillation; the propargyl alcohol is recovered in an azeotrope.

*Handling of High-Pressure Acetylene.* Acetylene can explode or even detonate if the pressure rises above 1.4 bar (→ Acetylene). In butynediol plants this deflagration is often ignited by copper acetylide that was accidentally deposited in the head space; if it dries up, it can explode even upon slight vibration. Because other sources of ignition can never be excluded special precautions are required when handling acetylene at pressures above 1.4 bar. When filling all equipment with metal packings and all pipes with bundles of thin tubes (max. diameter 12.5 mm) the transition from deflagration to detonation (50–100 fold increase in pressure) can be excluded. Explosion (10–12 fold increase in pressure), however, cannot completely be avoided; this is accounted for by planning a 12-fold pressure for the whole apparatus [1]. When planning such plants the technical rules for handling acetylene [7] and the publication of SARGENT [8] are recommended.

**Environmental Protection.** The off-gas, which mainly consists of acetylene and other hydrocarbons, is burnt. Waste-water problems arise when regenerating the catalyst. The water used for rinsing is treated biologically after traces of catalyst have been deposited. Spent catalyst must always be wet. Copper and bismuth can be recovered from the catalyst by acid treatment.

**Analysis.** Because of the high boiling points and the instability of the compounds involved, GC analysis is of limited value only. The determination of byproducts by titration or by fractional distillation (and subsequent GC analysis of the fractions) is time consuming but more reliable.

**Storage and Transportation.** The storage temperature should be kept below 40 °C and storage times longer than a few months should be avoided because the butynediol flakes tend to set up.

**Uses.** Approximately 95% of the butynediol is hydrogenated in aqueous solution to 1,4-butanediol and occasionally to 2-butene-1,4-diol. In addition, butynediol is an intermediate for the production of 3-chloro-2-butynyl N-(3-chlorophenyl)carbamate, a herbicide. Solvay reacts butynediol with alkylene oxides and obtains a flame retardant after subsequent addition of bromine to the triple bond. Considerable amounts of butynediol are added to electroplating baths (nickel and copper) as brightening agent. In steel inhibitor picklings, the diol acts as a corrosion inhibitor.

**Commercial Products.** Butynediol is traded as 34% aqueous solution by GAF and BASF (Golpanol BOZ liquid or Korantin BH liquid with hexamethylenetetramine being added to the latter). The anhydrous product is also traded as flakes. Products containing residues (GAF) are 97–98% pure. Distilled butynediol (BASF) is 99% pure: Butynediol pure cryst., Korantin BH solid, Golpanol pure solid.

# 1.2.  2-Butene-1,4-diol

**Physical Properties.** 2-Butene-1,4-diol [*110-64-5*], HOCH$_2$–CH=CH–CH$_2$OH, C$_4$H$_8$O$_2$, $M_r$ 88.1, *mp* 10 °C, *bp* 240 °C (at 1013 mbar), *bp* 140 °C (at 24 mbar), is a pale yellow liquid, which is soluble in water. The product should be distilled at reduced pressure because it deteriorates above 180 °C.

| | |
|---|---|
| Specific heat capacity | 2.71 J g$^{-1}$ K$^{-1}$ |
| Flash point | 127 °C |
| Ignition temperature | 335 °C |

Vapor pressure

| t, °C | 110.4 | 123.7 | 140 | 171.3 | 240 |
|---|---|---|---|---|---|
| p, mbar | 4.4 | 9.6 | 24 | 101.7 | 1013 |

**Chemical Properties.** Butenediol undergoes the typical reactions of alcohols, such as ester formation with anhydrides or acid chlorides, substitution of the hydroxyl groups by halogen, and the formation of polyesters with dicarboxylic acids. Typical reactions of the double bond are addition of halogen, epoxidation with hydrogen peroxide, copolymerization with vinyl esters, and the classic Diels-Alder reaction with dienes, e.g., with hexachlorocyclopentadiene.

**Production.** 2-Butene-1,4-diol is produced industrially by partial hydrogenation of 2-butyne-1,4-diol in aqueous solution and in the presence of a palladium catalyst, which generally is poisoned with zinc or other metals [9]. The main product of this synthesis is cis-2-butene-1,4-diol. Raney nickel can also be used as a catalyst for the hydrogenation [10]. Another method for the production of 2-butene-1,4-diol is the acetoxylation of butadiene followed by hydrolysis [11].

**Analysis.** The 2-butene-1,4-diol concentration in the product is determined by gas chromatography, the water content by the Karl Fischer method.

**Storage and Quality.** 2-Butene-1,4-diol must be stored in rust-free containers. Typical commercial specifications are: purity 98.5%, mp 10 °C, color light to slightly yellow, water content < 0.5%.

**Uses.** 2-Butene-1,4-diol is used for the production of endosulfan, chlorinated bicyclo-[2.2.1]heptene-(2)-bis(oxyalkylene-5,6) sulfite, an insecticide [12]; for the production of pyridoxine (vitamin B 6) [13]; and in mixtures with other compounds as bactericide.

**Economic Aspects.** Worldwide 4000–5000 t/a of 2-butene-1,4-diol are produced, most of which is used for the production of crop protection products and vitamins.

## 1.3. 1,4-Butanediol

**Physical Properties.** 1,4-Butanediol [*110-63-4*], HOCH$_2$CH$_2$CH$_2$CH$_2$OH, C$_4$H$_{10}$O$_2$, $M_r$ 90.12, is a colorless, almost odorless, hygroscopic liquid (see Table 1), readily soluble in water, alcohols, ketones, glycol ethers, and glycol ether acetates, less soluble in diethyl ether and esters, and not miscible with aliphatic and aromatic hydrocarbons and chlorinated hydrocarbons.

**Table 1.** Physical properties of 1,4-butanediol

| | | | | | |
|---|---|---|---|---|---|
| mp | 20.2 °C | | | | |
| bp | 230.5 °C (at 101.3 kPa) | | | | |
| Density $\varrho$ | 1.017 g/cm$^3$ (at 20 °C); 1.0154 g/cm$^3$ (at 25 °C) | | | | |
| Critical temperature $t_c$ | 446 °C | | | | |
| Critical pressure $p_c$ | 41.2 bar | | | | |
| Vapor pressure: $t$, °C | 60 | 100 | 140 | 180 | 200 |
| $p$, kPa | ca. 0.031 | 0.47 | 4.08 | 21.08 | 41.5 |
| Heat of fusion $\Delta H_f$ | 16.3 kJ/mol ± 5% | | | | |
| Heat of vaporization: $t$, °C | 131.4 | | 193.2 | 215.6 | 230.5 |
| $\Delta H_v$, kJ/mol | 68.2 | | 59.4 | 57.8 | 56.5 |
| Specific heat capacity: $t$, °C | 20 | | 50 | 100 | 150 |
| $c$, J g$^{-1}$ K$^{-1}$ | 2.2 ± 2% | | 2.46 ± 2% | 2.9 ± 3% | 3.33 ± 4% |
| Specific heat capacity of a 50% aq. solution: $t$, °C | 20 | | 50 | 75 | 100 |
| $c$, J g$^{-1}$ K$^{-1}$ | 3.4 ± 2% | | 3.56 ± 2% | 3.69 ± 2% | 3.82 ± 3% |
| Heat of combustion $\Delta H_c$ | 2585 kJ/mol | | | | |
| Thermal conductivity: $t$, °C | 30 | | 50 | 70 | 100 |
| $\lambda$, W m$^{-1}$ K$^{-1}$ | 0.2100 | | 0.2091 | 0.2083 | 0.2069 |
| Thermal conductivity of a 50% aq. solution: $t$, °C | 20 | | 50 | 100 | 150 |
| $\lambda$, W m$^{-1}$ K$^{-1}$ | 0.3601 | | 0.3694 | 0.3886 | 0.3984 |
| Viscosity $\eta$ | 91.56 mPa · s (at 20 °C); 71.5 mPa · s (at 25 °C) | | | | |
| Refractive index $n_D$ | 1.4460 (at 20 °C); 1.4446 (at 25 °C) | | | | |
| Dielectric constant $\epsilon$ | 31.4 | | | | |
| Flash point | 134 °C | | | | |

**Chemical Properties.** 1,4-Butanediol is readily cyclized in acid medium to give tetrahydrofuran. Dehydrogenation in the presence of copper–zinc–aluminum catalysts gives butyrolactone.

1,4-Butanediol reacts with monocarboxylic acids to give diesters [14]. Esterification with dicarboxylic acids and their derivatives leads to partially crystalline, linear, thermoplastic polymeric esters [15], [16].

At ca. 200 °C, 1,4-butanediol reacts with ammonia or an amine over nickel or cobalt catalysts and in the presence of hydrogen to give pyrrolidine or pyrrolidine derivatives [17], [18]. Phosgene reacts with 1,4-butanediol at −5 °C to give the bis(chloroformate) of butanediol [19], [20]. Acrylonitrile adds to 1,4-butanediol at 20–100 °C and in the presence of catalytic amounts of alkali to give 1,4-bis(2-cyanoethoxy)butane [21]. Like other alcohols, 1,4-butanediol can be vinylated, giving the divinyl ether [22].

**Production.** 1,4-Butanediol is made on a large industrial scale by continuous hydrogenation of the 2-butyne-1,4-diol [23] over modified nickel catalysts [24], [25]. The one-stage flow process is carried out at 80–160 °C and 300 bar according to Figure 3.

An aqueous solution of 2-butyne-1,4-diol (30–50%), together with carbon monoxide-free hydrogen and recycled reaction mixture, which acts as medium for dissipation of heat, is lead over a reduced nickel–copper–manganese catalysts on silica gel strands. The initial temperature in the reactor is 80 °C; the temperature must not

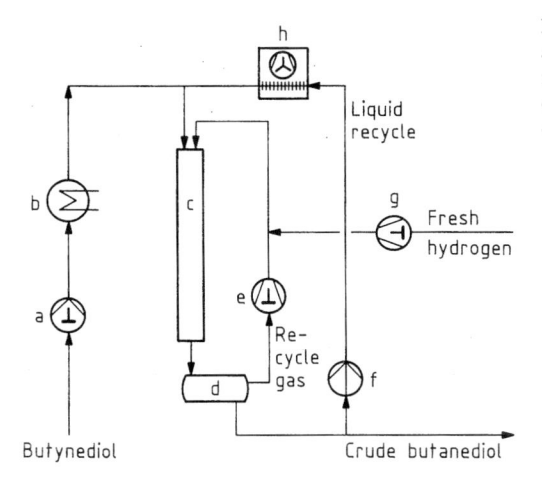

**Figure 3.** Hydrogenation of 2-butyne-1,4-diol
a) Piston pump; b) Preheater; c) Hydrogenation reactor; d) Gas–liquid separator; e) Recycle compressor; f) Canned motor pump; g) Hydrogen compressor; h) Air cooler

exceed 170 °C. In order to obtain a better distribution of the liquid, hydrogen is also circulated.

The raw product contains methanol, propanol, and butanol as byproducts as well as traces of 2-methyl-1,4-butanediol, hydroxybutyraldehyde, acetals, and triols. The reactor effluent is worked up to pure 1,4-butanediol by fractional distillation [26].

The hydrogenation of butynediol can also be carried out in two process steps. In the first stage, for example, at 40 bar, mainly 2-butene-1,4-diol is obtained, which is then completely hydrogenated at 300 bar in the second stage.

A further variant is the low-pressure hydrogenation at ca. 20 bar over a suspended Raney nickel catalyst, followed by hydrogenation at 120–140 °C and 140–210 bar on a fixed-bed contact [24].

*Other Processes.* 1,4-Butanediol can also be made from raw materials other than acetylene. Allyl alcohol and synthesis gas ($CO + H_2$) can be converted to 1,4-butanediol according to a process invented by Daicel [27], [28].

Mitsubishi uses a three-step process [29], [30]: (1) the catalytic reaction of butadiene and acetic acid yields 1,4-diacetoxy-2-butene [*18621-75-5*]; (2) subsequent hydrogenation gives 1,4-diacetoxybutane; and (3) hydrolysis leads to 1,4-butanediol:

$$CH_2=CH-CH=CH_2 + 2\,CH_3COOH + 1/2\,O_2$$
$$\longrightarrow CH_3OOC-CH_2-CH=CH-CH_2-COOCH_3$$
$$\xrightarrow{H_2,\ H_2O} 2\,CH_3COOH + HO-CH_2CH_2CH_2CH_2-OH$$

BASF also has patented a similar process, in which acetic acid first adds to butadiene; the product then isomerizes to 1,4-diacetoxy-2-butene [31], [32].

According to a process practiced by Toyo Soda [33] chlorine first adds to butadiene to form a mixture of 1,4-dichloro-2-butene [*764-41-0*] and 3,4-dichloro-1-butene [*760-23-6*]. This mixture reacts with sodium acetate to form 1,4-diacetoxy-2-butene, which is subsequently hydrogenated directly to 1,4-butanediol.

In a patent described by Shell butadiene reacts with a peroxide to diperoxybutene, which is then converted to 1,4-butanediol by hydrogenation [34].

1,4-Butanediol also is a starting material for the production of *tetrahydrofuran* [*109-99-9*], which can be made directly from butadiene according to a process developed by Chevron [35]: 1,2-epoxy-3-butene is made from butadiene and a peroxide and subsequent isomerization yields 2,5-dihydrofuran [*1708-29-8*], which is hydrogenated to tetrahydrofuran:

$$CH_2=CH-\overset{O}{CH-CH_2} \longrightarrow \underset{O}{\bigcirc} \xrightarrow{H_2} \underset{O}{\bigcirc}$$

Other products derived from 1,4-butanediol are butyrolactone and tetrahydrofuran. In addition, both 1,4-butanediol and its derivatives are used in the textile, leather, photographic, food, and pharmaceutical industries.

The worldwide capacity for 1,4-butanediol is ca. 235 000 t/a. European producers are BASF and GAF/Chemische Werke Hüls; producers in the United States are Du Pont, GAF, and BASF Wyandotte. Whereas all these producers use acetylene as starting material, a plant at Mitsubishi Chemical Industries has been making 1,4-butanediol from butadiene since 1982.

**Storage and Quality.** 1,4-Butanediol can be stored indefinitely. The product is noncorrosive and therefore can be transported in cast iron containers. When it is stored for longer periods, storage tanks of steel or aluminum are necessary in order to avoid traces of iron in the product. In this case, a cover of dry nitrogen also is recommended. Typical commercial specifications are: purity 99.5 – 99.8 %, *mp* 19.9 – 20.0 °C.

**Uses and Economic Aspects.** 1,4-Butanediol is a versatile intermediate for the chemical industry. The most important area of application is the production of polyurethanes and poly(butylene terephthalate). Among the polyurethanes produced from 1,4-butanediol, cellular and compact elastomers are of prime importance. Poly(butylene terephthalate) is processed particularly to plastic materials and hot-melt adhesives, but is used also for the production of plastic films and fibers.

# 2. Other Butanediols

## 2.1. 2,3-Butanediol

2,3-Butanediol [513-85-9], 2,3-butylene glycol, exists in three stereoisomeric forms: Previously, 2,3-butanediol was obtained by bacterial fermentation of hexoses and pentoses. By pyrolysis of the diacetate very pure butadiene can be obtained, which has been used in the United States for the production of synthetic rubber. Today butenes from crack gases are the raw material.

```
        CH₃                      CH₃
         |                        |
    H-C-OH                   HO-C-H
         |                        |
    HO-C-H                   H-C-OH
         |                        |
        CH₃                      CH₃

2S,3S-2,3-Butanediol      2R,3R-2,3-Butanediol
dextrorotatory            [24347-58-8]
                          levorotatory

                  CH₃
                   |
              H-C-OH
                   |
              H-C-OH
                   |
                  CH₃

            (R,S)-2,3-Butanediol
                [5341-95-7]
              optically inactive
```

**Physical Properties.** The 2,3-butanediols, $C_4H_{10}O_2$, $M_r$ 90.12, are colorless and odorless, strongly hygroscopic, oily liquids or crystals with a sweet taste. They are miscible with water and easily soluble in low molecular mass alcohols and ketones. For further physical data, see Table 2.

**Chemical Properties** [36]. The dehydrogenation of 2,3-butanediol yields acetoin [513-86-0] and diacetyl [431-03-8]. Dehydration leads chiefly to 2-butanone. The oxidation, e.g., with periodate, gives acetaldehyde; the reaction can be used for analytical determination. 2,3-Butanediol forms cyclic esters, acetals, and ketals. With diisocyanates, polyurethanes are obtained.

**Production.** After removal of butadiene and isobutene from crack gases, a $C_4$ hydrocarbon fraction, called $C_4$ raffinate II, is obtained, which contains approximately 77% butenes and 23% of a mixture of butane and isobutane. By chlorohydrination of this fraction with a solution of chlorine in water and subsequent cyclization of the chlorohydrins with sodium hydroxide, a butene oxide mixture of the following composition is obtained:

55% *trans*-2,3-butene oxide, 30% *cis*-2,3-butene oxide, 15% 1,2-butene oxide.

Hydrolysis of this mixture (50 bar, 160–220 °C, reaction enthalpy $\Delta H = -42$ kJ/mol) yields a mixture of butanediols which are separated by vacuum fractionation. In order

**Table 2.** Physical properties of 2,3-butanediols

|  | (R,S)-2,3-Butanediol | 2R,3R-2,3-Butanediol | 2S,3S-2,3-Butanediol | Racemate (R,R- and S,S-2,3-Butanediol) |
|---|---|---|---|---|
| mp, °C | 35.5–36.5 | 19 |  | 7.6 |
| bp (976 mbar), °C | 181.7 | 179–180* | 179–182* | 176.7 |
| bp (13.16 mbar), °C | 83.5 | 77.3–77.4 |  | 75.3–75.6 |
| $d_4^{25}$ | 0.9939 | 0.9873 | 0.9872 |  |
| $n_D^{25}$ | 1.43719 | 1.43095 | 1.4306 | 1.43109 |
| $[\alpha]_D^{25}$ |  | –13.16°* | +11.8°* |  |
| viscosity (35 °C), mPa · s | 65.6 | 21.8 |  |  |
| mp of bis(4-nitrobenzoates), °C | 193–193.5 | 143–143.5 |  | 128–128.5 |

* Theoretically bp and absolute amount of rotation of enantiomers should be the same.

to avoid the formation of polyethers during the hydrolysis, an excess of water must be used. The fractionation of the butanediols is easier than that of the butene oxides. By this reaction sequence, *meso*-2,3-butanediol is obtained from *trans*-2-butene via *trans*-2,3-butene oxide; the racemic mixture of R,R- and S,S-2,3-butanediol is formed analogously from *cis*-2-butene via *cis*-2,3-butene oxide. For a discussion of stereochemistry of this reaction, see also [37].

**Analysis, Storage.** The butanediol mixture is analyzed best by GC (polyethylene glycol as stationary phase).

Storage and transportation do not provide any problems. However, the hygroscopicity of the product requires some precautions.

**Uses.** 2,3-Butanediol (at least 80% meso isomer, the rest racemic mixture) is used as a cross-linking agent for naphthalene-1,5-diisocyanate in the production of specific hard-rubber products (Vulkollan). Derivatives of 2,3-butanediol are important as insecticides (Sapecron: acetal with N-methylcarbamate of salicylic aldehyde) and as intermediates in the pharmaceutical industry. 2,3-Butanediols have some interest as humectants and in the synthesis of polymers and plasticizers.

## 2.2. 1,3-Butanediol

1,3-Butanediol [*107-88-0*], $M_r$ 90.12, bp 207.5 °C (at 1013 mbar), bp 103–104 °C (at 10 mbar), $d_4^{20}$ 1.0053, $n_D^{20}$ 1.4410, is miscible with water and ethanol. The molecule has one center of chirality, and the data given above refer to the racemate. (R)-(–)-1,3-Butanediol [*6290-03-5*] has a specific rotation (ethanol) of $[\alpha]_D^{25}$ = –18.8°.

1,3-Butanediol is produced as an intermediate in the manufacture of butadiene from acetaldol. This process has, however, been abandoned by most companies. The compound is mainly used as a component of special polyester resins.

## 2.3. 1,2-Butanediol [38]

**Physical Properties.** 1,2-Butanediol[*584-03-2*], $C_4H_{10}O_2$, $M_r$ 90.12, *mp* −50 °C, *bp* 195 – 196.9 °C (at 101.3 kPa), is a colorless liquid, $d_4^{20}$ 1.0023, $n_D^{20}$ 1.4382. The diol is miscible with water in all proportions, readily soluble in alcohols, slightly soluble in ethers and esters, and insoluble in hydrocarbons. The dynamic viscosity at 20 °C is 73 mPa · s and the flash point 107 °C.

**Chemical Properties.** 1,2-Butanediol is a typical glycol and readily forms acetals and ketals; with dicarboxylic acids or anhydrides polyesters are formed and with diisocyanates polyurethanes.

**Production.** 1,2-Butanediol is synthesized by addition of water to 1,2-epoxybutane [*106-88-7*]:

$$CH_2\text{-}CH\text{-}CH_2\text{-}CH_3 \xrightarrow{H_2O} CH_2\text{-}CH\text{-}CH_2\text{-}CH_3$$
$$\underset{O}{\phantom{CH_2}} \qquad\qquad \underset{OH\ \ OH}{\phantom{CH_2}}$$

The hydration reaction is exothermic ($\Delta H = -93$ kJ/mol). A 10 to 20-fold molar excess of water is used to suppress polyether formation. The reaction is carried out either without a catalyst at 160 – 220 °C and 10 – 30 bar or in the presence of catalysts below 160 °C and only slightly above atmospheric pressure. Sulfuric acid or strongly acid ion exchange resins are used as catalysts [38]. Depending on the excess of water, selectivity is 70 – 92 %. Higher ethers of 1,2-butanediol are formed as byproducts.

**Uses.** The field of application for 1,2-butanediol is not very broad. It is used mainly as a solvent and intermediate.

# 3. Toxicology

*1,2-Butanediol and 1,3-Butanediol.* The median lethal dose ($LD_{50}$) of 1,2-butanediol is 16 g/kg (rat, oral) [39]; the corresponding value of 1,3-butanediol is 29.6 g/kg [40], [41]. Repeated administration of 1,3-butanediol to rats (2 a, 10 wt % in the feed) and to dogs (2 a, 3 wt % in the feed) did not reveal any toxic effects [42]. In high doses, 1,3-butanediol has a narcotic effect with a specific depressive effect on the central nervous system [43], [44]. 1,3-Butanediol is approved as a food additive by the U.S. Food and Drug Administration [45].

*2,3-Butanediol* has an $LD_{50}$ value of 8.9 g/kg (mouse, oral) [40].

*1,4-Butanediol*, however, with an $LD_{50}$ value of 1.78 g/kg [39] or 1.5 g/kg [46] (both rat, oral) should be classified as harmful according to the EC Council Directive. The observed depressive effect on the central nervous system is attributed to the metabolite $\gamma$-hydroxybutyric acid.

None of the butanediol isomers either irritate or sensitize the skin; however, they cause strong (1,2-butanediol) to slight (1,3- and 1,4-butanediol) eye irritation [39].

*2-Butene-1,4-diol* with an $LD_{50}$ value of about 860 mg/kg (rat, oral) is harmful. Eye and skin irritation are not observed, even after prolonged contact. Rats could inhale air saturated with 2-butene-1,4-diol at 20 °C for 8 h without showing any symptoms [46]. Repeated intake by rats (5% in the feed) led to death after 7–11 days; 30% in the feed led to the animals' death after four days [47].

*1-Butene-3,4-diol* with an $LD_{50}$ value of approximately 1.5 g/kg (rat, oral) is also harmful. An 8-h inhalation of air saturated with the compound at 20 °C by rats does not lead to any effects. The animals' skin was not irritated after contact for several hours; however, irritation to the eyes may occur.

*2-Butyne-1,4-diol* is to be classified as "toxic" having an $LD_{50}$ value of about 100 mg/kg (rat, oral). Strong irritation to the skin – especially after contact for several hours – and to the eyes were observed. An 8-h inhalation of air saturated with the compound at 20 °C by rats did not lead to any adverse effects [46].

Neither TLV nor MAK values have been established for the compounds cited above.

# 4. References

[1] W. Reppe: *Chemie und Technik der Acetylen-Druckreaktionen,* 2nd ed., Verlag Chemie, Weinheim 1952.
[2] C. J. S. Appleyard, J. F. C. Gortshore: Manufacture of Butynediol at IG Ludwigshafen, Bios-Report 367, OTS-Report PB 28556, US Department of Commerce.
[3] D. L. Fuller, A. O. Zors, H. M. Weir: The Manufacture of Butynediol from Acetylene and Formaldehyde, Fiat-Report no. 296, 1946, OTS-Report PB 80334 US Department of Commerce.
[4] BASF, DE-AS 2040501, 1970 (G. Boettger, H. Hoffmann, W. Reiß, L. Schuster, H. Toussaint).
[5] BASF, DE-AS 2421407, 1974 (J. Dehler, H. Hoffmann, W. Reiß, R. Schnur, S. Winderl, P. Zehner).
[6] W. Reppe, *Justus Liebigs Ann. Chem.* **596** (1955) 8.
[7] *Technische Regeln Acetylen,* Carl Heymanns Verlag, Köln.
[8] H. B. Sargent, *Chem. Ing.* 1957 (Feb.) 250.
[9] BASF, DE 2431929, 1981.
[10] GAF, DE-AS 1139832, 1961.
[11] K. Takehira, H. Mimoun, I. Sérée de Roche, *J. Catal.* **58** (1979) 155–169.
[12] Hoechst, DE-AS 1015797, 1954.
[13] R. A. Firestone, E. E. Harris, W. Reuter, *Tetrahedron* **23** (1967) no. 2, 943–955.
[14] I.G. Farbenind., FR 889079, 1943.
[15] Wingfoot Corp., GB 630992, 1949.
[16] J. Nelles, O. Bayer, W. Tischbein, F. Baehren, US 2417513, 1947.
[17] W. Reppe, C. Schuster, E. Weiss, US 2421650, 1947.
[18] J. H. Paden, H. Adkins, *J. Am. Chem. Soc.* **58** (1936) 2487.
[19] I.G. Farbenind., FR 905141, 1945.

[20] US Department of Commerce: *The Continuous Production of 1,4-Butanediol Dichlorocarboxylic Ester*, PB 58619.
[21] L. H. Smith: *Synthetic Fiber Development in Germany*, Textile Research Institute, New York 1964.
[22] I.G. Farbenind./BASF, DE 679607, 1939.
[23] I.G. Farbenind./BASF, DE 858094, 1937.
[24] GAF, US 3449445, 1967 (F. E. Wetherill).
[25] BASF, DE 2917018, 1979 (K. Baer, W. Reiss, W. Schroeder, D. Voges).
[26] S. H. H. Chow, J. D. Verbsky, US 3852164, 1974.
[27] *Chem. Ind. (Düsseldorf)* **32** (1980) 275.
[28] M. Tamura, S. Kumano, *Chem. Econ. Eng. Ref.* **12** (1980) no. 9, 32.
[29] Mitsubishi Chem., DE-OS 2345160, 1974; DE-OS 2424539, 1974; DE-OS 2504637, 1975; DE-OS 2510088, 1975; DE-OS 2510089, 1975; DE-OS 2505749, 1975.
[30] Y. Tanabe, *Hydrocarbon Process.* **60** (1981) no. 9, 187.
[31] BASF, DE-OS 2444004, 1976.
[32] BASF, DE-OS 2454768, 1976.
[33] Toyo Soda, US 3720704, 1973.
[34] Shell Oil, US 4384146, 1983.
[35] Chevron Research, US 3932468, 1976.
[36] *Beilstein*, **1 (3)** 2178, 2180, 2181, 2183; **1 (4)** 2524, 2525.
[37] C. E. Wilson, H. J. Lucas, *J. Am. Chem. Soc.* **58** (1936) 2396.
[38] K. Szafraniak, J. Myszkowski, A. Zielinski, W. Pyc, PL 79662, 1978; *Chem. Abstr.* **92** (1980) 22031 n;*Przem. Chem.* **52** (1973) no. 11, 744–746; *Chem. Abstr.* **80** (1974) 47385 w.
[39] V. K. Rowe et al.: *Patty's Industrial Hygiene and Toxicology*, vol. **2 C**, Wiley Interscience Publ., New York 1982, p. 3874.
[40] L. Fischer et al., *Z. Gesamte Exp. Med.* **115** (1949) 22.
[41] A. Loeser, *Pharmazie* **4** (1949) 263.
[42] R. A. Scala et al., *Toxicol. Appl. Pharmacol.* **10** (1967) 160.
[43] G. D. Frye et al., *J. Pharmacol. Exp. Ther.* **216** (1981) 306.
[44] G. S. Stoewsand et al., *Proc. Int. Congr. Nutr. 7th 1966* **4** (1967) 1082–7.
[45] W. R. Hewitt et al., *Toxicol. Appl. Pharmacol.* **64** (1982) 529.
[46] BASF, unpublished results 1959–1981.
[47] H. Schlüssel, *Naunyn Schmiedebergs Arch. Exp. Pathol. Pharmakol.* **221** (1954) 67.

# Butanols

HEINZ-DIETER HAHN, Ruhrchemie Aktiengesellschaft, Oberhausen, Federal Republic of Germany (Chaps. 2–8)

GEORG DÄMBKES, Ruhrchemie Aktiengesellschaft, Oberhausen, Federal Republic of Germany (Chaps. 2–8)

NORBERT RUPPRICH, Hoechst Aktiengesellschaft, Frankfurt/Main, Federal Republic of Germany (Chap. 9)

| | | |
|---|---|---|
| 1. Introduction . . . . . . . . . . . . . 951 | 8. | Economic Aspects . . . . . . . . . 963 |
| 2. Physical Properties . . . . . . . . 952 | 9. | Toxicology and Occupational Health . . . . . . . . . . . . . . . . . 964 |
| 3. Chemical Properties . . . . . . . 954 | 9.1. | 1-Butanol . . . . . . . . . . . . . . . 964 |
| 4. Production . . . . . . . . . . . . . 956 | 9.2. | 2-Butanol . . . . . . . . . . . . . . . 965 |
| 5. Quality Requirements and Control . . . . . . . . . . . . . . . . 960 | 9.3. | 2-Methyl-1-propanol . . . . . . . 966 |
| 6. Storage and Transportation . . 961 | 9.4. | 2-Methyl-2-propanol . . . . . . . 966 |
| 7. Uses . . . . . . . . . . . . . . . . . . 961 | 10. | References . . . . . . . . . . . . . . 967 |

# 1. Introduction

Butanols (butyl alcohols) are aliphatic saturated $C_4$ alcohols ($C_4H_9OH$, $M_r$ 74.12). There are four structural isomers of the alcohols: two primary, one secondary, and one tertiary; as there is an asymmetric C atom in the secondary alcohol, there are two stereoisomers of 2-butanol.

| Formula | Nomenclature | Common name |
|---|---|---|
| $CH_3CH_2CH_2CH_2OH$ | 1-butanol [71-36-3] | n-butanol |
| $CH_3CHCH_2OH$<br>$\quad\vert$<br>$\quad CH_3$ | 2-methyl-1-propanol [78-83-1] | isobutanol |
| $CH_3CH_2\overset{*}{C}HCH_3$<br>$\qquad\quad\vert$<br>$\qquad\quad OH$ | 2-butanol [78-92-2] | sec-butanol |
| $\quad\ CH_3$<br>$\quad\ \vert$<br>$CH_3-C-OH$<br>$\quad\ \vert$<br>$\quad\ CH_3$ | 2-methyl-2-propanol [75-65-0] | tert-butanol |

In addition to the systematic IUPAC nomenclature, other, nonsystematic names for the various butanols are in common use.

1-Butanol (*n*-butanol) occurs in nature in compound form. It also occurs, sometimes in high concentrations, in fusel oils obtained by fermentation. The first industrial production of 1-butanol, around 1912, was based on the discovery of the bacterium *Clostridium acetobutylicum Weizmann*, which causes carbohydrates to ferment to give mainly acetone and 1-butanol.

The continually increasing demand for 1-butanol could only be met by developing new manufacturing processes, such as the hydrogenation of crotonaldehyde formed by the aldolization of acetaldehyde; the Reppe synthesis (propylene carbonylation); and, in particular, the process by which most 1-butanol is now produced, the hydrogenation of *n*-butyraldehyde, easily obtainable by the hydroformylation of propylene.

2-Methyl-1-propanol (isobutanol) occurs in natural products as well as in fusel oils (74% of the total alcohol in the product of molasses fermentation), from which it can be isolated. The isobutyl oil synthesis, a reaction related to methanol synthesis from CO and $H_2$, first made it possible to produce larger quantities of 2-methyl-1-propanol. Nowadays, nearly all 2-methyl-1-propanol is produced by the oxo synthesis (propylene hydroformylation). In the last few years this straightforward production process and the attractive prices have led to a considerable expansion of the market for isobutanol.

The secondary and tertiary butanols are obtained by hydration of the corresponding unsaturated hydrocarbons, the butenes, whereas dehydration of the butanol yields the corresponding butene.

Thus, 2-butanol (*sec*-butanol), the simplest alcohol with an asymmetric carbon atom, can be obtained by the acid-catalyzed hydration of 1- and 2-butene whereas hydration of 2-methylpropene (isobutene) yields 2-methyl-2-propanol (*tert*-butanol). The ready availability of 1- and 2-butene and 2-methylpropene from olefins derived from petroleum cracking has made the commercial application of these reactions possible. Most 2-butanol produced is further processed to 2-butanone (methyl ethyl ketone).

The rapid development of 2-methyl-2-propyl methyl ether as an additive for low-lead gasolines has increased the availability of 2-methyl-2-propanol, from which it can be derived as byproduct.

# 2. Physical Properties

With the exception of 2-methyl-2-propanol (*mp* 25.6 °C) butanols are colorless liquids. Butanols exhibit a characteristic odor; their vapors have an irritant effect on mucous membranes and a narcotic effect in higher concentrations.

All butanols are completely miscible with common organic solvents. Only 2-methyl-2-propanol is completely miscible with water. The major characteristic physical properties of butanols are compiled in Table 1.

**Table 1.** Physical properties of butanols

| | 1-Butanol | 2-Methyl-1-propanol | 2-Butanol | 2-Methyl-2-propanol |
|---|---|---|---|---|
| $mp$, °C | − 89.3 | − 107.9 | − 114.7 | + 25.6 |
| $bp$, °C | 117.7 | 107.9 | 99.5 | 82.55 |
| Density $d_4^{20}$ | 0.8098 | 0.8027 | 0.8065 | 0.7867 |
| Refractive index $n_D^{20}$ | 1.3991 | 1.3959 | 1.39719 | 1.3841 |
| Viscosity at 20 °C, mPa · s | 3.0 | 4.0 | 4.2 (15 °C) | 3.3 (30 °C) |
| Specific heat, J g$^{-1}$ K$^{-1}$ | 2.437 at 30–80 °C | 2.5263 at 30–80 °C | 2.81 at 20 °C | 3.035 |
| Heat of vaporization, J/g | 591.64 | 578.83 | 562.75 | 535.78 |
| Heat of fusion, J/g | 125.2 | | | 91.61 |
| Heat of combustion, kJ/g | 36.111 | 35.981 | | 35.588 |
| Critical pressure, hPa | 48.4 | 48 | | |
| Critical temperature, °C | 287 | 265 | 265 | 235 |
| Surface tension at room temp., mN/m | 22.3 | 23.0 | 23.5 | |
| Dielectric constant at room temperature | 17.8 | 18.8 | 15.5 | 11.4 |
| Evaporation number (ether = 1) | 33 | 24 | | |
| Solubility in water | | | | |
| at 20 °C, wt% | 7.7 | 8.5 | 12.5 | miscible |
| at 30 °C, wt% | 7.08 | 7.5 | 18 | miscible |
| Solubility of water in butanol | | | | |
| at 20 °C, wt% | 20 | 15 | | miscible |
| at 30 °C, wt% | 20.62 | 17.3 | 36.5 | miscible |
| Rotation $[\alpha]_D^{20}$ | | | (+) (S)-2-butanol + 13.73° (−) (R)-2-butanol − 13.79° | |

See Chapter 6 for safety data, such as flash points, ignition temperatures, and lower explosive limits.

For further physical properties of butanols see also [1]. The thermodynamic properties of all butanols have been examined in detail; cf. [2], [3] for 1-butanol; [4] for 2-methyl-1-propanol; [5]–[7] for 2-butanol; [8], [9] for 2-methyl-2-propanol.

References [10]–[12] give an outline of liquid–vapor equilibria for the distillation of butanols. H. A. Rizk and N. Youssef [13] as well as G. E. Rajala and J. Crossley [14] have investigated the dielectric properties of butanol in detail. D. Davis [15] has set up nomograms for the densities of aqueous solutions of 1-butanol, 2-methyl-1-propanol, and 2-methyl-2-propanol.

Further physical properties of the systems comprising water with the various butanols have been described by several authors; cf. [16]–[18] for 1-butanol/water, [19], [20] for 2-butanol, and [21] for 2-methyl-2-propanol.

For the phase equilibria of the ternary systems isobutyraldehyde/2-methyl-1-propanol (isobutanol)/water and 2-butanol/2-methyl-2-propanol (tert-butanol)/water cf. [22] and [23].

Table 2 shows the composition of some binary azeotropic mixtures. For other azeotropes see [24]–[26].

**Table 2.** Azeotropic mixtures

| Compound A | Compound B | bp, °C comp. B | bp, °C azeotrope | wt% butanol in azeotrope |
|---|---|---|---|---|
| 1-Butanol | water | 100 | 92.7 [26] | 57.5 [26], 55.5 [24] |
|  |  |  | 93.0 [24] | 63.0 [25] |
| 1-Butanol | acetic acid 1-butyl ester | 126.2 | 117.6 | 67.2 |
| 1-Butanol | 2-methyl-1-propanol | 99.5 |  | no azeotrope |
| 1-Butanol | cyclohexane | 81.4 | 79.8 | 10.0 |
| 1-Butanol | toluene | 110.6 | 105.6 | 27.0 |
| 2-Methyl-1-propanol | water | 100 | 89.8 | 67 [26], 70 [24] |
| 2-Methyl-1-propanol | acetic acid 2-methylpropyl ester | 116.5 | 107.6 | 92/95 [26], 55 [24] |
| 2-Methyl-1-propanol | cyclohexane | 81.4 | 78.1 | 14.0 |
| 2-Methyl-1-propanol | toluene | 110.6 | 101.2 | 44.5 |
| 2-Butanol | water | 100 | 87.0 [26], 88.5 [24] | 73.2 [26], 68 [24] |
| 2-Butanol | acetic acid 2-methylpropyl ester | 112.2 |  | no azeotrope |
| 2-Butanol | 2-butanone | 79.6 |  | no azeotrope |
| 2-Butanol | cyclohexane | 81.4 | 76.0 | 18.0 |
| 2-Butanol | toluene | 110.6 | 95.3 | 55.0 |
| 2-Methyl-2-propanol | water | 100 | 79.9 | 88.24 |
| 2-Methyl-2-propanol | cyclohexane | 81.4 | 71.3 | 37.0 |

# 3. Chemical Properties

The chemical properties of butanols are mainly those of primary, secondary, and tertiary alcohols. In particular the reactive primary alcohols serve as starting materials for a wide range of reactions. The following deserve special mention:

**Dehydration.** The catalytic dehydration of alcohols for the preparation of olefins is a well-established reaction [27]. Of the butanols 2-methyl-2-propanol (*tert*-butanol) undergoes this reaction most readily. Even with dilute sulfuric acid water is split off on warming, leading to the formation of very pure 2-methylpropene (isobutene) [115-11-7] [28]. As 2-methylpropene is readily obtainable from the $C_4$-fractions of olefins from cracked petroleum this reaction is no longer of industrial importance, however. See Chap. 4 for the manufacture of pure 2-methylpropene. The dehydration of 2-methyl-1-propanol (isobutanol) similarly leads to the formation of 2-methylpropene.

The dehydration of 2-methyl-1-propanol to 2-methylpropene is carried out in the presence of $\gamma$-$Al_2O_3$ catalyst at about 300–350 °C with practically quantitative conversion and a 2-methylpropene selectivity of over 90%. In an analogous manner 1-butanol can be dehydrated to a mixture of 1-butene [106-98-9] and *cis*- and *trans*-2-butene [624-64-6], [590-18-1] [29], [30].

Selective dehydration of 1-butanol in mixtures with 2-methyl-1-propanol can be carried out with calcium zeolite molecular sieves [31], due to the difference in the spatial structure of the two butanols.

Also, at lower temperature dibutyl ether [*142-96-1*] can be formed from 1-butanol in the presence of dehydration catalysts [30]. The branched butanols show little tendency to undergo this reaction.

**Oxidation.** As primary and secondary alcohols 1-butanol, 2-methyl-1-propanol, and 2-butanol can be dehydrogenated to the corresponding carbonyl compounds. Dehydrogenation can be carried out even at low reaction temperatures with oxidizing agents such as manganese(IV) oxide in sulfuric acid, nitric acid, chromic acid, or selenium dioxide [32].

Even without oxidizing agents the use of suitable catalysts (e.g., Cu) at higher temperatures enables the alcohols to be dehydrogenated to aldehyde or ketone until the thermodynamic equilibrium has been established [33]–[36]. 2-Butanol dehydrogenates to 2-butanone (methyl ethyl ketone) [*78-93-3*] at about 350 °C [37].

As a tertiary alcohol 2-methyl-2-propanol cannot be selectively dehydrogenated; splitting of the molecule occurs under extreme conditions.

Primary alcohols can be oxidized to carboxylic acids under other reaction conditions. In addition to the standard oxidizing agents, this oxidation can be accomplished by the reaction of the alcohol with alkali hydroxide at 275 °C. In this reaction, which is based on the observations of GUERBERT, the alkali salt of butyric acid [*107-92-6*] is formed together with hydrogen and 2-ethylhexanol [*104-76-7*] (cf. also [38]).

The Koch-Haaf reaction can be used to prepare trimethylacetic acid [*75-98-9*] from 2-methyl-1-propanol and 2-methyl-2-propanol by the addition of carbon monoxide in the presence of sulfuric acid [39]. With readily available 2-methyl-1-propanol (isobutanol) from the oxo synthesis, trimethylacetic acid can be obtained in 89% yield [40].

**Alkylation.** Butanols can be employed in a wide variety of alkylation reactions. *N*-alkyl-, *N,N*-dialkyl-, or *N,N,N*-trialkylamines are obtainable with ammonia and amines. The use of isomeric mixtures of alcohols leads to the corresponding mixed amines. Ring alkylation of aromatic hydrocarbons proceeds with butanol in the presence of Friedel-Crafts catalysts [41].

**Esterification.** Butanols can be converted in the usual manner into butyl esters with inorganic and organic acids. The reaction is generally carried out in the presence of acid catalysts. In certain cases excess of acid which is to be esterified can serve as the catalyst. The rate of ester formation greatly depends on the structure of the carboxylic acid and of the alcohol. The primary butanols react more rapidly than 2-butanol, which in turn reacts more quickly than 2-methyl-2-propanol. For the formation of esters with acid chlorides and acid anhydrides, as well as other esterification processes, see [42].

# 4. Production

Of the many available processes for the preparation of butanols, the following have achieved industrial importance:

*1-Butanol:*
  propylene hydroformylation (oxo synthesis)
  Reppe synthesis
  crotonaldehyde hydrogenation

*2-Methyl-1-propanol:*
  propylene hydroformylation
  catalytic hydrogenation of carbon monoxide
  homologization reaction

*2-Butanol:*
  *n*-butene hydration

*2-Methyl-2-propanol:*
  2-methylpropene hydration
  byproduct of propylene oxide and methyl *tert*-butyl ether production

**Oxo Synthesis.** The most important process for the manufacture of 1-butanol and 2-methyl-1-propanol is propylene hydroformylation with subsequent hydrogenation of the aldehydes formed.

In the oxo reaction (hydroformylation) carbon monoxide [630-08-0] and hydrogen [1333-74-0] are added to a carbon–carbon double bond in the liquid phase in the presence of catalysts (hydrocarbonyls or substituted hydrocarbonyls of Co, Rh, or Ru) [43]. In the first reaction step aldehydes are formed with one more C-atom than the original olefins. For olefins with more than two C-atoms, isomeric aldehyde mixtures are normally obtained. In the case of propylene these consist of 1-butanal and 2-methylpropanal.

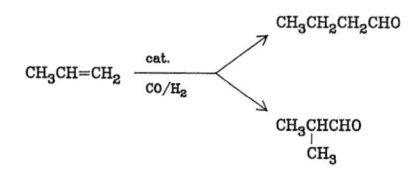

$$CH_3CH=CH_2 \xrightarrow[CO/H_2]{cat.} \begin{array}{l} CH_3CH_2CH_2CHO \\ \\ CH_3CHCHO \\ \phantom{CH_3CH}|\\ \phantom{CH_3CH}CH_3 \end{array}$$

There are several variations of the hydroformylation process, the differences being in the reaction conditions (pressure, temperature) as well as the catalyst system used. The classic high-pressure process exclusively used until the beginning of the 1970s operates at pressures of 20–30 MPa (200–300 bar) $CO/H_2$ and temperatures of 100–180 °C. The catalyst is Co. It leads to about 75% 1-butanol and about 25% 2-methyl-1-propanol.

The new process developments of the past few years have led to a clear shift in the range of products. The processes operating at relatively low pressures (1–5 MPa, 10–50 bar) use modified Rh-catalysts. The isomeric ratios achieved are about 92:8 [44] or 95:5 [45] 1-butanol to 2-methyl-1-propanol. However, by the use of unmodified Rh the percentage of 2-methyl-1-propanol can be increased to about 50%.

Catalytic hydrogenation of the aldehydes leads to the formation of the corresponding alcohols.

As only primary alcohols can be obtained via the oxo synthesis, it is not possible to produce 2-butanol and 2-methyl-2-propanol by this process.

**Reppe Process.** 1-Butanol and 2-methyl-1-propanol can also be produced on a commercial scale by the carbonylation of propylene [115-07-1] developed by REPPE. In this process, developed in 1942, [47], [48], olefins, carbon monoxide, and water are made to react under pressure in the presence of a catalyst (tertiary ammonium salt of polynuclear iron carbonyl hydrides). The difference between this process and the classic Co-catalyzed hydroformylation is that at low temperature (about 100 °C) and low pressure (0.5–2 MPa, 5–20 bar) alcohols are formed directly from the olefin.

As with the oxo synthesis the carbon monoxide can be added to both C-atoms of the double bond which means that, when propylene is used, 1-butanol and 2-methyl-1-propanol are obtained in a ratio of 86:14.

$$CH_3CH=CH_2 + 3\ CO + 2\ H_2O \xrightarrow{90\%} \begin{array}{l} CH_3CH_2CH_2CH_2OH \\ + 2\ CO_2 \\ CH_3 \\ | \\ CH_3CHCH_2OH \end{array}$$

The catalyst is sensitive both to air and to higher temperatures at which, in the presence of water and $CO_2$, it decomposes into iron carbonate. To achieve adequate reaction rates the catalyst, carbonyl triferrate, must be present in concentrations of $\gg 10\%$ in the reaction solution; this is attained by the presence of dissolving agents (N-alkylpyrrolidine); see [49], [50], [51] for details of the process.

The Reppe process has not been as successful as propylene hydroformylation with Co catalysts despite favorable *n/iso* ratios in the reaction products and milder reaction conditions [52]. This can be attributed to the more expensive process technology.

**Hydrogenation of Crotonaldehyde** [4170-30-3]. Until the mid 1950s the manufacture of 1-butanol based on acetaldehyde was the preferred process. With the development of the oxo synthesis, however, it has lost its importance even in Japan and the USA and is no longer in use.

The individual steps of the process are as follows:
aldol condensation

$$2\ CH_3CHO \longrightarrow CH_3CH(OH)CH_2CHO$$

splitting off of water

$$CH_3CH(OH)CH_2CHO \longrightarrow CH_3CH=CHCHO + H_2O$$

hydrogenation

$$CH_3CH=CHCHO + 2\,H_2 \longrightarrow CH_3CH_2CH_2CH_2OH$$

Acetaldehyde is aldolized to acetaldol at normal temperature and pressure in the presence of alkaline catalysts. With conversions of about 60% the acetaldol yield is about 95%. Unreacted acetaldehyde can be distilled off and recycled. The removal of water and formation of aldehyde is brought about by acidification of the acetaldol with acetic acid or phosphoric acid and subsequent distillation, whereby the crotonaldehyde is obtained almost quantitatively as the top product.

Various gas- and liquid-phase processes have proved their value for the hydrogenation of crotonaldehyde to 1-butanol. Copper catalysts are particularly useful. About 1000 kg of 1-butanol can be obtained from 1350 kg of acetaldehyde [53].

The future economic importance of the crotonaldehyde hydrogenation process will depend to a large extent on developments in the raw materials market. As "crude oil" raw material becomes scarcer and more expensive as a basis for the oxo synthesis, the alternative route based on ethanol from fermentation will become increasingly competitive. The biomass substrate for ethanol production is almost inexhaustible. The ethanol is first dehydrogenated to form acetaldehyde, from which the synthesis can proceed.

Such a process route is an increasingly interesting alternative for tropical countries with large supplies of cheap biomass as well as for the more developed countries of the third world which do not have their own oil resources [54].

**Catalytic Hydrogenation of Carbon Monoxide.** In addition to the $CO/H_2$-based methanol synthesis developed commercially in the 1920s, other analogous processes were developed for the manufacture of higher alcohols. Compared with the methanol synthesis, however, these processes found only limited application. Nowadays they have no commercial importance.

The most significant process for the production of 2-methyl-1-propanol was the BASF isobutanol oil synthesis. Reaction products with about 50% methanol [67-56-1] and 11–14% 2-methyl-1-propanol as well as various other components were obtained from CO and $H_2$ at temperatures of about 430 °C and pressures of about 30 MPa (300 bar) in the presence of a KOH-modified methanol synthesis catalyst [55], [56]. BASF ceased its isobutyl oil synthesis in 1952 when 2-methyl-1-propanol and its secondary product 2-methylpropene could be obtained more economically using the oxo synthesis or via a petrochemical route. Further comprehensive experiments to obtain the desired butanol from the cheap raw materials methanol and ethanol with the homologization reaction have so far proved fruitless. Butanols have so far only been obtained in small amounts as byproducts. A review is given in [57].

**Olefin Hydration.** 2-Butanol (*sec*-butanol) and 2-methyl-2-propanol (*tert*-butanol) can be obtained by the acid-catalyzed addition of water to 1- or 2-butene [*106-98-9*], [*107-01-7*] and 2-methylpropene [*115-11-7*] respectively. The water adds at the double bond in accordance with the Markownikoff rule. The difference in olefin reactivity can be used for the selective manufacture of 2-butanol or 2-methyl-2-propanol from technical $C_4$-fractions where the two alcohols are prepared in turn and isolated. 2-Methylpropene can be completely converted to 2-methyl-1-propanol with 65% sulfuric acid. Hydration of *n*-butene, on the other hand, requires 75–80% $H_2SO_4$ [58]. Because of the tendency of the olefins to polymerize readily, the requisite reaction temperatures must not be exceeded.

In addition to this indirect hydration another process variant is known to industry in which the olefins are hydrated directly at elevated temperatures (about 200–300 °C) and higher pressures (10–35 MPa, 100–350 bar) [59], [60].

$$CH_3CH=CHCH_3 + H_2O \underset{}{\overset{cat.}{\rightleftarrows}} CH_3CH_2\underset{OH}{C}HCH_3$$

Various catalysts have been described for direct hydration, such as tungsten oxide [61] and aqueous aluminum hydroxide gel suspensions [62]. For the hydration of 1-butene and 2-butene the best results are obtained at temperatures between 220 and 300 °C or in the case of 2-methylpropene, at 190–245 °C [63]. Direct hydration of the olefins has not yet established itself as a serious competitor to indirect hydration. For process improvement see [64].

**Other Processes.** 2-Methyl-2-propanol (*tert*-butanol) is formed during the manufacture of propylene oxide from 2-methylpropane [*75-28-5*] in an amount of about 1.2 t per ton of propylene oxide [27, pp. 834, 1005]. In an intermediate stage *tert*-butyl hydroperoxide [*75-91-2*] is formed from isobutane. Another process, in which 2-methyl-2-propanol is obtained as a by-product, is the splitting of methyl *tert*-butyl ether to highly pure 2-methylpropene or 2-methyl-2-propanol. Similarly the methyl *tert*-butyl ether synthesis with slight traces of water in the feedstock (methanol and $C_4$-olefins) produces 2-methyl-2-propanol as a byproduct [65].

In the literature a number of other methods for the preparation of butanol are cited which are generally of no significance other than for laboratory preparation. A comprehensive review is to be found in B. CORNILS and E. ZILLY [38].

**Table 3.** ASTM Specifications for butanols

| | 1-Butanol | 2-Methyl-1-propanol | 2-Butanol | ASTM Standard |
|---|---|---|---|---|
| ASTM-Standard: | D 304-80 | D 1719-81 | D 1007-80 | |
| Sampling | | | | D 268-80 |
| Specific gravity | 0.810–0.813 | 0.802–0.804 | 0.807–0.809 | D 268-80 |
| Color (platinum–cobalt scale) | < 10 | < 10 | < 10 | D 1209 |
| Distillation range | 117.7 °C within a range of 1.5 °C | 107.9 °C within a range of 2 °C | 98.0–101.0 °C | D 1078 |
| Nonvolatile matter | < 0.005 g/100 mL | < 0.005 g/100 mL | < 0.005 g/100 mL | D 1353 |
| Odor | characteristic, nonresidual | characteristic, nonresidual | characteristic, nonresidual | D 1296 |
| Water (Karl Fischer) | < 0.1 wt% | < 0.2 wt% | < 0.5 wt% | D 1364 |
| Miscibility with vol% heptane (99%) at 20 °C | 19 | 19 | 19 | D 1476 |
| Acidity (as free acetic acid) | < 0.005 wt% = 0.047 mg KOH/g sample | < 0.003 wt% = 0.028 mg KOH/g sample | < 0.002 wt% = 0.019 mg KOH/g sample | D 1613 |

# 5. Quality Requirements and Control

On an industrial scale the four different butanols are generally marketed as pure substances and not as isomer mixtures. Because of their many applications both as commodities and as materials for the manufacture of special fine chemicals a high standard of purity is demanded for the butanols.

The quality is mainly controlled by determining the physical and chemical properties of the alcohols, such as density, refractive index, color, water content, hydroxyl value, acid value, carbonyl value, and distillation range. The characteristic values are normally determined according to the ASTM [66], DIN [67], or the BS methods [68].

Table 3 lists the standard specifications which are stipulated by the American Society for Testing and Materials for 1-butanol, 2-butanol, and 2-methyl-1-propanol.

The B. S. I. (British Standards Institution) has issued specifications for 1-butanol (BS 508: 1966) and 2-butanol (BS 1993: 1968).

The best way to verify isomer purity and detect traces of impurities is by gas-chromatographic analysis. Such analyses can be carried out rapidly and with good reproducibility, on a routine basis, enabling the use of a high sampling frequency.

Separation can be effected by the use of both polar and non-polar packed columns or appropriate capillary columns.

For general purity analyses a thermal conductivity detector system is used; trace (impurity) analysis is carried out using a flame ionization detector system. Experience has shown that nitrogen, hydrogen, and helium are suitable carrier gases.

**Table 4.** Safety data for butanols

|  | UN-No. | Flash point, °C | Ignition limits in air, vol% | Ignition temperature, °C | Danger class, IMDG-code | MAK[a], ppm | TLV[b], ppm | TWA[c], ppm |
|---|---|---|---|---|---|---|---|---|
| 1-Butanol | 1120 | + 34 | 1.4 – 11.3 | 380 | 3.3 | 100 | 50 | 100 |
| 2-Methyl-1-propanol (isobutanol) | 1212 | + 28 | 1.7 – 10.9 | 430 | 3.3 | 100 | 75 | 100 |
| 2-Butanol (*sec*-butanol) | 1121 | + 24 |  | 390 | 3.3 | 100 | 150 | 150 |
| 2-Methyl-2-propanol (*tert*-butanol) | 1122 | + 11 | 2.3 – 8 | 470 | 3.2 | 100 | 150 | 100 |

[a] For the Federal Republic of Germany.
[b] Threshold Limit Value.
[c] Time-weighted average concentration.

The degree of purity of commercial 1-butanol, 2-methyl-1-propanol, and 2-butanol is at least 99 wt %. 2-Methyl-2-propanol is supplied both as a product of at least 99.8 wt % purity and as an azeotropic mixture with 11 to 12 wt % water.

In addition to characteristic value determination and gas-chromatographic analysis a number of special analyses and tests may be carried out depending on the intended application, such as esterification tests with phthalic anhydride and color tests with concentrated sulfuric acid or $KMnO_4$.

# 6. Storage and Transportation

Butanols can be stored or dispatched in untreated mild steel or enamelled steel drums provided that ingress of moisture is prevented. Stainless steel containers are also suitable. The butanols are transported in rail and road tank cars, and in drums as well as in tanker vessels and containers. It is important to observe the national and international safety regulations relating to aviation, ocean-going and inland waterway shipping, rail and road transport as well as the particular safety requirements for butanols.

Some important data relating to the safe handling, storage, and transport of butanols are compiled in Table 4.

# 7. Uses

**1-Butanol.** 1-Butanol is used principally in the field of surface coating. In the USA, for instance, approx. 85 % of 1-butanol production finds an outlet in this sector of the market. It is used either directly as a solvent for varnishes or it is converted into derivatives which then serve as solvents or monomer components.

1-Butanol cannot be used directly as a solvent in the production of nitrocellulose lacquers; in admixture with toluene, ethanol, or various esters, however, it is an excellent thinner.

The addition of 5–10% of 1-butanol is also valuable in overcoming "blushing" (unwanted white opacity) which may arise when large quantities of thinners, particularly volatile solvents, are used.

1-Butanol is also useful for regulating the viscosity and improving the flow properties of varnishes and for the prevention of "streaking" in paints and lacquers based on spirit-soluble gums and resins.

Although not itself a solvent for substances such as polystyrene and chlorinated rubber, 1-butanol can be successfully used in quantities up to 20% as a diluent for the commonly used solvents for these substances, which are mainly the esters of the saturated carboxylic acids; in particular, the acetates.

The acrylic ester of 1-butanol has become increasingly important during the last decade. It has become an essential component of latex paints since such coatings are not only robust but can also be produced relatively cheaply. The 1-butyl esters of phthalic, adipic, sebacic, oleic, azelaic, stearic, and phosphoric acids are used as plasticizers and additives for surface coatings. The most important of these products is di-1-butyl phthalate (DBP). Consumption is, however, stagnating and in some countries it decreased slightly within the last years.

In the USA in particular, but also in the Federal Republic of Germany and Belgium, 1-butanol is used for the manufacture of butylamines.

1-Butanol has numerous applications in the plastics and textile sector. It is used as a coagulation bath for spinning acrylic fibers and in the dyeing of poly(vinyl alcohol) fibers.

**2-Methyl-1-propanol.** The fields of application of 2-methyl-1-propanol (isobutanol) closely resemble those of 1-butanol. Indeed, because of its lower price it can often replace 1-butanol and is accordingly used as a solvent, diluent, and additive for nitrocellulose and synthetic resins; wetting agent; cleaner additive; and component of printing inks and related products. Its capacity for dissolving resins (ketone, phthalate, urea, and melamine–formaldehyde resins) is of commercial importance. As a solvent for phenol–formaldehyde resins, however, 2-methyl-1-propanol is not as effective as 1-butanol.

Several esters exhibit excellent solvent properties; in particular, isobutyl acetate is a very good solvent for fats, chlorinated rubber, polystyrene, and coumarone resins. The esters of phthalic acid, adipic acid, some dicarboxylic acids as well as phosphoric acid are used as plasticizers, particularly for PVC and its copolymers and for cellulose derivatives.

The isobutyl esters of 2,4-dichloro- and 2,4,5-trichlorophenoxyacetic acid exhibit herbicidal activity. For the anti-freezing properties of 2-methyl-1-propanol in gasoline see [72].

Also of interest is the use of 2-methyl-1-propanol as an extracting agent in the recovery of ammonium phosphate [73].

**2-Butanol.** Practically all of the 2-butanol produced is dehydrogenated to 2-butanone (MEK or methyl ethyl ketone) which has found considerable application because of its good solvent properties and its favorable boiling point (79.57 °C). 2-Butanol itself is also used as a solvent: in particular, when mixed with aromatic hydrocarbons it is especially suitable as a solvent for alkyd resins and ethylcellulose lacquers. The ability of 2-butanol to dissolve both water and oils is applied in the manufacture of brake fluids and cleaning agents.

2-Butanol is converted into amines by treatment with ammonia. The acetate is recommended as a good solvent for nitrocellulose; the xanthate is used in ore flotation. 2-Butanol also finds application in the manufacture of perfumes, dyestuffs, fruit essences, and wetting agents.

**2-Methyl-2-propanol.** Consumption of 2-methyl-2-propanol is small in comparison with the other butanols, although its use appears to be growing. The solvent sector is again the major field of application. 2-Methyl-2-propanol is also used as an agent for introducing the *tert*-butyl group into organic compounds (e.g., *tert*-butyl-phenol for the preparation of oil-soluble resins and antioxidants, trinitro-*tert*-butyltoluene as artificial musk) and as a starting material for the preparation of peroxides (polymerization catalysts).

2-Methyl-2-propanol competes with 2-propanol as a gasoline additive to prevent carburetor freezing. The methylether can be used as an antiknock agent in gasoline. Its use as a fuel additive has increased explosively in recent years: 175 000 t in 1982; 400 000 t in 1983; for 1990 the estimated consumption is 400 000 – 800 000 t.

# 8. Economic Aspects

Both in the industrialized countries and worldwide, there is considerable excess capacity for the manufacture of 1-butanol and 2-methyl-1-propanol, the two major products in this group of substances (Table 5).

Plant utilization was approx. 60 – 70 % in 1983.

The uses can vary considerably from one region to another. As well as direct use as a sol vent the main areas of application are in plasticizers, butyl acetate, acrylic esters, and glycol esters (see Table 6).

**Table 5.** Butanol capacities in the leading producer countries in 1984

|  | 1-Butanol/*2-methyl-1-propanol, 1000 t | 2-Butanol, 1000 t |
|---|---|---|
| United States | 700 | |
| F.R. of Germany | 300 | 60 |
| Western Europe |  | 422** |
| Japan | 206 | 146 |

\* Any 1-butanol production plant also can be used for the manufacture of 2-methyl-1-propanol;
\*\* 1982

**Table 6.** Butanol consumption by end use, %

|  | Japan (1982) | USA (1980) |
|---|---|---|
| Plasticizers | 13 | 8 |
| Solvents | 18 | 17 |
| Butyl acetate | 20 | 9 |
| Butyl acrylate | 24 | 29 |
| Glycol ethers and esters | 13 | 23 |
| Miscellaneous | 12 | 14 |

# 9. Toxicology and Occupational Health [74]–[78]

The main effects of exposure to excessive concentrations of all butanols are irritation of the mucous membranes and depression of the central nervous system. Animal studies have shown low acute oral, dermal, and inhalation toxicity. There is insufficient data to make a reliable assessment of the long-term effects in animals. The limited amount of documented experience in humans suggests that excessive exposure can cause irritation, headache, nausea, fatigue, and dizziness.

## 9.1. 1-Butanol

The lowest median lethal dose ($LD_{50}$) published in the literature is 790 mg/kg (rat, oral) and 4200 mg/kg (rabbit, dermal) [74]. Inhalation of high concentrations of 1-butanol by experimental animals results in pulmonary and eye irritation, incoordination, and narcosis. The lowest $LC_{50}$ published in the literature is 8000 ppm (rats, inhal., 4 h) [74]. Application of drops of 1-butanol into the rabbit eye produces severe corneal irritation. 1-Butanol has a slight to moderate irritating effect on rabbit skin [75]. 1-Butanol was found to be non-mutagenic in the Ames test [75].

Conjunctivitis and keratitis, sometimes accompanied by small intracorneal vacuoles, were reported among workers exposed to 1-butanol vapors. Intracorneal vacuoles proved to be reversible after exposure had ceased [76]. In a ten-year-study, workers exposed to 1-butanol at concentrations of 200 ppm and above developed corneal inflammation associated with a burning sensation, blurring of the vision, lacrimation, and photophobia. The mean erythrocyte count was slightly lowered. There were only rare complaints of irritation when the concentration was 100 ppm [75].

Further data indicate that eye irritation may be expected when the concentration of 1-butanol is above 50 ppm [76]. According to another study humans exposed to 1-butanol showed mild irritation already at 25 ppm, followed by headaches at 50 ppm [77]. Workers exposed to 80 ppm experienced a greater hearing loss (hypoacusia) in comparison with an unexposed group [77]. Dermatitis of the fingers and hands upon exposure to 1-butanol is also reported [76].

*Exposure limits:*
  TLV (ceiling limit) 50 ppm (skin) [77],
  OSHA air standard (TWA) 100 ppm [74],
  MAK value 100 ppm.

## 9.2.   2-Butanol

The lowest $LD_{50}$ published in the literature is 4400 mg/kg (rat, oral) [77]. 2-Butanol, like the other butanols, is a depressant to the central nervous system. The lowest lethal concentration published in the literature is 10670 ppm for mice exposed for 225 min; repeated exposure of mice to 5330 ppm produced narcosis without death; and no acute signs of intoxication were observed in mice exposed to 1650 ppm for 420 min [75].

2-Butanol caused severe corneal injury when applied to the rabbit eye but was not found to be irritating to the rabbit skin [75].

According to one report industrial exposure to about 100 ppm has not resulted in any difficulties [77]. Another study concludes that irritation of the eyes, nose, and throat, headache, nausea, fatigue, and dizziness have been experienced from excessive exposure to 2-butanol [75].

*Exposure limits:*
  TLV 100 ppm (TWA); 150 ppm (STEL) [77],
  OSHA air standard 150 ppm (TWA),
  MAK value 100 ppm.

## 9.3. 2-Methyl-1-propanol

The lowest published $LD_{50}$ is 2460 mg/kg (rat, oral) [75]. A dermal $LD_{50}$ of 4250 mg/kg in the rabbit indicates a low toxicity due to local application [75]. The lowest $LC_{50}$ for 2-methyl-1-propanol is 8000 ppm (rats, inhal., 4 h) [77]. The compound produces narcosis in mice that had been intermittently exposed to 6400 ppm for a total of 136 h. The narcotic dosage is followed by slight organic changes in the liver and kidneys. Repeated exposure of mice to 2125 ppm for 9.25 h did not cause any adverse effect [76].

Application of 2-methyl-1-propanol to the rabbit eye causes moderate to severe irritation. A moderate irritation was observed after 24-h exposure of the rabbit skin [74].

The results of long-term studies in which a small number of rats received the material either orally or subcutaneously suggest that 2-methyl-1-propanol may have a carcinogenic potential [75]. Exposure of *E. coli* bacteria resulted in mutagenesis [74].

In humans, there was no evidence of eye irritation during repeated 8-h exposure to levels in the order of 100 ppm [75]. Industrial exposure to mixed vapors of butyl acetate and 2-methyl-1-propanol has caused vacuolar keratitis in several workers but it is not known which of the substances was responsible [76]. Application of 2-methyl-1-propanol to human skin resulted in slight erythema and hyperemia [76].

*Exposure limits:*
TLV 50 ppm (TWA), 75 ppm (STEL) [77],
OSHA air standard 100 ppm (TWA) [74],
MAK value 100 ppm.

## 9.4. 2-Methyl-2-propanol

The lowest $LD_{50}$ is reported to be 3500 mg/kg (rat, oral) [74]. The acute effect of the bad-smelling 2-methyl-2-propanol is that of a narcotic. Prolonged contact with rabbit skin failed to cause irritation [75]. The testing of the compound by *NTP* (National Toxicology Programm, U.S. Department of Health and Human Services) for carcinogenesis bioassay began in June 1983 [74].

Following application to human skin there was no reaction other than slight erythema and hyperemia. Irritating effects, headache, nausea, fatigue, and dizziness are reported as symptoms of excessive exposure [75].

*Exposure limits:*
TLV 100 ppm (TWA), 150 ppm (STEL) [77],
OSHA air standard 100 ppm (TWA) [74],
MAK value 100 ppm.

# 10. References

[1] R. W. Gallant, *Hydrocarbon Process.* **46** (1967) no. 1, 183–189. R. W. Gallant: *Physical Properties of Hydrocarbons,* vol. **1**, Gulf Publ. Comp., Houston, Texas, 1968, pp. 84–91.
[2] H. A. Gundry, A. J. Head, G. B. Lewis, *Trans. Faraday Soc.* **58** (1962) 1309–1312.
[3] C. Mosselman, H. Dekker, *Recl. Trav. Chim. Pays-Bas* **88** (1969) no. 3, 257–265.
[4] J. F. Counsell, E. B. Lees, J. F. Martin, *J. Chem. Soc. A* 1968, no. 8, 1819–1823.
[5] N. S. Berman, Doctoral Thesis (1962), Univ. of Texas, University Microfilms, Ann Arbor, Michigan; *Chem. Eng. Prop.* **59** (1963) no. 1, 85.
[6] R. J. L. Andou, J. E. Connett, J. F. Counsell, E. B. Lees, J. F. Martin, *J. Chem. Soc. A* 1971, no. 4, 661–664.
[7] D. Ambrose, C. H. S. Sprake, *J. Chem. Soc.* 1971, no. 9, 1261–1262.
[8] E. T. Beynon, Jr., J. J. McKetta, *J. Phys. Chem.* **67** (1963) 2761–2765.
[9] E. T. Beynon, University Microfilms, Ann Arbor, Michigan, Order no. 65-8025, 75 pp.; *Diss. Abstr.* **26** (1965) no. 2, 930; *Chem. Abstr.* **63** (1965) no. 12, 16196.
[10] W. B. Kay, W. E. Donham, *Chem. Eng. Sci.* **4** (1955) 1–16.
[11] M. Hirata, *Kagaku To Seibutsu* **2** (1964) 311–316; *Chem. Abstr.* **64** (1966) 12203.
[12] E. P. Ageev, G. M. Pančenkov, V. A. Spivak, *Zh. Fiz. Khim.* **41** (1967) no. 5, 1155–1157.
[13] H. A. Rizk, N. Youssef, *Z. Phys. Chem. Frankfurt/M* 58 (1/4) (1968) 100–113; *Z. Phys. Chem. Leipzig* NF 244 (1970) no. 3/4, 165–172.
[14] G. E. Rajala, J. Crossley, *Can. J. Chem.* **49** (1971) no. 22, 3617–3622.
[15] D. Davis, *Ind. Chem.* **39** (1963) no. 2, 75.
[16] J. J. Kipling, *J. Colloid Sci.* **18** (1963) 502–511.
[17] D. Hessel, G. Geiseler, *Z. Phys. Chem. Leipzig* **229** (1965) no. 314, 199–209.
[18] V. M. Komarov, B. K. Kricevcov, *Zh. Prikl. Khim. Leningrad* **39** (1966) 2604–2606.
[19] A. I. Alzybejewa, W. P. Beloussow, N. W. Owtracht, A. G. Moratschewski, *J. Phys. Chem. (Moscow)* **38** (1964) 1242–1247.
[20] G. Schneider, C. Russo, *Ber. Bunsenges. Phys. Chem.* **70** (1966) no. 9/10, 1008–1014.
[21] H. Arm, *Helv. Chim. Acta* **45** (1962) 1803–1806.
[22] A. G. Moračevskij, N. A. Smirnova, R. V. Lyzlova, *Zh. Prikl. Khim. Leningrad* **38** (1965) 1212–1267.
[23] A. G. Moračevskij, Z. P. Popovic, *Zh. Prikl. Khim. Leningrad* **38** (1965) 2129–2131.
[24] *Handbook of Chemistry and Physics,* 55th ed., CRC Press, Inc., Cleveland, Ohio, 1974/75.
[25] Hoechst AG/Ruhrchemie AG: *Oxo Synthesis Products (Aldehydes, Alcohols, Acids),* 2nd ed., 1971.
[26] L. H. Horsley, W. S. Tamplin: "Azeotropic Data III," *Adv. Chem. Ser.* **116** (1973).
[27] F. Asinger: *Die Petrochemische Industrie,* Akademie-Verlag, Berlin 1971, p. 302.
[28] B. C. Gates, J. S. Wisnouskas, H. W. Heath, *J. Catal.* **24** (1972) no. 2, 320–327.
[29] H. Pines, W. O. Haag, *J. Am. Chem. Soc.* **82** (1960) 2471.
[30] H. Knözinger, R. Köhne, *J. Catal.* **5** (1966) 264–270.
[31] P. B. Weisz, *Erdöl Kohle Erdgas Petrochem.* **18** (1965) no. 7, 525–527.
[32] L. Bexten, J. Weber in: *Methodicum Chimicum,* vol. **5** (1975) p. 277.
[33] E. Buckley, J. D. Cox, *Trans. Faraday Soc.* **63** (1967) 895.
[34] A. Amariglio, X. Duval, *Bull. Soc. Chim. Fr.* 1966, 2100–2106.
[35] K. J. Miller, J. L. Wu, *J. Catal.* **27** (1972) no. 1, 60–63.
[36] U. R. Rao, R. Kumar, N. R. Kuloor, *Ind. Eng. Chem. Process Des. Dev.* **8** (1969) 9.
[37] *Hydrocarbon Process. Pet. Refin.* **44** (1965) no. 11, 234.

[38] B. Cornils, E. Zilly in: *Methodicum Chimicum,* vol. **5** (1975) p. 84.
[39] H. Koch, W. Haaf, *Justus Liebigs Ann. Chem.* **618** (1958) 251–266.
[40] Ruhrchemie AG, DE-OS 1793369, 1968 (J. Falbe, B. Cornils, J. Meis, H. Tummes, J. Weber).
[41] L. M. Kozlova, I. A. Romaan, *Izv. Akad. Nauk Latv. SSR Ser. Khim.* 1968, no. 6, 727–730.
[42] *Methodicum Chimicum,* vol. **5** (1975) p. 637.
[43] J. Falbe: *Carbon Monoxide in Organic Synthesis,* Springer Verlag, Berlin-Heidelberg-New York 1970.
[44] UCC, DE-AS 2062703, 1970. Johnson Matthey Ltd., DE-AS 1939322, 1969.
[45] RP-Industries, DE 2627354, 1976. Stamicarbon, EP 40891, 1980.
[46] *Ullmann,* 3rd ed., supplement vol., p. 87.
[47] W. Reppe, H. Vetter, *Justus Liebigs Ann. Chem.* **582** (1953) 133.
[48] N. V. Kutepow, H. Kindler, *Angew. Chem.* **72** (1960) 802.
[49] H. Kindler, K. Eisfeld, *Ind. Chim. Belge* **32** (1967) 650.
[50] *Ullmann,* 3rd ed., vol. **5,** p. 134.
[51] BASF, DE-AS 1114796, 1960; DE-AS 1114797, 1960.
[52] *Hydrocarbon Process.* **50** (1971) no. 11, 138.
[53] Kyowa Hakko Kogyo, *Hydrocarbon Process.* **48** (1969) no. 11, 160.
[54] W. Swodenk, *Chem. Ing. Tech.* **55** (1983) 683.
[55] *Winnacker-Küchler,* 3rd ed., vol. **3**, pp. 349, 386.
[56] H. Friz, *Chem. Ing. Tech.* **40** (1968) 999.
[57] H. Bahrmann, W. Lipps, B. Cornils, *Chem. Ztg.* **106** (1982) 249.
[58] H. Kropf, *Chem. Ing. Tech.* **38** (1966) 837.
[59] ICI, US 2815391, 1953.
[60] M. Sittig: *Chemicals from $C_4$-Hydrocarbons,* Noyes Development Corp., Park Ridge 1966, pp. 73–79.
[61] ICI, US 3164641, 1959.
[62] Montecatini, US 3164641, 1959.
[63] Gulf Research & Development Co., US 3006970, 1958.
[64] Jap. Oil, DE-OS 1944392, 1968 (T. Horie).
[65] CWH AG, DE 2629769, 1976 (F. Obenaus et al.); Hydrocarbon Process. 61 (1982) no. 3, 121–123. *Hydrocarbon Process.* **60** (1981) no. 8, 101–106.
[66] *Annual Book of ASTM Standards,* Section 6, vol. 06.03: "Paint-Fatty Oils and Acids, Solvents, Miscellaneous; Aromatic Hydrocarbons; Naval Stores," American Society for Testing and Materials, Philadelphia 1984.
[67] *DIN-Taschenbuch 30,* Beuth-Vertrieb GmbH, Berlin-Köln-Frankfurt 1971.
[68] *British Standard Yearbook,* British Standards Institution, Inc., Royal Charter, London 1971.
[69] Hoechst AG: *"Lösemittel Hoechst,"* VOC L, Frankfurt 1984. Hoechst AG: *"Hoechst Solvents, Manual for Laboratory and Industry,"* 5th ed., Frankfurt 1975.
[70] G. Sorbe, *GIT-Fachz. Lab.* **16** (1972) no. 5, 598.
[71] K. Nabert, G. Schön: *Sicherheitstechnische Kennzahlen brennbarer Gase und Dämpfe,* 2nd ed., Deutscher Eichverlag, Berlin 1963.
[72] T. Hammerich, H. Schildwächter, *Erdöl Kohle Erdgas Petrochem.* **18** (1965) no. 12, 972–976.
[73] *Chem. Eng. Int. Ed.* **79** (1972) 32–36.
[74] R. J. Lewis, R. L. Tatken (eds.): *Registry of toxic effects of chemical substances,* National Institute for Occupational Safety and Health *(RTECS ONLINE data base)* DIMDI, 10. 01. 84.
[75] V. K. Rowe, S. B. McCollister in G. D. Clayton, F. E. Clayton (eds.): *Patty's Industrial Hygiene and Toxicology,* 3rd ed., vol. **2 C**, Wiley Interscience, New York 1982, pp. 4571-4588.

[76] E. Browning: *Toxicity and Metabolism of Industrial Solvents,* Elsevier, Amsterdam-London-New York 1965, pp. 342–355.
[77] American Conference of Governmental Industrial Hygienists: *Documentation of Threshold Limit Values for substances in workroom air,* Cincinnati, Ohio 1980, vol. **4,** pp. 54, 55, 235.
[78] A. I. Cederbaum, G. Cohen, *Biochem. Biophys. Res. Commun.* **97** (1980) no. 2, 730–736.

# 2-Butanone

WILHELM NEIER, Deutsche Texaco AG, Moers, Federal Republic of Germany
GUENTER STREHLKE, Deutsche Texaco AG, Moers, Federal Republic of Germany

1. Introduction . . . . . . . . . . . . . 971
2. Physical Properties . . . . . . . . 972
3. Chemical Properties . . . . . . . 973
4. Production . . . . . . . . . . . . . 975
4.1. Catalytic Dehydrogenation of sec-Butyl Alcohol (SBA) in the Gaseous Phase . . . . . . . . . . . 976
4.2. Liquid-Phase Oxidation of n-Butane . . . . . . . . . . . . . . . . 977
4.3. Direct Oxidation of n-Butenes (Hoechst-Wacker Process) . . . 978
5. Quality, Storage, Transportation . . . . . . . . . . . 978
6. Uses . . . . . . . . . . . . . . . . . . 979
7. Economic Aspects . . . . . . . . 980
8. Toxicology . . . . . . . . . . . . . 980
9. References . . . . . . . . . . . . . 981

# 1. Introduction

2-Butanone [78-93-3], methyl ethyl ketone, MEK, is the second link in the homologous series of aliphatic ketones and, next to acetone, the most important commercially produced ketone. 2-Butanone is produced primarily by dehydrogenation of 2-butanol, analogous to the production of acetone by dehydrogenation of gaseous isopropyl alcohol on copper, zinc, or bronze catalysts at 400–550 °C. At 80–95% sec-butyl alcohol conversion, MEK selectivity is > 95%. In some cases MEK can be produced in the same facilities as acetone. Butenes (dehydration) and higher ketones (autocondensation) are byproducts. In 1983, 700 000 t of MEK were produced worldwide.

2-Butanone is produced by ARCO (US), AKZO (NL), Biochimica (ES), BP (GB), Carbochlor (Argent.), Celanese (US, CA, Mexico), Esso (GB, US), Maruzen (JP), PCUK (FR), Shell (FR, NL, US, CA), Texaco (FRG), Tonen (JP), UCC (US), Sasol (ZA). Interest in MEK as a solvent for paints and adhesives has been growing in recent years. In general, MEK is considered to be a competitor for ethyl acetate, especially as a low-boiling solvent. It has broad application as a solvent for nitrocellulose, cellulose acetate – butyrate, ethylcellulose, acrylic resins, vinylacetates, and vinylchloride – vinylacetate copolymer (based on synthetic surface-coating preparation). It is favored as a

**Table 1.** Physical data of 2-butanone

| | |
|---|---|
| $M_r$ | 72.11 |
| $mp$, °C | – 86.9 |
| $bp$, °C | 79.6 |
| Relative density, | |
| $\quad d_4^{20}$ | 0.8045 |
| $\quad d_{20}^{20}$ | 0.80615 |
| Refractive index, $n_D^{20}$ | 1.3788 |
| Evaporation number (ether=1) | 2.6 |
| Critical temperature, °C | 262.45 |
| Critical pressure, MPa | 4.15 |
| Critical density, $\varrho$, g/L | 0.270 |
| Dynamic viscosity at 20 °C, mPa · s | 0.323 |
| Surface tension at 20 °C, mN/m | 24.6 |
| Molar heat $c_p$ at 23.8 °C, J mol$^{-1}$ K$^{-1}$ | 160.8 |
| Heat of fusion, J/mol | 7456 |
| Heat of vaporization at 79.6 °C, J/mol | 31.2 |
| Combustion enthalpy at constant pressure, 25 °C, kJ/mol | 2444.3 |
| Thermal conductivity, W m$^{-1}$ K$^{-1}$ | |
| $\quad$ at 0 °C | 0.150 |
| $\quad$ 20 °C | 0.145 |
| $\quad$ 50 °C | 0.137 |
| Solubility at 20 °C | |
| $\quad$ 2-butanone in water, mass fraction, % | 27.5 |
| $\quad$ water in 2-butanone, mass fraction, % | 12.5 |
| Flash point (DIN 51755), °C | –1 |
| Explosion limits in air at 20 °C, 101.3 KPa, | |
| $\quad$ lower, volume fraction, % | 1.8 |
| $\quad$ upper, volume fraction, % | 11.5 |
| Ignition temperature, °C | 505 |
| Ignition class (VED) | G 1 |
| Explosion class (VED) | 1 |
| Electric conductivity at 20 °C, Ohm$^{-1}$ cm$^{-1}$ | $5 \times 10^{-8}$ |
| Dipole moment, Debye | 3.18 |
| Dielectric constant of the liquid at 20 °C | 15.45 |

lacquer solvent because of its low viscosity, high solids concentration, and great diluent tolerance.

Moreover, MEK can be used as an activator for oxidative reactions, as a selective extractant, as a special solvent for dewaxing mineral oil fractions, and as a chemical intermediate.

## 2. Physical Properties

Physical data of 2-butanone are shown in Table 1.

2-Butanone, methyl ethyl ketone, MEK, $CH_3COCH_2CH_3$, is a relatively mobile, colorless liquid. Its typical odor resembles that of acetone. The compound is only partially water-miscible, whereas it is completely miscible with most organic solvents. 2-

**Table 2.** Binary azeotropic mixtures containing MEK

| Second component | Mass fraction of MEK | bp at 1013 hPa, °C |
|---|---|---|
| Water | 88.7 | 73.4 |
| Benzene | 37.5 | 78.4 |
| n-Hexane | 29.5 | 64.3 |
| n-Heptane | 73.0 | 77.0 |
| Cyclohexane | 40 | 72 |
| 1,3-Cyclohexadiene | 40 | 73 |
| Methanol | 30.0 | 63.5 |
| Ethanol | 60.9 | 74.0 |
| Isopropyl alcohol | 68.0 | 77.5 |
| tert-Butyl alcohol | 73 | 77.5 |
| Ethyl acetate | 18.0 | 77.0 |
| Methyl propionate | 52 | 79.25 |
| Propyl formate | 55 | 79.45 |
| Chloroform | 96.0 | 79.65 |
| Carbon tetrachloride | 71.0 | 73.8 |
| Carbon disulfide | 15.3 | 45.85 |
| Propyl mercaptan | 75 | 55.5 |
| Thiophene | 55.0 | 76.0 |
| Ethyl sulfide | 20 | 77.5 |

Butanone forms binary and ternary azeotropic mixtures in combination with water and several other organic solvents (see Table 2).

2-Butanone does not form a binary azeotropic mixture with toluene, m-xylene, n-butanol, isobutanol, sec-butyl alcohol, 4-methyl-2-pentanol, allyl alcohol, acetic acid methyl ester, acetic acid isopropyl ester, acetic acid n-butyl ester, acetic acid isobutyl ester, 1,1-dichloroethane, isobutyl chloride, propyl bromide, ethyl iodide, trichloroethylene, dichlorobromomethane, isobutyl bromide, formic acid, and acetic acid.

2-Butanone forms ternary azeotropes with water/benzene and water/carbon tetrachloride [3].

# 3. Chemical Properties

Under normal conditions and in the absence of atmospheric oxygen MEK is stable. Care must be taken after prolonged storage because peroxides may form in the presence of oxygen [4].

2-Butanone is unsaponifiable and, unlike esters, does not form corrosive products upon hydrolysis. It is heat and light stable. It decomposes only after prolonged UV exposure (yielding ethane, methane, carbon monoxide, ethylene, and diacetyl) [5].

Diacetyl [431-03-8] is formed by oxidation with air in the presence of special catalysts [6].

Methyl ethyl ketone peroxide [19393-67-0], a polymerization catalyst, is formed by oxidation with a 30% solution of hydrogen peroxide [7]. Nitric acid and other strong oxidants oxidize MEK to a mixture of formic and propionic acids [7].

sec-Butyl alcohol [78-92-2] is obtained by catalytic reduction with hydrogen [7]. It can also be formed by electrolytic reduction in sodium acetate solution or by reduction with ammonium amalgam or lithium aluminum hydride. 3,4-Dimethyl-3,4-hexanediol is obtained by electrolytic reduction in an acidic medium or by reduction with magnesium amalgam [7].

Methyl ethyl ketone forms addition products with hydrogen cyanide as well as with sodium and potassium hydrogen sulfites. In an alkaline medium MEK condenses with aldehydes to form higher unsaturated ketones. Condensation with formaldehyde to form methyl isopropenyl ketone [563-80-4], an intermediate for further syntheses, is of particular interest. During base-catalyzed autocondensation in the liquid phase and during gase-phase condensation on alkalinized copper catalysts, the carbonyl group reacts with the methyl group, whereas during acid-catalyzed condensation the methylene group in α-position to the carbonyl group is attacked [8].

$$2\ CH_3COCH_2CH_3 \xrightarrow{Base} CH_3CH_2-\underset{\underset{CH_3}{|}}{\overset{\overset{OH}{|}}{C}}-CH_2COCH_2CH_3$$

$$2\ CH_3COCH_2CH_3 \xrightarrow{Acid} CH_3CH_2-\underset{\underset{CH_3}{|}}{C}=\overset{\overset{CH_3}{|}}{C}-CO-CH_3 + H_2O$$

Methyl ethyl ketone and citral [5392-40-5] condense to form methylpseudoionone that can be cyclized to methylionone, a compound used for producing synthetic violet perfume.

During condensation with low-molecular aldehydes (during base-catalyzed and acid-catalyzed aldolization) the α-position of the carbonyl group is first occupied [9], [10].

$$CH_3COCH_2CH_3 + RCHO \xrightarrow{H^+} CH_3COC=CH_2 + H_2O$$
$$\qquad\qquad\qquad\qquad\qquad\qquad\quad |$$
$$\qquad\qquad\qquad\qquad\qquad\qquad HCR$$

Thus, the base-catalyzed aldolization with less than amounts of formaldehyde yields 2-methyl butane-1-ol-3-one [9] and exhaustive hydroxymethylation, with reduction of the carbonyl group (crossed Cannizzaro reaction), produces desoxyanhydroeneaheptite [11].

$$CH_3COCH_2CH_3 + 6\ CH_2O + 1/2\ Ca(OH)_2 \longrightarrow$$

$$\text{(cyclic product)} + 1/2\ Ca(HCOO)_2 + H_2O$$

When MEK is reacted with primary and secondary alcohols, higher ketones are obtained. Reaction with sec-butyl alcohol gives ethyl amyl ketone [106-68-3] [12].

Methyl ethyl ketone reacts with polyoxy compounds or epoxides to form cyclic products.

Amyl nitrite [110-46-3] attacks the $CH_2$ group in α-position to the carbonyl group and yields the monooxime of diacetyl.

$$CH_3COCH_2CH_3 + C_5H_{11}ONO \xrightarrow{HCl} CH_3\underset{\underset{NOH}{\|}}{C}OCCH_3$$

The keto group reacts with amino groups with elimination of water. In combination with hydroxylamine [7803-49-8], methyl ethyl ketoxime, an antiskinning agent, is formed.

Condensation of MEK with aliphatic esters and anhydrides gives $\beta$-diketones.

Phenols react with MEK to form oxyphenylene compounds. In combination with phenol, 2,2-hydroxyphenyl butane is obtained, a homolog of hydroxyphenyl propane (Bisphenol A [80-05-7]), an important material for the production of synthetic resins.

Methyl ethyl ketone can be halogenated in the $\alpha$-position. Methyl ethyl ketone reacts with Grignard compounds to form tertiary alcohols. With acetylene in the presence of sodium amide 3-methyl-1-pentyn-3-ol [77-75-8] is formed. N-Methyl-formyl-aminobutane is obtained from MEK plus N-methylformamide. The Reformatzky reaction produces the $\beta$-oxyester from monobromine-substituted esters [12].

# 4. Production

Most MEK (88%) is produced today by dehydrogenation of sec-butyl alcohol (SBA). SBA can easily be produced by hydration of n-butenes (from petrochemically produced $C_4$ raffinates) in a two-step process (catalyst: sulfuric acid), or in a single-step process by direct addition of water, acidic ion-exchange resins being used as a catalyst [13]. The remaining 12% MEK is produced by processes in which liquid butane is catalytically oxidized, giving both acetic acid and MEK [14].

The direct oxidation of n-butenes (Hoechst-Wacker process, Maruzen process, [15], [16]) has not been generally accepted, because of undesired byproducts.

The sec-butylbenzenehydroperoxide route giving phenol and MEK by acid-catalyzed splitting [17] is uneconomical. The autoxidation of liquid sec-butyl alcohol, giving MEK and hydrogen peroxide [18], and the catalytic oxidative hydration of gaseous n-butenes [19] are also uneconomical. The oxidation of n-butenes with ethylbenzenehydroperoxide to form butylene oxides, and subsequent hydration and formation of ketones appears to be moderately attractive [20]. Styrene, n-butanol, and MEK are obtained in coupled production.

## 4.1. Catalytic Dehydrogenation of sec-Butyl Alcohol (SBA) in the Gaseous Phase

The catalytic dehydrogenation of SBA is an endothermic reaction (51 kJ/mol). The equilibrium constant for SBA can be calculated as follows [21]:

$\log K_p = -2.790 \times T^{-1} + 1.51 \log T + 1.865$

($T$ = reaction temperature, in K)

The MEK concentration in the reaction mixture increases with the temperature and reaches its maximum at approx. 350 °C [22].

Copper [23], zinc [24], or bronze [25] are used as catalysts in gas-phase dehydrogenation. The latter two require high dehydrogenation temperatures (400 °C). Dehydration of SBA to butenes takes place as a side reaction on zinc oxide. Platinum on alumina [26], copper or chromium [27], [28] as well as copper, and zinc on alumina [26] are recommended as dehydrogenation catalysts for aqueous SBA.

Commercially used catalysts are reactivated by oxidation, after 3 to 6 months use. They have a life expectancy of several years. Catalyst life and alcohol conversion are impaired by contamination with water, butene oligomers, and di-sec-butyl ether [29].

Deutsche Texaco developed a process in which practically anhydrous sec-butyl alcohol has been produced since 1983 by direct hydration of n-butene, catalyzed by acidic ion-exchange resin (Figure 1); it is then dehydrogenated on a copper-based precipitating catalyst at 240–260 °C under normal pressure [13], [23]. The LHSV (4 L/L of catalyst · h), the conversion (90–95%), and the catalyst life until reactivation becomes necessary (3–4 months) render the process economically attractive.

sec-Butyl alcohol is dehydrogenated in a multi-tube reactor. The reaction heat (51 kJ/mol) is supplied by heat transfer oil. The reaction products leave the reactor as a gas and are split into liquid crude MEK and hydrogen on cooling. The hydrogen is purified by further cooling. The reaction is highly selective. Autocondensation to higher ketones (e.g., 5-methyl-3-heptanone) is much lower in comparison to acetone. In addition, the dehydration to butenes on copper catalysts is for the most part prevented.

During reaction and subsequent treatment, practically no waste disposal problems arise.

Table 3 lists further processes for the production of MEK by gas-phase hydrogenation of sec-butyl alcohol.

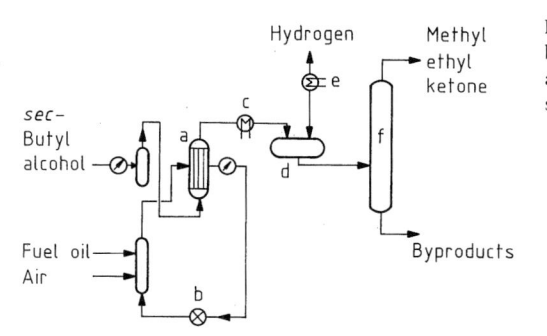

**Figure 1.** Gaseous-phase dehydrogenation of sec-butyl alcohol (Deutsche Texaco AG process)
a) Reactor; b) Oil circulation heating; c) Condenser; d) Separator; e) Refrigerator; f) Distillation

**Table 3.** Further processes for producing methyl ethyl ketone by gase-phase dehydrogenation of sec-butyl alcohol

| Company | Catalyst | $H_2O$ content, vol% | Temperature, °C | Pressure, MPa | Conversion, % | Selectivity, mol% | Yield, mol% | Reference |
|---|---|---|---|---|---|---|---|---|
| Standard Oil | $ZnO/Bi_2O_3$ | 0 | 400 | 0.1–0.3 | | | 80 | [30] |
| Esso Research & Eng. | $ZnO/Na_2CO_3/Al_2O_3$ | 0 | 413 | | 96–97 | 97 | 93–94 | [31] |
| Maruzen Oil | bronze | | 390 | 0.3 | 80 | 99 | 79.2 | [25] |
| Knapsack Griesheim | $CuO/CrO$ | 0 | 270–320 | | 88–93 | 93–96 | | [32] |
| Toyo Rayon | $CuO/NaF/SiO_2$ | | 300 | | 96 | 100 | | [33] |
| Ruhrchemie | 60% Cu, $Cr_2O_3$, MgO, 12% $SiO_2$, 10% $H_2O$ | | 260 | | 90 | 100 | 90 | [27] |
| Veba-Chemie | 22% Cu, 8% $BaCrO_4$, 2% $Cr_2O_3$, 0.5% $Na_2O$, 61% $SiO_2$ | 90.4 | 180 | | 57.3 [a] | 63 [b] | 97.8 [c] | [28] |
| Shell-Chemie | 0.05% $Pt/Al_2O_3$ | 0 | 358 | 0.6 | 92.5–93.5 | 96.3 | 90 | [29] |
| Shell-Chemie | 5% Cu, 5% $Cr/Al_2O_3$ | 0 | 286 | 0.6 | 81–85 | 86–92 | 73–77 | [29] |

[a] Relative to sec-butyl alcohol.
[b] Relative to di-sec-butyl ether.
[c] Total yield.

## 4.2. Liquid-Phase Oxidation of *n*-Butane

2-Butanone is a byproduct in the liquid-phase oxidation of *n*-butane to acetic acid. Autoxidation of *n*-butane takes place in liquid phase according to a radical mechanism yielding MEK as an intermediate and acetic acid as the end product. The continuous plug flow process developed by Union Carbide allows the partial collection of MEK intermediate [34]. MEK and acetic acid (mass ratio 0.15–0.23:1.0) are obtained by non-catalyzed liquid-phase oxidation at 180 °C and 5.3 MPa (53 bar) with remixing. Continuous oxidation under plug flow conditions at 150 °C, 6.5 MPa (65 bar), and a residence time of 2.7 min forms MEK and acetic acid at mass ratios of up to 3:1 [34]. Celanese uses acetic acid as a solvent and cobalt acetate and sodium acetate as homogeneously dissolved catalysts [35]. It is a batch process performed at

160–165 °C and 5.7 MPa (57 bar). MEK and acetic acid are obtained at a mass ratio of 0.4 : 1.0.

## 4.3. Direct Oxidation of *n*-Butenes (Hoechst-Wacker Process)

In the direct oxidation of *n*-butenes according to the Hoechst-Wacker process, oxygen is transferred in a homogeneous phase onto *n*-butenes using a redox salt pair, $PdCl_2$/ 2 CuCl [36], [37]. The salt pair is subsequently reoxidized.

$$n-C_4H_8 + PdCl_2 + H_2O \longrightarrow CH_3COC_2H_5 + Pd + 2\ HCl$$

$$Pd + 2\ CuCl_2 \rightleftharpoons PdCl_2 + 2\ CuCl$$

*n*-Butenes can be converted into the following reaction products (conversions of up to 95 % are attained):

| Reaction products: | Selectivities, mol% |
|---|---|
| MEK | 86 |
| *n*-Butyraldehyde | 4 |
| Chlorinated products | 6 |
| Carbon dioxide | 1 |

The main disadvantages are: formation of chlorinated butanones and *n*-butyraldehyde and corrosion caused by free acids.

The Maruzen process is similar [38], [39]. Oxygen is transferred by an aqueous solution of palladium sulfate and ferric sulfate.

Other processes employing the same oxygen transfer principle were developed by Consortium für Elektrochemie [40] and Eastman Kodak [41]. For further processes, see [42] – [45].

# 5. Quality, Storage, Transportation

Today it is possible to produce high-purity MEK (DIN 53247, ASTM D 740, BS 1940).

The present sales specifications of Deutsche Texaco AG are listed in Table 4.

A stable dilute potassium permanganate solution indicates high purity (permanganate time according to ASTM D 1363).

Storage life of MEK is limited. Carbon steel containers (ST 3529) are suitable for short-term storage and transportation. Stainless steel (316 SS) or containers with a tin lining are recommended for long-term storage. Once autocatalysis has started, it continues even if storage is continued in inert containers. During long-term storage

**Table 4.** Sales specifications for MEK

| | | |
|---|---|---|
| Purity, wt% | GC | min. 99.7 |
| $d_4^{20}$ | (DIN 51757) | 0.804–0.806 |
| $d_{20}^{20}$ | | 0.805–0.807 |
| Boiling range, °C | DIN 51751 | within 0.5 |
| | ASTM D 1078 | incl. 79.6 |
| Acidity (as acetic acid), mg KOH/g | | 0.0001 |
| Water content, wt%, | DIN 51777 | |
| | ASTM D 1364 | max. 0.1 |
| Color (Pt–Co, APHA), | ASTM D 1209 | max. 10 |
| Nonvolatile matter, wt% | | max. 0.002 |

the formation of peroxide must be prevented. Since MEK is somewhat hygroscopic, water is absorbed from the air.

The following regulations for transportation of MEK must be observed [46].

IMDG-Code: D 3308, Kl. 3.2 UN no. 1193, RID, ADR, ADNR: class 3, Rn 301, 2301, and 6301 respectively, no. 1 a, category Kl n; European Council, Yellow Book 78/79: no. 606-002-00-3; European Communities: Guideline/D VgAst, no. 606-002-00-3; UK: Blue Book, Fla.L. IMDG-Code E 3080; USA: CFR 49, 172.101, Fla.L.; IATA RAR: art. no. 726 Fla.L.

# 6. Uses

MEK is an important solvent with properties similar to those of acetone. MEK has the following advantages in comparison to other solvents with comparable rates of evaporation: very high power of dissolution, high ratio of dissolved matter to viscosity, miscibility with a large number of hydrocarbons without impairing the solids content or viscosity, favorable volume/mass ratio due to its low density.

The following natural substances, plastics, and resins can be dissolved in MEK: rosin, ester resins, pentaerythritol ester resins, Congo ester, dammar (dewaxed), nitrocellulose, low-molecular cellulose acetate, cellulose acetobutyrate, cellulose acetostearate, methylcellulose, epoxy resins, nearly all alkyd and phenolic resins, polyvinylacetate, vinylchloride/acetate mixed polymerizates, vinylchloride/vinylidene chloride mixed polymerizates, coumarone–indene resins, sulfonamide resins, cyclohexanone resins, acrylic resins, polystyrene, chlorinated rubber, polyurethane.

Cellulose triacetate, high-molecular cellulose acetate, poly(vinylchloride), poly(vinylbutyral), polysulfide rubber cannot be dissolved in MEK. Shellac is only partially soluble.

Other areas of application are production of synthetic leather, transparent paper, printing inks, aluminum foil lacquers; degreasing of metal surfaces; extraction of fats, oils, waxes, natural resins; dewaxing of mineral oils [47].

**Table 5.** Sales survey (FRG, 1979)

|  | % |
|---|---|
| Paints, lacquers, printing inks, aluminum foil lacquers | 40 |
| Coating and printing of plastics | 20 |
| Chemical industry (incl. sound carrier), pharmaceutical industry | 13 |
| Adhesives | 11 |
| Miscellaneous | 16 |

In contrast to its uses as a solvent, use as a chemical feedstock is of minor importance despite the great number of possible reaction; however, condensation with formaldehyde to obtain methyl isopropenyl ketone, autocondensation to form ethyl amyl ketone, and mixed condensation with acetone to obtain methyl amyl ketone are of interest. Methyl ethyl ketoxime, used as an antiskinning agent in lacquers, is of minor importance. Methyl ethyl ketone peroxide is used as a polymerization initiator for unsaturated polyesters. Diacetyl serves as a butter flavorer. The perfume industry reacts MEK with citral to obtain perfume components such as methylpseudoionone. Since 1962 MEK is permitted as an alcohol denaturant in the Federal Republic of Germany (by decree of *Bundesmonopolverwaltung* in Offenbach).

# 7. Economic Aspects

Table 5 shows the 1979 sales (in %) for the various markets in the Federal Republic of Germany.

The worldwide consumption of MEK in 1979 amounted to 642 000 t (Western Europe 166 000 t, Eastern Europe 18 000 t, North America 315 000 t, Central and South America 34 000 t, Asia, Australia, Oceania 100 000 t, Africa 9000 t).

MEK is also available under the following names:

| German: | Ethylmethylketon, Acetonersatz, β-Ketobutan, 2-Butanon |
| English: | Butanone, 2-butanone, MEK, methylacetone, meetco |
| French: | Méthyléthylcétone, butane-2-one, ethylméthylcétone, MEC |

# 8. Toxicology

The odor threshold of MEK is 10 ppm; both, MAK and TLV are established at 200 ppm. The inhalation of MEK vapor has narcotic effects. The vapor irritates the eyes and the nasal and pharyngeal mucous membranes [46]. Frequent and prolonged contact with liquid MEK causes skin moisture loss and slight irritation [48]. Sensitive persons may develop dermatoses [49]. Liquid MEK temporarily irritates the eye and corneas [48].

MEK is usually absorbed through the respiratory tract. It may also be absorbed by the skin, but the cutaneous $LD_{50}$ in rabbits is above 8 mL/kg [49]. The MEK metabolism has been studied in guinea pigs. MEK is both reduced to 2-butanol and oxidized to 3-hydroxy-2-butanone. However, unlike 2-hexanone and $n$-hexane, which are further oxidized to form neurotoxic 2,5-hexanedione after oxidation of the $\omega$-1 C atom, the hydroxybutanone is not further oxidized, but converted to 2,3-butanediol [50].

Animal tests have shown that the neurotoxic effect of 2-hexanone may be potentiated by simultaneous administration of MEK [51], [52]. The chronic inhalation of 200 ppm (MAK, TLV) does not seem to be harmful [53].

Even workroom concentrations of 500–700 ppm over an extended period of time do not cause permanent damage.

The $LD_{50}$ (oral, rat) is 2500–3400 mg/kg [54], [55].

Toxic concentrations for water organisms [46]: average lethal concentration for fish: 5600 mg/L. Maximum permissible concentrations for *Pseudomonas putida*: 1150 mg/L, for *Scenedesmus quadricanda*: 4300 mg/L, for*Microcystis aeruginosa*: 120 mg/L. For small crabs (*Daphnia magma*) the $LC_0$ is 2500 mg/L, the $LC_{50}$ is 8890 mg/L, and the $LC_{100} > 10000$ mg/L.

# 9. References

General References

*Beilstein* **1**, 668, **1(1)**, 347, **1(2)**, 726, **1(3)**, 2770, **1(4)**, 3243.
F. Asinger: *Die Petrolchemische Industrie*, Akademie Verlag, Berlin 1971, p. 1086.
J. Mellan: *Ketones*, Chemical Publ., New York 1968.
Winnacker-Küchler: *Organische Technologie II*, vol. **6**, Hanser Verlag, München 1982, pp. 77, 81.
*Römpps Chemie Lexikon*, 8th ed., vol. **1**, p. 541.
G. Hommel: *Handbuch der gefährlichen Güter*, Springer Verlag, Berlin 1980.
*Kirk-Othmer*, 3rd ed., vol. **13**, p. 894.
*Ullmann*, 4th ed., vol. **14**, p. 193 ff.

Specific References

[1] VDI-Wärmeatlas 1983, VDI-Verlag Düsseldorf.
[2] A. Weissberger: *Organic Solvents*, vol. **II**, 3rd ed., Wiley-Interscience, New York 1970.
[3] L. H. Horsley: *Azeotropic Data*, Advances in Chemistry Series 116, Am. Chem. Soc., Washington D.C. 1973.
[4] O. L. Mageli, J. R. Kolczynski, Lucidol Division, Wallace & Tiesman Inc., *Encycl. Polym. Sci. Technol.* 1964–1972, vol. **9**, 831; N. A. Milas, *J. Am. Chem. Soc.* **81** (1959) 5824.
[5] W. Davis Jr., *Chem. Rev.* **40** (1947) 240–244.
[6] *Beilstein* (Syst. No. 95) **1**, 769, **1(1)**, 397, **1(2)**, 824, **1(3)**, 3098.
[7] *Beilstein* **1**, 668, **1(1)**, 347, **1(2)**, 726, **1(3)**, 2770, **1(4)**, 3243.

[8]   A. T. Nielsen, *Organic Reactions*, vol. **16**, J. Wiley & Sons, New York 1968, p. 20.
[9]   Rheinpreußen AG, CA 605368, 1955 (W. Grimme, J. Wöllner).
[10]  Rheinpreußen AG, DE 1198814, 1963 (J. Wöllner, F. Engelhardt).
[11]  J. R. Roach, H. Wittcoff, S. E. Miller *J. Am. Chem. Soc.* **69** (1947) 2651.
[12]  Technical Brochure LB 11, Deutsche Shell, Frankfurt (Main) 1976.
[13]  Erdöl Informations-Dienst A. M. Stahmer, vol. 37, no. 28 (1984).
[14]  Union Carbide, US 3196182, 1965 (R. N. Cox).
[15]  Hoechst, US 3215734, 1965 (E. Katzschmann).
[16]  Maruzen Oil, JP 46-2010, 1971.
[17]  Union Carbide, DE-OS 2300903, 1973 (F. P. Wolf).
[18]  N. V. de Bataafsche Petroleum Maatschappij, DE 935503, 1955.
[19]  Stamicarbon, NL 69, 016990, 1971 (J. W. Geus, J. H. Kruit, P. P. Nobel).
[20]  SRI-Report, PEP-Review no. 80-3-1 (1981), Menlo Park, California.
[21]  H. J. Koll et al., *J. Am. Chem. Soc.* **67** (1945) 1084.
[22]  C. Padovani et al., *Riv. Combust.* **5** (1951) 81.
[23]  Rheinpreußen AG, DE 1147933, 1958 (R. Langheim, H. Arendsen).
[24]  Esso, US 2885442, 1959 (W. J. G. McCullock, I. Uirshenbaum).
[25]  Maruzen Oil, Technical Brochure, Aug. 1969.
[26]  Shell, DE-OS 2028350, 1970 (B. Stouthamer, A. Kwantes).
[27]  Ruhrchemie AG, DE-OS 2347097, 1973 (W. Rottig, C. O. Frohning, H. Liebern).
[28]  Veba-Chemie, DE-OS 1913311, 1969 (W. Ester, W. Heitmann).
[29]  Shell, DE 2028350, 1970 (B. Stouthamer, A. Kwantes).
[30]  Standard Oil, US 2436970, 1948 (V. F. Mistretta).
[31]  Esso, US 2835706, 1958 (C. E. Cordes).
[32]  Knapsack-Griesheim, DE 1026739, 1958 (W. Opitz, W. Urbanski).
[33]  Toyo Rayon, JP 43-3163, 1968.
[34]  Union Carbide, US 3196182, 1965.
[35]  Celanese, US 2704294, 1955 (C. S. Morgan, N. C. Robertson).
[36]  Consortium f. Elektrochemie, GB 878777, 1961.
[37]  J. Smidt et al., *Erdöl Kohle Erdgas Petrochem.* **16** (1963) 560.
[38]  Maruzen Oil, JP 46-2010, 1971.
[39]  Maruzen Oil, DE 1951759, 1970 (N. Irinchijma, H. Taniguchi).
[40]  Consortium f. Elektrochemie, US 3080425, 1963 (J. Smidt, W. Hafner, R. Jira).
[41]  Eastman Kodak, GB 1099348, 1968 (H. J. Hagemeyer, F. C. Canter).
[42]  Union Carbide, DE-OS 2300903, 1973 (F. P. Wolf).
[43]  Union Carbide, DE-OS 2421168, 1974 (B. J. Argento, E. A. Rick).
[44]  F. Conssenant, US 2829165, 1958.
[45]  C. S. Cronan, *Chem. Eng. Int. Ed.* **67** (1960) Feb. 8, 63.
[46]  G. Hommel: *Handbuch der gefährlichen Güter*, Springer Verlag, Berlin-Heidelberg-New York 1980.
[47]  W. P. Gee et al., *Refine. Nat. Gasoline Manuf.* **15** (1936) no. 6, 205.
[48]  H. F. Smyth Jr., *Am. Ind. Hyg. Ass. Q.* **17** (1956) 129.
[49]  G. D. Clayton, F. E. Clayton (eds.): *Patty's Industrial Hygiene and Toxicology*, 2nd rev. ed., vol. **2**, Wiley Interscience, New York 1962.
[50]  Di Vincenzo et al., *Toxicol. Appl. Pharmacol.* **36** (1976) 511.
[51]  K. Saida et al., *J. Neuropath. Exp. Neurol.* **35** (1976) 207.
[52]  S. N. Ducket, *Experientia* **30** (1974) 1283.

[53] M. T. Okawa, *U.S. Nat. Techn. Inform. Serv. P.B. Rep.* 1973, no. 229 166/46 A, Avail NTIS, Gov. Rep. Announce (US), 74 (10) 51 (1974).
[54] E. T. Kimura, M. E. Donn, W. P. Dodge, *Toxicol. Appl. Pharmacol.* **19** (1971) 699.
[55] N. J. Sax: *Dangerous Properties of Industrial Materials*, 3rd ed., Van Nostrand Reinhold, New York 1969, p. 495.

# Butenes

FRITZ OBENAUS, Hüls AG, Marl, Federal Republic of Germany
WILHELM DROSTE, Hüls AG, Marl, Federal Republic of Germany
JOACHIM NEUMEISTER, Hüls AG, Marl, Federal Republic of Germany

| | | | | | |
|---|---|---|---|---|---|
| 1. | Introduction | 985 | 7. | Analysis | 996 |
| 2. | Physical Properties | 986 | 8. | Storage and Transportation | 996 |
| 3. | Chemical Properties | 986 | 9. | Uses and Economic Data | 997 |
| 4. | Resources and Raw Materials | 990 | | | |
| 5. | Upgrading of Butenes | 991 | 10. | Toxicology | 1001 |
| 6. | Quality Specifications | 995 | 11. | References | 1001 |

## 1. Introduction

Butenes are unsaturated olefinic hydrocarbons, $C_4H_8$, $M_r$ 56.1080. There are four isomers:

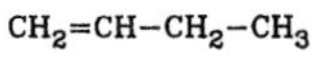

$CH_2=CH-CH_2-CH_3$

**1** 1-Butene

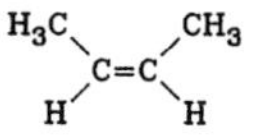

$\begin{array}{c} H_3C \\ \phantom{H_3}C=C \\ H \end{array}\begin{array}{c} CH_3 \\ \\ H \end{array}$

**2** cis-2-Butene

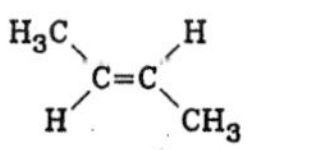

$\begin{array}{c} H_3C \\ \phantom{H_3}C=C \\ H \end{array}\begin{array}{c} H \\ \\ CH_3 \end{array}$

**3** trans-2-Butene

$\begin{array}{c} CH_3 \\ | \\ CH_2=C-CH_3 \end{array}$

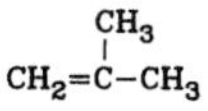

**4** 2-Methylpropene, Isobutene

"Butylenes", the older name for "butenes", is still used today; **4** is frequently referred to as "isobutylene". The designation "n-butenes" refers to mixtures of **1**, **2**, and **3**.

All the butenes, which do not exist as natural products, have been known for more than 100 years, but remained in very limited use and importance. Scarce availability has been for long the reason. However with the growth of cracking processes in crude oil refining and for ethylene production, butenes have been obtained as coproducts in huge quantities. Since then the complicated nature of the raw $C_4$ streams, containing besides the four butene isomers also the two butane isomers and multiple unsaturated $C_4$ hydrocarbons, has been the main barrier for specific chemical use of the butenes.

Until today availability of butene mixtures exceeds by far worldwide the production of the single butene isomers and their derivatives. But during the last decade the sharp price increase of hydrocarbons accelerated the development of economic separation processes, which opened for the butenes access to appropriate upgraded use.

## 2. Physical Properties

Butenes are colorless, flammable gases at room temperature and atmospheric pressure. They are completely miscible with alcohols, ethers, and hydrocarbons [1]. Butenes are only slightly water soluble and water is only slightly butene soluble [2]. Important physical properties are summarized in Table 1. Other thermodynamic properties and transport regulations are reported in [3]–[5].

## 3. Chemical Properties

Butenes behave as typical olefins. The main reactions are acid-catalyzed addition reactions, isomerization, and polymerization. The four isomers, including two isomers with nonterminal double bonds and one branched-chain olefin, show differences in their chemical behavior: while the 2-butenes as the lowest olefins with nonterminal double bonds generally show a minor chemical activity, isobutene, the lowest branched-chain olefin, exhibits higher reactivity, especially in addition and polymerization reactions. This difference in reactivity permits, e.g., the separation of the other isomers through selective reaction of isobutene. The differences in activity (isobutene ≫ 1-butene > 2-butenes) are due to different electron densities, polarities, and steric effects.

**Hydration.** Acid-catalyzed hydration of butenes is one of the commercially most important processes. Both gas- and liquid-phase processes are used. For kinetic and mechanistic studies see references [6], [7] (→ Butanols). Isobutene yields tert-butyl alcohol (TBA). Sulfuric acid (45 wt%) is commonly used as a protonating agent [8]. This process has also been recently carried out in the presence of sulfonated styrene–divinylbenzene ion-exchangers [9]. The latter liquid-phase process has fewer side reactions, such as formation of di- and tri-isobutene, less equipment corrosion, and fewer environmental problems. Hydration is commercially used for separating isobutene from mixed butenes.

Hydration of isobutene-free n-butenes to form sec-butyl alcohol (SBA) is catalyzed under more severe conditions usually by sulfuric acid [10] or more recently by acid ion-exchange resins [11].

**Etherification.** The acid-catalyzed addition of alcohols to butenes yields alkyl butyl ethers (→ Ethers, Aliphatic). Reaction of isobutene with methanol, yielding methyl tert-butyl ether (MTB or MTBE), is of technical importance. Liquid phase and ion exchange

**Table 1.** Physical properties of butenes

|  |  | 1-Butene [106-98-9] | cis-2-Butene [590-18-1] | trans-2-Butene [624-64-6] | Isobutene [115-11-7] |
|---|---|---|---|---|---|
| Melting point (101.3 kPa) | °C | − 185.35 | − 138.92 | − 105.53 | − 140.34 |
| Boiling point (101.3 kPa) | °C | − 6.25 | + 3.72 | + 0.88 | − 6.90 |
| Critical temperature | °C | 146.45 | 162.43 | 155.48 | 144.75 |
| Critical pressure | MPa | 4.02 | 4.20 | 4.10 | 4.00 |
| Critical density | g/cm$^3$ | 0.240 | 0.234 | 0.238 | 0.239 |
| Density of liquid at 25 °C | g/cm$^3$ | 0.5888 | 0.6154 | 0.5984 | 0.5879 |
| Density of gas at 0 °C, 101.3 kPa | kg/m$^3$ | 2.582 | 2.591 | 2.591 | 2.582 |
| Vapor pressure at |  |  |  |  |  |
| 0 °C | kPa | 127.3 | 87.9 | 98.4 | 130.3 |
| 20 °C | kPa | 252.9 | 181.2 | 199.7 | 257.0 |
| 40 °C | kPa | 457.4 | 337.5 | 367.8 | 462.8 |
| 60 °C | kPa | 766.7 | 579.6 | 626.0 | 774.3 |
| 80 °C | kPa | 1207.8 | 931.3 | 999.2 | 1219.0 |
| 100 °C | kPa | 1807.1 | 1416.4 | 1512.1 | 1824.7 |
| Vapor pressure (Antoine equation constants)* |  |  |  |  |  |
| temp. range | °C | − 82 to + 13 | − 73.4 to + 23 | − 76 to + 20 | − 82 to +12 |
| A |  | 6.84290 | 6.86926 | 6.86952 | 6.84134 |
| B |  | 926.10 | 960.10 | 960.80 | 923.20 |
| C |  | 240.00 | 237.00 | 240.00 | 240.00 |
| Heat of vaporization at saturation pressure |  |  |  |  |  |
| at 25 °C | J/g | 358.7 | 394.5 | 380.3 | 366.9 |
| at bp | J/g | 390.6 | 416.2 | 405.6 | 394.2 |
| Isobaric specific heat at 25 °C |  |  |  |  |  |
| gas in ideal state | J kg$^{-1}$ K$^{-1}$ | 1528 | 1408 | 1566 | 1589 |
| liquid at 101.3 kPa | J kg$^{-1}$ K$^{-1}$ | 2299 | 2250 | 2276 | 2336 |
| Enthalpy of formation $\Delta H_f^0$ at 25 °C, 101.3 kPa | kJ/mol | − 0.04 | − 6.91 | − 11.1 | − 16.9 |
| Free enthalpy of formation $\Delta G_f^0$ at 25 °C, 101.3 kPa | kJ/mol | 71.38 | 65.98 | 63.10 | 58.11 |
| Heat of combustion [to H$_2$O (liquid) and CO$_2$ (gas)] at constant pressure and 25 °C | kJ/mol | − 2719.1 | − 2712.3 | − 2708.1 | − 2702.3 |
| Ignition temperature (DIN 51794) | °C | 384 | 325 | 325 | 465 |
| Flammability limits in air at 20 °C, 101.3 kPa |  |  |  |  |  |
| lower | vol% | 1.6 | 1.7 | 1.7 | 1.8 |
| higher | vol% | 9.3 | 9.7 | 9.7 | 8.8 |

* $\log_{10} p = A - B/(t+C)$, where $p$ is in mm Hg and $t$ in °C; to convert mm Hg to kPa, divide by 7.528.

resins as catalysts are commonly used [12]. The etherification of *n*-butenes requires more severe conditions and is of no commercial importance.

**Halogenation.** 1-Butene reacts with halogens at room temperature to give 1,2-dihalogenbutane and reaction with 2-butenes gives 2,3-dihalogenbutane. Allylic substitution occurs at higher temperatures (> 200 °C). This reaction is of no practical importance. Isobutene readily reacts with chlorine at low temperatures forming methallyl chloride ClCH$_2$C(CH$_3$)=CH$_2$ [13].

**Hydroformylation.** Hydroformylation of butenes in the presence of cobalt or rhodium catalysts gives valeric aldehydes and the corresponding amyl alcohols. $n$-Pentanol and 2-methylbutanol are formed from $n$-butenes, while isobutene gives only 3-methylbutanol [14].

**Hydrocarboxylation.** The catalyzed reaction of butenes with carbon monoxide and water yields carboxylic acids. Particularly isobutene readily forms pivalic acid $(CH_3)_3CCOOH$ in the presence of strong acid (Koch reaction [15]).

**Isomerization.** With increasing temperatures the following reactions take place:

$$cis\text{-2-butene} \rightleftharpoons trans\text{-2-butene} \quad (1)$$
$$1\text{-butene} \rightleftharpoons 2\text{-butenes} \quad (2)$$
$$n\text{-butenes} \rightleftharpoons \text{isobutene} \quad (3)$$

While *cis/trans*-isomerization (Eq. 1) occurs at ambient temperatures in the presence of catalysts, carbon rearrangement (Eq. 3) requires temperatures of about 450 °C. A large number of catalysts, e.g., Lewis acids, Brønsted acids, metal oxides, and zeolites, are effective. A review of catalysts and equilibrium constants as a function of temperature is given in [16].

**Polymerization, Oligomerization.** The two types of reactions depend on the catalyst system used:

a) Coordination catalysts (Ziegler-Natta)
b) Liquid or solid Brønsted and Lewis acids

For a review see [17].

Only 1-butene yields high-molecular material such as isotactic or atactic poly-1-butene using Ziegler-Natta catalysts. Isotactic polymers [18] and copolymers with ethylene or propene are produced industrially.

Acidic substances induce poly- or oligomerization by forming carbenium ions [19]. Molecular mass is dependent on many reaction variables, especially temperature: increasing temperature generally results in lower molecular masses.

Polymerization of highly pure isobutene to polyisobutene occurs in an inert solvent at temperatures between $-10$ and $-100$ °C [20]. Copolymerization with 1–3% isoprene under similar conditions yields butyl rubber. Polymerization of isobutene in butene mixtures in the presence of $AlCl_3$ between $-10$ and $+80$ °C yields polybutenes with a few percent $n$-butene and isobutane as chain terminators. The molecular masses lie between 300 and 2500. About 80–95% of the isobutene is usually converted to polymer [21].

Oligomerization of isobutene to dimers and trimers is done by extracting isobutene from mixed butenes using 65–70 wt% sulfuric acid and subsequent heating to 100 °C (cold acid process). An alternative process uses acid ion-exchange resins instead of sulfuric acid (Bayer). The resulting isobutene oligomers, mainly 2,2,4-trimethylpentenes, contain isomeric *n*-butene cooligomers, which increase with higher isobutene conversions. A good part of the accompanying 1-butene is isomerized to 2-butenes [22].

*n*-Butenes are commercially oligomerized by various homogeneous or heterogeneous catalysts. The Dimersol X Process (IFP), which is commercially applied in Japan (Nissan Oil), uses homogeneous Ziegler-Natta catalysts and forms, in addition to higher oligomers mainly 3-methylheptenes [23]. The Hüls/UOP Octol Process is an example of a heterogeneous catalyzed process which oligomerizes *n*-butenes by different mechanisms depending on the catalyst. The dimers are either mainly 3,4-dimethylhexenes or 3-methylheptenes containing considerable amounts of *n*-octenes and 3,4-dimethylhexenes. A commercial plant is on stream in the Federal Republic of Germany (Hüls).

Highly branched oligomers for fuel are produced by cooligomerizing complete butene streams alone or in the presence of propene to polymer gasoline (Cat Poly; UOP Catalytic Condensation Process) [24]. The process uses phosphoric acid on kieselguhr.

**Dehydrogenation.** Dehydrogenation of *n*-butenes is important in the USA for the production of 1,3-butadiene. The reaction is carried out discontinuously at 600–700 °C over a chromium/alumina catalyst (Catadiene process, Houdry [25]). Catalytic oxidative dehydrogenation processes in the presence of air and steam running continuously at lower temperatures were later developed (Oxo-D process, Petrotex [26]).

**Oxidation.** Air oxidation of isobutene [27] or *tert*-butyl alcohol [28] over complex mixtures of transition metal oxides gives methacrolein, which is oxidized in a second reactor to methacrylic acid. The selectivities are between 70–80% at 80% conversion. Two plants in Japan are in operation (Asahi Glass, Mitsubishi Rayon).

Oxidation of isobutene with ammonia and oxygen to produce methacrylonitrile (ammoxidation) is an extension of SOHIO's propene-based acrylonitrile technology.

*n*-Butenes and mixed butenes can be oxidized over $V_2O_5/P_2O_5$-catalysts to maleic anhydride with selectivities of about 50–60 mol%. The temperatures are between 350 and 450 °C. Any isobutene present in the feed is rapidly oxidized to CO, $CO_2$, and water.

Three plants, one in the Federal Republic of Germany (Bayer [29]) and the others in Japan (Mitsubishi Chemical [30], Nichiyu Chemical) are presently on stream.

Catalytic gas-phase oxidation of *n*-butenes in the presence of steam yields acetic acid [31]. Higher yields can be achieved by liquid-phase oxidation of *sec*-butyl acetate, which is formed in a separate reaction of *n*-butenes with acetic acid [32].

**Disproportionation (Metathesis).** The Phillips Triolefin Process, which was developed for the production of ethylene and *n*-butenes from propylene, can be reversed to react 2-butenes with ethylene to yield propene. The most common catalyst systems are tungsten, molybdenum, or rhenium complexes; for a review see [33].

**Prins Reaction.** The acid-catalyzed reaction of isobutene with formaldehyde is the most important synthetic route for isoprene production. Several plants using two-step processes (IFP, Bayer, Kuraray) via intermediate 4,4-dimethylmetadioxane formation are operating in USSR, Japan, and Romania [34].

**Alkylation.** Friedel-Crafts alkylation of aromatics, primarily phenol, *p*-cresol, and catechol, with isobutene in the presence of acid catalysts yields *tert*-butylaromatics. The formation of the thermodynamically favored para-substituted products can be minimized by varying the reaction time, amount of catalyst, and temperature [35].

Alkylation for motor fuel involves reacting mixed butenes with excess isobutane in the liquid phase over sulfuric or hydrofluoric acid catalyst [36]. The main product, aside from numerous other isomers, is 2,2,4-trimethylpentane (isooctane).

# 4. Resources and Raw Materials

Butenes are rarely made purposely. Almost all butenes are mainly byproducts of gasoline and ethylene manufacture. The operation of these processes is governed by the demand for the major products, while the quantity and composition of the butene streams are left to vary.

Petroleum refining is the largest butene source in the world. Most butenes are generated in catalytic crackers, which convert vacuum gas oils into gasoline and medium distillates. Thermal refining processes (visbreaking, thermal cracking, coking) also byproduce butenes, but their volume is inconsequential compared to the quantity from cat crackers.

The butane–butene ("B–B") streams from catalytic cracking may account for 10 to 13 wt% of fresh feed depending on cracking severity and the catalyst. The butene portion may vary somewhat, but generally comprises about 50% of the total weight. This is illustrated together with a breakdown of the butenes in Table 2.

Even though steam crackers account worldwide for only a small percentage of the total butene production, the products from these facilities are more important for chemical use than butenes produced in refineries. The yield of pyrolysis $C_4$ in olefin plants is highly dependent on the type of feedstock. Liquid feedstocks, i.e., naphtha and gas oil, are converted up to 12 wt% to pyrolysis $C_4$, while light hydrocarbon feedstocks, i.e., ethane and propane yield only roughly 2 or 4 wt%, respectively. More severe cracking conditions favor ethylene formation and lower the higher olefin and butene yields [37]. Typical composition of pyrolysis $C_4$ obtained by naphtha steam cracking under medium severity is shown in Table 2.

Catalytic dehydrogenation of *n*-butane is currently only of some importance in Eastern Europe as a source in *n*-butenes, which usually serve as intermediates in the butadiene production. Several years ago, *n*-butenes obtained by dehydrogenation were

**Table 2.** Typical composition (in wt%) of $C_4$ fractions from fluid catalytic cracking (FCC) with zeolite catalyst and from naphtha steam cracking (St.Cr.) under medium severity

| Component | FCC | St.Cr. |
|---|---|---|
| Isobutane | 37 | 2 |
| Isobutene | 15 | 26 |
| 1-Butene | 12 | 14 |
| 1,3-Butadiene | < 0.5 | 43 |
| $n$-Butane | 13 | 6 |
| $trans$-2-Butene | 12 | 5 |
| $cis$-2-Butene | 11 | 4 |

also a major source of butadienes in the USA, but were replaced by the increased amounts of butadiene from steam cracking.

Dehydrogenation of isobutane is used today extensively only in the USSR. Increasing availability of butanes as condensate from natural gas or gas from crude-oil production (field butanes) in countries far from the consumers has generated plans for new dehydrogenation plants primarily for the production and conversion of isobutene to methyl *tert*-butyl ether (MTB) [38]. Three newly developed catalytic processes – Houdry's Catofin [39], UOP's Oleflex [40], and Phillip's STAR [41] – are offered for commercial use.

*tert*-Butyl alcohol (TBA), made by Arco (formerly Oxirane) as a coproduct in its propene oxide plants in the USA and the Netherlands [42], is another source of isobutene.

Ethylene oligomerization processes for production of alpha-olefins with 6–20 carbon atoms yield about 10 wt% 1-butene as a byproduct which can be isolated in high purity. Three plants in the USA (Ethyl Corp., Gulf, Shell) and one in Japan (Mitsubishi Chemical) produce 1-butene via this route. The Alphabutol process of IFP for selective dimerization of ethylene to 1-butene has been recently developed and is now being offered for commercial use [43].

Worldwide butene availability, especially as crude mixtures, is enormous. Estimated world primary production of butenes (1984) is given in Table 3. A breakdown into particular areas and sources shows remarkable differences. More than half of the total is generated in North America, but the production of the more valuable butenes from steam cracking is highest in Western Europe.

# 5. Upgrading of Butenes

Upgrading production of pure butene isomers, their derivatives, or a concentrate of the *n*-butenes, with the above mentioned exceptions, always requires separation from the $C_4$ mixtures. The pyrolysis $C_4$ mixture from steam cracking after removal of butadiene, which is recovered in almost all cases by extractive distillation, forms a butene-rich stream known as raffinate-I (Table 4). This is the preferred petrochemical feed-

**Table 3.** Estimated world primary production of butenes in 1984 ($10^3$ t)

| Area | Source | Sum of butenes | Isobutene | $n$-Butenes |
|---|---|---|---|---|
| North America | St.Cr. | 1130 | 590 | 540 |
| | Ref. | 13600 | 4420 | 9180 |
| | Dehydro | 30 | – | 30 |
| | TBA | 250 | 250 | – |
| | $C_2H_4$ | 40 | – | 40 |
| | Total | 15050 | 5260 | 9790 |
| South America and Caribbean | St.Cr. | 220 | 120 | 100 |
| | Ref. | 2300 | 710 | 1590 |
| | Dehydro | 60 | – | 60 |
| | Total | 2580 | 830 | 1750 |
| Western Europe | St.Cr. | 1750 | 920 | 830 |
| | Ref. | 2350 | 730 | 1620 |
| | TBA | 50 | 50 | – |
| | Total | 4150 | 1700 | 2450 |
| Eastern Europe | St.Cr. | 620 | 330 | 290 |
| | Ref. | 930 | 320 | 610 |
| | Dehydro | 650 | 50 | 600 |
| | Total | 2200 | 700 | 1500 |
| Asia and Australia | St.Cr. | 1060 | 550 | 510 |
| | Ref. | 1680 | 530 | 1150 |
| | Dehydro | 40 | – | 40 |
| | $C_2H_4$ | 10 | – | 10 |
| | Total | 2790 | 1080 | 1710 |
| Africa and Middle East | St.Cr. | 60 | 30 | 30 |
| | Ref. | 1260 | 390 | 870 |
| | Total | 1320 | 420 | 900 |
| World | St.Cr. | 4840 | 2540 | 2300 |
| | Ref. | 22120 | 7100 | 15020 |
| | Dehydro | 780 | 50 | 730 |
| | TBA | 300 | 300 | – |
| | $C_2H_4$ | 50 | – | 50 |
| | Total | 28090 | 9990 | 18100 |

stock. Where demand exceeds supply, the more diluted "B–B streams" from cat crackers are also processed.

Butenes cannot be separated by mere distillation. Their boiling points are too close together, especially those of isobutene and 1-butene, being the isomers which have established substantial outlets for high purity qualities. Extractive distillation is sometimes used to separate butenes from butanes, but cannot separate butene isomers. Molecular sieve adsorption processes offer potential for isolation of pure butenes [44], but are not yet in commercial use.

**Isobutene Separation.** All industrially used butene separation processes are based on the higher chemical reactivity of isobutene, the key substance in the butene mixtures. The isobutene derivatives are easily separated and can be split back during

**Table 4.** Typical composition (in wt%) of the $C_4$ fraction from naphtha steam cracking after (1) butadiene extraction (raffinate-I), (2) conversion of isobutene to MTB * (raffinate-II), (3) isobutane removal and selective hydrogenation * (*n*-butene concentrate)

| Component | Raff.-I | Raff.-II | *n*-Butene conc. |
|---|---|---|---|
| Isobutane | 3 | 6 | < 0.05 |
| Isobutene | 45 | < 0.1 | < 0.1 |
| 1-Butene | 25 | 45 | 48 |
| 1,3-Butadiene | < 0.5 | < 0.8 | < 0.001 |
| *n*-Butane | 11 | 19 | 20 |
| *trans*-2-Butene | 9 | 17 | 18 |
| *cis*-2-Butene | 7 | 13 | 14 |

* Related to the Hüls process (Fig. 1)

a subsequent step to produce pure isobutene or be recovered for particular use. In early stages several similar processes for isobutene separation by reactive extraction with sulfuric acid (45 to 70 wt%) were applied [45]. Isobutene can be isolated in more than 99 wt% purity as *tert*-butanol (TBA) or dimer/trimer mixture depending on the acid concentration and type of acid regeneration. Only the latter is formed by the so-called "cold acid process", in which higher acid concentrations are added. This process is still widely used to reduce isobutene to < 0.2 wt% in the spent $C_4$ mixture.

The difficulty and high cost of the traditional processes to separate isobutene has been the main factor in retarding growth of butenes for specific chemical use. Within the last decade new processes have been introduced that are superior form the view points of economics, environment and energy conservation.

The etherification of isobutene with methanol in the $C_4$ mixtures to form methyl *tert*-butyl ether (MTB) is the key step in these processes. This very selective reaction takes place in the liquid phase between 40 and 100 °C using a macroporous cation exchange resin as a catalyst. Hüls was the first to modify its raffinate-I based MTB production in 1978 to achieve more than 99.9% isobutene conversion [46]. All other $C_4$ components in the co-produced raffinate-II remain unchanged (Table 4). The broad acceptance of MTB in recent years as a valuable octane booster opened an unlimited outlet for all surplus isobutene. This, together with the ease of complete isobutene conversion to MTB, makes the corresponding *n*-butenes available, especially as raffinate-II at low cost for specific use. Even though several processes for splitting MTB by reversing synthesis to high-purity isobutene are available [47], none are in commercial use.

Another new process introduced by Hüls in 1981 selectively hydrates isobutene to *tert*-butanol (TBA) under non-corrosive liquid-phase conditions. A cation exchanger acts as catalyst under conditions similar to the MTB process. Equilibrium and incomplete phase miscibility limits the conversion to TBA to about 90%. Due to the different equilibriums, decomposition of TBA is more favored than that of MTB. The Hüls process decomposes in a second step well below 150 °C heterogeneously catalyzed in liquid phase TBA to high-purity isobutene (Table 5) [9], [48].

***n*-Butene Concentration.** Once isobutene content has been reduced to the raffinate-II level shown in Table 4, recovery of high-purity 1-butene, which is worldwide in growing demand, is possible by fractionation. However, a preceding hydrotreating step is necessary to convert selectively butadiene and traces of acetylenes to achieve the desired product quality (Table 5) [49].

2-Butene is the preferred feedstock for butadiene production by dehydrogenation and is therefore sometimes concentrated. It is of no commercial importance as a high-purity product. Because 1-butene reacts in most cases to the same products as obtained from 2-butene (hydration, esterification, oxidation, oligomerization) the use of 2-butene in mixed *n*-butene concentrates has proved to be a convenient feed for chemical production.

An example of *n*-butene concentrate composition is illustrated in Table 4. Higher concentration can be obtained by removing butanes through extractive distillation or the more recently commercialized molecular sieve adsorption method [50]. Lower concentrations, especially *n*-butane mixtures in the bottom of 1-butene fractionation, are also useful for producing *n*-butene derivatives.

**Complementing Process Integration.** Whether a butene upgrading process is economical or not depends largely on suitable process combinations. A rather simple combination is e.g. the production of MTB in front of an alkylation of FCC $C_4$'s, whereby the contained isobutene is higher upgraded and alkylation of the remaining butenes becomes independant of the availability of additional isobutane. A steady outlet of the products to the gasoline pool is in such cases generally no problem. More sophisticated systems are required to properly upgrade the more highly valued raffinate-I butenes. Continuous processing of the entire $C_4$ stream requires that varying demand for different isomers or derivatives be compensated by flexible production schedules that allow production of alternate products. This leads to highly integrated process systems. Figure 1 shows the Hüls scheme for raffinate-I processing.

Raffinate-I is fed in a variable ratio to the synthesis of TBA and MTB. Primary product of the TBA synthesis is a TBA/water mixture that is either decomposed to give high-purity isobutene or purified to give TBA/water azeotrope and dehydrated TBA. The raffinate-I not required for TBA/isobutene is conjointly used with remaining $C_4$ compounds from TBA synthesis in MTB production. Whereas TBA is more easily converted to pure isobutene, MTB is the more economic gasoline improver and according to its thermodynamically favored synthesis isobutene conversion is accomplished for the total stream, resulting in virtually isobutene-free raffinate-II (Table 4).

Isobutane is separated together with water from the raffinate-II by fractionation. Selective hydrogenation of the bottom product leads to *n*-butene concentrate with a composition shown in Table 4. Fractionation of the *n*-butene concentrate yields polymerization grade 1-butene. The 2-butene/*n*-butane mixture remaining after 1-butene separation is like any surplus of *n*-butene concentrate suitable for the production of di-*n*-butenes through oligomerization.

**Table 5.** Isobutene and 1-butene composition (in wt%) in high-purity commercial products (Hüls)

| Component | Isobutene | 1-Butene |
|---|---|---|
| Isobutene | 99.98 | 0.15 |
| 1-Butene | 0.005 | 99.70 |
| 2-Butenes | 0.01 | < 0.01 |
| Butanes | 0.005 | 0.15 |
| 1,3-Butadiene | < 0.001 | < 0.001 |
| *tert*-Butanol | < 0.0005 | < 0.0001 |
| Water | < 0.003 | < 0.001 |
| Sulfur | < 0.0001 | < 0.0001 |

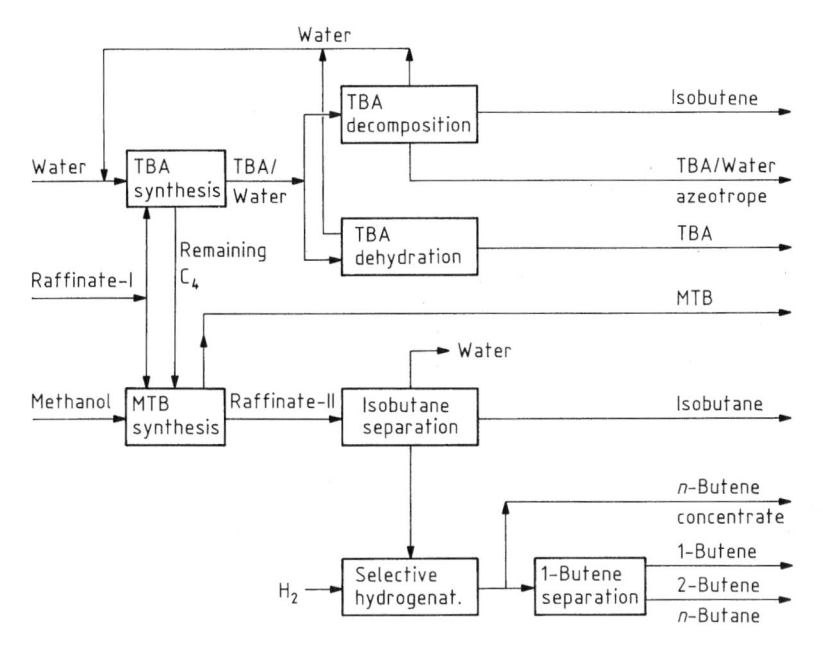

**Figure 1.** Hüls integrated raffinate-I processing

# 6. Quality Specifications

Due to different production processes and applications, no standard quality specifications for the butenes are available. The two isomers which have become commodities as high purity products show in general a purity greater than 99 wt%. Table 5 illustrates the typical isobutene and 1-butene composition as obtained through the Hüls processes.

## 7. Analysis

Analysis of pure butenes and butene mixtures containing *n*-butane, isobutane, 1,3-butadiene, small amounts of $C_3$- and $C_4$-acetylenes, and diolefins are generally performed by gas–liquid chromatography. A routine GLC-analysis using a packed column is described in the ASTM Method D-1717.

An $Al_2O_3$ capillary column (Chrompack) permits the simultaneous determination of all $C_1$- to $C_5$-components combined with improved detection (usually $\leq$ 5 ppm).

The water content of butenes often plays an important role in further processing. It can be determined either by Karl Fischer titration [51] or more precisely by a commercial on-line hygrometer (e.g., Panametrics).

All other impurities are determined by general standard methods.

## 8. Storage and Transportation

Butenes are usually stored and transported as pressurized liquids.

Most countries have specific legislation pertaining to the erection and operation of tank farms for LPG as butenes.

The following regulations pertaining to the transport of dangerous goods must be observed while transporting butenes:

*GGVE/GGVS:*
  class 2, number 3 b

*RID/ADR:*
  class 2, number 3 b

*ADNR:*
  class 2, number 6, category F

*IMDG-Code/GGVSee:*
  class 2, UN-No. 1012 (*n*-butenes)
  class 2, UN-No. 1055 (isobutene)

*IATA-DGR/ICAO-Code:*
  class 2 (3) UN-No. 1012 (*n*-butenes)
  class 2 (3) UN-No. 1055 (isobutene)

The transport containers for butenes (steel cylinders, steel drums, road tankers, pressure gas tank wagons and tank containers) must conform to these regulations in respect to construction materials, design, fittings, and marking. They must be approved for use at not less than 1 MPa (10 bar) test pressure.

# 9. Uses and Economic Data

The estimated world butene consumptions in 1984 are shown according to regions in Table 6. More than half of the butenes produced worldwide are utilized as alkylate and polymer gasoline. One third is used without any conversion as fuel, mainly as fuel gas or blendstock for gasoline. Only about 15% of the butenes are converted to specific chemicals. Even the greater part of the isobutene transformed to pure MTB is ultimately burned as a gasoline component.

Table 6 shows some remarkable regional differences. In North America, Western Europe, Eastern Europe, and Japan, most butenes are converted into higher value gasoline components and chemical products, whereas in industrially less developed regions, butenes are usually directly used without further conversion for their calorific value.

*Alkylate gasoline with butenes.* In North America nearly 80% of butenes are converted into alkylate gasoline, whereas alkylate gasoline plays a lesser role in other regions. However, since more than half of all butenes are produced in North America, almost 50% of the butenes used worldwide are converted into alkylate gasoline, the largest outlet for butenes.

*Polymer Gasoline.* The oligomerization of butenes, in most cases mixed with propene, gives highly branched $C_6$–$C_{12}$-olefins [24]. It is the oldest method for reconverting the low boilers obtained from catalytic cracking into higher molecular mass gasoline components. This was an important process, particularly in the 1940s, for the production of aviation fuel. Today, it has been largely replaced by alkylation, which gives higher octane rating products.

*Straight Fuel.* Butenes are also used directly for such diverse applications as a blendstock in gasoline, LPG, and as plant fuel. The least demand exists in North America, where paraffinic light hydrocarbons are readily available from natural gas liquids and butenes are more highly valued as alkylation feedstock. Consumption is higher in areas as Europe, where LPG production is more or less restricted to refineries.

*Methyl tert-Butyl Ether (MTB).* MTB is almost exclusively admixed to carburetor fuels as an octane booster. It showed the highest growth rate of all butene derivatives during the last decade [52]. The current endeavors toward a further reduction in the lead alkyl content of gasoline, particularly in the USA and Western Europe, further increase the importance of MTB. Since the isobutene in raffinate-I is already largely utilized, isobutene from catalytic cracking or other sources – upstream of alkylation – is increasingly used as feedstock for the production of MTB [53].

*tert-Butanol (TBA).* Large quantities of TBA from isobutane are obtained as a byproduct of propene oxide production by ARCO [42]. It is predominantly used in its impure form as blending component for gasoline [54] or to produce pure isobutene (see Table 3).

Smaller quantities of pure TBA are obtained by direct hydration of isobutene-containing $C_4$ cuts. This process, presently only in use in Japan and the Federal

**Table 6.** Estimated world butene consumptions in 1984 ($10^3$ t)

| Area | Product | Sum of butenes | Isobutene | $n$-Butenes |
|---|---|---|---|---|
| North America | Alkylate gasoline | 11860 | 3690 | 8170 |
| | Polymer gasoline | 390 | 120 | 270 |
| | Straight fuel uses | 1220 | 410 | 810 |
| | MTB | 500 | 500 | – |
| | Di-/triisobutene | 30 | 30 | – |
| | Butyl rubber | 250 | 250 | – |
| | Polybutene | 250 | 250 | – |
| | 1,3-Butadiene | 60 | – | 60 |
| | 1-Butene in polyolef. | 180 | – | 180 |
| | 2-Butanol (SBA) | 250 | – | 250 |
| | Special chemicals | 210 | 100 | 110 |
| | Total | 15200 | 5350 | 9850 |
| South America and Caribbean | Alkylate gasoline | 260 | 90 | 170 |
| | Polymer gasoline | 30 | 10 | 20 |
| | Straight fuel uses | 2100 | 665 | 1435 |
| | MTB | 20 | 20 | – |
| | Polybutene | 5 | 5 | – |
| | 1,3-Butadiene | 60 | – | 60 |
| | 1-Butene in polyolef. | 10 | – | 10 |
| | 2-Butanol (SBA) | 15 | – | 15 |
| | Special chemicals | 30 | 10 | 20 |
| | Total | 2530 | 800 | 1730 |
| Western Europe | Alkylate gasoline | 900 | 280 | 620 |
| | Polymer gasoline | 180 | 60 | 120 |
| | Straight fuel uses | 1710 | 380 | 1330 |
| | MTB | 370 | 370 | – |
| | TBA | 10 | 10 | – |
| | Di-/triisobutene | 120 | 120 | – |
| | Butyl rubber | 190 | 190 | – |
| | Polybutene | 130 | 130 | – |
| | 1-Butene in polyolef. | 50 | – | 50 |
| | 2-Butanol (SBA) | 200 | – | 200 |
| | Special chemicals | 180 | 100 | 80 |
| | Total | 4040 | 1640 | 2400 |
| Eastern Europe | Alkylate gasoline | 280 | 100 | 180 |
| | Straight fuel uses | 910 | 190 | 720 |
| | MTB | 90 | 90 | – |
| | Butyl rubber | 60 | 60 | – |
| | Isoprene | 250 | 250 | – |
| | 1,3-Butadiene | 600 | – | 600 |
| | Special chemicals | 10 | 10 | – |
| | Total | 2200 | 700 | 1500 |
| Asia and Australia | Alkylate gasoline | 360 | 140 | 220 |
| | Polymer gasoline | 90 | 30 | 60 |
| | Straight fuel uses | 1960 | 730 | 1230 |
| | MTB | 10 | 10 | – |
| | TBA | 40 | 40 | – |
| | Di-/triisobutene | 20 | 20 | – |
| | Butyl rubber | 50 | 50 | – |
| | Polybutene | 30 | 30 | – |
| | Isoprene | 20 | 20 | – |
| | 1,3-Butadiene | 40 | – | 40 |

**Table 6.** (continued)

| Area | Product | Sum of butenes | Isobutene | $n$-Butenes |
|---|---|---|---|---|
| | 1-Butene in polyolef. | 30 | – | 30 |
| | 2-Butanol (SBA) | 100 | – | 100 |
| | Special chemicals | 40 | 10 | 30 |
| | Total | 2790 | 1080 | 1710 |
| Africa and Middle East | Alkylate gasoline | 130 | 35 | 95 |
| | Straight fuel uses | 1190 | 385 | 805 |
| | 1-Butene in polyolef. | 10 | – | 10 |
| | Total | 1330 | 420 | 910 |
| World | Alkylate gasoline | 13790 | 4335 | 9455 |
| | Polymer gasoline | 690 | 220 | 470 |
| | Straight fuel uses | 9090 | 2760 | 6330 |
| | MTB | 990 | 990 | – |
| | TBA | 50 | 50 | – |
| | Di-/triisobutene | 170 | 170 | – |
| | Butyl rubber | 550 | 550 | – |
| | Polybutene | 415 | 415 | – |
| | Isoprene | 270 | 270 | – |
| | 1,3-Butadiene | 760 | – | 760 |
| | 1-Butene in polyolef. | 280 | – | 280 |
| | 2-Butanol (SBA) | 565 | – | 565 |
| | Special chemicals | 470 | 230 | 240 |
| | Total | 28090 | 9990 | 1 8100 |

Republic of Germany, serves also for the isolation of pure isobutene via TBA. That portion is not included in the figures in Table 6. This pure TBA is in Japan predominantly used for the production of methyl methacrylate [28]. Other countries use TBA as a solvent for the production of *tert*-butyl peroxides or as a stabilizer for chlorinated hydrocarbons among others.

*Di- and Triisobutenes.* The decreased use of sulfuric acid extraction for isobutene separation has resulted in di- and triisobutene production for chemical use only.

Diisobutene is mainly used for the production of isononanol and octylphenol, which are required as intermediates for plasticizers and detergents, respectively.

*Butyl Rubber.* More than half of the 1 million tons of high-purity isobutene produced in 1984 was converted to butyl rubber. Butyl rubber, which is made by copolymerizing high-purity isobutene with small amounts of isoprene, is used mainly as a tire inner liner because of its excellent air impermeability.

Also included are the smaller amounts of so-called polyisobutene, which does not contain any comonomers and is used in films, among others [20].

*Polybutene.* Polybutene is a polymerization product of isobutene obtained directly from raffinate-I or the catalytic cracker $C_4$ cut and contains therefore about 2% $n$-butene. The largest portion is in the 1000 molecular mass range and is used as a precursor for lube oil additives [55]. Polymers with higher molecular masses are used as caulks and sealants.

*Isoprene.* Isoprene can be directly extracted from the steam cracker $C_5$ cut or synthesized by reacting formaldehyde with pure isobutene. Even though this process is only used in Eastern Europe, especially in the USSR, and in one plant in Japan, isoprene production via isobutene is the second largest outlet of pure isobutene.

*Butadiene.* Although butadiene production via *n*-butene dehydrogenation is the single largest *n*-butene consumer for chemical use, less than 15 % of worldwide butadiene production in 1984 was based on dehydrogenation of *n*-butenes. Even in the United States, where dehydrogenation was once dominating, it is presently only used to augment the supply when butadiene from extractive distillation of steam cracker $C_4$ cut is in short supply. Butadiene production by two-step dehydrogenation via *n*-butane is used in countries with an ample butane supply, such as the USSR and Mexico.

*1-Butene in Polyolefins.* High-purity 1-butene demands have multiplied during the last 10 years and are expected to continue to grow at this rate [56]. Most is used as a comonomer for linear low-density polyethylene (LLDPE), which contains up to 10 %. However, 1-butene is also used as a comonomer for modifying high-density polyethylene (HDPE), which contains up to 4 %.

Shell in the United States uses about 20000 t of high-purity 1-butene for the production of poly-1-butene per year.

*sec-Butanol (SBA).* SBA made by hydration is after butadiene quantitatively the second most important product from *n*-butenes. Raffinate-II is a typical feed material. Almost all the SBA is further converted by catalytic dehydrogenation into methyl ethyl ketone (MEK), which is mainly used as a solvent and diluent in the surface-coating and plastic industries.

*Special Chemicals.* Butenes are used to produce various special chemicals.

Isobutene derivatives include, among others, phenols, such as 4-*tert*-butylphenol, 2,6-di-*tert*-butyl-4-methylphenol (BHT), 2- or 3-*tert*-butyl- 4-hydroxyanisole (BHA), most of which are used as stabilizers. Other isobutene derivatives worth mentioning are *tert*-butylamine (intermediate for lubricating oil additives), *tert*-butylmercaptan (gas odorant), pivalic acid, 3-methylbutyric acid (isovaleric acid), 3-methyl-1-butanol, methallyl chloride, and triisobutylaluminum.

With a few exceptions, such as *n*-butene 1,2-oxide (stabilizer for chlorinated hydrocarbons) *n*-butene derivatives are generally derived from mixtures (raffinate II or *n*-butene concentrate). These include oligomers (intermediates for plasticizers, detergents, and lube oil additives), maleic anhydride, the oxo derivatives *n*-valeric acid, and 2-methylbutyric acid, and the corresponding alcohols 1-pentanol and 2-methyl-1-butanol, which are primarily used as solvents.

# 10. Toxicology

A specific toxic action of the butenes is not known. They have, however, a narcotic action at higher concentrations causing vomiting, giddiness, and intoxication. Butene gas in high concentrations, as built up by rapid evaporation, displaces atmospheric oxygen so that there is a danger of suffocation. The symptoms, however, disappear rapidly in fresh air.

Skin contact with the liquid causes frost bite due to the high negative heat of vaporization. Toxicological data are available only for isobutene.

Acute toxicity of isobutene:

$LC_{50}$, inhalation, rats: 620 g/m$^3$     4 h
$LC_{50}$, inhalation, mice: 415 g/m$^3$     2 h [57]

Chronic toxicity of isobutene:

Isobutene shows little chronic toxicity because rats inhaling high doses (up to 8000 ppm) for 90 days show no evidence of any toxicological effects. Evidence of mutagenic or cell transforming activity (Ames Test, Mouse Lymphoma Assay, Cell Transformation Study) is also absent [58].

# 11. References

[1] K. Smeykal, H. Lütgert, *Chem. Tech. Leipzig* **14** (1962) 202–207; *Chem. Abstr.* **57** (1962) 10103 h.
[2] W. F. Hoot, A. Azarnoosh, J. J. McKetta, *Pet. Refin.* **36** (1957) no. 5, 255–256.
[3] ASTM Committee D-2: Physical Constants of Hydrocarbons $C_1$ to $C_{10}$, ASTM Data Series Publication DS 4 A, American Society for Testing and Materials, Philadelphia 1971.
[4] Technical Data Handbook-Petroleum Refining, American Petroleum Institute, New York 1970.
[5] R. W. Gallant: *Physical Properties of Hydrocarbons*, vol. **1**, Gulf Publishing Comp., Houston 1968, pp. 11, 157.
[6] F. Axel, H. J. Warnecke, H. Langemann, *Monatsh. Chem.* **113** (1982) 541–546; *Chem. Abstr.* **97** (1982) 91333 c.
[7] M. Katsuno, *Kogyo Kagaku Zasshi* **44** (1941) 275–282; *Chem. Abstr.* **44** (1950) 7752 b.
[8] H. Kröper, K. Schlömer, H. M. Weitz, *Hydrocarbon Process.* **48** (1969) no. 9, 195–198.
[9] *Chem. Eng. N.Y.* **90** (1983) no. 25, 60.
[10] *Hydrocarbon Process.* **54** (1975) no. 11, 118.
[11] Deutsche Texaco AG, EP 0051164, 1982 (W. Neier, W. Webers, R. Ruckhaber, G. Osterburg, W. Ostwald).
[12] F. Obenaus, W. Droste, *Erdöl Kohle Erdgas Petrochem.* **33** (1980) 271–275.
[13] W. Reeve, D. H. Chambers, C. S. Prickett, *J. Am. Chem. Soc.* **74** (1952) 5369.
[14] B. Cornils: "Hydroformylation. Oxo Synthesis, Roelen Reaction" in J. Falbe (ed.): *New Synthesis with Carbon Monoxide*, Springer Verlag, Berlin-Heidelberg-New York 1980, p. 149.

[15] H. Bahrmann: "Koch Reactions" in J. Falbe (ed.): *New Synthesis with Carbon Monoxide,* Springer Verlag, Berlin-Heidelberg-New York 1980, p. 372.
[16] V. R. Choudhary, *Chem. Ind. Dev.* **8** (1974) no. 7, 32–41; *Chem. Abstr.* **82** (1975) 5131 w.
[17] D. Stadermann, H. Hartung, G. Heublein, (Part I) *Chem. Tech. Leipzig* **35** (1983) no. 4, 169–174; *Chem. Abstr.* **98** (1983) 180218 g; (Part II) *Chem. Tech.Leipzig* **35** (1983) no. 6, 290–296; *Chem. Abstr.* **99** (1983) 38819 f.
[18] P. W. de Leeuw, C. R. Lindegreen, R. F. Schimber, *Chem. Eng. Prog.* **76** (1980) no. 1, 57–63.
[19] A. Onopchenko, B. L. Cupples, A. N. Kresge, *Ind. Eng. Chem. Prod. Res. Dev.* **22** (1983) no. 2, 182–191.
[20] H. Güterbock: *Polyisobutylen und Isobutylen-Mischpolymerisate,* Springer Verlag, Berlin 1959, p. 62.
[21] D. Mark, A. R. Orr, *Pet. Refin.* **35** (1956) no. 12, 185–186.
[22] G. Scharfe, *Hydrocarbon Process.* **52** (1973) no. 4, 171–173.
[23] J. F. Boucher, G. Follain, D. Fulop, J. Gaillard, *Oil Gas J.* **80** (1982) no. 13, 84–86.
[24] P. C. Welnert, P. Egloff, *Pet. Process.* **3** (1948) 585–586, 589–590, 592–593.
[25] R. G. Craig, J. M. Duffalo, *Chem. Eng. Prog.* **75** (1979) no. 2, 62–65.
[26] L. M. Welch, R. J. Croce, H. F. Christmann, *Hydrocarbon Process.* **57** (1978) no. 11, 131–136.
[27] D. J. Hucknell: *Selective Oxidation of Hydrocarbons,* Academic Press, London-New York 1974, pp. 104–111.
[28] T. Hasuike, H. Matsuzawa, *Hydrocarbon Process.* **58** (1979) no. 2, 105–107.
[29] G. Lenz, *Chem. Anlagen + Verfahren* **76** (1976) no. 7, 27–29;*Chem. Abstr.* **86** (1977) 5860 r.
[30] *Eur. Chem. News* **25** (1974) no. 630, 30.
[31] Chemische Werke Hüls, DE 1279011, 1968 (R. Brockhaus) (= US 3431297).
[32] Chemische Werke Hüls, DE 3019932, 1981 (R. Brockhaus, H. Rademacher, B. Scholz)
[33] R. Streck, *Chem. Ztg.* **10** (1975) 397–413; *Chem. Abstr.* **84** (1976) 73132 p.
[34] M. Hellin, M. Davidson, *Inf. Chim.* **206** (1980) 163–184; *Chem. Abstr.* **94** (1981) 139101 t.
[35] S. H. Patinkin, B. S. Friedmann: "Alkylation of Aromatics with Alkenes and Alkanes" in G. A. Olah (ed.): *Friedel-Crafts and Related Reactions,* vol. II, Interscience Publishers, Inc., New York 1964, Part 1, p. 1.
[36] J. Weitkamp, S. Maixner, *Erdöl Kohle Erdgas Petrochem.* **36** (1983) 523–529.
[37] B. Schleppinghoff, *Erdöl Kohle Erdgas Petrochem.* **27** (1974) 240–245.
[38] *Eur. Chem. News* **42** (1984) no. 1128, 18.
[39] R. G. Craig, E. A. White, *Hydrocarbon Process.* **59** (1980) no. 12, 111–113.
[40] V. V. Vera, F. Imai, *Hydrocarbon Process.* **61** (1982) no. 4, 171–174.
[41] F. M. Brinkmeyer, D. F. Rohr, M. E. Olbrich, L. E. Drehmann, *Oil Gas J.* **81** (1983) no. 13, 75–78.
[42] *Eur. Chem. News* **37** (1981) no. 1004, 23.
[43] D. Commereuc, Y. Chauvin, J. Gaillard, J. Léonard et al., *Hydrocarbon Process.* **63** (1984) no. 11, 118–120.
[44] A. J. de Rosset, J. W. Priegnitz, D. J. Korous, D. B. Broughton, Presentation to the ACS 175th National Meeting at Anaheim, March 12–17, 1978.M. S. Adler, D. R. Johnson, Presentation to the AIChE 85th National Meeting at Philadelphia, June 4–8, 1978.
[45] Chemische Werke Hüls, DE 2362115, 1975 (W. Zerrweck) (= US 3979474).
[46] *Oil Gas J.* **77** (1979) no. 1, 76–77.
[47] V. Fattore, M. M. Mauri, G. Oriani, G. Paret, *Hydrocarbon Process.* **60** (1981) no. 8, 101–106.
[48] Chemische Werke Hüls, EP-A 0082937, 1983 (F. Obenaus, B. Greving, H. Balke, B. Scholz) (= US 4423271)

[49] Chemische Werke Hüls, DE-OS 3143647, 1983 (F. Obenaus, F. Nierlich, O. Reitemeyer, B. Scholz) (= US 4517395).
[50] *Oil Gas J.* **82** (1984) no. 31, 109.
[51] E. Scholz, *Fresenius Z. Anal. Chem.* **303** (1980) 203–207.
[52] I. Steel, Unsaturated C4's – A West European Producer's View, Presentation to the ECMRA Meeting at Oslo, Oct. 11–13, 1982.
[53] R. N. Davis, R. S. Andre, *Proc. Refin. Dep. Am. Pet. Inst.* **59** (1980) 59–64; *Chem. Abstr.* **93** (1980) 242.2345.
[54] E. G. Guetens, Jr., J. M. DeJovine, G. J. Yogis, *Hydrocarbon Process.* **61** (1982) no. 5, 113–117.
[55] Chem. Systems, Synthetic Lubricants, Report No. 77–6 (1978) 173.
[56] Buten-1 LLDPE, *Chem. Week* **129** (1981) Dec. 16, 30–34.
[57] B. B. Shugaer, *Arch. Environ. Health* **18** (1969) no. 6, 878–882; *Chem. Abstr.* **71** (1970) 47775 n.
[58] Hazleton Laboratories Europe Ltd., Report no. 2916–13/11 (November 1982); unpublished.

# Butyrolactone

Hans Jochen Mercker, BASF Aktiengesellschaft, Ludwigshafen, Federal Republic of Germany (Chaps. 2-7)

Heinz Kieczka, BASF Aktiengesellschaft, Ludwigshafen, Federal Republic of Germany (Chap. 8)

1. Introduction . . . . . . . . . . . . . 1005
2. Physical Properties . . . . . . . . 1005
3. Chemical Properties . . . . . . . . 1006
4. Production . . . . . . . . . . . . . 1007
5. Quality Specifications and Analysis . . . . . . . . . . . . . . . 1008
6. Storage and Transportation . . 1008
7. Uses . . . . . . . . . . . . . . . . . . 1009
8. Toxicology and Occupational Health . . . . . . . . . . . . . . . . 1009
9. References . . . . . . . . . . . . . 1010

## 1. Introduction

$\gamma$-Butyrolactone or $\gamma$-hydroxybutyric acid lactone [96-48-0], $C_4H_6O_2$, $M_r$ 86.09, became industrially available in the 1940s as a result of the work of W. Reppe and his colleagues at BASF.

To date, the Reppe process (see Chap. 4) has remained the most important process for the manufacture of butyrolactone. Butyrolactone is important as an intermediate in the manufacture of pyrrolidone derivatives and as a solvent for polymers and agrochemicals.

Methods of synthesis of historical and preparative interest are summarized in *Beilstein* [1] and in *Houben-Weyl* [2]. In the mid-1960s butyrolactone was discovered in nature, namely, as a normal metabolic product in animals [3].

## 2. Physical Properties

Butyrolactone, *mp* −43.53 °C, *bp* 206 °C (at 101.3 kPa), is a colorless, slightly hygroscopic liquid having a faint odor. Butyrolactone is miscible in all proportions with water, alcohols, esters, ethers, ketones, and aromatic hydrocarbons. It has limited miscibility with linear and cyclic aliphatic hydrocarbons. It is an excellent solvent for numerous polymers and a selective solvent for lower hydrocarbons. The main physical properties of $\gamma$-butyrolactone are listed below.

| Heat of solution in water | 2500 J/mol |
|---|---|
| Critical data: $p_{crit.}$ | 3.35 MPa (33.5 bar) |
| $T_{crit.}$ | 436.5 °C |
| Specific heat $c_p$ (l) | 1680 J kg$^{-1}$ K$^{-1}$ at 25 °C |
| | 1850 J kg$^{-1}$ K$^{-1}$ at 100 °C |
| | 2200 J kg$^{-1}$ K$^{-1}$ at 200 °C |
| Specific heat $c_p$ (g) | 1275 J kg$^{-1}$ K$^{-1}$ at 100 °C |
| | 1575 J kg$^{-1}$ K$^{-1}$ at 200 °C |
| | 1820 J kg$^{-1}$ K$^{-1}$ at 300 °C |
| Heat of vaporization | 535 kJ/kg at 206 °C |

Vapor pressure as the function of temperature:

| $t$, °C | 20 | 60 | 100 | 140 | 180 | 206 |
|---|---|---|---|---|---|---|
| $p$, kPa | 0.15 | 1.07 | 5.25 | 19.05 | 55.3 | 101.3 |

| Evaporation number (DIN 53170) | 216 |
|---|---|
| Flash point (DIN 51758) | 104 °C |
| Ignition temperature (DIN 51794) | 455 °C |
| Heat of combustion at constant volume or constant pressure | 234 kJ/g |
| Density | $d_4^0$ 1.15; $d_4^{20}$ 1.13; $d_4^{40}$ 1.11 |
| Refractive index $n_D^{20}$ | 1.4352 |
| Viscosity (20 °C, DIN 53015) | 1.9 mPa s |
| Surface tension | 44.6 × 10$^{-5}$ N/cm |
| Dielectric constant | 39.1 (20 °C) |
| Thermal conductivity (25–65 °C) | 0.276 J m$^{-1}$ s$^{-1}$ K$^{-1}$ |

# 3. Chemical Properties

In aqueous solution there is an equilibrium between the lactone and free hydroxybutyric acid, which is 100% to the lactone side at 0 °C and about 80% at 100 °C. In the presence of one mole of alkali per mole of lactone, equilibrium is displaced 100% to the hydroxybutyric acid side; therefore, the butyrolactone content can be determined by titration with a base. At pH 7 the compound is extremely stable. Butyrolactone reacts with bases, hydrogen halides, and alcohols (in the presence of acids), undergoing ring cleavage and giving derivatives of γ-hydroxybutyric acid. It reacts with ammonia, amines, carbonyl compounds, and halogens in the α position without ring cleavage. For example, butyrolactone condenses with ethyl acetate [141-78-6] to give α-acetobutyrolactone, an intermediate in the production of vitamin $B_1$:

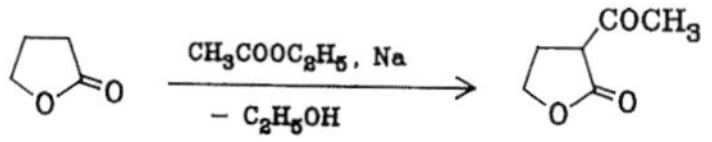

Cleavage of the lactone ring gives derivatives of γ-substituted butyric acid. For example, butyrolactone reacts with sodium sulfide to give the rubber additive thiodibutyric acid [5152-99-8]:

$$2 \;\text{[butyrolactone]} + Na_2S \longrightarrow S(CH_2CH_2CH_2COONa)_2$$

Reaction with phenols gives phenoxybutyric acids [6303-58-8]:

$$\text{[butyrolactone]} + \text{[PhOH]} \longrightarrow \text{Ph–O}(CH_2)_3COOH$$

Replacement of the ring oxygen by nitrogen yields pyrrolidone derivatives, a reaction often used in industry. For example, butyrolactone reacts with methylamine [74-89-5] to give N-methylpyrrolidone [872-50-4]:

$$\text{[butyrolactone]} + CH_3NH_2 \longrightarrow \text{[N-methylpyrrolidone]} + H_2O$$

The chemistry of butyrolactone, and its use in a variety of syntheses, was investigated, above all, by W. REPPE and colleagues [4], [5]. Reactions, generally typical of lactones, are described in *Houben-Weyl* [2].

Homopolymerization of butyrolactone can be effected only under very high pressure (2000 MPa = 20 000 bar) [6].

# 4. Production

**Dehydrogenation of Butane-1,4-diol** [110-63-4] (→ Butanediols, Butenediol, and Butynediol). The most important process for preparing butyrolactone is the endothermic dehydrogenation of butane-1,4-diol in the gas phase.

$$\text{HO–}(CH_2)_4\text{–OH} \longrightarrow \text{[butyrolactone]} + H_2 \quad \Delta H = +61.6 \text{ kJ/mol}$$

Preheated butanediol vapor is introduced into a hot stream of circulating hydrogen and passed at atmospheric pressure over the catalyst, copper on pumice, at 240 °C. The yield of butyrolactone is about 90%. The reaction takes place via γ-hydroxybutyraldehyde [25714-71-0] [7].

The hydrogen split off requires only simple purification before reuse. The crude butyrolactone isolated from the circulating gas stream has a low content of byproducts, including butane-1,4-diol and butyric acid [107-92-6], from which butyrolactone is separated by distillation.

The apparatus can be constructed of steel. However, where parts of the apparatus come into contact with hot crude product containing butyric acid, they must be made of stainless steel. BASF [8] and General Aniline and Film Corp. (GAF) use the Reppe process. Liquid-phase dehydrogenation at ≈ 200 °C is not industrially significant.

**Hydrogenation of Maleic Anhydride** [108-31-6]. In the preparation of butyrolactone by hydrogenating maleic anhydride, molten maleic anhydride is fed into a preheated circulating stream of hydrogen and passed under a pressure of 6–12 MPa (60–120 bar) at 160–280 °C over a nickel-containing catalyst [9].

$$\text{(maleic anhydride)} + 3\,H_2 \longrightarrow \text{(butyrolactone)} + H_2O$$

$$\Delta H = -211 \text{ kJ/mol}$$

The reaction takes place via succinic anhydride [108-30-5] and can, by choice of the conditions, be continued to produce tetrahydrofuran [109-99-9]. The excess hydrogen is washed with water and recycled to the synthesis. Byproducts contained in the butyrolactone are separated out of the circulating gas: propanol [71-23-8], butanol [71-36-3], propionic acid [79-09-4], and butyric acid [107-92-6]. The butyrolactone is separated from these by distillation.

Because of the acids present, both the synthesis apparatus and the distillation apparatus must be made of stainless steel. The Japanese manufacturers Nippon Hydrofuran [10] and Dainippon Hokkaido [11] use this process.

**Manufacturers.** Butyrolactone is manufactured by BASF (Ludwigshafen, Federal Republic of Germany), GAF (Grasselli, N.J., United States), Nippon Hydrofuran (Mizushima, Japan), and Dainippon Hokkaido (Hokkaido, Japan).

**Other Processes.** Processes via tetrahydrofuran [12], olefins [13], butynediol [110-65-6] [14], or butadiene [106-99-0] [9] or by carbonylation [15] are not industrially important.

# 5. Quality Specifications and Analysis

Technical-grade butyrolactone is about 99.7% pure, the remainder being $\approx$ 0.1% of butyric acid, $\approx$ 0.1% of butane-1,4-diol, and $\approx$ 0.1% of water.

Butyrolactone is usually analyzed by gas chromatography. Its purity can also be determined by hydrolysis with 1 M NaOH and back-titration. The result obtained includes any butyric acid present. The butyric acid is determined separately by direct titration with alkali. The water content is determined by titration with the Karl Fischer reagent.

# 6. Storage and Transportation

Almost all metallic materials, except zinc, are resistant to butyrolactone [16]; in steel vessels, the lactone becomes somewhat yellow on prolonged storage. Surprisingly, the compound attacks concrete vigorously.

Butyrolactone does not fall under the "VbF West German Order on Flammable Liquids." It is not a dangerous material in the sense of the West German Road (GGVS/ADR), Railway (GGVE/RID), Air (IATA-DGR), Inland Water (ADNR), and Ocean (IMOG) Traffic Regulations; DOT identification NA 9188, DOT hazard classification ORM-E.

# 7. Uses

The main use of butyrolactone is as an intermediate in the synthesis of N-methylpyrrolidone (NMP) [17], pyrrolidone [18], herbicides (e.g., MCPB = $\gamma$-2-methyl-4-chlorophenoxybutyric acid) [19], growth regulators (e.g., $\alpha$-(4-methylbenzylidene)-$\gamma$-butyrolactone [*5418-24-6*] [20]), $\alpha$-acetobutyrolactone (a vitamin $B_1$ intermediate) [21], and the rubber additive thiodibutyric acid [22]. Butyrolactone is used as a solvent for polymers [23] and as a polymerization catalyst [24]; in hairwave compositions [25] and sun lotions [26]; and in pharmaceuticals [27]. It is also used in printing inks, e.g., for ink-jet printing [28], as an extractant in the petroleum industry [29], as a stabilizer for chlorohydrocarbons [30] and phosphorus-based pesticides [31], and as a nematocide [32].

# 8. Toxicology and Occupational Health

$\gamma$-Butyrolactone exhibits no particular acute toxicity. The acute oral $LD_{50}$ in rats is about 1580 mg/kg [33]. Data in the literature vary between 800 and 1600 mg/kg (rat, mouse) and 500 and 700 mg/kg (guinea pig) [34]. The dermal $LD_{50}$ in guinea pigs is given as 5600 mg/kg [34]. $\gamma$-Butyrolactone acts as a narcotic following both oral and intraperitoneal administration. This central nervous effect results from the action of $\gamma$-hydroxybutyric acid, a metabolite of $\gamma$-butyrolactone.

In rabbits, $\gamma$-butyrolactone does not cause skin irritation; however, it does irritate the eye mucosa [33]. The compound can be absorbed through the skin, and the uptake of larger amounts produces a narcotic effect. $\gamma$-Butyrolactone does not cause skin sensitization in guinea pigs.

Inhalation (8 h) of air saturated with $\gamma$-butyrolactone at room temperature produces no symptoms in rats [33].

Administration of $\gamma$-butyrolactone in drinking water for four weeks at a daily dosage of 3 g/kg leads to depression of the central nervous system in rats. Apart from a slight reduction in body weight gain, the animals showed no further health effects [35].

*Carcinogenicity.* The tumor incidence in male mice that from an age of eight weeks were given lifelong dermal treatment three times weekly with a solution of 10% $\gamma$-

butyrolactone in benzene was not higher than that of control animals treated with benzene alone. Also the survival time of the animals was unaffected by the administration of the substance [36]. The twice weekly lifelong dermal treatment of male and female mice from the age of four weeks with a 1% solution of $\gamma$-butyrolactone in acetone does not lead to increased tumor incidence or to reduction of the average survival time in comparison with control animals treated with acetone alone [37].

Neither the lifelong oral administration of 1000 mg of $\gamma$-butyrolactone per kg feed, nor the lifelong administration of 2 mg $\gamma$-butyrolactone per 0.1 mL of water via stomach tube to male and female mice from an age of four weeks lead to an increased tumor incidence or to a reduction in the median survival time in comparison with untreated control animals [37].

On the basis of these results and in agreement with the literature [34], no carcinogenic potential is attributed to $\gamma$-butyrolactone.

## 9. References

[1] *Beilstein*, **7**, 234, **7 (1)**, 130, **7 (2)**, 286, **17/5 (3/4)**, 4159.
[2] *Houben-Weyl*, **6/2**, 561–852.
[3] G. Quadbeck, *Dtsch. Med. Wochenschr.* **90** (1965) 403.
[4] W. Reppe, *Justus Liebigs Ann. Chem.* **596** (1955) 163–224; *Chem. Ing. Tech.* **22** (1950) 361–373; *Neue Entwicklungen auf dem Gebiet der Chemie des Acetylens und Kohlenoxids*, Springer Verlag, Berlin-Göttingen-Heidelberg 1949, pp. 45–47.
[5] S. Kano, S. Shibuya, T. Ebata, *Heterocycles* **14** (1980) 661–711.
[6] F. Korte, W. Glet, *J. Polym. Sci. Part B* **4** (1966) 685–689.
[7] S. Oka, *Bull. Chem. Soc. Jpn.* **35** (1962) 986–989.
[8] Fiat Final Report no. 945 (24. Oct. 1946).
[9] Mitsubishi Petrochemical, DE-OS 1593073, 1966; DE-OS 1901870, 1969 (T. Asano, J. Kanetaka).
[10] J. Kanetaka, T. Asano, S. Masumune, *Ind. Eng. Chem.* **62** (1970) 24–32.T. Yoshimura, *Chem. Eng. N.Y.* **76** (11. Aug. 1969) 70–72; *Chem. Week* **104** (1969) 63–72.S. Minoda, M. Miyajima, *Hydrocarbon Process.* **49** (1970) no. 11, 176–178.
[11] S. Ushio, *Chem. Eng. N.Y.* **78** (27. Dec. 1971) 24–26.
[12] Quaker Oats, US 3074964, 1961 (A. P. Dunlop, E. Sherman). H. Hara, JP-Kokai 7887347, 1978.
[13] Toa Nenryo Kogyo K.K., JP-Kokai 75154237, 1975 (Y. Okumura, Y. Nagashima). Standard Oil, US 4238357, 1980 (Th. Haase, F. A. Pesa).
[14] Y. Shvo, Y. Blum, *J. Organomet. Chem.* **238** (1982) C 79–C 81.
[15] Texaco, US 3061614, 1958 (W. M. Sweeney, J. A. Patterson).
[16] Dechema, Werkstoff Tabelle 323 (1955).
[17] I.G. Farbenind., DE-AS 694043, 1938 (C. Schuster, A. Seib).
[18] BASF, GB 1312463, 1973 (W. Himmele, E. Hofmann, H. Hoffmann).
[19] *Chem. Age London* **71** (1954) 1239.
[20] S. Huneck, K. Schreiber, DD 112884, 1975.
[21] W. Reppe, *Justus Liebigs Ann. Chem.* **596** (1955) 163–164.

[22] BASF, DE 917247, 1952 (W. Reppe, H. Friedrich).
[23] Merkblatt Butyrolacton, BASF July 1964.
[24] Mitsubishi Petrochemical, JP 8129881, 1981.
[25] Shiseido, DE-OS 2421248, 1974 (K. Ono, K. Torii, T. Ozawa).
[26] BASF, EP-A 44970, 1982 (F. Thoemel, W. Hoffmann, D. Degner).
[27] W. Klunk, A. C. McKeon, *Science* **217** (1982) 1040–1042. W. Klunk, D. F. Covey, *Mol. Pharmacol.* **22** (1982) 431–437. Grissmann Chemicals, GB 2028653, 1980 (E. Diethalm).
[28] Whittaker Corp., DE-OS 2936241, 1980 (I. R. Mansukhani).
[29] Esso Research, US 3013962, 1958 (Ch. N. Kimberlin Jr., W. J. Mattox). P. Alessi, I. Kikic, *Int. Congr. Scand. Chem. Eng. Proc.* 1980, 366–373.
[30] Du Pont, US 2958712, 1958 (F. W. Starks).
[31] Dow Chemical, JP-Kokai 7986618, 1979.
[32] GAF, US 3086907, 1958 (F. A. Hessel).
[33] BASF, unpublished results (1960).
[34] *IARC Monographs on the Evaluation of the Carcinogenic Risk of Chemicals to Man:* γ-Butyrolacton, vol. **11**, Lyon 1976, 231–239.
[35] M. C. Nowycky, R. H. Roth: "Chronic γ-Butyro lactone (GBL) Treatment: A Potential Model of Dopamine Hypoactivity," *Naunyn Schmiedebergs Arch. Pharmacol.* **309** (1979) 247–254.
[36] B. L. Van Duuren et al.: "Carcinogenicity of Epoxides, Lactones, and Peroxy Compounds," *J. Nat. Canc. Inst.* **31** (1963) 41–55.
[37] G. Rudali et al.: "A propos de l'action cancérigène de la γ-butyrolactone chez les souris," *C.R. Acad. Sc. Paris, Série D,* **282** (1976) 799–802.

# Caprolactam

JOSEF RITZ, BASF Aktiengesellschaft, Ludwigshafen, Federal Republic of Germany (Chaps. 2–8)
HUGO FUCHS, BASF Aktiengesellschaft, Ludwigshafen, Federal Republic of Germany (Chaps. 2–8)
HEINZ KIECZKA, BASF Aktiengesellschaft, Ludwigshafen, Federal Republic of Germany (Chap. 9)
WILLIAM C. MORAN, BASF Corporation, Freeport, Texas, United States (Chaps. 2–8)

| | | | | | |
|---|---|---|---|---|---|
| 1. | History | 1013 | 4.2. | Photooximation | 1026 |
| 2. | Physical Properties | 1014 | 4.3. | Production from Toluene | 1028 |
| 3. | Chemical Properties | 1015 | 4.4. | Recovery from Nylon-6 Waste | 1031 |
| 4. | Production | 1015 | 4.5. | Former Commercial Routes | 1031 |
| 4.1. | Processes via Cyclohexanone | 1017 | 4.6. | New Routes in Development | 1032 |
| 4.1.1. | Cyclohexanone Oxime | 1017 | 4.7. | Environmental Protection | 1036 |
| 4.1.2. | Beckmann Rearrangement to Caprolactam | 1018 | 5. | Quality Specifications | 1037 |
| 4.1.3. | Improvement of the Oximation Process | 1020 | 6. | Analysis | 1037 |
| 4.1.4. | Gas-Phase or Catalytic Rearrangement of Cyclohexanone Oxime | 1024 | 7. | Storage and Transportation | 1039 |
| | | | 8. | Economic Aspects | 1039 |
| | | | 9. | Toxicology | 1040 |
| 4.1.5. | Modified Oximation and Beckmann Rearrangement | 1025 | 10. | References | 1041 |

# 1. History

ε-Caprolactam (2-oxohexamethylenimine, hexahydro-1$H$-azepin-2-one) has been known since the 19th century. S. GABRIEL and T. A. MAAS synthesized caprolactam in 1899 by cyclization of ε-aminocaproic acid. O. WALLACH synthesized caprolactam by Beckmann rearrangement of cyclohexanone oxime. Commercial interest increased in 1938, when P. SCHLACK of IG Farbenindustrie produced the first spinnable polymer by polycondensation of caprolactam. Since then, caprolactam has gained importance. Large-scale industrial production has increased rapidly. In 1986, world production may reach 3 million metric tons.

## 2. Physical Properties

Caprolactam [105-60-2], $C_6H_{11}ON$, is a white, hygroscopic, crystalline solid with a characteristic odor.

Its basic properties are summarized below:

| | |
|---|---|
| $M_r$ | 113.16 |
| mp, °C | 69.2 |
| bp, °C | |
|   at 101.3 kPa | 268.5 |
|   at 6.7 kPa | 174 |
|   at 1.3 kPa | 134 |
|   at 0.4 kPa | 111 |
| Density, kg/L | |
|   at 120 °C | 0.9829 |
|   at 100 °C | 0.9983 |
|   at 80 °C | 1.0135 |
| Viscosity, mPa · s | |
|   at 120 °C | 2.93 |
|   at 100 °C | 4.87 |
|   at 80 °C | 8.82 |
| Specific heat, kJ kg$^{-1}$ K$^{-1}$ | |
|   at 150 °C | 2.345 |
|   at 80 °C | 2.135 |
| Heat of fusion, kJ/kg | 123.5 |
| Heat of polycondensation, kJ/kg | 138 |
| Heat of vaporization, kJ/kg | |
|   at 268 °C | 481 |
|   at 168 °C | 574 |
|   at 105 °C | 628 |
| Vapor pressure, kPa | |
|   at 268 °C | 101.3 |
|   at 168 °C | 5.3 |
|   at 105 °C | 0.25 |
| Flash point, °C | 139.5 |
| Ignition temperature, °C | 375 |
| Lower explosion limit | 1.4 vol% at 135 °C |
| Upper explosion limit | 8.0 vol% at 180.5 °C |
| Thermal conductivity coefficient, kJ m$^{-1}$ h$^{-1}$ K$^{-1}$, at 76–183 °C | 0.5 |
| Coefficient of volume expansion, K$^{-1}$, at 80–90 °C | 0.00104 |

Caprolactam is soluble in polar and aromatic solvents and slightly soluble in high molecular mass aliphatic hydrocarbons. The solubility of caprolactam in various solvents is summarized in Table 1.

**Table 1.** Solubility of caprolactam (wt%)

| Solvent | Solubilities at | | | |
|---|---|---|---|---|
| | 20 | 30 | 40 | 50 °C |
| Water | 82 | 86.5 | 90 | 93.5 |
| Toluene | 26 | 36.5 | 51 | 66.5 |
| Ethyl acetate | 24.2 | 33.3 | 48.5 | 66.2 |
| Methyl ethyl ketone | 34.6 | 45.7 | 59.2 | 72.9 |
| Cyclohexanone | 34.6 | 42.2 | 54.5 | 68.2 |
| Cyclohexane | 2 | 2.5 | 7 | 18.5 |

## 3. Chemical Properties

Polymerization is caprolactam's most important chemical property. The ring is hydrolyzed at 260–270 °C. Linear polymer chains are formed by polycondensation [1]. Caprolactam also reacts directly by polyaddition with the polymer chains. These reactions lead to an equilibrium between the polymer and caprolactam favoring a 90% conversion to polymer. Caprolactam can also be polymerized by anionic polymerization at low moisture contents, preferably less than 100 ppm. A catalyst and cocatalyst system is necessary. The reaction temperature is lower than in hydrolytic polymerization.

Caprolactam can undergo all cyclic amide reactions, such as oxidation, hydrolysis, N-alkylation, phosphogenation, and nitration.

Molten caprolactam absorbs atmospheric oxygen. Oxygen reacts at 75 °C with caprolactam to produce very small amounts of peroxide. Above 100 °C adipic acid imide is formed [2].

This reaction is catalyzed by traces of heavy metals.

Caprolactam is heat stable in the absence of oxygen [3].

Caprolactam is quantitatively converted to ε-aminocaproic acid by hydrolysis with aqueous acids or alkalis [4]. Treatment with gaseous methanol in the presence of a dehydrating catalyst gives N-methyl-ε-caprolactam, a versatile solvent [5]. Caprolactam reacts with phosgene to give chloroformic acid ε-caprolactim ester. Nitration gives nitrocaprolactam, and subsequent reduction amino-caprolactam. L-Lysine is formed from amino-caprolactam by hydrolysis and resolution of the racemate [6]. Reactions with other chemical reagents are of no industrial importance.

## 4. Production

Many methods for caprolactam production have been developed. All commercial processes are based on benzene or toluene from BTX extract streams. Figure 1 summarizes the various processes.

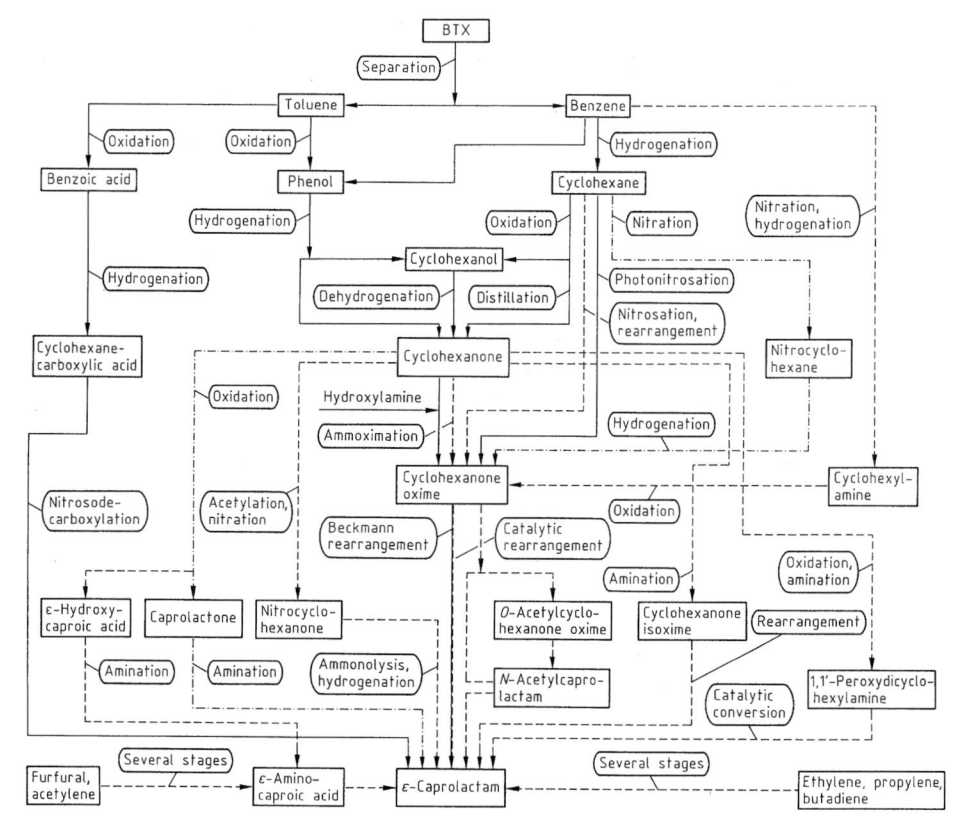

**Figure 1.** Routes to caprolactam
—— routes in commercial use; ——— routes once in commercial use; ······· routes in laboratory development

Large-scale industrial processes are, without exception, multistage processes in which ammonium sulfate and sometimes organic compounds are formed as byproducts.

Cost-efficiency depends a great deal on the byproducts. Large quantities of ammonium sulfate are undesirable in the manufacture of caprolactam. The profitability of a process is dependent on this.

Large-scale industrial processes employ cyclohexanone, cyclohexane, or toluene as starting materials.

About 90% of the caprolactam is produced by using the conventional cyclohexanone process. Cyclohexanone is obtained by catalytic oxidation of cyclohexane with air, or by hydrogenation of phenol and dehydrogenation of the cyclohexanol byproduct. The conversion of cyclohexanone to cyclohexanone oxime followed by Beckmann rearrangement gives caprolactam. The yield from cyclohexanone approaches 98%.

About 10% of caprolactam is produced by photonitrosation of cyclohexane or by nitrosation of cyclohexanecarboxylic acid in the presence of sulfuric acid; in the latter process cyclohexanecarboxylic acid is produced from toluene.

## 4.1. Processes via Cyclohexanone

### 4.1.1. Cyclohexanone Oxime

In caprolactam processes using cyclohexanone, water containing cyclohexanone oxime is formed and isolated as an intermediate.

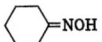

**Physical Properties.** Cyclohexanone oxime is a white crystal with a slightly pungent odor. Typical properties are listed below:

| | |
|---|---|
| $M_r$ | 113.16 |
| *mp*, °C | |
|   water content 0 wt% | 89.4 |
|   water content 3 wt% | 74.6 |
|   water content 5 wt% | 68.4 |
|   water content 7 wt% | 65.7 |
| *bp*, °C | |
|   at 101.3 kPa | 209.8 |
|   at 6.7 kPa | 130 |
|   at 1.3 kPa | 96 |
|   at 0.4 kPa | 74 |
| Density, kg/L | |
|   at 100 °C | 0.969 |
|   at 90 °C | 0.976 |
| Viscosity, mPa · s | |
|   at 100 °C | 4.34 |
|   at 90 °C | 6.11 |
| Specific heat, kJ kg$^{-1}$ K$^{-1}$ at 90–120 °C | 1.766 |
| Heat of fusion, kJ/kg | 98.8 |
| Heat of rearrangement, kJ/kg | 1815 |
| Heat of vaporization, kJ/kg | |
|   at 204 °C | 490.3 |
|   at 168 °C | 499.5 |
|   at 105 °C | 540 |
| Vapor pressure, kPa | |
|   at 204 °C | 101.3 |
|   at 168 °C | 28.0 |
|   at 105 °C | 2.0 |
| Flash point, °C | 100 |
| Ignition temperature, °C | 285 |

Cyclohexanone oxime can be distilled without decomposition only under reduced pressure. For unknown reasons, violent decomposition may occur [7]. Traces of acid can initiate spontaneous rearrangement.

**Chemical Properties.** The Beckmann rearrangement to caprolactam is the most important industrially used reaction. The catalytic reduction of cyclohexanone oxime with hydrogen gives cyclohexylhydroxylamine and cyclohexylamine [7]. Hydrolysis with mineral acids gives cyclohexanone and the corresponding hydroxylamine salt.

**Production.** The reaction of cyclohexanone with hydroxylamine is the best known method for the production of cyclohexanone oxime.

$$\text{C}_6\text{H}_{10}=\text{O} + \text{NH}_2\text{OH} \xrightarrow{-\text{H}_2\text{O}} \text{C}_6\text{H}_{10}=\text{NOH}$$

Hydroxylamine is used in its sulfate or phosphate form.

Some manufacturers produce hydroxylamine by the modified Raschig process using ammonium salts as starting materials. Oximation is then carried out with aqueous hydroxylammonium sulfate solution containing ammonium sulfate. The sulfuric acid liberated is neutralized with ammonia to form ammonium sulfate. The amount of ammonium sulfate formed during oximation is about 2.7 tons per ton of cyclohexanone oxime [8].

The BASF process (Figure 2) for the production of hydroxylammonium sulfate solution generates about 0.7 tons of ammonium sulfate per ton of cyclohexanone oxime during the oximation step. A similar process has been developed by Inventa [9]. BASF and Inventa obtain the hydroxylammonium sulfate solution by hydrogenation of nitric oxide over a platinum catalyst in the presence of dilute sulfuric acid. The hydroxylammonium sulfate solution is reacted with cyclohexanone and ammonia in an oxime reactor.

This reaction is conducted with thorough mixing at 85–90 °C in a weak acidic solution.

Cyclohexanone oxime is obtained as a moist melt. It is separated from the aqueous ammonium sulfate solution in a separator drum. The purified aqueous ammonium sulfate solution is free of substances that would be undesirable in fertilizer- or technical-grade salt uses.

## 4.1.2. Beckmann Rearrangement to Caprolactam

All caprolactam manufacturers use sulfuric acid or oleum as a rearrangement medium. The rearrangement is more complete in concentrated sulfuric acid. Excess sulfur trioxide further increases the speed of the rearrangement [10].

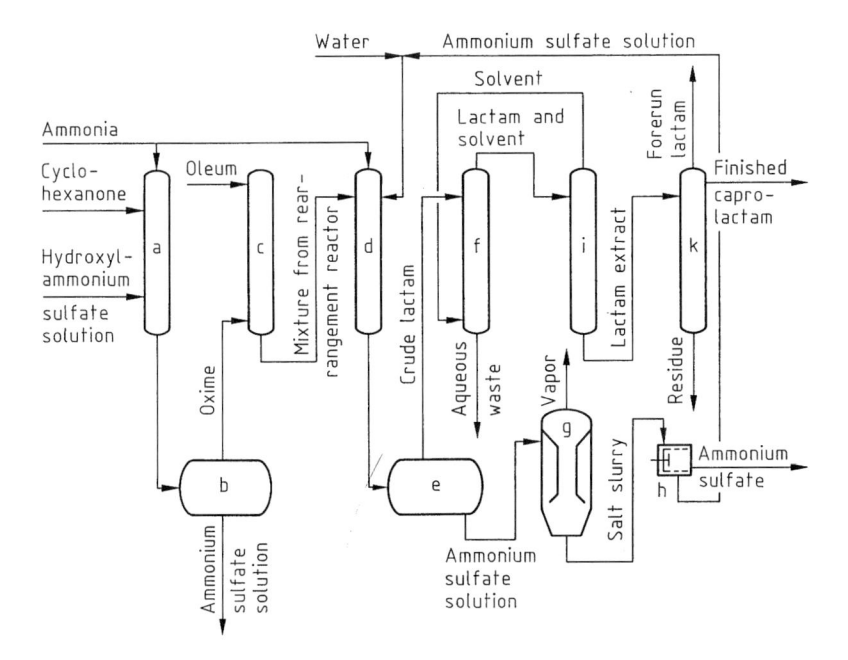

**Figure 2.** BASF caprolactam production
a) Oximation; b) Oxime separation; c) Rearrangement; d) Neutralization; e) Crude-lactam separation; f) Extraction; g) Crystallization; h) Centrifuge; i) Solvent distillation; k) Lactam distillation

The rearrangement, which includes opening of the cyclohexyl ring, is a very rapid, highly exothermic reaction.

Since the Beckmann rearrangement is highly exothermic, molten cyclohexanone oxime and concentrated oleum (27%) are introduced simultaneously into a relatively large amount of already rearranged product.

Molar ratios of cyclohexanone oxime and oleum may range from 1 to 1.05 [11]. Sulfur trioxide binds the water in the moist cyclohexanone oxime to form sulfuric acid and catalyzes the rearrangement. Cooling reduces the heat of reaction. The reaction gives the sulfate of caprolactam in excess sulfuric acid. It is hydrolyzed by neutralization with ammonia or ammonia water. The reactions take place according to the following equations:

The rearrangement mixture is removed from the reactor at the rate it is formed. It is neutralized in ammonia and water to prevent precipitation of solid ammonium sulfate. The reaction product is then separated from the lighter phase above the concentrated ammonium sulfate solution as 70% aqueous crude caprolactam. The crude caprolactam is removed in a separator drum and fed to an extractor.

Caprolactam is extracted with solvents such as benzene, toluene [12], or chlorinated hydrocarbons. The virtually caprolactam-free aqueous solution is discharged into a waste treatment plant or incinerated.

Concentrated ammonium sulfate solution is further evaporated and crystallized.

The caprolactam-containing solvent is fed into a distillation column. Solvent-free caprolactam is further distilled under reduced pressure. Very pure caprolactam is obtained in molten form. It is either used as such or solidified in a flaker. A total of 4.5 tons of ammonium sulfate per ton of caprolactam is obtained from both oximation and neutralization with the Raschig process. The BASF or Inventa process produces a total of 2.5 tons of ammonium sulfate per ton of caprolactam in oximation and neutralization.

Caprolactam yields can be improved and undesirable byproduct formation reduced by multi-stage rearrangement and the adjustment of reaction conditions in each stage. Acid concentration, free sulfur trioxide concentration, amount of oxime added, reaction temperature, and degree of mixing are critical variables at each stage [13].

Some processes differ from the BASF process with regard to purification of crude caprolactam.

In the Allied Chemical process, solvent extraction is followed by crystallization from aqueous caprolactam solution.

In the DSM process, the crude caprolactam is first extracted with a solvent and then reextracted with water. Before being distilled, the caprolactam extract is subjected to a number of physicochemical purifications, such as hydrogenation and ion exchange treatment [9].

Polimex developed a different approach by combining crystallization from aqueous caprolactam solution and oxidizing treatment. This physical and chemical purification process is followed by distillation [14].

Inventa licenses caprolactam plants that utilize crystallization as a final purification process.

## 4.1.3. Improvement of the Oximation Process

Oximation processes which form no or little ammonium sulfate have been developed by DSM [15], [16] and BASF [17]. These processes comprise a combination of hydroxylamine and oxime manufacture.

**Hydroxylamine–Phosphate Oxime Process (HPO Process of DSM; Fig.3)** The process proceeds according to the following equation:

$$C_6H_{10}=O + HNO_3 + 3\,H_2 \longrightarrow C_6H_{10}=NOH + 3\,H_2O$$

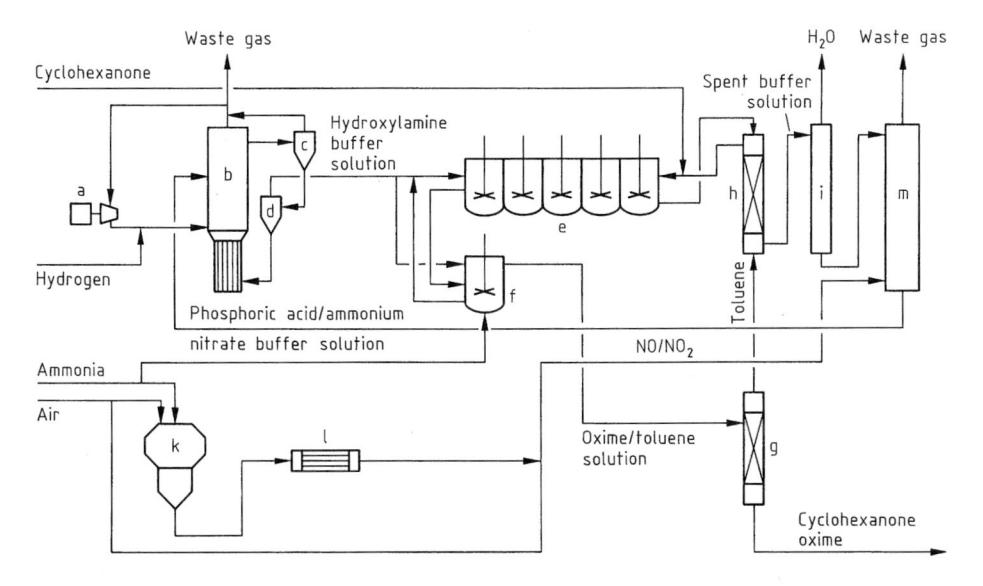

**Figure 3.** DSM HPO hydroxylamine and cyclohexanone oxime production
a) Compressor; b) Hydroxylamine generator; c) Separation; d) Filtration; e) Oximation; f) Neutralization; g) Solvent distillation; h) Extraction; i) Toluene stripping; k) Ammonia combustion; l) Condenser; m) Decomposition and absorption column

Oximation is conducted in a hydroxylamine phosphoric acid buffer solution. After separation of the oxime, the remaining aqueous ammonium phosphate buffer solution is recycled to hydroxylamine synthesis and concentrated. The reactions involved are described below:

1) Reduction of the phosphoric acid/ammonium nitrate buffer solution with hydrogen, and formation of hydroxylammonium phosphate:

$$NH_4NO_3 + 2\ H_3PO_4 + 3\ H_2 \longrightarrow$$
$$[NH_3OH]^+ [H_2PO_4]^- + NH_4H_2PO_4 + 2\ H_2O$$

Palladium on graphite or alumina carrier is used as a catalyst.

2) Oxime formation:

$$[NH_3OH]^+ [H_2PO_4]^- + NH_4H_2PO_4 + 2\ H_2O + \bigcirc\!\!=\!\!O \longrightarrow$$
$$\bigcirc\!\!=\!\!NOH + H_3PO_4 + NH_4H_2PO_4 + 3\ H_2O$$

3) After the oxime has been separated, the nitrate ions consumed are replaced by addition of 60% nitric acid.

$$H_3PO_4 + NH_4H_2PO_4 + 3\ H_2O + HNO_3 \longrightarrow$$
$$2\ H_3PO_4 + NH_4NO_3 + 3\ H_2O$$

The hydroxylamine formed by catalytic hydrogenation of nitrate ions reacts with free phosphoric acid in the buffer solution at pH 1.8 to form a hydroxylammonium phosphate solution. The reaction takes place in special columns; the unreacted hydrogen is separated from the catalyst suspension in a separator, and is recycled to the reaction via a compressor.

After the catalyst has been filtered and recycled, the hydroxylamine buffer solution reacts with cyclohexanone in the oximation reactor to produce cyclohexanone oxime. Toluene is used as the solvent and phosphoric acid is liberated.

Oximation takes place in a cascade of mixers and separators with a countercurrent process at pH 2. Conversion is 98%.

The remaining 2% of unreacted cyclohexanone is oximated at pH 4.5 with about 3% of the mainstream hydroxylamine. Ammonia is added. The toluene, containing about 30% oxime, is separated from the aqueous buffer solution. The cyclohexanone oxime is freed from toluene and fed into the Beckmann rearrangement process. The distilled solvent is recycled to the oximation process [18]. To avoid poisoning of the catalyst, residual cyclohexanone and oxime are extracted from the separated buffer solution. Toluene is used as the extracting agent and the process is carried out in a packed pulsed column. The exhausted buffer solution still contains dissolved toluene, which is stripped with steam. In addition, the excess ammonium ions formed as a reduction byproduct in the buffer solution must be removed. This is achieved with nitrous gases from the combustion of ammonia. This reaction proceeds according to the following equation:

$$2\ NH_4H_2PO_4 + NO + NO_2 \longrightarrow 2\ N_2 + 3\ H_2O + 2\ H_3PO_4$$

The reaction takes place in a decomposition column, and the nitrogen generated passes into the vent gas. The excess nitrogen oxides are absorbed in a downstream column and recycled for hydroxylamine synthesis.

**Oximation without Neutralization or Acidic Oximation (BASF).** The reactions involved are summarized by the following basic formula [17]:

$$\bigcirc\!=\!O + NO + 3/2\ H_2 \longrightarrow \bigcirc\!=\!NOH + H_2O$$

1) Nitric oxide is catalytically hydrogenated in an ammonium hydrogen sulfate solution:

$$NO + 3/2\ H_2 + (NH_4)HSO_4 \longrightarrow (NH_3OH)(NH_4)SO_4$$

Platinum on graphite is used as catalyst.

2) Cyclohexanone oxime is formed by reacting cyclohexanone with ammonium hydroxylammonium sulfate:

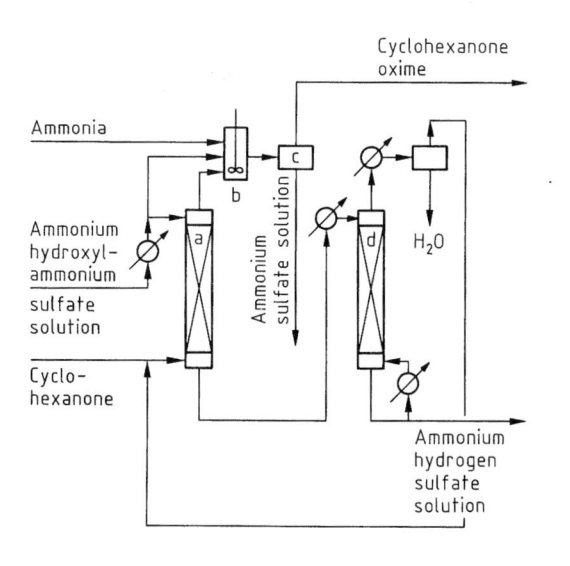

**Figure 4.** BASF acidic oximation
a) Oximation; b) After-oximation; c) Ammonium purge; d) Stripping column

$$\bigcirc=O + (NH_3OH)(NH_4)SO_4 \longrightarrow$$

$$\bigcirc=NOH + H_2O + (NH_4)HSO_4$$

Where as the sulfuric acid set free in BASF's classic process is neutralized with ammonia for separation of cyclohexanone oxime, further neutralization of ammonium hydrogen sulfate is not required for cyclohexanone oxime recovery. Hence, the ammonium hydrogen sulfate is recycled directly into hydroxylamine production (Fig. 4).

After filtration and separation of the catalyst, the ammonium hydroxylammonium sulfate solution is fed into the oximation process. Oximation takes place in a special column by means of countercurrent flow of cyclohexanone and hydroxylamine solution with 97–98% conversion.

Temperature is maintained above the melting point of cyclohexanone oxime.

Complete conversion of cyclohexanone is achieved with classic after-oximation.

Some ammonium sulfate is produced was byproduct in both hydroxylamine synthesis and after-oximation. The ammonium level can be kept constant with a purge to after-oximation. The corresponding sulfate deficit is made up by addition of sulfuric acid.

The aqueous phase of oximation contains ammonium hydrogen sulfate as well as traces of cyclohexanone and cyclohexanone oxime. The recycled ammonium hydrogen sulfate solution must be free of residual carbon. This is achieved in a stripping tower, where residual cyclohexanone oxime is decomposed and cyclohexanone stripped. The resulting ammonium hydrogen sulfate solution is virtually free of organic material and recycled to hydroxylamine production.

Acidic oximation gives 0.1 ton ammonium sulfate per ton of cyclohexanone oxime.

## 4.1.4. Gas-Phase or Catalytic Rearrangement of Cyclohexanone Oxime

Interest in heterogeneous catalytic rearrangement of cyclohexanone oxime was prompted by economic considerations in order to avoid the ammonium sulfate generated in Beckmann rearrangement.

In 1938, Du Pont proposed a process in which the cyclohexanone oxime vapors are passed over water-eliminating catalysts such as oxides of silicon, aluminum, titanium, or magnesium, preferably in the presence of ammonia, under atmospheric or reduced pressure at 250–350 °C.

In the early 1950s, Leuna-Werke demonstrated surface catalysis [19]. This work led to worldwide activity on catalytic rearrangement by such companies as BASF, Bayer, Solvay, BP Chemicals, Mitsubishi, Asahi, and Honshu.

An additional catalytic rearrangement process was developed in the USSR.

Figure 5 shows the process developed and proven by BASF in a large-scale pilot plant. The catalyst system contains boric acid. It is advantageous to have the catalyst in a fluidized bed during reaction and regeneration. A fluidized flow system is also possible [20].

The catalyst can be transported by means of a pneumatic conveyor from the fluidized-bed reactor to the catalyst regenerator and back. In the catalyst regenerator, the catalyst is regenerated with air at 700–900 °C [20]–[22]. Dust is removed from the stack gases in a cyclone. The catalyst can be cooled to the reaction temperature before recycling.

The fluidized-bed reactor can be operated under atmospheric, reduced, or elevated pressure by introducing a certain amount of inert gas as a fluidizing agent to the bottom of the reactor.

The cyclohexanone oxime can be injected into the 275–375 °C fluidized bed as a gas or liquid via nozzles. It can be introduced as a solid with the aid of an inert-gas stream [23]–[25]. The pre-evaporation of cyclohexanone oxime is particularly advantageous.

The recovery of rearrangement heat supplies the heat of fusion and evaporation.

Catalyst dust deposited in the cyclone can be reused for the production of fresh catalyst. The hot vapors from the cyclone can be condensed on a surface condenser or in a column [26]. The inert gas required for fluidization can be circulated. Crude caprolactam obtained by catalytic rearrangement (yield up to 96%, based on cyclohexanone oxime) is less pure and the impurities are different from those in caprolactam produced by Beckmann rearrangement. It is therefore purified by combined extraction, crystallization, and distillation methods, as described in references [27]–[37].

However, caprolactam has not yet been produced catalytically on an industrial scale.

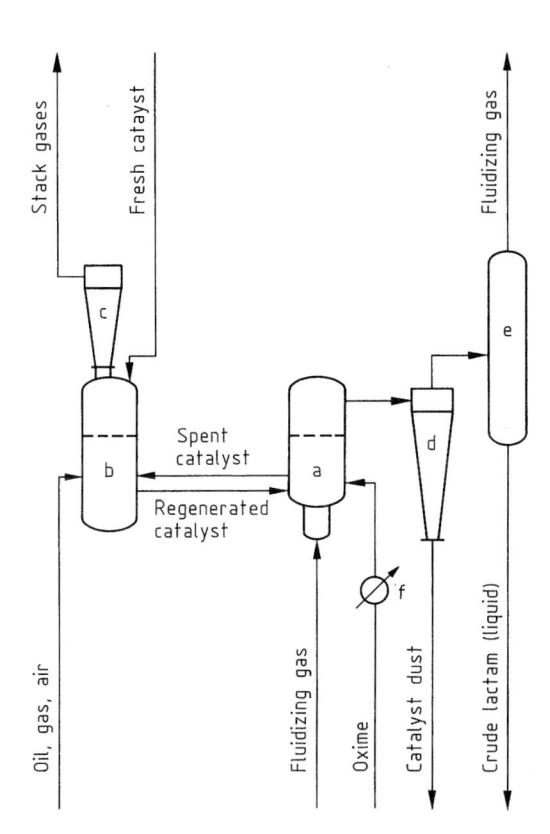

**Figure 5.** Catalytic oxime rearrangement
a) Fluidized-bed reactor; b) Catalyst regenerator; c) Cyclone; d) Cyclone; e) Condensation; f) Pre-evaporation

## 4.1.5. Modified Oximation and Beckmann Rearrangement

It has been proposed that cyclohexanone be reacted with hydroxylammonium nitrate instead of hydroxylammonium sulfate. In this case, ammonium nitrate is formed as byproduct in oximation [38].

Another process bypasses ammonium salts as byproducts in oximation. The acid obtained during the reaction of cyclohexanone with hydroxylammonium salts is removed by electrodialysis [39].

The Beckmann rearrangement of cyclohexanone oxime can be carried out in polyphosphoric acid [40] instead of oleum. The rearrangement mixture is neutralized in an organic solvent in which caprolactam and polyphosphoric acid are soluble. The monoammonium dihydrogen phosphate and diammonium monohydrogen phosphate formed during neutralization with ammonia are separated.

It has also been proposed that the rearrangement of cyclohexanone oxime be carried out in acetic acid over acidic ion exchangers [41]. Mixtures of oleum and phosphoric acid, sulfur trioxide in liquid sulfur dioxide, thionyl chloride, phosgene, hydrofluoric acid in acetic anhydride, and chlorosulfonic acid in chlorinated hydrocarbons can also be used [42].

Attempts have been made to neutralize the mixture obtained from rearrangement in oleum with metal oxides, such as calcium oxide, instead of ammonia [43]. It is proposed that the resulting calcium sulfate is then decomposed thermally into recyclable calcium oxide and sulfur dioxide. Diluting the rearrangement reaction mixture with water and extracting the caprolactam with a solvent such as phenol is another method for avoiding the production of ammonium sulfate [44].

Another approach is to dilute the rearrangement mixture with aqueous ammonium sulfate or ammonium dihydrogen phosphate resulting in splitting caprolactam sulfate into caprolactam and sulfuric acid and then extracting caprolactam [45]. The resulting sulfuric acid solution containing ammonium salts is used in phosphoric acid production. It has been proposed for economic reasons to decompose ammonium sulfate from rearrangement back into ammonia and sulfur dioxide containing gas streams [46].

## 4.2. Photooximation

In the 1950s, Toray industries developed a photochemical process for the production of caprolactam (Fig. 6).

Photonitrosation (PNC) converts cyclohexane to cyclohexanone oxime dihydrochloride followed by Beckmann rearrangement.

The cyclohexane is reacted with nitrosyl chloride to give cyclohexanone oxime hydrochloride:

1) Preparation of nitrosylsulfuric acid from nitrous gases obtained from combustion of ammonia, and sulfuric acid:

$$2\ H_2SO_4 + NO + NO_2 \longrightarrow 2\ NOHSO_4 + H_2O$$

2) Preparation of nitrosyl chloride by reaction with hydrogen chloride:

$$NOHSO_4 + HCl \longrightarrow NOCl + H_2SO_4$$

3) Photochemical reaction:

$$\bigcirc + NOCl + HCl \xrightarrow{h\nu} \bigcirc\!\!=\!NOH \cdot 2\ HCl$$

The industrial photonitrosation process is based on the development of efficient photoreactors. Toray designed an immersion lamp with a high radiation efficiency and capacity as well as long life. In order to remove the short-wave radiation below 365 nm, which contributes to tar formation on the lamps, either an absorbent is added to the cooling water, or the light source is surrounded by a glass filter [47].

The cyclohexanone oxime produced by photonitrosation of cyclohexane is separated as the dihydrochloride in the presence of excess hydrogen chloride. This compound exists in the form of oil droplets and forms a heavier lower phase in cyclohexane; this lower phase is subjected to Beckmann rearrangement with excess sulfuric acid or oleum

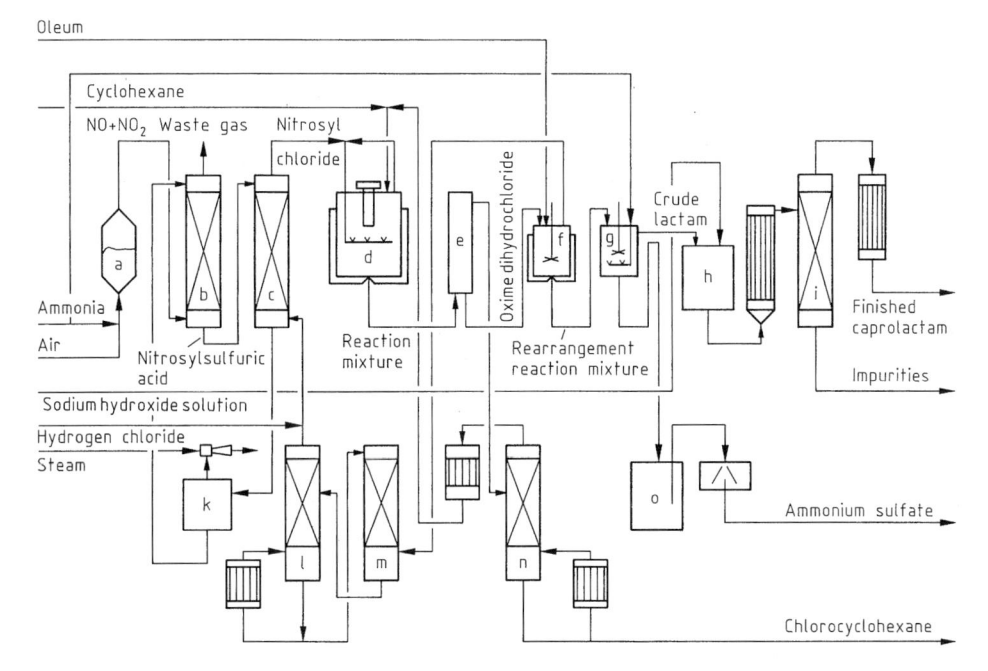

**Figure 6.** Toray PNC caprolactam production
a) Ammonia combustion; b) Nitrosylsulfuric acid generator; c) Nitrosyl chloride generator; d) Photonitrosation; e) Cyclohexane/cyclohexanone oxime separation; f) Rearrangement; g) Neutralization; h) Chemical treatment; i) Drying and lactam distillation; k) Dewatering of sulfuric acid; l) Hydrogen chloride regenerator; m) Hydrogen chloride recovery; n) Cyclohexane recovery; o) Ammonium sulfate recovery

to give caprolactam. Hydrogen chloride is set free and recycled. The rearrangement reaction mixture is neutralized with ammonia water to give crude lactam and ammonium sulfate.

*PNC process description:* Nitrous gases are produced in an ammonia burner and cooled. They then react with sulfuric acid to form nitrosylsulfuric acid. This is brought into contact with excess hydrogen chloride to give nitrosyl chloride. The remaining sulfuric acid is dehydrated and then either recycled into the nitrosyl chloride production stage, or circulated. The gaseous nitrosyl chloride/hydrogen chloride mixture is passed through liquid cyclohexane. The conversion to oxime dihydrochloride is carried out in a photoreactor with actinic light from cooled mercury immersion lamps. This product separates at the bottom as a heavy oil. In order to prevent the deposit of oxime salt and resinous coating, the walls of the lamp cooler are periodically washed with concentrated sulfuric acid [48]. The thermal energy emitted from the light source must be removed by cooling water. Since nitrosyl chloride, hydrogen chloride, and oxime dihydrochloride are extremely corrosive, the photoreactor is made of titanium [49] or lined with enamel or PVC.

Unreacted nitrosyl chloride is removed from the photoreactor with the excess hydrogen chloride and recycled into the photochemical reaction. The oily oxime dihydrochloride is then rearranged with oleum. Unreacted cyclohexane is purified during cyclohexane regeneration and reused. Hydrogen chloride is liberated during Beckmann rearrangement if carried out under the same conditions as with pure oxime which is then absorbed during the hydrogen chloride regeneration stage in dilute hydrochloric acid. The solution is concentrated in the hydrogen chloride recovery stage and recycled into nitrosyl chloride production. The mixture from the rearrangement reaction is neutralized with ammonia water and the aqueous crude caprolactam is purified. This includes a chemical process, drying, and distillation. In cyclohexane regeneration, a small chlorocyclohexane residue is obtained. The ammonium sulfate solution is crystallized by evaporation. In the PNC Toray process, 1.55 tons of ammonium sulfate are produced per ton of caprolactam [50].

## 4.3. Production from Toluene

Snia Viscosa developed a toluene-based process in 1960 [51] (Fig. 7).
Caprolactam is obtained in three steps:

1) Catalytic oxidation of toluene with air to benzoic acid:

$$\text{C}_6\text{H}_5-\text{CH}_3 + 3/2\ \text{O}_2 \rightarrow \text{C}_6\text{H}_5-\text{COOH} + \text{H}_2\text{O}$$

2) Hydrogenation of benzoic acid to cyclohexanecarboxylic acid:

$$\text{C}_6\text{H}_5-\text{COOH} + 3\ \text{H}_2 \rightarrow \text{C}_6\text{H}_{11}-\text{COOH}$$

3) Nitrosodecarboxylation of cyclohexane-carboxylic acid to caprolactam in the presence of oleum:

$$\text{C}_6\text{H}_{11}-\text{COOH} + \text{NOHSO}_4 \xrightarrow{\text{Oleum}} \text{caprolactam} + \text{H}_2\text{SO}_4 + \text{CO}_2$$

During the last stage, nitrosocyclohexane and cyclohexanone oxime are formed in situ. Cyclohexanone oxime immediately undergoes rearrangement to caprolactam.

*Snia process description:* The oxidation of toluene with air is carried out in liquid phase using a cobalt catalyst at 160–170 °C and 0.8–1 MPa (8–10 bar) pressure. The yield is well above 90% of theory. The gases from the oxidation reactor contain mainly nitrogen with small amounts of oxygen, carbon dioxide and carbon monoxide. They are cooled to 7–8 °C in order to recover unreacted toluene. The water accumulated during the reaction is removed in a separator drum and toluene is recycled to the reactor. In addition to toluene and cobalt catalyst, the liquid reaction product contains about 30% benzoic acid and also various intermediates and byproducts.

Most of the toluene is removed by distillation, and the remaining concentrated solution is given into a rectification column. The lower-boiling intermediates and the remaining toluene are removed at the top and reused. Benzoic acid in vapor form is

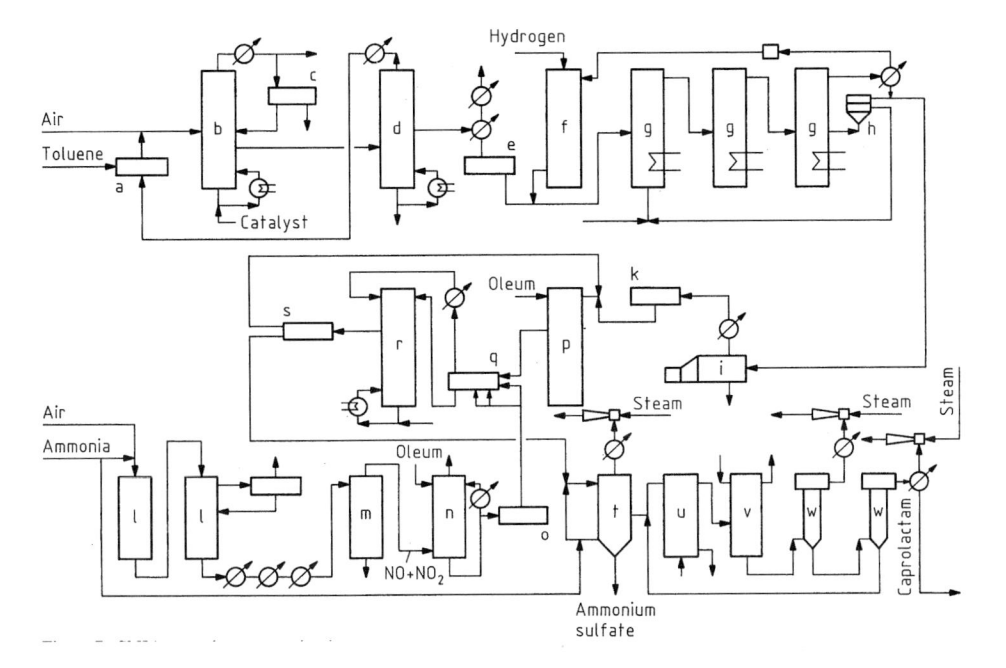

**Figure 7.** SNIA caprolactam production
a) Toluene tank; b) Oxidation; c) Separation; d) Rectification; e) Benzoic acid tank; f) Benzoic acid/hydrogen mixture; g) Benzoic acid hydrogenation; h) Removal of catalyst; i) Cyclohexanecarboxylic acid distillation; k) Cyclohexanecarboxylic acid tank; l) Ammonia combustion; m) Separation; n) Nitrosylsulfuric acid generator; o) Nitrosylsulfuric acid tank; p) Cyclohexanecarboxylic acid/oleum mixture; q) Rearrangement; r) Hydrolysis; s) Solution of cyclohexanecarboxylic acid in cyclohexane; t) Neutralization and ammonium sulfate crystallization; u) Solvent extraction; v) Water extraction; w) Lactam distillation

removed from the rectification column as a side stream. The high-boiling byproducts leave the column as a residue. They include the cobalt salt, which is reprocessed. The benzoic acid obtained is suitable for hydrogenation without any further purification.

Hydrogenation of benzoic acid to cyclohexanecarboxylic acid is carried out in liquid phase in the presence of a palladium on graphite catalyst. A series of stirred reactors are used with a temperature of about 170 °C and pressure of 1–1.7 MPa (10–17 bar). Conversion is 99.9% with a commensurate yield. The catalyst is centrifugally separated from the liquid reaction product, mixed with benzoic acid, and reused in hydrogenation. Cyclohexanecarboxylic acid is distilled under reduced pressure. Any catalyst still present remains in the residue. A 73% nitrosylsulfuric acid solution in sulfuric acid is used for the nitrosation of cyclohexanecarboxylic acid. It is obtained by absorption of nitrous gases (from ammonia combustion) in concentrated sulfuric acid or oleum. Cyclohexanecarboxylic acid is premixed with oleum at ambient temperature and fed into a multistage nitrosation reactor. Exactly defined amounts of cyclohexanecarboxylic acid and nitrosylsulfuric acid are reacted at each stage, ensuring the complete reaction of the nitrosating agent. The conversion rate of the cyclohexanecarboxylic acid is maintained at about 50%. One mole of carbon dioxide per mole of caprolactam is set free along with small amounts of other gases such as nitric oxide, carbon monoxide, sulfur dioxide, and nitrogen. In order to maintain the reaction and remove the reaction heat, this process is carried out in boiling cyclohexane under atmospheric pressure. Depending on the organic acid, less than 10% of the yield is lost.

Sulfonic acids are byproducts. During the reaction, 3.3 moles of sulfuric acid per mole of caprolactam are consumed. Nitrosation with subsequent rearrangement is the most important part of the process [52].

Products from the nitrosation/rearrangement step are hydrolyzed with water at low temperatures. Cyclohexane extracts unreacted cyclohexanecarboxylic acid. The final residues are extracted from the sulfuric acid solution, which has been diluted with water, by countercurrent extraction using cyclohexane as the solvent. The recovered cyclohexanecarboxylic acid is freed from the solvent and recycled to the process. The acidic caprolactam solution containing excess sulfuric acid is then fed into the neutralization stage, where it is neutralized with ammonia. Neutralization is carried out directly in a crystallizer under reduced pressure. Two liquid layers are formed, a saturated ammonium sulfate solution and a concentrated aqueous caprolactam solution. The caprolactam solution is purified in several stages. It is first separated from the water-soluble byproducts by toluene extraction. An aqueous caprolactam solution is then obtained by countercurrent extraction of the caprolactam–toluene solution with water. The toluene-soluble byproducts remain in the organic layer. Water is removed from the aqueous caprolactam solution and the product distilled to give pure caprolactam.

Crystallization of the separated saturated ammonium sulfate solution gives 4.1 tons of ammonium sulfate per ton of caprolactam.

Research centered on reduction of ammonium sulfate production:

Cyclohexanecarboxylic acid is treated with oleum in the first version of this process to produce pentamethylene ketene.

$$\text{C}_6\text{H}_{11}\text{-COOH} \xrightarrow[-\text{H}_2\text{O}]{\text{Oleum}} (\text{CH}_2)_5\text{C}=\text{CO} \xrightarrow[-\text{CO}_2]{+\text{NOHSO}_4} \text{caprolactam}$$

The pentamethylene ketene is then nitrosodecarboxylated to caprolactam in the presence of oleum. Ammonium sulfate is decreased to 2 tons per ton of caprolactam by this route [53].

The second version produces no ammonium sulfate as a result of a modification in the separation procedure: Caprolactam dissolved in sulfuric acid is extracted by diluting the solution with small amounts of water, which is then extracted with an alkylphenol. The remaining sulfuric acid is subjected to thermal cracking, which destroys the impurities, and the $SO_2$ formed is recycled into the process. Aside from not generating ammonium sulfate, this version is also advantageous in that impurities do not present waste disposal problems [54].

## 4.4. Recovery from Nylon-6 Waste

Solid nylon-6 waste is depolymerized in a kettle operation with the aid of a cracking catalyst and superheated steam [55].

The mixture of steam and caprolactam leaving the kettle is fed to an evaporation column, where the mixture is condensed and concentrated under atmospheric pressure. The concentrate is purified by adding an oxidizing agent that converts impurities to readily removable compounds. The caprolactam is purified through distillation.

The washwater obtained from extraction of nylon-6 chips contains oligomers of caprolactam as well as caprolactam. The oligomers are separated after preconcentration in a thin-film evaporator and then depolymerized in the same manner as the solid waste. The resulting aqueous caprolactam solution is purified chemically. Subsequent distillation gives water, a fraction containing low-boiling compounds, and caprolactam as the principal product.

Oligomers can also be depolymerized by other methods [56], [57]. The washwater of nylon-6 chip production is concentrated at a sufficiently high temperature to maintain the oligomers in solution.

The caprolactam-oligomer solution is fed into a fixed- or fluidized-bed reactor. The oligomers are cracked at 275–350 °C.

A special aluminum oxide catalyst is used and a yield of 95%, relative to the oligomers in the feed, is obtained. After cracking, the caprolactam is purified by conventional methods [57].

## 4.5. Former Commercial Routes

Union Carbide has developed a process for the manufacture of caprolactam based on the following reaction sequence [58]:

$$\text{cyclohexanone} + CH_3C(O)\text{-OOH} \longrightarrow \varepsilon\text{-caprolactone} + CH_3COOH$$

$$\xrightarrow{NH_3} \text{caprolactam} + H_2O$$

This reaction can be carried out by two methods:

Cyclohexanone is reacted with peracetic acid in an anhydrous medium, such as acetone. With an excess of cyclohexanone, ε-caprolactone is obtained in a yield of about 90%.

Oxidation can also be carried out in situ. Cyclohexanone is reacted with air at 25–50 °C with the simultaneous introduction of acetaldehyde. An excess of cyclohexanone is also used in this case. Manganese, cobalt, platinum, palladium, vanadium, and zirconium salts and their oxides on carriers are examples of suitable oxidation catalysts. The acetic acid formed as a byproduct during the oxidation (about 1.1 kg per kg of lactam) is separated from the ε-caprolactone by distillation. The conversion of caprolactone to caprolactam with ammonia can either be carried out in anhydrous or aqueous medium, preferably at 350–425 °C at increased pressure. An excess of ammonia is used. The caprolactam formed during the ammonolysis of caprolactone in aqueous medium is isolated by extraction. The total yield is 65–70%.

Union Carbide operated a plant with a capacity of about 20 000 t/a using this method.

Several companies have developed processes for the production of caprolactam using nitrocyclohexane as a starting product. Nitrocyclohexane is converted to cyclohexanone oxime by partial reduction with hydrogen in the presence of silver oxide, zinc oxide, or chromium oxide catalysts in an aqueous phase [59], [60]. The yield can be increased by using lead-modified platinum and palladium catalysts [61]. The reduction is carried out under increased pressure at 100–150 °C. The byproducts include cyclohexylamine.

Du Pont operated a caprolactam plant based on catalytic reduction of nitrocyclohexane to cyclohexanone oxime with a capacity of about 25000 t/a. Rearrangement was carried out in the conventional manner. The plant was shut down.

The catalytic conversion of nitrocyclohexane to caprolactam is a modified method never used industrially. The reaction is carried out over borophosphoric acid catalysts in the gas phase at 150–450 °C [62].

## 4.6. New Routes in Development

Several attempts are being made to develop new routes using cheaper raw materials or more economical reaction steps or minimizing byproducts and waste.

**Hydrogen Peroxide Processes.** Inventa describes a process for the production of cyclohexanone oxime by oxidation of cyclohexylamine with hydrogen peroxide in the presence of catalysts and stabilizers [63]. Salts of acids of group VI B metals are used as catalysts. The yield based on cyclohexylamine is 98%. An adduct of 1 mol of cyclohexylamine and 1 mol of cyclohexanone oxime is formed as an intermediate, which is hydrolyzed to cyclohexylamine and cyclohexanone oxime.

USSR researchers succeeded in catalytically converting cyclohexanone with hydrogen peroxide in the presence of ammonia in aqueous solution to cyclohexanone oxime [64]. Toa Gosei developed a process based on this approach [65]. The reaction is carried out with an excess of ammonia at 10–30 °C under atmospheric pressure. Tungsten and tin compounds are used as catalysts. The cyclohexanone oxime yield is said to be well above 90%. The cyclohexanone oxime is obtained from the reaction mixture by extraction with organic solvents. About 95% of the catalyst can be recovered from the aqueous solution. Catalyst still in solution is adsorbed on carbon.

Ammonium sulfate byproduct is formed by both the Inventa and Toa Gosei process during rearrangement of the cyclohexanone oxime.

Another hydrogen peroxide process known as the 1,1′-Peroxydicyclohexylamine Process has been demonstrated by BP Chemicals [66] and Degussa [67]:

$$2 \, \text{C}_6\text{H}_{10}\text{=O} \xrightarrow{H_2O_2} \text{(OH)(HO)C}_6\text{H}_{10}\text{-O-O-C}_6\text{H}_{10} \xrightarrow[-H_2O]{NH_3}$$

$$\text{C}_6\text{H}_{10}(\text{NH})\text{-O-O-C}_6\text{H}_{10} \xrightarrow{cat.} \text{caprolactam} + \text{C}_6\text{H}_{10}\text{=O}$$

Cyclohexanone is reacted with hydrogen peroxide in a molar ratio of 2 : 1 in the presence of an organic solvent, such as methanol, ethanol, or isopropanol, and a stabilizer, such as sodium ethylenediaminetetraacetate, to give 1,1′-dihydroxydicyclohexyl peroxide.

The peroxide reacts with ammonia to form 1,1′-peroxydicyclohexylamine. Salts of organic acids are used as accelerators. The reaction temperature for the first stage is 25 °C, while the second stage reacts at about 25 – 45 °C; the reaction can also be carried out in a single stage. 1,1′-Peroxydicyclohexylamine is converted to caprolactam and cyclohexanone in the presence of catalysts such as lithium halides; molten caprolactam is used as a solvent. The reaction mixture is then distilled at 100 – 115 °C and 1.3 – 2 kPa (13 – 20 mbar). Cyclohexanone is recycled and caprolactam is purified in a conventional manner.

This process proceeds without any byproduct formation.

**Oxidation of Cyclohexylamine with Elemental Oxygen.** Allied proposed the oxidation of cyclohexylamine to cyclohexanone oxime with elemental oxygen [68].

Cyclohexylamine vapor is brought into contact with oxygen. Helium is used as a diluent to keep cyclohexylamine and oxygen concentration outside the explosion limits. This gas is passed over silica gel catalyst at temperatures between 120 – 250 °C and adequate pressure. A yield of about 11% per pass of the reaction mixture over the catalyst is reported.

The cyclohexanone oxime obtained can be processed to caprolactam in the usual way.

**Direct Oximation of Cyclohexanone.** Allied introduced a process for ammoximation of cyclohexanone [69].

Ammonia, air, and cyclohexanone react in the presence of silica, alumina, or gallia catalysts at temperatures between 50 and 500 °C. A solvent can be employed to remove cyclohexanone oxime from the catalyst.

A 40% yield of cyclohexanone oxime is obtained.

**Bis(nitrosocyclohexane) Process.** Monsanto described a process for the production of cyclohexanone oxime using electrical discharge [70]. In this process, cyclohexane is reacted with either nitric oxide, alkyl nitrites, or nitroalkanes to form bis(nitrosocyclohexane). If the reaction is carried out at about 100 °C, the bis(nitrosocyclohexane) undergoes in situ rearrangement to give cyclohexanone oxime.

Pressure, temperature, ratio of reactants, residence time, and current density are important prerequisites for obtaining a good yield. Other companies have developed similar processes.

**ε-Hydroxycaproic Acid – ε-Caprolactone Process.** Caprolactam production via ε-hydroxycaproic acid and ε-caprolactone has long been known [71].

Teijin described a process for ε-hydroxycaproic acid production by oxidation of cyclohexanone with air in the presence of such catalysts as azoisobutyronitrile or cyclohexyl peroxide. Substantial amounts of adipic acid are formed as byproduct. The yield of ε-hydroxycaproic acid and adipic acid is about 80% [72].

Several companies have attempted to produce caprolactam from ε-hydroxycaproic acid. ε-Hydroxycaproic acid and its esters, including ε-caprolactone, can be converted, either in the liquid or in the gas phase, to caprolactam using ammonia.

Suitable solvents for carrying out the reaction in the liquid phase are water, alcohols, chloroform, or dioxane. Nickel, cobalt [73], copper, palladium, platinum [74], etc. are suitable catalysts. The reaction is carried out between 140 and 475 °C and usually under pressure. The yield of caprolactam after three cycles is 96 – 98 % [75].

In the gas-phase process, ε-caprolactone or the esters of ε-hydroxycaproic acid together with hydrogen, ammonia and water are passed over catalysts at temperatures of about 120 – 320 °C. The catalysts are titanium dioxide, alumina, or silica and copper or nickel [76], or copper oxide/chromium oxide [77]. The caprolactam yield is 90% or higher.

The production of ε-caprolactone from hexane-1,6-diol, and its conversion to caprolactam, has also been described [78].

**2-Nitrocyclohexanone Process.** Another process forming no byproducts has been proposed by Techni-Chem [9], [79].

Cyclohexanone is reacted with excess acetic anhydride in the presence of ketene at 140 °C to give cyclohexenyl acetate. The catalysts are sulfuric or hydroiodic acid. The amount of ketene is such that the acetic acid formed during acetylation of cyclohexanone with acetic anhydride is reconverted to acetic anhydride with ketene. The reaction mixture is removed from the residue by flash distillation under reduced pressure.

Unreacted cyclohexanone and acetic anhydride are separated and recycled. Cyclohexenyl acetate is converted to 2-nitrocyclohexanone in an exothermic reaction at 30 – 50 °C in excess acetic anhydride and concentrated nitric acid (acetyl nitrate). Excess acetyl nitrate is removed under reduced pressure, absorbed by acetic anhydride, and recycled to the nitration stage. 2-Nitrocyclohexanone, acetic anhydride, and the resulting acetic acid are separated. Acetic acid is recycled into the ketene generator and acetic anhydride is returned to the nitration stage.

2-Nitrocyclohexanone is then quantitatively cracked with excess aqueous ammonia at 40–60 °C to give the ammonium salt of ε-nitrocaproic acid. The aqueous solution of the ammonium nitrocaproate is hydrogenated in the presence of a Raney nickel catalyst at 100 °C and 2 MPa (20 bar). The conversion to ε-aminocaproic acid is virtually quantitative. After the removal of catalyst and ammonia, the 5–25% aqueous ε-aminocaproic acid solution is converted to caprolactam at 300 °C and 10 MPa (100 bar). Caprolactam is then extracted from unreacted ε-aminocaproic acid. ε-Aminocaproic acid is returned to the cyclization process. The process is illustrated by the following equation:

$$\text{cyclohexanone} + CH_2C=O \xrightarrow{(CH_3CO)_2O} \text{1-acetoxycyclohexene (OCOCH}_3\text{)}$$

$$\xrightarrow[HNO_3]{(CH_3CO)_2O} \text{2-nitrocyclohexanone} + CH_3COOH$$

$$\xrightarrow{H_2O/NH_3} NO_2-(CH_2)_5-COOH \xrightarrow[\text{cat.}]{H_2} NH_2-(CH_2)_5-COOH \rightarrow \text{caprolactam (C=O, NH)}$$

Teijin described a process for the manufacture of caprolactam from 2-nitrocyclohexanone. In this process, ring cleavage, hydrogenation, and cyclization are carried out in one step: 2-Nitrocyclohexanone is reacted with hydrogen in the presence of water and ammonia at 150–300 °C over hydrogenation catalysts [80].

**3,3-Pentamethyleneoxaziridine Process.** Leuna-Werke developed a process for the production of caprolactam by heating 3,3-pentamethyleneoxaziridine (cyclohexanone isoxime) [81].

This cyclohexanone isoxime is obtained by reacting excess cyclohexanone in aqueous ammonia solution in the presence of a water-immiscible organic solvent, with hypochlorite solution at − 10 to + 50 °C.

The cyclohexanone isoxime solution is then heated to 250–350 °C in an organic solvent. The distillate contains caprolactam. The conversion of 3,3-pentamethyleneoxaziridine to caprolactam can also be carried out in sulfuric acid.

**Catalytic Deacetylation of N-Acetylcaprolactam.** Kanebo discussed the preparation of caprolactam by treating O-acetylcyclohexanone oxime in steam over silica–alumina, silica–magnesia, or silica–zirconia catalysts. The reaction is carried out at 150–450 °C and reduced pressure [82].

N-Acetylcaprolactam is converted to caprolactam by reaction with cyclohexanone oxime in the presence of a catalyst. Caprolactam and O-acetylcyclohexanone oxime are separated by distillation.

O-Acetylcyclohexanone oxime is recycled to rearrangement [83].

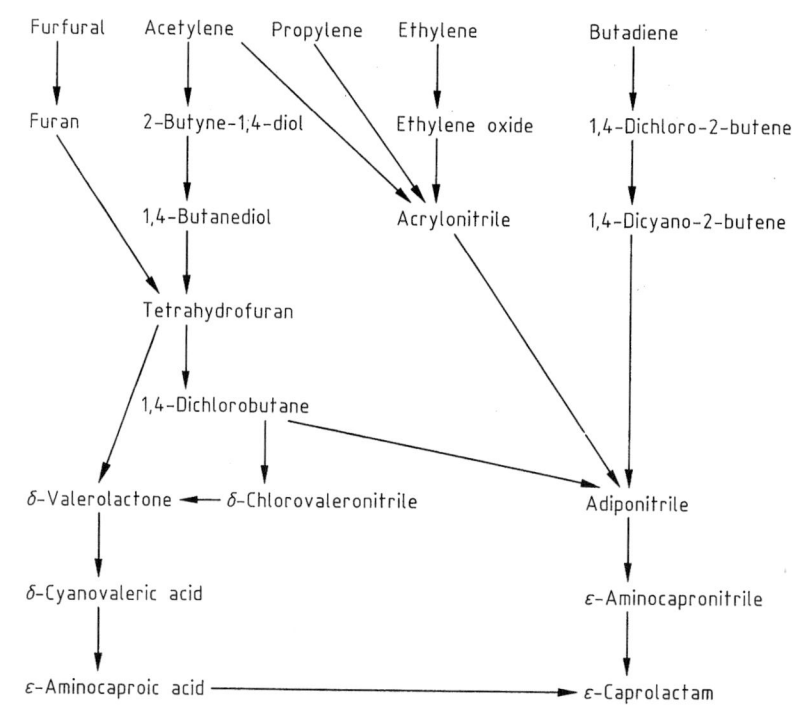

**Figure 8.** Routes for caprolactam production from furfural, acetylene, propylene, ethylene, and butadiene

**Acetylene, Ethylene, Propylene, Butadiene, or Furfural based Processes.** Figure 8 summarizes potential routes for the production of caprolactam.

However, none of these processes is used commercially.

## 4.7. Environmental Protection

Atmospheric and liquid wastes from caprolactam production require disposal methods that take local and legal restrictions into consideration:

1) Crude caprolactam extraction produces a vent gas stream containing traces of organic solvent which must be reduced to acceptable levels by physical methods such as carbon adsorption.
2) In all processes using Beckmann rearrrangement heavy bottoms in crude caprolactam extraction containing ammonium sulfate and other sulfur compounds are formed. They can be incinerated to produce sulfuric acid.
3) The residue of finished caprolactam distillation can be incinerated.

**Table 2.** Standard specifications of caprolactam (BASF)

| | |
|---|---|
| Solidification point | min. 69 °C |
| Moisture content | max. 0.05% |
| Permanganate absorption number | max. 5 |
| Volatile bases | max. 0.5 meq/kg |
| Content of free bases | max. 0.1 meq/kg |
| Content of free acids | max. 0.05 mg/kg |
| Oxime content | max. 5 ppm |
| Color number of 50% aqueous solution | max. 5 APHA |
| Absorbance 290 nm (50% aqueous solution; cuvette length: 1 cm) | max. 0.05 |

Aqueous solutions of caprolactam should be clear in all concentration ranges

# 5. Quality Specifications

Industrial caprolactam is 99.9 – 99.94% pure. Main contaminant is usually water amounting to 0.04 – 0.1%.

High purity is requested by the users who operate processes sensitive to quality fluctuations.

Standard specifications for polymerization and subsequent processing to fibers and plastics are listed in Table 2.

The most important quality criteria are the ppm amounts of impurities which can be oxidized with potassium permanganate, as well as the basic impurities which are present in free or volatile form.

Quality specifications and the methods used for determining these values vary from manufacturer to manufacturer.

# 6. Analysis

*Appearance and Solubility* Caprolactam melt is clear. A clear solution is obtained by dilution with water in all concentrations.

*Permanganate Numbers* Two different approaches are used.

1) Visual comparison with standard solution: 1 mL of 0.01 N potassium permanganate solution is added to 100 mL of aqueous caprolactam solution at 20 °C. 0.01 N is equivalent to $c$ (1/5 $KMnO_4$) 0.01 mol/L. The time taken for the color to change to that of the standard solution is measured. Typical values are shown in Table 3.
2) Photometric method: The absorbance of a 1 wt% caprolactam solution (50 mL or 100 mL aqueous solution) is measured 250 or 600 s after the addition of 0.01 N potassium permanganate (1 mL or 2 mL) at a specific wavelength and compared to water treated with permanganate.

The extinction times 100 is referred to as the Klett-Summerson value or permanganate absorption number (Table 4).

**Table 3.** Permanganate numbers of caprolactam

| Method | Concentration of the caprolactam/water solution | Standard solution composition in 1 L of water | Permanganate number in s |
|---|---|---|---|
| 1 | 1 wt% | 2.5 g $Co(NO_3)_2 \cdot 6H_2O$ + 0.01 g $K_2Cr_2O_7$ | 10 000 to 40 000 |
| 2 | 1 wt% | 3 g $CoCl_2 \cdot 6H_2O$ + 2 g $CuSO_4 \cdot 5H_2O$ | 3 600 to 4 000 |
| 3 | 3 wt% | 3 g $Co(NO_3)_2 \cdot 6H_2O$ + 0.012 g $K_2Cr_2O_7$ | 5 000 to 10 000 |

**Table 4.** Permanganate absorption number of caprolactam

| Method | Wave length | Time, s | Extinction × 100 |
|---|---|---|---|
| 1 | 410 nm | 250 | Klett-Summerson value: max. 7 |
| 2 | 420 nm | 600 | permanganate absorption number: max. 5 |

Permanganate numbers can be converted into permanganate absorption numbers. A permanganate number of 10 000 s (method 3/ Table 3) corresponds to a permanganate absorption number of 4 (method 2/Table 4).

*Volatile Bases*

Volatile bases are separated by steam distillation in the presence of aqueous sodium hydroxide, collected in a receiver, and determined by titration.

Methods differ in the caprolactam concentrations and the amounts of sodium hydroxide solution used. The amounts of volatile bases are expressed either in meq/kg or in mg of $NH_3$/kg.

*Color Number*

The color number is expressed in APHA units, and corresponds to the platinum – cobalt number, also referred to as Hazen color number or degree. It is used to determine the absorbance or extinction of a 50 wt% aqueous caprolactam solution. It is measured in comparison to standard solution or by photometric determination of the extinction compared to distilled water. The latter measurement is carried out in a 5-cm cell at wavelengths of 390 and 465 nm. The standard value amounts to 5 – 10 APHA units.

*Solidification Point*

The solidification point is determined by melt analysis. This analysis is related to the moisture content of caprolactam. The standard value amounts to 68.8 – 69.0 °C (corrected), corresponding to a moisture content of 0.04 – 0.1% water.

The moisture content itself is determined by titration with Karl-Fischer solution.

*Free Bases or Acids*

The free base or acid content is measured by acidimetric titration to pH 7.0 of an aqueous caprolactam solution.

*Residual Cyclohexanone Oxime*

Cyclohexanone oxime is measured colorimetrically after hydrolysis to hydroxylamine, oxidation, diazotization, and formation of a colored diazonium compound or other methods.

**Table 5.** Caprolactam capacities in 1984 (kt/a)

| | |
|---|---|
| Western Europe | 845 |
| USA | 505 |
| Latin America | 115 |
| Asia | 615 |
| Japan | 455 |
| Eastern Europe | 800 |
| World caprolactam capacity | 2880 |

*Absorbance at 290 nm*

The absorbance of a 50% aqueous solution of caprolactam is measured at a wavelength of 290 nm and corrected to a cuvette length of 1 cm.

The International Organisation for Standardization (ISO) is attempting to standardize analytical methods for caprolactam. Methods for "Determination of crystallizing point – ISO 7060-1982", "Determination of colour of 50% aqueous caprolactam solution – ISO DIS 8112-1983", and "Determination of absorbance at a wavelength of 290 nm – ISO 7059-1982" have been finalized and recommended.

# 7. Storage and Transportation

Molten caprolactam is stored and shipped in stainless steel or aluminum containers at 75–80 °C in the absence of moisture and under a blanket of nitrogen containing less than 10 ppm of oxygen.

Flaked caprolactam is supplied in polyethylene bags for general purposes or in aluminum coated paper bags for anionic polymerization. Flake caprolactam should be stored in dry warehouses and protected against direct sunlight. It is advisable to keep the ambient temperature below 45 °C (wherever possible) and the relative humidity below 65%. In areas with high temperatures and humidity, an additional outer reinforced paper bag can be used for protection.

# 8. Economic Aspects

The United States, Japan, the Soviet Union, Belgium, the Netherlands, Italy, and Germany are the main producers of caprolactam. With the exception of France and the United Kingdom virtually all industrialized nations possess caprolactam facilities of their own.

Recent world capacity is summarized in Table 5.

Some 90% of all caprolactam produced is processed to filament and fiber. Nearly 10% is used for the production of plastics. Only small quantities are used for chemical syntheses.

# 9. Toxicology

Caprolactam is of relatively low acute toxicity: $LD_{50}$ 1155 mg/kg (rat, oral) [84] resp. 1660 mg/kg (rat, oral); the symptoms of acute intoxication are tonoclonic convulsions. Rabbits and cats are more sensitive to caprolactam than rats [85].

Repeated epicutaneous application (50% ether solution; 10 times) and intracutaneous injection (0.1% in physiological NaCl solution) did not cause local irritation or sensitization to the skin of guinea pigs [85]. Other animal experiments, however, indicated weak sensitization [86]. Repeated instillation of 3–4 drops of a 5% or 10% solution into the conjunctival sac of the rabbit eye caused no irritation [87].

At the workplace caprolactam is usually inhaled as a vapor or dust. Repeated inhalations (7 times, 7 h) of dust containing 118–261 mg/m$^3$ was tolerated by guinea pigs without any symptoms except for occasional coughing [88]. In a subchronic inhalation study rats were exposed to relatively high concentrations of caprolactam (vapor: 500 mg/m$^3$; aerosol: 120–150 mg/m$^3$) 45 times for 4 h per day: the genital cycle of female rats was prolonged, the formation of follicles was retarded, and the reproduction of corpus luteum was stimulated [89].

Caprolactam was not teratogenic after oral administration of 1/10 of the $LD_{50}$ (116 mg/kg) to rats from day 6 to 15 of gestation [90].

Effects on kidneys after chronic exposure are being discussed as the most important toxic effect [91].

Short-term screening for mutagenic/carcinogenic potential on bacterial and mammalian cells (Ames test, CHO test, cell transformation test) was negative for caprolactam [92].

When caprolactam was administered in the feed to mice (7500 and 15 000 ppm) and rats (3750 ppm) over a period of two years no carcinogenic effect was observed; only the body weight gain was retarded with increasing concentrations [93].

Long-term experience in the handling of caprolactam has demonstrated that skin and eye irritation may occur. Skin sensitization is apparently very rare [94]. Long-term exposure of female workers to caprolactam (concentration: "a multiple of 10 mg/m$^3$") led to irregularities of menstruation and disorders during the course of pregnancy [89]. Another publication, however, reports no adverse health effects at workplace concentrations of 40–60 mg/m$^3$ [95].

Exposure levels are established at 1 mg/m$^3$ (TLV for dust) and 20 mg/m$^3$ (5 ppm; TLV for vapor) in the United States and at 25 mg/m$^3$ (MAK) in the Federal Republic of Germany.

# 10. References

General References

Houben-Weyl, **10/1**, 897–1016; **10/4**, 1–308, 449–472; **11/2**, 529–564.

M. Sittig: Chemical Process Monograph no. 21, *Caprolactam and Higher Lactams,* Noyes Development Corp., Park Ridge, N.J. 1966.

"Caprolactam Production: A Survey of Current Technology," *Eur. Chem. News* (1976) April 30; "Process Survey: Caprolactam," Eur. Chem. News (1969) May 2.

J. J. McKetta et al. in *Encyclopedia of Chemical Processing and Design,* vol. **6**, Marcel Dekker, New York – Basel 1978, 72–95.

*Kirk-Othmer,* 3rd ed., vol. **18**, 425–436.

W. Rösler et al.: "Entwicklungstendenzen und technologische Fortschritte bei der Produktion von ε-Caprolactam," *Chem. Tech. Leipzig* **30** (1978) no. 2, 67–73.

C. van de Moesdijk: "De Industriele Bereiding van Caprolactam, Het Monomer van Nylon 6," *PT-Procestech.* **36** (1981) no. 3, 147–153.

Specific References

[1] H. Hopff et al.: *Die Polyamide,* Springer Verlag, Berlin 1954, pp. 7, 23.
[2] A. Rieche et al., *Z. Chem.* **3** (1963) 443–452.
[3] A. Rieche et al., *Kunststoffe* **57** (1967) 49–52.
[4] *Houben-Weyl,* **11/2**, 565.
[5] K. Wehner et al., *Chem. Tech. Leipzig* **33** (1981) 193.
[6] *Eur. Chem. News* (1966) June 3, 36.
[7] *Houben-Weyl,* **10/4**, 236.
[8] A. H. Jubb, *Educ. Chem.* **8** (1971) 23–25.
[9] *Eur. Chem. News* (1969) May 2, Caprolactam Supplement.
[10] O. Wichterle et al., *Collect. Czech. Chem. Commun.* **16** (1951) 591–602.
[11] H. Hopff et al.: *Die Polyamide,* Springer Verlag, Berlin 1954, pp. 61–62.
[12] BASF, DE 1194863, 1961.
[13] Allied Chemical Corporation, US 3914217, 1975.
[14] A. Krzysztoforski et al., *Chem. Prod.* (1977) Dec. 8.
[15] J. Damme et al., *Chem. Eng. N.Y.* **79** (1972) July 10, 54–55. A. H. de Roij et al., *Chemtech* May 1977, 309–315.
[16] Stamicarbon, DE-OS 2106385, 1971.
[17] BASF, DE 2508247, 1975.
[18] Stamicarbon, DE 2029114, 1970.
[19] W. Dawydoff, *Chem. Tech. Leipzig* **7** (1955) 647–655.
[20] BASF, DE 1195318, 1961.
[21] BASF, DE-AS 1242620, 1961.
[22] Solvay et Cie, BE 695573, 1967.
[23] BASF, DE 1227028, 1962.
[24] BASF, EP 17945, 1980.
[25] Bayer, DE 2641414, 1976.
[26] BASF, DE-OS 1445549, 1964.
[27] Tore K. K., JP 24783, 1967.

[28] Inventa AG, CH 317460, 1953; CH 484142, 1967.
[29] Kanegafuchi Boseki K. K., JP 26300, 1966.
[30] V. Kaleb, *Sb. Vys. Sk. Chem. Technol. Praze K. Chem. Inz.* Chemische Industrie (Czech) 19/44, 4, 162 (1969).
[31] S. Stempel et al., Chemische Industrie (Czech) 19/44, 10467 (1969).
[32] Zimmer Aktiengesellschaft, DE-OS 1620755, 1965.
[33] Chemicke Zavody Na Slovensko, GB 666717, 1949.
[34] Bayer, DE-OS 1930218, 1969.
[35] Allied Chemical Corporation, US 3406167, 1965.
[36] VEB Chemieanlagenbau Erfurt, DE-OS 2035859, 1970.
[37] Stamicarbon, DE 953168, 1954.
[38] BASF, DE-OS 2100034, 1971.
[39] BASF, DE-OS 2062436, 1970.
[40] BASF, DE-OS 1545617, 1965.
[41] Inventa AG, CH 394212, 1962.
[42] *Houben-Weyl*, **8**, 669–670; **11/2**, 529–564.
[43] Toyo Rayon, DE-OS 1916149, 1969.
[44] Bayer, GB 1021709, 1964.
[45] Stamicarbon, DE-OS 1620468, 1965; 2129657, 1971.
[46] Stamicarbon, DE-OS 2704561, 1977.
[47] G. A. Turner, *Chem. Process. (London)* **15** (1969) 4–6.
[48] Toyo Rayon, DE-AS 1150975, 1960.
[49] Y. Ito et al., *Ann. N.Y. Acad. Sci.* 155 (1969) 618–624.
[50] *Hydrocarbon Process.* **62** (1983) Nov., 85.
[51] Snia Viscosa, IT 603606, 1960; 604795, 1960; 608873, 1960.
[52] M. Taverna et al., *Hydrocarbon Process.* **49** (1970) 137–145.
[53] G. Siali, *Hydrocarbon Process.* **53** (1974) July, 124; *Hydrocarbon Process.* **54** (1975) Jan., 83.
[54] A. Hcath, *Chem. Eng. N.Y.* **81** (1974) no. 15, 70.
[55] R. Conrad, *Chem. Ing. Tech.* **45** (1973) 1510.
[56] SU 176680, 1964.
[57] BASF, DE 3030735, 1980.
[58] Union Carbide, US 3000880, 1959; 3025306, 1960; 3064008, 1960.
[59] C. Grundmann et al., *Angew. Chem.* **62** (1950) 556–560.
[60] Stamicarbon, DE-AS 1012910, 1953.
[61] Du Pont, DE-AS 1073490, 1957.
[62] Du Pont, US 2634269, 1951.
[63] Inventa AG, DE 939808, 1952.
[64] D. L. Lebedew et al., *J. Gen. Chem. USSR (Engl. Transl.)* **30** (1960) no. 92, 1631.
[65] Toa Gosei Chem. Ind., DE-AS 1274124, 1967.
[66] BP Chemicals Ltd., DE-OS 1695503, 1967; 1770477, 1968; 1803872, 1968.
[67] Degussa, DE-OS 2003269, 1970; 2004440, 1970.
[68] Allied Chemical Corporation, EP-A 43445, 1981.
[69] Allied Chemical Corporation, US 4163756, 1979; 4281194, 1981; 4225511, 1980.
[70] Monsanto, DE-OS 2027810, 1970.
[71] Anorgana GmbH, DE 935544, 1953.
[72] Teijin Ltd., JP 7210368, 1967; 7211736, 1969; 7211413, 1967.
[73] Teijin Ltd., JP 6614983, 1964.

[74] Société d'Electrochimie d'Ugine, FR 1411872, 1964.
[75] Stamicarbon, NL 6412868, 1964.
[76] Teijin Ltd., DE-OS 2111216, 1971.
[77] Kanegafuchi-Boseki K. K., FR 1506874, 1966.
[78] W. Reppe: *Neuere Entwicklungen auf dem Gebiet der Chemie des Acetylens und Kohlenoxyds*, Springer Verlag, Berlin – Göttingen – Heidelberg 1949, 38.
[79] The Techni-Chem Comp., DE-OS 1940809, 1969.
[80] Teijin Ltd., DE-OS 1931121, 1969.
[81] VEB Leuna Werke "Walter Ulbricht", DE-OS 1961474, 1969; 1961473, 1969; 2055165, 1970.
[82] *Chem. Eng. News* **51** (1973) no. 15, 114.
[83] Kanebo Co. Ltd., DE-OS 2307302, 1973.
[84] G. Bornmann, A. Löser, *Arzneim. Forsch.* **9** (1959) 9.
[85] BASF, unpublished results (1965/66).
[86] T. P. Ivanova et al., *Farmakol. Toksikol. (Kiev)* **8** (1973) 178.
[87] F. Hohensee, *Faserforschung* **1** (1951) 299.
[88] M. W. Goldblatt et al., *Br. J. Ind. Med.* **11** (1954) 1.
[89] A. P. Martynova et al., *Gig. Tr. Prof. Zabol.* **11** (1972) 9.
[90] BASF, unpublished results (1978).
[91] M. A. Friedman et al., *Food Cosmet. Toxicol.* **18** (1980) 39.
[92] E. J. Greene et al., *Environ. Mutagen.* **1** (1979) 399.
[93] National Toxicology Program (USA), Report 1982, NIH Publ. 81–1770, NTP-80-26.
[94] H. Janson, *Zentralbl. Haut Geschlechtskr.* **26** (1959) 37.
[95] Z. Zwierzchowski et al., *Med. Pr.* **18** (1967) no. 4, 357.

# Carbamates and Carbamoyl Chlorides

*Indvidual keywords:* → *Dithiocarbamic Acid and Derivatives*

PETER JÄGER, BASF Aktiengesellschaft, Ludwigshafen/Rh., Federal Republic of Germany (Chaps. 2–4)

COSTIN N. RENTZEA, BASF Aktiengesellschaft, Ludwigshafen/Rh., Federal Republic of Germany (Chaps. 2–4)

HEINZ KIECZKA, BASF Aktiengesellschaft, Ludwigshafen /Rh., Federal Republic of Germany (Chap. 5)

| | | | | |
|---|---|---|---|---|
| 1. | Introduction .............. 1045 | 4. | Carbamoyl Chlorides ....... 1048 |
| 2. | Salts of Carbamic Acid...... 1045 | 5. | Toxicology and Occupational Health................. 1053 |
| 3. | Esters of Carbamic Acids .... 1046 | 6. | References............... 1055 |

## 1. Introduction

Carbamic acid, $H_2NCOOH$, the half amide of carbonic acid, does not exist as the free acid but forms numerous stable metal salts, esters, halides, amides (i.e., urea and urea derivatives), and simple or mixed anhydrides. A large number of derivatives result from replacement of the hydrogen atoms on the nitrogen atom by organic radicals.

## 2. Salts of Carbamic Acid

The salts of carbamic acid (carbamates) are relatively unstable. In aqueous solution they hydrolyze slowly at room temperature, and more rapidly on heating, to the corresponding metal carbonates and ammonia. Heating of the anhydrous solid produces cyanates from alkali-metal carbamates, and cyanamide from alkaline earthmetal carbamates.

Of the unsubstituted salts only *ammonium carbamate* is made industrially. It is present in technical-grade ammonium carbonate (hartshorn, sal volatile) and is an intermediate, which is not isolated, in the industrial synthesis of urea.

# 3. Esters of Carbamic Acids

Carbamic acid esters are referred to as carbamates, in the same way as the metal salts, or by the trivial name urethane.

**Physical Properties.** The unsubstituted and the *N*-phenyl carbamates are generally crystalline, whereas the simple mono and di N-alkyl substituted compounds are liquids at ambient temperature. The physical data of the simple esters are shown in Table 1.

**Preparation.** There is no outstanding synthetic route to the esters. Depending on the carbamate to be prepared and the availability of the starting materials, one of the following methods has an economic advantage.

During the preparation of N-monosubstituted esters temperatures of 60–80 °C should not be exceeded, in order to prevent decomposition into isocyanate and alcohol.

*Alcoholysis of Carbamoyl Chlorides:*

$$R^1R^2N-COCl + R^3OH \xrightarrow{-HCl} R^1R^2N-COOR^3$$

In order to obtain good yields the reaction should be carried out in the presence of a base, such as a tertiary amine or pyridine, or in the presence of an acylation catalyst, such as 4-dimethylaminopyridine [1].

*Aminolysis of Chloroformates:*

$$2\,R^1R^2NH + ClCOOR^3 \longrightarrow R^1R^2N-COOR^3 + R^1R^2NH \cdot HCl$$

Since the second mole of amine neutralizers the hydrogen chloride no other base is required; however, acylation catalysts may be used if necessary. The reaction can also be carried out with tertiary amines with the formation of an alkyl chloride [2].

*Reaction of Urea or Urea Nitrate with Alcohols in the Presence of Heavy Metals* [3]:

$$H_2N-CO-NH_2 + ROH \xrightarrow{ZnCl_2} H_2N-COOR + NH_3$$

All the lower unsubstituted carbamates (e.g., methyl or ethyl carbamate) are prepared by this method.

*Reaction of Alcohols with Isocyanates:*

$$R^1NCO + R^2OH \longrightarrow R^1NH-COOR^2$$

Alkyl or aryl isocyanates yield monosubstituted carbamates, whereas isocyanic acid, HNCO, yields unsubstituted esters [4]. For the substituted isocyanates the order of reactivity is *prim.* alkyl > *sec*-alkyl > *tert*-alkyl > phenyl. Lewis bases such as triethylamine and Lewis acids such as dibutyltin dilaurate [5] are useful catalysts.

**Table 1.** Physical properties of simple carbamates

| Substituent | Formula | $M_r$ | CAS reg. no. | mp, °C | bp, °C at 101.3 kPa | Density, g/cm³ | State at 20 °C |
|---|---|---|---|---|---|---|---|
| Methyl | $H_2N-COOCH_3$ | 75.07 | [598-55-0] | 54 | 177 | 1.136 (56 °C) | solid |
| Ethyl | $H_2N-COOC_2H_5$ | 89.09 | [51-79-6] | 48.1 | 185 | 1.0599 (48.2 °C) | solid |
| n-Propyl | $H_2N-COOC_3H_7$ | 103.12 | [627-12-3] | 60 | 196 | | solid |
| Isopropyl | $H_2N-COOCH(CH_3)_2$ | 103.12 | [1746-77-6] | 92.4 | 181 (94.6 kPa) | 0.995 (66 °C) | solid |
| n-Butyl | $H_2N-COOC_4H_9$ | 117.15 | [592-35-8] | 54 | 204 | | solid |
| Isobutyl | $H_2N-COOCH_2CH(CH_3)_2$ | 117.15 | [543-28-2] | 67 | 207 | 0.947 (77.7 °C) | solid |
| n-Pentyl | $H_2N-COOC_5H_{11}$ | 131.17 | [638-42-6] | 55.5–62.5 | 116.7 (1.73 kPa) | | solid |
| n-Hexyl | $H_2N-COOC_6H_{13}$ | 145.20 | [2114-20-7] | 65 | | | solid |
| Phenyl | $H_2N-COOC_6H_5$ | 137.14 | [622-46-8] | 143 | | 1.079 (60 °C) | solid |
| Methyl N-methyl | $CH_3NH-COOCH_3$ | 89.09 | [6642-30-4] | | 158 (101.9 kPa) | 1.065 | liquid |
| Methyl N-ethyl | $C_2H_5NH-COOCH_3$ | 103.12 | [6135-31-5] | | 165 (102.4 kPa) | | liquid |
| Methyl N,N-dimethyl | $(CH_3)_2N-COOCH_3$ | 103.12 | [7541-16-4] | | 131 | 1.012 (15 °C) | liquid |
| Ethyl N,N-dimethyl | $(CH_3)_2N-COOC_2H_5$ | 117.15 | [687-48-9] | | 147 | 0.972 (15 °C) | liquid |
| 1-Naphthyl N-methyl | $CH_3NHCOOC_{10}H_7$ | 201.22 | [63-25-2] | 142 | (decomp.) | | solid |

*Transamidation of Simple Carbamates with Higher Amines:*

$$R^1NH_2 + H_2N-COOR \xrightarrow{Cat.} R^1NH-COOR + NH_3$$

Zinc acetate is commonly used as the catalyst [6].

*Transesterification of Lower Alkyl Carbamates with Higher Alcohols:*

$$R^1R^2N-COOR^3 + R^4OH \xrightarrow{Cat.} R^1R^2N-COOR^4 + R^3OH$$

Aluminum isopropylate is commonly used as the catalyst [7].

*Aminolysis of Dialkyl Carbonates:*

$$R^1O-COOR^1 + HNR^2R^3 \longrightarrow R^2R^3N-COOR^1 + R^1OH$$

*Rearrangements of ω-Alkoxyalkyl- and ω-Phenoxyalkylcarbamoyl Chlorides [8]:*

$$R^1O(CH_2)_n N{<}^{COCl}_{R^2} \longrightarrow Cl(CH_2)_n N{<}^{COOR^1}_{R^2}$$

*Oxidative Amination of Carbon Monoxide:*

$$R^1NH_2 + CO + R^2OH + 1/2\,O_2 \xrightarrow{Cat.} R^1NH-COOR^2 + H_2O$$

Preferred catalysts are iron compounds [9] in the presence of small amounts of a noble metal.

*Reductive Carbonylation of Organic Nitro Compounds:*

$$C_6H_5NO_2 + CO + R^1OH \xrightarrow{Cat.} C_6H_5NH-COOR^1$$

The usual catalyst is palladium together with an iron compound [10].

**Uses.** Carbamates are used mainly as crop protection agents and pharmaceuticals, with the greater proportion being used as insecticides, herbicides, and fungicides. For examples, see p. 1052.

Of the simple carbamates, the methyl ester and the ethyl ester (urethane) still have a certain importance, especially as starting materials for the preparation of higher carbamates. The condensation product of ethyl carbamate and formaldehyde is used to give a crease-resistant finish to textiles [11].

# 4. Carbamoyl Chlorides

Carbamoyl chlorides, also known as carbamic acid chlorides, are industrially important as intermediates because of their ready availability and of their reactivity.

**Table 2.** Physical properties of some carbamoyl chlorides

| N-Substituents | Formula | $M_r$ | CAS reg. no. | mp, °C | bp, °C | Density |
|---|---|---|---|---|---|---|
| | $H_2N-COCl$ | 79.49 | [463-72-9] | ca. 50 | 61–2 | |
| Methyl | $CH_3NH-COCl$ | 93.51 | [6452-47-7] | 90 | 93–4 | |
| Dimethyl | $(CH_3)_2N-COCl$ | 107.54 | [79-44-7] | −33 | 167 | 1.17 |
| Ethyl | $C_2H_5NH-COCl$ | 107.54 | [41891-13-8] | | 92.3 | |
| Diethyl | $(C_2H_5)_2N-COCl$ | 135.59 | [88-10-8] | −32 | 186 | 1.073 |
| n-Propyl | $n\text{-}C_3H_7NH-COCl$ | 121.57 | [41891-16-1] | | 109 (2.67 kPa) | |
| Di-n-propyl | $(n\text{-}C_3H_7)_2N-COCl$ | 163.65 | [27086-19-7] | −21 | 100–4 (1.6 kPa) | 1.023 |
| Diisopropyl | $(iso\text{-}C_3H_7)_2N-COCl$ | 163.65 | [19009-39-3] | 56–7 | | |
| Di-n-butyl | $(n\text{-}C_4H_9)_2N-COCl$ | 191.70 | [13358-73-1] | | ca. 245 | 0.991 |
| Di-sec-butyl | $(sec\text{-}C_4H_9)_2N-COCl$ | 191.70 | [36756-72-6] | −42–6 | 151 (0.6 kPa) | 1.009 |
| Diisobutyl | $(iso\text{-}C_4H_9)_2N-COCl$ | 191.70 | [38952-42-0] | −18 | 105–9 (0.13 kPa) | 0.985 |
| Di-n-hexyl | $(n\text{-}C_6H_{13})_2N-COCl$ | 247.81 | [27086-21-1] | | 162–4 (0.13 kPa) | |
| Di-n-octyl | $(n\text{-}C_8H_{17})_2N-COCl$ | 303.92 | [27086-22-2] | | 200–2 (0.13 kPa) | |
| Methoxy-methyl | $CH_3O\diagdown N-COCl$ / $CH_3\diagup$ | 123.54 | [30289-28-2] | | 86 (5.0 kPa) | |
| Hexamethylene | ⟨ring⟩N−COCl | 161.63 | [27817-35-2] | −66 | 135 (2.67 kPa) | 1.159 |
| Oxydiethylene | O⟨ring⟩N−COCl | 149.58 | [15159-40-7] | ca. −50 | 116 (2.0 kPa) | 1.32 |
| Phenyl | $C_6H_5NH-COCl$ | 155.58 | [2040-76-8] | 59 | | |
| Diphenyl | $(C_6H_5)_2N-COCl$ | 231.68 | [83-01-2] | 85 | 138 (0.13 kPa) | |

**Physical Properties.** The simple carbamoyl chlorides are colorless liquids or solids, usually with pungent odor. The physical properties of the most important carbamoyl chlorides are given in Table 2.

**Chemical Properties.** Carbamoyl chloride, $H_2N-COCl$, is unstable; it may be stabilized by Lewis acids (e.g., aluminum chloride) as adducts and in this form can be used in the Friedel-Crafts acylation of arenes [12]. Since many of the monosubstituted carbamoyl chlorides also deteriorate on storage, the corresponding isocyanates are used instead. Disubstituted carbamoyl chlorides are stable and can be stored for prolonged periods.

In their chemical behavior the carbamoyl chlorides resemble the carboxylic acid chlorides. Their most important reaction is the nucleophilic displacement of the chlorine atom.

Since carbamoyl chlorides are less reactive than carboxylic acid chlorides, acid acceptors or acylation catalysts such as 4-dimethylaminopyridine are generally used in reactions involving the liberation of hydrogen chloride. Table 3 gives an overview of the reactions of carbamoyl chlorides.

Hydrolysis of carbamoyl chlorides yields the corresponding substituted ammonium chloride and carbon dioxide instead of the unstable carbamic acid. This reaction is used to remove traces of carbamoyl chloride vapor from exhaust gases or to render residues harmless.

**Table 3.** Chemical reactions of carbamoyl chlorides

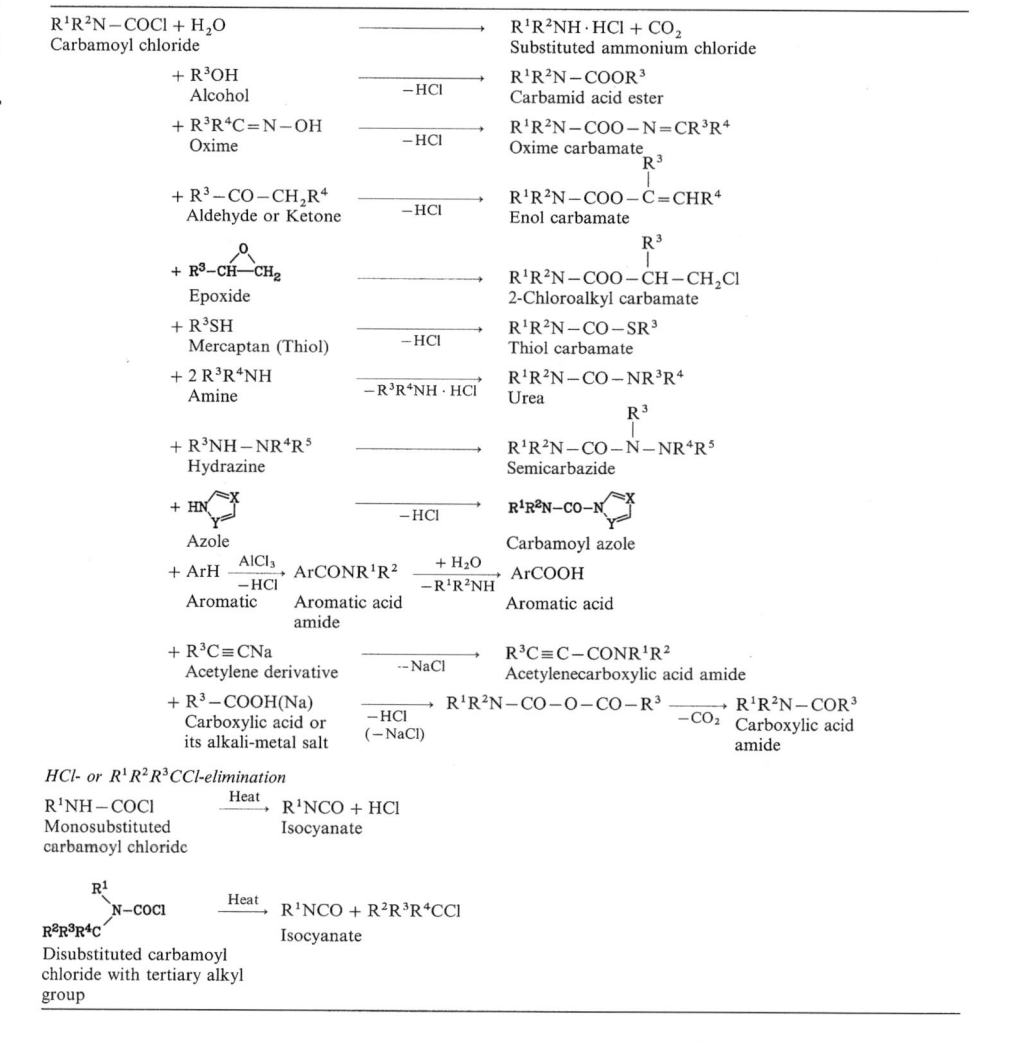

| Reactant | Conditions | Product |
|---|---|---|
| $R^1R^2N-COCl + H_2O$ (Carbamoyl chloride) | → | $R^1R^2NH \cdot HCl + CO_2$ (Substituted ammonium chloride) |
| $+ R^3OH$ (Alcohol) | $-HCl$ | $R^1R^2N-COOR^3$ (Carbamid acid ester) |
| $+ R^3R^4C=N-OH$ (Oxime) | $-HCl$ | $R^1R^2N-COO-N=CR^3R^4$ (Oxime carbamate) |
| $+ R^3-CO-CH_2R^4$ (Aldehyde or Ketone) | $-HCl$ | $R^1R^2N-COO-\underset{\underset{R^3}{|}}{C}=CHR^4$ (Enol carbamate) |
| $+ R^3-\overset{O}{\overset{|}{CH}}-CH_2$ (Epoxide) | | $R^1R^2N-COO-\underset{\underset{R^3}{|}}{CH}-CH_2Cl$ (2-Chloroalkyl carbamate) |
| $+ R^3SH$ (Mercaptan (Thiol)) | $-HCl$ | $R^1R^2N-CO-SR^3$ (Thiol carbamate) |
| $+ 2 R^3R^4NH$ (Amine) | $-R^3R^4NH \cdot HCl$ | $R^1R^2N-CO-NR^3R^4$ (Urea) |
| $+ R^3NH-NR^4R^5$ (Hydrazine) | | $R^1R^2N-CO-\underset{\underset{R^3}{|}}{N}-NR^4R^5$ (Semicarbazide) |
| $+ HN\overset{X}{\diagdown}$ (Azole) | $-HCl$ | $R^1R^2N-CO-N\overset{X}{\diagdown}$ (Carbamoyl azole) |
| $+ ArH$ (Aromatic) | $\xrightarrow[-HCl]{AlCl_3}$ ArCONR$^1$R$^2$ (Aromatic acid amide) $\xrightarrow[-R^1R^2NH]{+H_2O}$ | ArCOOH (Aromatic acid) |
| $+ R^3C\equiv CNa$ (Acetylene derivative) | $-NaCl$ | $R^3C\equiv C-CONR^1R^2$ (Acetylenecarboxylic acid amide) |
| $+ R^3-COOH(Na)$ (Carboxylic acid or its alkali-metal salt) | $\xrightarrow[(-NaCl)]{-HCl}$ $R^1R^2N-CO-O-CO-R^3$ $\xrightarrow{-CO_2}$ | $R^1R^2N-COR^3$ (Carboxylic acid amide) |

*HCl- or $R^1R^2R^3$CCl-elimination*

| | | |
|---|---|---|
| $R^1NH-COCl$ (Monosubstituted carbamoyl chloride) | $\xrightarrow{Heat}$ | $R^1NCO + HCl$ (Isocyanate) |
| $\underset{R^2R^3R^4C}{\overset{R^1}{\diagdown}}N-COCl$ (Disubstituted carbamoyl chloride with tertiary alkyl group) | $\xrightarrow{Heat}$ | $R^1NCO + R^2R^3R^4CCl$ (Isocyanate) |

**Production.** In industry phosgene, $COCl_2$, is usually the starting material.

$$COCl_2 + R^1R^2NH \longrightarrow R^1R^2N-COCl + HCl$$

$R^1$, $R^2$ = alkyl, alkoxy, aryl

At low reaction temperatures (cold phosgenation) the hydrogen chloride liberated removes half the amine as the ammonium salt $R^1R^2NH \cdot HCl$. Complete conversion of valuable amines into the corresponding carbamoyl chlorides is achieved by adding a base, e.g., pyridine.

The same is achieved by the use of higher temperatures (about 90–100 °C, hot phosgenation) in which case the hydrogen chloride is removed from the reaction mixture as a gas. Crystallization of the substituted ammonium chloride during the course of the reaction can lead to difficulties with stirring or to blockage of the reactor. Sometimes, however, the amine salt is used as the starting material. Phosgenation is carried out at room temperature because the salt, unlike the free amine, does not react with the carbamoyl chloride already formed to produce a substituted urea:

$$R^1R^2NH + R^1R^2N-COCl \longrightarrow R^1R^2N-CO-NR^1R^2 + HCl$$

To repress urea formation, excess phosgene is used together with an appropriate amount of solvent as a diluent. Good mixing must also be employed to prevent any local excess of amine.

The solvent used should be a good solvent for phosgene, e.g., toluene, chlorobenzene, or ethyl acetate [15], and should be easily separable from the carbamoyl chloride, i.e., the boiling point difference must be great enough. This separation is not required when the carbamoyl chloride serves as the reaction medium [21].

The amine also reacts in the gas phase with phosgene, if necessary in a diluent inert gas (e.g., nitrogen), at temperatures above 250 °C [22]. Carbamic acid chloride can be prepared in this way from ammonia and phosgene at 400–450 °C [12].

Phosgene also reacts under mild conditions (50–60 °C) with tertiary amines containing at least one lower alkyl group (preferably methyl or ethyl) to give alkyl halides as byproducts [23]:

$$COCl_2 + R^1R^2NR^3 \longrightarrow R^1R^2N-COCl + R^3Cl$$

$R^3$ = methyl or ethyl

Carbamoyl chlorides can also be obtained by the addition reaction of phosgene to the C=N double bond of azomethines [24]–[26]. α-Chloro- or α,β-unsaturated carbamoyl chlorides are formed, depending on the reaction conditions:

$$R^1N=C(R^2)-CH(R^3)(R^4) + COCl_2 \longrightarrow R^1N(COCl)-CCl(R^2)-CH(R^3)(R^4) \longrightarrow R^1-N(COCl)-C(R^2)=C(R^3)(R^4) + HCl$$

In the laboratory, carbamoyl chlorides may be prepared by the addition of hydrogen chloride to isocyanates:

$$RNCO + HCl \longrightarrow RNH-COCl$$

or by the chlorination of disubstituted formamides with sulfur dichloride, $SCl_2$, sulfuryl chloride, $SO_2Cl_2$ [27], phosphorus trichloride, $PCl_3$, or thionyl chloride, $SOCl_2$ [28]:

$$R^1R^2N-CHO \longrightarrow R^1R^2N-COCl$$

or by the action of phosphorus oxychloride, $POCl_3$, or phosphorus pentachloride, $PCl_5$, on carbamic acid esters [29]:

$$R^1R^2N-COOR^3 \longrightarrow R^1R^2N-COCl$$

as well as by carbonylation of chloramines [30]:

$$R^1R^2N-Cl + CO \longrightarrow R^1R^2N-COCl$$

**Uses.** As is apparent from the reactions shown in Table 3, carbamoyl chlorides are intermediates with many uses. Of primary industrial importance are the reactions with alcohols, phenols, and oximes to give carbamic acid esters; with thiols (mercaptans) to give thiocarbamates; with amines and hydroxylamines to give substituted ureas; and with imidazoles and triazoles to give carbamoyl azoles. All these are important classes of compounds in crop protection. The following crop protection materials are examples of products derived from carbamoyl chlorides:

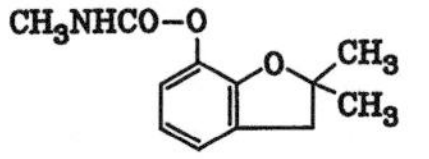

Carbofuran (insecticide)

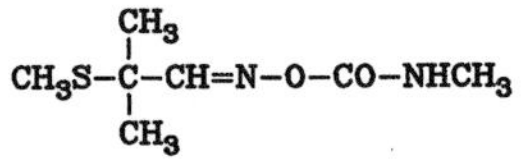

Temik (insecticide)

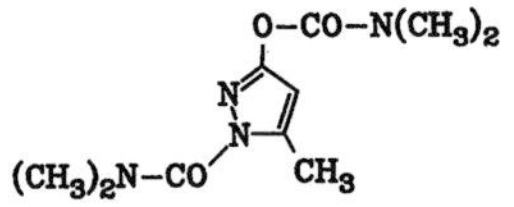

Dimetilan (insecticide)

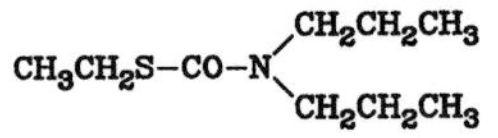

Eptam (herbicide)

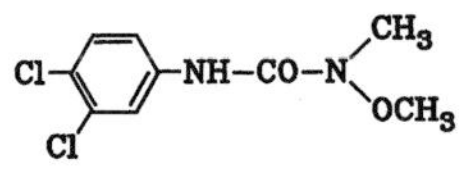

Linuron (herbicide)

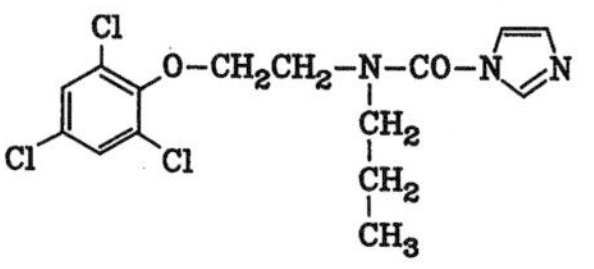

Sportak (fungicide)

**Handling and Transportation.** The industrially important lower carbamoyl chlorides in particular are chemically and biologically (see Chap. 5) active substances. The selection of suitable apparatus and corrosion resistant materials (e.g., glass, polytetrafluoroethylene, glass-lined steel, or, if necessary, tantalum or Hastelloy) requires much care and due attention must also be given to the various regulations existing in different countries. In the Federal Republic of Germany, for example, dimethylcarbamoyl chloride and N-chloroformylmorpholine (morpholinocarbamoyl chloride) are specifically mentioned in Appendix II No. 1 of the regulations dealing with dangerous materials [31] as well as in Appendix II of the regulations dealing with noxious emissions [32].

Most of the details about technical safety data are to be found in [33]–[35].

# 5. Toxicology and Occupational Health

**Ethyl Carbamate.** There is little data available on the acute effects of this compound: the median lethal dose ($LD_{50}$) upon oral administration in mice is 2500 mg/kg [36].

The most important toxicological effect is its *carcinogenicity*. Ethyl carbamate has proved to be carcinogenic in mice, rats, and hamsters upon oral, inhalative, subcutaneous, or intraperitoneal administration [37]. Lung tumors, lymphomas, hepatomas, melanomas, tumors of the blood vessels, and skin tumors (upon treatment of the skin) have occurred in the tests. The carcinogenic effects were observed upon administration of the following doses: lifelong treatment of rats with 1 ppm ethyl carbamate in drinking water; lifelong treatment of hamsters with 0.2% ethyl carbamate in drinking water; two treatments (once a week) of the skin of mice with 120 mg ethyl carbamate, followed by eighteen applications (once a week) of 0.3 mL of a 0.5% preparation in croton oil; inhalative treatment of mice (20–60 min per d) with aerosols containing varying amounts of ethyl carbamate, for a maximum of 14.5 weeks. Even short-term treatments, for example, a ten-day treatment of young mice with 0.4% ethyl carbamate in drinking water, led to the development of tumors. Further investigations have confirmed the types of effects described here.

Ethyl carbamate has been classified by the MAK commission as being a carcinogen in animal experiments and is listed in group III A/2.

**Diethylcarbamoyl Chloride.** The median lethal dose ($LD_{50}$) upon oral administration to rats is about 200 mg/kg [38]. The compound is irritating to skin and eyes. The inhalation of an atmosphere saturated with its vapors at 20 °C is lethal to rats after 3 h [39]. Diethylcarbamoyl chloride has proved to be mutagenic on *E. coli* strains in mutagenicity tests [40]. It is weakly mutagenic in high doses in the Ames test [41]. In view of these results and on account of the structural similarity to dimethylcarbamoyl chloride, a carcinogenic potential cannot be ruled out. The MAK commission has listed this substance in group III B.

**Dimethylcarbamoyl Chloride.** The $LD_{50}$ is 1170 mg/kg upon oral administration to rats and 350 mg/kg upon intraperitoneal administration to mice. The compound is irritating to the skin and eyes. All the test animals survived an 8-min inhalation test in an atmosphere saturated with vapors of the substance at 20 °C. Longer exposure times led to death. No sensitization was observed in animal experiments [42].

Dimethylcarbamoyl chloride proved to be a mutagen in numerous short-term tests for mutagenicity. They included the Ames test [43], [44], cell transformation test [45], sister chromatid exchange [46], [48], and micronucleus test [47]. The compound caused damage to the DNA of bacteria, fungi, and plants, but not in mammalian cells in vitro. It was mutagenic in bacteria, fungi, and plants, as well as in mammalian cells in vitro;

however, it was not mutagenic in insects. Chromosomal aberrations were observed in fungi and plants, and in mammalian cells in vitro and in vivo, but not in humans. A review is given in [49].

Dimethylcarbamoyl chloride is carcinogenic in animal experiments. Mice developed tumors at the site of application upon percutaneous treatment with 2 mg substance in 0.1 mL acetone, three times weekly, for 492 d. Its tumorigenic effect was confirmed in another dermal application study. The subcutaneous application of dimethylcarbamoyl chloride to mice (weekly injection of 5 mg substance in 0.05 mL tricapryline over a period of 26 weeks) also caused local tumors and — as with percutaneous treatment — reduced the survival rate. Intraperitoneal treatment of mice (weekly dose of 1 mg substance in 0.05 mL tricapryline over a period of 450 d) led to increased incidences of lung tumors [50]. Lifelong inhalation of 1 ppm of dimethylcarbamoyl chloride over 6 h every day led to squamous cell carcinomas of the nasal mucosa in 51% of the hamsters employed. In a comparable study using more sensitive rats, 96% of the animals developed tumors of the respiratory tract [51]. The MAK commission has listed this substance in group III A/2; in the LV list dimethylcarbamoyl chloride is in group A 2.

**Diphenylcarbamoyl Chloride.** The $LD_{50}$ in rats after a single oral administration is above 2000 mg/kg. It is not irritating to the skin and eyes but caused sensitization in animal experiments [52].

**Morpholinocarbamoyl Chloride.** Its acute toxicity in rats is relatively small. The $LD_{50}$ after oral administration of the substance in peanut oil is about 3500 mg/kg.

The compound is only weakly mutagenic in the Ames test. However, in view of the instability of the compound in aqueous solutions, it is doubtful whether it is only a weak mutagen.

Weekly subcutaneous injection of rats with 300 mg/kg in peanut oil over more than 140 d led to tumors around the site of injection after 140 d. This treatment caused the death of all the animals within 340 d. As many as 58 of the 60 test animals in the test developed fibrosarcomas at the site of injection, and two animals developed adenocarcinomas of the mamma. Weekly subcutaneous treatment with 100 mg/kg in peanut oil over 140 d led to the death of all the rats within 335 d. As many as 59 of the 60 animals showed a sarcoma at the site of injection; in contrast, there were only three incidences of tumors among the sixty control animals treated with peanut oil alone. Morpholinocarbamoyl chloride has been listed in group III A/2 by the MAK commission [53].

# 6. References

General References

*Beilstein* **3**, 20–31; **3 (1)**, 9–15; **3 (2)**, 18–27; **3 (3)**, 39–65; **3 (4)**, 37–80.
*Houben-Weyl*, **8**, 111–118, 137–149; **E (4)**, 36–64, 142–192.
P. Adams, F. Baron: "Esters of Carbamic Acid," *Chem. Rev.* **65** (1965) 567–602.
G. Scheuerer: "Carbamate als Agrarchemikalien," *Fortschr. Chem. Forsch.* **9** (1967/68) 254–294.

Specific References

[1] G. Höfle, W. Steglich, H. Vorbrüggen, *Angew. Chem.* **90** (1978) 602–615; *Angew. Chem. Int. Ed. Engl.* **17** (1978) 569–583.
[2] M. Matzner, R. P. Kurkjy, R. J. Cotter, *Chem. Rev.* **64** (1964) 645.
[3] A. Paquin, *Z. Naturforsch.* **1** (1946) 518–523.
[4] Ota-Seiyaku Co., JP 4716433, 1972.
[5] T. Francis, M. P. Thorne, *Can. J. Chem.* **54** (1976) 24.
[6] BASF, EP 18583, 1981.
[7] W. M. Kraft, *J. Am. Chem. Soc.* **70** (1948) 3569.
[8] R. Banks, R. F. Brookes, D. H. Godson, *J. Chem. Soc. Perkin Trans. 1* 1975, 1836–1840.
[9] BASF, DE-OS 2910132, 1979.
[10] Bayer, DE-OS 2819826, 1978.
[11] R. Arceneaux, J. Fricke, Jr., J. Reid, G. Gautreaux, *Am. Dyest. Rep.* **50** (1961) 849–853.
[12] H. Hopf, H. Ohliger, *Angew. Chem.* **61** (1949) 183–185.
[13] B. F. Filipasic, R. Patarcity, *Chem. Ind. (London)* 1969, 166–167.
[14] F. Boberg, G. Schultze, *Chem. Ber.* **88** (1955) 275–280.
[15] The Boots Co., GB 1469772, 1973.
[16] G. A. Olah, J. Olah: *Friedel-Crafts and Related Reactions*, vol. **III**, Interscience, New York 1964, pp. 1262–1267.
[17] R. Epsztein, *C. R. Hebd. Séances Acad. Sci.* **240** (1955) 989–990.
[18] SU 229489, 1967 (A. Zalikin, Y. Sterpikheev).
[19] J. Lawson, Jr., J. Croom, *J. Org. Chem.* **28** (1963) 232–235.
[20] J. Tilley, A. Sayigh, *J. Org. Chem.* **28** (1963) 2076–2079.
[21] Ciba-Geigy, DE-OS 2206365, 1972.
[22] R. Slocombe, E. Hardy, J. Saunders, R. Jenkins, *J. Am. Chem. Soc.* **72** (1950) 1888–1891.
[23] H. Babad, A. G. Zeiler, *Chem. Rev.* **73** (1973) 75–91.
[24] BASF, DE-OS 1901542, 1969.
[25] Bayer, DE-OS 2146069, 1971.
[26] H. Kiefer, *Synthesis* 1972, 39–42.
[27] U. Hasserodt, *Chem. Ber.* **101** (1968) 113–120.
[28] N. Schindler, W. Plöger, *Chem. Ber.* **104** (1971) 969–971.
[29] O. Schmidt, *Ber. Dtsch. Chem. Ges.* **36** (1903) 2479.
[30] T. Saegusa, T. Tsuda, Y. Isegawa, *J. Org. Chem.* **36** (1971) 857–860.
[31] Verordnung über gefährliche Arbeitsstoffe (Arb-StoffV), publ. 11th February 1982.
[32] Zwölfte Verordnung zur Durchführung des Bundes-Immissionsschutzgesetzes (Störfall-Verordnung) – 12. BImSchV.

[33] G. Hommel: *Handbuch der gefährlichen Güter,* Springer Verlag, Berlin – Heidelberg – New York 1983.
[34] *Encyclopedia of Occupational Health and Safety,* International Labour Organisation, Geneva 1983.
[35] *Handbook of Toxic and Hazardous Chemicals,* Noyes Publications, Park Ridge, New Jersey, 1981.
[36] *Arzneim. Forsch.* **9** (1959) 595.
[37] *IARC Monogr. Eval. Carcinog. Risk Chem. Man* **7** (1974) 111–140.
[38] BASF, unpublished results, 1977.
[39] BASF, unpublished results, 1970.
[40] J. F. Finklea, National Institute for Occupational Safety and Health; Current Intelligence Bulletin no. 12, July 7, 1976.
[41] BASF, unpublished results, 1978.
[42] W. Hey et al., *Zentralbl. Arbeitsmed. Arbeitsschutz* **24** (1974) 71–77.
[43] J. F. Finklea, National Institute for Occupational Safety and Health; Current Intelligence Bulletin no. 11, July 7, 1976.
[44] J. McCann et al., *Proc. Nat. Acad. Sci. USA* **72** (1975) 979–983.
[45] C. Heidelberger, *Mutat. Res.* **114** (1983) 283–385.
[46] E. L. Evans et al. in P. J. de Serres, J. Ashby (eds.): *Progress in Mutation Research. Evaluation of Short-Term Tests for Carcinogenesis,* vol. **1,** Elsevier, Amsterdam 1981, Chap. 49.
[47] B. Kirkhart in F. J. de Serres, J. Ashby (eds.): *Progress in Mutation Research. Evaluation of Short-Term Tests for Carcinogenesis,* vol. **1,** Elsevier, Amsterdam 1981, Chap. 67.
[48] A. T. Natarajan et al. in F. J. de Serres, J. Ashby (eds.): *Progress in Mutation Research. Evaluation of Short-Term Tests for Carcinogenesis,* vol. **1,** Elsevier, Amsterdam 1981, Chap. 50.
[49] *IARC Monogr. Suppl.* **4** (1982) 118.
[50] *IARC Monogr. Eval. Carcinog. Risk Chem. Man* **12** (1976) 77–84.
[51] A. R. Sellakumar et al., *J. Environ. Pathol. Toxicol.* **4** (1980) 107–115.
[52] BASF, unpublished results, 1967.
[53] Deutsche Forschungsgemeinschaft (ed.): *Maximale Arbeitsplatzkonzentration (MAK) und Biologische Arbeitsstofftoleranzwerte (BA),* VCH Verlagsges., Weinheim 1985.

# Carbazole

GERD COLLIN, Rütgerswerke AG, Duisburg-Meiderich, Federal Republic of Germany

HARTMUT HÖKE, Rütgerswerke AG, Castrop-Rauxel, Federal Republic of Germany

1. Physical Properties ........ 1057
2. Chemical Properties........ 1057
3. Production .............. 1058
4. Analysis ................ 1058
5. Uses and Economic Aspects .. 1058
6. Derivatives .............. 1059
7. Toxicology............... 1059
8. References............... 1060

Carbazole (dibenzopyrrole) [86-74-8], $C_{12}H_9N$, was first found in coal tar by C. GRAEBE and C. GLASER in 1872.

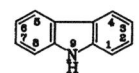

## 1. Physical Properties

$M_r$ 167.21; mp 246 °C, bp at 101.3 kPa 354.8 °C; $\varrho$ at 18 °C 1.1035 g/cm$^3$; plates or tables; sublimable; readily soluble in acetone, slightly soluble in ether and alcohol, barely soluble in chloroform, acetic acid, carbon tetrachloride, and carbon disulfide, soluble in concentrated sulfuric acid; heat of fusion 176.3 kJ/kg; heat of combustion 3.719× 10$^4$ kJ/kg at 25 °C.

## 2. Chemical Properties

The N-hydrogen of carbazole which is a secundary amine can be substituted with alkali metals. During halogenation, nitration, and sulfonation, substitution initially takes place at the 3- and 6-positions; the 1,3,6- and 1,3,6,8-derivatives are formed under more rigorous conditions.

Hydrogenation yields 1,2,3,4-tetrahydro-, hexahydro-, or dodecahydrocarbazole. Oxidation with chromates gives 3,3′-dicarbazyl, with permanganate 9,9′-dicarbazyl. Carboxylation with alkali and carbon dioxide gives, depending on the temperature, carbazole-3- or carbazole-1-carboxylic acid. Alkylation yields N-alkylcarbazoles.

# 3. Production

High-temperature coal tar contains an average of 1.5% carbazole. Carbazole is obtained as a co-product in the production of anthracene. Due to its higher boiling point and better solubility, it can be separated from anthracene by extraction with pyridine, ketones, benzene/methanol, N-methylpyrrolidone, dimethylacetamide, dialkyl sulfoxides, or dialkylformamides, or by azeotropic distillation with ethylene glycol (→ Anthracene). The chemical separation of carbazole by means of potassium hydroxide or concentrated sulfuric acid fusion has become uneconomical. Pyridine mother liquors obtained from the preparation of pure anthracene yield pure carbazole by means of concentration and recrystallization from chlorobenzene.

Carbazole can be synthesized by converting cyclohexanone azine to octahydrocarbazole and subsequent dehydrogenation [1], by reductive cyclization of 2-nitrobiphenyl [2], [3], or by dehydrogenation and cyclization of diphenylamine [4]–[6], o-aminobiphenyl [7]–[9], or N-cyclohexylideneaniline [10]. These synthetic processes are not yet of industrial significance, because sufficient quantities of carbazole can be prepared from coal tar.

# 4. Analysis

Gas–liquid chromatography is suitable for the quantitative determination of carbazole in complex mixtures, such as in coal tar or coal-tar fractions [11]. Thin-layer chromatography can also be used to identify carbazole along with other compounds [12]. Purity is determined by UV spectroscopy. As little as 0.005% carbazole in anthracene can be detected using phosphorescence spectroscopy. Small quantities of phenanthrene do not interfere with the results [13].

# 5. Uses and Economic Aspects

Carbazole is obtained from coal tar in amounts of several thousand t/a, primarily for dye synthesis. Co-products of anthracene recovery representing the raw material of carbazole are up to 100 000 t/a.

The blue sulfur dye Hydron Blue is commercially synthesized by condensing carbazole with p-nitrosophenol and subsequent sulfurization.

The insecticide Nirosan, 1,3,6,8-tetranitrocarbazole, has been commercially available since 1939. N-Vinylcarbazole is polymerized to give poly(vinylcarbazole) (Luvican, Polectron) [14], which is thermally and chemically stable and has a high softening temperature and low dielectric loss and photoconductivity. These properties make it useful in electrical industry and for electrostatic copying methods [14]–[16]. Carbazole reacts with phenols and formaldehyde in the presence of acidic catalysts to give novolaks. These can be cured with hexamethylenetetramine to form highly heat–resistant polymers [17]. Concrete plasticizers are obtained by cocondensation of carbazole with phenols and formaldehyde and subsequent sulfonation [18]. Thermal condensation of 3,6-diaminocarbazole with dicarboxylic acid gives elastic and thermally stable polyamides [19].

# 6. Derivatives

*N-Ethylcarbazole* [86-28-2], $C_{14}H_{13}N$; $M_r$ 195.27; *mp* 68 °C; by ethylation of carbazole–potassium with diethyl sulfate [20].

*1,3,6,8-Tetranitrocarbazole* [4543-33-3], $C_{12}H_5O_8N_5$; $M_r$ 347.20; *mp* 312 °C; by nitration of carbazole with a mixture of nitric and sulfuric acid.

*N-Vinylcarbazole* [1484-13-5], $C_{14}H_{11}N$; $M_r$ 193.25; *mp* 65 °C; from carbazole–potassium with ethylene oxide or vinyl chloride, or from carbazole and acetylene [14].

# 7. Toxicology

The $LD_{50}$ of carbazole ranges from 200 mg/kg (intraperitoneal, mice) to more than 5000 mg/kg (oral, rats) [21], [22]. Due to this fact it can be considered as non-toxic. Folliculitis and comedos were reported by workers with carbazole contact [23]. In rats and rabbits, carbazole is glucuronized and excreted in the urine [24]. There is no evidence of carcinogenicity for carbazole.

# 8. References

General References

*Beilstein*, **20**, 433; **20(1)**, 162; **20(2)**, 279; **20(3)**, 3824.

H. J. V. Winkler: *Der Steinkohlenteer und seine Aufarbeitung*, Verlag Glückauf, Essen 1951, pp. 181–187.

H.-G. Franck, G. Collin: *Steinkohlenteer*, Springer Verlag, Berlin – Heidelberg – New York 1968, pp. 56–58, 180–182.

*Erzeugnisse aus Steinkohlenteer*, Rütgerswerke, Frankfurt 1958.

J. A. Joule in: A. R. Katritzky (ed.), *Advances in Heterocyclic Chemistry, vol.* **35**, Academic Press, Orlando 1984, pp. 83–198.

Specific References

[1] BASF, DE 1158518, 1958 (M. Seefelder, H. Maisack).
[2] Allied Chem. & Dye Corp., US 2508791, 1948 (M. S. Larrison).
[3] M. Tashiro, JP-Kokai 80122760, 1980.
[4] BASF, DE 937590, 1952 (H. Friedrich, O. Stichnoth, H. Waibel).
[5] American Marietta, US 2921942, 1956 (H. M. Grotta).
[6] Schering, DE 2418503, 1974 (B. Akermark, L. Eberson, E. Jonsson, E. Pettersson).
[7] Monsanto, US 2479211, 1949 (C. Conover).
[8] Ciba, DE 889592, 1951 (D. Porret).
[9] H. Suhr, U. Schöch, G. Rosskamp, *Chem. Ber.* **104** (1971) 674–676.
[10] Snam Progetti, NL 7115166, 1971 (M. M. Mauri, P. A. Moggi, U. Romano).
[11] H. D. Sauerland, *Brennst. Chem.* **45** (1964) 55–56.
[12] E. Sawicki, H. Johnson, K. Kosinski, *Microchem. J.* **10** (1966) 72–102.
[13] M. Zander, *Angew. Chem.* **76** (1964) 922; *Angew. Chem. Int. Ed. Engl.* **3** (1965) 755.
[14] J. M. Pearson, M. Stolka: *Poly(N-vinylcarbazole)*, Polymer Monographs no. **6**, Gordon and Breach Science Publishers, New York – London – Paris 1981.
[15] W. Klöpffer, *Kunststoffe* **61** (1971) 533–539.
[16] U. König, V. Stepanek, *Kunststoffe* **69** (1979) 223–227.
[17] Rütgerswerke, DE 2033015, 1970 (J. Omran, H.-G. Franck, M. Zander).
[18] Rütgerswerke, DE 3210458, 1982 (J. Omran, M. Zander).
[19] Rütgerswerke, DE 2125128, 1971 (J. Omran, H.-G. Franck, M. Zander).
[20] Farbwerke Hoechst, DE 2132961, 1971 (T. Papenfuhs).
[21] N. J. Sax: *Dangerous Properties of Industrial Materials*, Van Nostrand Reinhold Comp., New York – London 1979, p. 468.
[22] E. Eagle, A. J. Carlson, *Pharm. Exp. Ther.* **99** (1950) 450–57.
[23] L. Jirásek, Z. áva, *Prakt. Lek.* **35** (1955) 34–37.
[24] S. R. Johns, S. E. Wright, *J. Med. Chem.* **7** (1964) 158–161.

# Carbohydrates

*Individual keywords:* → *Cellulose Esters,* → *Cellulose Ethers,* → *Gluconic Acid,* → *Lactose and Derivatives*

JOCHEN LEHMANN, Institut für Organische Chemie und Biochemie, Universität Freiburg, Freiburg, Federal Republic of Germany

KNUT RAPP, Südzucker AG, Grünstadt-Obrigheim, Federal Republic of Germany

| | | |
|---|---|---|
| 1. Introduction .............. 1061 | 6.1. | Reactions of the Hydroxyl Groups ................. 1070 |
| 2. Monosaccharides .......... 1063 | | |
| 3. Oligosaccharides .......... 1066 | 6.2. | Reactions of the Carbonyl Group .................. 1074 |
| 4. Polysaccharides .......... 1068 | | |
| 5. Nomenclature ............. 1069 | 6.3. | Other Reactions ........... 1077 |
| 6. Reactions of Carbohydrates .. 1070 | 7. | References ................ 1082 |

# 1. Introduction

The term "carbohydrates" describes a major group of naturally occurring chemical compounds of widely varying molecular mass. The low molecular mass carbohydrates comprise a family of more or less sweet-tasting, water-soluble, colorless products with closely related physical, chemical, and physiological properties. They are often generically called "sugars," reflecting their relationship to "sugar," which is the familiar sucrose.

The high molecular mass carbohydrates are called *polysaccharides*. Their physical properties differ significantly from those of low molecular mass. They do not taste sweet and many do not dissolve in water.

All carbohydrates are composed of units of *monosaccharides* or their derivatives. Carbohydrates are usually classified by the number of monosaccharide units in their molecules. A monosaccharide contains an uninterrupted chain of a limited number (five and six are most common) of carbon atoms. In oligosaccharides (disaccharide, trisaccharide, tetrasaccharide, etc.) and in polysaccharides, monosaccharides are linked together through glycosidic oxygen atoms.

*Oligosaccharides* are considered to be low molecular mass carbohydrates. There is, however, no definite borderline between the oligosaccharides and the polysaccharides.

In general, naturally occurring oligosaccharides rarely consist of more than 5, and polysaccharides of less than 100, monosaccharide units.

Most monosaccharides, oligosaccharides, and polysaccharides have the common molecular formula $C_n(H_2O)_m$ (from which the common name "carbohydrates" is derived), in which $n$ is either equal to $m$ (monosaccharides) or a little larger than $m$ (oligo- and polysaccharides). Oligomers are formed by condensation (formal elimination of water) of monosaccharide units, and conversely oligo- and polysaccharides can be degraded to monosaccharides by hydrolysis. The whole biomass on earth has been estimated to be about 90% carbohydrates, which means that the molecular formula of the total organic matter on earth is only slightly different from that of the carbohydrates.

Carbohydrates have important functions: primarily they supply chemical energy to the living cell and serve as raw materials for the construction of other natural products. Carbohydrates are mainly formed in the photochemical carbon dioxide fixation in green plants, a complicated biochemical process which provides for the efficient conversion of solar energy into chemical energy. The uses of carbohydrates for the supply of primary energy and as raw materials in synthetic organic chemistry represent an important facet of present and future technology.

Carbohydrates are industrially utilized to a considerable extent. Glucose [50-99-7] is the starting material for the production of vitamin C (ascorbic acid). Maltose [69-79-4], a product of the enzymatic degradation of starch, can be reduced to maltitol, which is used as a low-calorie sweetener. A similar sweetener is isomalt (palatinit) [64519-82-0], a mixture of the disaccharide alcohols $O$-α-D-glucopyranosyl-(1 → 1)-D-mannitol [20942-99-8] and $O$-α-D-glucopyranosyl-(1 → 6)-D-glucitol [534-73-6], which are obtained by reduction of isomaltulose (palatinose) [13718-94-0], a product of enzymatic isomerization of sucrose. Fructose [57-48-7] itself is used as a sweetener by diabetics.

Many nonionic detergents are derived from carbohydrates. Used in soap or shampoo, they are less irritating than their ionic counterparts and can easily be degraded biologically, therefore causing no environmental problem. Polysaccharides are produced and used in a wide variety of applications. These include substitutes for blood serum (dextran), adhesives (cellulose and starch basis), wrapping material (cellophane), additives to food and beverages, pharmaceuticals and cosmetics (starch, pectic substances, plant and microbial gums), and drilling fluids in oil fields. Cellulose in the form of wood or cotton is one of the most important industrial raw materials.

*Production.* Natural carbohydrates are obtained exclusively from biological sources. Total syntheses are impractical, and even the simple 3-epimerization of D-glucose into another monosaccharide, D-allose, which is not found in nature, is tedious. Chemical syntheses of oligosaccharides by coupling suitable derivatives of monosaccharides are extremely difficult and only possible on a small scale. The chemical synthesis of a polysaccharide is almost impossible. One possibility is the polymerization of 1,6-anhydro sugar derivatives, such as substituted levoglucosans, which yields (1 → 6)-α-D-glucopyranans.

# 2. Monosaccharides

Monosaccharides are either polyhydroxy aldehydes (-oses, aldoses) or polyhydroxy ketones (-uloses, ketoses). The carbon skeleton in most monosaccharides is linear (not branched). In almost all ketoses the carbonyl group is located at carbon atom 2. Most monosaccharides have either six (hexoses and hexuloses) or five (pentoses and pentuloses) carbon atoms. Monosaccharides with fewer carbon atoms (trioses, tetroses, and tetruloses) or more carbon atoms (heptoses, octoses, heptuloses, octuloses, etc.) rare. The most important and abundant monosaccharides are hexoses: D-glucose, D-mannose, and D-galactose. D-Fructose is a hexulose. Ubiquitous pentoses are L-arabinose, D-xylose, and D-ribose.

**Stereoisomerism.** Monosaccharides (as polyhydroxy aldehydes or ketones) possess several asymmetric carbon atoms. Hexoses have four, hexuloses and pentoses three, and pentuloses two chirality centers. The existence of several asymmetric carbon atoms in one molecule gives rise to families of stereoisomers. For a hexose with four asymmetric carbon atoms, there exist $2^4$ stereoisomers, eight of which are enantiomers (mirror images) of the other eight. Hexuloses exist in four pairs of enantiomers, pentuloses and tetroses in two. The diastereomeric monosaccharides (diastereomers are stereoisomers that are not mirror images of each other) have trivial names. In Figure 1, D-hexoses, D-hexuloses, D-pentoses, and D-tetroses are represented as open-chain Fischer projections. In the Fischer projection all carbon atoms form a bow with the convex part pointing toward the viewer. According to the Fischer convention, the orientation of the hydroxyl group attached to the highest numbered asymmetric carbon atom determines whether a monosaccharide belongs to the D- or the L-series. If this group is positioned on the right of the carbon chain, assignment is to the D-series, and vice versa. Most naturally occurring monosaccharides have the D-configuration. The chirality of carbohydrates makes this class of natural compounds an ideal source for the preparation of chiral synthons (intermediates for chemical synthesis), which are needed for the syntheses of pharmaceuticals, agrochemicals, pheromones, etc. [26], [27].

**Pyranoses and Furanoses.** Monosaccharides form cyclic intramolecular hemiacetals if their carbon skeleton permits. The relative stabilities of these hemiacetals depend on the ring size. In general, six-membered rings (pyranoses) are most common, whereas only a few five-membered rings (furanoses) are stable. Smaller rings are unstable, and larger ones can only be detected in aqueous equilibrium mixtures. Structures of the cyclic hemiacetals are represented by the Haworth projection formulas of the D-series of hexopyranoses and hexofuranoses shown in Figure 2. The rings are positioned in a horizontal plane with the ring oxygen atom away from the viewer. Because of cyclization, the carbonyl carbon, which is called the *anomeric* carbon atom, becomes asymmetrically substituted. In the Haworth projection formula of a D-monosaccharide, the hydroxyl group attached to this carbon atom is positioned above or below the ring

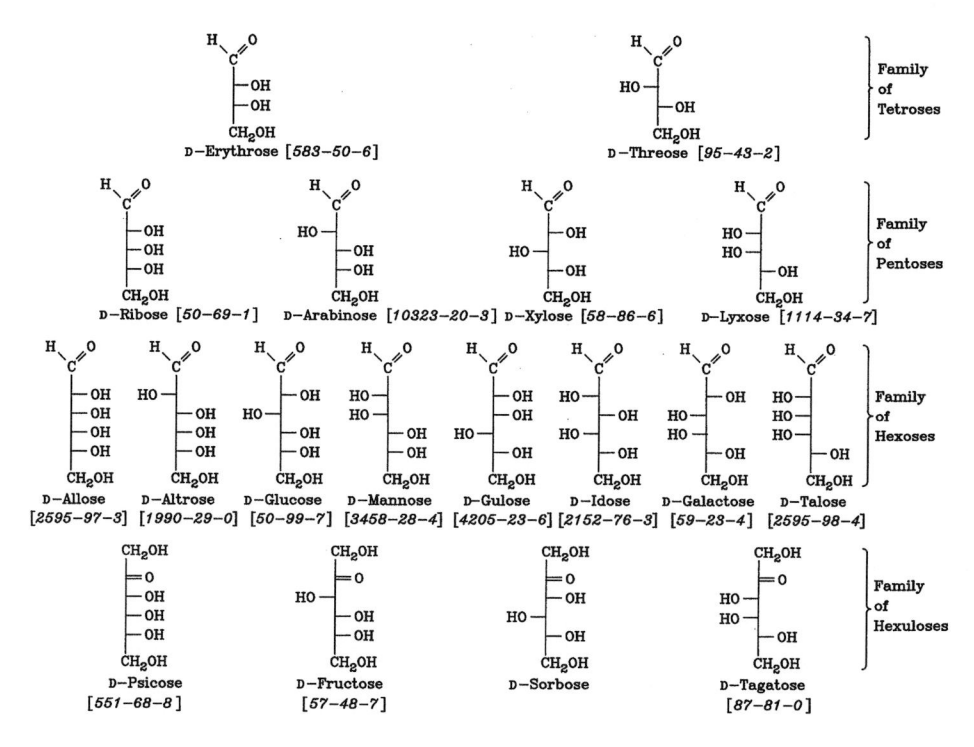

**Figure 1.** Fischer projection formulas

plane, and the monosaccharide is assigned the β- or α-*anomeric configuration*, respectively. This definition is reversed in the L-monosaccharide series.

Like corresponding alicyclic compounds with cyclopentane or cyclohexane structures, furanose and pyranose rings are not planar but are puckered in order to achieve minimum deformation of the tetrahedral bond angle as well as minimum nonbonded interaction among the ring substituents. The following formula shows the most stable conformation of β-D-glucopyranose [492-61-5], i.e., all substituents of the tetrahydropyran ring are in an equatorial position.

The formation of cyclic hemiacetals in solution is a dynamic process, which depends on pH, temperature, solvent, and other factors, such as the substituents of the sugar molecule. Thus, glucose forms approximately a 2:1 mixture of the β-pyranoid and α-pyranoid forms, whereas fructose at 22 °C in water is a 72:20:5:3 mixture of the β-pyranoid, β-furanoid, α-furanoid, and α-pyranoid forms, respectively; the composition at 80 °C is 55:31:10:4 [28].

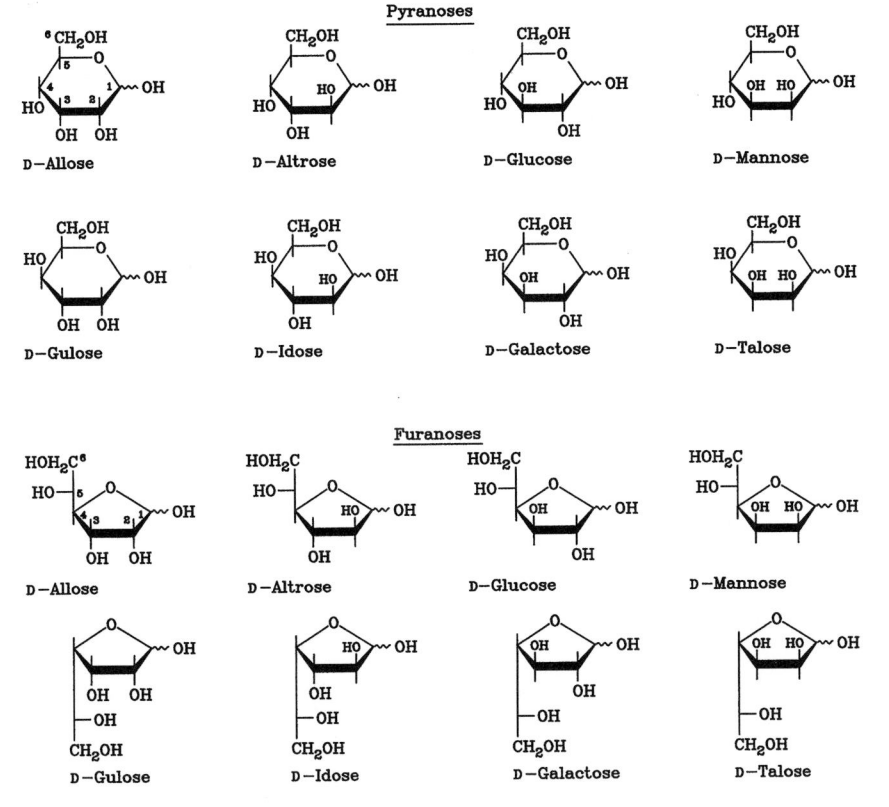

**Figure 2.** Haworth projection formulas of hexoses
Vertical lines with no substituent designation stand for "H" (hydrogen).

This feature makes the chemistry of (keto-) sugars very difficult, because one sugar often yields several products (five- and six-membered rings; $\alpha$ and $\beta$ forms). On the other hand, a single sugar may allow the synthesis of linear (open chain), furanoid, and pyranoid derivatives as chiral intermediates.

**Structural Variations of Monosaccharides.** Not all monosaccharides have the molecular formula $C_n(H_2O)_n$ or the typical unbranched carbon skeleton. If a hydroxyl group in a typical monosaccharide is formally replaced by hydrogen or by an amino group, deoxymono-saccharides or aminodeoxymonosaccharides, respectively, are obtained. Some of these derivatives are quite common in nature. For example, N-acetyl-2-amino-2-deoxy-D-glucopyranose (N-acetyl-D-glucosamine [1398-61-4]), is a building block of the polysaccharide chitin, a constituent of the hard shells of crustaceans and other arthropods. It is also found in fungi and yeasts.

```
        H    O
         \\ //
          C
    H ──┼── NHCOCH₃
   HO ──┼── H
    H ──┼── OH
    H ──┼── OH
         CH₂OH
```
N-Acetyl-D-glucosamine

Another common derivative is the deoxy monosaccharide 6-deoxy-L-galactopyranose (L-fucose [2438-80-4]).

```
        H    O
         \\ //
          C
   HO ──┼── H
    H ──┼── OH
    H ──┼── OH
   HO ──┼── H
          CH₃
```
L-Fucose

On the other hand, branched-chain monosaccharides such as streptose and apiose are rare.

```
        H    O                H    O
         \\ //                  \\ //
          C                     C
    H  ──┼── OH           H  ──┼── OH
   O=C  ──┼── OH        HOH₂C ──┼── OH
    H                              CH₂OH
   HO  ──┼── H
          CH₃
```
Streptose       Apiose

There are some unusual monosaccharides in which hydroxyl groups are replaced by sulfur-containing groups; an example is 6-deoxy-D-glucose-6-sulfonic acid (sulfoquinovose), a constituent of the plant sulfolipid found in chloroplast membranes.

```
        H    O
         \\ //
          C
    H ──┼── OH
   HO ──┼── H
    H ──┼── OH
    H ──┼── OH
         CH₂SO₃H
```
Sulfoquinovose

# 3. Oligosaccharides

Formal condensation of two or more monosaccharides with expulsion of one or more molecules of water and formation of acetal bridges yields oligosaccharides. Theoretically the number of possible isomers increases with the number of monosaccharide units (a disaccharide composed of two hexopyranoses can have 5120 distinguishable isomeric forms). In reality nature produces only a limited number of oligosaccharides. However, on the cellular level the importance and variety of the oligosaccharide part of glycoproteins or glycolipids is very high (cell–cell interaction, substrates for enzymes, ligands for receptors, proteins, antigenic properties of cells, cancer-determined changes of cell surface carbohydrates). Because of their different reactivity reducing and nonreducing oligosaccharides exist. In nonreducing oligosaccharides the anomeric hydroxyl groups (lactol groups) are all blocked by acetal formation. The abundant disaccharides α,α-trehalose [99-20-7] and sucrose [57-50-1], the trisaccharide raffinose [512-69-6], and the tetrasaccharide stachyose [10094-58-3] (the latter two derived from sucrose) are nonreducing.

Trehalose is widespread in the lower species of the plant kingdom (fungi, young mushrooms, yeasts, lichens, and algae). In baker's yeast it accounts for as much as 15% of the dry mass. In the animal kingdom this disaccharide circulates in the metabolic cycle of insects like glucose does in the mammalian cycle. Trehalose occurs as a storage carbohydrate in eggs, larvae, and pupae of insects. The α,α-trehalose 6,6'-diester of mycolic acid is the so-called cord factor [29].

There are only very few naturally occurring oligosaccharides with a free anomeric hydroxyl group, which therefore possess reducing properties. The most important example is lactose [63-42-3], an ingredient of the milk of mammals (→ Lactose and Derivatives).

Lactose

The reducing gluco-oligosaccharides maltose, maltotriose, cellobiose, and isomaltose are enzymatic degradation products of the polysaccharides amylose, cellulose, and amylopectin; therefore, they cannot be regarded as native oligosaccharides.

Sucrose is produced in crystalline form at a level of about $10^8$ t/a, primarily from sugar cane and sugar beets. The quantity of sucrose synthesized by all plants on earth is estimated to be $150 \times 10^9$ t/a.

# 4. Polysaccharides

Polysaccharides are high molecular mass carbohydrates. Like oligosaccharides, they are formed by linking monosaccharides together through glycosidic bonds. The molecular mass range of polysaccharides is wide, between approximately five thousand for inulin (polyfructose) and several million for glycogen (polyglucose) or gum arabic (polygalactose with side chains formed by other sugars). The general name for a polysaccharide is glycan. There are homo- and heteroglycans, depending on whether they are composed of one or more kinds of monosaccharides. Very few heteroglycans contain more than five different monosaccharides. In polysaccharides the monosaccharide units can be arranged in either a linear or a branched fashion. The branched polymers have either a comblike or a treelike structure (Fig. 3). Regularly repeating units of two or more monosaccharides are typical of heteroglycans. Unlike proteins, polysaccharides are not genetically determined and therefore are usually not monodisperse, which means that not all molecules of a given glycan have the same structure. They can differ in the degree of polymerization (polymolecular), in proportions of monosaccharides, in proportion of linkage types, or in distribution of side chains (polydisperse). Many polysaccharides contain monosaccharides that have been chemically modified after the polymer has been formed. Among these modifications are methyl ester formation with D-galacturonic acid (pectic acid), O-acetylation (alginic acid), O- and N-sulfation (heparin), and 3,6-anhydro ring formation by intramolecular etherification (agarose). Only a limited number of monosaccharides are regularly found in polysaccharides. These are mainly the pentoses D-xylose and L- and D-arabinose; the hexoses D-glucose, D- and L-galactose, D-mannose, D- and L-fucose, and L-rhamnose; the N-acetylated monosaccharides N-acetyl-D-glucosamine and N-acetyl-D-galactosamine; the uronic acids D-glucuronic acid, D-galacturonic acid, and D-mannuronic acid; and the ulosonic acid N-acetylneuraminic acid. The uronic acids L-iduronic acid (in heparin)

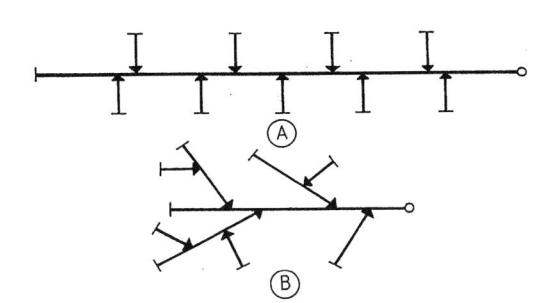

**Figure 3.** Segments of two possible arrangements of chains in comblike (or herringbone) structure (A) or multiply branched treelike structure (B) Reproduced by permission [9]
⊢——: nonreducing termini of chains; —→⊢: junctions at branch points; ——○: reducing ends

and L-guluronic acid (in alginic acid) are formed by enzymatic 5-epimerization from D-glucuronic acid and D-mannuronic acid, respectively, after the polysaccharide chain has been biosynthetically assembled.

The physical properties of polysaccharides are very different. The globular and lower molecular mass polysaccharides like inulin or amylose are water-soluble. Some, especially such linear fibrous high molecular mass polysaccharides as cellulose and chitin, are completely insoluble in water. Many polysaccharides only swell in water and at the most form colloidal, often highly viscous or gelling, solutions. Intermolecular association among segments of the polysaccharide chains by salt bridges (intermolecular calcium diuronates) or intermolecular formation of double helices gives rise to the widely meshed three-dimensional networks of polysaccharide chains, which are responsible for the gelling of very dilute solutions of pectic acid and carrageenan.

# 5. Nomenclature

According to common practice, trivial names are used for monosaccharides and for many naturally occurring oligosaccharides. With the development of carbohydrate chemistry, however, and ever-increasing numbers of newly defined compounds, it has become necessary to introduce a semisystematic nomenclature which has been approved by IUPAC (International Union of Pure and Applied Chemistry) [30] and IUB (International Union of Biochemistry) [31]. This nomenclature is based on the classical names for monosaccharides which appear, written in italics, as a "configurational prefix." For example, D-*xylo*, L-*arabino*, D-*gluco* refer to the distribution of asymmetric carbon atoms along a carbon chain of any length with the configuration of the corresponding monosaccharide.

The chain length is given by the root, such as tetrose, pentose, hexose, or heptose, if the monosaccharide is an aldose, or tetrulose, pentulose, hexulose, heptulose, if it is a ketose. In ketoses the position of the oxo group is indicated by the position number. D-Fructose is systematically named D-*arabino*-2-hexulose. If more than four asymmetric carbon atoms are present in a carbon chain, the configurational prefix is formed by combination, as exemplified by D-*glycero*-D-*gluco*-heptose. The larger part (D-*gluco*) is always closest to the carbonyl group. Methyl or methylene groups, or carbon atoms that carry an amino group instead of a hydroxyl group, are designated by the prefix "deoxy." D-Glucosamine therefore has the systematic name 2-amino-2-deoxy-D-*gluco*-hexose.

The ring size is indicated by a suffix: pyranose for six-membered rings, furanose for five-membered rings, and pyranulose for six-membered ketose rings. The six-membered cyclic hemiacetal of D-fructose is named D-*arabino*-2-hexopyranulose. The symbol α or β for the anomeric configuration is always written together with the configurational symbol D or L (α-D, β-D, α-L, β-L).

Names of oligosaccharides are formed by combining the monosaccharide names, usually the trivial names. The nonreducing disaccharide sucrose (formula 3) is β-D-fructofuranosyl-α-D-glucopyranoside. The endings "yl" and "ide" describe the fructose part as the aglycone and the glucose part as the glycone in this "glycoside." It is thus clearly indicated that both sugars are glycosidically linked by their anomeric hydroxyl groups. In reducing oligosaccharides the reducing monosaccharide is the root, and all attached monosaccharide units are named as substituents. The disaccharide lactose (formula 3) is therefore named *O*-β-D-galactopyranosyl-(1 → 4)-D-glucopyranose. Position numbers and arrows indicate a β-configurated glycosidic bond between the anomeric hydroxyl group (carbon atom 1) of D-galactose (glyconic part) and the hydroxyl group at carbon atom 4 of D-glucose (aglyconic part).

Some linear homopolysaccharides in which the anomeric configuration, the ring size of the monosaccharide unit, and the positions of glycosidic linkages are known can be named as glycans, for example, amylose as (1 → 4)-α-D-glucopyranan.

# 6. Reactions of Carbohydrates

## 6.1. Reactions of the Hydroxyl Groups

**Esters of Inorganic Acids.** Phosphoric acid esters of monosaccharides are important metabolites and are present in every organism. D-Glucose 1-phosphate [59-56-3], D-glucose 6-phosphate [56-73-5], and D-fructose 1,6-diphosphate [488-69-7] are key intermediates in both biological carbohydrate degradation and synthesis.

Glucose 1-phosphate

Monosaccharide phosphates are constituents of many nucleotides; for example, adenosine triphosphate (ATP), nicotinamide adenine dinucleotide (NAD$^+$), and flavine adenine dinucleotide (FAD) all contain D-ribose 5-phosphate. Polymeric phosphate esters of D-ribose and 2-deoxy-D-ribose form the backbone of ribonucleic acid (RNA) and deoxyribonucleic acid (DNA).

Monosaccharide 1-phosphates, in contrast to other monosaccharide phosphates, such as 6-phosphates, are acetal esters and therefore not very stable in aqueous mineral acids. They hydrolyze to give the free monosaccharide and phosphoric acid. Under basic conditions the preferred reaction of monosaccharide 1-phosphates is condensation to cyclic diesters. Most naturally occurring 1-phosphates are $\alpha$-anomers. Because of a strong anomeric effect, they are thermodynamically more stable than the $\beta$-anomers. $\alpha$-1-Phosphates are conveniently prepared from per-O-acetylated monosaccharides or oligosaccharides by heating with anhydrous phosphoric acid [38], [39]. Phosphorylation of hydroxyl groups other than the anomeric groups is carried out by acylation of suitably blocked derivatives with diphenyl or dibenzyl chlorophosphate in pyridine. The phosphate group can be unblocked by catalytic hydrogenolysis [40]. Sulfate groups are present in many biologically important polysaccharides, such as heparin and chondroitin sulfate. Sulfated monosaccharides can be prepared from suitable monosaccharide derivatives by reaction with chlorosulfuric acid in pyridine [41].

The nitrate esters of cellulose (nitrocellulose) contain as many as three $ONO_2$ groups per glucose unit. The product with about 13% nitrogen is the well-known guncotton (C. F. SCHÖNBEIN, 1846; whereas celluloid is nitrocellulose containing about 10% nitrogen plasticized with camphor; it is one of the oldest known plastics. This plastic was once the principal photographic and movie film, but has been supplanted by other films because of its high flammability.

**Esters of Organic Acids.** Acetate esters of cellulose are manufactured on a large scale. Acetylation can be performed with acetic anhydride, in pyridine or in a neutral solvent with sulfuric acid catalyst. The degree of acetylation determines the solubility; the triacetate (3.0-acetate) is soluble in chloroform, the 2.5-acetate in acetone, and the 0.7-acetate in water. These esters have many uses as such, and are also used as synthesis intermediates, using the acetyl group to protect hydroxyl groups under neutral or weakly acidic conditions. The acetyl group can be removed by transesterification with methanol, using sodium methoxide catalyst and distilling off the methyl acetate [42]. Mixed esters such as cellulose acetate propionate and cellulose acetate butyrate are widely used in the production of lacquers, films, and plastics ($\rightarrow$ Cellulose Esters). Other organic esters, such as benzoates and carbonates, are prepared from the corresponding acyl chlorides in pyridine. Sulfonate groups such as $p$-toluenesulfonyl (tosyl) and methanesulfonyl (mesyl) can be introduced by using the acid chlorides, and provide excellent leaving groups in nucleophilic substitution reactions. The very reactive trifluoromethanesulfonyl group is frequently used as a leaving group in nucleophilic substitution and elimination reactions, and is introduced by using trifluoromethanesulfonic anhydride.

Polysaccharide esters in which the carbohydrate portion is the acid component occur in the plant kingdom in fruits, roots, and leaves. For example, pectins are high molecular mass polygalacturonic acids joined by $(1 \rightarrow 4)$-$\alpha$-glycosidic links, in which some of the carboxylic acid groups are esterified with methanol. In the production of fruit juices the formation of methanol, which can be liberated through the action of pectinesterase, should be avoided.

Pectins in which 55–80% of the carboxyl groups are esterified are called high-methoxyl pectins (HM-pectins), and have the important property of gelling at very low concentrations ($\approx$ 0.5%) in water in the presence of sugars and acid. Low-methoxyl (LM, < 50% of the carboxyl groups esterified) pectins form gels with divalent cations such as $Ca^{2+}$; 0.5% of a low-methoxyl pectin can bind 99.5% of the water in the gel matrix. These pectins can be used as gelling agents in the production of jellies from fruit juices. Undesired coagulation is a problem with low-methoxyl pectins, depending on the amount of calcium ions present. This problem has been avoided by the introduction of "amidated" pectins, formed by treatment of high-methoxyl pectins with ammonia, which converts some of the methyl ester groups into galacturonamide groups. These amidated pectins form gels at lower concentrations of $Ca^{2+}$ ions, and are stable over a wider pH range than the high-methoxyl pectins.

**Ethers.** In carbohydrate chemistry methyl and benzyl ether blocking groups are frequently used. Methyl ethers are useful in methylation analysis for structural investigations of polysaccharides. Benzyl ethers are widely used in preparative carbohydrate chemistry because, unlike O-methyl groups, O-benzyl groups are easily removed by hydrogenolysis. Triphenylmethyl ether groups are used for the selective blocking of primary hydroxyl functions.

The ethers are conveniently prepared from alkyl bromides, iodides, or sulfates in polar aprotic solvents such as dimethylformamide or dimethyl sulfoxide. Agents for deprotonation of the hydroxyl group and for binding the mineral acids liberated include alkali hydroxides or hydrides and barium or silver oxide. With high molecular mass carbohydrates, quantitative deprotonation is best carried out with sodium or potassium methylsulfinyl methanide (the conjugate base of dimethyl sulfoxide). The "naked" anions thus formed are easily alkylated [43]. Carbohydrates containing base-labile groups can be methylated with diazomethane–boron trifluoride etherate [44]. Trimethylsilyl ethers, although extremely sensitive to base as well as to acid-catalyzed hydrolysis, are often used in analytical and preparative carbohydrate chemistry. The pertrimethylsilyl ethers of monosaccharides and small oligosaccharides are relatively volatile, highly lipophilic, and thermostable, and therefore, ideal derivatives for gas chromatographic analysis. The trimethylsilyl ethers are rapidly formed in pyridine by using a mixture of hexamethyldisilazane and trimethylchlorosilane [45].

Cellulose ethers are generally manufactured by the Williamson synthesis: reaction of sodium cellulose (prepared by treating cellulose with 20 to > 50% sodium hydroxide) with an organic halide such as chloromethane or sodium monochloroacetate. The latter reagent produces sodium carboxymethylcellulose (NaCMC), which is widely used, for example, as a thickening agent in foods. Worldwide production of NaCMC is in the range of several hundred thousand tons per year ($\rightarrow$ Cellulose Ethers).

**Acetals.** Cyclic acetals are readily formed from adjacent hydroxyl groups of carbohydrates. The reactions normally are carried out in the reagent aldehyde or ketone as solvent with an electrophilic catalyst ($H_2SO_4$ or $ZnCl_2$). Acetal formation under these conditions is thermodynamically controlled and usually very specific. Aldehydes form predominantly 1,3-dioxane derivatives, and ketones predominantly 1,3-dioxolane derivatives:

1,2: 3,4–Diisopropylidene–
α–D–galactopyranose

5,6–Ethylidene–
α–D–galactopyranose

Acetal formation is often used to block the reactivity of hydroxyl groups or to provide volatile derivatives for analytical procedures. Commonly used acetals are the benzylidene derivatives (from benzaldehyde) and isopropylidene derivatives (from acetone). Both types of cyclic acetals are easily hydrolyzed under mild acidic conditions (aqueous acetic acid).

**Intramolecular Ethers.** Formal dehydration of carbohydrates to yield anhydro derivatives can be achieved under mild conditions by intramolecular nucleophilic displacement of sulfonyl ester groups or halides in the presence of nonnucleophilic bases such as diazabicycloundecene. The intramolecular reactions are especially favored when epoxides (1,2-anhydro compounds) or five- or six-membered rings (1,4- and 1,5-anhydro compounds) can be formed. Intramolecular condensation with participation of the anomeric center (intramolecular glycosides) can be achieved by heating phenyl glycosides in basic solution or by pyrolysis of homoglycans (cellulose or starch). The condensation of hexopyranosyl compounds usually takes place between C-1 and C-6 to form glycosans:

1,6–Anhydro–D–glucopyranose
(Levoglucosan)

If the molecule does not contain suitable nucleophilic groups, sulfonyl esters or halides of carbohydrate derivatives can be converted into unsaturated hydroxy compounds (glycals or glycosenides) under basic conditions. The intramolecular ethers and the unsaturated derivatives, especially enol ethers, are valuable starting materials for the synthesis of deoxy sugars and aminodeoxy sugars.

**Oxo Compounds.** Hydroxyl groups in suitably protected carbohydrate derivatives can be converted into carbonyl groups by oxidation with dimethyl sulfoxide and electrophilic reagents such as dicyclohexylcarbodiimide, $P_2O_5$, or acetic anhydride. For example, 1,2:5,6-diisopropylidene-α-D-glucofuranose can be oxidized to 1,2:5,6-diisopropylidene-α-D-ribofuranos-3-ulose:

1,2: 5,6–Diisopropylidene–
α–D–ribofuranos–3–ulose

The same product can also be obtained by oxidation with catalytic amounts of ruthenium dioxide and sodium metaperiodate in aqueous solution [46]. Oxo compounds of this kind are especially useful for the preparation of branched-chain sugars. Additions to the carbonyl group are often highly stereospecific. Thus, reduction of 1,2:5,6-diisopropylidene-α-D-ribofuranos-3-ulose with $NaBH_4$ yields almost quantitatively 1,2:5,6-diisopropylidene-α-D-allose, making D-allose, which is not a naturally occurring hexose, easily accessible.

## 6.2. Reactions of the Carbonyl Group

**Reduction.** If the carbonyl group in carbohydrates is not protected by glycoside formation, it can be reduced under mild conditions. Reduction can be achieved with $NaBH_4$ in aqueous solution. Alkali-sensitive carbohydrates can be reduced with $NaBH_3CN$ in acetic acid.

Industrial production of alditols (monosaccharide and oligosaccharide alcohols) involves high-pressure hydrogenation with Raney nickel catalyst. Hydrogenation of glucose yields the sugar alcohol sorbitol, produced at a level of 500 000 t/a worldwide. *Sorbitol* [50-70-4] has a sweet taste and is used in foods for diabetics. It is also the precursor for the synthesis of ascorbic acid (vitamin C), with about 20 % of the annual production going to this use. Sorbitol is used as a humectant in cosmetic and pharmaceutical formulations and in foodstuffs. It is also used as an alcoholic component in the preparation of rigid polyurethane foams. Fatty acid esters of monoanhydrosorbitol (1,4-sorbitan) are widely used as emulsifiers and nonionic surfactants. The mono- and dinitrate esters of 1,4:3,6-dianhydrosorbitol (isosorbide [652-67-5]) are coronary vasodilators.

*Mannitol* [87-78-5], the second most important hexitol, is prepared by hydrogenation of the fructose portion of invert sugar [36], which yields a mixture of mannitol and sorbitol. In contrast to sorbitol, mannitol is not hygroscopic; world production in 1980 was 10 000 t. Mannitol is used in the manufacture of dry electrolytic capacitors and synthetic resins; in the pharmaceutical industry as a diluent for solids and liquids and in the preparation of the vasodilator mannitol hexanitrate; in the food industry as an anticaking and free-flow agent, and as a lubricant, stabilizer, and nutritive sweetener.

**Oxidation.** Controlled stoichiometric oxidations of carbohydrates to yield glyconic acids or derivatives thereof are limited to aldoses. Such oxidations can be carried out almost quantitatively either enzymatically by dehydrogenases or oxidases or chemically with bromine or iodine in buffered solution. The latter reaction is used for quantitative determination of aldoses and their derivatives. Oxidations in basic solution, especially with heavy metal ions ($Cu^{2+}$, $Hg^+$, $Fe^{3+}$), are usually ill-defined and therefore unstoichiometric. Today quantitative sugar determinations based on such oxidations (Tollens' reagent, Fehling's solution) are only of historic interest. Because of the very different reaction rates for anomers the reaction with bromine or iodine can be applied to determine quantitatively the amount of anomers in equilibrium mixtures, for instance, in the case of glucose.

**Chain Elongation.** Monosaccharides with more than six carbon atoms and monosaccharides labelled with $^{13}C$ or $^{14}C$ can be prepared by extension of the carbon chain of common aldoses. Classical methods for chain extension are the cyanohydrin synthesis [47] and the nitromethane synthesis [48]:

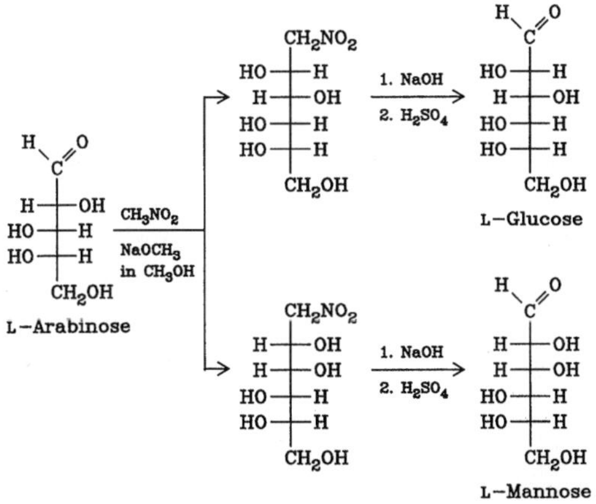

The former reaction has recently been improved by the discovery of a method for direct conversion of a cyano group into an aldehyde group [49]. Chain elongation of free sugars in aqueous solution has been achieved by reaction with stable phosphoranes. Thus, the carbon chain of D-glucose can be extended by reaction with ethoxycarbonylmethylenetriphenylphosphorane to yield ethyl 2,3-dideoxy-D-*gluco*-octenoate [50]. Such products can be converted into C-glycosyl compounds by intramolecular addition reactions.

**Thioacetals.** Under strongly acidic conditions monosaccharides react with thiols to yield thioacetals. These open-chain compounds can be used to prepare monosaccharide derivatives with a free carbonyl group, such as 2,3,4,5,6-penta-*O*-acetyl-D-glucose:

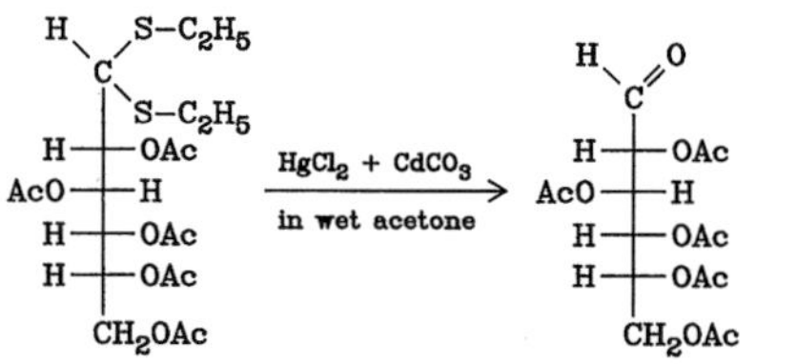

**N-Glycosyl Derivatives.** In buffered, weakly acidic solutions such nitrogen bases as amines, hydrazines, or hydroxylamine react with monosaccharides and reducing oligosaccharides to give condensation products. Like the free sugars, these condensation products exist in solution as equilibrium mixtures of isomers. Generally the pyranosyl derivatives are thermodynamically more stable. Crystalline condensation products therefore tend to have a six-membered pyranosyl ring. *N*-Glycosyl derivatives made from stronger bases may undergo the Amadori rearrangement. Thus, in the presence of oxalic acid at 70–80 °C, *N*-butyl-D-glucosylamine is converted into 1-(butylamino)-1-deoxy-D-fructose:

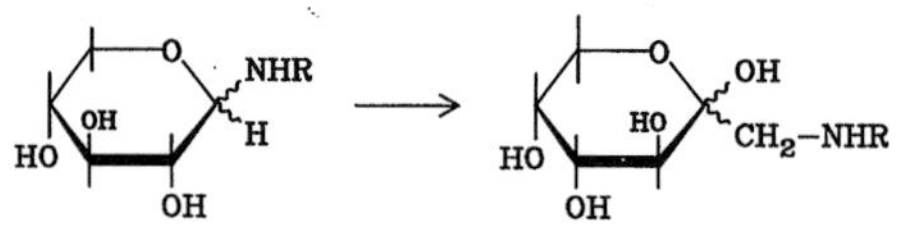

where R = *n*-C$_4$H$_9$

The formation of osazones from monosaccharides or reducing oligosaccharides and phenylhydrazines is also based on the Amadori rearrangement.

**O-Acetylglycosyl Halides.** Per-*O*-acetylated monosaccharides can be converted smoothly into glycosyl halides by dissolving them in cold solutions of the hydrogen halide in glacial acetic acid (acetates of acid-sensitive oligosaccharides may undergo cleavage of glycosidic bonds). The product, mainly with the α-configuration, is formed within a few hours:

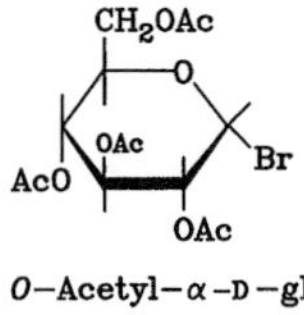

*O*–Acetyl–α–D–glucopyranosyl–
bromide(α–D–*O*–acetobromoglucose)

The fluorides are quite stable and can be deacetylated to the unsubstituted glycosyl fluorides. The iodides are extremely unstable. The bromides are commonly used for glycosylation reactions.

**Glycosides.** Almost all glycoside syntheses are carried out with *O*-acetyl-α-glycosyl bromides. In the presence of acid-binding reagents (silver oxide, silver carbonate,

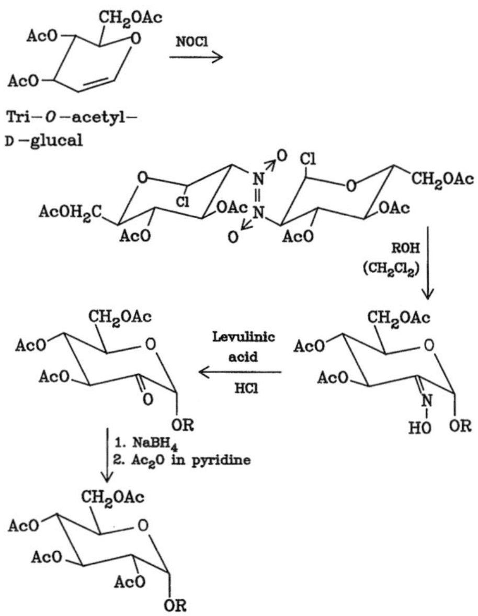

**Figure 4.** Stereoselective glycoside formation

mercury cyanide, or organic bases), they react with alcohols to form glycosides [51], predominantly 1,2-*trans*-glycosides (with most monosaccharides this is the β-anomer). Use of soluble heavy metal salts (e.g., silver perchlorate) and solvents with good solvating properties (acetonitrile, nitromethane) favors the formation of 1,2-*cis*-glycosides. These isomers can be obtained almost exclusively by reaction of acetylated glycals with nitrosyl chloride to give the dimeric 2-nitrosopyranosyl chlorides, which in the presence of an alcohol and dichloromethane form the α-glycosides [52]. Removal of the oximino group by transoximation with acetaldehyde or levulinic acid frees the 2-oxoglycosides, which can be reduced stereoselectively to 1,2-*cis*-glycosides with NaBH$_4$ (Fig. 4).

Glycosides derived from a monosaccharide and a volatile aliphatic alcohol (methanol or ethanol) are best prepared by boiling the sugar in the alcohol with a small amount of acid catalyst (Fischer synthesis). The use of acidic ion-exchange resins is convenient; the resins can be removed by filtration after termination of the reaction. The Fischer synthesis generally yields α-glycosides (anomeric effect) [53].

## 6.3. Other Reactions

**Hydrolysis.** The hydrolysis of disaccharides like sucrose and lactose, polysaccharides like starch, and hemicellulosic materials is of great importance in the food and fermentation industries [32]. For example, the approximately 1:1 mixtures of glucose and fructose (*invert sugar* from hydrolysis of sucrose) and of glucose and galactose (from

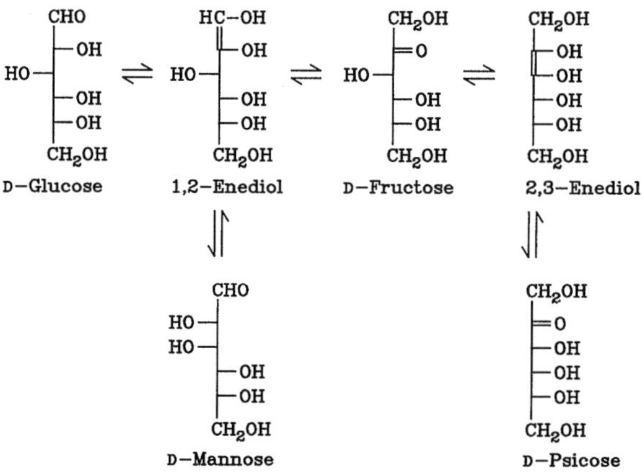

**Figure 5.** Lobry de Bruyn–van Ekenstein rearrangement

hydrolysis of lactose) are used as liquid sweeteners in food and beverages. The latter can substitute for lactose, the carbohydrate in milk and milk products, which cannot be digested by many nonwhite adults because they lack the splitting enzyme lactase [33].

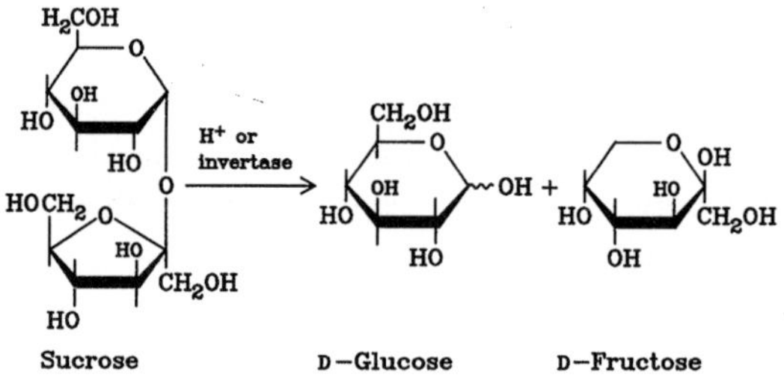

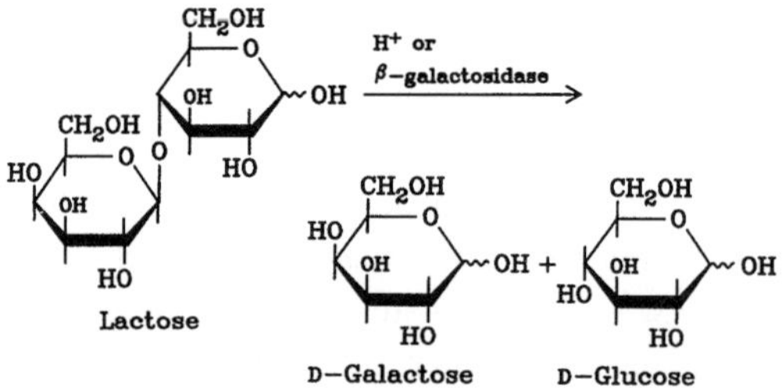

Hydrolysis of sucrose, starch, and cellulose provides monosaccharides that can be fermented to yield several important chemicals, particularly ethanol.

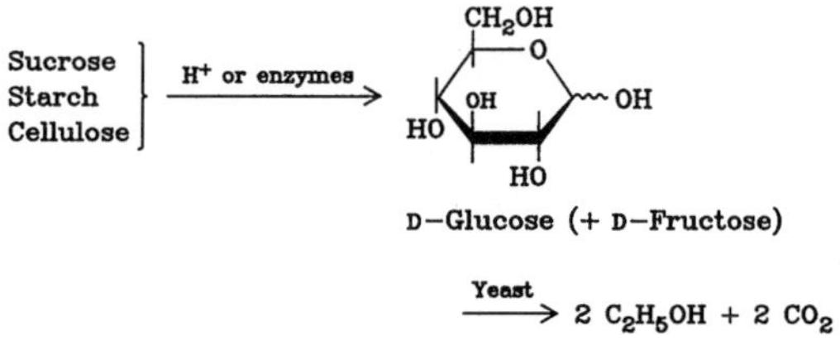

Because these starting materials grow by photosynthesis in amounts of $(1-2) \times 10^{11}$ t/a, they are valuable sources of renewable energy.

Hydrolysis of oligo- and polysaccharides is generally carried out in aqueous acid or by means of enzymes. Hydrolysis of starch yields the liquid sweetener glucose syrup (corn syrup). From the $5.1 \times 10^6$ t of starch produced in the United States in 1982, about $2 \times 10^6$ t was consumed as liquid sweetener [34].

Agricultural wastes (cereal straws and brans) contain pentosans, polysaccharides based on D-xylose and L-arabinose. The conversion of these into monosaccharides by acid hydrolysis and the subsequent dehydration to furfural [98-01-1] and levulinic acid is a well-known industrial process:

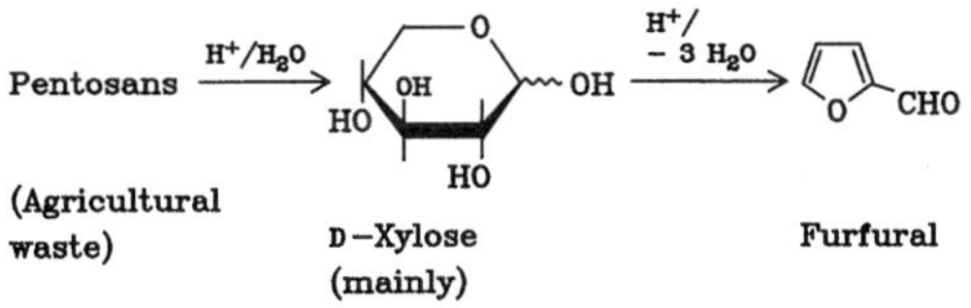

Pentosans (Agricultural waste) $\xrightarrow{H^+/H_2O}$ D-Xylose (mainly) $\xrightarrow{H^+/-3 H_2O}$ Furfural

Furfural is a widely used industrial chemical; worldwide production in 1982 was $200 \times 10^3$ t ($\rightarrow$ Furan and Derivatives).

The trisaccharide raffinose, which is a storage carbohydrate in many plants such as sugar beets, beans, and trees, can be cleaved enzymatically with α-galactosidase into sucrose and galactose. This reaction is used in the beet sugar industry to increase the yield of sucrose, as well as to improve the digestibility of food from leguminous plants. Raffinose can be fermented partially by baker's yeast to form melibiose.

**Isomerization.** Monosaccharides and reducing oligosaccharides undergo enediol rearrangements, especially under alkaline conditions (Lobry de Bruyn-van Ekenstein rearrangement; Fig. 5). From the two aldoses, glucose and mannose, one ketose, fructose, is obtainable. These three sugars are present in an equilibrium mixture; their concentrations depend on the pH: in neutral aqueous solution the ratio of glucose : fructose : mannose is 4 : 3 : 1; in strong alkali it is 2 : 2 : 1 [35].

The 1,2-enediol is assumed to be the common intermediate. From the 2,3-enediol the keto sugar psicose is derived. After benzilic acid rearrangement the 2,3-enediol is converted into the saccharinic acid derivative shown below; this compound is structurally related to the nucleoside sugar D-ribose.

$$\text{D-Fructose} \xrightarrow{\text{Ca(OH)}_2} \text{2,3-Enediol}$$

$$\xrightarrow{-\text{H}_2\text{O}} \xrightarrow{\text{Benzilic acid rearrangement}} \text{2-}C\text{-Methyl-}\text{D-ribonolactone}$$

One disadvantage of chemical isomerizations is the formation of such degradation products as lactic acid. However, these reactions offer the possibility to make lactulose from lactose or (iso-)maltulose from (iso-)maltose.

Enzymatic isomerizations are also industrially important. For example, glucose syrup (from starch) is isomerized to high-fructose corn syrup (HFCS) containing about 42 % fructose by passing glucose syrup over immobilized enzymes. The fructose content can be increased by chromatography of high-fructose corn syrup on ion-exchange resins.

Fructose is sweeter than glucose and can be used by diabetics.

The world production of corn syrup and high-fructose corn syrup in 1980 was $10^7$ t, with a selling price about 90 % that of sucrose [36]. The production of total corn sweeteners in the United States in 1985 was $6.8 \times 10^6$ t, including $4.4 \times 10^6$ t of high-fructose corn syrup (dry basis). The production of HFCS is expected to double over the next five years.

**Decomposition.** High temperatures lead to decomposition (dehydration) of carbohydrates with darkening (caramelization). This can be used to produce the caramel color, e.g., the color of cola beverages. Decomposition in the presence of amino acids (Maillard reaction) is responsible for many color- and flavor-forming reactions, such as baking bread and roasting meat or coffee.

The controlled decomposition of hexoses yields furan derivatives. The dehydration of fructose gives 5-hydroxymethylfurfural [67-47-0].

**Nucleophilic Reactivity.** The carbonyl groups in all carbohydrates are normally masked by acetal or hemiacetal formation; therefore, nucleophilic reactions of carbohydrates involve primarily the hydroxyl groups, which can be acylated, alkylated, and acetalized (see Section 6.1).

**Electrophilic Reactivity.** In monosaccharides and reducing oligosaccharides, the reactions of the potentially electrophilic hemiacetal group are typical. Thus, transglycosylation, nucleophilic substitution of an anomeric hydroxyl group (lactol group), is a typical reaction in carbohydrate chemistry. In solution the hemiacetal form of carbohydrates is in equilibrium with a minor amount of the free carbonyl form. Accordingly, carbohydrates undergo typical carbonyl reactions with good nucleophiles (addition, addition with subsequent elimination, and enolization followed by isomerization or other reactions).

**Degradation of Monosaccharides.** The two most important methods for degrading monosaccharides are (1) the stepwise degradation via bissulfonyl derivatives prepared by oxidation of monosaccharide dithioacetals and (2) the oxidative periodate degradation. Degradations serve analytical purposes and can also be used for the preparation of shorter-chain monosaccharides. According to D. L. MacDonald and H. O. L. Fischer [37] D-arabinose can be prepared from D-mannose diethyl dithioacetal: the dithioacetal (mercaptal) is oxidized by perpropionic acid to the disulfone, which can be cleaved by aqueous ammonia into D-arabinose and the insoluble bis(ethylsulfonyl)methane.

Sodium metaperiodate cleaves carbon chains bearing vicinal hydroxyl groups to yield carbonyl compounds. Unsubstituted monosaccharides are completely degraded into one-carbon units. Thus, formaldehyde is formed from the hydroxymethyl group, and all other carbon atoms yield formic acid. Partially blocked derivatives are degraded accordingly. For example, 2,3-isopropylidene-L-rhamnitol yields 2,3-isopropylidene-L-erythrose when oxidized by periodate:

L-Erythrose is obtained by mild acid hydrolysis of the isopropylidene acetal.

# 7. References

General References

[1] J. Lehmann: *Chemie der Kohlenhydrate,* Thieme Taschenbuch der Organischen Chemie, Reihe B, vol. **10,** Thieme, Stuttgart 1976.

[2] J. Stanek, M. Cerny, J. Kocourek, J. Pacak: *The Monosaccharides,* Academic Press, New York-London 1963.

[3] J. Stanek, M. Cerny, J. Pacak: *The Oligosaccharides,* Academic Press, New York-London 1965.

[4] R. W. Bailey: *Oligosaccharides,* Pergamon Press, London 1965.

[5] J. F. Stoddart: *Stereochemistry of Carbohydrates,* Wiley-Interscience, New York-London-Sydney-Toronto 1971.

[6] S. Hanessian: *Deoxy Sugars,* Am. Chem. Soc. Publications, Washington, D.C., 1968.

[7] W. Pigman, D. Horton: *The Carbohydrates - Chemistry and Biochemistry,* **4** vols., Academic Press, New York-London 1971–1980.

[8] R. W. Jeanloz: *The Amino Sugars,* **3** vols., Academic Press, New York-London 1965–1969.

[9] G. O. Aspinall (ed.): *The Polysaccharides,* **2** vols., Academic Press, New York-London 1982–1983.

[10] R. L. Whistler, J. N. B. Miller: *Methods in Carbohydrate Chemistry,* Academic Press, New York-London, published regularly since 1962.

[11] R. S. Tipson, D. Horton: *Advances in Carbohydrate Chemistry and Biochemistry,* Academic Press, New York-London.

[12] Specialist Periodical Reports, *Carbohydrate Chemistry,* Chemical Society London.

[13] S. J. Angyal: "Zusammensetzung und Konformation von Zuckern in Lösung", *Angew. Chem.* **81** (1969) 172–182; *Angew. Chem. Int. Ed. Engl.* **8** (1969) 157.

[14] B. Capon: "Mechanisms in Carbohydrate Chemistry," *Chem. Rev.* **69** (1969) 407–498.

[15] R. U. Lemieux, J. C. Martin: "Applications of Empirical Rules for Optical Rotation to Problems of Conformational Analysis," *Carbohydr. Res.* **13** (1970) 139–161.

[16] J. S. Brimacombe: "Some Recent Neighbouring Group Participation and Rearrangement Reactions of Carbohydrates," *Fortschr. Chem. Forsch.* **14** (1970) 367–388.

[17] R. J. Ferrier: "Newer Observations on the Synthesis of O-Glycosides," *Fortschr. Chem. Forsch.* **14** (1970) 389–429.

[18] H. Simon, A. Kraus: "Mechanistische Untersuchungen über Glykosylamine, Zuckerhydrazone, Amadori-Umlagerungsprodukte und Osazone," *Fortschr. Chem. Forsch.* **14** (1970) 430–471.

[19] H. Paulsen, H. Behre, E.-P. Herold: "Acyloxonium-Ion-Umlagerungen in der Kohlenhydratchemie," *Fortschr. Chem. Forsch.* **14** (1970) 472–525.

[20] M. Cerny, J. Stanek: "1,6-Anhydroaldohexopyranosen," *Fortschr. Chem. Forsch.* **14** (1970) 526–555.

[21] F. W. Lichtenthaler: "Branched-Chain Aminosugars and Aminocyclanols via Dialdehyde-Nitroalkane Cyclization," *Fortschr. Chem. Forsch.* **14** (1970) 556–577.

[22] J. S. Brimacombe: "Synthese von Antibiotica-Zukkern,"*Angew. Chem.* **83** (1971) 261–274; *Angew. Chem. Int. Ed. Engl.* **10** (1971) 236.

[23] H. Grisebach, K. Schmid: "Chemie und Biochemie verzweigtkettiger Zucker," *Angew. Chem.* **84** ( 1972) 192–206; *Angew. Chem. Int. Ed. Engl.* **11** (1972) 159.

[24] G. Wulff, G. Röhle: "Ergebnisse und Probleme der O-Glykosidsynthese," *Angew. Chem.* **86** (1974) 173–187; *Angew. Chem. Int. Ed. Engl.* **13** (1974) 157.

[25]   D. A. Rees, E. J. Welsh: "Sekundär- und Tertiärstruktur von Oligosacchariden in Lösungen und Gelen," *Angew. Chem.* **89** (1977) 228–239; *Angew. Chem. Int. Ed. Engl.* **16** (1977) 214.

Specific References

[26]   A. Vassala in R. Scheffold (ed.): *Modern Synthetic Methods,* Verlag Salle + Sauerländer, Frankfurt 1980, pp. 173–267.
[27]   S. Hanessian: *Total Synthesis of Natural Products; The "Chiron" Approach, X;* Organic Chemistry Series, vol. **3**, Pergamon Press, Oxford 1983.
[28]   B. Schneider, F. W. Lichtenthaler, G. Steinle, H. Schiweck, *Liebigs Ann. Chem.* 1985, 2443–2453.
[29]   C. K. Lee (ed.): *Developments in Food Carbohydrates* part 2, Applied Science Publishers, London 1980, pp. 1–89.
[30]   IUPAC Commission on Nomenclature of Organic Chemistry and IUPAC-IUB Commission on Biochemical Nomenclature, *Eur. J. Biochem.* **21** (1971) 455–477.
[31]   IUPAC-IUB Joint Commission on Biochemical Nomenclature, *Pure Appl. Chem.* **53** (1981) 1901–1905.
[32]   J. Szejtli: *Säurehydrolyse glycosidischer Bindungen,* VEB Fachbuchverlag, Leipzig 1976.
[33]   N. Kretschmer in D. M. Paige, T. M. Bayless (eds.): *Lactose Digestion,* Johns Hopkins University Press, Baltimore-London 1981, p. 3.
[34]   N. E. Lloyd, W. J. Nelson in R. L. Whistler, J. N.B. Miller, E. F. Paxchall (eds.): *Starch Chemistry and Technology,* 2nd ed., Academic Press 1984, p. 611.
[35]   A. P. G. Kieboom, H. van Bekkum, *J. R. Neth. Chem. Soc.* **103** (1984) 1.
[36]   P. J. Picard in G. G. Birch, K. J. Parker (eds.): *Nutritive Sweeteners,* Applied Science Publishers, London-New Jersey 1982, pp. 145 ff.
[37]   D. L. MacDonald, H. O. L. Fischer, *J. Am. Chem. Soc.* **74** (1952) 2087–2090.
[38]   D. L. MacDonald, *J. Org. Chem.* **27** (1962) 1107–1109.
[39]   A. H. Olavesen, E. A. Davidson, *J. Biol. Chem.* **240** (1965) 992–996.
[40]   C. E. Ballou, D. L. MacDonald, *Methods Carbohydr. Chem.* **2** (1963) 270–272.
[41]   E. G. V. Percival, *J. Chem. Soc.* 1945, 119–123.
[42]   G. Zemplen, E. Pascu, *Chem. Ber.* **62** (1929) 1613–1614.
[43]   H. E. Conrad, *Methods Carbohydr. Chem.* **6** (1972) 361–364.
[44]   J. O. Deferrari, E. G. Gros, J. O. Mastronardi, *Carbohyd. Res.* **4** (1967) 432–434.
[45]   C. C. Sweely, R. Bentley, M. Makita, W. W. Wells, *J. Am. Chem. Soc.* **85** (1963) 2497–2507.
[46]   D. Horton, L. G. Tindall, Jr., *Carbohyd. Res.* **15** (1970) 215–232.
[47]   N. K. Richtmyer, *Methods Carbohydr. Chem.* **1** (1962) 160–167.
[48]   J. C. Sowden, *Methods Carbohydr. Chem.* **1** (1962) 132–135.
[49]   H. P. Albrecht, D. P. Repke, J. G. Moffatt, *J. Org. Chem.* **38** (1973) 1836–1840.
[50]   N. K. Kochetkov, B. A. Dmitriev, *Dokl. Akad. Nauk SSSR* **151** (1963) 106.
[51]   K. Igarashi, *Adv. Carbohyd. Chem. Biochem.* **34** (1977) 243–283.
[52]   R. U. Lemieux, T. L. Nagabhutan, *Methods Carbohydr. Chem.* **6** (1977) 487–496.
[53]   G. N. Bollenback, *Methods Carbohydr. Chem.* **2** (1963) 326–328.

# Carbonic Acid Esters

Hans-Josef Buysch, Bayer AG, Krefeld, Federal Republic of Germany

| | | | | |
|---|---|---|---|---|
| 1. | Introduction ............. 1085 | 6. | Quality Specifications....... 1089 |
| 2. | Physical Properties ........ 1085 | 7. | Analysis ................ 1089 |
| 3. | Chemical Properties........ 1086 | 8. | Storage and Transportation .. 1089 |
| 4. | Production ............. 1087 | 9. | Uses ................... 1089 |
| 5. | Environmental Protection and Toxicology............... 1088 | 10. | Economic Aspects ......... 1091 |
| | | 11. | References............... 1091 |

## 1. Introduction

The monoesters of carbonic acid can only be isolated as salts, simple anhydrides, or mixed anhydrides with carboxylic acids [1]. In the past, they have aroused little industrial interest. On the other hand, the symmetrical and unsymmetrical diesters obtained from aliphatic and aromatic hydroxy compounds are widely used, especially as intermediates and solvents.

## 2. Physical Properties

The industrially important dialkyl carbonic acid esters are all colorless liquids, and most of them have pleasant odors. The aromatic esters, on the other hand, are solids. 2-Oxo-1,3-dioxolane (ethylene carbonate) and 2-oxo-4-methyl-1,3-dioxolane (propylene carbonate) dissolve readily in water, while the dimethyl ester of carbonic acid (dimethyl carbonate) and the diethyl ester of carbonic acid (diethyl carbonate) are sparingly soluble in water. The others, including aromatic carbonates, are nearly insoluble in water. They are soluble in many organic solvents, particularly polar solvents, such as esters, ketones, ethers, alcohols, and aromatic hydrocarbons. The lower aliphatic carbonates form azeotropic mixtures with several organic solvents [2]. The five-membered cyclic carbonates, especially 2-oxo-1,3-dioxolane, are extensively used as solvents for aromatics, polymers, and saltlike compounds [3].

Table 1. Properties of industrially important carbonic acid esters

| Ester | mp, °C | bp, °C/p, kPa | $d_4^{20}$ | $n_D^{20}$ | Flash point, °C |
|---|---|---|---|---|---|
| Dimethyl carbonate [616-38-6] | 4 | 90.2/101.3 | 1.073 | 1.3687 | 14 (closed cup) |
| Diethyl carbonate [105-58-8] | −43 | 125.8/101.3 | 0.9764 | 1.3843 | 33 (closed cup) |
| Diallyl carbonate [15022-08-9] | | 97/8.13 | | 1.4280 | |
| Diethylene glycol bis(allyl carbonate) [142-22-3] | − 4 | 160/0.27 | 1.143 | 1.4503 | 177 (open cup) |
| Diphenyl carbonate [102-09-0] | 78.8 | 302/101.3 | $1.1215_4^{87}$ | | 168 (closed cup) |
| Ethylene carbonate [96-49-1] | 39 | 248/101.3 | $1.3218_4^{39}$ | $1.4158^{50}$ | |
| Propylene carbonate [108-32-7] | −48.8 | 242/101.3 | $1.2069_{20}^{20}$ | 1.4189 | |

The physical properties of industrially important carbonic acid esters are presented in Table 1.

# 3. Chemical Properties

The ester structure and a strong tendency to form carbon dioxide play dominant roles in the chemical properties of carbonic acid esters. Hydrolysis, which becomes more difficult with increasing molecular mass of the ester, leads to the formation of hydroxy compounds and carbon dioxide. Apart from massive steric effects, transesterification between carbonic acid esters and hydroxy compounds adheres to the following rule: the more nucleophilic hydroxy compound displaces the less nucleophilic compound, and if both hydroxy compounds have the same, or nearly the same, nucleophilicity, then the less volatile compound displaces the more volatile one. Hence, even with lower-mass alcohols, diaryl carbonates form dialkyl carbonates and the less nucleophilic phenols, but lower-mass dialkyl carbonates react with higher-mass alcohols to give the higher-mass dialkyl carbonates; the lower-boiling alcohol is removed by distillation. In this case, transesterification occurs in steps with the intermediate formation of a mixed ester. The reaction of dialkyl carbonates with phenols to give diaryl carbonates, contrary to the above rule, can be achieved by using suitable catalysts [4]. It proceeds, however, at a relatively slow rate.

Carbonic acid esters react with primary and secondary amines to give urethanes and ureas [5]. At higher temperature and in the presence of catalysts, aliphatic carbonates lose carbon dioxide and act as alkylating agents, e.g., for amines [6] and phenols [7]. Rapid oxyethylation can be achieved in this way by using five-membered cyclic carbonates [8]. In the absence of compounds to be alkylated, these cyclic carbonates can be broken down to oxiranes and carbon dioxide, which is the reverse of the formation reaction [9]. Open-chain carbonic acid esters can be catalytically broken down at high temperatures to form alcohols, carbon dioxide, and olefins [10]. Dialkyl carbonates undergo Claisen condensation with aliphatic carboxylic esters to form malonates [11].

# 4. Production

The phosgenation of hydroxy compounds is currently the most important method employed to produce carbonic acid esters [12]. Diethyl carbonate, one of the most frequently used aliphatic carbonates, is produced by passing gaseous phosgene in boiling ethanol, in a molar ratio of 1 part of phosgene : 2.5 – 4 parts of ethanol, into a glass or enamel apparatus consisting of a heatable reaction vessel with a reflux condenser and a distillation setup. The reaction gives essentially quantitative yields ($\geq 99\%$) and also forms hydrogen chloride. Excess ethanol is removed from the reaction mixture by distillation, and diethyl carbonate is obtained as the residue.

To recover high purity ($\geq 99.8\%$) diethyl carbonate that is free of chlorine ($\leq 10$ mg/kg Cl), the raw product is distilled. If necessary, a base, such as an alkali carbonate or an alkaline earth carbonate, is used in the distillation. The yields, based on phosgene, are $\geq 95\%$ of the theoretical value.

Dimethyl carbonate is formed by the reaction of methanol with phosgene or methyl chloroformate in the presence of a concentrated sodium hydroxide solution [13]. This two-phase reaction produces very high yields of a pure product. Alkali-catalyzed transesterification can be used to obtain higher-mass dialkyl carbonates from lower-mass ones. Unsymmetrical aliphatic or aromatic – aliphatic carbonic acid esters, e.g., 9-oxo-2,5,8,10-tetraoxatridec-12-enoic 2-propenyl ester [diethylene glycol bis(allyl carbonate)] [14], are produced by condensing a chloroformic acid ester with the appropriate hydroxy compound in the presence of a base, such as sodium hydroxide, pyridine, or dimethylaniline [15], [16]. Aromatic carbonates are also synthesized by the phosgenation of alkaline phenolates in a two-phase system [17]. Trialkylamines or tetraalkylammonium salts are used as catalysts [18]. A series of specific carbonates have been synthesized in a similar manner [19].

High-temperature phosgenation of phenol, even in the absence of alkali, results in the cleavage of hydrogen chloride and essentially quantitative yields of carbonate, if an N-heterocyclic compound is used as catalyst [20].

Carbonic acid ester synthesis starting with carbon dioxide and carbon monoxide, but without phosgene, has recently attracted attention. For example, alcohols react with ethylene glycol carbonate to give essentially quantitative yields of dialkyl carbonates [21], [22]. Ethylene glycol carbonate is made from ethylene oxide [23] and carbon dioxide under pressure. Other oxiranes react in an analogous manner. Reagents such as alkyl iodides and ammonium and phosphonium compounds are used as catalysts [24] – [28].

The direct catalytic reaction of alcohols with carbon dioxide to form dialkyl carbonates succeeds only in the presence of hydrophilic agents, e.g., zeolites [29]. Indeed, excellent conversions are achieved by means of a special catalysis, using ethylene oxide as a water acceptor [30].

At temperatures greater than 200 °C, high-boiling alcohols catalytically displace ammonia from urea to form carbonic acid esters [31] – [33].

Carbonic acid esters can also be prepared by the reaction of salts of carbonic acid esters with alkyl halides [34]. High yields of carbonic acid esters can be obtained by oxidation, particularly air oxidation of hydroxy compounds and carbon monoxide in the presence of complex palladium [35], [36] or copper [37]–[40] catalysts. Dimethyl carbonate will be produced by this method [41], whereby an azeotropic mixture with methanol is formed and separation is subsequently required [42]–[44].

An analogous synthesis of diaryl carbonates using carbon monoxide is unattractive because of the high cost of the catalyst [45]–[47]. Cyclic carbonates with a ring size of more than five members [48], especially the six-membered cyclic carbonates [49], can be readily formed either by the urea route [50] or by depolymerizing relatively low molecular mass polycarbonates.

# 5. Environmental Protection and Toxicology

The safety precautions taken when working with phosgene must be taken during the normal production of carbonic acid esters. The exhaust gases are subjected to hydrolysis and then burned. Waste water is freed of all undesired substances, e.g., phenol, by extraction and desorption.

Although carbonic acid esters are not generally considered dangerous working substances, they must enter neither the atmosphere nor the water supply. Vapors must be removed by adequate exhaust systems and spilled liquids must be absorbed onto an appropriate material, which is subsequently burned.

All important carbonic acid esters are flammable substances. In particular, the short-chain, aliphatic esters can be easily ignited because of their low flash points. Carbonic acid esters form explosive mixtures with air.

Diethyl carbonate has an $LD_{50}$ (subcutaneous) of 8500 mg/kg (rat), the TDLo is 500 mg/kg (mouse). Dimethyl carbonate, a potential methylating agent, should be handled with care, in spite of the fact that its $LD_{50}$ (oral) is 13 g/kg and its $LD_{50}$ (intraperitoneal) is 1600 mg/kg (rat). Diphenyl carbonate can be considered a relatively nontoxic substance. Its TDLo is 28 g/kg (mouse). The same applies to ethylene glycol carbonate for which the $LD_{50}$ is 10 g/kg (rat, oral) and to propylene glycol carbonate for which the $LD_{50}$ is 29 g/kg (rat, oral). Diethylene glycol bis(allyl carbonate) is more toxic with an $LD_{50}$ of 349 mg/kg (rat, oral) and more irritating to the skin, especially wet skin. Its $LD_{50}$ is 3038 mg/kg (rabbit, dermal) [51].

# 6.  Quality Specifications

Carbonic acid esters are isolated in very pure form by distillation. Indeed, dialkyl carbonates of $\geq 99.6\%$ purity are obtained as colorless liquids, free of water and containing less than 0.1% phosgene and hydrogen chloride. Diphenyl carbonate of 99.9% purity, recovered in the molten state, has a color index of 20 – 30 (APHA) and a maximum phenol content of 0.05%. The diethylene glycol bis(allyl carbonate) on the market is more than 94% pure and has a maximum APHA color index of 30.

# 7.  Analysis

Gas chromatography has proven to be the simplest and most efficient method of analysis applicable to practically all impurities in carbonic acid esters. Titrations are used to detect water and traces of hydrogen chloride. Chloroformate can be detected by titration or gas chromatography [52].

# 8.  Storage and Transportation

Because carbonic acid esters are not corrosive they can be stored, in the absence of moisture, in stainless steel tanks for at least one year. Diphenyl carbonate can be stored in the molten state at 150 °C for approximately one month. However, allyl carbonates can only be stored for a limited time because of their tendency to polymerize. The permissible period of storage depends primarily on the temperature. Dialkyl carbonates are transported in steel barrels or tank cars. Diphenyl carbonate is delivered as flakes in polyethylene sacks or in the molten state in tank cars.

# 9.  Uses

Salts of carbonic acid monoesters can be used to carboxylate compounds such as ketones [53], [54]. Diethyl carbonate is an excellent solvent in the mid-boiling range. Diethyl carbonate and dimethyl carbonate are predominantly used as intermediates in the production of urethanes and ureas [55] – [58], other aliphatic carbonic acid esters, e.g., allyl carbonic acid esters [59], diaryl carbonates [60], aliphatic [61] – [63] and aromatic [64] polycarbonates, cyclic carbonic esters [48], [49], [65], and oxetanes [66]. Transesterification with N-acylated 2-aminoethanols followed by elimination of carbon dioxide leads to polymerizable N-vinyl compounds [67]. Dimethyl carbonate is seldom used as a solvent, but it is increasingly encountered as a methylating agent because it is better than dimethyl sulfate [68]. It is used in the synthesis of O-methylphenols [7], [69] and N-methylanilines, enabling the methylation of sterically hindered phenols and anilines [70], [71].

Diphenyl carbonate is used solely as an intermediate. It is important for the synthesis of lower-mass aliphatic monoisocyanates, which starts with the corresponding ureas or allophanates [72], and for the preparation of aliphatic and aromatic polycarbonates by means of transesterification [62].

Dicarboxylic acids react with diphenyl carbonate to eliminate carbon dioxide and form diphenyl esters, which are the starting materials for the production of high molecular mass aromatic polyesters [73], [74].

The addition of diphenyl carbonate at the end of polycondensation of polyamides and polyesters causes an increase in viscosity [75]. Plastics having an excellent optical quality are obtained by polymerizing diallyl esters of carbonic acid such as diethylene glycol bis(allyl carbonate) [76]–[78]. Allyl carbonates can be used to modify other vinyl polymers [79], [80].

Mixed aromatic and aliphatic carbonic acid esters are used to produce a softer poly(vinyl chloride) [81]. They also serve as aids in the dyeing of textiles [82]. Aliphatic carbonates possessing ether groups are used as lubricants and hydraulic fluids [63]. Carbonic acid esters active as herbicides are customarily obtained from alkynols [83], and those active as acaricides, fungicides, and insecticides are obtained from dinitroalkylphenols [84], [85]. Dihalogenocyanophenyl alkyl carbonates [86], [87] and alkyl phenyl carbonates possessing a fluorinated aliphatic group can also be used to make pesticides [88]. Similarly, cyclic carbonic esters can be used in many different ways. Five-membered cyclic carbonates are polar and, for this reason, are excellent solvents, e.g., for polyacrylonitrile [3]. They are also important for the extractive separation of mixtures [89] and as additives for hydraulic fluids [90]. They are often used in oxyalkylation reactions of OH, SH, or aromatic NH groups in place of oxiranes, which are more difficult to handle [3], [8]. They readily react with carboxylic acids to form 2-hydroxyethyl esters [91], e.g., bis(2-hydroxyethyl) terephthalate, which condenses to polyethylene terephthalate. Catalytic transesterification is a very useful reaction. For example, ethylene glycol carbonate reacts with aliphatic hydroxy compounds to yield ethylene glycol, the corresponding carbonic acid esters [20], [21], and polycarbonates [92]. Primary and secondary aliphatic amines are converted into urethanes [93] or ureas [94].

Selective, catalytic cleavage of five-membered cyclic carbonates yields oxiranes and carbon dioxide [95], [96]. This reaction has become more interesting since the successful catalytic oxidation of inexpensive olefins and carbon dioxide to form 2-oxo-1,3-dioxolanes [95], [97]. Methods for the polymerization of five-membered cyclic carbonates have not yet been worked out; the reaction is not reproducible and usually yields polyether polycarbonates because of the partial elimination of carbon dioxide [98]. The hydrolysis of five-membered cyclic carbonates produces high yields of very pure 1,2-diols [24].

In comparison, six-membered cyclic carbonates, e.g., 5,5-dimethyl-2-oxo-1,3-dioxane [*3592-12-9*], easily undergo catalyzed polymerization to form high molecular mass carbonates. In the presence of suitable bis(oxo)dioxanes, cross-linked polymers can also be obtained [49]. Some carbonic acid esters are suitable for use as flavor and fragrance sources, e.g., 4-isopropyl-5,5-dimethyl-2-oxo-1,3-dioxane [*32368-14-2*], which is used to enhance the aroma of foods, perfumes, and cosmetics and provides the product with an aroma like that of coumarin or tobacco [99].

Some special carbonic acid esters have acquired importance as protecting groups in peptide synthesis [100].

## 10. Economic Aspects

Large amounts of carbonic acid esters do not reach the market, but are utilized as intermediates and are further processed. Hence, exact production figures are not available. However, the worldwide production and consumption is estimated to be tens of thousands of tons per year. The kilogram price is two to five dollars. The importance of carbonic acid esters is shown not so much by these figures as by the fact that they are used in the synthesis of high-quality products.

## 11. References

[1]  *Houben-Weyl*, **E4,** 64–112.
[2]  *Beilstein*, **3 (III)** , 4, 6.
[3]  Jefferson Chem. Comp. Techn. Bulletin (1960).
[4]  Bayer, EP 880, 1977 (H. Krimm, H.-J. Buysch, H. Rudolph).
[5]  N. Yamazaki, T. Iguchi, F. Higashi. *J. Polym. Sci. Polym. Chem. Ed.* **17** (1979) no. 3, 835–841.
[6]  Anic S.p.A., DE-OS 3007196, 1979 (G. Jori, U. Romano).
[7]  Anic S.p.A., GB-A 2026484, 1978 (G. Jori, U. Romano).
[8]  Mellon Inst. of Ind. Res., US 2448767, 1942 (W. W. Carlson).
[9]  Union Carbide Corp., EP-A 47474, 1980 (C. H. McMullen).
[10] National Distillers & Chem. Corp., DE-OS 3331929, 1982 (C. Blewett).
[11] H. Krauch, W. Kunz: *Reaktionen der organischen Chemie,*5th ed.,  Hüthig Verlag, Heidelberg 1976, pp. 370–372. Mallinckrodt Chem. Works, US 2454360, 1948 (V. H. Wallingford, A. H. Homeyer).
[12] H. Babad, A. G. Zeiler, *Chem. Rev.* **73** (1973)  no. 1, 81.
[13] Bayer, EP 21211, 1979 (H.-J. Buysch, H. Krimm, H. Böhm).
[14] PPG Industries, US 4273726, 1979 (S. Altuglu).
[15] Pennwalt Corp., DE-OS 2926354, 1978 (J. R. Angle, U. D. Wagle, D. C. Reid).
[16] M. Matzner, R. R. Kurkjy, R. J. Cotter, *Chem. Rev.* **64** (1964) 645.
[17] R. S. Hanslick, W. F. Bruce, A. Mascitti, *Org. Synth.* **33** (1953) 74.
[18] Bayer, DE 1101386, 1956 (L. Bottenbruch, H. Schnell).

[19] R. E. Stenseth, R. M. Schisla, J. W. Baker, *J. Chem. Eng. Data* **9** (1964) 390–397.
[20] Bayer, DE-OS 2447348, 1974 (H.-J. Buysch, H. Krimm).
[21] Bayer, EP 1082, 1977 (H.-J. Buysch, H. Krimm, H. Rudolph).
[22] Bayer, EP 1083, 1977 (H.-J. Buysch, H. Krimm, H. Rudolph).
[23] W. J. Peppel, *Ind. Eng. Chem.* **50** (1958) 767–770.
[24] Union Carbide Corp., US 4314945, 1977 (C. H. McMullen, J. R. Nelson, C. B. Ream, J. A. Sims Jr.).
[25] N. Ryoki, N. Akira, M. Harno, *J. Org. Chem.* **45** (1980) no. 19, 3735–3738.
[26] Halcon SD Group Inc., DE-OS 3244456, 1981 (R. J. Harvey, H. M. Sachs).
[27] Standard Oil Co., EP-A 69494, 1981 (J. E. Rinz, C. Paparizos, D. R. Harrington).
[28] BASF, DE 1169459, 1960 (W. Münster, E. Dreher).
[29] Bayer, EP-A 85347, 1982 (J. Genz, W. Heitz).
[30] Bayer, DE 2748718, 1977 (H.-J. Buysch, H. Krimm, H. Rudolph).
[31] Bayer, EP 13957, 1979 (W. Heitz, P. Ball).
[32] Bayer, EP 13958, 1979 (W. Heitz, P. Ball).
[33] BASF, EP 41622, 1981 (W. Harder, F. Merger, F. Towae).
[34] Tokuyama Soda Co., DE-OS 2838701, 1977 (S. Kazuo, S. Toshiaki).
[35] F. Rivetti, U. Romano, *J. Organomet. Chem.* **174** (1979) 221–226.
[36] Rohm, EP 90977, 1982 (E. Baumgartner, G. Schröder, S. Besecke).
[37] Shell Int. Res., EP-A 71286, 1981 (E. Drent).
[38] U. Romano, R. Tesei, M. M. Mauri, P. Rebora, *Ind. Eng. Chem. Prod. Res. Dev.* **19** (1980) no. 3, 396–403.
[39] Bayer, DE-OS 3016187, 1981 (G. Stammann, R. Becker, J. Grolig, H. Waldmann).
[40] General Electric Co., US 4360477, 1981 (J. E. Hallgren, G. M. Lucas).
[41] *Chem. Eng. (N.Y.)* **90** (1983) no. 9, 15.
[42] Bayer, EP 894, 1977 (H.-J. Buysch, H. Krimm, H. Rudolph).
[43] BASF, DE-OS 2706684, 1977 (W. Himmele, K. Fischer, G. Kaibel, K. Schneider, R. Irnich).
[44] Anic S.p.A., DE-OS 2607003, 1975 (U. Romano).
[45] General Electric Co., DE-OS 2738437, 1977 (J. E. Hallgren).
[46] General Electric Co., DE-OS 2815501, 1978 (A. J. Chalk).
[47] J. E. Hallgren, G. M. Lucas, R. O. Matthews, *J. Organomet. Chem.* **204** (1981) no. 1, 135–138.
[48] Bayer, DE-OS 3103140, 1981 (H. Krimm, H.-J. Buysch).
[49] Bayer, US 4440937, 1981 (H. Krimm, H.-J. Buysch).
[50] Bayer, DE-OS 3103137, 1981 (H. Krimm, H.-J. Buysch).
[51] Registry of Toxic Effects of Chemical Substances, NIOSH, Washington, USA, 1980.
[52] B. J. Gudzinowicz, *Anal. Chem.* **37** (1965) no. 8, 1051–1053.
[53] Montedison S.p.A., DE-OS 2429627, 1973 (E. Alneri, G. Bottaccio, M. V. Carletti, G. Lana).
[54] Montedison S.p.A., DE-OS 2612577, 1975 (G. Bottaccio, G. P. Chiusoli, E. Alneri, M. Marchi, G. Lana).
[55] Dow Chem., EP-A 65026, 1980 (A. E. Gurgiolo).
[56] Bayer, EP-A 48371, 1980 (H.-J. Buysch, H. Krimm, W. Richter).
[57] Bayer, EP-A 48927, 1980 (H.-J. Buysch, H. Krimm, W. Richter).
[58] Hoechst, DE 2541741, 1975 (H. Loewe, J. Urbanietz, D. Düwel, R. Kirsch).
[59] Anic S.p.A., EP-A 35304, 1980 (U. Romano, G. Jori).
[60] Bayer, EP 879, 1977 (H. Krimm, H.-J. Buysch, H. Rudolph).
[61] Soc. Nationale des Poudres et Explosifs, US 4005121, 1975 (J.-P. G. Senet).
[62] Bayer, DE 1031512, 1956 (H. Schnell, G. Fritz).

[63] Dow Chem., US 3657310, 1967.
[64] Bayer, DE-OS 3017419, 1980 (H. Krimm, H.-J. Buysch, H. Rudolph).
[65] Bayer, DE-OS 3103135, 1981 (H. Krimm, H.-J. Buysch).
[66] Bayer, DE 950850, 1954 (H. Schnell, K. Raichle, W. Biedermann).
[67] Bayer, EP-A 87659, 1982 (H. Krimm, H.-J. Buysch, P. M. Lange, R. Klipper).
[68] Company Publication of Chimica Saline, Enichimica Subsidiary, Milan, Italy.
[69] BASF, DE-OS 2807762, 1978 (F. Merger, F. Towae, L. Schroff).
[70] PPG Industries, EP-A 104598, 1982 (R. B. Thompson).
[71] PPG Industries, EP-A 104601, 1982 (K. J. Sienkowski).
[72] BASF, EP 50739, 1980 (V. Schwendemann, D. Mangold).
[73] Bayer, GB 958798, 1960.
[74] Sumitomo Chem. Co., JP-Kokai 80100399, 1979; *Chem. Abstr.* **94** (1981) 653386.
[75] General Electric Co., DE-OS 2359260, 1973 (D. W. Fox, D. E. Floryan).
[76] Company Publication of PPG Industries, Chemicals Group, One Gateway Centre, Pittsburgh, PA 15222, USA.
[77] PPG Industries, US 4398008, 1981 (M. S. Misura).
[78] PPG Industries, EP 80339, 1981 (C. W. Eads, J. C. Crano).
[79] PPG Industries, DE 3149499, 1980 (J. C. Crano, R. L. Haynes).
[80] Air Products & Chem. Inc., BE 856911, 1977 (D. D. Dixon, F. L. Herman).
[81] Stauffer Chem. Corp., US 4403056, 1980 (L. S. Giolito).
[82] Bayer, US 3630664, 1967 (K. Fuhr, J. Nentwig, H. Rudolph, J. Romatowski).
[83] Chemagro Corp., US 3348939, 1964 (D. W. Gier).
[84] M. Pianka, P. Sweet, *J. Sci. Food. Agric.* **19** (1968) 676–681.
[85] M. Pianka, *J. Sci. Food Agric.* **17** (1966) 47–56. Fabriek van Chem. Prod., US 3198824, 1961.
[86] Boehringer, DE-OS 2054225, 1970 (H. M. Becher, R. Sehring).
[87] Boehringer, DE-OS 2060825, 1970 (H. M. Becher, R. Sehring).
[88] PPG Industries, US 4022609, 1970 (D. E. Hardies, J. K. Rinehart).
[89] Union Carbide Corp., US 2837585, 1958 (J. R. Anderson, J. V. Murray, C. H. Young). Phillips Petroleum Co., US 4115206, 1977 (T. P. Murtha).
[90] Dow Chem., FR 1572282, 1967 (R. J. Nankee, J. R. Avery, J. E. Schrems).
[91] Bayer, DE-AS 1280240, 1966 (A. Böckmann, H. Vernaleken, L. Bottenbruch, H. Rudolph, H. Schnell).
[92] PPG Industries, DE-AS 1302067, 1958 (H. C. Stevens).
[93] S. Morrell, *Eur. Rubber J.* **164** (1982) no. 2, 39.
[94] BASF, EP-A 2526, 1977 (G. Hamprecht, K. Fischer, O. Woerz).
[95] Texaco Dev. Corp., US 4374259, 1979.
[96] Texaco Dev. Corp., GB 2092127, 1981.
[97] Atlantic Richfield Co., DE-OS 3031289, 1980 (J.-L. Kao, G. A. Wheaton, H. Shalit, M. N. Sheng).
[98] K. Soga, S. Hosada, Y. Tazuke, S. Ikeda, *J. Polym. Sci. Polym. Lett. Ed.* **14** (1976) 161–165.
[99] Internat. Flavors and Fragrances Inc., US 4402985, 1981 (R. M. Boden, M. Licciardello).
[100] *Houben-Weyl*, **15/1**, 46–907.

# Carboxylic Acids, Aliphatic

*Indvidual keywords:* → *Acetic Acid;* → *Acrylic Acid and Derivatives;* → *Adipic Acid;* → *Chloroacetic Acids;* → *Citric Acid;* → *Crotonaldehyde and Crotonic Acid;* → *Dicarboxylic Acids, Aliphatic;* → *Ethylenediaminetetraacetic Acid and Related Chelating Agents;* → *Fatty Acids;* → *Formic Acid;* → *Glyoxylic Acid;* → *Hydroxycarboxylic Acids, Aliphatic and Oxocarboxylic Acids; Lactic Acid;* → *Maleic and Fumaric Acid;* → *Malonic Acid and Derivatives;* → *Mercaptoacetic Acid and Derivatives;* → *Methacrylic Acid and Derivatives;* → *Nitrilotriacetic Acid; Oxalic Acid;* → *Propionic Acid and Derivatives;* → *Sorbic Acid;* → *Tartaric Acid*

WILHELM RIEMENSCHNEIDER, Hoechst Aktiengesellschaft, Frankfurt, Federal Republic of Germany

| | | | | |
|---|---|---|---|---|
| 1. | Introduction | 1096 | 10.1. | Butyric Acids ............ 1107 |
| 2. | Physical Properties | 1096 | 10.2. | Valeric Acids ............ 1108 |
| 3. | Chemical Properties | 1099 | 10.3. | Octanoic Acids........... 1109 |
| 4. | Natural Sources | 1100 | 10.4. | 2-Ethylhexanoic Acid ....... 1109 |
| 5. | Production | 1100 | 10.5. | Nonanoic Acids ........... 1110 |
| 5.1. | Saturated Monocarboxylic Acids | 1101 | 10.6. | Isodecanoic Acid .......... 1110 |
| 5.1.1. | Aldehyde Oxidation | 1101 | 10.7. | Pivalic Acid............. 1111 |
| 5.1.2. | Carboxylation of Olefins, Koch Process | 1102 | 10.8. | Versatic Acids and Neo Acids 1111 |
| | | | 10.9. | Propiolic Acid ........... 1111 |
| 5.1.3. | Oxidation of Alkanes | 1103 | 11. | Trade Names, Economic Aspects ................ 1112 |
| 5.1.4. | Alkali Fusion of Alcohols | 1103 | | |
| 5.2. | Unsaturated Monocarboxylic Acids | 1103 | 12. | Toxicology and Occupational Health ................. 1112 |
| 6. | Environmental Protection | 1104 | 13. | Derivatives ............. 1114 |
| 7. | Quality Specifications and Analysis | 1105 | 13.1. | Acyl Halides ............ 1114 |
| | | | 13.2. | Anhydrides ............. 1115 |
| 8. | Storage and Transportation | 1105 | 13.3. | Lactams................ 1115 |
| 9. | Uses | 1105 | 13.4. | Halogenated Carboxylic Acids. 1116 |
| 10. | Specific Aliphatic Carboxylic Acids | 1107 | 14. | References............... 1117 |

# 1. Introduction

Aliphatic carboxylic acids have the general formula

$$R-C\underset{OH}{\overset{O}{\lessgtr}}$$

where R is H or a straight-chain or branched-chain alkyl group. The first three members of this homologous series, formic, acetic, and propionic acid, are exceptionally important and are discussed in separate articles. Similarly, acids containing 12 or more carbon atoms are described separately under → Fatty Acids. The unsaturated acids acrylic, methacrylic, crotonic, and sorbic acid are also treated separately, as are chloroacetic and oxo- and hydroxycarboxylic acids.

The continuing scientific importance of the $C_4 - C_{11}$ carboxylic acids arises from their functions in the metabolism of plants and animals (see Chap. 4). The great commercial value of these acids, and particularly of their salts and esters, is based on their synthetic utility (see Chap. 9).

Numerous aliphatic carboxylic acids were first obtained from natural sources in the 19th century along with many other natural products. However, these natural sources, which usually yielded straight-chain acids containing an even number of carbon atoms, have since been replaced by large-scale synthetic operations (see Chaps. 5 and 10).

# 2. Physical Properties

The properties of the important saturated and unsaturated aliphatic monocarboxylic acids are listed in Table 1 which also includes typical values of commercial mixtures.

Carboxylic acids are usually colorless. The $n$- and isoalkane carboxylic acids up to $C_9$ are liquids, as are the highly branched acids up to $C_{13}$, except for pivalic acid.

The volatility and characteristic odor of carboxylic acids decrease with increasing molecular mass. The odors of butyric and valeric acids are generally perceived as unpleasant; higher homologs and commercial mixtures have fainter smells. The lower saturated carboxylic acids are miscible with water; solubility decreases rapidly from $C_5$ up. Higher carboxylic acids are almost insoluble in water but miscible with most organic solvents. They are partially associated in polar solvents. In aqueous solution, carboxylic acids behave as weak acids; their strength changes only slightly with increasing molecular mass. The dissociation constants are between 1 and $2 \times 10^{-5}$.

Table 1. Physical properties of $C_4-C_{13}$ monocarboxylic acids

| Systematic name | CAS registry number | Trivial name | Molecular formula | $M_r$ | $fp$ or $mp$, °C | $bp^a$, °C | $\varrho^b$, g/cm³ | $n_D^{20\,c}$ | $\eta^d$ | $S^e$ | $A^f$ |
|---|---|---|---|---|---|---|---|---|---|---|---|
| **Straight-chain acids** | | | | | | | | | | | |
| n-Butanoic acid | [107-92-6] | n-butyric acid | $CH_3CH_2CH_2COOH$ | 88.11 | −5.5 | 164.0 | $0.9629^{15}$ | 1.4003 | 1.6 | | |
| n-Pentanoic acid | [109-52-4] | n-valeric acid | $CH_3(CH_2)_3COOH$ | 102.1 | −34.5 | 186.3 | $0.9437^{15}$ | 1.4105 | 2.19 | 4 | 14 |
| n-Hexanoic acid | [142-62-1] | caproic acid | $CH_3(CH_2)_4COOH$ | 116.2 | −3.9 | 205.3 | $0.9313^{15}$ | 1.4188 | | | |
| n-Heptanoic acid | [111-14-8] | enanthic acid | $CH_3(CH_2)_5COOH$ | 130.2 | −7.5 | 223.0 | $0.9222^{15}$ | 1.4245 | | | |
| n-Octanoic acid | [124-07-2] | caprylic acid | $CH_3(CH_2)_6COOH$ | 144.2 | 16.3 | 239.3 | $0.9088^{20}$ | 1.4280 | | | |
| n-Nonanoic acid | [112-05-0] | pelargonic acid | $CH_3(CH_2)_7COOH$ | 158.2 | 12.5 | 253–254 | $0.9068^{18}$ | | | 0.026 | |
| n-Decanoic acid | [334-48-5] | capric acid | $CH_3(CH_2)_8COOH$ | 172.3 | 31.3 | 268.4 | $0.8858^{40}$ | 1.4285 | | | |
| n-Undecanoic acid | [112-37-8] | undecylic acid | $CH_3(CH_2)_9COOH$ | 186.3 | 29.3 | 284.6 | $0.9905^{25}$ | $1.4202^{70}$ | | | |
| n-Dodecanoic acid | [143-07-7] | lauric acid | $CH_3(CH_2)_{10}COOH$ | 200.4 | 44 | 225* | $0.8690^{50}$ | $1.4304^{50}$ | | | |
| **Branched-chain acids** | | | | | | | | | | | |
| 2-Methylpropanoic acid | [79-31-2] | isobutyric acid | $(CH_3)_2CHCOOH$ | 88.11 | −46.1 | 154.7 | $0.9529^{15}$ | 1.3964 | 1.1 | 21 | 37 |
| 2-Methylbutanoic acid | [116-53-0] | α-methylbutyric acid | $CH_3CH_2\underset{\underset{CH_3}{\mid}}{C}HCOOH$ | 102.1 | ca. −80 | 177 | $0.938^{20}$ | | 2.1 | 4.5 | 66 |
| 3-Methylbutanoic acid | [503-74-2] | isovaleric acid | $(CH_3)_2CHCH_2COOH$ | 102.1 | −37 | 176.5 | $0.9308^{15}$ | 1.4053 | 2.45 | 2.5 | 11 |
| 2,2-Dimethylpropanoic acid | [75-98-9] | pivalic acid | $(CH_3)_3COOH$ | 102.1 | 35 | 163.7 | $0.905^{50}$ | | | 22 | |
| 2-Ethylbutanoic acid | [88-09-5] | α-ethylbutyric acid | $CH_3CH_2\underset{\underset{C_2H_5}{\mid}}{C}HCOOH$ | 116.2 | −15.3 | 193 | $0.923^{20}$ | | 3.3 | 1.8 | 3.6 |
| 2-Methylpentanoic acid | [97-61-0] | isocaproic acid | $CH_3(CH_2)_2\underset{\underset{CH_3}{\mid}}{C}HCOOH$ | 116.2 | | 196 | $0.921^{20}$ | 1.414 | 2.9 | 1.5 | 2.8 |
| 2-Ethylhexanoic acid | [149-57-5] | α-ethylcaproic acid | $CH_3(CH_2)_3\underset{\underset{C_2H_5}{\mid}}{C}HCOOH$ | 144.2 | | 220 | $0.908^{20}$ | 1.425 | 7.8 | 0.2 | 1.2 |
| 2-n-Propylpentanoic acid | [99-16-1] | di-n-propylacetic acid | $(CH_3CH_2CH_2)_2CHCOOH$ | 144.2 | ca. −60 | 229 | $0.906^{20}$ | 1.430 | 8.1 | 0.2 | 1.2 |
| **Commercial mixtures** | | | | | | | | | | | |
| Isomeric $C_8$ acids | [25103-52-0] | isooctanoic acid | $C_7H_{15}COOH$ | 144.2 | ca. −80 | 225–238 | 0.912–0.919 | | 7.5 | 0.4 | 1.8 |

Table 1. continued

| Systematic name | CAS registry number | Trivial name | Molecular formula | $M_r$ | fp or mp, °C | $bp^a$, °C | $\varrho^b$, g/cm³ | $n_D^{20\,c}$ | $\eta^d$ | $S^e$ |
|---|---|---|---|---|---|---|---|---|---|---|
| Isomeric $C_9$ acids | [26896-18-4] | isononanoic acid | $C_8H_{17}COOH$ | 158.2 | ca. −70 | 232−245 | 0.895−0.902 | 1.430 | 11.2 | 0.3 | 1.3 |
| Isomeric $C_{13}$ acids | [25448-24-2] | isotridecanoic acid | $C_{12}H_{25}COOH$ | 214.4 | ca. −46 | 280−315 (90%) | 0.897−0.902 | 1.446 | 49.0 | <0.1 | 0.3 |
| Mixtures of $C_9$−$C_{11}$ acids | [71700-95-3] | versatic acid 911 | | | 140−162** | 0.92 | 1.447 | | | |
| **Unsaturated carboxylic acids** | | | | | | | | | | |
| 2,2-Dimethylpropenoic acid | [541-47-9] | 2,2-dimethylacrylic acid | $(CH_3)_2C=CHCOOH$ | 100.1 | 69 | 199 | $1.066^{24}$ | | 1.5 | |
| Propynoic acid | [471-25-0] | propiolic acid | $CH\equiv CCOOH$ | 58.1 | 9 | 144 (decomp.) | $1.139^{15}$ | | | |

\* At 1.33 kPa.
\*\* At 270 kPa.

$^a$ at 1013 mbar (DIN 53171)
$^b$ (DIN 51 757)
$^c$ (DIN 51 423, Bl. 2)
$^d$ Viscosity, mPa s (DIN 51 550)
$^e$ Solubility in water at 20 °C, wt %
$^f$ Absorption of water at 20 °C, wt %

# 3. Chemical Properties

The carboxyl group is the target of most reactions [1], [2] and is hardly affected by the carbon chain. However, the carboxyl group activates the $\alpha$-position, which is also an important reaction site.

Alkali metal salts of carboxylic acids are readily obtained by neutralization. They undergo only limited hydrolysis in aqueous solutions, which are nearly neutral. Salts of the lower carboxylic acids are soluble in water. Alkali metal salts are poorly soluble in organic solvents; heavy metal salts, on the other hand, are soluble in organic solvents and can be prepared from the alkali metal salts. Zinc salts can be produced by treating the metal with the acids at high temperature. Esters are formed by an acid-catalyzed reaction between carboxylic acids and alcohols or olefins. Polyols such as glycerol or cellulose may react completely or partially. In the second case mixtures of esters are formed. Because of steric hindrance, esterification becomes more difficult with increasing chain length or branching. Addition of carboxylic acids to acetylene gives vinyl esters capable of polymerization.

In the presence of halogenating compounds, e.g., thionyl halides, phosphorus halides, or phosphorus oxyhalides, the highly reactive, unstable acyl halides, a valuable class of synthetic intermediates, are formed. With ammonia or amines the acyl halides give amides, with alcohols esters, and with carboxylic acids anhydrides. These derivatives can also be obtained directly from the carboxylic acids.

Ketenes are obtained from the lower carboxylic acids or anhydrides by pyrolysis; the higher homologs are obtained by dehydrochlorination of the acyl chlorides.

In the presence of such catalysts as aluminum oxide, thorium oxide, or boron phosphate, monocarboxylic acids give ketones with loss of carbon dioxide and water:

$$2\ RCOOH \xrightarrow{Cat.} R\underset{O}{\overset{\parallel}{C}}R + CO_2 + H_2O$$

Reactions involving an unsubstituted carbon chain usually require more severe conditions than reactions of the carboxyl group. Halogenation often produces mixtures but sites of chain-branching are attacked preferentially. For example, isobutyric acid gives mostly the $\alpha$-chloro derivative. Nitric acid oxidation gives dicarboxylic acids with some chain degradation. Permanganate oxidation produces oxalic and acetic acids under alkaline and acidic conditions, respectively.

## 4. Natural Sources

Low concentrations of aliphatic acids are widely distributed in nature. The free acids occur in plants, e.g., formic acid in stinging nettles and pine needles. Butyric acid is found in vegetable oils and in animal fluids, such as sweat, tissue fluids, and milk fat. Isobutyric acid is found in carob bean, isovaleric acid in valerian root. Other acids occur in fruit.

In animals, free carboxylic acids are important metabolites in the breakdown of carbohydrates, fats, and proteins, in the so-called tricarboxylic acid cycle (citric acid cycle). Nearly all naturally occurring acids have an even number of carbon atoms, which suggests a common synthetic pathway utilizing two-carbon units.

Esters occur more frequently than free acids and are important constituents of perfumes and essential oils. Esters of glycerol with various aliphatic acids are found in animals and plants. However, natural sources of carboxylic acids up to $C_{11}$ have little or no commercial importance, because they frequently occur in low concentration only and their isolation is difficult.

## 5. Production [1], [2]

The four important commercial methods of production are (1) aldehyde oxidation, (2) carboxylation of olefins (Koch synthesis), (3) paraffin oxidation, and (4) alkali fusion of alcohols. The hydrolysis of natural oils such as castor and soybean oil for heptanoic acid and undecanoic acid is less important.

At present, biochemical processes are of limited importance. Butyric acid is produced from sugar or starch by *Bacillus butylicus,* from methanol by *Butyribacterium methylotrophicum* [3], or from CO or $CO_2$ by *Eubacterium limosum* [4]. However, commercial interest in biochemical methods may increase in the near future.

Similarly, catalytic air oxidation of alcohols has not yet attracted the interest of industry. Examples are the oxidation of 3-methyl-1-butanol to isovaleric acid (82% yield), of 2-ethyl-1-hexanol to 2-ethylhexanoic acid (84% yield), and of 1-decanol to *n*-decanoic acid (89% yield) [5].

Although hydroformylation of methanol gives excellent yields of acetic acid, higher alcohols give much poorer yields. Thus, 1-heptanol gave only a 66% yield of 1-octanoic acid with a ruthenium catalyst [6]. Homologation with palladium complexes and iodides gave similar results [7].

Carboxylic acids are also obtained by careful oxidation at low temperature and low concentration with one of the following agents: ozone, nitric acid, chlorine, chromic acid, permanganate, or periodate. However, these reagents are used especially when other functional groups are present in the molecule, and the reactions are largely limited to laboratory operations.

High-pressure hydrocarboxylation is a recent development that appears to offer commercial possibilities. An example is the conversion of propylene into isobutyric acid (main product) and *n*-butyric acid in the presence of a palladium catalyst [8].

## 5.1. Saturated Monocarboxylic Acids

### 5.1.1. Aldehyde Oxidation

The principal method for the commercial production of $C_4$–$C_{13}$ carboxylic acids is catalytic oxidation of the corresponding aldehydes:

$$RCHO + 1/2\ O_2 \longrightarrow RCOOH$$

This highly exothermic reaction liberates between 250 and 300 kJ/mol.

The starting materials are produced catalytically from olefins with CO and $H_2$ by the oxo process, which was developed in 1938 by Ruhrchemie AG, Oberhausen, in Germany [9]. Virtually any olefin can be used in this process, giving access to a wide range of aldehydes oxidizable to the corresponding carboxylic acids. Other aldehydes can be produced via aldol condensation, increasing the range of accessible carboxylic acids. Because of the complex nature of the olefinic raw materials, the higher carboxylic acids obtained ($C_8$ and higher) are usually mixtures of branched-chain products.

The aldehydes are oxidized by air or oxygen with or without a catalyst. Liquid-phase oxidation without solvent is preferred, although vapor-phase oxidation is feasible. Side reactions are suppressed when the temperature is only slightly increased. Thorough dispersion of oxygen in the liquid medium is important, especially in the absence of catalysts [10].

Effective catalysts are the salts of metals that appear in more than one oxidation state, e.g., silver, cerium, cobalt, chromium, copper, iron, manganese, molybdenum, nickel, and vanadium. For liquid-phase oxidation, these metals are used as soaps or naphthenates, or in the form of complexes, such as $K_3[Fe(CN)_5 \cdot H_2O]$ [11]. In vapor-phase oxidation, oxides are predominantly used, supported on silicates, diatomaceous earth, silica gel, quartz, or alumina. Catalysts containing manganese or copper or both can be reclaimed by precipitation with oxalic acid [12].

Where side reactions or decomposition may occur during oxidation, as in the case of 2-ethylhexanal, selectivity is improved by using the alkali or alkaline earth salts of weak acids [13].

The following conditions are optimal for the oxidation of isobutyraldehyde to isobutyric acid: (1) temperature 50 °C, residence time ca. 3 h, and 0.02 g catalyst per mole of isobutyraldehyde; (2) temperature 60 °C, residence time 50 min, and 0.06 g catalyst per mole of isobutyraldehyde. The catalyst in each case is a mixture of cobalt, manganese, and copper acetates (1:1:1); oxygen is passed through in slight excess. Conversion is 100%, selectivity 90% [14]. A low pressure gives isobutyric acid with more desirable properties [15].

Peroxycarboxylic acids appear to be intermediates in both uncatalyzed and catalyzed liquid-phase oxidation:

$$R-\underset{\underset{O}{\|}}{C}-H \xrightarrow{O_2} R-\underset{\underset{O}{\|}}{C}-OOH$$

Aldehyde oxidation is usually carried out in stainless steel, but glass and enamelled vessels are also used (see Chap. 8).

## 5.1.2. Carboxylation of Olefins, Koch Process

Carboxylic acids are produced directly from olefins at high pressure as follows [16], [17]:

$$R^1R^2C=CR^3R^4 + CO + H_2O \rightarrow R^1R^2CH-\underset{\underset{R^4}{|}}{\overset{\overset{R^3}{|}}{C}}-COOH$$

$$\Delta H = \text{ca.} -150 \text{ kJ/mol}$$

This eliminates the oxo-aldehyde step.

Reppe carbonylation, limited commercially to the conversion of ethylene into propionic acid, is carried out in the presence of metal carbonyls. The Koch carbonylation employs proton catalysis and is usually accompanied by double bond and structure isomerization to give branched, predominantly tertiary acids [18]. Mixtures are mainly produced; the proportion of tertiary acids depends on reaction conditions. For example, at 80 °C, a CO pressure of 10 MPa (100 bar), and prolonged residence time, all butene isomers are converted mainly into pivalic acid (trimethylacetic acid).

In addition to olefins (except for ethylene) and cycloalkenes, alcohols and alkyl halides may also serve as starting materials. Strong mineral acids, such as $H_2SO_4$, HF, or $H_3PO_4$, alone or in combination with $BF_3$, or $SbF_5$, serve as catalysts; anhydrous conditions are required to give high yields.

The commercial process employs two stages. In the first stage, using a stirred vessel cascade, CO is added to the olefin in the presence of the catalyst, at 20–80 °C and 5–10 MPa (50–100 bar); the preferred catalyst is $H_3PO_4 \cdot BF_3$. In the second stage water is added. The reaction mixture separates into a product phase containing the carboxylic acid and into a second phase that contains the catalyst which can then be recycled. The crude acids are washed with sulfuric acid, hydrogen carbonate, and citric acid, and are then distilled. Tertiary carboxylic acids are formed at high conversion rates with 85–95% selectivity. Carboxylic acids derived from the dimerized olefins are produced as byproducts. The use of alcohol instead of water in the second stage produces esters.

The Koch process is employed commercially by Shell, Exxon (Enjay) [19], [20], Kuhlmann, and Du Pont. In addition to isobutene which gives pivalic acid, diisobutene and $C_6$–$C_8$ and $C_8$–$C_{10}$ cuts are also used as raw materials. On the basis of these olefins an economical and widespread production has been developed. The products are marketed as Versatic Acids (Shell), Neo Acids (Exxon), and CeKanoic Acids (Kuhlmann). The Koch synthesis is conducted in Hastelloy-C or lined pressure vessels (see Chap. 8).

### 5.1.3. Oxidation of Alkanes

The oxidation of paraffin waxes, polyethylenes, or polypropylenes with oxygen or air produces mixtures of straight- and branched-chain monocarboxylic acids ranging from formic acid to acids with the same number of carbon atoms as the starting material.

The reaction takes place above 100 °C in a melt or an aqueous suspension. The pressure is slightly above atmospheric; a catalyst may be employed. The heat of reaction must be effectively removed.

This method is not selective; it produces large amounts of $CO_2$ and modest yields of products which are inseparable mixtures of acids, but can be used as such.

In this way, Rhône-Poulenc produces butyric acid in large amounts as a byproduct of naphtha oxidation.

A variant of alkane oxidation is the vapor-phase cracking of high-molecular alkanes in the presence of $CO_2$ at 500–700 °C with very short residence times [21], [22].

### 5.1.4. Alkali Fusion of Alcohols

This method is still used occasionally to produce carboxylic acids of intermediate chain length in good yields:

$RCH_2OH + NaOH \longrightarrow RCOONa + 2\,H_2$

Oxo alcohols from $C_8$ are possible starting products; their carbon skeleton is preserved. The process is carried out with NaOH or NaOH–KOH mixtures at a temperature of 250–300 °C (ca. 2 MPa) in stainless steel or copper pressure vessels [23]; [24].

## 5.2. Unsaturated Monocarboxylic Acids

Acrylic, methacrylic, crotonic, and sorbic acids are discussed in separate articles; other unsaturated carboxylic acids with less than 12 carbon atoms are not of commercial importance. However, the following two processes are employed on a small scale:

**Table 2.** Biodegradation of sodium carboxylates

| Acid | Degradation, in 10–15 days, % | BOD$_5$ [b] |
|---|---|---|
| Isobutyric | >95 | 1160 |
| Isovaleric | >95 | 1140 |
| 2,3-Dimethylbutyric | >96 | <10 |
| 2,2-Dimethylpentanoic | >95 | <10 |
| Isoheptanoic | 95 | 1350 |
| Isooctanoic [c] | 96 | 650 |
| Isononanoic | 95 | 125 |
| Isodecanoic | 79 | 690 |
| Tridecanoic | 48 | 40 |
| Dimethylacrylic | 96 | 590 |

[a] Straight-chain monocarboxylates are easily degraded.
[b] Biochemical oxygen demand: mg O$_2$ per liter of waste-water; sum over 5 days.
[c] Mixture.

*Dimethylacrylic acid* is obtained in excellent yields from acetone and ketene:

$$\begin{array}{c} H_3C \\ \phantom{H_3}C=O \\ H_3C \end{array} + CH_2=C=O \longrightarrow \begin{array}{c} H_3C \\ \phantom{H_3}C=CHCOOH \\ H_3C \end{array}$$

Homologous ketones and ketenes react similarly.

*Propiolic acid*, acetylenecarboxylic acid, H–C≡C–COOH, is produced by oxidizing propargyl alcohol at a lead anode in the presence of sulfuric acid [25]. The temperature is kept below 20 °C; the electrolytes are separated from each other by membranes. 2-Butyne-1,4-diol gives acetylenedicarboxylic acid, HOOC–C≡C–COOH.

# 6. Environmental Protection

The strong and unpleasant odors of many of these acids call for special protective measures. For example, the odor threshold of *n*-butyric acid is $8.8 \times 10^{-13}$ g/L. These acids are readily removed from waste gas emanating from plants and storage tanks by an alkaline wash. Mixtures with aldehydes, e.g., from oxo syntheses, are removed by burning in a waste-gas flare or in a muffle furnace.

The degrees of degradation and BOD$_5$ values in a biological treatment plant have been determined for sodium carboxylates in wastewater using the Test for Biological Degradation from OECD-Guidelines 302 B (see Table 2).

# 7. Quality Specifications and Analysis

Manufacturers' specifications for the European market are given in Table 3. Specifications for the world market are similar.

Titration with aqueous alkali is satisfactory for many purposes; the preferred indicators are bromobenzene green, bromophenol blue, or phenolphthalein. The less soluble higher acids are titrated in aqueous methanol or ethanol or with methanolic KOH. Lauric acid is a suitable calibration standard.

Titration against a glass–calomel electrode also gives excellent results; a graph is used to calculate the acid concentration [26].

Mixtures can be analyzed with great precision by gas chromatography; higher acids are first converted into methyl esters. The stationary phase in the GC column may contain Carbowax 20M impregnated with terephthalic acid, or polyglycol esters such as type Sp $1200/H_3PO_4$ (10 : 4%). Helium or nitrogen is used as carrier gas.

Paper or thin-layer chromatography [27] and spectrophotometric methods [28] are also used.

# 8. Storage and Transportation

Most aliphatic carboxylic acids are corrosive and flammable; flammability properties are given in Table 4. Carboxylic acids must be stored and transported in corrosion-resistant containers. Stainless steel and aluminum are used, but aluminum is subject to attack at certain concentrations. Polyethylene-lined metal drums are suitable for the lower acids. For acids with more than six carbon atoms, drums coated with acid-resistant paint are used. The higher-melting acids, e.g., lauric acid, are transported in paper bags or fiber drums lined with polyethylene.

# 9. Uses

Carboxylic acids may undergo many reactions and have a wide variety of applications [1].

Esters, particularly those of branched-chain acids, are widely employed as solvents. The lower esters are good solvents for nitrocellulose and other cellulose derivatives. Esters have low polarity and low water solubility. Large quantities are used in perfumes [29]. They are the principal constituents of many essential oils. Because of their flavor and aroma, the lower esters are also known as fruit ethers.

# Carboxylic Acids, Aliphatic

**Table 3.** Manufacturers' specifications for commercial aliphatic carboxylic acids

| Acid | Content, wt% min. | Boiling range, 2–97% at 101.3 kPa, °C | Density at 20 °C, g/cm³ | Refractive index $n_D$ at 20 °C | Hazen Color Index, max. APHA | Water content, wt%, max. | Freezing or melting temperature, °C |
|---|---|---|---|---|---|---|---|
| n-Butyric | 99.5 | 161–164 | 0.956–0.959 | 1.398–1.399 | 15 | 0.1 | −5 |
| n-Valeric | 99 | 184–188 | 0.937–0.940 | 1.408–1.409 | 20 | 0.1 | −32 |
| Isobutyric | 99.4 | 151–155 | 0.946–0.950 | 1.393–1.394 | 15 | 0.1 | −46 |
| 2-Methylbutyric | 99.3 | 176 | 0.935 | 1.406 | 5 | 0.05 | <−60 |
| Isovaleric | 98.5 | 175–177 | 0.925–0.927 | 1.403–1.404 | 15 | 0.2 | ca. −26 |
| 2-Methylpentanoic | 99 | 196 | 0.921 | 1.414 | ca. 10 | 0.1 | <−60 |
| 2-Ethylbutyric | 99 | 192–196 | 0.921–0.924 | 1.413–1.414 | 15 | 0.2 | −13 |
| 2-Ethylhexanoic | 99.2 | 226–230 | 0.905–0.907 | 1.425–1.426 | 15 | 0.1 | −60 |
| Di-n-propylacetic | 99.5 | 225–229 | 0.903–0.906 | 1.424–1.426 | 20 | 0.1 | <−60 |
| Isooctanoic * | 99 | 225–238 | 0.913–0.917 | 1.428–1.431 | 20 | 0.1 | <−60 |
| Isononanoic ** | 99 | 233–245 | 0.897–0.901 | 1.429–1.431 | 20 | 0.1 | <−60 |
| Isodecanoic * | 99 | 252–269 | 0.903–0.906 | 1.436–1.438 | 35 | 0.1 | ca. −60 |
| Isotridecanoic * | 99 | | 0.897–0.902 | 1.446–1.447 | 50 | 0.2 | −46 |

\* Mixture of isomers.
\*\* Ca. 90% 3,5,5-trimethylhexanoic acid.

Table 4. Flammability properties of aliphatic carboxylic acids

| Acid | Flash point, °C (Abel-Pensky) | Ignition temperature, °C |
|---|---|---|
| n-Butyric | 69 | 425 |
| Isobutyric | 59 | 420 |
| n-Valeric | 86 | 360 |
| 2-Methylbutyric [b] | 77 | 495 |
| 3-Methylbutyric | 74 | 440 |
| 2-Methylpentanoic | 95 | 390 |
| 2-Ethylbutyric | 87 | 345 |
| 2-Ethylhexanoic | 114 | 320 |
| 2-n-Propylpentanoic | 113 | 350 |
| Isooctanoic [c] | 118 | 440 |
| Isononanoic | 120 | 405 |
| Isodecanoic | 128 | 355 |
| Isotridecanoic | 150 | 340 |
| Dimethylacrylic [b] | 93 | 465 |

[a] Ignition group G 2 according to VDE 0165, unless otherwise stated.
[b] Ignition group G 1.
[c] Mixture of isomers.

Carboxylic acids are incorporated into plastics. Much of the acids produced from oxo-process aldehydes is used for the manufacture of alkyd resins. Long-chain carboxylic acid groups increase solubility in organic solvents, compatibility with softeners, suppleness, and toughness.

Copper salts of branched-chain acids are important preservatives. Copper and zinc salts are used as fungicides and insecticides. Cobalt, lead, manganese, and iron salts are employed as drying agents for paints and printing inks. Because of their good miscibility with solvents and binding agents oxo-process acids give drying agents with a higher metal content than that afforded by naphthenic acids.

Carboxylic acids are also used as intermediates for textile chemicals, dyes, drugs, plastics, and agricultural chemicals. Amides and other derivatives serve as corrosion inhibitors, detergents, flotation aids, and oil additives.

# 10. Specific Aliphatic Carboxylic Acids

## 10.1. Butyric Acids

*n-Butyric acid* is produced commercially by the liquid-phase oxidation of n-butyraldehyde with oxygen. In the continuous process conversion is 97–98% and selectivity over 95%. The product, a colorless mobile liquid, is purified by vacuum distillation. Its penetrating and persistent odor is pungent when concentrated, and rancid or sweaty when diluted. The acid is miscible at room temperature with water and most organic solvents. It is steam-volatile and forms an azeotrope with water (18.5 wt% of butyric acid) boiling at 99.4 °C. Butyric acid esters are used to manufacture lacquers, plastics, and perfumes. Cellulose butyrates are more soluble in organic solvents than cellulose acetate and more miscible with resins and softeners.

Cellulose acetate butyrate is used for plastics with good physical stability and melting behavior. These plastics are suitable for molding. Being highly resistant to humidity, heat, and light, cellulose acetate butyrate gives excellent cable coatings and fishing tackle (→ Cellulose Esters).

Glycerol butyrates and N-arylbutyramides serve as softeners in cellulose-based plastics. The esters of monoalcohols have a pleasant, fruity aroma and are used in perfumes.

*Isobutyric acid* (2-methylpropanoic acid) is produced almost exclusively by the oxidation of isobutyraldehyde, a cheap byproduct of the hydroformylation of propene. This oxidation is less selective than that of *n*-butyraldehyde. Isobutyric acid is a colorless, clear liquid with a characteristically pungent and unpleasant odor. Above 26 °C, isobutyric acid and water are completely miscible. At 20 °C the aqueous phase contains 22 wt% of isobutyric acid and the organic phase contains 55 wt% of water. Miscibility with most organic solvents is also good. The water–isobutyric acid azeotrope boils at 99.3 °C and contains 21 wt% of the acid.

Isobutyric acid is valuable as a polar solvent [30], as starting material for diisopropyl ketone [31], and as a herbicide [32]. Its salts are used as textile chemicals, tanning compounds, stabilizers, catalysts, and preservatives. The esters are used in perfumes, as solvents, and as special oils for aircraft turbines.

## 10.2. Valeric Acids

*n-Valeric acid*, *n*-pentanoic acid, is a clear liquid with a characteristic, unpleasant odor. It is produced by the oxidation of *n*-valeraldehyde with oxygen. Its miscibility with water is limited, but it is fully miscible with common organic solvents.

Many esters of *n*-valeric acid with the lower alcohols are used as solvents and in aromas and perfumes [33]; esters with higher alcohols are used as softeners [34]. The ethyl ester is a lubricant. The sodium and calcium salts and some derivatives are used as rodenticides or fungicides.

*2-Methylbutyric acid* is produced by oxidation of 2-methylbutanal. Its properties and uses resemble those of *n*-valeric acid.

*Valeric acid.* A mixture of *n*-valeric acid (65%) and 2-methylbutyric acid (35%) is obtained directly by oxidation of the valeraldehyde mixture produced by oxo synthesis from *n*-butene. In its properties and uses it resembles *n*-valeric acid.

*Isovaleric acid*, 3-methylbutyric acid, is a clear liquid with a highly unpleasant odor. It is produced by oxidation of isovaleraldehyde, which is obtained from isobutene. It is sparingly soluble in water (ca. 2.5%), but readily soluble in organic solvents. It is used to make fungicides, rodenticides, and sedatives, narcotics, and other drugs. The esters are used as solvents and softeners and in perfumes [29]. The ammonium salts in particular are known for their rodenticidal properties.

## 10.3. Octanoic Acids

*n-Octanoic acid*, caprylic acid, is produced from oxo aldehydes. It is used in hydraulic fluids, machining oils, flotation agents [35], and as a wood preservative [36].

*Isooctanoic acid* is produced by the oxidation of the corresponding aldehyde. The commercial product is a mixture of methyl-branched isomers. It is a clear colorless liquid with a weak odor, practically insoluble in water, but miscible with most organic solvents. It is used in alkyd resins for high-quality lacquers.

Esters of isooctanoic acid, especially those of glycols and polyglycols, and amides are used as plasticizers for PVC polymers. Esters with glycols are widely employed in lubricating oils.

The lithium, magnesium, and aluminum salts are used in lubricants and the heavy metal salts in binding agents. Other derivatives are used as corrosion inhibitors, detergents, flotation aids, oil additives, hydraulic fluids, and preservatives. Uses are similar to those of 2-ethylhexanoic acid.

## 10.4. 2-Ethylhexanoic Acid

2-Ethylhexanoic acid is produced by oxidation of 2-ethyl-1-hexanol or 2-ethylhexanal; the latter is obtained in 95 % yield by hydrogenation of 2-ethyl-2-hexenal, which is itself formed by the aldol condensation of *n*-butyraldehyde (→ Aldehydes, Aliphatic and Araliphatic). In this way propene becomes the source of 2-ethylhexanoic acid. Its accessibility and utility have made it one of the acids with the highest production capacities worldwide. It is of particular importance in the production of alkyd resins used for baking enamels because of their exceptional stability even at temperatures over 200 °C and their good weathering and aging stability.

Esters of 2-ethylhexanoic acid, especially those of diglycols, triglycols, and polyethylene glycol, are used as plasticizers and stabilizers for PVC polymers. Peroxides of 2-ethylhexanoic acid are catalysts in the production of low-density polyethylene.

The lead, cobalt, manganese, and zinc salts are soluble in hydrocarbons. Such solutions serve as drying agents for lacquers and enamels. The iron, nickel, and cobalt salts are used as stabilizers for silicones. The copper salt is a fungicide for marine and other applications.

The acid and its derivatives are also used in the manufacture of lubricants, detergents, flotation aids, and corrosion inhibitors, as catalysts for polyurethane foaming, for solvent extraction [37], and for dye granulation [38]. Versatic Acid 10 (Shell) is used similarly.

## 10.5. Nonanoic Acids

*Pelargonic acid,* n-nonanoic acid, is manufactured via the oxo process and from oleic acid in 96% yield by double-bond cleavage with ozone, nitric acid [39], or chlorine [40].

Like isononanoic acid, it is used in the manufacture of lubricating oils, peroxides, perfumes, and catalysts for alkyd resins.

*Isononanoic acid,* produced by the oxidation of isononyl aldehyde, is a mixture of isomers, containing more than 90% 3,5,5-trimethylhexanoic acid. The starting aldehyde is obtained by the hydroformylation of diisobutene. The acid is also obtained directly from diisobutene by the Koch synthesis in 83 – 86% sulfuric acid at 20 °C and a CO pressure of 7 MPa (70 bar); the main product in this case is 2,2,4,4-tetramethylpentanoic acid.

Isononanoic acid is a clear, colorless liquid with a weak odor. It is almost insoluble in water, but absorbs about 1.6% water at 20 °C. It is miscible with most organic solvents. It is used for the same applications as isooctanoic acid; its higher molecular mass is sometimes advantageous.

Isononanoic acid is used in alkyd resins for high-quality lacquers and coatings. As such, it provides excellent mechanical properties, surface hardness, color stability up to 200 – 230 °C, excellent gloss, and no yellowing.

Isononanoic acid esters are good stabilizers. Glycol and polyglycol esters are used in hydraulic fluids; they are also suited as lubricants under extreme conditions, e.g., in jet aircrafts, because of their good viscosity/temperature coefficient over a wide range. Metal salts find application in lubricating greases and as oxidative drying agents for lacquers. The copper salt is a fungicide. Derivatives are used as corrosion inhibitors, detergents [41], flotation aids, and catalysts.

Isononanoic acid can be replaced by Versatic Acid 10 (Shell) and Cardura E (Shell) or by CeKanoic 9 (Kuhlmann).

## 10.6. Isodecanoic Acid

Isodecanoic acid is produced by oxidizing the mixture of aldehydes obtained from tripropylene in the oxo process. It is used in drying agents, PVC stabilizers, and alkyd resins; the copper salt is a wood preservative [42]. Isodecanoic acid may be replaced by versatic acid 10 or neodecanoic acid.

## 10.7. Pivalic Acid

Pivalic acid, trimethylacetic acid, is produced from isobutene in ca. 95% yield by the Koch synthesis at 40–60 °C at a CO pressure of 5 MPa (50 bar) and with HF as catalyst [43]–[45]. Isobutanol or *tert*-butyl alcohol also serve as starting material [46].

Chain branching confers unusual properties on pivalic acid. The esters are very resistant to hydrolysis; some are heat-resistant lubricants. The poly(vinyl)esters form highly reflective lacquers with good weathering properties. The acid is used in cosmetics and for analgesic, anti-inflammatory, and antispasmodic preparations. It can replace *n*-valeric acid in various applications.

## 10.8. Versatic Acids and Neo Acids

These two acids, designated by trade mark (see Chap. 11), are prototypes of a group of highly branched carboxylic acids, mostly of 5–13 carbon atoms. They are produced on a large scale by the Koch synthesis; each product is a mixture of branched isomers and acids with varying numbers of carbon atoms. The name (e.g., Neononanoic Acid) or a numerical suffix (e.g., Versatic Acid 10) ordinarily refers to the number of carbon atoms in the constituent of highest molecular mass.

All these acids are liquids with a weak odor, miscible with organic solvents, but only slightly soluble in water. They can be vacuum distilled and are stable to about 250 °C. The exceptional thermal stability and the resistance to oxidation and hydrolysis are caused by the bulky tertiary alkyl groups of these acids or esters. They are therefore suitable as components of synthetic lubricants or hydraulic fluids. The vinyl esters are used as comonomers and plasticizers in vinyl chloride resins. The glycidyl esters are alkyd resin modifiers. The metal salts are drying agents.

## 10.9. Propiolic Acid

Propiolic acid, acetylenecarboxylic acid, is produced electrochemically from propargyl alcohol and used in agriculture and in the control of nematodes [47].

**Table 5.** Trade names and manufacturers of carboxylic acids

| Chemical name | Trade name | Manufacturer |
|---|---|---|
| Valeric acid mixture | Neopentanoic Acid 5 | Exxon |
| Valeric acid mixture | Versatic Acid 5 | Shell |
| Octanoic-, nonanoic-, and decanoic acid mixtures with isomers | Neodecanoic Acid 10 | Exxon |
| | Versatic Acid 10 | Shell |
| | Cardura E | Shell |
| Isooctanoic acid mixture | CeKanoic 8 | Kuhlmann |
| Isononanoic acid mixture | CeKanoic 9 | Kuhlmann |
| Isodecanoic acid mixture | CeKanoic 10 | Kuhlmann |
| Nonanoic and undecanoic acid mixture | Versatic Acid 911 | Shell |

# 11. Trade Names, Economic Aspects

Most carboxylic acids are sold under their chemical names; exceptions are listed in Table 5. The 1980 consumption figures and 1983/84 production capacities for the most important products with a known capacity of more than 50 t/a are given in Table 6 [48].

# 12. Toxicology and Occupational Health

The saturated aliphatic carboxylic acids of 4–11 carbon atoms are not known as human poisons [49]; large doses are harmful when applied locally. The lower homologs are corrosive. With increasing chain length and decreasing water solubility, they become less irritating. They can be detected even in low concentrations by the pungent and sweaty odor of the lower homologs and the paraffinic odor of the higher ones. Inhalation of high concentrations can cause breathing difficulties and lung edema. Carboxylic acids are classified as dangerous substances; see also [50]. Chlorination increases acid strength and toxicity. Protective goggles and gloves must be worn when handling carboxylic acids to prevent contact with skin or clothing. If contact occurs, the affected area should be liberally rinsed with water and the clothing changed. Chlorocarboxylic acids require special precautions.

The lethal doses $LD_{50}$ for rats are given in Table 7. Neither MAK values nor TLVs have been established, indicating that harmful concentrations are not likely to arise in the workplace.

**Table 6.** Economic aspects of aliphatic monocarboxylic acids

| | Manufacturers [a]; production capacities (1983/1984), t/a | | | 1980 Consumption, t | | |
|---|---|---|---|---|---|---|
| | Western Europe | USA | Japan | Western Europe | USA | Japan |
| n-Butyric acid | Hoechst [b]<br>Hüls<br>BASF<br>Rhône-Poulenc | Eastman; max. 50 000<br>Celanese; 5 000 | Daicel; 250 | 10 000 | 11 800 | 90 |
| Isobutyric acid | Hoechst<br>BASF<br>Hüls<br>Rhône-Poulenc | Eastman; 500 | | < 1 000 | | |
| n-Pentanoic acid (n-valeric acid and iso acids) | Hoechst<br>BASF | Eastman<br>UCC<br>Exxon (as neopentanoic acid) | Daicel | 1 000 | 6 800 | 50 |
| n-Hexanoic acid (n-caproic acid) | | | | | 550 | |
| 2-Ethylhexanoic acid | Hoechst<br>BASF/Beroxo<br>Berol-Kemi<br>CdF<br>Romania | Eastman; 25 000<br>UCC | Toyo Gosei<br>Daisan Kasei<br>Sanken Kako | 10 000 | 14 100 | 1 900 |
| Heptanoic acid (enanthic acid) | Kuhlmann<br>Atochem | | | 11 000 | 10 900 | very little |
| Octanoic acid (caprylic acid) | Hoechst<br>Kuhlmann | Procter & Gamble; 5000 | | | 1 000 | |
| Nonanoic acid (pelargonic acid) | Unilever<br>Akzo<br>Henkel<br>Hoechst | Givaudan<br>Celanese<br>Emery | | 15 000 | 9 800 | very little |
| Decanoic acid (capric acid) | Shell | Procter & Gamble; 500<br>Ashland; 500 | | | 500 | |
| Pivalic acid | BASF<br>Shell | Exxon<br>Shell | | | 6 800 | 110 |
| Versatic Acids | Shell | Shell | | 20 000 | | |
| Neo Acids | | Exxon; 15 000 | | 5 000 | 6 800 | 110 |
| Lauric acid | | Ashland; 5000<br>Capital City; 5000<br>Procter & Gamble; 500<br>Humko Sheffield; 500<br>Glymo Chem.; 500 | | | | |
| 2-Chloropropanoic acid | | Dow; max. 25 000 | | | 500 | |

[a] With total capacities over 50 t/a.
[b] Total carboxylic acid production capacity, 30 000 t/a [51]

**Table 7.** Toxicity of aliphatic carboxylic acids

| Acid | $LD_{50}{}^a$, mg/kg |
|---|---|
| n-Butyric | 2940 |
| n-Valeric | 1120 |
| Isobutyric | 280 |
| Isovaleric | 2000 |
| Pivalic | 900 |
| 2-Methylpentanoic | 2040 |
| n-Hexanoic | 3000 |
| 2-Ethylbutyric | 2200 |
| 2-Ethylhexanoic | 3000 |
| Di-n-propylacetic [b] | 670 |
| n-Nonanoic | 3200 |
| Isooctanoic [c] | 1.41 |
| Isononanoic [d] | 3135 |
| n-Decanoic | 129 |
| Isodecanoic [c] | 3.73 |
| Dimethylacrylic | 3650 |

[a] Rats, oral.
[b] 2-n-Propylpentanoic acid [51].
[c] Mixture of isomers.
[d] Ca. 90% 3,5,5-trimethylhexanoic acid.

# 13. Derivatives

## 13.1. Acyl Halides

Only the acyl chlorides have commercial importance. They are obtained from carboxylic acids and their salts by halogenation with $PCl_5$, $POCl_3$, $SOCl_2$, $SO_2Cl_2$, or $COCl_2$. Alternatively, olefins are treated with hydrochloric acid and carbon monoxide at high pressure in the presence of palladium catalysts:

$$CH_2=CH_2 + CO + HCl \longrightarrow CH_3CH_2COCl$$

Vinylacetyl chloride [1470-91-3], $CH_2=CH-CH_2COCl$, can be produced from carbon monoxide and allyl chloride; acid bromides can be obtained by the same method.

Isobutyryl fluoride [430-92-2], $(CH_3)_2CHCOF$, can be prepared in 95% yield from isopropyl fluoride [420-26-8], $(CH_3)_2CHF$, using CO and anhydrous HF at 0–80 °C and 1–5 MPa (10–50 bar) [52].

Acid chlorides have a pungent odor. They are thermally stable but easily hydrolyzed.

## 13.2. Anhydrides

Anhydrides of the lower carboxylic acids are mobile-to-oily liquids. Viscosity increases with molecular mass. The higher anhydrides are solids. From $C_4$ to $C_7$ the unpleasant odor resembles that of the carboxylic acids; the higher homologs possess little or no odor. In water, the lower members hydrolyze readily to the corresponding carboxylic acid; the higher members require prolonged heating.

Anhydrides can be obtained by the reaction of acyl chlorides with carboxylic acids or their salts or directly by oxidation of aldehydes with oxygen. An important synthesis of the higher homologs makes use of mixed anhydrides with acetic acid. These are formed from the acid by reaction with acetic anhydride or ketene. They react with a second mole of acid to give the symmetrical anhydride and one mole of acetic acid that is removed by a distillation; the removal may be accelerated by passing a stream of inert gas through the mixture. The process may be carried out in one step; yields are over 90%:

$$RCOOH + CH_2=C=O \longrightarrow RCOOCOCH_3$$
$$RCOOCOCH_3 + RCOOH \longrightarrow (RCO)_2O + CH_3COOH$$

Another method is the addition of acetic anhydride to $\alpha$-olefins using di-*tert*-butyl peroxide as the catalyst. The higher anhydrides are used mainly in the acylation of cellulose, e.g., propionate and butyrate ($\rightarrow$ Cellulose Esters).

## 13.3. Lactams

Lactams are internal amides of aminocarboxylic acids; the lactams of 3-, 4-, and 5-aminocarboxylic acids are more stable than the acids themselves. They are produced by the reaction of ammonia with chlorocarboxylic acids, hydroxycarboxylic acids, or lactones. The principal industrial process is the Beckmann rearrangement of ketoximes, used to produce caprolactam and laurolactam ($\rightarrow$ Caprolactam).

High yields of pyrrolidones and piperidones (five- and six-membered lactams, respectively) are obtained by the reaction of unsaturated amines with carbon monoxide in the presence of a cobalt carbonyl catalyst [53].

The most important reaction of lactams is polymerization to produce polyamide fibers (Nylon) and plastics.

The principal lactams of commercial importance are caprolactam (BASF, Du Pont, Toyo Rayon, Snia Viscosa, Union Carbide, Allied Chemical), pyrrolidone, and *N*-methylpyrrolidone (BASF); the last is used as a solvent and extractant, for example, for the removal of olefins and butadiene from crack gas mixtures (BASF process) or of aromatics from petroleum (Lurgi Arosolvan process). *N*-Vinylpyrrolidone polymers are used in cosmetics.

## 13.4. Halogenated Carboxylic Acids

Mono-, di-, and trichloroacetic acid are produced commercially on a large scale and are discussed in a separate article. The higher chloro- and fluorocarboxylic acids are important as intermediates and agents for pest control.

*Chlorocarboxylic acids* are produced by liquid- and gas-phase chlorination at ca. 300–400 °C; mixtures usually result. Liquid-phase reaction is accelerated by UV light. Tertiary carbon atoms react preferentially, but chlorinating agents such as $POCl_3$, $PCl_5$, and $SO_2Cl_2$ favor α-substitution. In addition, chlorination of aldehydes in aqueous solution produces α-chloroisobutyric and other α-chloro acids [54]. Addition of HCl to α,β-unsaturated acids gives β-chloro acids. For example, a high yield of 3-chloropropionic acid [107-94-8] is produced from concentrated HCl and acrylic acid at 60–100 °C [55]. Halogen addition yields α,β-dihalopropionic acid.

The most important chlorinated aliphatic acid, 2,2-dichloropropionic acid [75-99-0] [56], can be made by chlorination of propionic acid, propionaldehyde, or propanol (→ Propionic Acid and Derivatives) [57]. It is a colorless liquid, bp 186–189 °C, $d_4^{23}$ 1.389.

The sodium salt, $CH_3CCl_2COONa$, is known as the herbicide Dalapon [127-20-8] (Dow Chemical), a hygroscopic powder which decomposes at 166.5 °C and is soluble in water to the extent of 90 g/100 mL at 25 °C. The calcium and magnesium salts are also highly soluble in water. The oral toxicity ($LD_{50}$) of Dalapon in rats is 6600–9300 mg/kg. It is a soil herbicide and a selective, systemic leaf herbicide and is used against perennial grasses [58]. However, its importance in Europe is decreasing.

Less important products are 3-chlorobutanoic acid [1951-12-8] and 3-chloro-2,2-dimethylpropanoic acid [13511-38-1] (Kodak).

The Koch synthesis with unsaturated halohydrocarbons appears likely to become more important [59]. For example, chloropivalic acid [13511-38-1] can be obtained in ca. 70% yield from methallyl chloride using a $BF_3$ catalyst.

$$ClCH_2-\underset{\underset{CH_2}{\|}}{\overset{\overset{CH_3}{|}}{C}} + CO + H_2O \xrightarrow{Cat.} ClCH_2-\underset{\underset{CH_3}{|}}{\overset{\overset{CH_3}{|}}{C}}-COOH$$

The chlorine of chloroacetic acid and other α-chloro acids may be replaced, e.g., by amino groups. β-Chloro acids can be converted into unsaturated acids by eliminating HCl; γ- and δ-chloro acids yield lactones.

*Fluorocarboxylic Acids.* Perfluorocarboxylic acids have recently gained in importance; the lower homologs are used as intermediates and the higher ones, from about $C_6$, in surfactants. These compounds are discussed in a separate article (→ Fluorine Compounds, Organic).

# 14. References

General References

[1] Chem. Technology Rev. no. 9, *Fatty Acids Synthesis and Applications,* Noyes Data Corp., Park Ridge, N.J., 1973.
[2] L. F. Hatch, S. Matar: "From Hydrocarbons to Petrochemicals, Part 12, Chemicals from $C_4$'s," *Hydrocarbon Process.* **57** (1978) no. 8, 153–165.

Specific References

[3] Wisconsin Alumini Res., US 4 425 432, 1984 (J. G. Zeikus, L. H. Lynd).
[4] University Patents Inc., US 4 377 638, 1983 (M. P. Bryant, B. Genthner).
[5] S. Miyazaki, Y. Suhara, *J. Am. Oil. Chem. Soc.* **55** (1978) 536–538.
[6] Texaco, DE 2 930 280, 1979 (J. F. Knifton); US 4 334 092, 1980 (J. F. Knifton).
[7] Texaco, US 4 334 093, 1980 (J. F. Knifton); 4 334 094, 1980 (J. F. Knifton).
[8] Standard Oil, EP-A 33 422, 1981 (F. A. Pesa, T. A. Haase); EP-A 52 419, 1982 (F. A. Pesa, T. A. Haase).
[9] B. Cornils in J. Falbe (ed.): *New Syntheses with Carbon Monoxide,* Springer Verlag, Berlin-Heidelberg-New York 1980, pp. 1–181.
[10] Ruhrchemie, DE 1 083 800; DE 1 154 454, 1959.
[11] Ruhrchemie, DE-OS 2 931 154, 1981 (B. Cornils, W. Devin, J. Weber).
[12] Celanese, US 4 246 185, 1981 (F. Wood, Jr.)
[13] BASF, DE 950 007, 1951.
[14] L. Nowakowski, M. Kaledkowka, *Zh. Prikl. Khim. (Leningrad)* **57** (1984) no. 11, 2569–2574; *Chem. Abstr.* **102** (1985) 61 767 b.
[15] *Chem. Eng. (N.Y.)* **84** (1977) no. 12, 110–115.
[16] H. Koch et al., *Brennst. Chem.* **36** (1956) 321; K. E. Möller, *Brennst. Chem.* **47** (1966) 15.
[17] H. Bahrmann in J. Falbe (ed.): *New Syntheses with Carbon Monoxide.* Springer Verlag, Berlin-Heidelberg-New York 1980, pp. 372–408.
[18] *Kirk-Othmer,* 3rd ed., vol. **4,** pp. 863–871.
[19] W. J. Ellis, C. Ronning, Jr., *Hydrocarbon Process.* **44** (1965) 139.
[20] *Chem. Process. (Chicago)* **45** (1982) no. 11, 14.
[21] Chevron, US 4 016 185, 1977 (J. Wilkes).
[22] BP Chemicals, GB 1 563 202, 1980 (K. R. Dobson, M. Lynn).
[23] Chemie Linz, GB 1 373 275, 1974.
[24] Daisan Kasei, JP 75/96 517, 1975 (T. Nishida, H. Maruyama, S. Ushirode).
[25] P. Jäger, H. Hannebaum, H. Nohe (BASF), *Chem. Ing. Tech.* **50** (1978) no. 10, 787–789.
[26] L. Lykken, *Anal. Chem.* **22** (1950) 398–505.
[27] *Methodicum Chimicum,* vol. **1,** G. Thieme Verlag, Stuttgart 1973, p. 1010.
[28] A. R. Baldwin, H. E. Longenecker, *Oil Soap (Chicago)* **22** (1945) 151–153.
[29] D. L. J. Opdyke: "Monographs on Fragrance," *Food Cosmet. Toxicol.* **16** (1978) Suppl. 1, 839–841; **17** (1979) Suppl., 735–741, 841–843; **19** (1981) Suppl. 2, 237–245.
[30] Hoechst, DE 1 352 351, 1964; DE 1 443 826, 1962.
[31] H. J. Arpe, J. Falbe, *Brennst. Chem.* **48** (1967) 69.
[32] P. Luberg, U. Härter, *Arzneim. Forsch.* **20** (1970) 1874.
[33] Ch. Morel, *Soap Perfum. Cosmet.* **23** (1950) 732, 932.

[34] Hoechst, DE 912 397, 1951.
[35] US Depart. of the Interior, US 4 192 737, 1980 (P. Thompson, J. Huiatt, D. Seidel).
[36] B. D. Hager, US 4 061 500, 1977 (B. D. Hager).
[37] Rustenburg Refiners, GB 2 065 092, 1981 (R. Grant).
[38] Ciba Geigy, DE 2 921 646, 1979 (J. A. Macphersan, J. A. Stirling, J. R. Robert, J. M. McCrae).
[39] V. W. Advani, B. Y. Rao, *J. Oil. Technol. Assoc. India (Bombay)* **8** (1976) no. 2, 27–30; *Chem. Abstr.* **86** (1977) 74 857 g.
[40] Nippon Oil, JP-Kokai 80/77675, 1980.
[41] BASF, DE 2 754 359, 1979 (R. Baur, K. Oppenländer, D. Stickigt).
[42] Australia Commonwealth Sc. and Res. Org. AU 77/9923, 78/35331 (D. F. McCarthy).
[43] Shell, DE 1 924 151, 1969.
[44] Ruhrchemie, DE 1 793 369, 1968.
[45] S. Pawlenko, *Chem. Ing. Tech.* **40** (1968) 52.
[46] Mitsubishi Gas, JP-Kokai 80/12870, 1980; 81/97245, 1981.
[47] Kanesho Co., JP 83/52089, 1983.
[48] T. Gibson, O. Kamatari, J. Wahlen: *Chemical Economics Handbook,* Marketing Res. Report "Oxo Chemicals", SRI-International, Dec. 1981.
[49] D. W. Fasset in *Patty,* 2nd ed., vol. **II,** p. 1962; D. Guest, G. Katz, B. D. Astill, in *Patty,* 3rd ed., vol. **2 C,** 4901–4987, 1982.
[50] TSCA Toxic Substances Control test, Issue 1978.
[51] *Chem. Mark. Rep.*1977, March 7, 3.
[52] Ashland Oil, US 4 451 670, 1984 (D. Grote, R. Grimm, R. Norton).
[53] J. Falbe, H. J. Schulze-Steinen, F. Korte, *Chem. Ber.* **98** (1965) 1923.
[54] Continental Oil, US 3 661 986, 1969; US 3 661 986, 1969.
[55] Mitsui Toatsu Chem., JP 83/124 738, 1983.
[56] R. Wegler, L. Eue: *Chemie des Pflanzenschutzes und der Schädlingsbekämpfungsmittel,* vol. **2,** Springer Verlag, Berlin 1970, pp. 263–264
[57] L. de Buyck et al., *Bull. Soc. Chim. Belg.* **89** (1980)no. 6, 441–458; *Chem. Abstr.* **94** (1981) 15 141 m.
[58] Brit. Crop Prot. Council: *The Pesticide Manual,* 7th ed., London 1983, pp. 154–155.
[59] K. E. Müller, *Brennst. Chem.* **47** (1966) 14.

# Carboxylic Acids, Aromatic

*Individual keywords:* → Benzoic Acid; → Cinnamic Acid; → Hydroxycarboxylic Acids, Aromatic; → Naphthalene Derivatives; → Phthalic Acid and Derivatives; → Salicylic Acid; → Terephthalic Acid and Dimethyl Terephthalate; and → Isophthalic Acid

FREIMUND RÖHRSCHEID, Hoechst Aktiengesellschaft, Frankfurt/Main, Federal Republic of Germany

| | | |
|---|---|---|
| 1. | Introduction | 1120 |
| 2. | Physical and Chemical Properties | 1120 |
| 3. | Production | 1125 |
| 3.1. | Carboxylation with Carbon Dioxide | 1125 |
| 3.2. | Dicarboxylic Acids Formed by Rearrangement | 1126 |
| 3.3. | Side-Chain Chlorination of Methylbenzenes | 1126 |
| 3.4. | Oxidation with Nitric Acid | 1127 |
| 3.5. | Oxidation with Molecular Oxygen in the Vapor Phase | 1127 |
| 3.6. | Oxidation with Molecular Oxygen in the Liquid Phase | 1128 |
| 3.6.1. | Solvent-Free Oxidation | 1128 |
| 3.6.2. | Oxidation in Acetic Acid | 1128 |
| 4. | Polycarboxylic Acids and Anhydrides | 1129 |
| 4.1. | Trimellitic Acid and Trimellitic Anhydride | 1129 |
| 4.2. | Pyromellitic Acid and Pyromellitic Dianhydride | 1131 |
| 4.3. | Other Benzenepolycarboxylic Acids and Anhydrides | 1131 |
| 4.4. | Naphthalic Acid and Naphthalic Anhydride | 1132 |
| 4.5. | 3,4,9,10-Perylenetetracarboxylic Acid and Perylene Dianhydride | 1132 |
| 4.6. | 2,6-Naphthalenedicarboxylic Acid | 1133 |
| 4.7. | 1,4,5,8-Naphthalenetetracarboxylic Acid and 1,8:4,5-Naphthalenetetracarboxylic Dianhydride | 1133 |
| 5. | Economic Aspects | 1134 |
| 6. | Toxicology | 1134 |
| 7. | References | 1134 |

# 1. Introduction

The production of virtually all industrially important aromatic carboxylic acids involves either the partial oxidation of aromatic alkyl groups or the partial oxidative degradation of the aromatic ring with atmospheric oxygen. The oxidation of the methyl groups of xylene exemplifies the former case.

$$H_3C-C_6H_4-CH_3 + 3\,O_2 \longrightarrow HOOC-C_6H_4-COOH + 2\,H_2O$$

The partial oxidation of naphthalene is an example of the oxidation of the ring.

$$2\,C_{10}H_8 + 9\,O_2 \longrightarrow 2\,C_6H_4(CO)_2O + 4\,CO_2 + 4\,H_2O$$

Such oxidizing agents as dichromate, permanganate, hypochlorite, chlorine, and nitric acid, which were used a few decades ago, have been replaced in almost all cases by atmospheric oxygen. The advantage of air oxidation is that only carbon dioxide and water are formed as the main byproducts. The progress made in modern oxidation techniques has gone hand in hand with the development of the synthesis of benzoic acid and of the two major products phthalic acid and terephthalic acid.

# 2. Physical and Chemical Properties

The physical properties of the most important aromatic carboxylic acids are given in Table 1. The monocarboxylic acids of benzene and naphthalene are, without exception, crystalline solids melting at temperatures above 100 °C. The melting points of the dicarboxylic and polycarboxylic acids are usually higher than those of the monocarboxylic acids. The acidity of benzoic acid ($pK_a$ 4.17) and of the naphthoic acids ($\alpha$: $pK_a$ 3.7, $\beta$: $pK_a$ 4.15) is a little greater than that of acetic acid ($pK_a$ 4.74). The acidity is enhanced by electron-attracting substituents, especially those in the ortho position. Hence, in aromatic polycarboxylic acids, the acidity of the first carboxyl group is increased stepwise by the addition of further carboxyl groups (phthalic acid: $pK_a$ 3.00, pyromellitic acid: $pK_a$ 1.92).

In general, the chemistry of the carboxyl group of aromatic carboxylic acids does not differ from that of the carboxyl group of aliphatic carboxylic acids. Both form salts, esters, anhydrides, acid amides, and acid halides ($\rightarrow$ Carboxylic Acids, Aliphatic). For this reason, only the most important reactions and some special features are discussed in detail here.

Table 1. Aromatic carboxylic acids

| Acid | Common name | Empirical formula | $M_r$ | Chemical structure | mp, °C | bp, °C | Solubility in 100 g water, g | |
|---|---|---|---|---|---|---|---|---|
| | | | | | | | at 25 °C | at 100 °C |
| Benzenecarboxylic acid [65-85-0] | benzoic acid | $C_7H_6O_2$ | 122.12 | (C6H5-COOH) | 122.374 | 249 | 0.34 | 2.19 (75 °C) |
| 2-Hydroxybenzenecarboxylic acid [69-72-7] | salicylic acid | $C_7H_6O_3$ | 138.12 | (2-OH-C6H4-COOH) | 162 | sublimes at 135 °C (650 Pa) sublimes at 150 °C (1930 Pa) | 0.22 | 2.2 (80 °C) |
| trans-3-Phenyl-2-propenoic acid [621-82-9] | trans-cinnamic acid | $C_9H_8O_2$ | 148.16 | (C6H5-CH=CH-COOH) | 134 | 300 (101.3 kPa) | 0.055 | |
| 1,2-Benzenedicarboxylic acid [88-99-3] | phthalic acid | $C_8H_6O_4$ | 166.13 | (1,2-(COOH)2-C6H4) | 210 (forms anhydride) | | 0.7 | 19.0 |
| 1,3-Benzenedicarboxylic acid [121-95-5] | isophthalic acid | $C_8H_6O_4$ | 166.13 | (1,3-(COOH)2-C6H4) | 348 (sublimes) | | 0.012 | 0.25 |
| 1,4-Benzenedicarboxylic acid [100-21-0] | terephthalic acid | $C_8H_6O_4$ | 166.13 | (1,4-(COOH)2-C6H4) | > 300 (sublimes) | | 0.0019 | 0.035 |
| 1,2,4-Benzenetricarboxylic acid [528-44-9] | trimellitic acid | $C_9H_6O_6$ | 210.14 | (1,2,4-(COOH)3-C6H3) | 238 (forms anhydride) | | 2.1 | 60 |
| 1,3,5-Benzenetricarboxylic acid [554-95-0] | trimesic acid | $C_9H_6O_6$ | 210.14 | (1,3,5-(COOH)3-C6H3) | 368 | | 0.27 | 6.4 |
| 1,2,3-Benzenetricarboxylic acid [569-51-7] | hemimellitic acid | $C_9H_6O_6$ | 210.14 | (1,2,3-(COOH)3-C6H3) | 197 (forms anhydride) | | very soluble | |

**Carboxylic Acids, Aromatic**

**Table 1.** continued

| Acid | Common name | Empirical formula | $M_r$ | Chemical structure | $mp$, °C | $bp$, °C | Solubility in 100 g at 25 °C |
|---|---|---|---|---|---|---|---|
| 1,2,4,5-Benzenetetracarboxylic acid [89-05-4] | pyromellitic acid | $C_{10}H_6O_8$ | 254.15 | (structure) | 280 (forms dianhydride at 290 °C and 1.7 kPa) | | 1.5 > 30 |
| 1,2,3,4-Benzenetetracarboxylic acid [476-73-3] | mellophanic acid | $C_{10}H_6O_8$ | 254.15 | (structure) | 244–246 (decomp.) | | soluble very soluble |
| 1,2,3,5-Benzenetetracarboxylic acid [479-47-0] | prehnitic acid | $C_{10}H_6O_8$ | 254.15 | (structure) | 263–266 | | soluble very soluble |
| Benzenepentacarboxylic acid [1585-40-6] | | $C_{11}H_6O_{10}$ | 298.16 | (structure) | 238–239 | | soluble very soluble |
| Benzenehexacarboxylic acid [517-60-2] | mellitic acid | $C_{12}H_6O_{12}$ | 342.17 | (structure) | 290 (in closed capillary) 285 decomp. | | soluble |
| 1,8-Naphthalenedicarboxylic acid [518-05-8] | naphthalic acid | $C_{12}H_8O_4$ | 216.20 | (structure) | forms anhydride > 60 °C | | sparingly soluble |
| 2,6-Naphthalenedicarboxylic acid [1141-38-4] | | $C_{12}H_8O_4$ | 216.20 | (structure) | > 340 | | soluble in aqueous ethanol |
| 1,4,5,8-Naphthalenetetracarboxylic acid [128-97-2] | | $C_{14}H_8O_8$ | 304.22 | (structure) | forms dianhydride > 80 °C | | moderately soluble |

Table 1. continued

| Acid | Common name | Empirical formula | $M_r$ | Chemical structure | $mp$, °C | $bp$, °C | Solubility in 100 g water, g | |
|---|---|---|---|---|---|---|---|---|
| | | | | | | | at 25 °C | at 100 °C |
| 4,9,10-Perylenetetracarboxylic acid [81-32-3] | | $C_{24}H_{12}O_8$ | 428.36 | | forms dianhydride | | very soluble | |

The esterification and amidation reactions of bifunctional carboxylic acids that yield polyesters and polyamides are of considerable industrial importance. The most important polyester, poly(ethylene terephthalate), is prepared by a two-stage reaction. Terephthalic acid or dimethyl terephthalate reacts with ethylene glycol at 200 °C in the molten state to give diglycol terephthalate:

$$ROOC-C_6H_4-COOR + 2\ HOCH_2CH_2OH \xrightarrow{200\ °C} HOCH_2CH_2OOC-C_6H_4-COOCH_2CH_2OH + 2\ ROH$$

where R = H or CH$_3$

In the next stage, conducted under vacuum conditions at approximately 280 °C and in the presence of an antimony catalyst, polycondensation occurs which releases ethylene glycol.

$$n\ HOCH_2CH_2OOC-C_6H_4-COOCH_2CH_2OH \xrightarrow[280\ °C]{Sb^{3+}} {+\!\!\{CO-C_6H_4-COOCH_2CH_2O\}_n\!\!+} + n\ HOCH_2CH_2OH$$

Among the polyamides, the condensation polymers of terephthalic acid and *p*-phenylenediamine (Kevlar) and of isophthalic acid and *m*-phenylenediamine (Nomex) are of practical importance because they can be converted into flameproof, temperature-resistant fibers.

$$+\!\!\{NH-C_6H_4-NH-CO-C_6H_4-CO\}+\!\! \quad \text{Kevlar}$$

Aromatic polycarboxylic acids readily undergo cyclization to give anhydrides if a strainless five- or six-membered ring can be formed. The reaction is carried out by heating at low pressure or by boiling with acetic anhydride.

The five-membered cyclic anhydrides react readily with water, alcohol, or ammonia. They undergo only partial esterification with alcohols, but the product can be completely esterified at a higher temperature.

$$\text{phthalic anhydride} + ROH \rightarrow \text{o-}C_6H_4(COOH)(COOR) \xrightarrow[\Delta]{ROH} \text{o-}C_6H_4(COOR)_2 + H_2O$$

They react with ammonia or primary amines to form acid amides, which when heated form imides.

$$\text{phthalic anhydride} + RNH_2 \rightarrow \text{o-}C_6H_4(COOH)(CONHR) \xrightarrow{\Delta} \text{phthalimide-NR} + H_2O$$

The acid amides or acid imides can be converted to the *o*-aminocarboxylic acids by a sodium hypochlorite–sodium hydroxide solution; e.g., phthalimide can be converted to anthranilic acid. This is an example of the Hofmann degradation.

The six-membered cyclic anhydrides, which have the structure of naphthalic anhydride, form readily and are resistant to hydrolysis. They react easily with ammonia or primary amines to give imides.

The reaction with *o*-phenylenediamine yields imidazoles. The alkaline salts of aromatic dicarboxylic acids can undergo carboxyl group rearrangements at high temperature to form thermodynamically more stable structures (Henkel reaction).

The copper-catalyzed oxidative decarboxylation of benzoic acid to form phenol is of considerable commercial importance. In the process developed by Dow Chemical, a mixture of air and steam at 220–250 °C and ca. 200 kPa is passed into molten benzoic acid. The resultant phenol leaves the reactor as a vapor. A selectivity of approximately 85% is achieved using a copper benzoate catalyst.

# 3. Production

## 3.1. Carboxylation with Carbon Dioxide

Specific carboxylic acids can be prepared by introducing a carboxyl group directly into the aromatic ring. This is usually carried out by reacting carbon dioxide with an aryl Grignard reagent or an aryl lithium compound. However, this reaction is of limited use.

In the Kolbe–Schmitt reaction, carbon dioxide at approximately 0.5 MPa (5 bar) reacts at 130–180 °C with the alkali salt of a phenol or naphthol to give a hydroxycarboxylic acid (→ Salicylic Acid, → Hydroxycarboxylic Acids, Aromatic).

## 3.2. Dicarboxylic Acids Formed by Rearrangement

Procedures involving carboxyl group rearrangements are known as the Henkel I and Henkel II procedures. They are named after the company that discovered them [1].

In the Henkel I procedure, the potassium salt of a dicarboxylic acid undergoes rearrangement to give its thermodynamically more stable form. For example, dipotassium phthalate is converted into dipotassium terephthalate by Zn–Cd-catalyzed rearrangement under carbon dioxide at 1.5 MPa (15 bar) and 430–440 °C [2].

The Henkel II procedure is based on the disproportionation of aromatic carboxylic acids. For example, potassium benzoate can be converted into dipotassium terephthalate and benzene with a 95 % selectivity by using the same reaction conditions as described above for the Henkel I procedure [3]. Although these methods have been applied by Teijin (Henkel I) and Mitsubishi Chemical (Henkel II) to produce terephthalic acid, they are used today only in the production of special carboxylic acids, e.g., 2,6-naphthalenedicarboxylic acid [1141-38-4].

## 3.3. Side-Chain Chlorination of Methylbenzenes

For the preparation of acid chlorides, the side-chain chlorination of methyl-substituted aromatic compounds is often highly advantageous [4]. Indeed, photochlorination of toluenes and xylenes at 80–140 °C in glass, lead, or nickel reactors forms trichloromethyl-substituted aromatic compounds. The reaction must be carried out with complete exclusion of any trace of iron. The product in turn reacts with one equivalent of water or carboxylic acid over an iron(III) chloride or zinc chloride catalyst to yield the acid chloride.

where X = H, Br, Cl, F, $CCl_3$, or aryl

The acid chloride can be hydrolyzed with water to form the corresponding carboxylic acid.

## 3.4. Oxidation with Nitric Acid

The oxidation of methyl-substituted aromatic compounds to the corresponding carboxylic acids requires long heating (> 160 °C) with dilute nitric acid [5]. These conditions are chosen to reduce nitration. A supply of air at 3–4 MPa can lower the consumption of nitric acid. The reaction is catalyzed by ammonium vanadate. Even if the oxidation conditions are optimum, a marked nitration of the aromatic compound occurs. This method was at one time very important, and is still used by AB Bofors Nobel Kemi to oxidize nitrotoluenes, chloronitrotoluenes, and chlorotoluenes.

$$\text{4-NO}_2\text{-C}_6\text{H}_4\text{-CH}_3 + 2\,\text{HNO}_3 \longrightarrow \text{4-NO}_2\text{-C}_6\text{H}_4\text{-COOH} + 2\,\text{NO} + 2\,\text{H}_2\text{O}$$

## 3.5. Oxidation with Molecular Oxygen in the Vapor Phase

The vapor-phase air oxidation of alkyl-substituted aromatic compounds or of multinuclear aromatic compounds to form aromatic carboxylic acids (e.g., *o*-xylene or naphthalene to phthalic anhydride) proceeds at 350–500 °C over a vanadium pentoxide catalyst. The catalyst is usually modified with titanium dioxide or sodium sulfate and is supported on a silica gel or silicon carbide carrier [6], [7]. The oxidation can take place in a fixed-bed reactor or in a fluidized-bed reactor. In the former case, the catalyst is firmly held in place in a multitubular reactor with 10 000 or more tubes of about 1.5 cm inner diameter. Large amounts of heat are generated by the reaction, necessitating the use of a eutectic molten salt for cooling the tubes. The hydrocarbon reactant and air are passed into the reactor as a gaseous mixture. This procedure is used by Chemische Fabrik von Heyden, Veba Chemie, Koppers, BASF, Monsanto, and Rhône-Poulenc, and is suitable for the oxidation of *o*-xylene (at 380–400 °C, molar yield 78%), naphthalene (at 400 °C, molar yield approximately 80%), acenaphthene, and durene.

In 1944, the oxidation procedure using a continuous fluidized-bed reactor was introduced on a large industrial scale in the United States. It was used by Badger–Sherwin-Williams for the production of phthalic anhydride from naphthalene (see chapter 1). In this process liquid naphthalene is injected into the fluidized solid catalyst at 350–380 °C, using preheated air to maintain the flow. The main advantages of this technique are better control of the reaction heat and minimization of flammability problems. The yields obtained by it are lower than those in the fixed-bed process, especially when the starting material is *o*-xylene. The Badger Company, United Coke–Foster Wheeler, and ICI use the fluidized-bed reactor for the oxidation of naphthalene.

The gases leaving the reactor are passed through switch condensers where the product is almost quantitatively deposited on the cooled tubes. Hot oil is in turn passed through the tubes, and the product is melted off. The exhaust gas must be scrubbed before being discharged into the atmosphere. The vapor-phase oxidation gives molar yields of up to 80% and is suitable for the production of maleic anhydride, benzoic acid, phthalic anhydride, naphthalic anhydride, and pyromellitic anhydride.

## 3.6. Oxidation with Molecular Oxygen in the Liquid Phase

Many carboxylic acids are destroyed by the high temperatures required in the gas-phase oxidation technique or they are just not sufficiently volatile. These acids can be effectively produced by catalytic oxidation in the liquid phase. The starting materials are alkyl-substituted aromatic compounds, and the reaction is carried out either in a column or in a pressure reactor with effective mixing. Although acetic acid is generally the most useful solvent, oxidation can also occur in the absence of solvent. Cobalt or manganese salts are commonly used as catalysts.

### 3.6.1. Solvent-Free Oxidation

A process developed by Snia Viscosa involves the oxidation of toluene to benzoic acid at 165 °C and 1 MPa (10 bar) and uses cobalt carboxylates as catalysts. A 20–30% conversion and a selectivity of 90% are achieved [8]–[10].

The Dynamit Nobel–Hercules procedure is used to make dimethyl terephthalate (DMT). This joint oxidation of $p$-xylene and the methyl ester of toluic acid at 150–170 °C and 400–800 kPa (4–8 bar) yields toluic acid and the monomethyl ester of terephthalic acid. This mixture is converted by pressure esterification with methanol at 200–250 °C to toluic acid methyl ester and DMT. The latter is isolated by distillation and recrystallized from methanol. The toluic acid methyl ester is then returned to the oxidation reactor along with fresh xylene [11].

### 3.6.2. Oxidation in Acetic Acid

Oxidation without addition of a solvent is in many ways disadvantageous because the resulting carboxylic acids have high melting points, and, therefore, only limited conversion rates can be attained. However, 100% conversion is possible when working in acetic acid. The resultant carboxylic acids usually precipitate as crystals, e.g., terephthalic acid and $p$-nitrobenzoic acid.

The main catalysts are soluble cobalt and manganese carboxylates. However, only in a few cases do these metal ions alone give a satisfactory degree of conversion. To facilitate oxidation of chlorotoluene, *p*-nitrotoluene, or toluic acid to the corresponding carboxylic acids, the following additional requirements must be met.

In the Amoco procedure, the addition of bromide to cobalt acetate and manganese acetate in acetic acid produces a highly reactive catalyst. *p*-Xylene is oxidized at 180–210 °C and 2–3 MPa (20–30 bar) using 0.1 wt% cobalt to a very pure terephthalic acid with a yield of over 95% [12]–[14]. This catalyst is also suitable for the preparation of tricarboxylic and tetracarboxylic acids of benzene [15]. The reactors must be coated with a resistant material (enamel, titanium, nickel-chromium-molybdenum alloys such as Hastelloy-C4) because bromide ion is highly corrosive.

Carboxylic acids can be efficiently obtained by the *cooxidation* of alkyl-substituted aromatic compounds with reagents (acetaldehyde, butane, 2-butanone) which form peroxy radicals during the course of the reaction. Two procedures for the synthesis of terephthalic acid are based on this principle. The liquid-phase oxidation procedure operated by Eastman Kodak uses acetaldehyde [16], [17]. In the Toray process, a modification of the Kodak method, paraldehyde is employed instead of acetaldehyde. Butane and 2-butanone have been used as cooxidizing agents by Mobil Oil and Olin Mathieson [18]. These methods can also be efficiently applied to the oxidation of other alkyl-substituted aromatic compounds.

# 4. Polycarboxylic Acids and Anhydrides

## 4.1. Trimellitic Acid and Trimellitic Anhydride

Trimellitic acid crystallizes from water as needles (Table 1). It dissolves to the extent of 1.0 g in 100 g of acetic acid at 25 °C [19]. Trimellitic anhydride reacts with acetic acid and water and is hygroscopic (Table 2).

Trimellitic anhydride alone is of industrial importance. Its preparation is best achieved by the liquid-phase catalytic oxidation of 1,2,4-trimethylbenzene (pseudocumene) with air (Amoco method) [15], [20], [21]. An alternative process involves oxidation with 7% nitric acid at 170–190 °C and 2 MPa, but it is inefficient compared with the air oxidation method [22].

The largest amount of trimellitic anhydride is used to make plasticizers for poly(vinyl chloride). These plasticizers have a lower volatility than phthalate plasticizers. Polyimides and polyesters of trimellitic anhydride have a high thermal resistance and are used in the production of wire enamels, coatings, and baking varnishes. The anhydride is an important curing agent in the production of epoxy resins. Polyester resins of

**Table 2.** Anhydrides of aromatic carboxylic acids

| Compound | Common name | Empirical formula | $M_r$ | Chemical structure | mp, °C | bp, °C |
|---|---|---|---|---|---|---|
| 1,3-Isobenzofurandione [85-44-9] | phthalic anhydride | $C_8H_4O_3$ | 148.12 | | 131.2 | 215 (440 Pa) 284.5 (101.3 kPa) |
| 1,3-Dihydro-1,3-dioxo-5-isobenzofuran-carboxylic acid [552-30-7] | trimellitic anhydride | $C_9H_4O_5$ | 192.13 | | 168 | 390 |
| 1,3-Dihydro-1,3-dioxo-4-isobenzofuran-carboxylic acid [3786-39-8] | hemimellitic anhydride | $C_9H_4O_5$ | 192.13 | | 196 | |
| Benzo[1,2-c:3,4-c']-difuran-1,3,4,6-tetrone [4435-60-3] | mellophanic dianhydride | $C_{10}H_2O_6$ | 218.12 | | 198 sublimes | |
| Benzo[1,2-c:4,5-c']difuran-1,3,5,7-tetrone [89-32-7] | pyromellitic dianhydride | $C_{10}H_2O_6$ | 218.12 | | 285 | 390 |
| Benzo[1,2-c:3,4-c':5,6-c'']-trifuran-1,3,4,6,7,9-hexone [4253-24-1] | mellitic trianhydride | $C_{12}O_9$ | 288.13 | | 310 (decomp.) | sublimes at 200 °C (400–500 Pa) |
| Naphtho[1,8-c,d]pyran-1,3-dione [81-84-5] | naphthalic anhydride | $C_{12}H_6O_3$ | 198.18 | | 276 | 422; sublimes at 215 °C (440 Pa) |
| [2]Benzopyrano[6,5,4-def][2]benzopyran-1,3,6,8-tetrone [81-30-1] | 1,8:4,5-naphthalenetetra-carboxylic dianhydride | $C_{14}H_4O_6$ | 268.18 | | > 300 | sublimes at 320 °C (400 Pa) |
| Perylo[3,4-c,d:9,10-c',d']dipyran-1,3,8,10-tetrone [128-69-8] | perylene dianhydride | $C_{24}H_8O_6$ | 392.33 | | > 300 | |

trimellitic anhydride are widely used as paint resins for water-based and solvent-based coatings.

## 4.2. Pyromellitic Acid and Pyromellitic Dianhydride

Pyromellitic acid crystallizes from water as a dihydrate, triclinic plates, mp 242 °C (Table 1). Pyromellitic dianhydride is very hygroscopic (Table 2).

Pyromellitic acid was originally made by Du Pont by oxidizing 1,2,4,5-tetramethylbenzene (durene) with nitric acid. Today, the only method employed industrially is the vapor-phase oxidation of durene, trimethylisopropylbenzene, or dimethyldiisopropylbenzene to the dianhydride [23]. The oxidation conditions are similar to those used for $o$-xylene. The yield is approximately 60 %. Most of the applications require very pure pyromellitic dianhydride. Thus, the raw substance must be freed from all impurities. Because of the high boiling point, purification by distillation is expensive and involves a large loss of material. Therefore, methods have been developed for the purification of pyromellitic dianhydride. Inert gases are passed through the raw product at a temperature above the dew point of the impurities (175–275 °C) [24], [25]. Complexes of dioxane or xylene are formed, isolated, and decomposed [26], and the impurities are dissolved and removed [27]–[29]. The only commercial producer is Chemische Werke Hüls AG with a capacity of 500 t/a.

The major part of the pyromellitic dianhydride produced is used in the production of polyimides that have excellent electrical and physical properties and a high heat resistance. Polyimide resins can be used for films, fibers, molding compounds, varnishes, and wire coatings. They are resistant to a constant temperature of 300 °C and a transient temperature of 500–600 °C. The trade names of these Du Pont products are Kapton (condensation product of pyromellitic dianhydride and 4,4′-diaminophenyl ether), Pyre-M 11, Pyralin, and Vespel.

## 4.3. Other Benzenepolycarboxylic Acids and Anhydrides

1,3,5-Benzenetricarboxylic acid, trimesic acid, is formed by the oxidation of 1,3,5-trimethylbenzene (mesitylene). This is carried out on an experimental scale (4.5 t/a) by the American Bio-Synthetic Corp. The applications are the same as those of trimellitic anhydride.

1,2,3-Benzenetricarboxylic acid (hemimellitic acid), hemimellitic anhydride, 1,2,3,4-benzenetetracarboxylic acid (mellophanic acid), mellophanic dianhydride, 1,2,3,5-benzenetetracarboxylic acid (prehnitic acid), benzenepentacarboxylic acid, benzenehexacarboxylic acid (mellitic acid) and mellitic trianhydride can all be obtained either by the air oxidation of the corresponding polymethylbenzenes using the Amoco process or by oxidation with nitric acid. The preparation of mellitic acid can also be achieved by the nitric acid oxidation of certain types of brown coal. None of these compounds is of commercial importance.

## 4.4. Naphthalic Acid and Naphthalic Anhydride

Naphthalic acid crystallizes as needles (Table 1). Naphthalic anhydride crystallizes from ethanol as needles (Table 2).

Naphthalic acid is prepared by the gas-phase oxidation of acenaphthene with air at 450–550 °C over a vanadium pentoxide catalyst. Yields of approximately 85 % are obtained. The mass-to-mass ratio of air to acenaphthene should be 80–100 to 1 [30]. The oxidation of acenaphthene can also be achieved in the liquid phase in acetic acid at 100–150 °C in the presence of cobalt, manganese, and bromide ion [31].

Naphthalic anhydride is primarily converted into imides or imide derivatives. Naphthalimide [81-83-4] (mp 300 °C) is formed by heating the anhydride in concentrated aqueous ammonia. Naphthalic anhydride is the starting material in the production of imidazole dyes. Imide derivatives are employed as optical brighteners for synthetic resins and fibers. Most naphthalimide is converted into derivatives of 3,4,9,10-perylenetetracarboxylic acid.

## 4.5. 3,4,9,10-Perylenetetracarboxylic Acid and Perylene Dianhydride

When heated, deep red 3,4,9,10-perylenetetracarboxylic acid (Table 1) yields red perylene dianhydride (Table 2).

When naphthalimide is heated to 220 °C with caustic potash [1310-58-3] and sodium acetate [127-09-3], the potassium salt of the leuco form of perylenetetracarboxylic diimide is formed. This product is subsequently oxidized in aqueous solution to form perylene diimide [81-33-4], $C_{24}H_{10}N_2O_4$, $M_r$ 390.36, Perylene Bordeaux. The diimide in turn forms perylene dianhydride on hydrolysis with concentrated sulfuric acid at 215 °C [32], [33].

The diimide and the dianhydride are starting materials for the synthesis of aliphatic- and aromatic-substituted diimides. The latter constitute an important class of red dyes and pigments used in the production of color-fast plastics and coatings. Approximately 100 t/a of the dyes and 1000 t/a of the pigments are manufactured worldwide.

## 4.6. 2,6-Naphthalenedicarboxylic Acid

2,6-Naphthalenedicarboxylic acid, is prepared by using the Henkel procedure starting with the potassium salt of 1,8-naphthalenedicarboxylic acid or 2-naphthoic acid [34], [35].

The salt hydrolyzes readily to form the corresponding acid. Polycondensation of 2,6-naphthalenedicarboxylic acid with diols yields highly ductile and heat-resistant polyesters, which are used in the manufacture of films and fibers [36]. The product is suitable for magnetic tapes and electrical insulation materials.

## 4.7. 1,4,5,8-Naphthalenetetracarboxylic Acid and 1,8:4,5-Naphthalenetetracarboxylic Dianhydride

1,4,5,8-Naphthalenetetracarboxylic acid readily loses water when heated above 80 °C to form 1,8:4,5-naphthalenetetracarboxylic dianhydride.

The tetracarboxylic acid is obtained by the oxidative degradation of pyrene from coal tar. The oxidation proceeds in a series of steps [37], [38].

A new, technically feasible, procedure is based on the reaction of acenaphthene with diketene in liquid HF to yield 1,2-dihydro-7-methyl-cyclopenta[cd]phenalen-5-one [42937-13-3], $C_{16}H_{12}O$ [39].

This compound in turn can be converted to the dianhydride by air oxidation in acetic acid in the presence of ammonium vanadate and nitric acid as catalysts [40].

Most of the naphthalenetetracarboxylic dianhydride is condensed with two equivalents of *o*-phenylenediamine to give a mixture of *cis*- and *trans*-1,4,5,8-naphthoylenebis(benzimidazole) (Indanthrene Scarlet GG), which when separated into the isomers yields Perinone Orange (trans), Indanthrene Brilliant Orange GR [*4424-06-0*], and Perinone Red (cis), Indanthrene Bordeaux RR [*4216-02-8*]. Both are used as pigments and vat dyes.

## 5. Economic Aspects

The benzene- and naphthalenepolycarboxylic acids are speciality items and are produced in comparatively small amounts. Amoco is the sole manufacturer of trimellitic anhydride in the United States; it has a production capacity of 25 kt/a. In 1983, the United States used 14 kt trimellitic acid ester as a special softener [41].

Chemische Werke Hüls is the sole supplier of pyromellitic anhydride. The production capacity is 500 t/a. The 1200–2000 t naphthalic anhydride produced in the EEC is all made by Chemische Fabrik Weyl (Germany). 1,4,5,8-Naphthalenetetracarboxylic acid is produced by Hoechst and by Toms River Chemical. The production capacity is less than 1 kt/a.

## 6. Toxicology

As a result of their acidity and their ability to dehydrate, the carboxylic acids and their anhy-drides may cause severe irritation to the pulmonary tract, skin, eyes, and mucous membranes. Some compounds have additional toxic properties. Exposure to trimellitic acid may result in noncardiac pulmonary edema and immunological sensitization [42]. Pyromellitic acid has an intraperitoneal $LD_{50}$ in mice of 300 mg/kg. Pyromellitic acid inhibits the enzyme aconitase [43]. Handling precautions include effective ventilation, protective clothing, and goggles if exposure to dust is expected.

## 7. References

[1]   B. Raecke, *Angew. Chem.* **70** (1958) 1–5.
[2]   Teikoku, GB 975113, 1964. Teijin, JP 4220525, 1967; JP 4420091, 1969.
[3]   Mitsubishi Chemical, JP 4719537, 1972; JP 4724549, 1972. Phillips Petroleum, DE 2230173, 1972 (Yu-Lin Wu, P. S. Hudson) (= US 3781341, 1973).

[4] *Winnacker-Küchler,* 4th ed., vol. **6,** pp. 165–169.
[5] E. Bengtsson, B. Holm, *Chem. Ztg.* **104** (1980) 349–351. *Houben-Weyl,* **4/1 a,** 660–670.
[6] H. Suter: *Phthalsäureanhydrid und seine Verwendung,* Steinkopff Verlag, Darmstadt 1972.
[7] M. S. Wainwright, N. R. Foster: "Catalysts, Kinetics, and Reactor Design in Phthalic Anhydride Synthesis," *Catal. Rev. Sci. Eng.* **19** (1979) no. 2, 211–292.
[8] Snia Viscosa, IT 657641, 1963 (L. Notarbartolo, G. Messina, V. Muench, R. Mattone); IT 803424, 1968 (G. Sioli, A. Salatini, S. Sanchioni).
[9] *Chem. Age (London)* **117** (1978) Sept. 29, 15.
[10] *Hydrocarbon Process.* **56** (1977) no. 9, 134; **58** (1979) no. 9, 145.
[11] E. Katzschmannn, *Chem. Ing. Tech.* **38** (1966) 1–10. *Hydrocarbon Process.* **64** (1985) no. 11, 129
[12] R. Landau, A. Saffer, *Chem. Eng. Prog.* **64** (1968) 20–26.
[13] Y. Ichikawa, Y. Takeuchi, *Hydrocarbon Process.* **51** (1972) no. 9, 103–108.
[14] Toray Industries, DE-OS 2749638, 1976 (H. Morimoto, Y. Tohda, H. Torigata, K. Nakaoka).
[15] N. E. Ockerbloom, *Hydrocarbon Process.* **51** (1972) no. 4, 114–118.
[16] Eastman Kodak, US 3240803, 1961 (B. Thompson, S. D. Neely).
[17] Toray Industries, JP-Kokai 7834738, 1978 (H. Torikata, H. Morimoto, H. Eguchi, K. Nakaoka).
[18] W. F. Brill, *Ind. Eng. Chem.* **52** (1960) 837.
[19] P. Stecher: *Trimellitic Anhydride and Pyromellitic Dianhydride,* Noyes Data, Park Ridge, N.J., 1971.
[20] *Trimellitic Anhydride. Data Sheet,* Amoco Chem. Corp., Chicago 1979. Standard Oil Indiana, US 3819659, 1974 (R. H. Baldwin, D. E. Burney, P. H. Towle, D. G. Micklewright); US 3931304, 1976 (G. L. Wampfler).
[21] Mitsubishi Gas Chem. Co., DE-OS 3044752, 1981 (T. Suzuki, K. Katahara, S. Naito, T. Tsuji).
[22] Bergwerksverband, *Hydrocarbon Process.* 50 (1971) no. 11, 214.
[23] Gelsenberg-Benzin AG, DE 1246707, 1963; DE 1518917, 1965; DE 1568682, 1966; DE 1618461, 1967. Veba-Chemie, DE-OS 1593536, 1966.
[24] Princeton Chem. Res. Inc., DE-AS 1468851, 1965 (J. F. McMahon).
[25] Hüls AG, DE-OS 2751979, 1979 (H. Haferkorn).
[26] Rhône-Poulenc, GB 1226695, 1971 (M. Balme, B. Rollet).
[27] Veba Chemie, DE-OS 1930011, 1969 (G. Ibing, K. Neubold).
[28] Gelsenberg-Benzin AG, DE-AS 1289039, 1965 (K. Peterlein).
[29] Princeton Chem. Res., US 3592827, 1971 (R. I. Bergmann).
[30] Rütgers Werke, BE 525660, 1956. BASF, US 3708504, 1973 (O. Kratzer, H. Suter, F. Wirth).
[31] Nippon Kayaku K. K., DE-OS 2846432, 1980 (R. Hasegawa).
[32] N. Woroshzow: *Grundlagen der Synthese von Zwischenprodukten und Farbstoffen,* 4th ed., Akademie Verlag, Berlin 1966, p. 924.
*Bios Final Report* no. 1484, p. 21.
[33] K. Venkataraman: *The Chemistry of Synthetic Dyes,* vol. 2, Academic Press, New York-London 1952.
[34] B. Raecke, H. Schirp, *Org. Synth. Coll.* **5** (1973) 813–816.
[35] I. I. Kiiko, *Sov. Chem. Ind. (Engl. Transl.)* **7** (1975) 818–821.
[36] *Chem. Week* **112** (1973) no. 17, 43.
[37] H. E. Fierz-David, L. Blangey: *Grundlegende Operationen der Farbenchemie,* 8th ed., Springer Verlag, Wien 1952, pp. 219–221.
[38] IG-Farbenind., DE 604445, 1934 (H. Vollmann, M. Corell).
[39] Hoechst, DE 2209692, 1972; DE 2262858, 1972 (K. Eiglmeier).
[40] Hoechst, DE 2343946, 1973 (F. Röhrscheid).

[41]  *Mod. Plast. Int.* **13** (1983) Nov., 35.
[42]  M. Sittig: *Hazardous and Toxic Effects of Industrial Chemicals,* Noyes Data, Park Ridge, N.J., 1979.
[43]  F. J. Letkiewiez, *US Environ. Prot. Agency NTIS* PB-248 (1975) 838.

# Cellulose Esters

KLAUS BALSER, Wolff Walsrode AG, Walsrode, Federal Republic of Germany (Chap. 1)

LUTZ HOPPE, Wolff Walsrode AG, Walsrode, Federal Republic of Germany (Chap. 1)

THEO EICHER, Stuttgart, Federal Republic of Germany (Chaps. 2.1 and 2.2)

MARTIN WANDEL, Bayer AG, Leverkusen, Federal Republic of Germany (Chaps. 2.2 and 2.4)

HANS-JOACHIM ASTHEIMER, Rhodia AG, Freiburg, Federal Republic of Germany (Chap. 2.3)

| | | |
|---|---|---|
| 1. | **Inorganic Cellulose Esters** | 1138 |
| 1.1. | **Esterification** | 1140 |
| 1.2. | **Cellulose Nitrate** | 1141 |
| 1.2.1. | Physical Properties | 1141 |
| 1.2.2. | Chemical Properties | 1143 |
| 1.2.3. | Raw Materials | 1145 |
| 1.2.4. | Production | 1149 |
| 1.2.4.1. | Cellulose Preparation | 1150 |
| 1.2.4.2. | Nitration | 1150 |
| 1.2.4.3. | Stabilization and Viscosity Adjustment | 1151 |
| 1.2.4.4. | Displacement and Gelatinization | 1153 |
| 1.2.4.5. | Acid Disposal and Environmental Problems | 1153 |
| 1.2.4.6. | Other Nitrating Systems | 1154 |
| 1.2.5. | Commercial Types and Grades | 1155 |
| 1.2.6. | Analysis and Quality Control | 1157 |
| 1.2.7. | Uses | 1159 |
| 1.2.8. | Legal Provisions | 1162 |
| 1.3. | **Other Inorganic Cellulose Esters** | 1163 |
| 1.3.1. | Cellulose Sulfates | 1163 |
| 1.3.2. | Cellulose Phosphate and Cellulose Phosphite | 1164 |
| 1.3.3. | Cellulose Halogenides | 1165 |
| 1.3.4. | Cellulose Borates | 1165 |
| 1.3.5. | Cellulose Titanate | 1165 |
| 1.3.6. | Cellulose Nitrite | 1166 |
| 1.3.7. | Cellulose Xanthate | 1166 |
| 2. | **Organic Esters** | 1167 |
| 2.1. | **Cellulose Acetate** | 1168 |
| 2.1.1. | Chemistry of Cellulose Esterification | 1168 |
| 2.1.2. | Raw Materials | 1169 |
| 2.1.3. | Industrial Processes | 1170 |
| 2.1.3.1. | Pretreatment | 1171 |
| 2.1.3.2. | Esterification | 1171 |
| 2.1.3.3. | Hydrolysis | 1173 |
| 2.1.3.4. | Precipitation and Processing | 1174 |
| 2.1.4. | Recovery of Reactants | 1175 |
| 2.1.5. | Properties of Cellulose Acetate | 1175 |
| 2.1.6. | Analysis and Quality Control | 1177 |
| 2.2. | **Cellulose Mixed Esters** | 1178 |
| 2.2.1. | Production | 1178 |
| 2.2.2. | Composition | 1178 |
| 2.2.3. | Properties | 1179 |
| 2.2.4. | Other Organic Cellulose Mixed Esters | 1180 |
| 2.2.5. | Uses | 1180 |
| 2.3. | **Cellulose Acetate Fibers** | 1181 |
| 2.3.1. | Properties | 1181 |
| 2.3.2. | Raw Materials | 1182 |
| 2.3.3. | Production | 1182 |
| 2.3.4. | Economic Aspects | 1183 |
| 2.4. | **Plastic Molding Compounds from Cellulose Esters** | 1184 |
| 2.4.1. | Physical Properties | 1184 |
| 2.4.2. | Polymer-Modified Cellulose Mixed Esters | 1192 |
| 2.4.3. | Chemical Properties | 1193 |
| 2.4.4. | Raw Materials | 1193 |
| 2.4.5. | Production | 1195 |
| 2.4.6. | Trade Names | 1195 |
| 2.4.7. | Quality Requirements and Quality Testing | 1195 |
| 2.4.8. | Storage and Transportation | 1196 |
| 2.4.9. | Uses | 1196 |
| 2.4.10. | Toxicology and Occupational Health | 1198 |
| 3. | **References** | 1198 |

# 1. Inorganic Cellulose Esters

**Definition.** Cellulose esters are cellulose derivatives which result by the esterification of the free hydroxyl groups of the cellulose with one or more acids, whereby cellulose reacts as a trivalent polymeric alcohol. Esterification can be carried out by using mineral acids as well as organic acids or their anhydrides with the aid of dehydrating substances. Cellulose nitrate [9004-70-0] is the most important and only industrially produced inorganic cellulose ester (abbreviation CN, according to DIN 7728, T 1, 1978). A comprehensive bibliography on inorganic cellulose esters may be found in [12] – [19].

**Historical Aspects** [20]. The nitric acid ester of cellulose is the oldest known cellulose derivative and is still the most important inorganic cellulose ester. The term "nitrocellulose" is still used, but it is not the precise scientific term for cellulose nitrate. Cellulose esters were first described and industrially used at a time when the structure of esters was unknown and information on the polymeric primary material cellulose was not yet available.

The nitration of polysaccharides with concentrated nitric acid had already been described in 1832. H. Braconnot obtained a white and easily inflammable powder when he transformed starch with nitric acid. The product obtained was xyloidine. Th.-J. Pelouze treated paper with nitric acid and obtained an insoluble product containing ca. 6% nitrogen which he called pyroxiline and which he provided for military use.

C. F. Schönbein and R. Böttger are considered to be the inventors of so-called guncotton (1845). They transformed cotton with a mixture of nitric and sulfuric acid into a highly nitrated product that could serve as a substitute for black powder. Production on an industrial scale was stopped in 1847 because its extremely rapid catalytic decomposition was the cause of numerous plant explosions. Production was legally prohibited in 1865.

The use of cellulose nitrate as an explosive brought new momentum to its further industrial and scientific development, as well as to its economic significance. F. Abel made a basic breakthrough in 1865 when he succeeded in developing a safe method of handling. He was able to achieve a better washing of the adhering nitrating acid and a hydrolytic decomposition of the unstable sulfuric acid ester by grinding the nitrated fibers in water. This process allowed this product to attain military importance for its use as gunpowder.

In 1875, A. Nobel phlegmatized nitroglycerine by mixing with cellulose nitrate and discovered blasting gelatin. In the 1880s, smokeless gunpowders were developed. Vieille developed Poudre B (blanche) and Nobel developed Ballistit, the first dibasic gunpowder from cellulose nitrate and nitrogycerol. Abel and Dewar developed a similar gunpowder called Cordit.

The discovery that fibrous products could be modified by, for example, dissolution in an alcohol/ether mixture (film for wound protection) or by gelatinization with softeners brought additional uses. Films made from camphor and castor oil were used in collodium photography as carriers for light-sensitive materials (Archer, 1851). Nitrofilms found increasing use in photography and cinematography until they were replaced by nonflammable cellulose acetate films.

The year 1869 is considered to be the beginning of the age of plastics. J. W. Hyatt discovered celluloid, the first thermoplastic synthetic material. It was originally used as a substitute for ivory in the production of billiard balls.

The practical use of cellulose nitrate as a raw material for lacquers began in 1882, when STEVENS suggested amyl acetate as a highly volatile solvent (Zaponlack). The nitro lacquers achieved importance at first after World War I (FLAHERTY, 1921), when new applications were being sought as a result of the sharply declining demand for gunpowder. Only after the possibility of depolymerization of cellulose nitrate by pressure boiling during production had become known during the 1930s it was possible to use cellulose nitrate in protective and pigmented lacquers. Thus, it became possible to use nitro lacquers for painting automobiles on the assembly line.

Cellulose xanthate [9032-37-5] is a cellulose ester obtained with the inorganic acid dithiocarbonic acid. CH. F. CROSS discovered this important alkali-soluble cellulose ester in 1891 while he was reacting cellulose with alkali and carbon disulfide. It represents the base of the viscose process introduced in 1894 by BEVAN and BEADLE for producing man-made cellulosic fibers (rayon, rayon staple) and cellophane.

Other cellulose esters with inorganic acids are presently only of theoretical interest and have not attained any industrial or economic importance.

**Present Significance.** Cellulose nitrate is still currently important, 150 years after its discovery. It is industrially produced in large quantities for diversified applications. The reasons for this are the relatively simple production process with high yields, its solubility in organic solvent systems and its excellent film-forming properties from such solutions (collodion cotton as a raw material for lacquers), compatibility and gelatinability with softeners and other polymers (thermoplastics), as well as inflammability (guncotton for explosives). Cellulose nitrate has maintained its importance as a raw material for the manufacture of protective and coating lacquers as well as blasting agents and explosives.

Densified products, colored or pigmented chips kneaded with softeners, as well as aqueous dispersion systems with a low solvent content, are available today. They facilitate transport and processing, secure existing application forms, open up new ones, and are becoming increasingly nonpolluting.

The viscose process with its essential intermediate cellulose xanthate will remain, because of the availability of a constantly regrowing raw material supply, an important source of textile fibers for years to come. Alternative processes are intensively being sought to reduce pollution by sulfurous decomposition gases resulting during manufacturing. Cellulose nitrite, the cellulose ester of nitrous acid, is the most prominent example.

Cellulose esters with other inorganic acids have been frequently described and investigated. Cellulose sulfate [9032-43-3] was of some interest because of its solubility in water, but never achieved any practical importance. Cellulose phosphate [9015-14-9], borate, and titanate show interesting properties such as fire retardation, but are not yet of any industrial significance.

## 1.1. Esterification

**Mechanism.** The alcoholic hydroxyl groups of cellulose are polar and can be substituted by nucleophilic groups in strongly acid solutions. The mechanism of esterification assumes the formation of a cellulose oxonium ion followed by the nucleophilic substitution of an acid residue and the splitting off of water. Esterification is in equilibrium with the reverse reaction; saponification can be inhibited largely by binding the resulting water.

$$\text{Cell–OH} + \text{H}^+ \longrightarrow \text{Cell–O}^+\!\!\begin{smallmatrix}H\\H\end{smallmatrix}$$

$$\text{X}^- + \text{Cell–O}^+\!\!\begin{smallmatrix}H\\H\end{smallmatrix} \longrightarrow \left[\text{X} \rightarrow \text{Cell} \rightarrow \text{O}^+\!\!\begin{smallmatrix}H\\H\end{smallmatrix}\right] \longrightarrow \text{X–Cell} + \text{O}\!\!\begin{smallmatrix}H\\H\end{smallmatrix}$$

**Course of Reaction.** The three functional hydroxyl groups on each anhydroglucose unit of cellulose are blocked by intermolecular and intramolecular hydrogen bonds and, therefore, are not freely accessible for the reaction partners. The supermolecular arrangement and microstructure within the cellulose fiber, whose intensity depends on the origin and previous history of the cellulose material, is determined by these hydrogen bonds. The accessibility to the reaction partners and the reactivity of the alcohol groups also depend on this structure.

Due to the fact that cellulose is insoluble in all common solvents, reactions to form derivatives are usually carried out in heterogeneous systems. As the reaction proceeds, new reactive centers are created so that ultimately almost all parts of the cellulose fibers are included and in special cases yield soluble derivatives which react to completion in a homogeneous phase.

Little information is available on the esterification process. The following two reaction types are under discussion:

– An intermicellar reaction, which initially consists of the penetration of the reaction partner into the so-called amorphous regions between the highly organized cellulose micelles and proceeds during the course of esterification from the surface to the innermost regions of the micelles. The reaction speed is determined by diffusion.
– An intramicellar or permutoid reaction, in which the reagent penetrates all regions including the micelles so that practically all cellulose molecules react almost simultaneously. The reaction speed is specified by adjustment of the esterification equilibrium.

Arguments for both mechanisms are based on X-ray analyses. The possibility exists that both reaction types occur and ultimately merge. This depends on the reaction conditions, especially the esterification mixture and the temperature.

The hydrogen bonds between the cellulose molecules are almost completely broken down during esterification. The introduction of ester groups separates the cellulose chains so completely that the fiber structure is either altered or completely destroyed. Whether the cellulose ester is soluble in a solvent or in water depends on the types of substituents added.

**Substitution.** The esterification reactions do not necessarily proceed stoichiometrically because of equilibrium adjustment. The maximal attainable substitution with a mean degree of substitution (DS) of 3 is generally not reached. A triester can only be obtained under carefully controlled conditions. The primary hydroxyl group on the C-6 atom reacts most readily, while the neighboring hydroxyl groups on the C-2 and C-3 atoms of the anhydroglucose ring react considerably slower due to steric hindrance.

Basically, esterification is possible with all inorganic acids. Limiting factors are the type and the size of the acid residue as well as the varying degree of acid-catalyzed hydrolysis, which can lead to a complete cleavage of the cellulose molecule as the result of statistical chain splitting.

## 1.2. Cellulose Nitrate

Summary monographs on cellulose nitrate in addition to those in the Reference list can be found in [21] and [22].

### 1.2.1. Physical Properties

Cellulose nitrate (CN) is a white, odorless, and tasteless substance. Its characteristics are dependent on the degree of substitution.

*Density.* The density of cellulose nitrate is dependent on its nitrogen content and, therefore, on the degree of substitution (Table 1).

The bulk density of commercially available CN types is between 0.25 and 0.60 kg/L for moistened CN cotton, 0.15 – 0.40 kg/L when converted to dry mass.

Cellulose nitrate chips, which contain at least 18% dibutyl phthalate in addition to cellulose nitrate, have a density of 1.45 g/cm$^3$ (measured at 20 °C in an air-comparison pycnometer). The bulk density is 0.3 – 0.65 kg/L.

*Specific Surface.* The laboratory apparatus described by S. Rossin [23] is best suited for the determination of the specific surface of cellulose nitrate, which is 1850 – 4700 cm$^2$/g, depending on the fineness of the cellulose nitrate.

The determination of the inner surface according to the BET method showed dependence on the molar mass (i.e., an inner surface area of 1.44 m$^2$/g would correspond to a molar mass of 180 000 g and a surface area of 2.41 m$^2$/g would correspond to a molar mass of 400 000 g).

**Table 1.** The density of cellulose nitrate in relation to its nitrogen content (degree of substitution)

| Nitrogen content, % | Degree of substitution (DS) | Density at 20 °C, g/cm$^3$ |
|---|---|---|
| 11.5 | 2.1 | 1.54 |
| 12.6 | 2.45 | 1.65 |
| 13.3 | 2.7 | 1.71 |

**Table 2.** Thermodynamic properties of some cellulose nitrates

| | | |
|---|---|---|
| Heat of formation | trinitrate | − 2.19 kJ/g |
| | dinitrate | − 2.99 kJ/g |
| | cellulose | − 5.95 kJ/g |
| Heat of combustion | trinitrate | − 9.13 kJ/g |
| | dinitrate | − 10.91 kJ/g |
| | cellulose | − 17.43 kJ/g |
| Specific heat | celluloid film (70% CN and 30% camphor) | 1.26 – 1.76 J g$^{-1}$ K$^{-1}$ |
| Thermal conductivity | celluloid film (70% CN and 30% camphor) | 0.84 kJ m$^{-1}$ h$^{-1}$ K$^{-1}$ |
| Heat of solution in acetone | CN with 11.5% N content | − 73.25 J/g |
| | CN with 14.0% N content | − 81.64 J/g |

It must, however, be noted that the degassing temperature was lowered from the usual 200 °C to 60 °C due to the fact that cellulose nitrate deflagrates at 180 °C. It is possible that complete desorption did not take place under these conditions.

*Thermodynamic Properties,* see [24, pp. 137–154]. The most important thermodynamic properties are listed in Table 2.

*Electrical Properties* [24, p. 136]. The following electrical properties were measured on cellulose nitrate containing 30 wt% camphor (celluloid):

| | |
|---|---|
| Dielectric constant | |
|   at 50–60 Hz | 7.0 – 7.5 |
|   at 10$^6$ Hz | 6.0 – 6.5 |
| Dissipation factor (tan $\delta$) | |
|   at 50–60 Hz | 0.09 – 0.12 |
|   at 10$^6$ Hz | 0.06 – 0.09 |
| Specific resistance | 10$^{11}$ – 10$^{12}$ $\Omega \cdot$ cm |

*Mechanical Properties* [25]. The stress–strain diagram of cellulose nitrate films shows the elongation and tensile strength to be dependent on the size of the molecule (expressed as a term of viscosity).

The higher the molecular mass of a CN, the more elastic is the film made from it. Films become more brittle and their tensile strength declines with decreasing molecular mass3.

*Optical Properties.* Cellulose nitrate films are optically anisotropic because of their microcrystalline structure. The colors change in polarized light in relation to the nitrogen content of the CN:

**Table 3.** Mechanical properties of CN lacquer films

| Type* | Elongation, % | Tensile strength, N/mm$^2$ |
|---|---|---|
| E 4  | 24–30 | 98–103 |
| E 6  | 23–28 | 98–103 |
| E 9  | 23–28 | 88–98 |
| E 13 | 20–25 | 88–98 |
| E 15 | 18–23 | 78–98 |
| E 21 | 12–18 | 78–88 |
| E 22 | 10–15 | 74–84 |
| E 24 | 8–12 | 69–78 |
| E 27 | 5–10 | 59–69 |
| E 32 | <5 | 39–49 |
| E 34 | <3 | 29–49 |

* According to DIN 53179: The E-type designation specifies the CN concentration (% in dry condition) in acetone which gives a viscosity of 400 ± 25 mPa · s.

| | |
|---|---|
| 11.4 % N | weakly red |
| 11.5–11.8 % N | yellow |
| 12.0–12.6 % N | blue to green |

The index of refraction is 1.51, and the maximal light transmission is achieved at 313 nm.

*Light Stability.* Exposure to sunlight, and especially to ultraviolet light, has a detrimental effect on cellulose nitrate film by causing it to become yellowish and brittle. Solvents, softeners, and resins can either promote or hinder yellowing.

## 1.2.2. Chemical Properties

The three hydroxyl groups of cellulose can be completely or partially esterified by nitrating acid. The varying degrees of nitration can be related to the following theoretical nitrogen contents:

| | |
|---|---|
| Cellulose mononitrate, $C_6H_7O_2(OH)_2(ONO_2)$: | 6.75 % N |
| Cellulose dinitrate, $C_6H_7O_2(OH)(ONO_2)_2$: | 11.11 % N |
| Cellulose trinitrate, $C_6H_7O_2(ONO_2)_3$: | 14.14 % N |

Cellulose nitrate with a nitrogen content between 10.8 and 12.6 % is a suitable raw material for lacquers, and CN with > 12.3 % N is suitable for explosives exclusively.

*Degree of Substitution – Nitrogen Content – Solubility.* The degree of substitution can be calculated from the nitrogen content of the various CN types (Fig. 1). The degree of substitution determines the solubility of cellulose nitrate in organic solvents. CN for lacquers can be classified according to its solubility in organic solvents as follows:

alcohol-soluble CN (A types)
  nitrogen content: appr. 10.9–11.3 %
  readily soluble in alcohols, esters, and ketones

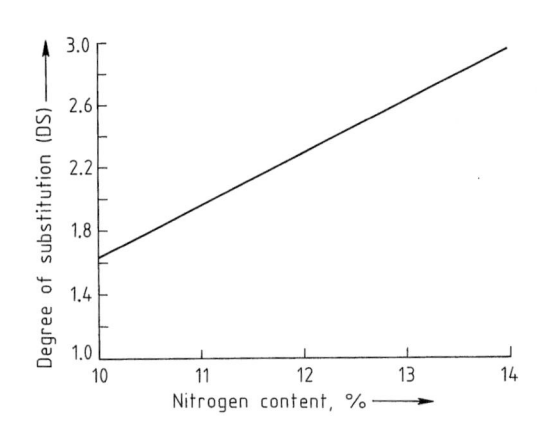

**Figure 1.** Variation of the degree of substitution with the nitrogen content of cellulose nitrate

moderately soluble CN (AM types)
nitrogen content: appr. 11.4–11.7%
soluble in esters, ketones, and glycol ethers with excellent blendability or compatibility with alcohol

CN soluble in esters (E types)
nitrogen content: 11.8–12.2% for lacquer cotton, up to 13.7% for guncotton
readily soluble in esters, ketones, and glycol ethers

*Intrinsic Viscosity – Degree of Polymerization* [24, pp. 85–121]. The mean number of anhydroglucose units in cellulose nitrate molecules is designated as the mean degree of polymerization (DP). The viscosity of the solution (at the same concentration in the same solvent) is generally considered to be a relative measure of the molecular mass. The molecular mass can be mathematically expressed as a function of the intrinsic viscosity (Staudinger–Mark–Houwink equation).

*Distribution of the Molecular Mass.* The starting material of cellulose nitrate is natural cellulose, the quality of which is subjected to annual growth cycles. It is, therefore, of great importance to have polymolecular data, such as the mean degree of polymerization and the distribution of the molecular mass, available in addition to viscosity, solubility behavior, and nitrogen values. These values are important, for example, in assessing the mechanical properties and aging processes of polymer products.

The isolation of the polymers according to their molecular mass can be achieved elegantly by gel permeation chromatography (GPC).

*Chemical Compatibility.* An every day use of cellulose nitrate is in nitro lacquers, where it is dissolved in organic solvents. In this solution, cellulose nitrate is extremely compatible with essential substances in the lacquer formulation such as alkyd resins, maleic resins, ketone resins, urea resins, and polyacrylates. A large number of softeners, such as adipates, phthalates, phosphates, and raw and saturated vegetable oils are compatible with cellulose nitrate.

*Chemical and Thermal Stability.* Cellulose nitrate, as a solid or in solution, should not be brought into contact with strong acids (degradation), bases (denitration), or organic amines (decomposition) since they all induce a destruction of cellulose nitrate. This may proceed very rapidly and lead to deflagration of the cellulose nitrate.

The ester bonds of cellulose nitrate, which can be broken by saponifying agents or by catalysis, are responsible for its physicochemical instability. This substance-specific property is dependent on the temperature, the specimen, and whether catalytically active decomposition products remain or are removed from the sample.

Another basic instability of cellulose nitrate is observed during the production process. Mixed sulfuric acid esters transmit a chemical instability to the nitrocellulose molecule. These mixed esters are destroyed in weakly acid water during the stabilization phase of production. The long reaction time required by this procedure can be considerably shortened by increasing the reaction temperature. The time required can be reduced to only a few hours by raising the temperature to 60–110 °C. Under these conditions the nitrate ester remains stable; the glucosidic bond of cellulose nitrate, on the other hand, is attacked. This property is used to advantage to specifically reduce the degree of polymerization of the cellulose nitrate.

Thermogravimetry, IR spectroscopy, and electron spectroscopy (ESCA) [26], [27] have recently been used to determine the extent of thermally induced and light-induced decomposition of cellulose nitrate. The reaction proceeds as follows:

$$\text{Cell} - \text{O} - \text{NO}_2 \longrightarrow \text{Cell} - \text{O} \cdot + \cdot \text{NO}_2$$

It is proceeded by a series of extremely exothermic oxidation reactions triggered by the $NO_2$ radical, which often leads to spontaneous deflagration. $NO_2$ is reduced to NO and in the presence of air $NO_2$ is reformed, thus initiating an autocatalytic chain reaction, at the end of which the gaseous reaction products $CO_x$, $NO_x$, $N_2$, $H_2O$, and HCHO are found.

By adding stabilizers such as weak organic bases (diphenylamine) or acids (phosphoric acid, citric acid, or tartaric acid), intermediary nitric oxides can be bound and the autocatalytic decomposition prevented.

Thermal decomposition does not occur at temperatures below 100 °C. The temperature (according to [28]) at which cellulose nitrate spontaneously deflagrates is used as a measure of its thermal stability. A well-stabilized lacquer cotton has a deflagration temperature of $\geq$ 180 °C. The deflagration temperature of plasticized cellulose nitrate chips with at least 18 wt% softener (i.e., dibutyl or dioctyl phthalate) is $\geq$ 170 °C.

The Bergmann–Junk test [29] and the warm storage test are additional methods for determining the stability of cellulose nitrate.

## 1.2.3. Raw Materials

**Cellulose.** Until the beginning of World War I, the only raw material available for nitration was cellulose obtained from cotton in the form of bleached linters (as flakes or crape). This was due to the high degree of purity ($\alpha$-cellulose > 98%), which allowed a high yield and products with good clarity and little yellowing.

Especially in times when linters were scarce, it was possible to produce gunpowder from wood celluloses, even unbleached, other cellulose fibers (annual plants), and even from wood if attention was given to the adequate disintegration of the raw materials. Lacquer types obtained from wood celluloses, especially from hardwood, gave dull and mat films and lacquers with inferior mechanical properties. This is due to the high content of pentosans, which is also nitrated but is easily split by hydrolysis in conventional nitrating acid systems and thus becomes insoluble.

The development of highly purified chemical-grade wood pulp by refinement with hot and cold alkali having $R_{18}$ values of 92–95% (see Table 16) allows this type of raw material to be used in the same manner as were linters, which currently are used only for the production of special and highly viscous CN types. The highly refined prehydrolyzed sulfate pulps with $R_{18}$ values of above 96% are especially well-suited for nitration. The viscosity range of CN products can be adjusted in advance by choosing an initial cellulose with an adequate DP. A low ash content, and above all a low calcium content, of the cellulose is important in preventing calcium sulfate precipitation during industrial nitration.

A comparative study on the nitrating behavior of linters and wood pulps [30] shows the morphological factors of the fibers (fiber length and distribution, cross-section form and thickness of the secondary wall, and fine structure including packing density, degree of crystallization, and lateral arrangement of the fibrils), the chemical composition of the cellulose (DP and polydispersibility) as well as the type, quantity, and topographic distribution of the accompanying hemicelluloses and lignin to be responsible for the nitrating capability of celluloses. These factors determine the swelling properties and thereby the uniformity of nitration, as well as the compressibility and the relaxation capacity of the fibers, which in turn influence the retention capacity of the fiber mass. Linters with a lower acid retention capacity of 110–130% are definitely superior to wood pulp (acid retention capacity of prehydrolyzed sulfate pulps up to 230%, of sulfite pulps up to 300%) in this respect. The suitability of a raw material for nitration can be tested by a specially developed machine that measures the compression and relaxation characteristics of a cellulose fiber pile.

Approximately 150 000 t (3.4%) of the annual worldwide production of $4.4 \times 10^6$ t of chemical-grade pulps are used for the production of cellulose nitrate.

**Industrial Nitrating Agents.** The so-called nitrating acid as developed by SCHÖNBEIN, the nitric acid/sulfuric acid/water system, is still the nitrating agent of choice for industrial purposes. The highest attainable degree of substitution using this system is at DS 2.7 13.4% N. This is achieved only when the nitric acid used is not hydrated and the molar ratio of nitric acid to sulfuric acid monohydrate is 1:2. The optimal nitrating mixture is as follows:

|             | $HNO_3$ | $H_2SO_4$ | $H_2O$ |
|-------------|---------|-----------|--------|
| molar ratio | 1       | 2         | 2      |
| wt%         | 21.36   | 66.44     | 12.20  |

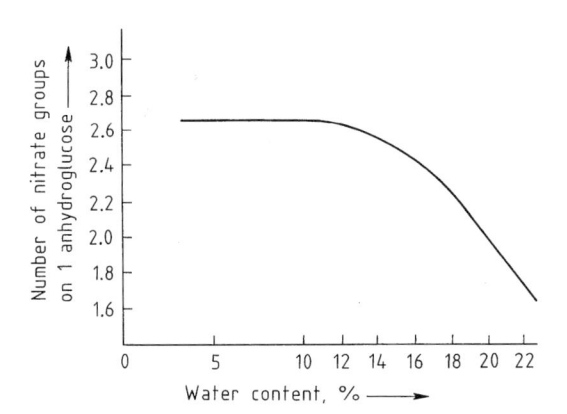

**Figure 2.** Dependence of the degree of esterification (DS) on the water content of the optimal nitrating mixture ($HNO_3 : H_2SO_4 = 1 : 2$)

Water plays a special role as far as the attainable degree of nitration is concerned. Below 12% water there is no increase in substitution, but a higher water content results in a drastic decline in the degree of nitration (Fig. 2).

It is assumed that increasingly hydrated nitric acid causes increased swelling and gelatinization of the cellulose so that the nitrating acid is no longer able to penetrate into the inner structures of the micelles.

The desired degrees of esterification can be adjusted by varying the nitrating acid mixture according to the CN types (Table 4), whereby in industrial processes the nitric acid content is kept nearly constant at 25–26%.

The ternary system $HNO_3/H_2SO_4/H_2O$ has been extensively investigated. The results are summarized in Figure 3.

The curves of the same degrees of substitution (% N) in relation to the nitrating acid composition are presented here. The cross-hatched band identifies those areas in which the cellulose material is extremely swollen and gelatinized. Three zones can therefore be differentiated in the phase diagram:

1) Area of technical nitration:

| | |
|---|---|
| nitric acid | 15–100% |
| sulfuric acid | 0–80% |
| water | 0–20% |

In this range, nitric acid is present in a nonhydrated form and induces true nitration (N content 10%). The industrially used range with 20–30% $HNO_3$, 55–65% $H_2SO_4$, and 8–20% water is also included in this area.

**Table 4.** Industrially used nitrating acid solutions

| CN type | Nitrating acid | | | N content, % | DS |
|---|---|---|---|---|---|
| | % HNO$_3$ | % H$_2$SO$_4$ | % H$_2$O | | |
| Lacquer cotton A | 25 | 55.7 | 19.3 | 10.75 | 1.90 |
| Celluloid cotton | 25 | 55.8 | 19.2 | 10.90 | 1.95 |
| Lacquer cotton AM | 25 | 56.6 | 18.4 | 11.30 | 2.05 |
| Dynamite cotton | 25 | 59.0 | 16.0 | 12.10 | 2.30 |
| Lacquer cotton E | 25 | 59.5 | 15.5 | 12.30 | 2.35 |
| Powder cotton | 25 | 59.8 | 15.2 | 12.60 | 2.45 |
| Guncotton | 25 | 66.5 | 8.5 | 13.40 | 2.70 |

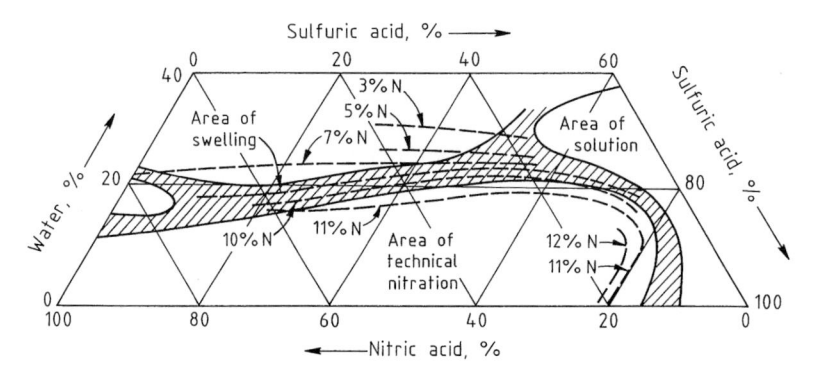

**Figure 3.** Composition of the nitrating mixtures and attainable N contents of cellulose nitrates

2) Area of solution:

| nitric acid | 0–10 % |
|---|---|
| sulfuric acid | 60–100 % |
| water | 0–40 % |

Little or no nitration takes place in this range. Cellulose is degraded to the point of complete dissolution in concentrated sulfuric acid.

3) Area of swelling:
Nitric acid is increasingly hydrated in this range of increasing water content. Nitration decreases rapidly.

A process developed in the United States, but less important, uses magnesium nitrate instead of sulfuric acid as a dehydrating agent [31]. Magnesium nitrate can bind water as its hexahydrate. The nitrating mixture consists of 45–94 % nitric acid, 3.3–34 % magnesium nitrate, and 2.7–21 % water; the ratio of magnesium nitrate : water is 1.2–2.2 : 1. A cellulose nitrate with an N content of 11.9 % was obtained, for example, with 64.5 % HNO$_3$, 19.5 % Mg(NO$_3$)$_2$, and 16 % H$_2$O. This nitrating system is appropriate for a continuous process, in which waste and washing acid are reprocessed in ion exchangers and the magnesium nitrate is recycled. Thus, acid and sulfate no longer pose a waste disposal problem.

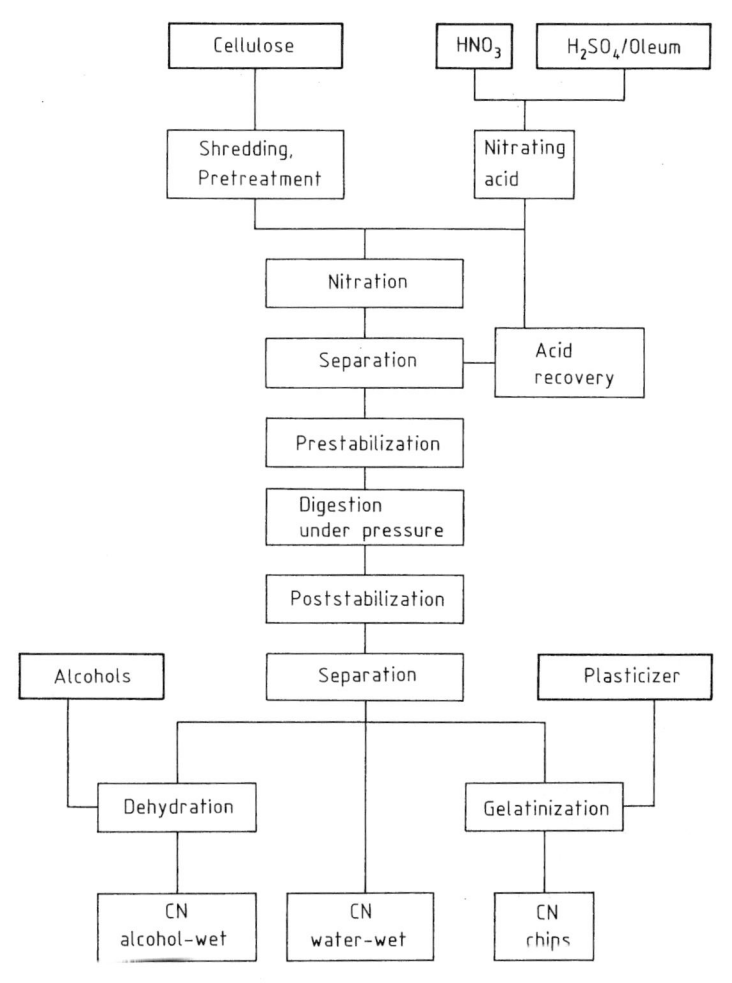

**Figure 4.** Flow diagram of cellulose nitrate production

## 1.2.4. Production

The flow diagram (Fig. 4) shows the industrial production of CN according to the mixed acid process. The viscosity of the end product is determined by the choice of the initial cellulose, and the degree of nitration is determined by the composition of the mixing acid. The final viscosity adjustment follows during the pressure boiling step (see Section 1.2.4.3).

## 1.2.4.1. Cellulose Preparation

Cotton linters with a moisture content of up to 7% are mechanically disintegrated homogeneously. Pressed pulp sheets must be appropriately shredded to obtain rapid and uniform nitration. Spruce or beech celluloses, preactivated with 20% sodium hydroxide (mercerization), were formerly used for this purpose in the form of crape papers with a mass per unit area of ca. 25 g/m$^3$. To avoid the costly transformation of the cellulose to paper sheets, a process was attempted to obtain a loose product resembling linters by direct disintegration of pulps to fibers. A moisture content of 50% proved to be optimal for nitration and washing out the acid. The required drying of the cellulose flakes before nitration proved to be disadvantageous.

The Stern shredder [32], in which the pulp sheets are torn rather than being cut into small elongated shreds to avoid compression at the edges, was a definite improvement. Currently, cellulose for nitration is used in the form of fluff, shreds, or chips. The packing density and compression behavior of the cellulose fibers in the fiber pile are decisive factors for the swelling and nitration kinetics, as well as the acid retention capability [30].

## 1.2.4.2. Nitration

Nitration on an industrial scale is still frequently carried out according to a batch process that was developed from a process described by Du Pont in 1922. The equipment is constructed of stainless steel. The adjusted and preheated nitrating acid reaches the stirring reactor that is charged with cellulose by means of a measuring system; a large excess of acid (1:20 to 1:50) is added to retain the ability of the reaction mixture to be stirred and to ensure that heat is carried off. The nitrating temperature is between 10 °C (dynamite type) and 36 °C (celluloid type). The total heat of reaction is estimated to be over 200 kJ per kg of CN, of which the enthalpy of formation of CN is about one-third.

Even though the reaction is nearly complete after ca. 5 min, the mixture remains in the reactor for about 30 min. The temperature must remain constant (cooling), since hydrolytic degradation processes that lead to considerable losses in yield begin at temperatures as low as 40 °C.

The theoretical yield of commonly used industrial types with a DS of 1.8–2.7 (10.4–13.4% N) is between 150 and 176% with respect to cellulose. The practical yield, however, is up to 15% lower and depends on the type and purity of the cellulose, as well as on the temperature and duration of nitration. Losses arise from the inevitable complete decomposition of cellulose to oxalic acid by way of oligo- and monosaccharides, whereby the nitric acid is reduced to nitrogen oxides, $NO_x$. In addition, mechanical losses during the subsequent separation process, due particularly to short fibers (cellulose from hard wood), must also be taken into consideration.

The reaction mixture is drained from the reactor into the centrifuge, where the excess acid is separated and removed at high speed and reprocessed for recycling. The mixture must remain moist so that it does not ignite and deflagrate.

The degree with which the product retains acid after separation is of economic importance because of the acid loss and the expense of the ensuing washings. Linters with an acid retention of 100–130% clearly surpass wood celluloses in this respect which, depending on the wood and cellulose type as well as its processing, can retain up to 3 times more adhering acid relative to CN [30].

The still acid-moist product is immediately placed into a great excess of water (consistency 1%) so that the adhering acid is displaced as rapidly as possible and the saponification of the CN is prevented.

Continuous nitrating processes, which are more economical, were developed in the 1960s [33], [34]; they ensure a more uniform product quality and are safer to handle. The nitrating system consists of two or more consecutively arranged straight-run vats or tube systems containing conveyers (screw conveyer or turbulence stirring apparatus) which forward the reaction mixture. The prepared cellulose is directed into this cascading equipment from storage bunkers over automatic weighing scales and continuously mixed with the added nitrating acid. It is important that the cellulose is rapidly added and immediately covered with acid. There it remains for 30–55 min. A newer process using a continuous loop-formed pressure reactor [35] requires the cellulose to remain only for 6–12 min. The reactant is then sent into a continuously operating special centrifuge, where the excess acid is separated and simultaneously taken up with water. The fact that the reactant remains only a few seconds reduces the risk of spontaneous deflagration and saponification.

Figure 5 shows schematically the continuous process according to Hercules [31].

The broken-up and preconditioned cellulose (a) is brought by way of the automatic scales (b) to the continuous reactor (c). The reaction product is centrifuged in a washing zone (d) and simultaneously washed by zones with water in a countercurrent. The product leaves the centrifuge almost free of acid, and the washed out acid can be recycled and reused almost without loss [36].

### 1.2.4.3. Stabilization and Viscosity Adjustment

The prestabilization step following the prewashing further purifies the product by means of repeated washing and boiling with water that contains 0.5–1% acid residue. The batch method requires large amounts of space, water, and energy; the required boiling time varies between 6 (celluloid type) and 40 h (guncotton). Automatic continuous processes have been developed in this case as well [37].

Most of the remaining sulfuric acid is removed during prestabilization, since it would promote the catalytic self-decomposition of CN. The sulfuric acid is bound by adsorption and esterified. A total sulfate content of 1–3% was found in weakly nitrated CN, of which 70–85% is in the form of the acidic sulfuric acid semiester, while highly

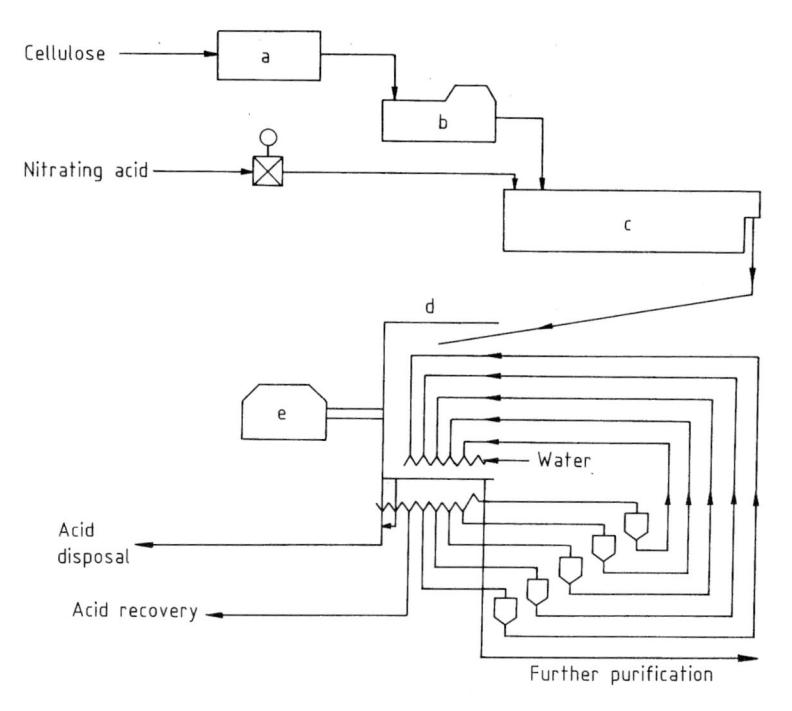

**Figure 5.** Continuous cellulose nitrate production according to Hercules
a) Preconditioning; b) Auto-matic scale; c) Reactor; d) Washing zone; e) Centrifuge

nitrated CN contains only 0.2–0.5 % total sulfate, of which 15–40 % is thought to exist as an ester. Semiesters can be easily saponified and washed out by boiling with water. It is not yet certain whether the so-called resistent sulfate content exists in the form of the neutral sulfuric acid ester or the physically adsorbed sulfuric acid.

The desired final viscosity of the CN is adjusted in the following process step, which is pressure boiling (digestion under pressure) in a consistency of 6–8 % at 130–150 °C, by means of specific degradation of the degree of polymerization. The remaining extremely low sulfuric acid content induces hydrolytic decomposition at this temperature and under pressure. The viscosity can, for example, be reduced to 1/10 of the initial viscosity within 3 h at 132 °C by using this process. This process made the development of high solid coating and protective nitro lacquers possible. The stabilization process of guncottons is accelerated by pressure boiling; dynamite wools are usually not pressure boiled.

Further product losses are due to chain degradation ranging from soluble cleavage products to oxalic acid. Nitrous gases ($NO_x$) are released by the reduction of nitric acid, which must be continuously drawn off to avoid decomposition of CN.

Pressure boiling can be achieved batchwise in autoclaves, as well as continuously in a tube reactor of 1500 m in length and 100 mm in diameter, e.g., with direct steam. A one-pot process in which prestabilization, pressure boiling, and poststabilization are carried out in one operation is described in [38].

During the stabilization process, the remaining sulfuric acid is almost completely removed by additional washing and boiling. While celluloid and lacquer types are finished in flaky, fibrous form, guncotton must be ground. This is done in grinding hollander engines at 12–15% consistency or continuously in a series of cone refiners, whereby the material is gradually concentrated from 3 to 10% between the various grinding steps. Sorting steps are inserted by hydrocyclones during the final washing processes. The last acid remnants in the fiber capillaries are removed during the grinding process by means of diffusion against water. Weak bases, sodium carbonate, or chalk are used to maintain a pH of 7. Stabilizers (organic acids) may be added during this step.

### 1.2.4.4. Displacement and Gelatinization

A water-wet CN cotton with a water content of 25–35% remains in the centrifuge after the final separation and is then packed into drums or PE sacks.

Water contained in celluloid and lacquer types is displaced by alcohols specified by the processors (ethanol, 2-propanol, *n*-butanol) in displacement presses or displacement centrifuges. Continuous processes prevail here also [39]. The resulting aqueous alcohols must be distilled to remove the water.

The water-wet CN cotton can be gelatinized with softeners such as phthalates in kneading aggregates and dried on drum or band driers for the production of CN chips [40], [41]. Colored chips are obtained by adding carbon black or pigments from which colored enamels can be produced without the use of ball mills or roller mills.

### 1.2.4.5. Acid Disposal and Environmental Problems

The nitrous gases formed during the nitrating and stabilizing side reactions are drawn off and washed out in trickling towers. The lower nitrogen oxides are regained after oxidation as 50–60% nitric acid.

The waste acids resulting from the first separation contain 2–3% more water and 3–4% less nitric acid than the initial mixture. They are circulated in a closed system and constantly regenerated with nitric acid and oleum. The acid that adheres to the product must also be replaced.

The proportion of adhering acid depends on the initial cellulose and the CN type. It ranges between 80% (guncotton) and up to 200–300% (lacquer types) with regard to CN and is removed with the water used for washing and boiling.

Aside from the economic aspects of acid loss in wastewater, environmental considerations are beginning to play an increasingly important role. While older manufacturing facilities using simple centrifuging to remove waste acid produced 300 m$^3$ of water per ton of CN, containing 0.5% acid and with a pH of 1, it has recently become possible to reduce the volume of wastewaters to a fraction of its previous volume by almost completely closing the cycles.

Before proceeding into the draining ditch, the wastewater must be separated from the hardly decomposable sludge consisting of cellulose and CN, and then be neutralized. The sulfate proportion can be reduced by calcium sulfate precipitation, while the nitrate proportion remains completely in the wastewater. Organic matter of communal sewage, for example, can be biologically decomposed without additional oxygen, whereby nitrates disappear almost completely as a result of biological denitrification.

The salt/acid process with magnesium nitrate (see Section 1.2.4.2) is more favorable with regard to the wastewater problem. Sulfates are completely absent, and magnesium nitrate is recycled and, therefore, causes no water pollution problems. The amount of wastewater can be reduced by 80% and the nitric acid requirement by 83% in comparison to the formerly used discontinuous processes.

### 1.2.4.6. Other Nitrating Systems

Numerous attempts have been made to improve nitration by the introduction of other nitrating systems, or at least to increase the degree of substitution. Further details may be obtained from [12], [14], [17], and [21]. Table 5 gives a summary of alternative nitrating systems, none of which was able to displace the ternary system $HNO_3/H_2SO_4/H_2O$ for industrial nitration.

Nitration with pure nitric acid is possible in principle. Esterification is not possible with acid concentrations below 75%. Acid concentrations less than 75% cause the formation of the unstable so-called Knecht compound, which has been described as either a molecular complex or an oxonium salt of the nitric acid. Cellulose nitrates with 5–8% N, which dissolve in excess acid, are formed at acid concentrations of 78–85%. Nitrogen contents of 8–10% are attained at concentrations between 85 and 90% $HNO_3$; these products have a strong tendency to gelatinize. Heterogeneous nitration without apparent swelling takes place at a $HNO_3$ concentration above 89%, and 13.3% N can be achieved with 100% $HNO_3$. Nitration can be increased to 13.9% N with 100% $HNO_3$ by addition of inorganic salts such as sulfates, acid phosphates, and particularly nitrates, preferably in a 15% concentration.

The nitric acid/phosphoric acid system is of special interest in a 1:1 ratio with 2.5% phosphorus pentoxide added, with which an almost completely nitrated product of great stability was achieved. The nitric acid/acetic acid/acetic anhydride system in a ratio of 2:1:1 gives highly nitrated and highly stable products in which the fiber structure remains intact largely. After extraction of these nitrates with water or alcohol, the theoretical degree of substitution of the trinitrate may be attained.

Nitrating systems which achieve a high degree of nitration without degradation of the cellulose chain are of special scientific interest. This process is known as polymer analogous nitration. After a critical examination of all known nitrating mixtures, the nitric acid/acetic acid/acetic anhydride system in a ratio of 43:32:25 at 0 °C [42] was recommended for determining the molecular mass of native celluloses of such solutions by using absolute methods and the intrinsic viscosity number [43]. The system

**Table 5.** Nitrating systems

| Nitrating system | Max. N content, % | Comments |
|---|---|---|
| $HNO_3/H_2SO_4/H_2O$ | 13.4 | Industrial nitration |
| $HNO_3 < 75\%$ | | "Knecht compound," unstable |
| 78–85% | 8 | Dissolution in the nitrating acid |
| 85–89% | 10 | Gelatinization |
| 90–100% | 13.3 | No swelling |
| $HNO_3$ + nitrates, sulfates, phosphates | 13.9 | |
| $HNO_3$ vapor | 13.75 | Slow reaction, stable nitrate |
| $HNO_3$ vapor + nitrogen oxides | 13.8 | |
| $N_2O_5$ | 14.12 | |
| $N_2O_5$ in $CCl_4$ | 14.14 | Trinitrate |
| $HNO_3$ in $CH_2Cl_2$ | 14.0 | |
| $HNO_3$ in nitromethane | 14.0 | Homogeneous reaction |
| $HNO_3$ + $H_3PO_4/P_2O_5$ | 14.04 | Rapid reaction without decomposition (polymer analogue) |
| | 14.12 | After extraction with methanol |
| $HNO_3$ + acetic acid/acetic anhydride | 14.08 | Great stability |
| | 14.14 | After extraction with ethanol |
| $HNO_3$ + propionic acid/butyric acid | 14.0 | |

anhydrous nitric acid in dichloromethane also allows the application of such polymer analogous reactions at temperatures between 0 and −30 °C [44]. Other authors [45] prefer the system nitric acid/phosphoric acid/phosphorus pentoxide.

**Nitration in the Laboratory.** Preparative cellulose nitration with $HNO_3/H_2SO_4$ nitrating acid to products with whatever N content up to 13.65% is desired, stabilization and stabilization tests, nitration with the nitric acid/phosphoric acid (< 13.9% N) and nitric acid acetic anhydride systems up to the trinitrate, denitration with hydrogen sulfide to cellulose II, the analytic determination of the N and sulfate content, and the solution of the CN and the viscosity determination of the solution are extensively described in [46].

## 1.2.5. Commercial Types and Grades

Cellulose nitrates receive, because of their fluffy structure and cottonlike appearance, the additional designation "cotton."

Two parameters are decisive for the industrial use of cellulose nitrate:

Nitrogen content (including the resulting solubility properties)
Viscosity

**Table 6.** Cellulose nitrate types

| Type | N content, % | Degree of substitution (DS) |
|---|---|---|
| Celluloid cotton | 10.5 – 11.0 | 1.82 – 1.97 |
| Alcohol-soluble lacquer cotton | 10.9 – 11.3 | 1.94 – 2.06 |
| Lacquer cotton moderately soluble in alcohol | 11.4 – 11.7 | 2.08 – 2.17 |
| Ester-soluble lacquer cotton | 11.8 – 12.2 | 2.20 – 2.32 |
| Powder cotton | 12.3 – 12.9 | 2.55 – 2.57 |
| Guncotton | 13.0 – 13.6 | 2.58 – 2.76 |

As seen in Table 6, cellulose nitrates with differing nitrogen contents have various applications. Cellulose nitrates for lacquers are available in numerous viscosities. It is possible to categorize all stages of viscosity according to the European norm (DIN 53179), but the viscosity of cellulose nitrates is primarily categorized by using the Cochius method and the British or American ball drop method (ASTM D 1343-69).

In addition to the so-called cotton types densified CN types are available. These may be obtained by either nitrating compressed cellulose or by subsequently compressing the fluffy cellulose nitrate. It is possible to almost double the bulk density by compression.

For safety reasons, the commercially available CN cotton types must be wetted with at least 25 wt% water or aliphatic alcohols. In addition to water, ethanol, $n$-butanol, and 2-propanol may also be used as wetting agent.

The largest manufacturers of cellulose nitrates are the following:

| | |
|---|---|
| Hercules Inc. | USA |
| Wolff Walsrode AG | FR Germany |
| Hagedorn | FR Germany |
| WNC Nitrochemie GmbH | FR Germany |
| Société Nationale des Poudres et Explosifs (SNPE) | France |
| Imperial Chemical Industries (ICI) | Great Britain |
| S.I.P.E. Nobel S.p.A. | Italy |
| Unión de Explosivos Río Tinto S.A. | Spain |
| Bofors | Sweden |
| Asahi | Japan |
| Daicel Chemical Industries, Ltd. | Japan |

Many countries in South America, Asia, and Eastern Europe maintain small CN production facilities. The total world capacity may be estimated to 150 000 t/a of dry cellulose nitrate.

**Other Commercial Types.** Also available, in addition to cellulose nitrate cotton types, are so-called cellulose nitrate chips, made from cellulose nitrate plasticized by gelatinizing softeners. For safety reasons, the softener content has been established at a minimum of 18 wt%. Chips are preferred in processes where alcohols interfere in the formulation of lacquers.

The dispersions of cellulose nitrate with softeners or resins in water manufactured by the Wolff Walsrode AG are other available forms. The solvent-free or low-solvent dispersions are not polluting and may be used in all areas in which cellulose nitrate lacquers also are used [55].

## 1.2.6. Analysis and Quality Control

The most important analytical characteristics relate to the determination of the N content and, thereby, the average degree of substitution (DS), as well as the viscosity of the solution as a measure of the average molecular mass or chain-length.

**Analytic Tests.** The most commonly used analytic procedures are summarized in [25], [46], [47], and [48].

*Dry content* is determined by careful drying of a small, thinly layered alcohol or water-wet sample at room temperature for 12–16 h, in a weighing glass at 100–105 °C for 1 h, or with compressed warm air at 60–65 °C for 0.5–1 h.

*Ash content* is determined by decomposing a dried sample with $HNO_3$ and incinerating the residue. Specifications require that the ash content should not be above 0.3%.

*N-content* is determined by reducing nitrates according to the following reaction (Schulze-Thiemann):

$$NO_3^- + 3\ FeCl_2 + 4\ HCl \longrightarrow 3\ FeCl_3 + Cl^- + 2\ H_2O + NO$$

or by the following reaction:

$$2\ NO_3^- + 4\ H_2SO_4 + 3\ Hg \longrightarrow 3\ HgSO_4 + SO_4^{2-} + 4\ H_2O + 2\ NO$$

The resulting NO is collected in a Du Pont nitrometer.

**Stability Tests** [25]. *Deflagration Temperatures:* Well-stabilized CN deflagrates at temperatures above 180 °C.

*Bergmann–Junk Test* [29]: A quantity of 2 g of dried CN is kept at a temperature of 32 °C for 2 h in a special apparatus for the elimination reaction, after which time the amount of the developed nitrous gases (after reduction to NO) is determined. CN is stable according to this test if no more than 2.5 cm$^3$ of NO per gram is measured.

*Warm Storage Test:* A quantity of 5 g of dried CN is stored in a glass-stoppered tube at 75 °C. Note is then made when the first nitric oxide (red-brown gas) becomes visible. Well-stabilized CN can be stored at 75 °C for at least 10 days.

*ASTM Stability Test* [48]: After storage at 134.5 ± 0.5 °C the time is noted in which the nitrous gases discolor methyl violet test paper.

**Viscosity.** *Viscosity according to DIN 53179:* If CN is dissolved in acetone in the appropriate concentration, CN solutions meeting this requirement show a apparent

**Table 7.** Characterization of cellulose nitrates according to DIN 53179

| A types | CN concentration, % (absol. dry) | AM types | CN concentration, % (absol. dry) | E types | CN concentration, % (absol. dry) |
|---|---|---|---|---|---|
| | | | | E 1440 | 4 |
| | | | | E 1160 | 7 |
| | | | | E 950 | 9 |
| | | | | E 840 | 12 |
| | | AM 760 | 14 | | |
| | | AM 750 | 15 | | |
| | | | | E 730 | 15 |
| | | AM 700 | 17 | | |
| | | | | E 620 | 21 |
| | | | | E 560 | 22 |
| | | | | E 510 | 24 |
| A 500 | 27 | AM 500 | 27 | | |
| A 400 | 30 | | | E 400 | 27 |
| | | | | E 375 | 32 |
| | | AM 330 | 36 | E 330 | 34 |

dynamic viscosity of $400 \pm 25$ mPa · s in the ball drop viscometer according to Höppler (ball no. 4) at 20 °C (Table 7).

*Cochius Viscosity* [25]: The viscosity of the various cellulose types is measured in commonly used solvent mixtures:

A and AM types: butanol/ethylene glycol/toluene/ethanol in the following proportions $1:2:3:4$

E types: butanol/butyl acetate/toluene in the following proportions $3:4:5$

Dried CN is dissolved in varying concentrations depending on the type and the time which an air bubble requires to rise 500 mm between two calibrations in a 7 mm Cochius tube at 18 °C is measured in seconds. The Cochius seconds are converted to absolute viscosity units mPa · s by multiplying with the factor 3.64 mPa.

*Ball drop method according to ASTM* [49]: Dried CN is dissolved according to its viscosity stage in 12.5, 20.0, or 25.0% ethanol/toluene/ethyl acetate according to [48]. The drop time of the balls with a diameter between 1/4 and 1/16 in. at 25 °C is given in seconds or converted into Pa · s. Figure 6 shows the relationship between the degree of polymerization and the technical viscosity (fall velocity of the balls in a 17.2% CN solution in acetone).

Comparative viscosity charts for converting the various viscosities and comparing the various types are found in [25].

**Solubility and Color.** The color and cloudiness of solutions produced according to [48] are tested visually. Consistency, appearance, and depth of color can be controlled according to [50].

*Dilution with Toluene.* Toluene is added to a 12.2% CN solution in butyl acetate at 25 °C until CN continuously precipitates. The dilution factor is noted. The dilution

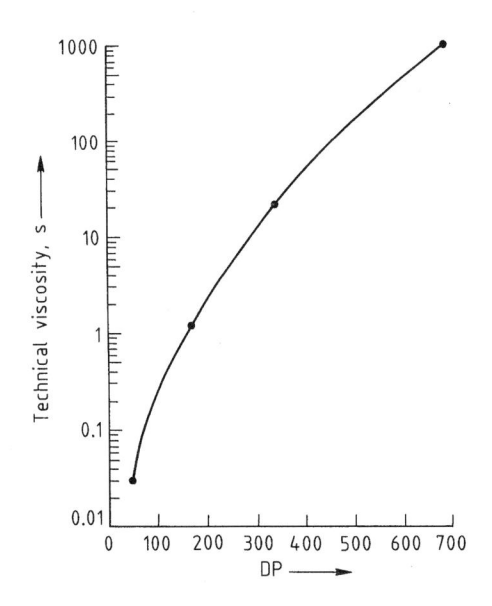

**Figure 6.** Degree of polymerization (DP) and technical viscosity ("ball drop" in seconds of a 12.2% CN solution with 12% N in acetone). DP = 170 × viscosity

ratio of CN solutions with other solvents and blending agents is determined according to [51].

**Film Test.** The solutions made according to [48] are diluted with an equal volume of butyl acetate and poured as a film onto a glass plate. The dried films are examined for undissolved particles, surface structure, transparency, and gloss.

## 1.2.7. Uses

**Explosives.** Explosives may be categorized according to their use:

blasting agents
propellants and shooting agents
detonating agents
igniting agents
pyrotechnical agents

Cellulose nitrates are used primarily as propellants and gun powder, whereby the following distinctions can be made: *monobasic powder*, which is based solely on cellulose nitrate; *dibasic powder*, which contains further energy carriers such as, for example, nitroglycerin or diglycol dinitrate in addition to cellulose nitrate; *tribasic powder*, which contains in addition to the components of the dibasic powder a third agent such as nitroguanidine.

The selection of the cellulose nitrate is of special importance. The types of cellulose nitrates that differ in the degree of nitration were standardized as follows:

| | | |
|---|---|---|
| CP I | (Collodium powder) also known as guncotton, nitrogen content: 13.3–13.5% |
| CP II | (Collodium) nitrogen content: 12.0– 12.7%, mostly 12.6% |
| PE | (Powder standard) nitrogen content: 11.5–12.0%, mostly 11.5% |

Aromatic amines, such as diphenylamine, are added to gunpowder as stabilizers. They are capable of binding the nitrous gases generated during the decomposition of the nitric acid ester. A mixture of ca. 80% highly nitrated gunpowder (13.4% N) and ca. 20% less-nitrated collodium cotton (12.5% N) is used for the production of the monobasic propellant powder. Since cellulose nitrate granules are easy to charge electrostatically, they are made conductive with a fine graphite coating.

The multibasic powders usually contain cellulose nitrate CP II. Mixtures of 40% PE cotton and 60% CP I are also used because they have the same energy content as CP II with 12.6% N.

The introduction of a third component to tribasic powder results in a lower heat of combustion in comparison to dibasic powder, thereby lengthening the life of the gun barrel.

Gunpowder is used in small-arms ammunition as well as large-caliber guns and tanks.

**Lacquers.** Cellulose nitrate lacquers are characterized by the outstanding film-forming properties of the physically drying cellulose nitrate. Moreover, cellulose nitrate is compatible with many other raw materials used in lacquers and can be used advantageously in combination with resins, softeners, pigments, and additives.

In addition to the nonvolatile lacquer components, the composition of the solvent mixture is decisive for the formation of a film.

The most important uses for nitro lacquers are as follows: wood lacquers (especially furniture lacquers), metal lacquers, paper lacquers, foil lacquers (also as hot sealing lacquers, e.g., cellophane, plastic, and metal foils), leather lacquers, adhesive cements, putties, and printing ink (for flexo and gravure printing).

The processes used for applying cellulose nitrate lacquers to substrates are as follows: spraying (compressed-air, airless, and electrostatic spraying), casting (for example, with a curtain coater), rolling (especially for the application of small amounts of lacquer), doctor knife coating, and dipping.

The casting and rolling processes are used for lacquering large, even areas. Irregularly shaped objects are sprayed. The choice of a suitable type of cellulose nitrate (e.g. completely or moderately soluble in alcohol, soluble in esters, degree of viscosity) is dependent on the lacquer type. A highly viscous cellulose nitrate type is used if elastic and thin applications are desired (e.g., leather). However, if hard and thick layers are desired, low-viscosity types are preferred.

The concentration or the degree of viscosity of the cellulose nitrate determine the viscosity of the lacquer solution. However, the formulation of the lacquer must be taken into consideration when the mode of application is chosen. For example, a highly viscous dipping lacquer cannot be sprayed or casted.

**Table 8.** CN lacquer cottons

| Ester-soluble type | Alcohol-soluble type |
|---|---|
| Possible use of alcohol in the lacquer formulation | Use of alcohol, especially ethanol, in any desired amount as a solvent |
| Good dilutability with aliphatic and aromatic hydrocarbons | Good dilutability with aromatic hydrocarbons |
| Very rapid solvent release | Rapid solvent release |
| Formation of hard films | Formation of films with thermoplastic properties |
| Attainment of good mechanical properties as far as the cold-check test, stretch, hardness, and tensile strength are concerned | Attainment of good mechanical properties; some special problems of lacquer production may be solved such as: Lacquers which can be diluted with ethanol in any desired manner (wood polishes) Odorless lacquers (printing inks) Gel dipping lacquers Hot sealing waxes (cellophane lacquers and aluminum foil lacquers) |

Furthermore, the striking differences between ester-soluble and alcohol-soluble types should be taken into consideration when nitro lacquers are formulated (Table 8).

For further information on the formulation of cellulose nitrate lacquers, see [52], [53].

**Dispersions.** Conventional cellulose nitrate lacquers contain between 60 and 90% organic solvents, which are released during drying. For economic and environmental reasons, it is desirable to substitute organic solvents by water. Aqueous cellulose nitrate/softener dispersions (e.g., Isoderm, Bayer AG; Coreal, BASF; Waloran N, Wolff Walsrode AG) are available for such absorbing substrates as leather [54]. Other aqueous cellulose nitrate dispersions for use on wood, foil, and metal have also been developed (Waloran N, special-types, Wolff Walsrode AG) [55]. The film forming process of water-insoluble cellulose nitrate requires a small amount of coalescents in the dispersion systems.

**Celluloid.** A special use of cellulose nitrate is in the production of celluloid [56]. Cellulose nitrate with a nitrogen content of 10.5–11.0% is mixed in a kneader with softeners, particularly camphor, and solvents (alcohols).

Normal celluloid contains ca. 25–30% camphor and 70–75% cellulose nitrate. Celluloid that contains 10–15% solvent can be formed into the desired articles in heated piston or screw presses (e.g., tubes and round and profile rods).

In the past decades, celluloid has been widely replaced by synthetic materials and thermoplastics. Celluloid is still of economic importance in the following areas: combs and hair ornaments, toilet articles, office supplies (drafting and measuring instruments), ping-pong balls, and various special uses.

## 1.2.8. Legal Provisions

**Toxicology and Industrial Safety.** Concentrated sulfuric acid, nitric acid, and nitrous gases formed during the production of cellulose nitrate are considered hazardous chemical products [57]:

1) Sulfuric acid
   5–15% EC-No. 016-020-01-5
   above 15 % EC-No. 016-020-00-8
2) Nitric acid
   20–70% EC-No. 007-004-01-9
   above 70% EC-No. 007-004-00-1
3) Nitrous gases
   EC-No. 007-002-00-0

They are subjected to the Arbeitsstoffverordnung (working substance regulation) [58] and must, therefore, be adequately labeled.

Concentrated nitric acid and mixed nitrating acids are oxidizing when brought into contact with organic materials [59]. The MAK values (maximum working place concentration) are as follows:

nitric acid vapors:   10 mL/m$^3$ (ppm);  25 mg/m$^3$
nitrogen oxides (NO$_2$): 5 mL/m$^3$  (ppm);  9 mg/m$^3$

Employees should be examined regularly for obstructive respiratory tract illnesses.

Cellulose nitrate is neither toxic nor hazardous to health [60]. Damping agents in CN and nitrous gases which may be formed during combustion and smoldering processes are potentially hazardous to health if inhaled.

Commercially available phlegmatized cellulose nitrate for the production of lacquer with less than 12.6% N contains at least 18% of a gelatinizing softener. According to the first paragraph in [58], cellulose nitrate is a hazardous substance and must be packaged and labeled accordingly. EEC regulations (1982) are similar.

Damping agents such as ethanol and 2-propanol are not subjected to these regulations; butanol belongs to category II d, but is not considered to be hazardous to health in a damped mixture of a maximum 35% concentration.

**Storage and Shipping.** Cellulose nitrate, especially guncotton, burns in air with a yellow flame and deflagrates if present in larger quantities, especially after rapid heating. An explosion can be caused by friction or a sharp impact. Dry CN has electrostatic charge. Friction, particularly on metals but also on plastics, can cause sparks which lead to a deflagration. Therefore, cellulose nitrate should be stored in a moist and cool place [60], [61]. Rooms in which cellulose nitrate is processed must be adequately protected according to the guidelines for protection from explosions.

Cellulose nitrate is subjected to the regulations governing explosives [62]. The transportation of phlegmatized cellulose nitrates proceeds according to the most recent versions of the hazardous materials regulation; see [25]. Wetted cellulose nitrate is shipped in thick-walled, galvanized, tightly closing iron or fiber drums which are adequately labeled.

Dried cellulose nitrate may not be shipped under any circumstances.

For further information on the properties, handling, storage, and transportation of hazardous goods, see also [63].

## 1.3. Other Inorganic Cellulose Esters

Summaries on the esterification products of cellulose with other inorganic acids may found in [12] – [19]. For recent publications from Soviet authors on the modification of cellulose, including esterification, see [64].

### 1.3.1. Cellulose Sulfates

Cellulose sulfates [9032-43-3] are the most frequently investigated of all other inorganic cellulose esters. The ability of concentrated sulfuric acid to dissolve cellulose, particularly in concentrations between 70 and 75%, has been known since 1819. After precipitation immediately following dissolution, the cellulose contains little or no bound sulfate. An almost homogeneous esterification takes place only if the cellulose is left in a sulfuric acid solution over a longer period of time. However, the ester yield is very poor. The major portions of the reaction products consist of hydrolytically split decomposition products with a maximum degree of substitution of 1.5.

In their free acidic form, cellulose sulfates are fairly unstable and easily saponified. A semiester was developed in 1953 in the United States [65] in an esterification mixture consisting of 1 mol of cellulose with 20 – 30% water, 3.5 – 15 mol of sulfuric acid, 0.3 – 1.0 mol of a primary or secondary $C_3$ – $C_5$ alcohol, and an inert volatile organic solvent such as toluene or carbon tetrachloride (reaction temperatures between – 5 and – 10 °C). The product was soluble in hot or cold water, yielded relatively stable, clear, and highly viscous solutions, and was recommended for use as a thickener for aqueous systems (emulsion paints and printing inks, printing pastes for textiles, and food products), as well as for fat- and oil-proof finish, and as paper glue. This product, however, is of no economic importance.

Numerous attempts have been made to find improved preparation methods for water-soluble cellulose sulfates stable to saponification. All currently known reaction systems are summarized in a tabular overview [65]. The reaction of cellulose with sulfuric acid in organic solvents, especially in lower-mass aliphatic alcohols, gives by way of a heterogeneous reaction fibrous and water-soluble cellulose sulfates with a maximum DS value of 1. More highly substituted products are obtained by reaction with sulfuric acid/acetic acid anhydride (up to a DS value of 2.8) or esterification with chlorosulfuric acid in pyridine or formamide. The reaction with $SO_3$ only or in various organic systems yields trisubstituted products. The reaction mechanism may be described as the addition of the strongly electrophilic $SO_3$ to the hydroxyl groups with the succeeding disintegration of the intermediately formed oxonium ion.

$$\text{Cell-OH} + SO_3 \longrightarrow \left[ \text{Cell} - \overset{+}{\underset{}{O}} - SO_3^- \right] \longrightarrow \text{Cell-O-SO}_3^- + H^+$$

Completely water-soluble, highly viscous sodium cellulose sulfate semiesters are obtained in homogeneous systems by the reaction of cellulose nitrite [67]. The intermediate that is formed and dissolved, cellulose nitrite, is obtained in the $N_2O_4$/dimethylformamide system and is at the same time transesterified by the $SO_3$/DMF complex. Uniformly substituted cellulose sulfate with a range of DS values between 0.3 and 2.0 and solution viscosities up to 7000 mPa · s (in 1% solution) can be obtained by using this process [68]. Such transesterified products can be cross-linked by metal ions to form highly effective thickening agents for aqueous media [69].

Such processes have recently been further developed [66] and make interesting novel fields of application accessible as a result of the rheological and gel-forming properties of the Na cellulose sulfate semiester.

Mixed esters such as cellulose acetate sulfates, cellulose acetate butyrate sulfates [70], cellulose acetate propionate sulfate [71], and ethyl cellulose sulfates [70], [72] are described in the patent literature.

Being polyelectrolytes, cellulose sulfates form salts and have ion-exchanging properties; thus, they have been recommended for use as cation exchangers [64, p. 65], [73], [74].

## 1.3.2. Cellulose Phosphate and Cellulose Phosphite

Reaction of cellulose with aqueous phosphoric acid gives the following unstable addition compound: $3\ C_6H_{10}O_5 \cdot H_3PO_4$, from which the cellulose can be regenerated unchanged by reaction with water. Cellulose phosphates [*9015-14-9*] with a low phosphorus content are obtained by reacting cellulose or linters with phosphoric acid in an urea melt [75]. Higher phosphorus contents and a lower degradation rate of the cellulose may be obtained with excess urea at reduced reaction time (ca. 15 min) and at high temperature (ca. 140 °C). Water-soluble cellulose phosphate with a high degree of substitution may be obtained from a mixture of phosphoric acid and phosphorus pentoxide in an alcoholic medium [76].

Phosphorylated cellulose fibers show increased swelling after partial hydrolytic degradation and transfer into the alkali salt form and were, therefore, suggested for use as adsorbents [77].

Cellulose phosphates with a 17% phosphorus content (this represents about 3/4 of the maximal possible substitution of triphosphate with 23% phosphorus) were already produced in 1933 by reacting cellulose with a mixture consisting of concentrated sulfuric acid and phosphoric acid in the presence of a weakly acidic catalyst [78].

Cellulose phosphites [*37264-91-8*] and cellulose phosphonates may be prepared by transesterification with alkyl phosphites. All cellulose esters containing phosphorus have fire-retarding properties [78] and have attracted some interest due to their ion-exchanging effect [74], [79], but are not yet industrially used.

### 1.3.3. Cellulose Halogenides

Various preparative methods are suitable for the synthesis of halogenated cellulose derivatives [64, p. 64]. Halogenation can be carried out by transesterification of such cellulose esters as tosylate, nitrate, and sulfate with hydrohalic acids [80]. Nucleophilic substitution proceeds considerably faster in homogeneous systems than in heterogeneous aqueous systems.

The Finkelstein transesterification process of cellulose nitrate with sodium iodide in anhydrous acetone leads to deoxyiodo cellulose. The reaction of cellulose with thionyl chloride, $SOCl_2$, in the presence of pyridine produced a monosubstituted, but strongly decomposed and unstable, hydrogen chloride ester.

Halogenation of cellulose improves its water-resistant and flame-resistant properties. Slight fluorination increases oil resistance and lowers the soiling potential of cellulose textiles [64]. Commercial applications are not yet known.

### 1.3.4. Cellulose Borates

The preparation of cellulose borate succeeded by means of transesterification of methyl and *n*-propyl borate with cellulose [64, p. 7]. The products with a maximum DS value of 2.88 are, however, extremely sensitive to hydrolysis and alcoholysis.

### 1.3.5. Cellulose Titanate

Cellulose can be reacted to cellulose titanates in a heterogeneous reaction system by reacting it with titanium tetrachloride in DMF or with chlorinated anhydrides, chlorinated ester anhydrides, and esters of the hypothetical orthotitanic acid $Ti(OH)_4$ [81]. Ethyl trichlorotitanate has been shown to be the most reactive. Esters with 16% titanium content are possible.

Cellulose esters with a titanium content between 3 and 5% do not burn or smolder. They possess considerable hydrolytic stability in neutral and weakly alkaline media, but are easily hydrolyzed at a low pH.

## 1.3.6. Cellulose Nitrite

The nitrite of cellulose came of scientific and possibly practical interest as a cellulose derivative in 1974 [67]. It is obtained by reacting cellulose with nitrosyl compounds such as dinitrogen tetroxide, $N_2O_4$ (corresponding to nitrosyl nitrate), or nitrosyl chloride, NOCl, in dimethylformamide or dimethylacetamide as a proton acceptor and solvent for the resulting ester. The reaction proceeds in a homogeneous phase to the trinitrite.

$$\text{Cell-OH} + \underset{\underset{O}{\|}}{N}-X \rightarrow \left[\text{Cell-O}-\underset{\underset{O}{\|}}{\overset{\overset{H}{|}}{\underset{+}{N}}}X^-\right] \rightarrow \text{Cell-O}-\underset{\underset{O}{\|}}{N} + H^+X^-$$

$$X = -O-NO_2 \text{ or } -Cl$$

Cellulose nitrite is extremely sensitive to hydrolysis. Chain degradation to a DP of 200 (level-off DP) was observed to take place within 3 h in the presence of water. The scientific and preparative importance of cellulose nitrite is based on its high reactivity, which may be used to produce many other cellulose esters, also mixed esters, by transesterification in a homogeneous phase [82]. Transesterification to stable cellulose sulfates has already been mentioned [67]. In this manner, water-soluble cellulose nitrates with a DS value of 0.5–0.6 may also be obtained [83].

Cellulose solutions produced under cold conditions (up to ca. 5 °C) in a $N_2O_4$/DMF system are relatively stable to degradation and can be produced, depending on the DP of the cellulose, up to a concentration of 14%. The cellulose can be regenerated in an unaltered form to cellulose II, with the result that this process has already been considered as an alternative to the environmentally detrimental viscose process [84]. Not only an attempt was made to achieve good mechanical textile properties from the regenerated fibers, but also to recycle the expensive solvent. An economic solution to the competition with the viscose process has not yet been found.

## 1.3.7. Cellulose Xanthate

Cellulose xanthate [*9032-37-5*], an important intermediary molecule for the production of regenerated cellulose according to the viscose process must also be considered as an ester of an inorganic acid, namely the nonexistent thiol–thion carbonic acid.

$$S=C\begin{matrix}\diagup SH \\ \diagdown OH\end{matrix}$$

The O ester of this compound with organic residues is the xanthic acid and the appropriate salts. Sodium cellulose xanthate is obtained by reacting alkali cellulose with carbon disulfide, which dissolves in dilute sodium hydroxide to an orange-yellowish, highly viscous solution, the so-called viscose.

$$\text{Cell-O}^-\text{Na}^+ + \text{C}\begin{smallmatrix}\diagup\text{S}\\\diagdown\text{S}\end{smallmatrix} \longrightarrow \text{Cell-O-C}\begin{smallmatrix}\diagup\text{S}^-\\\diagdown\text{S}\end{smallmatrix}\text{Na}^+$$

The regenerated cellulose is precipitated in the form of fibers (rayon, cord, and rayon staple), foils (cellophane), or tubes in precipitation baths containing sulfuric acid and salts. About $4 \times 10^6$ t of regenerated cellulose is presently produced worldwide by using the viscose process.

## 2. Organic Esters

Cellulose can theoretically form an unlimited number of organic acid esters because of its anhydroglucose units with three reactive hydroxyl groups each. Industrial possibilities are, however, drastically limited by the complex nature of the cellulose molecule. Highly esterified organic esters are, therefore, only produced from a few aliphatic fatty acids with up to four carbon atoms.

Cellulose acetate [*9004-35-7*], cellulose acetate propionate [*9004-39-1*], and cellulose acetate butyrate [*9004-36-8*], which have been known for quite some time and are in large-scale production, are especially important esters. Formic acid esters are, because of their instability, of no industrial importance.

Only cellulose acetate phthalate [*9004-38-0*] has found limited use as a tablet coating; all the other described cellulose esters such as cellulose palmitate, cellulose stearate, esters of unsaturated acids such as crotonic acid, or esters of dicarboxylic acids are not used industrially.

**Historical Aspects.** Cellulose acetate was first synthesized by P. SCHUTZENBERGER in 1865 by heating cellulose and acetic acid under pressure, whereby a product of very low molecularity was obtained. In 1879, A. P. N. FRANCHIMONT added sulfuric acid to the esterification process, which remains to this day the most frequently used catalyst. The limited solubility of cellulose acetate in less-expensive solvents and poor compatibility with the then-known softeners was a considerable obstacle for its industrial use. The problem was solved in 1904 when F. D. MILES and A. EICHENGRÜN simultaneously succeeded in synthesizing an acetone-soluble secondary acetate by partially hydrolyzing a primary triacetate.

During World War I, the less-flammable airplane paints based on cellulose secondary acetate reached considerable importance as a replacement for nitrocellulose. At almost the same time, the manufacture of foils, films, synthetic silk, and plastic masses developed.

An especially high number of publications and patents were achieved between 1920 and 1935. Ultimately, only a few processes proved to be industrially useful, most of which are still used today.

Even though the technology of cellulose ester production is considered for the most part to be complete, research continues in the fields of rationalization and improvement of production methods, leading to products with greater uniformity and improved properties, as well as in the development of new fields of application, especially with the introduction of cellulose acetate propionate and cellulose acetate butyrate.

## 2.1. Cellulose Acetate

### 2.1.1. Chemistry of Cellulose Esterification

The esterification reaction of the primary and secondary hydroxyl groups of cellulose does not basically differ from that of other alcohols. The peculiarities lie in the macromolecular structure of the cellulose molecule. The splitting of the molecule chain competes with the catalyzed esterification, but can be fairly well controlled under appropriate conditions. The speed and completeness of the reaction is dependent on the quality of the cellulose, whereas the different reactivities of the primary and secondary hydroxyl groups [83], [86] have little influence on industrial processes.

Acids, acid chlorides, and acid anhydrides are possible *esterification reagents* for the three hydroxyl groups in each glucose unit.

Esterification with free acids, with the exception of formic acid [87], whose esters are not stable, requires such high temperatures and catalyst concentrations that only low molecular mass products are obtained. Acid chlorides in pyridine, however, were suggested for use in the production of esters from higher fatty acids (lauric acid, stearic acid, and palmitic acid), without ever having attained any industrial significance. Attempt was made to manufacture cellulose acetate industrially with acetyl chloride and catalysts, but the process proved to be useless.

All industrial processes in current use, therefore, are based on acetic acid anhydride as a reactant, whereby theoretically 3 mol of anhydride per unit of glucose are used and 3 mol of acetic acid are formed.

Attempts to use ketene, which could be directly accumulated without incurring acetic acid, which must again be processed, did not lead to any results [88].

Numerous *catalysts* were suggested to accelerate the reaction. Only sulfuric acid and perchloric acid are of any practical importance. Zinc chloride, which is required in large amounts of 0.5 – 1 part per part of cellulose, is no longer used today because of the high reprocessing costs. Other mineral acids, however, are not sufficiently acidic in the water-free acetic acid – acetic acid anhydride system. The catalytic effect of sulfuric acid is primarily in the rapid and quantitative formation of acidic cellulose – sulfuric acid

**Table 9.** Analytical values obtained from bleached linters according to

| | |
|---|---|
| α-Cellulose | 99.7% |
| β-Cellulose | 0.2% |
| γ-Cellulose | 0.1% |
| Carboxyl groups | <0.02% |
| Total ash | 0.02% |
| Degree of polymerization | 1000–7000 |

esters (sulfoesters) [89], which are substituted by acetyl groups as the reaction progresses and the temperature rises.

Due to the topochemical character of the reaction, soluble cellulose esters with low acetic acid content can be obtained only from the triester stage by way of hydrolysis. Incompletely esterized and insoluble derivatives are found in addition to the triester before completion of the reaction.

Cellulose esters with low acetic acid content are produced by subsequent *hydrolysis* in a homogeneous system by adding water or dilute acetic acid to destroy the excess anhydride and possibly by adding sulfuric acid for acceleration and then again split off under controlled conditions (temperature, water content, and time) a certain number of acetyl groups without further breakdown of the cellulose chain.

## 2.1.2. Raw Materials

**Cellulose.** The production of high-quality cellulose esters requires that special attention be paid to the selection of the starting materials. The cellulose bases generally consist of highly purified cotton linters with an α-cellulose content of over 99% and celluloses from wood pulp which contain between 90 and 97% α-cellulose.

After the long layered spinnable cotton has been freed of the cotton seed by ginning, the remaining shorter fibers on the seed pod are usually removed with two cuts before the seeds go to the oil presses for further processing. The first cut gives about 4% longer linters relative to the entire cotton flower, which are preferentially processed to medicinal cotton, felt, paper, etc. The second cut gives about 8% shorter layered linters, which are best suited for further chemical processing.

The raw linters undergo mechanical cleaning by means of screening, pressure boiling in a 3–5% sodium hydroxide solution, and finally acid–alkaline bleaching. Special care should be taken during drying, since local overdrying of cellulose, the water content of which should lie between 3 and 8%, impairs the reactivity considerably. Table 9 shows analytical values of good linters [90].

For a long time, cellulose from wood pulp could only be used for the manufacture of lower-quality cellulose esters because of the 90–95% α-cellulose content. Celluloses with an α-cellulose content of 96% have been available for about 20 years. Due to special processing techniques, they give cellulose esters comparable to those produced from linters as far as tensile strength, color, clarity of the solutions, and light stability as well as thermal stability are concerned.

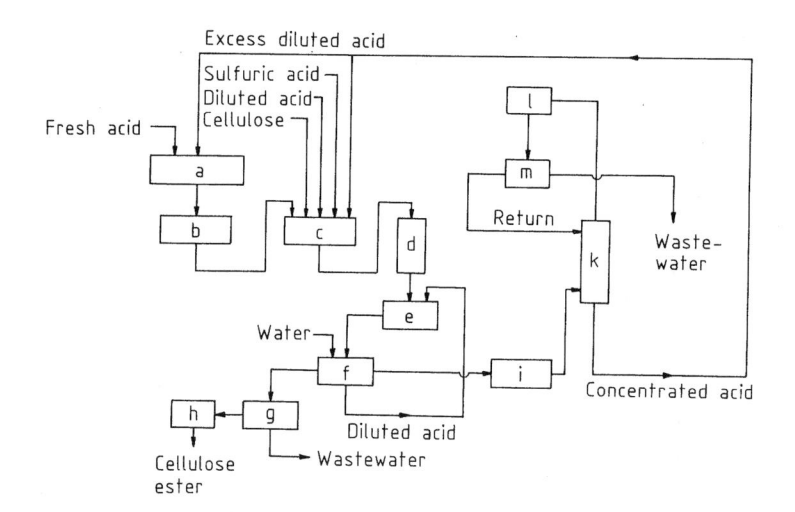

**Figure 7.** Flow chart for the production of cellulose esters according to [91]
a) Acid reconditioning; b) Acidanhydride; c) Esterification; d) Hydrolysis; e) Precipitation; f) Washing; g) Centrifuge; h) Drier; i) Evaporator; k) Azeotropic di-stillation; l) Cooler; m) Decanter

**Acetic Acid Anhydride.** Most manufacturers of cellulose acetate convert the resulting acetic acid to anhydride directly on the premises and adjust the concentrations as required for their process, generally between 90 and 95%.

## 2.1.3. Industrial Processes

Only a few of the proposed industrial processes for the manufacture of cellulose esters have attained industrial significance. Even when the fact that no two manufacturers use identical processes is taken into consideration, the following categories can be distinguished:

1) Acetylation in a homogeneous system (solution acetate)
   Use of glacial acetic acid as a solvent (glacial acetic acid process)
   Use of methylene chloride as a solvent (methylene chloride process)
2) Acetylation in a heterogeneous system (fiber acetate)

Whereas the triester that is formed during esterification according to the solution acetate process goes into solution and can subsequently be hydrolyzed to secondary acetate, fiber acetate is formed in the presence of nonsolvents, similar to the nitration of cellulose. This method does not permit hydrolysis.

A flow diagram of the entire process is shown in Figure 7.

## 2.1.3.1. Pretreatment

The cellulose is first dried to obtain a moisture content of 4–7%. Too little moisture would lower the reactivity of the cellulose, and too much moisture would lead to a higher acetic acid consumption and an extremely violent reaction start.

Acetic acid is generally used as a pretreatment reagent, which for some processes can contain small amounts of sulfuric acid to further improve the diffusion of the acetylating reagents. It is believed that the pretreatment process causes a partial swelling of the cellulose by splitting the hydrogen bonds so that the acetylating reagents can enter the fibers more rapidly.

A ratio of 30–100 parts of glacial acetic acid per 100 parts of cellulose, which is sprayed into the thoroughly mixed cellulose at temperatures of up to 50 °C in appropriate equipment such as mixing vats, is usually sufficient for the pretreatment process. Pretreatment, depending on the process and temperatures, takes from one to several hours.

## 2.1.3.2. Esterification

**Acetic Acid Process.** The acetylation mixture consists of glacial acetic acid as a solvent for triacetate, an excess of 10–40% glacial acetic acid anhydride and, depending on the process, 2–15% sulfuric acid with respect to the cellulose.

Esterification begins after the initial spontaneous reaction of the water contained in the cellulose with part of the anhydride; the semiliquid mass uniformly saturated with acetylation mixture forms a fiber pulp while developing temperatures of up to 50 °C and ultimately a highly viscous solution.

A decomposition of the cellulose chain takes place parallel to the strongly exothermic esterification. The chain degradation can be controlled by controlling the reaction temperature, e.g., by adding cooled solution portionwise and cooling the reaction vessel down to the desired temperature range. The use of glacial acetic acid as a solvent, in which the highly esterified triacetate is poorly dissolved, results in gel formation and precipitation of the triacetate. This is due to the formation of sulfate acetate, which is formed as an intermediate during the reaction. It is reesterified to pure acetic acid esters especially at higher temperatures.

After the reaction solution is free of fibers, the degradation of the cellulose ester molecule can continue until the desired viscosity is attained. The reaction is then stopped by adding water or dilute acetic acid, which destroys the excess anhydride.

Cooled kneaders are suitable reaction vessels in that they allow a rapid and uniform mixture and catalyst distribution through intensive mixing. This is important for controlling the reaction (Fig. 8).

**Methylene Chloride Process.** Using methylene chloride (*bp* 41 °C) as a solvent presents several advantages over acetic acid. Due to the fact that methylene chloride

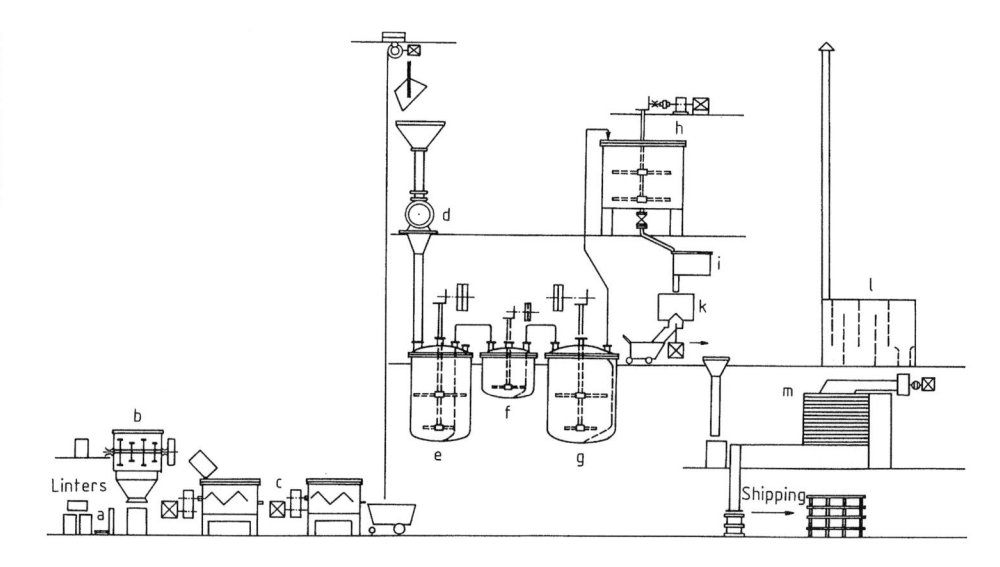

**Figure 8.** Cellulose acetate production by the kneader method according to [92]
a) Weighing scale; b) Sprinkling vat; c) Kneader; d) Mill; e) Rinsing vat; f) Stabilizing vat; g) Bleaching vat; h) Floater; i) Stock pan; k) Centrifuge; l) Dust chamber; m) Drier

is such an excellent solvent for primary triacetate, lower catalyst concentrations (1 % sulfuric acid) are required at higher esterification temperatures. Furthermore, due to its low boiling point, the heat of reaction can be removed by means of vaporization and return of the cooled methylene chloride. The reaction of highly viscous solutions can, thus, be better controlled. Finally, only a half to a third as much dilute acetic acid must be recycled compared with the glacial acetic acid process.

Table 10 shows typical examples for acetylation according to the glacial acetic acid and methylene chloride processes. Figure 9 shows a sceme for the production of cellulose acetate according to the methylene chloride process.

Acetylation according to the methylene chloride process is carried out largely in rotating drums (roll vats) or in horizontal containers with shovel-like stirrers on both sides. The problem of corrosion, which arises during esterification and especially during hydrolysis, has only been partially solved by using equipment constructed of bronze, high-alloy steels, or plates containing metals such as silver, titanium, or tantalum.

**Fiber Acetate.** Cellulose can be esterified maintaining its fiber structure by adding sufficient amounts of nonsolvents to the triacetate during acetylation. Carbon tetrachloride, benzene, or toluene can be used as nonsolvents [94], [95].

Temperature and catalyst concentrations are similar to those required for the solvation process. A large amount of liquid is required to keep the loose voluminous cellulose in suspension. Perchloric acid is preferred as a catalyst because of the great difficulty with which the sulfuric acid ester is split in this system.

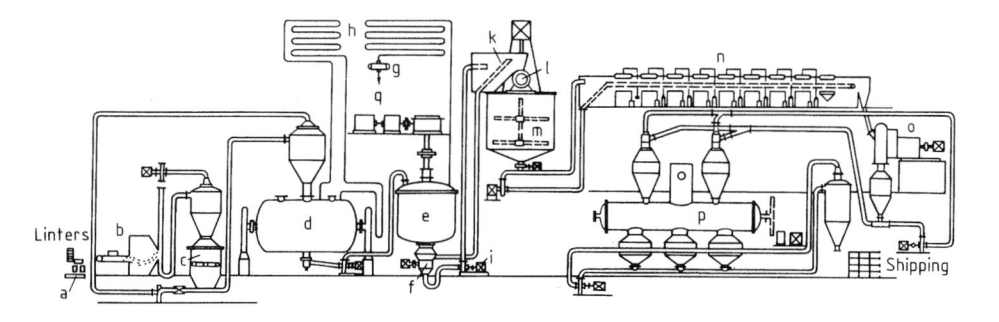

**Figure 9.** Cellulose acetate production according to the methylene chloride process [93]
a) Weighing scale; b) Bale opener; c) Sprinkling vat; d) Acetylator; e) Precipitating vat; f) Prebreaker; g) Vacuum vessel; h) Pipe cooler; i) Pump for viscous substances; k) Filter bath; l) Mill; m) Floater; n) Sprinkling line; o) Centrifuge; p) Vacuum shovel drier; q) To reprocessing of methylene chloride

**Table 10.** Acetylating preparations according to the glacial acetic acid process and the methylene chloride process

|  | Acetic acid process | Methylene chloride process |
|---|---|---|
| Cellulose | 700 kg | 3500 kg |
| Pretreatment | 700 kg glacial acetic acid | 1200 kg glacial acetic acid |
| Acetylation | 1900 kg anhydride | 10500 kg anhydride |
|  | 4800 kg glacial acetic acid | 14000 kg methylene chloride |
|  | 50 kg sulfuric acid | 35 kg sulfuric acid |

The hydrolysis of fiber acetate to acetone-soluble esters is not possible in a heterogeneous system. Its utilization is limited to special applications, such as the manufacture of foils and films from triacetate.

As shown in Figure 10, fiber acetate is produced by rotation in various directions and at various speeds in a perforated drum enclosed in a metal casing. The shaft of the drum is hollow so that liquid may be added during rotation [96].

**Continuous Processes.** In spite of numerous attempts, continuous acetylation and hydrolysis processes have achieved only limited industrial applications and are not used in the production of high-quality esters for, among others, plastic masses. Due to the varying reactivities of the different celluloses, considerable variations between batches are possible so that the development of a continuous process is seriously impeded.

### 2.1.3.3. Hydrolysis

After the esterification is completed, the process is interrupted by adding water or dilute acetic acid. Sufficient water must be added to decompose the excess anhydride and adjust the water content of the solution to 5–10% so that, aside from hydrolysis, no further decomposition of the molecular mass occurs, but at the same time the bound sulfuric acid is almost completely split off. The speed of hydrolysis depends on the

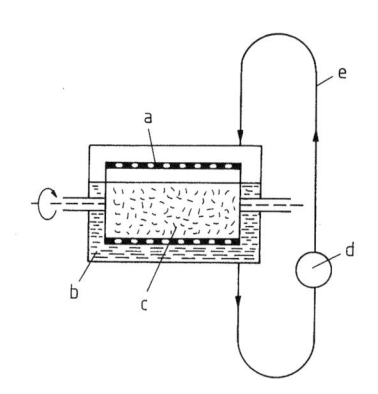

**Figure 10.** Acetylation equipment for cellulose triacetate fibers [92]
a) Perforated drum; b) Reaction solution; c) Cellulose fibers; d) Cooler for acetylating liquid; e) Acetylating liquid circulation

temperature, which – depending on the process – ranges between 40 and 80 °C, and on the amounts of sulfuric acid and water.

The course of hydrolysis is constantly monitored by checking the solubility of the secondary acetate. The reaction is terminated at the desired degree of substitution by neutralizing the catalyst, preferably with sodium acetate or magnesium acetate. The proportion of free hydroxyl groups in the ester should be maintained as reproducibly precise as possible because it determines the properties and the utilization of the ester.

### 2.1.3.4. Precipitation and Processing

The product precipitates in the form of flakes or powder during intense stirring after water or dilute acetic acid has been gradually added. The methylene chloride process requires that the methylene chloride be completely distilled before precipitation occurs. A part of the precipitating acid is added to the solution to just before the point of precipitation when the main portion of the acid is added. The process must be carefully monitored to achieve a product with an open structure which can be easily rinsed out. The precipitate can be broken down and thoroughly rinsed, whereby the dilute acid is brought back into the cycle. Rinsing is done primarily by using continuous methods based on the countercurrent principle (Fig. 9).

High-quality products for plastic masses are stabilized and bleached. Bound sulfuric acid remnants are removed by either boiling under pressure or heating in 1% mineral acids during stabilization.

After further rinsing and removal of excess water by suctioning or by thrust extraction, the product is carefully dried, preferably in a vacuum shovel drier, to a water content of $< 1-3\%$.

The cellulose acetate yield from a good process is at least 95% of the theoretical yield.

Major manufacturers of cellulose acetate are the following [103]: Bayer AG, Leverkusen; British Celanese Ltd., London; Celanese Corp. of America, New York; Courtaulds Ltd., Manchester; Daicel Ltd., Osaka, Japan; E. I. Du Pont de Nemours & Co., Inc., Wilmington; Eastman Kodak Comp., Kingsport; UCB Fabelta, Tubize, Belgium; Gevaert-Agfa, Antwerp; Hercules Powder Comp. Ltd., London; Rhodiaceta, Lyon; Soc. Rhodiatoce SpA, Milan; Rhône-Poulenc, Paris; VEB Orbitaplast, Eilenburg, GDR.

A number of products are further processed to fibers, films, or injection-molding compounds in the manufacturers' own facilities and never reach the market as a raw material.

## 2.1.4. Recovery of Reactants

The recovery of the incurred large amounts of acetic acid is a decisive factor for the profitability of a process.

**Recovery of Acetic Acid.** Depending on the process, 2 – 6 parts of 15 – 25 % dilute acetic acid per part of cellulose accumulate. They must be reprocessed to glacial acetic acid and acetic acid anhydride.

Only continuous processes consisting of a combination of extraction and azeotropic distillation are of current practical importance. The dilute acid is, for example, extracted with ethyl acetate in a countercurrent and is subsequently distilled, so that 99.8 % pure glacial acetic acid can be removed from the bottom of the column while the azeotropic ethyl acetate water is removed from the top.

**Recovery of Acetic Acid Anhydride.** Since only a portion of the accumulated glacial acetic acid is required for the acetylation process, the remainder must be converted to glacial acetic acid anhydride.

The ketene process developed by the Wacker Co. (→ Acetic Anhydride) is currently in general use [97]: Pure, almost anhydrous glacial acetic acid is continuously vaporized under a vacuum and is split to ketene in the presence of small amounts of the catalyst triethyl phosphate; ketene then reacts with glacial acetic acid to form the anhydride.

**Recovery of Methylene Chloride.** Methylene chloride can be recovered inexpensively because of its insolubility in water, which alllows its recovery from the raw solution without further processing steps in almost pure form.

## 2.1.5. Properties of Cellulose Acetate

Cellulose acetate and the other fatty acid esters are white, amorphous products that are commercially available as a powder or flakes. They are nontoxic, odorless, tasteless, and less flammable than nitrocellulose. They are resistant to weak acids and are largely stable to mineral and fatty oils as well as petroleum.

Some physical characteristics are given in Table 11.

Properties and applications of cellulose acetates are primarily determined by the following:

**Table 11.** Physical characteristics of cellulose acetate

| Characteristic | Triacetate | Secondary acetate |
|---|---|---|
| Density, g/cm$^3$ | 1.27 – 1.29 | 1.28 – 1.32 |
| Thermal stability, °C | > 240 | ca. 230 |
| Tensile strength of fibers, kg/mm$^2$ | 14 – 25 | 16 – 18 |
| Tensile strength of foils | | |
|    longitudinal, kg/mm$^2$ | 12 – 14 | 8.5 – 10 |
|    transverse, kg/mm$^2$ | 10 – 12 | 8.5 – 10 |
| Refractive index of fibers toward the fiber axis | | |
|    longitudinal | 1.469 | 1.478 |
|    transverse | 1.472 | 1.473 |
| Double refraction | −0.003 | +0.005 |
| Dielectric constant $\varepsilon$ | | |
|    50 – 60 Hz | 3.0 – 4.5 | 4.5 – 6.5 |
|    $10^6$ Hz | | 4.0 – 5.5 |
| Dielectric loss factor tan $\delta$ | | |
|    50 – 60 Hz | 0.01 – 0.02 | 0.007 |
|    $10^6$ Hz | | 0.026 |
| Specific resistance, $\Omega \cdot$ cm | $10^{13} - 10^{15}$ | $10^{11} - 10^{13}$ |
| Specific heat, J g$^{-1}$ K$^{-1}$ | | 1.46 – 1.88 |
| Thermal conduction, J m$^{-1}$ h$^{-1}$ K$^{-1}$ | | 0.63 – 1.25 |

1) viscosity of their solution
2) degree of esterification or the amount of bound acetic acid

*Viscosity* as an indicator for the degree of polymerization influences to a great extent the mechanical properties of the resulting fibers, films, or plastic masses, as well as their workability.

The *degree of esterification* primarily determines the solubility and compatibility with softeners, resin, varnish, etc., and ultimately also influences the mechanical properties.

The wide span of solubility properties of hydrolyzed cellulose acetate is shown in Table 12.

The compatibility of the plasticizer and the solubility in polar solvents increase with decreasing acetic acid content while the solubility in nonpolar solvents decreases. Moreover, a correlation between the incompatibility with nonsolvents such as water, alcohol, benzene, or toluene and a decreasing degree of esterification exists. Furthermore, a number of solvent combinations are known which are able to dissolve the cellulose acetate although each of the components is a nonsolvent.

Figure 11 shows a selection of solvents for the industrially interesting range of esterification (52 – 62% bound acetic acid). Detailed information on solubility and softener selection can be found in the literature and the manufacturers' information brochures.

**Table 12.** Solubility of cellulose acetate at various degrees of esterification

| Degree of esterification | Bound acetic acid | Chloroform | Acetone | 2-Methoxyethanol | Water |
|---|---|---|---|---|---|
| 2.8–3.0 | 60–62.5% | soluble | | | |
| 2.2–2.7 | 51–59% | | soluble | | |
| 1.2–1.8 | 31–45% | | | soluble | |
| 0.6–0.9 | 18–26% | | | | soluble |
| <0.6 | <18% | | | | |

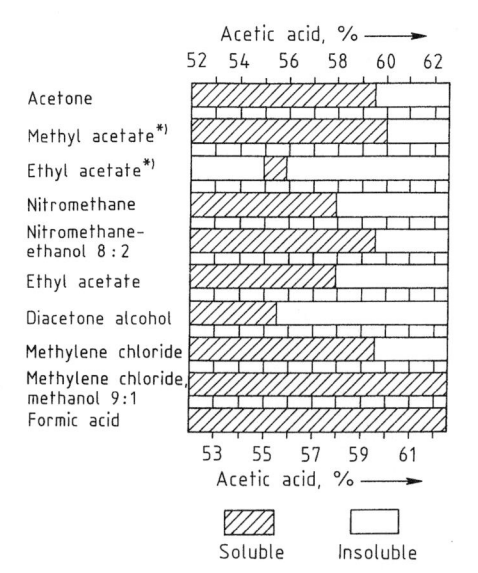

**Figure 11.** Solubility of cellulose acetate in various solvents (abridged according to [99])
\* Technical grade

## 2.1.6. Analysis and Quality Control

The *viscosity* is determined in practice by the usual methods. Along with the relative viscosity, the measurement of 15–20% solutions according to the ball drop method corresponding to ASTM 871-56 has been generally accepted.

The *acetic acid content* is generally determined by saponification of the ester with 0.5 N-potassium hydroxide and back-titration of the excess. The free acids from mixed esters are separated from the residue after saponification by means of distillation or extraction and by extraction or azeotropic distillation isolated and titrated. Recently, gas-chromatographic methods have found increasing popularity. A comprehensive presentation can be found in [100], [101].

Determination of the *free hydroxyl groups* in pure cellulose acetate is not necessary, since a precise analysis of bound acetic acid is possible. It is primarily used to characterize the mixed esters and can be carried out according to various methods. Complete esterification with acetic acid anhydride in pyridine and back-titration of the excess is a proven method [102]. The value is mostly given as a percentage of the hydroxyl or as the hydroxyl value (mg of KOH/g).

Additional quality control methods for unprocessed cellulose esters are the following: determination of temperature stability by heating to 220–240 °C and evaluation of discoloration and melting behavior, the determination of free acid as an indicator for the efficacy of the rinsing process, and determination of the ash content as well as clarity, color, and filterability of the solution.

## 2.2. Cellulose Mixed Esters

Apart from the long-established cellulose acetate, only cellulose mixed esters of acetic and propionic acid, cellulose acetate propionate [*9004-39-1*], or acetic and butyric acid, cellulose acetate butyrate [*9004-36-8*], have attained any notable importance.

Mention should also be made of cellulose acetate phthalate [*9004-38-0*], which is used in certain special fields of application.

Pure cellulose propionates [*9004-48-2*] and cellulose butyrates [*9015-12-7*] are difficult to produce and – like formates, palmitates and stearates, and esters of unsaturated acids and dicarboxylic acids – have attained no industrial importance [4].

### 2.2.1. Production

As far as the chemistry of the esterification reaction and the subsequent partial saponification are concerned, the basic description given in the chapter on cellulose acetate is also valid for mixed acids to a large extent.

The raw materials used are the same as for cellulose acetate, i.e., cotton linters or cellulose produced by special processes.

Pretreatment of the cellulose raw materials is similar to that used in the production of cellulose acetate. In practice, esterification takes place only in a homogeneous system and not, as is sometimes the case with cellulose acetate, in a heterogeneous system. The esterification mix consists of a mixture of anhydrides of acetic acid and propionic acid or of acetic acid and butyric acid. The reactivity of the aliphatic fatty acids decreases very rapidly as the chain-length increases.

Mixed esters consisting of propionic acid and butyric acid or of acetic acid, propionic acid, and butyric acid are not produced on an industrial scale.

### 2.2.2. Composition

The properties of the mixed esters are determined not only by their viscosity, but also in particular by the ratio of the two bound acids and by the content of free hydroxyl groups.

**Table 13.** Characteristic data of cellulose acetate propionate (Cellit PR)

|  | PR 900 | PR 800 | PR 500 |
|---|---|---|---|
| Acetyl content, % | 3.5 | 3.5 | 3.5 |
| Propionyl content, % | 45 | 45 | 45 |
| Hydroxyl content, % | 1.6 | 1.6 | 1.6 |
| Viscosity (DIN 53 015), mPa · s | 4700–7800 | 2200–3800 | 150–240 |
| Melting range, °C | 200–220 | 190–210 | 180–200 |

At present, cellulose acetate propionate and cellulose acetate butyrate flake is only produced by two manufacturers: Bayer AG, Leverkusen, in the Federal Republic of Germany, and the Eastman Kodak Co., Kingsport, Tennessee, in the United States.

Whereas pure cellulose acetates are clearly characterized by their viscosity and content of bound acetic acid, with mixed esters, data on the individual acids and possibly the free hydroxyl groups are also required.

## 2.2.3. Properties

As the degree of hydrolysis changes, the properties of cellulose mixed esters vary over a wide range from pure acetates to pure butyrates, with the propionates occupying a property-profile position between the cellulose mixed esters and pure acetates.

In the case of cellulose acetate butyrate, for example, if one considers the mixing range from pure cellulose acetate through the various mixing ratios to pure cellulose butyrate (the degree of esterification is adjusted in such a way that the esters in pure cellulose acetate contain 50–60% acetic acid, and those in pure butyrate 60–70% butyric acid), then the density varies from ca. 1.32 (cellulose acetate) to 1.16 (cellulose butyrate). The melting point is between ca. 300 °C (cellulose acetate) and 160 °C (cellulose butyrate), while the water absorption at 90% relative humidity varies from ca. 12% (cellulose acetate) to 1.5% (cellulose butyrate).

Solubility in various solvents (like acetone, methyl ethyl ketone, methyl isobutyl ketone, and various phthalate plasticizers) also varies over a wide range [104].

Typical examples of currently available cellulose acetate propionates, whose acetic acid and propionic acid levels are practically identical, are given in Table 13.

With cellulose acetate propionates, the acetic acid content varies from ca. 3 to 8%, and the propionic acid content from ca. 55 to 62%. With cellulose acetate butyrates, acetic acid contents of 19–23% and butyric acid contents between 43 and 47% are normal.

There are many more grades of cellulose acetate butyrate available than of cellulose acetate propionate.

Examples of cellulose butyrate grades are given in Table 14.

Table 14. Characteristic data of cellulose acetate butyrate (Cellit BP)

|  | BP 300 | BP 500 | BP 700/25 | BP 700/40 | BP 900 |
|---|---|---|---|---|---|
| Acetyl content, % | 14 | 14 | 15 | 14 | 15 |
| Butyryl content, % | 37 | 37 | 37 | 37 | 37 |
| Hydroxyl content, % | 1.2 | 1.2 | 0.8 | 1.2 | 0.8 |
| Viscosity (DIN 53015), mPa · s | 30–60 | 150–240 | 750–1500 | 750–1500 | 5000–8000 |
| Melting range, °C | 160–180 | 170–190 | 170–190 | 170–190 | 180–200 |

## 2.2.4. Other Organic Cellulose Mixed Esters

Cellulose acetate phthalates are cellulose mixed esters of minor industrial importance. They are produced from hydrolyzed cellulose acetate by reaction with an excess of phthalic anhydride in acetone or dioxane [106].

This produces esters of phthalic acid with a free carboxyl group. The products are used on a small scale as water- or alkali-soluble textile auxiliaries, tablet coatings, and antistatic agents in film coating.

## 2.2.5. Uses

The wide scope of variation for cellulose esters has led to the development of special grades for different fields of application.

Mixed esters based on cellulose acetate butyrate and cellulose acetate propionate are chiefly used in the production of molding plastics (see Section 2.4). The applications described in the following are of lesser importance.

**Films.** Triacetate is more widely used than cellulose acetate butyrate as an electrical insulating film, which is mainly produced by casting. Cellulose acetate butyrates with an acetic acid/butyric acid ratio between 2:1 and 1:1 are preferred.

The importance of cellulose mixed esters as film substrates in the photographic industry has greatly diminished as a result of the increasing use of polyester film.

**Surface coatings.** Cellulose ester lacquers, with their excellent lightfastness, gloss, low combustibility, and good thermal stability, coupled with their indifference to hydrocarbons, oils, and greases, became very quickly established in numerous fields of application.

One significant step forward was the introduction of cellulose acetate butyrates and cellulose acetate propionates. Both are particularly characterized by lower water absorption and good compatibility with extenders and, in the case of the low-viscosity grades, also allow the production of very high-solid lacquers. Recently, cellulose acetate butyrate has become particularly important as an extender for metal effect finishes [105].

**Table 15.** Physical properties of acetate fibers and tow

|  | Secondary acetate | Triacetate |
|---|---|---|
| Strength, cN/dtex | 1.0 – 1.5 | 1.0 – 1.5 |
| Stretch, % | 25 – 30 | 25 – 30 |
| Density, g/cm$^3$ | 1.33 | 1.30 |
| Moisture uptake, % (65 % relative humidity, 20 °C) | 6 – 6.5 | 4 – 4.5 |
| Water retention capability, % | 25 – 28 | 16 – 17 |
| Melting point, °C | 225 – 250 | decomposition at 310 – 315 |
| DP | 300 | 300 |

## 2.3. Cellulose Acetate Fibers

Cellulose acetate is the most important cellulose ester. It is primarily used for textile yarn and cigarette filter tow. The cellulose acetate is usually dissolved in a suitable organic solvent and spun by dry spinning. Secondary (2.5) acetate with an acetic acid content of 54 – 56 % is normally used, whereas only a small amount of cellulose triacetate is normally produced.

### 2.3.1. Properties

The viscosity and the filterability of the spinning solution (spinning dope) are particularly important in the production of cellulose acetate fibers. The spinning dope has a high viscosity, which depends on the degree of polymerization. The strength and stretch properties of the fibers also depend on the degree of polymerization as well as on the distribution of the acetate groups along the cellulose chain.

Because the fibers are produced by extruding the spinning dope through minute spinnerette holes, insoluble particles must first be removed from the spinning dope by filtration. These particles are primarily composed of very small, incompletely acetylated cellulose fibers, which will obstruct the spinnerett holes.

Secondary acetate and triacetate fibers have similar physical properties (Table 15). Their densities are lower than that of viscose rayon fibers and equal to that of wool. For textile yarns, the fibers should be as free of color as possible.

The chemical reactions of cellulose acetate are similar to those of organic esters. Cellulose acetate is hydrolyzed by strong acids and alkali; it is sensitive to strong oxidizing agents but not affected by hypochlirite or peroxide solutions.

Acetate fibers cannot be dyed under the same conditions as viscose rayon fibers because their swelling properties are different. Acetate fibers can only be dyed with water-disperse dyes at the boiling point of the medium usually in the presence of carriers. The carriers promote fiber swelling and enhance dye uptake by the fibers. The dyeing process coupled with the textile spinning operation assures color fastness. Triacetate fibers have better wash-and-wear properties than secondary acetate because of better dimensional stability and higher crease resistance.

Table 16. Typical properties of acetate wood pulps

| Characteristic* | Sulfite softwood pulp (conifer) | Sulfate hardwood pulp (deciduous) |
|---|---|---|
| $R_{10}$, % | 95 | 96 |
| $R_{18}$, % | 97 | 98 |
| Ash, % | 0.08 | 0.08 |
| Silica, % | 0.001 | 0.003 |
| Calcium, % | 0.006 | 0.008 |
| Pentosans, % | 1.2 | 1.2 |
| Moisture content, % | 6.5 | 6.5 |
| Apparent density, g/cm$^3$ | 0.45 | 0.5 |
| DP | 2300 | 1700 |

\* $R_{10}$ and $R_{18}$ are residues in 10 or 18% sodium hydroxide at 20 °C [108]

### 2.3.2. Raw Materials

Wood pulp produced from various softwood (conifer) or hardwood (deciduous) species is the cellulose source for the production of cellulose acetate fibers. The wood pulps are produced by the sulfite pulping process with hot alkali extraction or by the prehyrolized sulfate (Kraft) process with cold caustic extraction. The lignins and hemicelluloses are removed from the wood to give wood pulps with an α-cellulose content of over 96% (Table 16). High-purity cotton linters are no longer used in the production of cellulose acetate fibers for economic reasons.

For the production of high-quality cellulose acetate fibers the wood pulp must have (1) good swelling properties for uniform accessibility of the cellulose to the catalyst and the acetylation agent, (2) a uniform reactivity, and (3) a moderate acetylation rate for convenient process control. In addition, it must produce a fiber-free spinning solution which can easily be filtered.

### 2.3.3. Production

The general points discussed in Section 2.3.1 for the production of cellulose acetate also apply here. The sulfuric acid catalyst initially forms the cellulose sulfate ester. The sulfate groups are then replaced by acetyl groups as the acetylation proceeds. The sulfate ester contents is further reduced in the hydrolysis stage. However, any sulfate ester groups remaining at the end of the hydrolysis stage must be neutralized with an appropriate stabilizer, e.g., magnesium salts [109], [110]. Any "free" sulfate ester groups will affect the stability of the acetate because under the influence of heat and humidity they splitt off as sulfuric acid and degrade the fiber [111].

For secondary acetate spinning, acetone is used as the solvent. For triacetate, the solvent is 90% dichloromethane and 10% methanol or acetic acid (wet-spinning process). The viscosity of the spinning solution with a cellulose acetate concentration of 20–30% is between 300 and 500 Pa s at 45–55 °C. The spinning dope is filtered in

**Table 17.** Worldwide production of textile fibers and filter tow (1000 t)

| Fiber | 1983 | 1973 |
|---|---|---|
| Cotton | 14525 | 13718 |
| Wool | 1606 | 1432 |
| Silk | 55 | 43 |
| Full synthetic yarns | 11102 | 7640 |
| Cellulose yarns (rayon) | 764 | 964 |
| Acetate and triacetate fibers and filter tow | 608 | 610 |

several steps at 0.5 – 1.5 MPa and is then deaerated in large vessels for 24 h at constant temperature.

Dry spinning is used almost exclusively; wet spinning is occasionally used for triacetate only. The spinneretts for textile filament have between 20 and 100 holes and those for tow up to 1000. The fibers are formed by evaporating the solvent with a countercurrent of air at 80 – 100 °C in a 4- to 6-m spinning column. The fibers are then stretched while still plastic to increase their strength.

The spinning speeds range between 300 and 800 m/s depending on the titer (2 – 6 dtex). A core-skin structure is formed in triacetate fibers. The acetyl groups are distributed very regularly in cellulose triacetate compared to secondary acetate; therefore, crystallization occurs when triacetate fibers are heated at 180 – 200 °C (heat setting) [107], [111], [112]. This heat treatment, which enhances the wash-and-wear properties of triacetate textiles, requires several minutes at 180 °C or several seconds at 220 °C. Heating for shorter periods is not effective and longer heating periods lead to deterioration of the mechanical properties of the textile. Heat-setting reduces water retention to 10% and water absorption to 2.5%.

## 2.3.4. Economic Aspects

Secondary acetate and triacetate fibers for textiles and filter cigarette tow accout for 80% of all cellulose ester production. The balance is used for plastics and film. Secondary acetate and triacetate textile fibers have a small share (about 1%) of all textile fiber production. In recent years synthetic fibers have continued to expand, whereas the acetate and triacetate fiber production has declined slightly. The growth in cigarette filter tow production has offset this decline and the total acetate and triacetate fiber production has remained essentially unchanged (Table 17).

The weaker physical properties of acetate fibers and their higher production cost (almost twice as high than those of synthetic fibers) have contributed to the decline in acetate textile fiber production. Synthetic fibers are stronger, give better dimensional stability in textiles, and have better wash-and-wear properties. Acetate fibers have an advantage in comfort due to greater moisture absorption and, as a result, a major use of acetate textile fibers continues to be in coat linings.

## 2.4. Plastic Molding Compounds from Cellulose Esters

In the category "plastics made from natural materials," thermoplastics based on cellulose esters or cellulose mixed esters are still the most important [114].

As early as 1920, A. EICHENGRÜN developed thermoplastic cellulose ester molding compounds as a spraying and molding powder. Cellulose acetate and mixed esters are used in injection molding and extrusion; mixed esters are also used for fluidized-bed dip coating and rotational molding.

The use of inorganic cellulose esters (see 1.2.7) is continually decreasing in the plastics sector because of their high flammability.

### 2.4.1. Physical Properties

Like cellulose acetate, cellulose mixed esters can be plasticized at elevated temperatures by using plasticizers. This results in a large number of grades with property combinations found in no other type of thermoplastic.

Thermoplastic cellulose ester molding compounds are generally characterized by good transparency, high mechanical strength, and toughness; one particularly noteworthy feature is that the material reacts to mechanical stresses by exhibiting cold flow, so that the otherwise problematic insert molding of metal parts presents no problems and there is no risk of stress cracking. Light-stable, transparent material is available in a wide range of transparent, translucent, and opaque color shades. High surface gloss coupled with antistatic properties (i.e., electrical charges disperse rapidly and no annoying dust patterns form) ensures that moldings retain their attractive appearance for years. High surface elasticity ensures a good "natural feel" and imparts a "repolishing" effect to the material: this means that scratches disappear as the object is being used. The relatively low modulus of elasticity gives excellent damping of vibrations, so that the acoustic behavior is not affected by annoying resonance or ambient noise.

The individual cellulose esters generally differ in their mechanical properties and in their compatibility with plasticizers. As a rule, cellulose mixed esters contain higher-boiling plasticizers in amounts ranging from 3 to 25% whereas cellulose acetate contains 15–35% plasticizers. Heat distortion temperature increases as the plasticizer content is reduced. Mixed esters absorb considerably less water than cellulose acetates, with the result that parts produced from mixed esters retain their dimensional stability even in humid climates. Finally, cellulose acetate butyrates and (with certain restrictions) cellulose acetate propionates can also be treated with UV inhibitors to ensure serviceability of the moldings even during years of outdoor exposure [115]. In principle, it is possible to reinforce thermoplastic cellulose ester molding compounds with glass fibers [116].

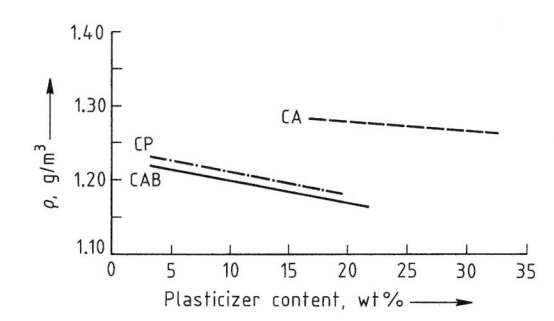

**Figure 12.** Density of cellulose acetate (CA), cellulose acetate propionate (CP), and cellulose acetate butyrate (CAB) as a function of the plasticizer content (determined in accordance with DIN 53479 or ISO/R 1183)

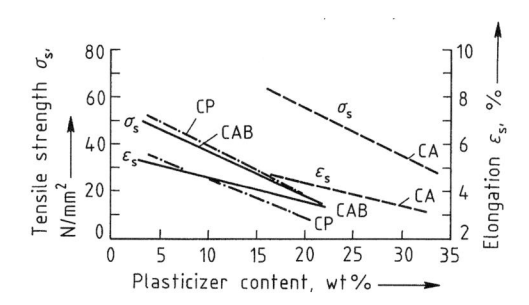

**Figure 13.** Tensile strength at yield $\sigma_s$ and elongation $\varepsilon_s$ of cellulose acetate, cellulose acetate propionate, and cellulose acetate butyrate as a function of the plasticizer content (determined in accordance with DIN 53455 or ISO/R 527; specimen no. 3, rate of deformation 25 mm/min)

Figures 12, 13, 14, 15, 16, 17, 18, 19, 20, 21, 22 show the physical properties as a function of the plasticizer content; the property levels may vary by as much as 15–20% in either direction, depending on the type of plasticizer and the relative viscosity of the cellulose ester. In all of the diagrams and tables featured here, the abbreviations used are as follows:

CA    = Cellulose acetate molding compound (acetic acid content > 55%)
CP    = Cellulose acetate propionate molding compound
CP*  = Polymer-modified cellulose acetate propionate molding compound
CAB  = Cellulose acetate butyrate molding compound
CAB* = Polymer-modified cellulose acetate butyrate molding compound

The indices (e.g., $CAB_{10}$) give the plasticizer content in percent by weight.

Table 18 shows the electrical properties of medium-hardness cellulose ester molding compounds; their shear moduli and damping properties are given in Figure 23. Figure 24 shows the position of the damping maxima as a function of the plasticizer content.

Long-term properties derived from the tensile creep test are shown in Figures 25, 26, 27, 28, 29. Time-to-failure curves, modulus of creep curves and isochronous stress–strain curves of slightly and highly plasticized grades of cellulose acetates, cellulose acetate propionates, and cellulose acetate butyrates are given here.

Figure 30 shows results of the dynamic fatigue test in the tensile pulsating range on a medium-plasticity cellulose acetate, a slightly plasticized cellulose acetate propionate, and a medium-plasticity cellulose acetate butyrate.

**Table 18.** Electrical properties of organic cellulose ester molding compounds

| Type of test | Unit | Test specification | Specimen | Cellulose acetate [129] | Cellulose acetate propionate [118] | Cellulose acetate butyrate [118] |
|---|---|---|---|---|---|---|
| Dielectric strength Ed (50 Hz, 0.5 kV/s) | | | | | | |
| dry | | VDE 0303 | Circular | 315 | 355 | 350 |
| 4 days at 80% rel. humidity | kV/cm | Pt. 2, | discs 95 mm | 290 | 330 | 330 |
| 24 h water immersion | | DIN 53481 | × 1 mm | 280 | 330 | 330 |
| Surface resistance $R_0$ | | | | | | |
| dry | | VDE 0303 | | $8 \times 10^{13}$ | $2 \times 10^{14}$ | $9 \times 10^{13}$ |
| 4 days at 80% rel. humidity | $\Omega$ | Pt. 3, | 150 × 15 × 4 mm | $3 \times 10^{12}$ | $1 \times 10^{13}$ | $9 \times 10^{12}$ |
| 24 h water immersion | | DIN 53482 | | $4 \times 10^{11}$ | $5 \times 10^{12}$ | $9 \times 10^{12}$ |
| Insulation resistance $R_a$ | | | | | | |
| dry | | VDE 0303 | | $5 \times 10^{15}$ | $5 \times 10^{15}$ | $5 \times 10^{15}$ |
| 4 days at 80% rel. humidity | $\Omega$ | Pt. 3, | 150 × 5 × 4 mm | $1 \times 10^{13}$ | $6 \times 10^{13}$ | $5 \times 10^{13}$ |
| 24 h water immersion | | DIN 53482 | | $7 \times 10^{11}$ | $2 \times 10^{13}$ | $2 \times 10^{13}$ |
| Volume resistivity $\varrho_D$ | | | | | | |
| dry | | VDE 0303 | Circular | $2 \times 10^{15}$ | $1 \times 10^{16}$ | $4 \times 10^{15}$ |
| 4 days at 80% rel. humidity | $\Omega$/cm | Pt. 3, | discs 95 mm | $2 \times 10^{12}$ | $5 \times 10^{13}$ | $6 \times 10^{13}$ |
| 24 h water immersion | | DIN 53482 | × 1 mm | $2 \times 10^{11}$ | $1 \times 10^{13}$ | $2 \times 10^{13}$ |
| Relative permittivity $\varepsilon_r$, dry | | | | | | |
| at 50 Hz | | VDE 0303 | Circular | 5.1 | 4.1 | 4.0 |
| at 800 Hz | | Pt. 4, | discs 95 and | 4.0 | 3.9 | 3.8 |
| at 1 MHz | | DIN 53483 | 80 mm × 1 mm | 4.1 | 3.6 | 3.4 |
| Dissipation factor tan $\delta$, dry | | | | | | |
| at 50 Hz | | VDE 0303 | Circular | 0.009 | 0.005 | 0.006 |
| at 800 Hz | | Pt. 4 | discs 95 and | 0.019 | 0.011 | 0.012 |
| at 1 MHz | | DIN 53483 | 80 mm × 1 mm | 0.050 | 0.026 | 0.028 |
| Tracking resistance | | VDE 030 | | | | |
| KB method | | Pt. 1/9.64 | | | | |
| Test solution A | | DIN 53480/6 | 20 × 15 mm | > 600 | > 600 | > 600 |

Figure 31 shows results of the alternating bending test on slightly and highly plasticized cellulose acetate, cellulose acetate propionate, and cellulose acetate butyrate molding compounds.

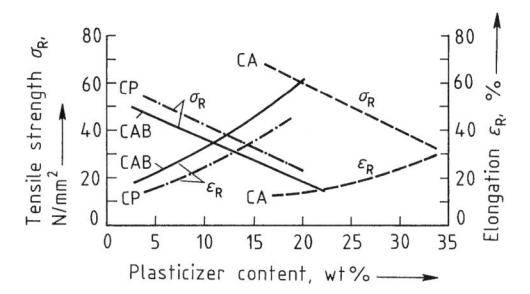

**Figure 14.** Tensile strength at break $\sigma_R$ and elongation $\varepsilon_R$ of cellulose acetate, cellulose acetate propionate, and cellulose acetate butyrate as a function of the plasticizer content (determined in accordance with DIN 53455 or ISO/R 527; specimen no. 3, rate of deformation 25 mm/min)

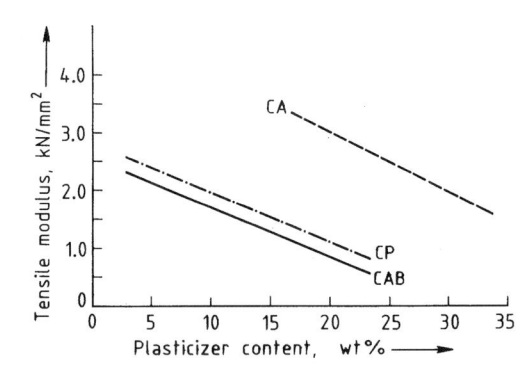

**Figure 15.** Tensile modulus of cellulose acetate, cellulose acetate propionate, and cellulose acetate butyrate as a function of the plasticizer content (determined in accordance with DIN 53455 or ISO/R 527)

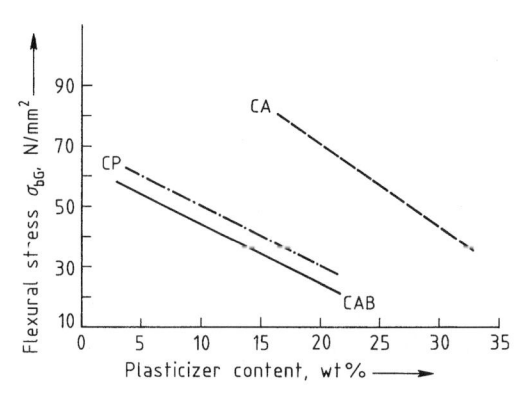

**Figure 16.** Flexural stress at a given strain $\sigma_{bG}$ of cellulose acetate, cellulose acetate propionate, and cellulose acetate butyrate as a function of the plasticizer content (determined in accordance with DIN 53452 or ISO/R 178; test specimen 4×10×80 mm, rate of deformation 2 mm/min)

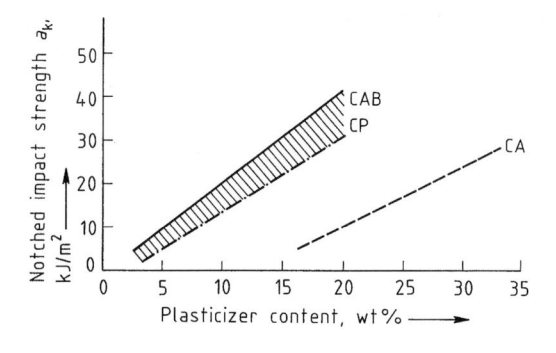

**Figure 17.** Notched impact strength $a_k$ of cellulose acetate, cellulose acetate propionate, and cellulose acetate butyrate as a function of the plasticizer content (determined in accordance with DIN 53453 or ISO/R 179; specimen no. 2)

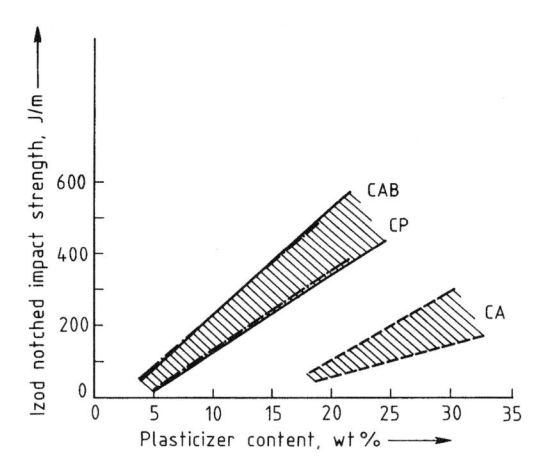

**Figure 18.** Izod notched impact strength of cellulose acetate, cellulose acetate propionate, and cellulose acetate butyrate as a function of the plasticizer content (determined in accordance with ASTM D 256, Method A, or ISO/R 180; test specimen 63.5×12.7×3.2 mm)

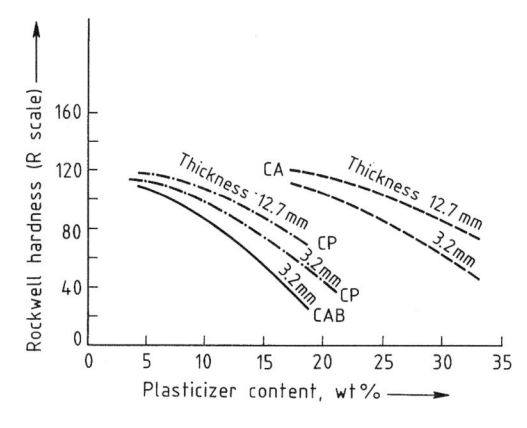

**Figure 19.** Rockwell hardness (R scale) of cellulose acetate, cellulose acetate propionate, and cellulose acetate butyrate as a function of the plasticizer content (determined in accordance with ASTM D 785)

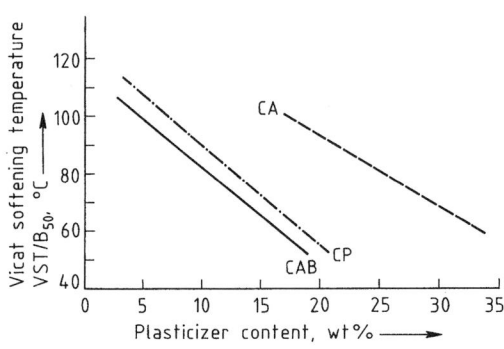

**Figure 20.** Vicat softening temperature $VST/B_{50}$ of cellulose acetate, cellulose acetate propionate, and cellulose acetate butyrate as a function of the plasticizer content (determined in accordance with DIN 53460/B or ISO/R 306; sheet 10×10×4 mm)

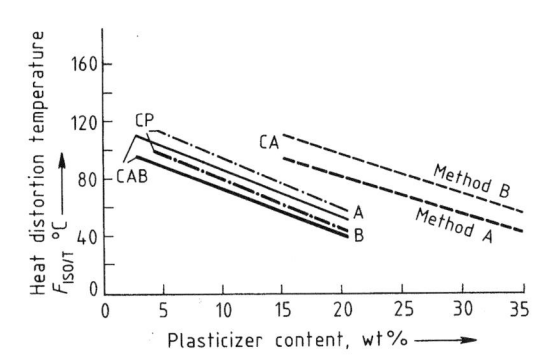

**Figure 21.** Heat distortion temperature $F_{\text{ISO}}$ of cellulose acetate, cellulose acetate propionate, and cellulose acetate butyrate as a function of the plasticizer content (determined in accordance with ASTM D 648, ISO/R 75, or DIN 53461; test specimen 12.7×12.7×120 mm)

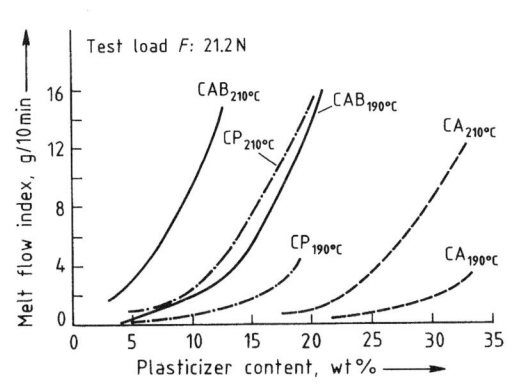

**Figure 22.** Melt flow index of cellulose acetate, cellulose acetate propionate, and cellulose acetate butyrate as a function of the plasticizer content (determined in accordance with DIN 53735 or ISO/R 1133)

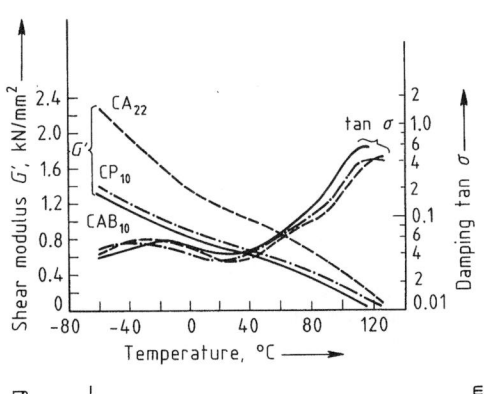

**Figure 23.** Shear modulus $G'$ and damping tan $\sigma$ of cellulose acetate$_{22}$, cellulose acetate propionate$_{10}$, and cellulose acetate butyrate$_{10}$ (determined in accordance with DIN 53445 or ISO/R 537)

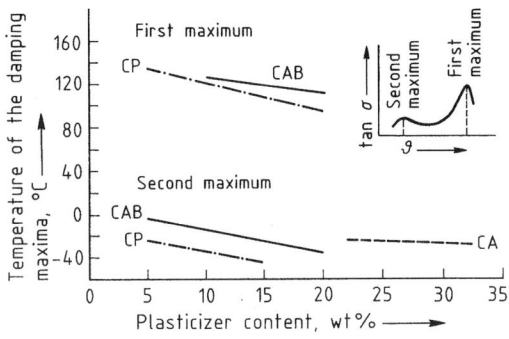

**Figure 24.** Temperature of the damping maxima of cellulose acetate, cellulose acetate propionate, and cellulose acetate butyrate as a function of the plasticizer content

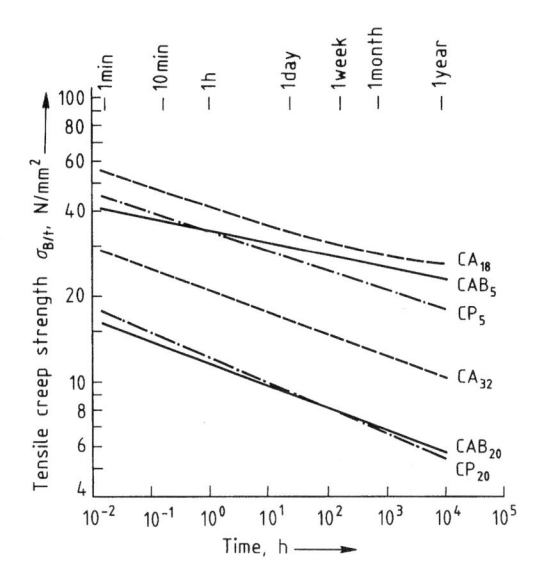

**Figure 25.** Tensile creep strength $\sigma_{B/t}$ of cellulose acetate, cellulose acetate propionate, and cellulose acetate butyrate (determined in accordance with DIN 53444 or ISO/R 899; test specimen no. 3)

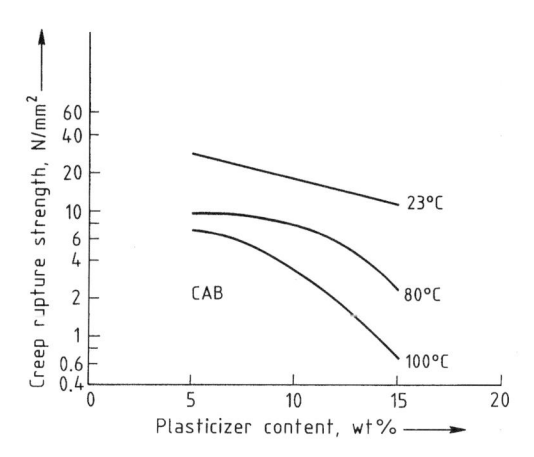

**Figure 26.** Creep rupture strength $\sigma_{B10_5}$ of cellulose acetate butyrate at 23 °C, 80 °C, and 100 °C as a function of the plasticizer content (determined in accordance with DIN 53444 or ISO/R 899)

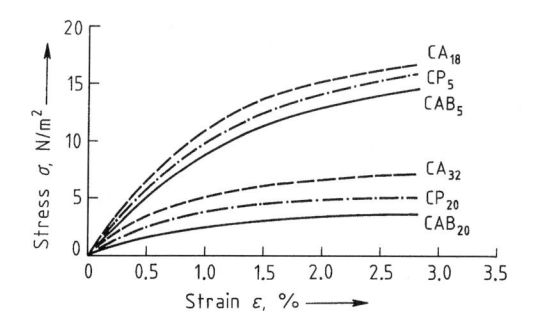

**Figure 29.** Isochronous stress–strain curves of cellulose acetate, cellulose acetate propionate, and cellulose acetate butyrate for 1000 h (determined in accordance with DIN 53444 or ISO/R 899)

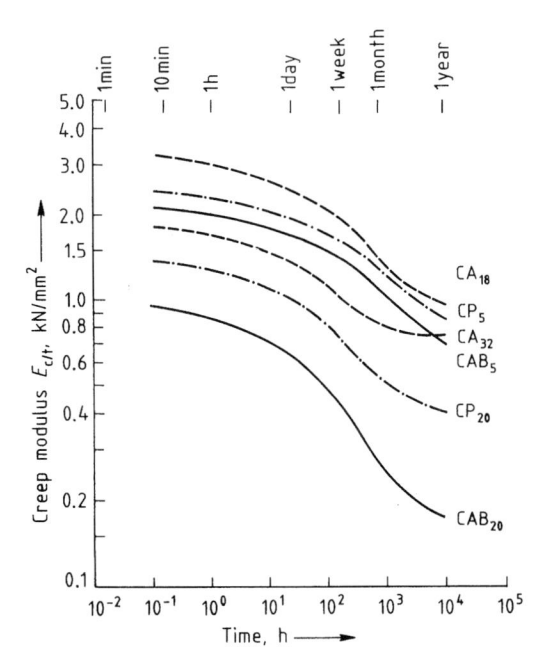

**Figure 27.** Creep modulus $E_{c/t}$ of cellulose acetate, cellulose acetate propionate, and cellulose acetate butyrate (determined in accordance with DIN 53444 or ISO/R 899)

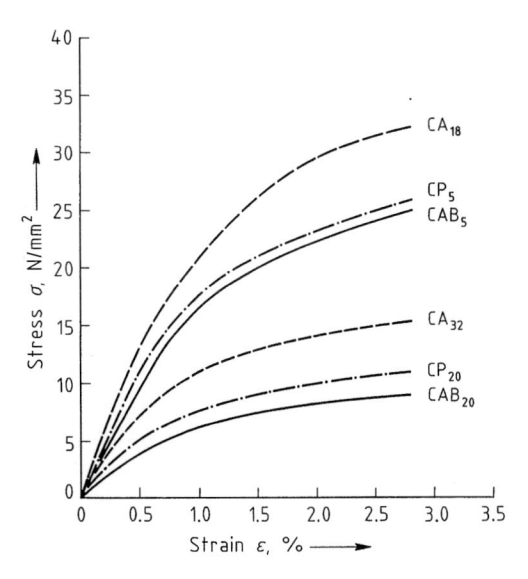

**Figure 28.** Isochronous stress–strain curves of cellulose acetate, cellulose acetate propionate, and cellulose acetate butyrate for 1 h (determined in accordance with DIN 53444 or ISO/R 899)

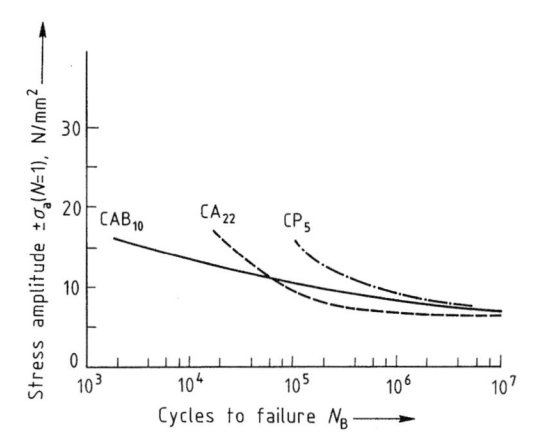

**Figure 30.** Dynamic fatigue test in the range of pulsating tensile stresses (number of load cycles) of cellulose acetate$_{22}$, cellulose acetate propionate$_5$, and cellulose acetate butyrate$_{10}$ (determined in accordance with DIN 50100; stress amplitude $\pm \sigma_a$ ($N = 1$) means stress amplitude under initial loading)

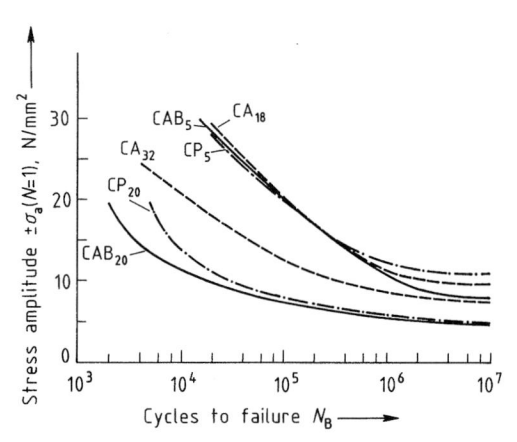

**Figure 31.** Dynamic fatigue test in the range of alternating flexural stresses (number of load cycles) of cellulose acetate, cellulose acetate propionate, and cellulose acetate butyrate (determined in accordance with DIN 50100; stress amplitude $\pm \sigma_a$ ($N = 1$) means stress amplitude under initial loading)

## 2.4.2. Polymer-Modified Cellulose Mixed Esters

In 1977, cellulose acetate butyrate molding compounds modified with ethylene–vinyl acetate without monomolecular plasticizers were introduced for the first time [117]. They have since become firmly established, particularly in the extrusion sector (automotive decorative trim, etc.). As far as the property pattern was concerned, the decisive factor was the matching of the ethylene–vinyl acetate component to the cellulose acetate butyrate used.

A polymer modification of cellulose acetate propionate without monomolecular plasticizers has also been successfully carried out [118]. In this case, however, a modifier of complex structure, based on an ethylene–vinyl acetate graft polymer, is necessary [119].

The main advantages of these polymer-modified cellulose mixed ester molding compounds over the former plasticized systems are that the existing combination of characteristic properties is retained, while the values for heat distortion temperature, creep behavior, and stiffness are considerably improved.

Another feature of these molding compounds is their freedom from plasticizer migration.

The polymer modifiers that are incorporated also delay the occurrence of crazing during long-term outdoor exposure, a fact which significantly increases the service life of, for example, external automotive decorative trim made from cellulose acetate butyrate molding compounds [117].

Some striking examples of the different properties of monomolecular-plasticized and polymer-modified cellulose acetate propionate and cellulose acetate butyrate are shown in Figures 27, 28, and 31.

Polymer modification has a similar effect on the tensile test, shear modulus, creep behavior, and hardness [118].

## 2.4.3. Chemical Properties

Thermoplastic cellulose ester molding compounds are resistant to white spirits, oil, and grease. Table 19 gives guide values for resistance to a range of substances, but thorough practical tests are recommended in each case.

## 2.4.4. Raw Materials

The following cellulose esters are used for the production of cellulose ester molding compounds:

Cellulose acetate propionate molding compounds,
   55–62% propionic acid content
   3–8% acetic acid content
Cellulose acetate butyrate molding compounds,
   43–47% butyric acid content
   19–23% acetic acid content
Cellulose acetate molding compounds,
   51.6–56.3% acetic acid content
   51.5–53.5% acetic acid content (for block acetate only)

Of the large number of plasticizers that are compatible with cellulose esters [6], the following have acquired industrial significance, either alone or in combination with one another:

*For cellulose acetate propionates and cellulose acetate butyrates:*
di-2-ethylhexyl phthalate, dibutyl adipate, di-2-ethylhexyl adipate, dibutyl azelate and dibutyl sebacate, dioctyl azelate, dioctyl sebacate, palmitates, stearates, etc.
*For cellulose acetates:*
dimethyl, diethyl, dibutyl, di-2-ethylhexyl, and di-2-methoxyethyl phthalate; triphenyl and trichloroethyl phosphate.

**Table 19.** Typical values* for the chemical resistance of organic cellulose ester molding compounds

| Solvent | Cellulose acetate (< 55% acetic acid) | Cellulose acetate (> 55% acetic acid) | Cellulose acetate propionate | Cellulose acetate butyrate |
|---|---|---|---|---|
| Water | + | + | + | + |
| Alcohols | – – | – – | – – | – – |
| Ethyl acetate | – – | 0 | 0 | 0 |
| Methylene chloride | – – | 0 | 0 | 0 |
| Acetone | 0 | 0 | 0 | 0 |
| Carbon tetrachloride | + | + | + – | + – |
| Trichloroethylene | + | + – | – – | – – |
| Perchloroethylene | + | + | + – | + – |
| Benzene | + | + – | – – | – – |
| Xylene | + | + | – – | – – |
| Petroleum spirit | + | + | + | + |
| Motor fuel mixture (high octane) | + | + | + – | + – |
| Mineral oil (paraffin) | + | + | + | + |
| Linseed oil | + | + | + | + |
| Turpentine oil | + | + | + – | + – |
| Lavender oil | + | + | – – | – – |
| Ether | + | + | + – | + – |
| Formalin | – – | – – | + – | + – |
| 2-Chlorophenol | 0 | 0 | 0 | 0 |
| Sulfuric acid, conc. | – | – | – | – |
| Sulfuric acid, 10% | + – | + – | + – | + |
| Hydrochloric acid, conc. | – | – | – | – |
| Hydrochloric acid, 10% | – | – | – | – |
| Nitric acid, conc. | – | – | – | – |
| Nitric acid, 10% | – | – | – | – |
| Caustic potash solution, 50% | – | – | – | – |
| Caustic potash solution, 10% | – | – | + – | + – |

\* Key to symbols: + = resistant; + – = resistant, but swells; – = not resistant; – – = not resistant, swells; 0 = soluble.

Ethylene–vinyl acetate copolymers have proven to be particularly suitable as polymer modifiers for cellulose acetate butyrate molding compounds, while graft polymers based on ethylene–vinyl acetate are preferred for cellulose acetate propionate molding compounds.

The *stabilizers* and *antioxidants* used for cellulose ester molding compounds include: alkali salts and alkaline-earth salts of sulfuric, acetic, and carbonic acid, tartaric acid, oxalic acid, citric acid, higher molecular mass epoxides, and phenolic antioxidants. In special cases these stabilizers and antioxidants can be complemented by others [120].

From the range of *ultraviolet absorbers* available, various benzophenones, benzotriazoles, salicylates, and benzoates are recommended for organic cellulose ester molding compounds [121].

*Processing auxiliaries* for cellulose ester molding compounds include zinc stearate, butyl stearate, and paraffin oil.

Numerous combinations of *dyes* can be used for coloring cellulose ester molding compounds [122]. The following groups of dyes have proven successful in practice: alkaline, acid, and substantive dyes (provided they are sufficiently soluble in the solvent); Zapon, Sudan, and Ceres dyes (provided they are sufficiently resistant to sublimation); and organic and inorganic pigments.

## 2.4.5. Production

Like cellulose acetates, cellulose acetate propionates and cellulose acetate butyrates are also thoroughly mixed at room temperature with plasticizers, stabilizers, antioxidants, dyes, and sometimes ultraviolet absorbers and processing aids. Plastification and homogenization are carried out at higher temperatures (between 150 and 210 °C, depending on the type and the degree of plasticization) in single- or twin-screw kneaders or roll mills. Depending on the type of equipment used, this results in granules in the form of pellets (bulk density 500–620 g/L) or cubes (bulk density 400–470 g/L).

There are no wastewater or waste gas problems associated with the production of thermoplastic cellulose ester molding compounds. The inevitable plasticizer vapors that occur during processing should be removed by exhaust ventilation.

## 2.4.6. Trade Names

The most important trade names of thermoplastic cellulose mixed ester molding compounds are as follows:

*Cellulose acetate propionates:* Cellidor CP (Bayer AG, Leverkusen, Federal Republic of Germany), Tenite Propionate (Eastman Chemical Products, Inc., Kingsport, United States).

*Cellulose acetate butyrates:* Cellidor B (Bayer AG, Leverkusen, Federal Republic of Germany), Tenite Butyrate (Eastman Chemical Products, Inc., Kingsport, United States).

*Cellulose acetate:* Acety (Daicel, Osaka, Japan), Cellidor S (Bayer AG, Leverkusen, FRG), Dexel (Courtaulds Chem. and Plastics, GB), Saxetat (VEB Eilenburg, GDR), Setilithe (Tubize Plastics, Belgium), Tenite Acetate (Eastman Chemical Products, Inc., Kingsport, United States).

## 2.4.7. Quality Requirements and Quality Testing

As has already been stated, the mechanical properties of thermoplastic cellulose ester molding compounds are dependent on the chain-length of the molecules, on the plasticizer or polymer modifier, the combination of plasticizers, and the content of plasticizer or polymer modifier.

With respect to cellulose acetate, determination of the viscosity and the viscosity ratio in a dilute solution [123], determination of the insoluble constituents [124], viscosity loss during molding [125], light absorption before and after heating [126], and the determination of constituents which are extractable with ethyl ether [127] are all standardized. The methods described for cellulose acetate are similarly applicable to cellulose mixed esters.

The manufacturers also carry out numerous in-house tests during the course of their quality control programs. These include determination of mechanical data, testing of purity, checking thermal stability, colorimetry, determination of flow properties (melt index, Brabender, extrusiometer), etc.

In the Federal Republic of Germany, cellulose ester molding compounds (CA, CP, and CAB) are standardized in accordance with DIN 7742, Parts 1 and 2, and in the United States in accordance with ASTM D 706 (cellulose acetate), D 707 (cellulose acetate butyrate), and D 1562 (cellulose acetate propionate).

## 2.4.8. Storage and Transportation

The transportation of thermoplastic cellulose ester molding compounds is not governed by the GGVS/ADR, GGVE/RID, GGVSee/IMDG code or DGR/ICAO regulations for the transportation of hazardous goods. The storage of thermoplastic cellulose ester molding compounds presents no problems. Even after 10 years in storage, no changes in composition have been detected. It is, however, recommended that thermoplastic cellulose ester molding compounds should be predried in accordance with the particular manufacturer's guidelines before processing.

## 2.4.9. Uses [128]

In the field of injection-molded eyeglass frames, cellulose acetate propionates have become increasingly important. This is due mainly to the fact that their dimensional stability is better than that of cellulose acetate, as a result of lower moisture absorption and greater stiffness and dimensional stability under heat [129].

Cellulose acetate propionate is also used for high-quality frames for sunglasses, protective goggles for industry, and sports goggles [129].

Cellulose acetate still dominates as sheet material for making ophthalmic frames. For this purpose, uni- or multicolored plates are either cast or made by extrusion.

Due to its high transparency, good impact resistance, and low level of light scattering, cellulose acetate propionate has become increasingly popular as a glazing for visors (for skiers, drivers, and workers in industry) and for sunglasses and sports goggles. These applications have been made possible by surface saponification of the plastic, which ensures permanent antifogging properties (the surface has very good wetting properties and excellent water absorption) [129].

Recently, cellulose acetate propionates with special infrared/ultraviolet-absorbing characteristics have become more important for welding goggles and certain types of sunglasses [129].

Where greater demands are placed on impact resistance, cellulose mixed esters as well as cellulose acetate are used (e.g., for the production of tool handles, hammer heads, and covers for handles of pliers, wrenches, etc.). Particularly important factors in this field of application are toughness, transparency, lightfastness, the "natural feel," the repolishing effect, and the absence of stress cracking. No problems occur when metal parts are insertion molded (the blades can even be driven "cold" into the handles), which brings obvious economic advantages.

With their minimal plasticizer migration, cellulose mixed esters are preferred over cellulose acetate for the packaging of toiletries. Other reasons for their use in this sector include the brilliance and depth of color, as well as the ability to produce special color effects.

Resistance to stress cracking and good toughness properties make cellulose mixed esters an excellent material for brushes, particularly toothbrushes, for which new designs require close spacing of the drill holes and a high bristle density. Cellulose mixed esters also satisfy the requirement for high tuft pull-out strength.

Cellulose acetate butyrate sheet is used for illuminated advertising signs, machine hoods, lamp covers, and dome lights. High light transmission, practically unrestricted choice of colors, antistatic characteristics, easy accurate processing, ease of joining (by simply gluing), good printing and coating properties, the absence of stress cracking, and finally, excellent mechanical strength are the main factors influencing the choice of this material. Cellulose acetate propionate is used in place of cellulose acetate for transparent, large-size seat shells.

Transparent, thin-section, and large-area hoods, lids, and covers with excellent toughness and good weathering resistance are injection-molded from specially modified cellulose acetate propionate.

Decorative trim made of cellulose acetate butyrate combined with aluminum foil [130] has been firmly established in industry for years. An aluminum foil is coated with the cellulose mixed ester and shaped in the crosshead die of an extruder [131]. With its practically unlimited scope for metal and wood effects, its elasticity, resistance to detergents, and simple fixing, this combination of materials has been used with great success in the automotive industry [132] in particular, as well as in the electrical, audio, and domestic appliance sectors [133].

Further applications for cellulose esters include lamp covers, toilet seats, writing and drawing instruments, tap handles, casings for disposable hypodermic syringes, transparent mouthwash spray attachments, handles, high-quality toys, instrument panel covers (glazing), and knife handle grips.

**Economic Facts.** With an apparently guaranteed supply of raw materials and a tremendous scope for variation of cellulose ester molding compounds and of their property combinations, this class of plastics should maintain its market significance in special areas of application for years to come.

### 2.4.10. Toxicology and Occupational Health

Cellulose acetate and cellulose propionate molding compounds comply with Recommendation XXVI of the Federal Health Authorities of the Federal Republic of Germany [134].

There are also various cellulose ester molding compounds on the market which satisfy the requirements of the U.S. Food and Drug Administration for plastics [135].

# 3. References

General References

[1] V. Stannet: *Cellulose Acetate Plastics*, Temple Press Ltd., London 1950.
[2] K. Thinius: *Analytische Chemie der Plaste*, Springer Verlag, Berlin-Göttingen-Heidelberg 1952.
[3] *Kirk-Othmer*, vol. **3**, pp. 357 ff.
[4] E. Ott, H. M. Spurlin, M. W. Grafflin: *Cellulose and Cellulose Derivatives*, 2nd ed., vol. **V**, part II, Interscience Publ., New York-London 1954.
[5] R. Houwink, A. J. Staverman: *Chemie und Technologie der Kunststoffe*, 4th ed., vol. **II/2**, Akademische Verlagsgesellschaft Geest & Porting KG, Leipzig 1963.
[6] K. Thinius:*Chemie, Physik und Technologie der Weichmacher*, VEB Deutscher Verlag für Grundstoffindustrie, Leipzig 1963.
[7] V. E. Yarsley, W. Flarell, P. S. Adamson, N. G. Perkins: *Cellulose Plastics*, Iliffe Book Ltd., London 1964.
[8] *Encyclopedia of Polymer Science and Technology*, vol. **3**, J. Wiley & Sons, New York 1965.
[9] R. Vieweg, E. Becker: *Kunstoff-Handbuch*, vol. **III**: "Abgewandelte Naturstoffe," Hanser Verlag, München 1965.
[10] P. B. Koslov, G. I. Braginski: *Chemistry and Technology of Polymers*, Iskustvo, Moscow 1965 (Russ.).
[11] H. Temming, H. Grunert: *Temming-Linters*, 2nd ed., P. Temming, Glückstadt 1972.

Specific References

[12] J. Barsha: "Inorganic Esters," in E. Ott, H. M. Spurlin, M. W. Grafflin (eds.): *Cellulose and Cellulose Derivatives*, 2nd ed., part II, Interscience Publ., New York-London 1954, pp. 713–762.
[13] E. D. Klug: "Cellulose Derivatives," in *Kirk-Othmer*, 2nd ed., vol. 4, J. Wiley & Sons, New York-London-Sydney-Toronto 1964, pp. 616–652.
[14] G. D. Hiatt, W. J. Rebell: "Esters," in N. M. Bikales, L. Segal (eds.): *Cellulose and Cellulose Derivatives*, part V, Wiley-Interscience, New York-London-Sydney-Toronto 1971, pp. 741–777.
[15] J. Honeyman: *Recent Advances in the Chemistry of Cellulose and Starch*, Heywood & Comp., London 1959.

[16]   E. Sjöström: *Wood Chemistry, Fundamentals and Applications,* Academic Press, New York-London 1981.
[17]   D. Fengel, G. Wegener: *Wood – Chemistry, Ultrastructure, Reactions,* De Gruyter, Berlin-New York 1983.
[18]   K. Balser: "Derivate der Cellulose," in W. Burchard (ed.): *Polysaccharide,* Springer Verlag, Berlin-Heidelberg-New York-Tokyo 1985, pp. 84–110.
[19]   L. C. Wadsworth, D. Daponte: "Cellulose Esters," in T. P. Nevell, S. H. Zeronian (eds.): *Cellulose Chemistry and its Applications,* Ellis Horwood Ltd., Chichester 1985, pp. 344–362. A. Revely: "A Review of Cellulose Derivatives and their Industrial Applications," in *Cellulose Chemistry and its Applications,* pp. 211–225.
[20]   Company Publication of Wolff Walsrode AG: *100 Jahre Collodiumwolle* (1979).
[21]   K. Fabel: *Nitrocellulose,* Enke Verlag, Stuttgart 1950.
[22]   F. D. Miles: *Cellulose Nitrate,* Oliver & Boyd, London 1955.
[23]   S. Rossin, *Mémorial des Poudres* **40** (1958) 457–471.
[24]   H. Temming, H. Grunert: *Temming-Linters,* 2nd ed. (engl.), Peter Temming AG, Glückstadt 1973.
[25]   Company Publication of Wolff Walsrode AG: *Walsroder Nitrocellulose,* no. 500/203/007 (1984).
[26]   J. Isler, D. Flegier: "The Self-Ignition Mechanism in Nitrocellulose," in J. F. Kennedy, G. O. Phillips, D. J. Wedlock, P. A. Williams (eds.): *Cellulose and its Derivatives,* Ellis Horwood, Chichester 1985, pp. 329–336.
[27]   A. H. K. Fowler, H. S. Munro: "Some Surface Aspects of the Thermal and X-Ray Induced Degradation of Cellulose Nitrates as Studied by ESCA," in [26] pp. 245–253.
[28]   Enclosure I of the International Rail Transport Regulations (RID), Appendix 1.
[29]   E. Berl, *Angew. Chem.* 1904, 982, 1018, 1074. E. Berl-Lunge: *Chemisch-technische Untersuchungsmethoden,* 8th ed., vol. **III,** Julius Springer, Berlin 1932, p. 1294.
[30]   P. Kassenbeck: "Faktoren, die das Nitrierverhalten von Baumwoll-Linters und Holzzellstoffen beeinflussen," report 3/83, ICT-Fraunhofer-Institut für Treib- und Explosivstoffe, Pfinztal-Berghausen 1983.
[31]   Hercules Powder, US 2776965, 1957; US 3063981, 1962.
[32]   Hercules Powder, US 2028080, 1936.
[33]   Hercules Powder, US 2950278, 1960.
[34]   Société Nationale des Poudres et Explosifs, FR 1394779, 1964; DE-OS 1246487, 1965; FR 1566688, 1968; DE-OS 1914673, 1969.
[35]   Société Nationale des Poudres et Explosifs, DE-OS 2813730, 1977.
[36]   Hercules Powder, US 2776944, 1957; US 2776964, 1957.
[37]   Wolff Walsrode, DE-OS 2727553, 1977; DE-OS 2727554, 1977.
[38]   Wasag Chemie, DE-OS 1771006, 1968.
[39]   Wolff & Co. Walsrode, DE-OS 1153663, 1961.
[40]   Wolff & Co. Walsrode, DE-OS 1203652, 1964.
[41]   Société Nationale des Poudres et Explosifs, DE-OS 2338852, 1973.
[42]   C. F. Bennett, T. E. Timell, *Sven. Papperstidn.* **58** (1955) 281–286.
[43]   D. A. J. Goring, T. E. Timell, *Tappi* **45** (1962) 454–460.
[44]   K. Thinius, W. Thümmler, *Makromol. Chem.* **99** (1966) 117–125.
[45]   M. Marx-Figini, *Makromol. Chem.* **50** (1961) 196–219.
[46]   J. W. Green in R. L. Whistler, J. W. Green, J. N. BeMiller (eds.): "Cellulose," *Methods in Carbohydrate Chemistry,* vol. **III,** Academic Press, New York-London 1963, pp. 213–237.

[47] W. J. Alexander, G. C. Gaul: "Cellulose Derivatives," in *Encyclopedia of Industrial Chemical Analysis*, vol. 9, J. Wiley & Sons, New York-London-Sydney-Toronto 1970, pp. 59–94.
[48] ASTM D 301-72 (1983): Soluble Cellulose Nitrate.
[49] ASTM D 1343-69 (1979): Viscosity of Cellulose Derivatives by Ball-Drop Method.
[50] ASTM D 365-79: Soluble Nitrocellulose Base Solutions.
[51] ASTM D 1720-79: Dilution Ratio in Cellulose Nitrate Solutions for Active Solvents, Hydrocarbons, Diluents and Cellulose Nitrates.
[52] A. Kraus: *Handbuch der Nitrocelluloselacke*, Westliche Berliner Verlagsgesellschaft Heenemann KG, Berlin, vol. **1**: 1955, vol. **2**: 1963, vol. **3**: 1961, vol. 4: 1966.
[53] H. Kittel: *Lehrbuch der Lacke und Beschichtungen*, vols. **1–8**, Heenemann GmbH, Stuttgart-Berlin 1971–1980.
[54] Bayer, DE-OS 2853578, 1978.
[55] Wolff Walsrode, DE-OS 3139840, 1981; DE-OS 3407932, 1984.
[56] *Ullmann*, 4th ed., **9** (1975) 179–183.
[57] Leaflet M 051 by the German Employers' Liability Insurance Association of the Chemical Industry: *Dangerous Chemical Substances*, Jedermann-Verlag Dr. Otto Pfeffer, Heidelberg 1984.
[58] *Verordnung über gefährliche Arbeitsstoffe* 11 Feb. 82, BGBl I, Carl Heymanns, Köln 1982, p. 144.
[59] Leaflet M 014 by the German Employers' Liability Insurance Association of the Chemical Industry: *Nitric Acid, Nitrogen Oxides, Nitrous Gases*, Jedermann-Verlag Dr. Otto Pfeffer, Heidelberg 1985.
[60] Leaflet M 037 by the German Employers' Liability Insurance Association of the Chemical Industry: *Nitrocellulose*, Jedermann-Verlag Dr. Otto Pfeffer, Heidelberg 1984.
[61] E. v. Schwartz: *Handbuch der Feuer- und Explosionsgefahr*, 5th ed., Feuerschutz-Verlag P. L. Jung, München 1958, p. 303.
[62] German Explosives Law, publ. 13 Sept. 1976, BGBl I, p. 2737; amended 10 Apr. 1981, BGBl I, Carl Heymanns, Köln 1981, p. 388.
[63] H. Dorias: *Gefährliche Güter*, Springer Verlag, Berlin-Heidelberg-New York-Tokyo 1984.
[64] Z. A. Rogovin, L. S. Galbraich, W. Albrecht (eds.): *Die chemische Behandlung und Modifizierung der Cellulose*, Thieme Verlag, Stuttgart-New York 1983.
[65] Hercules Powder Comp., US 2753337, 1953.
[66] B. Philipp, W. Wagenknecht: *Cellul. Chem. Technol.* **17** (1983) 443.
[67] R. G. Schweiger, *Tappi* **57** (1974) 86; *ACS Symp. Ser.* **77** (1978) 163; US 4138535, 1977.
[68] R. G. Schweiger, *Carbohydr. Res.* **70** (1979) 185.
[69] R. G. Schweiger, DE-OS 3025094, 1980.
[70] Eastman Kodak, US 3075962, 1963; US 3075963, 1963; US 3075964, 1963.
[71] Eastman Kodak, US 3068007, 1963.
[72] W. D. Slowig, M. E. Rowley, *Text. Res. J.* **38** (1968) 879.
[73] J. Pastyr, L. Kuniak, *Cellul. Chem. Technol.* **6** (1972) 249.
[74] J. B. Lawton, G. O. Phillips, *Text. Res. J.* **45** (1975) 4.
[75] K. Katsuura, T. Fujinami, *Kogyo Kagaku Zasshi* **71** (1968) 771.
[76] Eastman Kodak, US 2759924, 1956.
[77] Kimberly-Clark, US 3658790, 1970.
[78] National Chemical & Manufacturing, US 1896725, 1933.
[79] R. A. A. Muzzarelli, G. Marcotrigiano, C.-S. Liu, A. Frêche, *Anal. Chem.* **39** (1967) 1792.
[80] L. S. Sletkina, A. J. Poljakov, Z. A. Rogovin, *Vysokomol. Soedin* **7** (1965) 199; *Faserforsch. Textiltech.* **2** (1965) 299.

[81] D. A. Predvoditelev, M. S. Bakseeva; *Zh. Prikl. Khim.* (*Leningrad*) **45** (1972) 857; *Faserforsch. Textiltech.* **9** (1972) 361.
[82] D. C. Johnson: "Solvents for Cellulose," in *Cellulose Chemistry and its Applications,* Ellis Horwood, Chichester 1985, pp. 181–201.
[83] L. P. Clermont, F. Bender, *J. Polym. Sci. Part A-1* **10** (1972) 1669.
[84] R. B. Hammer, A. F. Turbak, *ACS Symp. Ser.* **58** (1977) 40. H. L. Hergert, *Tappi* **61** (1978) 63. ITT Rayonier, US 4056675, 1977.
[85] C. J. Malm, L. J. Tanghe, B. C. Laird, *J. Am. Chem. Soc.* **72** (1950) 2674.
[86] C. J. Malm, L. J. Tanghe, B. C. Laird, G. D. Smith, *J. Am. Chem. Soc.* **75** (1953) 80.
[87] J. Sakurada, *Kogyo Kagaku Zasshi* **35** (1932) B 123.
[88] Du Pont, US 1990483, 1935.
[89] C. J. Malm, L. J. Tanghe, B. C. Laird, *Ind. Eng. Chem.* **38** (1946) 77.
[90] H. Temming, H. Grunert: *Temming Linters,* 2nd ed., P. Temming, Glückstadt 1972.
[91] B. P. Rousse, *Encycl. Polym. Sci. Technol. 1964–1977,* **3**.
[92] *Ullmann,* 3rd ed., **5**, 187 ff.
[93] Bios Final Report No. 1600, Item No. 22, H. M. Stationary Office, London 1948.
[94] L. Ledderer, DE 200916, 1905.
[95] Boehringer, GB 363700, 1930.
[96] Bios Final Report No. 1850, Item No. 21, H. M. Stationary Office, London 1948.
[97] Cons. f. Elektrochem. Ind., DE 687065, 1933; US 2108829, 1934.
[98] Company Publication of Bayer AG, Leverkusen.
[99] Company Publication of Hercules Powder Comp., Wilmington, USA.
[100] E. Yarsley, W. Flavell, P. S. Adamson, N. G. Perkins: *Cellulosic Plastics,* Iliffe Book Ltd., London 1964.
[101] K. Thinius: *Analytische Chemie der Plaste,* Springer Verlag, Berlin-Göttingen-Heidelberg 1952.
[102] C. J. Malm, L. B. Genung, R. F. Williams, *Ind. Eng. Chem. Anal. Act.* **14** (1942) 935.
[103] H. Saechtling, W. Zebrowski: *Kunststoff-Taschenbuch,* 18th ed., C. Hanser, München 1971.
[104] C. J. Malm, C. R. Pordyce, H. A. Tanner, *Ind. Eng. Chem.* **34** (1942) 430. E. Ott, H. M. Spurlin, M. W. Grafflin: *Cellulose and Cellulose Derivatives,* vol. **V**, Part II, 2nd edn., Interscience Publ. Inc., New York-London 1954. W. Ballas, Internal communication, Bayer AG, Leverkusen.
[105] Bayer AG: *Cellit, Cellit BP and PR,* Brochures nos. KL 44172 and KL 44170, Leverkusen 1982.
[106] Eastman Kodak Comp.: Brochure on Cellulose Acetate Phthalate, Rochester.
[107] J. Corbiere, *Faserforsch. Textiltech.* **22** (1971) 71.
[108] Technical Association of Paper and Pulp Industry (TAPPI), DIN or ISO-standards.
[109] Celanese Corp. of America, US 2597156, 1952 (M. E. Martin, L. G. Reed).
[110] Eastman Kodak Co., US 2652340, 1953 (G. D. Hiatt, R. F. Willmans).
[111] E. Heim, *Chemiefasern* **16** (1966) 618.
[112] *Ullmann,* 4th ed., **9**, 213.
[113] *Textile Organon* **51** (1980) no. 6; **55** (1984) no. 5.
[114] W. Fischer, *Kunststoffe* **62** (1972) 653.
[115] Eastman Chem. Prod.: Specifications, T 58-3289, Kingsport, Tennessee 1958.
[116] W. Fischer, *Kunststoffe* **63** (1973) 292.
[117] W. Fischer, C. Leuschke, H. P. Baasch, *Kunststoffe* **67** (1977) no. 6, 348–252.
[118] C. Leuschke, M. Wandel, *Plastverarbeiter* **33** (1982) no. 9, 1095–1098.
[119] Bayer, DE-OS 2951800, 1979; DE-OS 2951747, 1979; DE-OS 2951748, 1979. K. Thinius: *Chemie, Physik und Technologie der Weichmacher,* VEB Deutscher Verlag für Grundstoffindustrie, Leipzig 1963.

[120] *Mod. Plast. Encycl.* 1970–1971, 840, 870.
[121] *Mod. Plast. Encycl.* 1970–1971, 876.
[122] *Mod. Plast. Encycl.* 1970–1971, 850.
[123] ISO/R 1157-1970.
[124] ISO/R 1598-1970.
[125] ISO/R 1599-1970.
[126] ISO/R 1600-1970.
[127] ISO/R 1875-1971.
[128] M. Wandel, C. Leuschke, *Kunststoffe* **74** (1984) no. 10, 589–592.
[129] Bayer AG: *Cellidor für die optische Industrie,* Company Publication KU 40029, Leverkusen 1984.
[130] J. Göller, H. Peters, W. Fischer, *Kunststoffe* **65** (1975) no. 6, 326–332.
[131] Glas Lab, DE-AS 1226291, 1955 (A. Shanok, V. Shanok, J. Shanok).
[132] Bayer AG: *Zierleisten und Profile in der Autoindustrie,* Company Publication KL 40025, Leverkusen 1979.
[133] Bayer AG: *Zierleisten in der Elektro- und Möbelindustrie,* Company Publication KL 40011, Leverkusen 1975.
[134] *Bundesgesundheitsblatt,* 19th ed. 1976, No. 6, Carl Heymanns Verlag, Köln-Berlin-Bonn-München.
[135] Eastman Kodak Comp.: Brochures on Tenite Propionate and Tenite Butyrate, nos. MB-33 C and MB-35 E, Rochester 1981 and 1983.

# Cellulose Ethers

Lothar Brandt, Hoechst AG, Niederlassung Kalle, Wiesbaden-Biebrich, Federal Republic of Germany

| | | |
|---|---|---|
| 1. | Introduction | 1204 |
| 1.1. | Chemical Structure | 1204 |
| 1.2. | Classification | 1205 |
| 1.3. | Principles of Synthesis | 1207 |
| 1.4. | History | 1208 |
| 1.5. | General Properties | 1208 |
| 1.5.1. | Solubility | 1208 |
| 1.5.2. | Viscosity | 1209 |
| 1.5.3. | Physical Appearance | 1209 |
| 1.5.4. | Stability | 1210 |
| 1.6. | Handling, Toxicology, and Ecology | 1210 |
| 2. | Production of Cellulose Ethers | 1211 |
| 2.1. | Raw Material | 1212 |
| 2.2. | Alkalization | 1212 |
| 2.3. | Etherification and Neutralization | 1213 |
| 2.4. | Workup | 1213 |
| 3. | Methyl Cellulose (MC) and Mixed Methyl Cellulose Ethers | 1214 |
| 3.1. | Classification and Production | 1214 |
| 3.2. | Properties | 1216 |
| 3.2.1. | Viscosity | 1216 |
| 3.2.2. | Gelation | 1218 |
| 3.2.3. | Solubility | 1219 |
| 3.2.4. | Other Properties | 1220 |
| 3.2.5. | Retarded Dissolution | 1221 |
| 4. | Ethyl Cellulose Ethers | 1222 |
| 4.1. | Ethyl Cellulose (EC) | 1223 |
| 4.2. | Mixed Ethyl Cellulose Ethers | 1223 |
| 5. | Hydroxyalkyl Cellulose Ethers | 1224 |
| 5.1. | Hydroxyethyl Cellulose (HEC) | 1224 |
| 5.1.1. | Production of Hydroxyethyl Cellulose | 1224 |
| 5.1.2. | Properties and Selected Uses | 1225 |
| 5.2. | Hydroxypropyl Cellulose | 1228 |
| 5.3. | Mixed Ethers | 1229 |
| 6. | Carboxymethyl Cellulose (CMC) | 1230 |
| 6.1. | Production | 1230 |
| 6.2. | Properties and Selected Uses | 1231 |
| 6.3. | Mixed Ethers with Carboxymethyl Groups | 1233 |
| 7. | Other Cellulose Ethers | 1233 |
| 7.1. | Ether Esters | 1233 |
| 7.2. | 2-(N,N-Diethylamino)ethyl Cellulose (DEAEC) | 1234 |
| 7.3. | Cyanoethyl Cellulose (CNEC) and Carboxyethyl Cellulose (CEC) | 1234 |
| 7.4. | Other Noncommercial Ethers | 1235 |
| 8. | Analysis | 1236 |
| 9. | Uses | 1238 |
| 10. | Economic Facts | 1243 |
| 10.1. | Trade Names and Suppliers | 1243 |
| 10.2. | Production Capacities | 1244 |
| 11. | References | 1245 |

# 1. Introduction [1]–[7]

Most cellulose ethers are water-soluble polymers; some types are also soluble in organic solvents. They are used as auxiliaries in a wide range of industrial fields. Their availability, economic efficiency, easy handling, low toxicity, and the great variety of types are reasons for a continuously expanding worldwide market.

## 1.1. Chemical Structure [8]–[12]

Cellulose ethers are derivatives of cellulose in which the hydroxyl groups are partially or completely replaced by ether substituents. Figure 1 shows the structural principle of cellulose ethers. Each $\beta$-D-anhydroglucose unit, the repeating unit of cellulose, is available for up to three ether groups in the C2-, C3-, and C6-positions of the unit.

The ether substituents are generally low molecular mass alkoxyl groups (one to four carbon atoms), which may be further substituted by such functional groups as carboxyl, hydroxyl, or amino. All substituents may be the same, but there may also be two or more different ether groups present in the same macromolecule. The substituents are randomly distributed along the chain with some preference for the C2- and C6-positions of each unit. The polymer ethers generally have no well-defined chemical structure, except for those that are totally etherified with a single type of group. These are only known as laboratory samples and have no commercial importance.

The amount of partial etherification is given for each substituent separately as the *DS value*, i.e., the *degree of substitution*. It ranges from 0 to 3 and is equivalent to the average of etherified hydroxyl groups per anhydroglucose unit.

Substitution by reactants that generate new free hydroxyl groups to be etherified is quantified by the *MS value*, the *molar substitution*. It gives the average number of moles of reactant added to one anhydroglucose unit. A typical reactant is ethylene oxide, yielding the hydroxyethyl (HE) substituent. In Figure 1 B the MS value for HE is 3.0.

The MS value has no upper limit. If the DS value is also known for such a substitution, the average chain length of ether side chains may be calculated from the ratio MS/DS.

Instead of specifying DS and MS values, some manufacturers characterize the substitution levels by the mass fractions (wt%) of the different ether groups (e.g., $-OCH_3$ or $-OC_2H_4OH$ for methyl or hydroxyethyl), based on the dry polymer. For the calculation of DS or MS from these mass fractions, see Chapter 8.

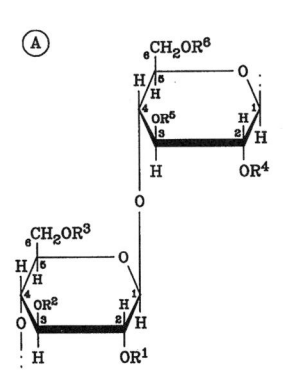

**Figure 1.** A) General structure of a section of two anhydroglucose units in a cellulose ether chain $R^1 - R^6 =$ H or organic substituent

B) Section of a carboxymethyl–hydroxyethyl cellulose chain with a degree of substitution (DS of carboxymethyl) of 0.5 and a degree of substitution (DS of hydroxyethyl) of 2.0; the molar substitution (MS of hydroxyethyl) is 3.0 in this case.

The amount of ether groups in this section corresponds to the average substitution levels, but in reality the substituent distribution along the chain is random

## 1.2. Classification [1]–[12]

Cellulose ethers are classified according to the chemical structures of the ether substituents. They may be grouped into anionic, cationic, or nonionic ethers. The nonionic ethers are further divided into water-soluble and organic-soluble products. *Single-substituted ethers* have only one kind of substituent, whereas *mixed ethers* have two or more different substituents in the polymer chain.

The main substituents of commercial products are listed in the upper part of Table 1. The lower part of Table 1 lists other well-known ether groups that either are no longer important commercially or have not yet become important commercially.

The order of substituent abbreviations for mixed ethers may either be alphabetical or follow the respective DS (MS) levels, i.e., HEMC or MHEC for 2-hydroxyethyl methyl cellulose with predominant methyl substitution.

Chemical Abstract Service registry numbers of commercial and other typical cellulose ethers are listed in Table 2.

**Table 1.** Ether substituents in cellulose ethers

| Abbreviation | Group | Structure | Ion activity | Abbreviations of commercial products[a] |
|---|---|---|---|---|
| CM or SCM | sodium carboxymethyl | $-CH_2-COO^-Na^+$ | anionic | CMC, CMHEC, (CMMC) |
| M | methyl | $-CH_3$ | none | MC, HEMC, HPMC, (HBMC, CMMC, EMC) |
| E | ethyl | $-CH_2-CH_3$ | none | EC, EHEC, (EMC) |
| HE | 2-hydroxyethyl | $-(CH_2-CH_2-O)_n-H$[b] | none | HEC, HEMC, CMHEC, EHEC, (HEHPC) |
| HP | 2-hydroxypropyl | $-(CH_2-CH-O)_n-H$[b]  \|  $CH_3$ | none | HPC, HPMC, (HEHPC) |
| HB | 2-hydroxybutyl | $-CH_2-CH-OH$  \|  $CH_2-CH_3$ | none | (HBMC) |
| TMAHP | 2-hydroxy-3-($N,N,N$-trimethylammonio)propyl, as chloride | $-CH_2-CH-CH_2$  \|  \|  $OH$  $N(CH_3)_3^+Cl^-$ | cationic | – |
| AE | 2-aminoethyl | $-CH_2-CH_2-NH_2$ | none[c] | – |
| DEAE | 2-($N,N$-diethylamino)ethyl | $-CH_2-CH_2-N(C_2H_5)_2$ | none[c] | (DEAEC) |
| B | benzyl | $-CH_2-C_6H_5$ | none | – |
| CNE | 2-cyanoethyl | $-CH_2-CH_2-CN$ | none | – |
| CE | sodium 2-carboxyethyl | $-CH_2-CH_2-COO^-Na^+$ | anionic | – |
| SE | sodium 2-sulfoethyl | $-CH_2-CH_2-SO_3^-Na^+$ | anionic | – |
| PM | monosodium phosphonomethyl | $-CH_2-PO_3H^-Na^+$ | anionic | – |

[a] In parentheses: minor products for special uses; ending "C" means "Cellulose." [b] Generally $n = 1, 2$, or 3. [c] Cationic in solutions of pH < 7.

**Table 2.** CAS registry numbers of cellulose ethers (for abbreviations of substituents, see Table 1)

| Single ethers | | Mixed ethers | |
|---|---|---|---|
| Abbreviation | CAS registry number | Abbreviation | CAS registry number |
| CMC[a] | [9004-32-4] | CMHEC[a] | [9088-04-4] |
| CMC[b] | [9000-11-7] | HEMC | [9032-42-2] |
| MC | [9004-67-5] | HPMC | [9004-65-3] |
| EC | [9004-57-3] | HBMC | [9041-56-9] |
| HEC | [9004-62-0] | CMMC[a] | [9088-05-5] |
| HPC | [9004-64-2] | EHEC | [9004-58-4] |
| DEAEC | [9013-34-7] | EMC | [9004-59-5] |
| BC | [9015-11-6] | HEHPC | [51331-09-0] |
| CNEC | [9004-41-5] | | |
| CEC[a] | [9032-39-7] | | |
| SEC[a] | [39277-57-1] | | |
| PMC[c] | [9069-33-4] | | |

[a] Sodium salt.
[b] Acid form (CMC-H).
[c] Disodium salt.

# 1.3. Principles of Synthesis

Etherification of cellulose proceeds under alkaline conditions; generally sodium hydroxide (NaOH) is used. Cellulose is first treated with aqueous caustic to yield swollen alkali cellulose, which is then etherified with the reagent. For the preparation of mixed ethers, the different reagents may be applied simultaneously or in subsequent reaction steps with intermediate purification, if necessary. There are four reaction types, summarized by the following equations (cellulose is abbreviated by Cell–OH):

$$\text{Cell–OH} + \text{R–X} + \text{NaOH} \longrightarrow \text{Cell–OR} + \text{H}_2\text{O} + \text{NaX} \quad (1)$$

$$\text{Cell–OH} + \text{R–CH–CH}_2\underset{\text{O}}{\diagdown\diagup} \xrightarrow{\text{OH}^-} \text{Cell–O–CH}_2\text{–CH–R} \quad (2)$$
$$\qquad\qquad\qquad\qquad\qquad\qquad\qquad\quad |$$
$$\qquad\qquad\qquad\qquad\qquad\qquad\qquad\text{OH}$$

$$\text{Cell–OH} + \text{CH}_2=\text{CH–Y} \xrightarrow{\text{OH}^-} \text{Cell–O–CH}_2\text{–CH}_2\text{–Y} \quad (3)$$

$$\text{Cell–OH} + \text{R–CHN}_2 \xrightarrow{\text{OH}^-} \text{Cell–O–CH}_2\text{–R} + \text{N}_2 \quad (4)$$

Equation (1) describes the Williamson etherification. R–X is an inorganic acid ester, X being halogen or sulfate. In industry, chlorides, R–Cl, are applied, e.g., methyl chloride, ethyl chloride, or sodium chloroacetate. Stoichiometric amounts of alkali are consumed in this type of reaction.

Equation (2) shows the alkali-catalyzed addition of an epoxide (e.g., R=H, $CH_3$, or $C_2H_5$) to the cellulosic hydroxyl with no alkali being consumed. This reaction may proceed further because new hydroxyl groups are generated during the reaction, leading to oligomeric alkylene oxide side chains:

$$\text{Cell–O}\!\!-\!\!\!\left(\text{CH}_2\text{–CH–O}\right)_n\!\!\text{H}.$$
$$\qquad\qquad\qquad\;\;|$$
$$\qquad\qquad\qquad\;\text{R}$$

The analogous addition of aziridine leads to aminoethyl ethers: Cell–O–$CH_2$–$CH_2$–$NH_2$. Acid-catalyzed epoxide addition is disadvantageous because epoxide polymerization is preferred and cellulose is hydrolyzed.

Equation (3) shows the addition of Cell –OH to an activated double bond in alkaline medium, Y being an electron-attracting substituent, such as CN, $CONH_2$, or $SO_3^-Na^+$. At present, this type is seldom applied industrially.

Equation (4), the etherification with diazoalkanes, has no industrial significance.

## 1.4. History [1], [4], [5], [8], [12]

The first cellulose etherification, methylation with dimethyl sulfate, was described by SUIDA in 1905. Nonionic alkyl ethers were claimed in patents of LILIENFELD (1912 and later), DREYFUS (1914), and LEUCHS (1920) as water-soluble or organic-soluble products. BUCHLER and GOMBERG benzylated cellulose in 1921, carboxymethyl cellulose was first prepared by JANSEN in 1918, and hydroxyethyl cellulose by HUBERT in 1920. Carboxymethyl cellulose was made commercially in the early 1920s in Germany. The industrial production of MC and HEC was established in the United States in 1937–1938 (for abbreviations of cellulose ethers, see Table 1). The production of water-soluble EHEC was initiated in Sweden in 1945. Since 1945 cellulose ether production has been expanding quickly in Western Europe, the United States, and Japan.

A great variety of mixed ethers have been developed with varying DS levels, viscosity levels, purity grades, and rheological properties. In the last few decades the main interest was aimed at optimizing production technology and at further diversifying types to obtain precisely tailored products for the customers' special use.

## 1.5. General Properties [8], [9], [11]–[13]

### 1.5.1. Solubility

The solubility of cellulose ethers in aqueous base, water, or organic solvents depends on the DS value and on the nature of the ether groups. Substances with DS below 0.1 are generally insoluble and differ from cellulose only in some physical and technical parameters such as tensile strength, surface potential, water absorption capacity, or dyeability. Such *modified celluloses* result from cellulose pretreatment in the textile and papermaking industries, but they are not commercially available as cellulose ethers.

Products become soluble in aqueous alkali, e.g., 5–8% NaOH, at DS grades ranging from 0.2 to 0.5, depending on the ether group. This solubility is retained also at higher DS levels for anionic and strongly hydrophilic nonionic types, but it disappears at higher DS levels if hydrophobic substituents predominate.

Most industrially produced cellulose ethers are soluble in water and/or organic solvents. Water solubility is obtained above a DS value of about 0.4 for anionic and above a DS (MS) value of about 1 for nonionic ethers. If hydrophobic ether groups predominate, the water solubility may again disappear at DS levels above 2. Within this range nonionic ethers are also soluble in such protic or polar aprotic solvents as lower aliphatic alcohols, ketones, or ethers. The more hydrophobic types additionally dissolve in chlorinated hydrocarbons, but rarely in purely aliphatic hydrocarbons.

Ethers containing only anionic groups are rarely soluble in organic solvents at all DS levels, except in such strongly polar aprotic solvents as dimethyl sulfoxide. In all cases, solubilities are somewhat greater with a lower molecular mass of the polymer.

The solubility of hydrophobic ethers in water is susceptible to higher temperature: the dissolved products undergo *thermal gelation* or coagulation but redissolve on cooling (see Section 3.2.2).

Although clear and even "brilliant" solubility is desirable for most uses, some commercial products form hazy solutions, which may contain undissolved particles or fibers. This is due to uneven substitution, either because of insufficient mixing of the reagents in the reaction vessel or because of strong irregularities in the cellulose chain (highly crystalline regions are hardly accessible for substitution). Impurities in the cellulose raw material, such as lignin, or the presence of crosslinking agents in the etherification reagent may also lead to insoluble residues.

## 1.5.2. Viscosity

Solutions of cellulose ethers are highly viscous, depending on concentration, temperature, average chain length of the macromolecule, and the presence of salts or other additives. The chain length of the original cellulose can be reduced in the manufacturing process by chemical means, resulting in lower viscosity of the final product (see Section 2.3).

The rheological behavior of a solution at a given concentration and temperature may be Newtonian, pseudoplastic, thixotropic, or even gel-forming, depending on the chain length, the substituent distribution, and the nature of the ether groups. The viscosities of 2% neutral, aqueous solutions at ambient temperature range from 5 to over $10^5$ mPa s.

## 1.5.3. Physical Appearance

Cellulose ethers are white or slightly yellowish solids. They are traded in granular form or as powders with moisture contents of up to ca. 10%. Apparent densities of powders range from 0.3 to 0.5 g/cm$^3$. Some fibrous products have apparent densities below 0.2 g/cm$^3$. Different purity grades are adjusted by the manufacturers according to the desired application. Pure products are odorless and tasteless. The crude grades may contain as much as 40 wt% sodium salts, e.g., NaCl. Additives may be incorporated to ensure stability, controlled dissolution, and easy handling.

In addition, commercial products may be blended with other water-soluble polymers, such as starch products, natural gums, or polyacrylamides, for defined rheological and other physical specifications.

## 1.5.4. Stability

Cellulose ethers are susceptible to cellulase-producing microorganisms. The enzyme preferentially attacks the unsubstituted anhydroglucose units, which results in hydrolytic chain cleavage and subsequent decrease in viscosity. Ether substituents protect the cellulosic backbone. Therefore, cellulose ethers become more stable with a rising DS value or with increasing uniformity of the substitution. In both cases, fewer unsubstituted anhydroglucose units are exposed to the hydrolytic enzyme.

Cellulose ethers are quite stable. They are not affected by air, moisture, daylight, moderate heat, and common pollutants. Strong oxidants generate peroxy and carbonyl groups, leading to further degradation under alkaline conditions. When alkaline solutions of cellulose are heated, the viscosity decreases distinctly. Strong acids degrade the chain by direct hydrolysis of the cellulosic acetal linkages. High-energy radiation destroys cellulose ethers as it does other organic polymers. Commercial products may contain biocides, buffer substances, or reducing agents to achieve nearly unlimited storage stability and unchanged viscosity under appropriate conditions. Solid cellulose ethers are thermally stable up to 80–100 °C. Higher temperature or prolonged heating, in some cases, causes crosslinking to form insoluble networks. The solid products are slightly degraded in the range of 130–150 °C; stronger degradation and browning occurs when they are heated at over 160–200 °C, depending on both the ether type and the heating conditions. Neutral aqueous solutions may be heated over prolonged periods without a viscosity loss after cooling to ambient temperature. An intermediate thermal gelation or coagulation does not affect the viscosity.

## 1.6. Handling, Toxicology, and Ecology

*Handling* [4], [5], [7]. Fine powders of cellulose ethers form explosive dusts in air as do natural polysaccharides or sawdust. Dry nonionic ethers undergo electrostatic charging similar to that of other organic polymers. When cellulose ethers are stored and handled, the general precautions concerning powdered organic polymers must be observed. The flammability is similar to that of cellulose. Spilling of solutions forms very slippery films that are difficult to remove.

*Toxicology* [7], [14], [15]. Cellulose ethers are generally nontoxic. Products causing health irritation may occur in research but are not made industrially. High-purity grades of most commercial products are approved as food additives and for use in cosmetic compositions. Toxic impurities or additives, such as mercury-containing biocides, do not permit such applications.

*Ecology* [7], [16]. The biodegradation of cellulose ethers by cellulase-producing microorganisms (see Section 1.5.4.) also occurs in wastewater and therefore prevents the accumulation of cellulose ethers. Glucose, glucose ethers, and ether oligomers result

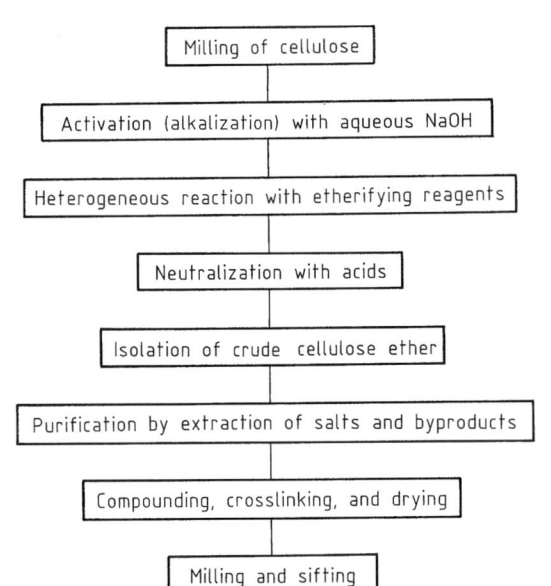

```
Milling of cellulose
        │
Activation (alkalization) with aqueous NaOH
        │
Heterogeneous reaction with etherifying reagents
        │
Neutralization with acids
        │
Isolation of crude cellulose ether
        │
Purification by extraction of salts and byproducts
        │
Compounding, crosslinking, and drying
        │
Milling and sifting
```

**Figure 2.** General scheme for the production of cellulose ethers

from enzymatic hydrolysis; they are further degraded to carbon dioxide and water in slow bioreactions. Biostable or toxic metabolites are not known.

Cellulose ethers have no fish toxicity and are poor nutrients for most microorganisms. Nevertheless, wastewater bacteria may adapt to enhanced cellulose ether degradation after some exposure. Products with high DS values appear to have very low biological oxygen demands under test conditions within short periods. Wastewaters of textile or papermaking plants have only low cellulose ether concentrations, and their total BOD or COD generally is low. Anionic ethers may be flocculated by using iron or aluminum salts to obtain insoluble and filterable residues. Wastewater may also be cleared of cellulose ethers and other soluble polymers by ultrafiltration.

# 2. Production of Cellulose Ethers

Figure 2 shows a general scheme for the production of cellulose ethers. The individual steps are discussed subsequently in the following sections. Special conditions for the production of specific ether types are treated in Chapters 3, 4, 5, 6, 7.

All known industrial processes use heterogeneous systems. Etherification of cellulose solutions is not economical because time–volume yield is much too low at the required cellulose concentration. Free choice of continuous or discontinuous processing is possible, either for each individual step or for the total process.

## 2.1. Raw Material [8], [9], [17]–[19]

For the production of cellulose ethers of high viscosity, *cotton linters* are used because of their high degree of polymerization ($\overline{DP}$ ca. 4000). For the high-viscosity types, degradation must be largely avoided by careful exclusion of oxygen; if necessary, antioxidants are added during the production process.

For the production of cellulose ethers with viscosities lower than 50 000 mPa s (2 % aqueous solution, ambient temperature), *sulfite-processed wood pulp* is more economical as the raw material. This pulp is similar to the "dissolving pulp" used for rayon fiber or film production. It is almost free of lignin, is highly purified, well-bleached, and has high α-contents of more than 86 %. *Sulfate-processed pulp* has only minor use for the production of cellulose ethers. The origin of the pulp is preferably coniferous waste wood because of its high $\overline{DP}$ and high α-contents, especially if it is made from knotty wood.

For some applications, hazy aqueous solutions of low viscosity and incomplete solubilities may be acceptable. The raw material for these cheaper types can be *beech pulp*, *hydrocellulose waste*, or even *sawdust*. Linters or pulp are employed either in commercial sheets or rolls or as milled and sifted powder.

## 2.2. Alkalization [8], [9], [17], [18], [20]

Traditional techniques use a steeping process in which the cellulose sheet is steeped in a bath of aqueous sodium hydroxide (at least 18 %). This causes swelling of the cellulose. The sheets are then pressed between rolls to the desired alkali and water content. Subsequently, the wet alkali cellulose sheet is shredded into fibers.

In modern processes, 30–70 % NaOH solution is sprayed onto dry cellulose powder in fast-turning, dry-mixing aggregates; the cellulose powder can also be impregnated with an inert organic solvent. Alternatively, the cellulose powder can be slurried in an organic solvent in normal stirred vessels before NaOH is added. Continuous alkalization in the steeping technique requires endless cellulose sheets provided from rolls. A stock container for cellulose powder is needed for the spraying technique. Shredding, mixing, or stirring devices must be chosen carefully and adjusted regularly to ensure a very uniform swelling and alkali distribution, the most important conditions for uniform etherification. Nonuniform alkalization causes severe loss in solubility, due to unetherified particles in the final product.

The alkali cellulose system ready for etherification must contain at least 0.8 mol of NaOH per mole of anhydroglucose to obtain uniformly substituted ethers, even when etherification reactions of type 2 or 3 (see Section 1.3) are employed. For Williamson reactions (type 1), more alkali is required, depending on the amount of reagent used and on the DS level desired. However, more than 1.5 mol of unconsumed NaOH per anhydroglucose unit is inconvenient and decreases the etherification yield because of reagent hydrolysis. About 5–20 mol of water per mole of anhydroglucose is needed to achieve sufficient swelling and good accessibility of the activated alkali cellulose. Again, a very high excess of water must be avoided because of reagent hydrolysis.

Lower viscosities in the final product can be adjusted while the process is in the alkalization step by exposing the alkaline material to air. In that case, well-defined conditions must prevail with regard to time, temperature, NaOH concentration, and the presence of catalytic amounts of iron, cobalt, or manganese salts, which catalyze the oxidative depolymerization. This procedure is called *ripening* or *aging* of alkali cellulose; it is best carried out in the absence of organic solvents.

## 2.3. Etherification and Neutralization
[8] – [10], [17]

In the case of lower alkyl chlorides or epoxides as etherifying reagents, the reaction is carried out in jacketed, agitated, stainless-steel or nickel-clad autoclaves, because pressure rises to about 3 MPa (30 bar) at higher temperature (50–120 °C). Addition of inert organic solvents lowers the pressure by dissolving the gaseous re-agent and facilitates the transport of reagent throughout the reaction mass. Such slurry processes are operated at pressures below about 0.3 MPa (3 bar). The reaction time varies with the reactivity of the etherifying agent from about 0.5 to 16 h. All processes, including simultaneous reaction of two different reagents to generate mixed ethers, may be accomplished in a single step or in several subsequent steps. Sometimes, single-substituent ethers are also produced in several steps with or without complete workup of the partially etherified intermediate products.

After the final reaction mass or slurry has been cooled, residual alkali is neutralized by acid, if necessary. Alkali-consuming Williamson reactions leave no or only small amounts of caustic, if an excess of etherifying reagent is applied. Hydrochloric, nitric, or acetic acid is generally used for neutralization; sulfuric or phosphoric acid is suitable for technical-grade production only, because the sodium salts of these acids are difficult to extract from water-soluble material. Addition of acetic acid to strong inorganic acid forms an acetate buffer at low excess of acid. This avoids local or overall acidification of the crude product, which could cause acid hydrolysis of the cellulosic chain or cross-linking in the case of carboxymethyl cellulose.

## 2.4. Workup [8], [9], [12]

Organic-soluble or thermogelling water-soluble ethers are purified by washing with hot water. Small soluble portions of hemicellulose ethers or degraded short-chain oligomers are extracted along with the salt into the wastewater; this reduces the yield to some extent, especially in the case of low-viscosity types. Predominantly hydroxyalkylated or carboxyalkylated products are often completely soluble in hot water. Salts are removed from these crude products step by step or continuously in countercurrent systems or by cascade extraction with mixtures of organic solvents containing a sufficient amount of water or methanol. Crude grades are only partially extracted or not at all.

The purified products are isolated in water- or in solvent-wet forms by centrifugal separation or filtration. Sometimes, the residual solvent must be removed by treatment with steam before drying for better solvent recycling. Solvents, lowboiling byproducts, and the excess of volatile reagents are generally recycled from the aqueous or organic liquors by distillation. Gaseous reagents are distilled at elevated pressure. In each case of solvent recycling, the exact azeotropic data of all components must be known. Weak crosslinking agents, such as glyoxal, are sometimes added to the wet mass of purified cellulose ether before drying to achieve controlled dissolution of the final product (see Section 3.2.5). Some processes involve wet milling before drying to obtain an easily transportable and uniformly wet material.

The product is dried in drum dryers or in other conventional drying equipment. Overheating or prolonged drying should be avoided because decreased solubility or thermal degradation of the product may occur. Therefore, cellulose ethers should not be dried exhaustively, and 1–10% of the water should remain in the product.

The material is subsequently milled under mild conditions. After sifting to obtain granular or powder types, the product may be blended with other substances before packing. Cellulose ethers are shipped in bags, drums, or containers. During transportation and storage, humidity must be excluded to prevent crusts from forming and to maintain a freely flowing product.

# 3. Methyl Cellulose (MC) and Mixed Methyl Cellulose Ethers

## 3.1. Classification and Production [11], [12], [21]–[23]

Methyl cellulose is a simplified term for cellulose ethers with predominant methyl substitution. This includes MC itself as well as the mixed ethers HEMC, HPMC, HBMC, EMC, and CMMC, which have properties similar to those of MC: thermal gelation in water below 100 °C like water-soluble MC and solubility in organic solvents like very highly substituted MC with DS value > 2.4.

Commercial MC is made in two types: (1) If the degree of substitution (DS) is 1.4–2.0, the product is soluble in cold water. (2) The lower substituted materials in the DS range from 0.25 to 1.0 are soluble in alkali. To dissolve an MC of the second type completely, 2–8% aqueous NaOH is adequate. Higher NaOH concentrations are required for lower DS values. Gelation temperatures of water-soluble MC drop with increasing DS. A typical methyl cellulose of DS 1.8 forms a gel at 54–56 °C. Methyl cellulose solutions may often contain insoluble particles because of nonuniform substitution. More clearly soluble types having higher gelation temperatures at equal DS values may be obtained by special alkalization techniques using NaOH–copper(II) complexes or benzyltrimethylammonium hydroxide as alkalizing agents. These reagents provide for better swelling of the alkali cellulose and enhanced accessibility of the macromolecule for etherification. However, these techniques are not used industrially for technical and economical reasons.

Mixed substitution is the more reasonable way to clear solutions and higher gelation temperatures. In addition, it allows many variations of the properties by adjusting the ratio of methyl to secondary substitution. Among the most important mixed ethers are *Hydroxypropyl methyl cellulose* (*HPMC*) and *hydroxyethyl methyl cellulose* (*HEMC*). They are made by additional action of propylene oxide or ethylene oxide on the alkali cellulose. Commercial HPMC, a very versatile class of products, has methyl DS values from 1.3 to 2.2 and hydroxypropyl MS values from 0.1 to 0.8. Hydroxyethyl methyl cellulose, manufactured mainly in Europe, has methyl DS values from 1.5 to 2.0 and hydroxyethyl MS values from 0.02 to 0.3. Hydroxybutylation with butylene oxide [106-88-7] to form hydroxybutyl methyl cellulose (HBMC) with a hydroxybutyl DS value of 0.1 leads to organic-soluble ethers for conventional methylation levels (DS 1.8). A British specialty is the water-soluble *ethyl methyl cellulose* (*EMC*) with methyl DS value of 0.9 and ethyl DS value of 0.4. Carboxymethylation to low DS values of about 0.05 produces *carboxymethyl methyl cellulose* (*CMMC*), a product with weakly anionic properties.

All secondary substitution, even at low substitution levels of the additional group, provides for clearer solutions compared with methyl cellulose. Gelation temperatures increase, especially when hydrophilic carboxymethyl or hydroxyethyl groups are introduced. A gelation temperature above 95 °C is undesirable, because it makes salt extraction with hot water, which is the usual purification step, impossible.

The production of methyl cellulose and related mixed ethers requires large amounts of alkali. For alkalization 35–60% aqueous NaOH is employed. The molar ratio of NaOH to anhydroglucose unit must be 3–4 to obtain water-soluble ethers with methyl DS values between 1.4 and 2.0. The viscosity level of the product is adjusted by adequate exposure to air during alkalization (aging). An excess of methyl chloride [74-87-3], depending on the amount of alkali, is used, thus converting all NaOH into NaCl. Methanol and dimethyl ether byproducts arise from reaction of methyl chloride with water. The resulting crude product is almost neutral and requires no or little acid for neutralization.

*Gaseous Methyl Chloride Process.* Methylation is started by warming alkali cellulose with part of the methyl chloride to 50 °C in a corrosion-resistant pressure vessel equipped with effective mixing or stirring devices. The exothermic etherification sustains a temperature of 60–100 °C for some hours, which is maintained by occasional heating or cooling. During the process some reagent evaporates with volatile byproducts. These are removed, condensed, and the methyl chloride is then recycled and replenished with fresh reagent to keep a constant concentration of gaseous methyl chloride in the reaction vessel. The blades and axis of the stirrer are hollow and have openings to recycle the reagent directly into the reaction mass.

*Liquid Methyl Chloride Process.* A continuous process uses liquid methyl chloride and needs reaction times of less than 1 h. It consists of slurrying the alkali cellulose under pressure in excessive reagent and then pumping the slurry through a partially heated reaction tube. This is followed by evaporation of volatile byproducts and excessive reagent. Other liquid-phase processes operate in the presence of an inert organic liquid, which reduces reaction pressure, facilitates heat transport, and sometimes requires smaller amounts of alkali by partly suppressing byproduct formation.

All these processes are equally suitable for the production of mixed ethers. The second reagent can be added before or after the methylation has begun. Optimized techniques comprise the gradual addition of reagents and temperature-programmed reaction steps.

Sodium chloride and such nonvolatile byproducts as glycols are removed by washing with hot water above the gelation temperature of the cellulose ether. The product is then dried in conventional equipment. Commercial methyl cellulose contains less than 1% NaCl; high-purity grades have less than 0.1% NaCl. Very low viscosity levels are obtained by subsequently treating the products with acid [24]. Controlled dissolving properties, mainly for powder-milled types, are adjusted by reaction with glyoxal (see Section 3.2.5).

## 3.2. Properties [11]–[13], [21]–[23], [25]

### 3.2.1. Viscosity

Figure 3 shows the influence of concentration on the viscosity of methyl cellulose solutions. Measuring viscosity by different methods may result in different apparent values, depending on the shearing force applied in the apparatus (Ubbelohde or Ostwald capillary viscometer, Hoeppler falling ball method, Ford beaker, etc.). This effect is caused by non-Newtonian flow or pseudoplasticity of methyl cellulose solutions. Pseudoplasticity becomes clearly obvious when the viscosity is determined in a rotary viscometer (e.g., made by Haake, Brookfield, Contraves, Epprecht, Fann). The apparent viscosities decrease at increasing rotational speeds (Fig. 4).

According to STAUDINGER the intrinsic viscosity $[\eta]$ is a fundamental viscosity value, related to the molecular mass ($\overline{M}$) by

$$[\eta] = K \times M^{\alpha}$$

$K$ and $\alpha$ are polymer-specific constants. The intrinsic viscosity $[\eta]$ is defined as the reduced viscosity $\eta_{\text{red}}$, extrapolated to zero concentration. The reduced viscosity values

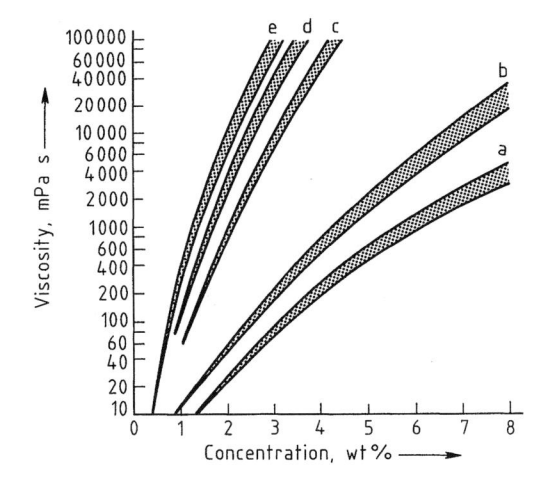

**Figure 3.** Viscosity vs. concentration for commercial types of hydroxyethyl methyl cellulose at 20 °C in aqueous solutions (Hoeppler viscometer) [12]
The fields indicate usual tolerances
Viscosity types: a) 20; b) 50; c) 1000; d) 4000; e) 10 000 (viscosities of 2% solutions, mPa s)

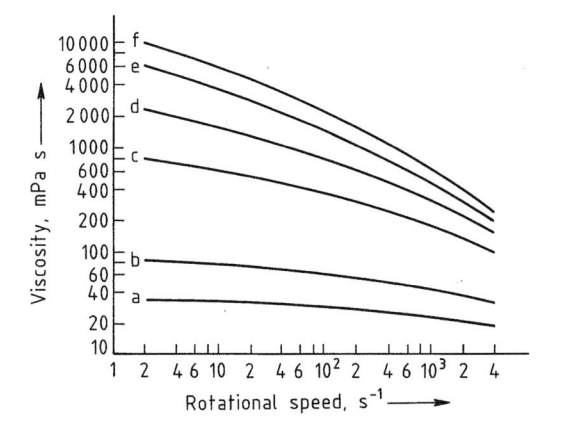

**Figure 4.** Viscosity vs. rotational speed for 2% hydroxyethyl methyl cellulose solutions of different viscosity types
a) 20; b) 50; c) 300; d) 1000; e) 4000; f) 10 000 (viscosities of 2% solutions measured at low shear in a Hoeppler viscometer, mPa s) [12]

$\eta_{red}$ can be calculated from $\eta_{red} = (\eta/\eta_0 - 1)/c$ by measuring the apparent viscosities $\eta$ at concentrations $c$, with $\eta_0$ being the apparent viscosity of the pure solvent.

The PHILIPOFF equation, relates the intrinsic to the apparent viscosity by $\eta/\eta_0 = (1 + [\eta] \cdot c/8)^8$. For long-chain, highly viscous cellulose ethers, this relation is better described by $\eta/\eta_0 = (1 + [\eta] \cdot c/9)^9$. These equations may be used for the determination of $[\eta]$ by a single measurement; cellulose ether concentrations should be about 0.2% in this case.

In cellulose ethers it is usually difficult to calculate $\overline{DP}$ and $\overline{M}$ by the Staudinger equation because $K$ and $\alpha$ are not known. Therefore, osmometric or light-scattering methods are preferred for the determination of the average molecular mass and degree of polymerization (Table 3).

Viscosities of methyl cellulose are constant within the pH range of 2 – 12. When the temperature is increased, the viscosity decreases. However, just before gelation occurs,

**Table 3.** Viscosities and molecular masses of methyl cellulose

| Apparent viscosity at 2% and 20 °C, mPa s | Intrinsic viscosity [$\eta$], dL/g | Average molecular mass $\overline{M}$* | Average degree of polymerization $\overline{DP}$* |
|---|---|---|---|
| 10 | 1.4 | 13000 | 70 |
| 40 | 2.05 | 20000 | 110 |
| 100 | 2.65 | 26000 | 140 |
| 400 | 3.90 | 41000 | 220 |
| 1500 | 5.7 | 63000 | 340 |
| 4000 | 7.5 | 86000 | 460 |
| 8000 | 9.3 | 110000 | 580 |
| 15000 | 11.0 | 120000 | 650 |
| 19000 | 12.0 | 140000 | 750 |

* Osmometric method.

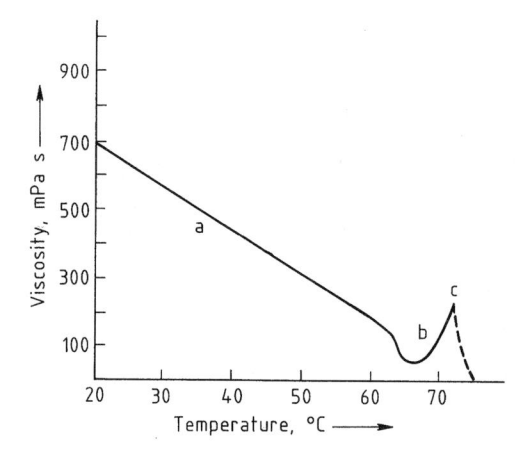

**Figure 5.** Viscosity vs. temperature of an aqueous methyl cellulose solution [12]
a) Normal decrease; b) Gelation interval; c) Coagulation

the viscosity increases in a small interval and then decreases sharply, due to flocculation of the polymer (Fig. 5).

## 3.2.2. Gelation

Gelation temperatures of various methyl cellulose types of a given substitution pattern decrease slightly with increasing viscosity of the solution (measured at ambient temperature); the viscosity increase may be caused either by increased concentration or by increased molecular mass (increasing intrinsic viscosity and average degree of polymerization, $\overline{DP}$).

Gelation temperatures are lowered by adding electrolytes; this effect depends on the kind of salt added. Some salts even prevent the dissolution of methyl cellulose at ambient temperature, or they cause flocculation when they are added in sufficient amount. Table 4 lists data for salt intolerance of HEMC. Both heat and the presence of electrolytes destroy hydrated structures that keep the polymer in solution by breaking hydrogen bonds between water and the polymer; this results in agglomeration. When

**Table 4.** Salt intolerance of hydroxyethyl methyl cellulose 2% aqueous solution at 20°C (viscosity: 300 mPas)

| Salt | Salt concentration causing coagulation, wt% |
|---|---|
| $MgSO_4$ | 4 |
| $Na_2CO_3$ | 4 |
| $Na_2SO_4$ | 5 |
| $Al_2(SO_4)_3$ | 5 |
| $Na_3PO_4$ | 5 |
| $(NH_4)_2SO_4$ | 6 |
| $FeSO_4$ | 7 |
| $CH_3CO_2Na$ | 8 |
| NaCl | 13 |
| KCl | 20 |
| $NH_4Cl$ | 25 |
| $CaCl_2$ | 30 |
| $NaNO_3$ | 30 |
| No coagulation in saturated solutions of $MgCl_2$, $FeCl_3$, and $CaSO_4$ | |

such strongly polar water-miscible organic solvents as alcohols or glycols are added, methyl cellulose solutions are stabilized and gelation temperatures increase because these solvents form stronger hydrogen bonds and more stable soluble complexes with the polymer. Methyl celluloses are incompatible with some specific additives, such as tannin, which form insoluble complexes even at low additive concentration. Most hydrocolloids are compatible with methyl celluloses, with slight gelling occurring in some cases. When isoviscous solutions of an anionic polymer, such as carboxymethyl cellulose, are mixed with nonionic methyl cellulose solutions, the viscosity increases due to synergic effects.

### 3.2.3. Solubility

Methyl cellulose becomes soluble in ethanol when the DS value is greater than 2.1. If the DS value exceeds 2.4, methyl cellulose is also soluble in such solvents as acetone or ethyl acetate. Above a DS value of 2.7, it is even soluble in some hydrocarbons (benzene, toluene). Commercial mixed ethers, such as HEMC and HPMC, are organic-soluble if the methyl DS value is in the range of 1.9–2.0 and the hydroxyalkyl MS value is about 0.3, especially in the case of HPMC. These types are also soluble in water. Good solubilities in mixtures of chlorinated hydrocarbons and ethanol or methanol are achieved by almost all types of water-soluble mixed methyl ethers. The viscosities in organic solutions depend on the kind of solvent or on the ratio of components in the case of mixtures (Fig. 6). Hydroxybutyl methyl cellulose is organic-soluble at a methyl DS value of 1.8–1.0 with only a little hydroxybutylation (MS value 0.05).

Methyl cellulose powders are best dissolved in water when the product is first slurried in hot water above the gelation temperature and is then cooled by adding

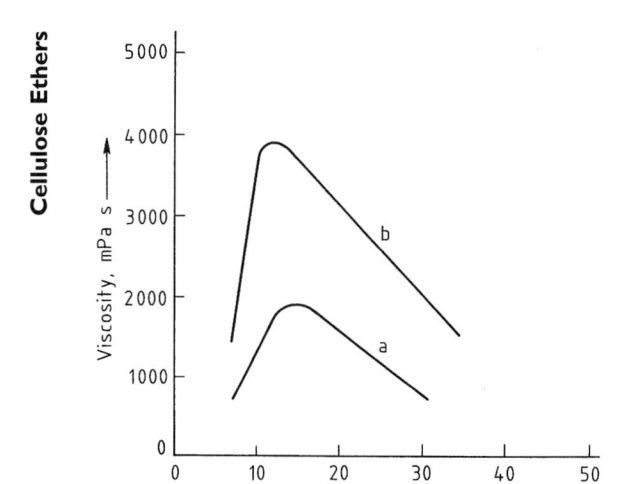

**Figure 6.** Relation between viscosity in organic solution and composition of solvent mixture ($CH_2Cl_2$–$CH_3OH$) for highly substituted hydroxyethyl methyl cellulose, 5 min after mixing Viscosities: a) 200 mPa s; b) 400 mPa s (viscosities of 2% aqueous solution at 20 °C, Hoeppler viscometer)

cold water to the well-stirred slurry. In this way, formation of lumps is avoided. Granular products may be added directly to cold water while the mixture is being stirred; the dissolution proceeds continuously, but somewhat more slowly in that case.

Methyl celluloses act as surfactants in aqueous systems. They reduce surface tension and support emulsification of two-phase layers. Foaming can be suppressed by usual defoaming agents such as silicon oil. The amphiphilic character is due to the presence of both hydrophilic OH and hydrophobic $OCH_3$ groups in the polymer.

### 3.2.4. Other Properties

Methyl celluloses have excellent water retention properties. This is used in cement and gypsum formulations, and also in water-based paints and in wallpaper adhesives, where the products' cohesiveness is also important. Water retention increases with an increasing amount of hydrophilic groups and increasing viscosity. Figure 7 shows such a relation between water retention and concentration.

Mixed hydroxyalkyl ethers are excellent film formers. The films have high water and oxygen permeability and are not affected by UV irradiation. They take up water reversibly when exposed to humid air; the amount corresponds to the relative humidity. Moreover, these ethers are oil-resistant and may be plasticized by common plasticizers. Water stability can be achieved by crosslinking with formaldehyde, phenols, or other resin-forming agents. Hydroxypropyl and hydroxybutyl methyl cellulose in the organic-soluble range are thermoplastic; in the presence of selected organic solvents at 120–190 °C, they can be extruded to sheets, which are useful for water-soluble, heat-sealable bags and envelopes. Low-viscosity types of methyl cellulose act as protective colloids in the emulsion polymerization of vinyl chloride.

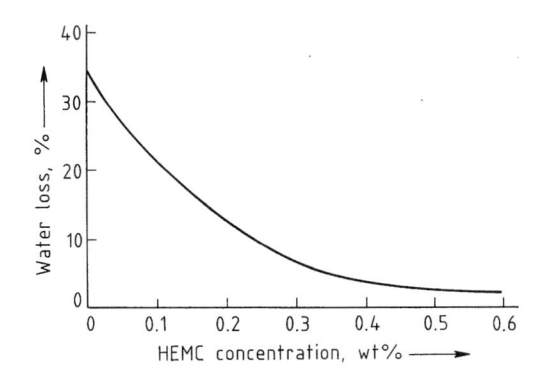

**Figure 7.** Water loss vs. concentration of hydroxyethyl methyl cellulose (dry basis) in a gypsum-based plaster [26]
HEMC type: 200 mPa s for 2% aqueous solution at 20 °C

## 3.2.5. Retarded Dissolution [27]

For most uses lump-free dissolution of the cellulose ether is absolutely mandatory. In some cases retarded dissolution may be desirable. This allows the addition of cellulose ether to a multicomponent formulation in water without an initial significant increase in viscosity. In this way stirring is still facilitated and such components as cements, paints, or adhesives may be added and thoroughly mixed. Viscosity then increases abruptly to the level desired in the final product.

This retardation is achieved at ambient temperature by incorporating glyoxal [107-22-2] as an aqueous solution of pH 4–5 into the water-wet material before it is dried. During drying, crosslinking takes place as the glyoxal reacts to form hemiacetals with the free hydroxyl groups of two polymer chains:

2 Cell – OH + OHC – CHO $\longrightarrow$ Cell – O – CHOH – CHOH – O – Cell

where Cell = cellulose ether chain.

This crosslinked product is insoluble in water and can be readily dispersed. During the hydration time, the period of time after dispersion in an aqueous layer, there is only a limited amount of swelling. Meanwhile, the hemiacetal structures slowly hydrolyze at a steady rate until the moment the very last bonds break. The chains are then liberated all at once, causing rapid dissolution and a rapid rise in viscosity. The rate of hydrolysis increases with temperature and with deviation from the pH range 4–5, the pH of maximum hemiacetal stability. The retardation is effectively suppressed in strongly acid solutions (pH < 2) and in alkaline solutions (pH > 9). Figures 8 and 9 illustrate this behavior at ambient temperature.

Sometimes an aqueous slurry of glyoxal-crosslinked product is prepared with pH ≈ 5. Viscosity development is then started by buffering the slurry to a slightly alkaline pH.

Glyoxal crosslinking is not only applied to methyl celluloses, but also to other cellulose ethers and to other OH-containing polymers, such as starch or natural gums. The retardation time is determined by the amount of glyoxal, the pH before drying, and the time and temperature of drying. Overheating or very low pH of the wet material

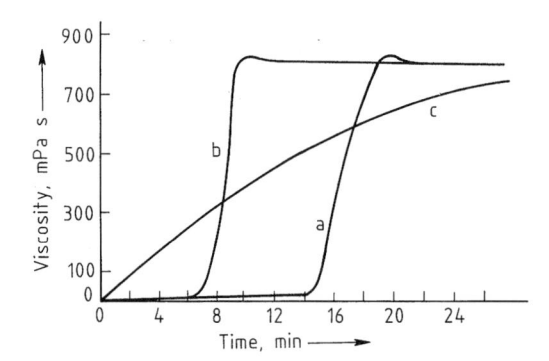

**Figure 8.** Viscosity of 2% aqueous solutions of hydroxyethyl methyl cellulose at 20 °C and pH 6 a) Powder, highly crosslinked with glyoxal (hydration time 15 min) b) Powder, weakly crosslinked (hydration time 7 min) c) Granular type, not crosslinked

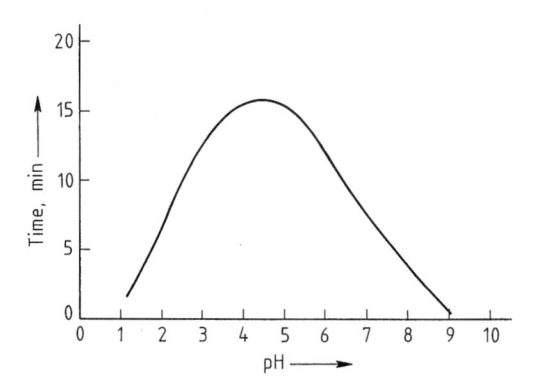

**Figure 9.** Hydration time vs. pH of a glyoxal-crosslinked hydroxypropyl methyl cellulose Maximum stability is at pH 4–5

causes irreversible crosslinking by the conversion of hemiacetals into acetals, and the cellulose ether then becomes completely insoluble.

Dialdehydes other than glyoxal, e.g., glutaraldehyde, are known to produce analogous effects, but are less efficient and not used industrially. Manufacturers' statements for retarding dissolution times generally are based on conditions of pH 6–7 and ambient temperature. Slight retardation at alkaline pH may be achieved by incorporation of borax instead of or in addition to glyoxal. Glyoxal-crosslinked products of any kind are not approved as food additives.

# 4. Ethyl Cellulose Ethers

The ethyl group of ethyl cellulose (EC) and mixed ethyl celluloses is more hydrophobic than the methyl group. Therefore, EC is industrially produced as a typically organic-soluble and water-insoluble type. Ethyl hydroxyethyl (EHEC) and ethyl methyl cellulose (EMC) are typical commercial mixed ethers. The types containing small amounts of ethyl groups are soluble in water.

## 4.1. Ethyl Cellulose (EC) [8], [9], [12], [28]

Ethyl cellulose is water-soluble from an ethyl DS value of 0.7 up to about 1.7. Above a DS value of 1.5, it is organic-soluble. Commercial EC that is made in the United States has DS values between 2.2 and 2.6. These types are soluble in mixtures of hydrocarbons and lower alcohols as well as in esters, ketones, or ethers of low molecular mass. Ethyl cellulose is a white or slightly yellowish powder. It is somewhat susceptible to autoxidation in air under strong light irradiation because of peroxide formation at the ethoxy groups. Antioxidants are used as stabilizing additives. Ethyl cellulose is highly bioresistant because of its high DS level. Various viscosity grades are available; the viscosity is generally standardized for 5 wt% solutions of EC in a mixture of toluene (80 vol%) and ethanol (20 vol%).

The production of EC using ethyl chloride [75-00-3] as the etherifying agent resembles that of MC, but higher temperatures of about 110 °C are required as well as longer reaction times of 8–16 h. Viscosity levels are adjusted by aging alkali cellulose before etherification or by controlled oxidation with air during the process. About one half of the ethyl chloride is consumed in side reactions, leading to ethanol and diethyl ether. These byproducts can be reconverted to ethyl chloride by high-temperature hydrochlorination in the liquid or gas phase in the presence of catalysts, e.g., $ZnCl_2$. The efficiency of the etherification, i.e., the fraction of the consumed ethyl chloride that is converted into cellulose ether groups, increases with the concentration of the NaOH used for alkalization; in general, solutions of 55–76% NaOH are applied. Greater efficiency is achieved in a two-stage process by adding solid alkali to the reaction mass after the initial amount of NaOH has been consumed. Salts are removed from the crude product by successive washing with water.

Ethyl cellulose is thermoplastic and can be extruded into plastic parts retaining their mechanical properties both in water and at low temperature. Sheet forming is possible by adding only small amounts of plasticizers. Ethyl cellulose is compatible with many waxes and resins, if the ethyl DS value does not exceed 2.6. The softening of commercial EC begins at about 130 °C; the flow temperatures range from 140 to 160 °C. Ethyl cellulose is also used in lacquers, hot-melt adhesives, and as a binder for tablets.

## 4.2. Mixed Ethyl Cellulose Ethers [29]

Berol in Sweden produces water-soluble *ethyl hydroxyethyl cellulose* (*EHEC*) of various viscosity levels. It has an ethyl DS value of 0.9 and a hydroxyethyl MS value of 0.8 or 2.0. Ethyl groups in these products are bound directly to the cellulosic hydroxyls, not to the hydroxylic end groups of hydroxyethyl or related oligomeric ether structures. The EHEC products have properties similar to those of HEMC or HPMC (see Chapter 3). They undergo thermal gelling in neutral aqueous solution at 65–70 °C. Some types are reversibly crosslinked with glyoxal. The overall high substitution (including both ether species) causes high bioresistance. Other types of EHEC with a higher ethyl DS value and a lower hydroxyethyl MS value are produced in the United States as organic-soluble products comparable to EC.

# 5. Hydroxyalkyl Cellulose Ethers
[8], [9], [12], [30]–[33]

Commercial hydroxyalkyl celluloses are hydroxyethyl cellulose (HEC), hydroxypropyl cellulose (HPC), and such mixed ethers as hydroxyethyl hydroxypropyl cellulose (HEHPC), carboxymethyl hydroxyethyl cellulose (CMHEC), and specialties with nitrogen-containing basic or cationic groups in the second substituent. Within this class, HEC is produced on the largest scale. Other mixed ethers containing minor amounts of hydroxyalkyl groups have been described in Chapters 3 and 4.

## 5.1. Hydroxyethyl Cellulose (HEC)

The important, nonionic HEC is produced by only a few companies in the world. The largest capacities exist in the United States, others in Great Britain, in the Netherlands, and in the Federal Republic of Germany (Hercules, Union Carbide, BP, and Hoechst).

Research samples of HEC are known to be soluble in water at MS levels of 1.3, or even less. However, commercial HEC is substituted at MS values between 1.7 and 3.0. The corresponding DS values for all these water-soluble types range from 0.8 to 1.2. Analysis of products that have been further hydroxyethylated (MS value > 3.0) has not revealed higher DS levels. The additional ether groups are then mainly formed at the hydroxyl groups of the hydroxyethyl substituents or oligomeric ether side chains already present in the molecule. The average length of side-chain ether groups, defined as MS/DS, is calculated to be 1.5–2.5 in commercial HEC. Longer chains make production and application more difficult, although water solubility is retained. The common HEC types are not clearly soluble in such organic solvents as acetone, ethanol, or lower ethers. However, at high MS values hydroxyethyl cellulose becomes partially soluble in methanol and in some mixtures of water and water-miscible organic solvents even if the latter is the major component of the mixture.

### 5.1.1. Production of Hydroxyethyl Cellulose

Hydroxyethyl cellulose (HEC) is synthesized by reaction of alkali cellulose with ethylene oxide [75-21-8] in a slurry process. Acetone, isopropyl alcohol, *tert*-butyl alcohol, 1,2-dimethoxyethane, or a mixture of lower hydrocarbons with minor amounts of lower alcohols is used as the medium for dispersion. Cellulose is alkalized with

0.8 – 1.5 mol of NaOH per mole of anhydroglucose either before adding the solvent or by pouring aqueous NaOH directly into the stirred slurry. The water content depends on the kind and amount of the solvent used. In isopropyl alcohol slurries, 10 – 12 mol of water per mole of anhydroglucose have been reported most favorable. After the required amount of ethylene oxide has been added, the slurry is heated at 30 – 80 °C. Hydroxyethylation takes place within 1 – 4 h. The desired MS value is adjusted by the amount of reagent applied. About 70 % of the ethylene oxide reacts with cellulose to form ether groups. The remainder forms glycols by reaction with water, or monoethers of glycols by reaction with the solvent alcohols. For MS levels higher than about 2.5, the ethylene oxide efficiency decreases to 50 % or less, because side reactions are then preferred.

For greater efficiency, the ethylene oxide may be added in two stages. A minor amount is first reacted to form an HEC with an MS value below 1.0, which is insoluble in water and can be isolated after neutralization and salt extraction. This product is further etherified by using the major part of the ethylene oxide under similar conditions; however, in the second stage only catalytic amounts of NaOH are added to suppress side reactions. The etherification in the second stage does not require much alkali because the HEC resulting from the first stage (MS value < 1) is sufficiently activated by the hydroxyethyl groups; these act as spacers and provide for an amorphous structure. The procedure may be simplified by partially neutralizing the excess NaOH of the first stage to catalytic remainders before adding the second batch of ethylene oxide; the intermediate product is not isolated. Low amounts of alkali in the second stage not only increase ethylene oxide efficiency, but also make substituent distribution more uniform, which increases the bioresistance of the resulting HEC.

After neutralization with nitric, hydrochloric, or acetic acid, salts are removed by repeated extractions with solvent mixtures containing water and/or methanol. Before the mixture is dried, glyoxal may be added to the wet HEC to obtain products with retarded dissolution (see Section 3.2.5). Decreased viscosity can be achieved by conventional aging of alkali cellulose (2.2). Crude HEC may also be degraded by adding appropriate amounts of hydrogen peroxide to the hot alkaline slurry before cooling and neutralization.

## 5.1.2. Properties and Selected Uses

Commercial HEC is available in viscosity grades from about 10 to 100 000 mPa s (2 % aqueous solution at ambient temperature). Viscosity decreases with increasing temperature, but no thermal gelling occurs (Fig. 10).

Most HEC products are soluble not only in water, but also in mixtures of water and water-miscible organic solvents at water contents above 40 %. The compatibility with electrolytes is higher than that of methyl celluloses and other nonionic cellulose ethers. Salts with monovalent anions generally do not precipitate HEC from solution, even at

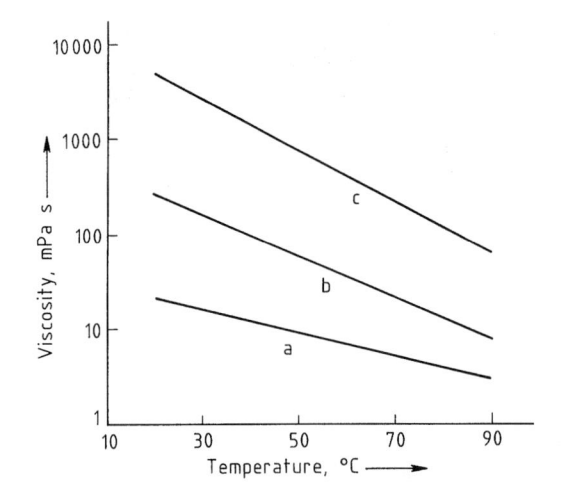

**Figure 10.** Viscosity of hydroxyethyl cellulose solutions vs. temperature
Types: a) 20; b) 300; c) 4000 (viscosities of 2% aqueous solutions at 20 °C, mPa s) [12]

**Table 5.** by electrolytes

| Salt | Salt concentration, causing coagulation, wt% | |
|---|---|---|
| | at 20 °C | at 75 °C |
| $Na_2SO_4$ | 13 | 8 |
| $MgSO_4$ | 12 | 7 |
| $Al_2(SO_4)_3$ | 9 | 7 |
| NaOH | 21 | 17 |
| $(NH_4)_2SO_4$ | 16 | 12 |
| $Na_2S_2O_3$ | 17 | 14 |
| $Na_2HPO_4$ | <5 | – |

No coagulation up to 30% salt at 20 °C with $NH_4NO_3$, $CaCl_2$, $MgCl_2$, NaCl, $NaNO_3$, $ZnCl_2$

high salt concentration or saturation. Electrolytes with divalent anions have limited compatibility with HEC at ambient and elevated temperature, as shown in Table 5.

However, higher temperatures may cause thermal gelling of solutions containing large amounts of chlorides or nitrates, especially at high MS values of the HEC.

Solutions of HEC are generally compatible with most water-soluble polymers including all other cellulose ethers, but they are incompatible with tannin, some phenols, and special complex salts, e.g., sodium molybdatophosphate, which form insoluble complexes with HEC. Solutions of HEC show pseudoplastic behavior similar to that of methyl celluloses. Generally, HEC is biodegraded by cellulolytic enzymes. Only very high and uniform substitution may protect the polymer against enzyme attack. Such products are declared "biostable," a property desired for paint formulations. Bioresistance is measured by comparing the viscosities of HEC solutions before and after enzymatic attack under defined conditions.

The surface tension of water is slightly decreased by HEC, but the surfactant effect of HEC is lower than that of methyl celluloses. Therefore, it causes only slight foaming, which can easily be suppressed by common antifoaming agents.

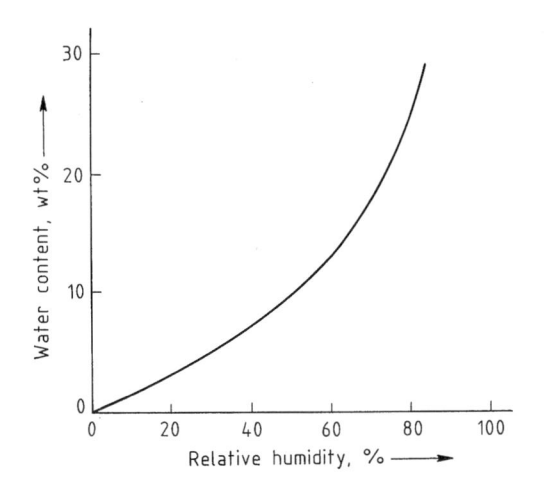

**Figure 11.** Maximum water take-up of an HEC film vs. relative humidity at ambient temperature [12]

Hydroxyethyl cellulose forms clear films, which can be plasticized by common agents such as glycerol, sorbitol, or polyglycols, present at levels of 5–30%. Films of HEC take up water from the air in a reversible manner, depending on the relative humidity (Fig. 11).

Films made from HEC are normally water-soluble. They can be made water-resistant temporarily or permanently by crosslinking with dialdehydes, or urea – or melamine – formaldehyde resins; these agents must be added in sufficient quantities to the HEC solutions before casting.

*A typical formulation of a water-resistant film* is 5 parts of HEC, 12 parts of 40 wt % glyoxal solution, and 83 parts of water. This solution is cast onto a plate and evaporated at 105 °C until dry and clear. The film resists softening for up to 1 h when soaked in water at 20 °C. Films can be made insoluble, even in boiling water, by adding 0.15 parts of $Al_2(SO_4)_3$ to the casting solution and by curing the dry film at 160 °C for 1 min.

Hydroxyethyl cellulose acts as a thickener, binder, stabilizer, film former, and protective colloid in a wide variety of uses. It is utilized in oil recovery and in cement formulations for its good water-retaining properties as well as in the textile and paper industries or in latex polymerization techniques. Although no toxic effects of HEC are known, it is not approved as a food additive, but it can be used in food-packaging adhesives or in paper and paperboard, either in contact with food or not.

## 5.2. Hydroxypropyl Cellulose [9], [11], [12], [32], [35], [36]

Hydroxypropyl cellulose is produced in the United States as a highly substituted nonionic cellulose ether with an MS level of about 4. The viscosities of commercial grades range from about 10 to 30 000 mPa s (2% aqueous solution, 25 °C). As for MC or HEC, the viscosity does not depend on pH in the range between 2 and 12. The products are soluble in cold water. They undergo thermal gelation at about 40 °C and precipitate completely at about 45 °C.

Hydroxypropyl cellulose is manufactured under pressure in a slurry process similar to that used for HEC, using propylene oxide [75-56-9] as the etherifying agent. The reaction medium may be hexane or liquid propylene oxide itself. The reaction of alkali cellulose with propylene oxide is slower than that with ethylene oxide and requires higher temperatures or longer reaction times. The crude product is purified like methyl cellulose by washing with hot water.

The high level of substitution and the more hydrophobic character of the hydroxypropyl and related oligomeric ether groups make commercial HPC soluble in polar organic solvents or mixtures with small amounts of water. Thus, HPC is soluble in methanol, ethanol, isopropyl alcohol – water (95:5), ethylene glycol monomethyl ether, chloroform, pyridine, dioxane, and cyclohexanone. It is poorly soluble in acetone, butanone, methyl acetate, isopropyl alcohol – water (99:1), dichloromethane, and ethylene glycol monobutyl ether. It is insoluble in aliphatic hydrocarbons, glycerol, benzene, toluene, tetrachloromethane, and vegetable oils. Aqueous HPC solutions are susceptible to strong electrolytes; precipitation occurs at ambient temperatures by adding 2 – 5% of $Na_2HPO_4$, $Na_2SO_4$, $Na_2CO_3$, $Al_2(SO_4)_3$, or $(NH_4)_2SO_4$ and about 10% of NaCl or $CH_3COONa$. Hydroxypropyl cellulose reduces the surface tension of water in a manner similar to methyl cellulose (43.0 mN/m at 0.2%), causing substantial foaming. It is a thermoplastic material and can be extruded at 160 – 180 °C without the need of a plasticizer. It takes up little water, not exceeding 12% at a relative humidity of 90%. Therefore, films or articles made from HPC are resistant to humidity and not tacky, although they are water-soluble. Highly concentrated solutions of HPC show liquid-crystal properties. Hydroxypropyl cellulose is approved as a food additive and is nontoxic and physiologically inert. Tests on humans have disclosed no evidence for skin irritation. Therefore, it is used as a stabilizer for whipped foods and in cosmetic emulsions. Its good film-forming properties make HPC convenient for coatings in the pharmaceutical and food industries.

## 5.3. Mixed Ethers

*Hydroxyethyl hydroxypropyl cellulose (HEHPC)*, an ether with mainly HP substitution, is produced in the United States as a material somewhat similar to HPC. The thermogelling properties of HEHPC in aqueous solution resemble those of methyl cellulose; i.e., gelling temperatures are higher than those of HPC [37].

Another commercial ether, also based on HPC, contains a minor amount of *aminoethyl groups* as secondary substituents (DS value of about 0.2), which may be introduced by using aziridine as the second etherifying reagent. The thermogelling properties of this ether resemble those of HPC in an aqueous medium from alkaline to neutral (gelation temperature 40–45 °C). At pH of 6 and less, the gelation temperature rises to about 95 °C, because the aminoethyl groups then form hydrophilic ammonium cations, which are stable in acid. When the product alone is dispersed in water, the pH rises to 9 because of the basic nature of the amino groups [38].

A mixed ether of some commercial importance is a *modified hydroxyethyl cellulose* with secondary substitution by cationic 3-trimethylammonium-2-hydroxypropyl ether groups, $(CH_3)_3N^+-CH_2-CHOH-CH_2-$, chloride being the counteranion. The DS value of the cationic groups of this product is 0.12–0.15, whereas the hydroxyethyl MS value is in the normal HEC range of about 1.9. The cationic group is introduced by using 2,3-epoxypropyltrimethylammonium chloride [*3033-77-0*], which is made according to the following scheme:

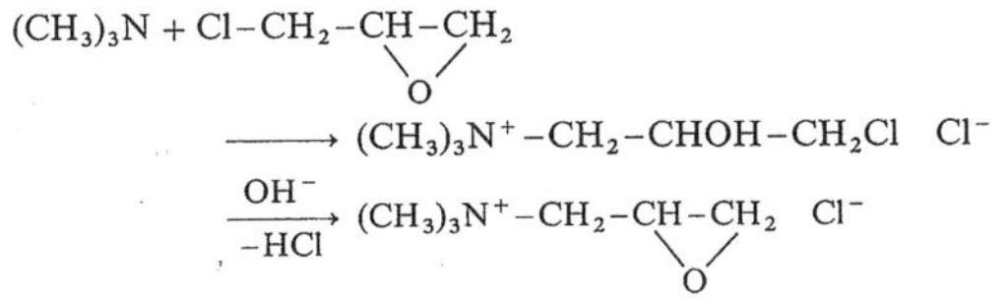

$$(CH_3)_3N + Cl-CH_2-CH-CH_2 \overset{}{\underset{O}{\diagdown\diagup}}$$

$$\longrightarrow (CH_3)_3N^+-CH_2-CHOH-CH_2Cl \quad Cl^-$$

$$\xrightarrow[-HCl]{OH^-} (CH_3)_3N^+-CH_2-CH-CH_2 \quad Cl^-\ \underset{O}{\diagdown\diagup}$$

The chlorohydrin intermediate is available commercially and may also be used for etherification. It reacts under the alkaline conditions via the epoxide. The resulting cationic mixed cellulose ether does not change its cationic character at any pH, because of the stable quaternary ammonium structure [39]. It has the same properties as HEC of comparable MS value, but moreover shows affinity to such polymers as keratin. It is, therefore, mainly used in hair-care products (shampoos, conditioners, or wave sets) as a thickener and hair-protecting agent. This modified HEC is available in different viscosity grades.

Another commercially important mixed ether, *carboxymethyl hydroxyethyl cellulose (CMHEC)*, is treated in Section 6.3.

# 6. Carboxymethyl Cellulose (CMC)

[8], [9], [11], [12], [40]–[43]

Sodium carboxymethyl cellulose is a water-soluble salt. It is manufactured by many companies throughout the world; large capacities exist in the United States and in Western Europe. Production of CMC is simpler than that of most other cellulose ethers because all reagents are solid or liquid and allow operation at atmospheric pressure. The etherifying reagent, sodium chloroacetate [*3926-62-3*] or chloroacetic acid [*79-11-8*], is easy to handle and very efficient. For this reason and because of its versatile properties as a thickener, film former, protective colloid, and water-retaining agent, CMC has become the largest industrial cellulose ether. Large quantities are produced in crude commercial grades without any refining for use in detergents, oil-drilling, and in the paper industry. High-purity grades are employed as food additives. Food-grade CMC is commonly known in the United States as *cellulose gum*. Viscosities range from 10 to over 50 000 mPa s for 2% aqueous solutions. The DS levels range from 0.3 to 1.2, but clear and fiber-free CMC solutions require a minimum DS value of about 0.5.

## 6.1. Production

For CMC production, alkali cellulose is generally made first by conventional techniques, using at least 0.8 mol of NaOH per mole of anhydroglucose. The Williamson reaction with sodium chloroacetate then requires at least stoichiometric amounts of alkali, but a slight excess of about 5% is required to complete the reaction in a reasonable period of time. If chloroacetic acid is employed, another mole of alkali is needed for its conversion into the sodium salt. Alternatively, the cellulose may first be mixed with a concentrated aqueous solution of the reagent by spraying or steeping, and alkali is then added to start etherification. There is no need for an inert solvent, but such solvents as acetone, isopropyl alcohol, or *tert*-butyl alcohol enhance the diffusion of alkali and reagent, which results in better reagent use and more uniform substitution, leading to better solubility in water. Solvents are employed either in small quantities to obtain an evenly wet reaction mass or in larger amounts for slurry processes.

The exothermic reaction requires little or even no initial heating, but sometimes cooling is needed to maintain the temperature between 25 and 70 °C. Shredders are suitable as reaction vessels; they are equipped with toothed, $\Sigma$-shaped blades and a cooling jacket. Rotary drums to tumble the mass during the reaction or double screw drives, which allow continuous processing of the wet reaction mass, are also suitable. For crude products it is also possible to transfer the well-mixed reaction mass to bins or wagons, in which the mass is shipped directly to the place of use, e.g., an oil-drilling station. The etherification is completed during transportation, and the crude, slightly alkaline product is ready for use on arrival. Slurry processes for higher qualities are

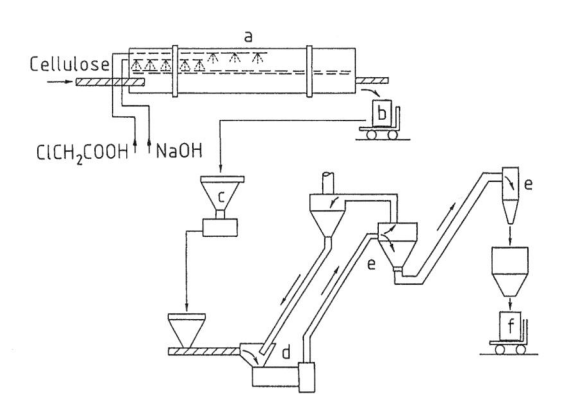

**Figure 12.** Continuous production of crude carboxymethyl cellulose using powdered cellulose, aqueous chloroacetic acid, and NaOH (Wyandotte Corp.) [12]
a) Rotary drum with spraying devices; b) Container for completing the reaction; c) Mill; d) Dryer; e) Cyclones; f) Transportation bin

carried out in simple vessels with heating and stirring. Figure 12 shows a schematic of a CMC manufacturing plant.

Crude products may contain up to 40 % salt (on dry basis). Etherification efficiency of the reagent ranges from 60 to over 80 %; the remainder is transformed into sodium glycolate by hydrolysis of chloroacetic acid. The reaction mass or slurry may be neutralized by hydrochloric or acetic acid. Decreased viscosities are adjusted by degradation of the product with hydrogen peroxide in the alkaline reaction mixture before neutralization. For purification, the salts, mainly sodium chloride, are extracted with methanol or methanol–water mixtures before drying.

## 6.2. Properties and Selected Uses

Most CMC solutions are highly pseudoplastic. Some of them are almost solid gels, which become fluid by vigorous stirring. They often show thixotropic behavior; i.e., shearing decreases the viscosity. This does not occur instantly but gradually, needing some constant shearing before the lower level is reached. After the shearing is stopped viscosity rises only slowly to the initial value. Highly and/or very uniformly substituted CMC products are less thixotropic.

In neutral CMC solutions, the molecules are arranged in uncoiled linear structures because of electrostatic repulsion of the anionic charges. Sodium carboxymethyl cellulose is the salt of a weak acid ($pK_a$ 4–5); at pH 5 about 90 % of the carboxyl groups are undissociated. If the pH is decreased further, precipitation occurs, depending on the conditions of acidification and on the DS value. Carboxymethyl cellulose with a DS value of 0.3–0.5 precipitates below pH 3; if the DS value is 0.7–0.9, precipitation occurs below pH 1. Repulsion between the carboxyl groups also decreases at high pH of about 12 because of the presence of alkali-metal cations, resulting in coiled chains and a drop in viscosity. Therefore, the viscosity of CMC solutions depends on pH; in particular, the highly viscous types exhibit a maximum at pH 6–7 (Fig. 13).

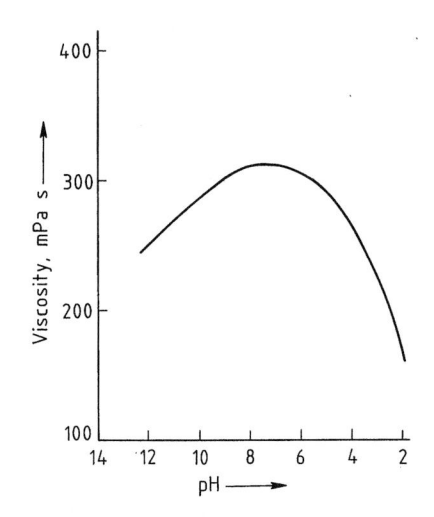

**Figure 13.** Viscosity of carboxymethyl cellulose vs. pH
(2% aqueous solution, 20 °C, DS value of 0.75) [12]

Carboxymethyl cellulose solutions are incompatible with heavy-metal salts (e.g., Cu, Ag, Pb), with trivalent cations (e.g., Al, Fe, Cr), and with cationic detergents and polymers. When these agents are added, CMC precipitates as an insoluble salt or complex. Such strong electrolytes as sodium salts decrease viscosity at low salt concentration by coiling the CMC chains, and sometimes enhanced thixotropy is observed. High salt concentrations may precipitate the cellulose ether. The addition of calcium salts leads to hazy solutions and may precipitate CMC of a low DS value. Generally, CMC becomes more soluble with increasing substitution (DS value of 1.2), which also makes the product more resistant to enzymatic degradation.

Carboxymethyl cellulose forms films which are unaffected by most organic liquids but become weak and flexible at high humidity, because they may absorb a large amount of water. Crosslinking with acid, alum, formaldehyde, or polyfunctional resins increases water resistance of CMC films or sheets. Slight crosslinking also occurs when solid CMC (neutral salt form) is heated at 80 °C for about 12 h. The few free acid groups present in this product then form esters with free hydroxyl groups of a neighboring chain. A high degree of ester crosslinking is achieved if a low pH is adjusted in the manufacturing process before drying. However, such partially crosslinked products are swellable in water and have a very high water-absorption capacity; this makes them useful for application in hygienic articles, e.g., diapers or tampons. The ester groups are hydrolyzed by alkali, thus rendering these derivatives alkali-soluble. Highly water-absorbing CMC can also be produced by adding small amounts of a second bifunctional reagent during etherification [44]–[46].

Carboxymethyl cellulose stabilizes aqueous suspensions of clay because of its high water retention; therefore, it is used by the oil industry to prepare stable drilling muds. It can protect textile surfaces, which makes it useful as a soil redeposition inhibitor in detergent compositions. Anionic dirt particles are repelled by the anionic charges of CMC. The affinity of CMC to cellulose makes it a useful warp sizing agent for yarns and an effective additive for coating paper and paperboards, where it improves surface and gloss ink printing characteristics. Pure sodium carboxymethyl cellulose is approved as food additive and applied as thickener and stabilizer.

## 6.3. Mixed Ethers with Carboxymethyl Groups [40]

*Carboxymethyl hydroxyethyl cellulose (CMHEC)* combines the excellent salt compatibility of HEC with the surface-protecting and suspension-stabilizing properties of CMC, especially at high-viscosity grades. Compared with CMC, the stability of CMHEC in acidic or strongly alkaline media is greatly improved.

Carboxymethyl hydroxyethyl cellulose can be produced by one- or two-stage reactions analogous to HEC or CMC slurry processes. The DS values of the two substituents can be adjusted to any suitable ratio. Even at a low carboxymethyl DS value (0.3) and low hydroxyethyl MS value (0.4), the products are clearly soluble in water. The increased stability of CMHEC against acids and electrolytes relative to that of CMC is caused by the hydroxyethyl groups that act as spacers and maintain a soluble hydrated structure. Therefore, CMHEC is not precipitated by aluminum salts, but the slight crosslinking action of the aluminum ions results in strong gelling in neutral solution; gelling can be suppressed by adding acid or alkali. All of these properties make CMHEC very useful in oil production, e.g., in drilling muds or completion fluids. The industrial importance of CMHEC has lately increased considerably.

*Carboxymethyl methyl cellulose (CMMC)*, a mixed ether with a low carboxymethyl DS value, is mentioned in Chapter 3. It is used as a binder and as an adhesive for tobacco sheets.

## 7. Other Cellulose Ethers

### 7.1. Ether Esters [12]

In addition to etherification, the OH groups of cellulose may also be esterified with carboxylic or dicarboxylic acids. Among the many possible combinations, *CMC acetate* and *HPMC phthalate* are typical examples.

Some ether ester derivatives of cellulose are suitable as tablet binders, tablet disintegrators, or tablet-coating materials in the pharmaceutical industry. Coatings of such ether esters contain free acid groups which make the product insoluble in neutral or acidic aqueous media but lead to soluble salt structures in slightly alkaline solutions. Therefore, drug release occurs only in the intestines. The free acid groups may be part of the ether component, e.g., acidic carboxymethyl, or part of the ester group, if cellulose is esterified with a bivalent acid, e.g., phthalic, phosphoric, or trimellitic acid.

## 7.2. 2-(N,N-Diethylamino)ethyl Cellulose (DEAEC) [12], [47]

Minor amounts of 2-(N,N-diethylamino)-ethyl cellulose are produced by the Williamson reaction of alkali cellulose with the hydrochloride of 2-chloroethyldiethylamine [869-24-9]. At low DS values (0.10 – 0.15), the product is insoluble in water. It is used as a weakly basic chromatographic material or as an anion exchanger to remove acids from solutions. Swelling in alkaline media may be prevented by crosslinking with epichlorohydrin during the synthesis. Commercial DEAEC products have an ion-exchanging capacity of about 0.7 mmol/g. The nonionic tertiary amino groups may be converted reversibly by acids into the corresponding ammonium salts. Alkylation of DEAEC in the presence of NaOH leads to the formation of quaternary ammonium groups, which results in strongly basic ion exchangers.

## 7.3. Cyanoethyl Cellulose (CNEC) and Carboxyethyl Cellulose (CEC) [8], [10], [48], [49]

Alkali cellulose reacts with acrylonitrile [107-13-1] under mild conditions (e.g., 30 °C, 1 – 2 % aqueous NaOH) to form cyanoethyl cellulose with little or no hydrolysis of the nitrile groups. Higher temperature and alkali concentration lead to undesired acrylonitrile polymerization and liberation of ammonia by nitrile hydrolysis. Highly substituted CNEC (DS value of 2.5 and more) can be produced by quick reaction of alkali cellulose in an excess of acrylonitrile. Although CNEC as such has little commercial significance, cyanoethylation to a low DS value is applied in the paper industry for improving sheet strength, heat resistance, and electrical resistance of some specialties.

A high degree of cyanoethylation leads to organic-soluble products with a remarkably high dielectric constant. The dielectric properties can be further improved by capping residual OH groups (e.g., by acetylation) to minimize dielectric losses. Such products may be employed as an embedding mass for pigments in electroluminescent devices.

When CNEC is heated in strongly alkaline solutions, elimination of acrylonitrile takes place in addition to nitrile hydrolysis. This results in less substituted *sodium carboxyethyl cellulose* (*CEC*). Highly substituted sodium CEC cannot be synthesized by the direct addition of sodium acrylate to alkali cellulose. It is best made by the addition of acrylamide [79-06-1] and subsequent hydrolysis under the alkaline conditions of the reaction. Carboxyethyl cellulose has properties similar to those of CMC, but is not produced industrially.

## 7.4. Other Noncommercial Ethers

*Benzyl cellulose* was manufactured in Germany [8], [9], [50]. It was aimed at substituting celluloid in photographic films, but it was not successful because of its incompatibility with other plastic materials, low tensile strength of films, and low light and heat stability. Etherification was effected with benzyl chloride [*100-44-7*] at about 130 °C, leading to products with a DS value of about 2.

*Sodium 2-sulfoethyl cellulose (SEC)* [10], [47], [51] is an interesting anionic ether. It is made by reaction of alkali cellulose with sodium 2-chloroethanesulfonate [*15484-44-3*] or sodium ethenesulfonate. Sulfoethyl cellulose is water-soluble if the DS value is greater than 0.3, and the compound is highly compatible with large amounts of alkali-metal salts. The addition of acid or of divalent or trivalent cations of heavy metals (e.g., Cu, Ag, Pb, Al, Fe, Cr, Ca, or Ba) does not precipitate SEC from aqueous solutions. Therefore, SEC has properties similar to those of CMC but is more stable.

*Phosphonomethyl cellulose (PMC)* [47], [52], a phosphorus-containing cellulose ether, can be synthesized by the reaction of chloromethane-phosphonic acid [*2565-58-4*] with alkali cellulose, either in the presence of a small amount of water and heating to 130 °C or in an organic slurry at 80–90 °C. After neutralization, the monosodium salt of PMC is obtained, which is water-soluble at a DS value as low as 0.15. If NaOH is added to pH 10, the acid groups are converted into disodium phosphonate. Solutions of PMC are very susceptible to acids and to cations which are not monovalent. The cellulose ether precipitates in the presence of minimal amounts of calcium or aluminum salts.

A more valuable product is the mixed ether *hydroxyethyl phosphonomethyl cellulose (HEPMC)* [53]. Its hydroxyethyl MS value is in the range of 1.6–2.5, and its phosphonomethyl DS value is below 0.15. That ether is highly soluble at any pH, like HEC, and does not coagulate, even when divalent or trivalent cations are added. However, it undergoes strong gelling, which can be adjusted by selecting specific cations for the various pH ranges: in the presence of aluminum or iron ions, gelling occurs only in the neutral range, whereas calcium or lead ions cause gelling of HEPMC solutions in alkaline media. Zirconium salts have no effect at pH greater than 3, but cause efficient gelling in highly acidic media of pH 0 and even less. The gelling characteristics of HEPMC are similar to those of CMHEC, but HEPMC is much more versatile with regard to pH variations.

A wide variety of other cellulose ethers have been prepared, but have not become commercial products. Some laboratory products containing three different common ether groups may become interesting in the future because they allow very precise control of rheological and other important properties. Careful evaluation of the optimum ratios of the different substitution levels is required in that case.

**Table 6.** Tests for the identification of cellulose ethers

| Test method | MC | HEMC | HPMC | HEC | CMMC |
|---|---|---|---|---|---|
| Tannin | + | + | + | + | + |
| Thermal gelation | + | + | + | – | + |
| Diphenylamine | + | + | + | – | + |
| Sodium nitroprussiate | – | + | + | + | – |
| Ninhydrin | – | – | + | – | – |

# 8. Analysis [12], [13], [54], [55]

Identification and characterization of cellulose ethers aims at the following:

1) Estimation of the cellulose ether content in an industrial product composition and isolation of the pure cellulose ether
2) Identification of the kind of ether groups and DS (MS) value analysis
3) Estimation of the average chain length ($\overline{DP}$) and of DP distribution
4) Analysis of the substituent distribution over the 2-, 3-, and 6-positions of the anhydroglucose; analysis of the uniformity of substitution along the chain and from particle to particle

**Isolation, Separation.** The isolation of cellulose ethers from a complex composition is mainly a problem of suitable extraction methods. The solvent of choice may be cold water for water-soluble types or an organic solvent, e.g., dichloromethane–methanol, for nonionic ethers, except HEC. After extraction other soluble compounds, including polymers, must be separated from the extract. Insoluble cellulose ethers must be made soluble first for extraction. For example, the insoluble aluminum salt of CMC in paper coatings must be converted into the soluble sodium salt by treatment with alkali.

Water-soluble cellulose ethers are then isolated from the aqueous extract by heat coagulation in the case of methyl celluloses, EHEC, and HPC; by coagulation with tannin in the case of nonionic cellulose ethers, including HEC; and by forming insoluble metal salts with copper or uranyl ions in the case of CMC. Coagulation by other additives may result in unspecific precipitation of mixtures, if other water-soluble polymers are present, e.g., starch derivatives, natural gums, or polyacrylamide.

Specific coagulation methods are generally sufficient for the identification of commercial cellulose ethers, if such other characteristics as foaming, solubility, or behavior in acid media are also determined. Color reactions with diphenylamine (methoxy groups), ninhydrin (hydroxypropyl groups), or sodium nitroprussiate (hydroxyalkyl groups) are more or less specific for polysaccharide ethers in general or for chemically similar groups of ether substituents. Table 6 gives a survey of some analytical tests [56].

**Substituent Analysis.** The DS (MS) value determination of nonionic ether groups is carried out by the Zeisel method, modified by Morgan for hydroxyalkyl derivatives; for further modifications, see ref. [57], [58].

The content of carboxymethyl groups of pure CMC is determined by gravimetry of the sulfated ash. Gravimetric, titrimetric, or even radiochemical CMC analysis may be performed by stoichiometric precipitation of the copper or uranyl salts.

The analytical result is in each case a mass fraction (wt %) of atoms or groups (e.g., Na or $CH_2COOH$ for carboxymethyl, $OCH_3$ for methyl, or $OC_2H_4OH$ for hydroxyethyl). The DS (MS) values are then calculated for single ethers from the following equation:

$$DS\,(MS) = P \cdot 162/(100 \cdot M_a - P \cdot M_d)$$

where $P$ is the mass fraction (in %) of the estimated unit or atom, $M_a$ is the molecular mass of that unit or atom, and $M_d$ is the difference of the molecular masses of a monosubstituted anhydroglucose unit (DS value of 1) and the unsubstituted one ($M = 162$). For a mixed ether, characterized by two mass fractions, the relation must be extended to the following:

$$DS_i = \frac{162\,P_i}{M_{ai} - M_{d1}/M_{a1} - P_2 \cdot M_{d2}/M_{a2}}$$

where $i = 1$ or 2, meaning the first or the second substituent.

As an example, the carboxymethyl DS value of sodium CMC is calculated from the mass fraction of sodium, $P_{Na}$, according to the following:

$$DS = 162 \cdot P_{Na}/(2300 - 80 \cdot P_{Na})$$

Each mass fraction of a group is related to pure and dry products; moisture and impurities must be taken into account.

The *phthalate ester method* for determining DS values of HEC in addition to MS estimation only seems to be valid up to an MS level of 1.5 [59].

**Degree of Polymerization.** The average chain-length ($\overline{DP}$) of cellulose ethers is related to the intrinsic viscosity. Light scattering and osmometric methods have been applied for measuring $\overline{DP}$. Recently, gel permeation chromatography has been developed as a promising method for evaluating molecular mass distribution. However, cellulose ethers must first be modified chemically before they are degraded to organic-soluble derivatives that can then be applied to commercial columns in common solvents [60].

**Substituent Distribution.** Unequal substitution of different particles is shown by testing the solubility. Undissolved residues of dilute aqueous solutions of cellulose ethers can be centrifuged. The undissolved particles are substituted to a much lower degree than the dissolved particles.

The pattern of substitution along the chain may roughly be characterized by total hydrolysis of the cellulose ether and subsequent chromatographic separation into unsubstituted glucose and the various substituted glucose ethers. Promising results, also significant for the substitution pattern at positions 2, 3, and 6, have been obtained by NMR analysis of the hydrolysates [61].

## 9. Uses [9], [12], [21]–[23], [25], [28]–[32],[40]–[42]

The most important properties of cellulose ethers determining their uses may be summarized as follows:

1) Thickening of aqueous or organic solutions
2) Stabilization of suspensions or emulsions
3) Water retention
4) Binding action
5) Action as protective colloid
6) Film formation
7) Adhesiveness

Further special properties include thermogelation, surfactant action, foam stabilization, thixotropy, ionic activity, and gelation by additives.

**Building Industry.** Nonionic water-soluble ethers are useful additives in construction materials because of their binding, suspension-stabilizing, and water-retaining activities. Methyl celluloses or EHEC are used in most cement- or gypsum-based formulations, such as masonry mortars, grouts, cement coatings, plasters, jointing compounds, and emulsion putties. The ethers improve the dispersion of sand or cement. They further intensify the adhesiveness, which is important for plasters, tile cement, and putties. Mixtures of HEMC or HPMC and polyacrylamides are used for mechanically applied plasters to achieve better workability. Hydroxyethyl cellulose (HEC) can be used in cements as a moisture-retaining agent as well as a retarder. For this field of use, HEHPC types may also become convenient. Cellulose ethers often are used in combination with gluconates in mortars as valuable set-retarding additives. In wallpaper glues, methyl celluloses or EHEC are used as thickeners and adhesives, most often combined with poly(vinyl acetate), CMC, and/or starch derivatives. In Southern Europe CMC is used as a substantial part of wallpaper glues.

Medium and highly viscous cellulose ether types are preferred in construction materials and wallpaper glues.

**Paints.** Methyl celluloses, EHEC, HEC, and nonthixotropic pure CMC are used in latex and distemper paints. They act as thickeners and as suspension aids for pigment particles. When the rheology is adjusted from pseudoplastic to nearly Newtonian flow, the workability of dripless and semigloss paints is enhanced. Direct addition of the thickener to the pigment grind is possible by choosing glyoxal-crosslinked products with retarded dissolution.

The organic-soluble types of methyl celluloses, EHEC, HBMC, and HPC are used in solvent-based paint removers to prevent evaporation of the solvent, e.g., a dichloromethane–alcohol mixture.

**Ceramics.** All kinds of water-soluble cellulose ethers can be applied in the production of ceramics as green strength binders exhibiting good burnout properties. Pure grades of the nonionic ethers cause no ash residue, while CMC causes only a small one. Methyl hydroxypropyl cellulose (MHPC) is applied with propylene glycol as a binder for the extrusion molding of ceramic capacitors and ferrite alumina porcelains. Carboxymethyl cellulose (CMC) is known as a plasticizer for electrical porcelain and for vitreous enamels.

**Textile Industry.** Cellulose ethers are useful warp sizing agents, and HEC and CMC are most commonly used. Crude grades of CMC can be applied for this purpose. Cellulose ethers can be desized after weaving without the need for enzymes because they are readily water-soluble. Carboxymethyl cellulose (CMC) is often applied in combination with starch; it causes only small wastewater problems from desizing because of its low BOD. Carboxymethyl cellulose is used in fabric sizing in laundering operations. It may also be used as a relatively permanent finish in the manufacture of fabrics, if the fabric is impregnated first and then treated with acid and heat. Hydroxyethyl cellulose (HEC) has also found application in textile finishes.

Both ethers, CMC and HEC (but also MC), are effective thickening agents in textile printing pastes. Hydroxyethyl cellulose is further used in carpet dyeing and in nonwoven binder formulations. Like CMC, it is suitable to thicken textile backcoatings.

**Paper Industry.** Carboxymethyl cellulose of low DS value is designed specifically as an internal additive, or beater size, to promote chemical hydration of the fibers: better hydration increases the dry strength. Higher substituted CMC or nonionic water-soluble cellulose ethers are applied with conventional equipment (size press, calender stack, off-machine coater) in paper size. Such coatings reduce wax consumption for waxed boards due to reduced penetration. Similarly, printing-ink consumption decreases and surface gloss improves. Smooth surfaces and improved resistance to grease and oil result from the good film-forming abilities of these coating materials.

**Oil Industry.** Drilling fluids or drilling muds are aqueous dispersions of clay, bentonite, and weighting agents, such as barite. Crude or pure-grade CMC is used as a colloidal thickener in carrying cuttings away from the bore hole to prevent settling. The CMC types used for drilling muds must fulfill certain specifications concerning salt compatibility, viscosity, and fluid-loss-inhibiting capacity, which are compiled by the Oil Companies Materials Association (OCMA). When porous formations are drilled through, fluid loss from drilling fluids can be minimized by adding CMC of an increased DS value (Fig. 14). Hydroxyethyl cellulose serves as an additive in oil well cements and as a thickener for fracturing fluids and heavy brines (e.g., containing

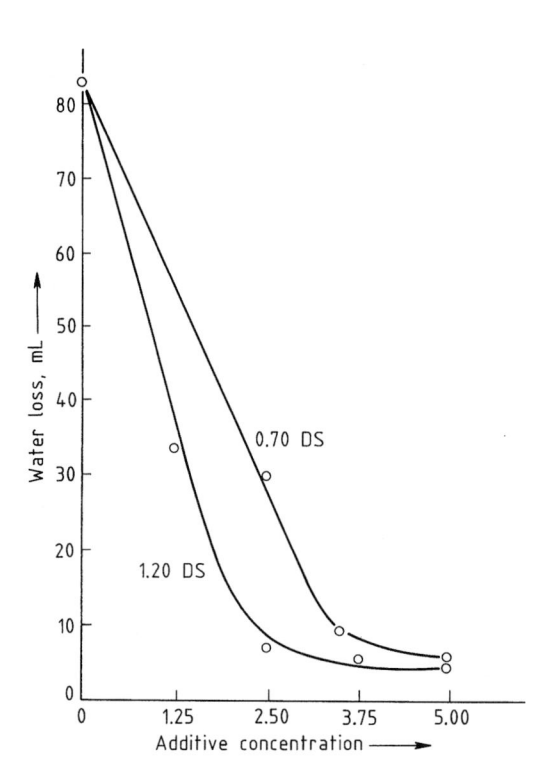

**Figure 14.** Effect of sodium carboxymethyl cellulose on the water loss of a saturated salt-water mud [42]

The unit of additive concentration is lb per barrel of mud (1 lb = 453.5 g; 1 barrel = 159 L)

CaBr$_2$, CaCl$_2$, or ZnCl$_2$), which are applied to workover and completion fluids. Another versatile agent in modern oil drilling techniques is CMHEC, which combines the compatibility of HEC with salts and the affinity of CMC to clays, due to its anionic groups.

**Detergents.** The detergent industry is the largest consumer of CMC. Mainly technical-grade CMC is used for detergent formulations. It acts as an inhibitor of soil redeposition for cotton fabrics after the soil is loosened by synthetic detergents. Detergent formulations contain 0.3–1% of CMC. Moreover, 2–5% of CMC (based on the mass of the builder) increases the soil-suspension ability in built soaps.

Hydroxyethyl cellulose is used as a thickener and protective colloid in liquid detergents and water-free hand cleaners. Methyl celluloses have become interesting for detergent formulations. They exhibit some soil-suspension action in fabrics based on synthetic fibers, where CMC is much less effective.

**Agriculture.** Cellulose ethers serve as suspending agents for solid pesticides in water-based sprays. The thickening action of HEC is used in spray emulsions to reduce drift. The ethers serve as spreader-stickers after application to bind the insecticide or fungicide to the plant leaves. Methyl celluloses or HEC are used in slurries for seed treatment to increase the seed coverage and to reduce exposure hazards caused by dusting of the protectant. Low-viscosity MC is applied in amounts of 25–50%, based on the mass of the dry protectant. Medium-viscosity HEMC is also added to agricultural dusts at a level of 6–12% of the dust to obtain better adhesion when the dust is wetted by rain or dew. Wettable powders are better dispersed when 0.5–2% of low-viscosity MC is added. Carboxymethyl cellulose is useful as a soil aggregant; its DS value should then be above 0.7 to avoid microbial deterioration.

**Polymerization.** Methyl celluloses, HPC, or HEC are used as suspension stabilizers and protective colloids in vinyl chloride polymerization for uniform particle distribution. Clearly soluble and highly purified types of low viscosity are required for this purpose. They are applied both in emulsion and in suspension techniques for the polymerization of styrene and vinyl acetate, as well as for copolymerization processes.

**Adhesives.** Cellulose ethers may be added to numerous adhesive formulations. Hydroxypropyl cellulose achieves thickening of solvent-based adhesives, whereas CMC and HEC are applied in water-soluble and emulsion-type resin adhesives. The methyl celluloses are especially convenient in the leather industry. Hides are pasted onto large frames of different materials by adhesive formulations containing MC, HEMC, or HPMC. The hides are then passed through a heating zone for drying; this causes no thinning, only gelling of the cellulose ether, which results in excellent adhesive bonds and fewer drop-offs during the subsequent tanning. Aqueous solutions of medium- or high-viscosity MC or HPMC, containing some plasticizer (0.2% *N*-acetylethanolamine) and 0.3% casein, are typical formulations.

**Cosmetics.** Carboxymethyl cellulose is used in dental impression materials and in toothpastes as a binding additive. All water-soluble cellulose ethers may serve as thickeners, stabilizers, suspending agents, and film formers in creams, lotions, or shampoos. Cationic, modified mixed ethers based on HEC are preferably used in haircare products (see Section 5.3). Methyl cellulose or HPMC provides a barrier against oil-soluble materials and is therefore used in protective creams against irritants ranging from tear gas to paint. Hydroxypropyl cellulose is used in alcohol-based hair dressings, perfumes, and colognes.

**Pharmaceutical Industry.** Pure grades of low-viscosity cellulose ethers are used for tablet coatings. Carboxymethyl cellulose is insoluble in the acidic environment of the stomach but soluble in the alkaline intestinal medium and can be applied as an enteric coating. Coatings based on HPC act as barriers to air and moisture. Methyl celluloses, CMC, or EHEC are further used as drug carriers, as tablet disintegrators, and as stabilizing agents for suspensions and emulsions. They are administered to humans as bulk laxatives, which are dispersed readily in sufficient amounts of water and are not metabolized. The suspending action of methyl celluloses or CMC is used for better dispersion of barium sulfate in X-ray diagnosis.

**Table 7.** European Community specifications for food-grade cellulose ethers

| E.C. number | Ether | Substitution levels | | Approximate molecular mass | pH of 1% solution | Max. content | |
|---|---|---|---|---|---|---|---|
| | | Group | Mass fraction, %[a] | | | Volatiles, wt% | Sulfated ash[a], wt% |
| E 461 | MC or HEMC | OCH$_3$ | 25–33 | 20 000–380 000 | 5–8 | 10 | 1.5 |
| | | OC$_2$H$_4$OH | 0–5 | | | | |
| E 463 | HPC | OC$_3$H$_6$OH | ≤4.6[b] | 30 000–1 000 000 | 5–8 | 10 | 0.5 |
| E 464 | HPMC | OCH$_3$ | 19–30 | 13 000–200 000 | 5–8 | 10 | 1.5 or 3[c] |
| | | OC$_3$H$_6$OH | 3–12 | | | | |
| E 465 | EMC | OCH$_3$ | 3.5–6.5 | 30 000–40 000 | 5–8 | 15 or 10[d] | 0.6 |
| | | OC$_2$H$_5$ | 14.5–19 | | | | |
| E 466 | Na CMC | OCH$_2$COONa[e] | 0.2–1.0[f] | 17 000–1 500 000 | 6–8.5 | 12 | 0.5[g] |

[a] On dry basis.
[b] MS value.
[c] For viscosities of 2% solutions above or below 50 mPa s.
[d] For fiber form or powder.
[e] May contain free acid groups.
[f] DS value.
[g] wt% NaCl + Na glycolate.

**Foods.** Methyl celluloses, HPMC, HPC, and CMC are approved as internal food additives in the United States, the European Community, and many other countries. In the European Community EMC and HEMC of low hydroxyethyl MS are further mentioned in the list of admitted thickening and gelling agents. Table 7 lists appropriate specifications. The toxicology of all substances mentioned has been fully evaluated by the Joint Expert Committee on Food Additives of the FAO/WHO (JECFA). The ADI (acceptable daily intake) of these products has been established at 25 mg/kg [62].

The U.S. Food Chemicals Codex lists CMC (cellulose gum), methyl celluloses, and HPC as food additives. Carboxymethyl cellulose is recognized as safe by the Code of Federal Regulations, Title 21, Part 182.1745, and can be used in standardized foods. Methyl cellulose is likewise classified under Title 21, Part 182.1480. Food-grade CMC may be admitted as kosher additive by Jewish authorities on request of the manufacturer.

All typical properties of cellulose ethers are utilized in foods. Cellulose ethers act as whipping aids (methyl celluloses or HPC in ice cream and whipped products), as gelling agents (jellies and puddings), as thickeners (fillings and dressings), as suspending agents (fruit juices and dairy products), or as protective colloids (emulsions and mayonnaises). The nonionic ethers exhibit a lack of syneresis of water at freezing temperatures, thus stabilizing frozen foods (mixtures of meat, fish, or fowl with vegetables). While the mixture is being fried, gelation of the cellulose ether holds these patties together and provides the moisture required to prevent burning of the vegetable components.

The ability of CMC to inhibit undesirable crystallization is used in ice creams as well as in such sugar products as syrup, icings, or glazes. Dipping batters for deep-fried, breaded foods may contain 1% of HPMC; gelation then occurs on contact with the hot fat, thus forming a barrier around the fried article, which greatly reduces fat absorption.

The food industry applies cellulose ethers in many other ways, generally in amounts of less than 0.5%.

**Miscellaneous.** Ethyl cellulose as an organic-soluble and widely compatible film-forming material is used in lacquers and coatings. It is suitable as an isolating lacquer for cables and wires and has a dielectric constant of $3.0-3.8$; its dielectric loss (tan $\delta$) at 1 kHz is in the range of $0.002-0.02$.

Plastic parts can be made from EC or HPC, the latter being a convenient material for water-soluble, blow-molded containers. Ethyl cellulose is further used in hot-melt adhesives. Methyl celluloses, EHEC, or CMMC are effective binders in tobacco sheets. Cellulose ethers, primarily CMC, are further used as binders for molding extrusion of pencils, in foundry cores, and in wrappings of welding electrodes. Low-viscosity aqueous solutions of cellulose ethers have a smaller friction resistance than water, which is desirable in fire-fighting to obtain an enhanced water current at equal pressure.

# 10. Economic Facts

## 10.1. Trade Names and Suppliers

The following list gives the trade names of different ether types. Manufacturing companies are named in parentheses; companies of the East Bloc are not included. The same general trade name of a company may comprise different ethers. Products having no special trade names are designated by "–." Viscosity types, purity grades, and DS (MS) levels are not specified; they are usually given by the manufacturers by additional alphanumeric codes.

*Methyl Cellulose* (*MC*): Celacol (Brit. Celanese, U. K.), Methocel A (Dow), Culminal (Henkel), Tylose M (Hoechst), Metolose SM (Shin-Etsu, Japan), Walocel M (Wolff-Walsrode), Marpolose (Matsumoto Yushi Seiyako, Japan).

*Hydroxyethyl Methyl Cellulose* (*HEMC*): Celacol (Celanese), Culminal (Henkel), Tylose MH (Hoechst), Walocel M (Wolff-Walsrode), Marpolose (Matsumoto).

*Hydroxybutyl Methyl Cellulose* (*HBMC*): Methocel HB (Dow).

*Hydroxypropyl Methyl Cellulose* (*HPMC*): Celacol (Celanese), Methocel, except A or HB (Dow), Culminal (Henkel), Metolose SH (Shin-Etsu), Marpolose (Matsumoto), Methofas (ICI), Walocel M (Wolff-Walsrode).

*Carboxymethyl Methyl Cellulose* (*CMMC*): Culminal (Henkel).

*Ethyl Hydroxyethyl Cellulose* (*EHEC*): Water-soluble: Bermocoll E (Berol, Sweden); organic-soluble: – (Hercules, USA and France).

*Ethyl Cellulose* (*EC*): Ethocel (Dow), – (Hercules).

*Ethyl Methyl Cellulose* (*EMC*): Edifas A (ICI).
*Hydroxypropyl Cellulose* (*HPC*): Klucel (Hercules).
*Aminoethyl-modified HPC:* Klucel 6 (Hercules).
*Hydroxyethyl Hydroxypropyl Cellulose* (*HEHPC*): Natrovis WSPD (Hercules).
*Methyl Hydroxypropyl Cellulose Phthalate*: – (Shin-Etsu).
*Hydroxyethyl Cellulose* (*HEC*): Natrosol (Hercules), Cellozise (UCC), Cellobond (BP, Netherlands and Brazil), Tylose H (Hoechst), Celacol (Celanese), – (Fuji, Japan).
*Cationic Modified HEC:* Ucare Polymer (UCC).
*Carboxymethyl Hydroxyethyl Cellulose* (*CMHEC*): – (Hercules), Tylose CHR (Hoechst).
*Carboxymethyl Cellulose* (*CMC*): Blanose (Hercules, Europe), Hercopac (Hercules, USA), Courlose (Celanese), Cellofas and Edifas (both ICI), Aku-CMC (Enka, Netherlands), Cellogen (Dai-Ichi, Japan), Tylose C (Hoechst), Relatin (Henkel), Walocel C (Wolff-Walsrode), Finnfix (Metsäliiton, Finland), Gabrosa (Montedison, Italy), Nymcel (Nyma, Netherlands), Cellufix (Svenska Cellulosa, Sweden), Cekol (Billerud Uddeholm Sweden), – (Daicel, Japan), – (Du Pont, USA), and other producers in several countries.

## 10.2. Production Capacities [63], [64]

The following approximate production data relate to the years 1982 and 1983.

*Carboxymethyl Cellulose.* Worldwide capacities total about 300 000 t/a. Most of that is located in the United States and in Western Europe. Japanese production is at about 30 000 t/a, ca. 55 % of which is exported to Southeast Asia. There are many suppliers, but only ca. ten of them produce pure CMC (ca. 120 000 t/a). Hercules is the leading company in the CMC market, especially in the United States; it produces about 40 000 t/a, 90 % of which is pure-grade material. The crude types and some semipurified types are applied in detergents, in the oil industry, and in warp sizes. The trend is going to purified grades, even in the U.S. oil industry. Very highly pure CMC of 99.5 % CMC content (dry basis) is used mainly in foods, pharmaceuticals, and cosmetics.

*Methyl Celluloses.* World capacities (except Japan) of water-soluble, nonionic, and thermogelling methyl celluloses total nearly 60 000 t/a. In the United States, Dow produces 15 000 t/a of MC and HPMC. Hoechst is Europe's main supplier with 16 000 t/a (MC, HEMC), whereas Henkel produces 9000 t/a (6000 in the Federal Republic of Germany, 3000 in Belgium). Capacities of both Dow – Europe and Wolff – Walsrode amount to 6000 t/a, and British Celanese supplies 4000 t/a.

About 20 000 t of the annual production is used in building materials, 13 000 t in adhesives (wallpaper glues), and 10 000 t in paints. A total of 6000 t is designed for polymerization, and another 5000 – 6000 t for tobacco, cosmetics, and other applications.

The Swedish Berol Co. produces about 8000 t/a of water-soluble EHEC, which is used mainly for the same purposes as the methyl celluloses.

*Hydroxyethyl Cellulose.* World capacities of HEC amount to about 56 000 t/a. There are 30 000 t in the United States (17 000 t Hercules, 13 000 t UCC), 22 000 t in Europe

(12 000 t Hercules – Europe, 8000 t BP, 2000 t Hoechst), 4000 t in Brazil (Union Carbide), and 500 t in Japan (Fuji).

About 40% of the HEC production is used in paints, 30% in the oil industry, and 20% in polymerization. Hercules puts emphasis on biostable materials (Natrosol B-types) of high MS values, whereas Union Carbide types aim at better salt compatibility by lower MS values. The new type of HEHPC (Hercules) is designed mainly for building materials.

# 11. References

The following list contains mainly books, chapters of books, and reviews citing numerous special reports and patents. Valuable information on commercial products is further available from the Company Publications of various suppliers (Section 10.1).

General References

[1] E. Ott, H. M. Spurlin, M. W. Grafflin (eds.): *High Polymers*, 2nd ed., vol. 5: "Cellulose and Cellulose Derivatives," parts I – III, Interscience, New York-London 1954.

[2] N. Bikales, L. Segal (eds.): *High Polymers*, vol. 5: "Cellulose and Cellulose Derivatives," parts IV and V, Wiley-Interscience, New York-London-SidneyToronto 1971.

[3] H. F. Mark, N. G. Gaylord, N. M. Bikales (eds.): *Encyclopedia of Polymer Science and Technology*, vols. **3,** 15, Wiley-Interscience, New York-London 1965.

[4] R. L. Davidson, M. Sittig (eds.): *Water-Soluble Resins,* 2nd ed., Van Nostrand Reinhold Co., New York 1968.

[5] R. L. Whistler, J. N. BeMiller (eds.): *Industrial Gums,* 2nd ed., Academic Press, New York-London 1973.

[6] H. Neukom, W. Pilnik (eds.): *Gelling and Thickening Agents in Foods,* Forster Publishing Ltd., Zürich 1980.

[7] R. L. Davidson (ed.): *Handbook of water-soluble gums and resins,* McGraw-Hill, New York 1980.

Specific References

[8] A. B. Savage, A. E. Young, A. T. Maasberg: *High Polymers*, part II, Interscience, New-York 1954, pp. 882 – 958.

[9] A. B. Savage in*High Polymers*, part V, Wiley —Interscience, New-York 1971, pp. 785 – 809.

[10] N. M. Bikales in *High Polymers*, part V, Wiley-Interscience, New-York 1971, pp. 811 – 833.

[11] E. D. Klug, H. E. t'Sas: *Gelling and Thickening Agents in Foods,* Forster Publishing Ltd., Zürich 1980163 – 174.

[12] L. Grosse in P. H. List, L. Hörhammer (eds.): *Hagers Handbuch der pharmazeutischen Praxis,* 4th ed., vol. **7 b,** Springer Verlag, Berlin-Heidelberg-New York 1977, pp. 111 – 154.

[13] L. Grosse, W. Klaus, *Seifen, Öle, Fette, Wachse* **100** (1974) 1 – 4.

[14] Food Chemicals Codex II, Food-Additive Regulation 121101, U.S. Pharmacopoeia XVIII.

[15] *WHO Tech. Rep. Ser.* **281** (1964) 82 – 88.

[16] M. G. Wirick, *J. Water Pollut. Control Fed.* **46** (1974) 512 – 521.

[17] H. M. Spurlin: *High Polymers*, part II, Interscience, New-York 1954, pp. 673 – 712.

[18] W. D. Nicoll, N. L. Cox, R. F. Conaway in*High Polymers*, part II, pp. 825 – 881.

[19]  L. Segal in *High Polymers*, part V, Wiley-Interscience, New-York 1971, pp. 719–739.
[20]  G. N. Richards in *High Polymers*, part V, Wiley-Interscience, New-York 1971, pp. 1007–1014.
[21]  A. B. Savage in *Encyclopedia of Polymer Science and Technology*, vol. **3,** Wiley-Interscience, New-York 1965, pp. 492–511.
[22]  G.Scheffler: *Water-Soluble Resins,* 2nd ed., Van Nostrand Reinhold Co., New York 1968 pp. 50–62.
[23]  G. K. Greminger jr. , A.B. Savage: *Industrial Gums,* 2nd ed., Academic Press, New York-London 1973. 619–647.
[24]  Dow Chemical, US 4061859, 1977 (W.-J. Cheng).
[25]  G. K. Greminger jr., K. L. Krumel:*Handbook of water-soluble gums and resins,* McGraw-Hill, New York 1980 3/1–3/25.
[26]  H. Kittel: *Lehrbuch der Lacke und Beschichtungen,* vol. **1**, part 3, Verlag W. A. Colomb, Oberschwandorf 1974.
[27]  Mo och Domsjö, US 2879268, 1959 (E. I. Jullander).
[28]  A. B. Savage in *Encyclopedia of Polymer Science and Technology,*, vol. **3**, pp. 475–492.
[29]  S. Lindenfels, E. I. Jullander: *Industrial Gums,* 2nd ed., Academic Press, New York-London 1973.673–693.
[30]  A. B. Savage in *Encyclopedia of Polymer Science and Technology,*, vol. 3, Wiley-Interscience, New-York 1965, pp. 511–519.
[31]  N. C. Eastman, J. K. Rose, M. Sittig (eds.): *Water-Soluble Resins,* 2nd ed., Van Nostrand Reinhold Co., New York 1968 pp. 63–90.
[32]  A. J. Desmarais in [5]pp. 649–672.
[33]  G. M. Powell: *Handbook of water-soluble gums and resins,* McGraw-Hill, New York 1980 12/1–12/22.
[34]  W. Hansi, W. Klaus, K. Mercator, *Dtsch. Farben Z.* **23** (1969) 305; **25** (1971) 493.
[35]  R. W. Butler, E. D. Klug: *Handbook of water-soluble gums and resins,* McGraw-Hill, New York 1980 13/1–13/17.
[36]  E. D. Klug in *Encyclopedia of Polymer Science and Technology,*, vol. 15, Wiley-Interscience, New-York 1965, pp. 307–314.
[37]  Hercules, US 3278521, 1966 (E. D. Klug).
[38]  Hercules, US 3431254, 1969 (E. D. Klug).
[39]  Union Carbide, US 3472849, 1969 (F. W. Stone, J. M. Rutherford).
[40]  E. D. Klug in *Encyclopedia of Polymer Science and Technology,* vol. **3,** Wiley-Interscience, New-York 1965, pp. 520–540.
[41]  G.S.Baird, K. Speicher:*Water-Soluble Resins,* 2nd ed., Van Nostrand Reinhold Co., New York 1968 pp. 99–180.
[42]  J. B. Batdorf, J. M. Rossman: *Industrial Gums,* 2nd ed., Academic Press, New York-London 1973. 695–729. J. B. Batdorf, J. M. Rossman in [5] pp. 695–729.
[43]  G.I. Stelzer, E. D. Klug:*Handbook of water-soluble gums and resins,* McGraw-Hill, New York 19804/1–4/28.
[44]  G. C. Tesoro, J. J. Willard in *High Polymers*, part V, pp. 835–875.
[45]  W. Klaus, P. W. Krause, H. Wurm, *Defazet-Aktuell* **26** (1972) 101–105.
[46]  Hoechst, US 4068068, 1978 (A. Holst, H. Lask, M. Kostrzewa).
[47]  J. D. Guthrie in*High Polymers*, part V, Wiley-Interscience, New-York 1971, pp. 1277–1291.
[48]  K. Ward jr., L. J. Bernardin in *High Polymers*, part V, Wiley-Interscience, New-York 1971, pp. 1169–1223.

[49] N. M. Bikales in *Encyclopedia of Polymer Science and Technology,*, vol. **3,** Wiley-Interscience, New-York 1965, pp. 540–541.
[50] D. J. Stannonis in *Encyclopedia of Polymer Science and Technology,*, vol. **3,** Wiley-Interscience, New-York 1965, pp. 542–549.
[51] Hercules, US 2580332, 1951 (V. R. Grassie).
[52] Hoechst, US 4379918, 1983 (L. Brandt, A. Holst).
[53] Hoechst, US 4358587, 4396433, 1982/83 (L. Brandt, A. Holst).
[54] L. Grosse, W. Klaus, *Fresenius Z. Anal. Chem.* **259** (1972) 195–203.
[55] G. Bartelmus, R. Ketterer, *Fresenius Z. Anal. Chem.* **286** (1977) 161–190.
[56] Henkel Company Publication: *"Culminal,"* Henkel KGaA, Düsseldorf.
[57] P. W. Morgan, *Ind. Eng. Chem. Anal. Ed.* **18** (1946) 500–504.
[58] R. U. Lemieux, C. B. Purves, *Can. J. Res. Sect. B* **25** (1947) 485–489.
[59] J. Quinchon, P. Pascal, *C. R. Hebd. Séances Acad. Sci.* **248** (1959) 225–228.
[60] H. G. Barth, F. E. Regnier, *J. Chromatogr.* **192** (1980) 275–293.
[61] A. Porfondry, A. S. Perlin, *Carbohydrate Res.* **57** (1977) 39–49.
[62] Richtlinie des Rates, 25. 07. 1978, I-7, pp. 16–19, in „Ausländisches Lebensmittelrecht, EG-Vorschriften", B. Behrs Verlag.
[63] *Chemical Economics Handbook,* Stanford Research Institute, Menlo Park, Nov. 1983.
[64] *Jpn. Chem. Week* **25** (1984) no. 1270, 4.